AF890345

Contributors

M. J Adams
Rothamsted Research, Harpenden Hertfordshire AL5 2JQ UK

O. J. Alabi
Research Fellow, International Institute of Tropical Agriculture, PMB 5320, Ibadan, Nigeria

Y. Antignus
Department of Virology, the Volcani Center, ARO, P. O. B. 6, Bet Dagan, 50250, Israel

G. I. Atiri
Department of Crop Protection & Environmental Biology, University of Ibadan, Ibadan, Nigeria

R. Ausher
Department of Crop Protection, Extension Service, Minstry of Agriculture and Rural Development, Israel

M. Bar-Joseph
Volcani Centre, Agricultural Research Organization, Bet Dagan, 50250, Israel

L.A. Calvert
Centro Internacional de Agricultura Tropical (CIAT), AA 6713, Cali, Colombia; E-mail: l.calvert@cgiar.org

J. Cohen
Dept. of Virology, Agricultural Research Organisation, Bet Dagan, Israel

G. Dahal
University of Minnesota, Department of Plant Pathology, St Paul, MN 55108, USA

D. Fargette
IRD, BP 64501, 34394. Montpellier cedex 5, France

C.M. Fauquet
International Laboratory for Tropical Agricultural Biotechnology (ILTAB)/Donald Danforth Plant Science Center, 975 N. Warson Road, St. Louis, MO 63132, USA

G. Fermin
Pacific Basin Agricultural Research Center, 99 Aupuni St., Suite 204. Hilo, HI 96720, USA

I.B.F. Fofana
International Laboratory for Tropical Agricultural Biotechnology (ILTAB) /Donald Danforth Plant Science Center, 975 N. Warson Road, St. Louis, MO 63132, USA; and Université de Cocody, UFR des Biosciences, Laboratoire de Génétique, 22 BP 582 Abidjan 22, Ivory Coast

S. Fuentes
Virology, Crop Protection Department, International Potato Center (CIP), Apartado 1558, Lima 12, Peru.

I.D. Garg
Central Potato Research Institute, Shimla, India

A. D. W. Geering
Agency for Food and Fibre Sciences, Department of Primary Industries, Queensland, 80 Meiers Road, Indooroopilly, Qld 4068, Australia

D. Gonsalves
Pacific Basin Agricultural Research Center, 99 Aupuni St., Suite 204. Hilo, HI 96720, USA

D.T. Gordon
Department of Plant Pathology, The Ohio State University, Wooster, Ohio, USA, E-Mail: gordon.4@osu.edu

L.S. Hagen
INRA-Station de Pathologie, Végétale, Domaine Saint Maurice, BP94, F-84143 Montfavet Cedex, France

R. O. Hampton
Research Plant Pathologist, U. S. Department of Agriculture, Agriculture Research Service, and Professor (Courtesy), Department of Botany and Plant Pathology, Oregon State University, Corvallis, OR 97331-2902, USA

D. Hanold
Waite Campus, University of Adelaide, Glen Osmond, South Australia

N. A. Harrison
Fort Lauderdale Research and Education Center, University of Florida, IFAS, Fort Lauderdale, Florida 33314-7719, USA

Monique Henry
CIMMYT, Mexico

J. H. Hill
Dept. of Plant Pathology, Iowa State University, Ames, Iowa.50011, USA.
E-mail: johnhill@iastate.edu

Jackie Hughes
Virologist, International Institute of Tropical Agriculture, (IITA), Ibadan, Nigeria (Address for correspondence: c/o L W Lambourn & Co Ltd., 26 Dingwall Road, Croydon, CR9 3EE, UK.

Mireille Jacquemond
INRA-Station de Pathologie, Végétale, Domaine Saint Maurice, BP94, F-84143 Montfavet Cedex, France

M.J. Jeger
Department of Agricultural Sciences, Imperial College, Wye, Ashford, Kent TN25 5AH, UK

K. Kahn
International Laboratory for Tropical Agricultural Biotechnology (ILTAB)/Donald Danforth Plant Science Center, 975 N. Warson Road, St. Louis, MO 63132, USA

S.M. P. Khurana
Central Potato Research Institute, Shimla, India

H. Koganezawa
National Agricultural Research Center for Western Region, Nishifukatsu, Hiroshima 721-8514, Japan

G. Konate
INERA, Laboratoire de Virologie, Station de Kamboinse, 01 BP 476 Ouagadougou, Burkina-Faso

N.K. Kulkarni
Department of Plant Pathology, University of Agricultural Sciences, GKVK, Bangalore-560 065, India

S.G. Kumari
Virology Laboratory, Germplasm Program, International Center for Agricultural Research in the Dry Areas (ICARDA), P.O. Box 5466, Aleppo, Syria

J.P. Labouisse
CIRAD-CP, Centre Agronomique de Recherche et de Formation du Vanuatu, Saraoutou, Espiritu Santo, Vanuatu

H. Lecoq
INRA, Station de Pathologie Végétale, Domaine Saint Maurice, BP 94, 84143 Montfavet cedex, France

R. F. Lee
University of Florida, Citrus Research and Education Center, Lake Alfred, FL 33850, USA

B. E. L. Lockhart
University of Minnesota, Department of Plant Pathology, St Paul, MN 55108, USA

G. Loebenstein
Dept. of Virology, Agricultural Research Organisation, Bet Dagan, Israel
E-mail : Gad-talma@barak.net.il

K.M. Makkouk
Virology Laboratory, Germplasm Program, International Center for Agricultural Research in the Dry Areas (ICARDA), P.O. Box 5466, Aleppo, Syria

V.G. Malathi
Advanced Centre for Plant Virology, Indian Agricultural Research Institute, NewDelhi 110012, India

Alija Manadilova
Institute of Molecular Biology and Biochemistry, Almaty, Kazakhstan

B. Mandal
Advanced Centre for Plant Virology, Indian Agricultural Research Institute, NewDelhi 110012, India

F. J. Morales
Virology Reseach Unit, International Center for Tropical Agriculture, Palmira, Colombia

J.P. Morin
CIRAD-CP, Montpellier, France

V. Muniyappa
Department of Plant Pathology, University of Agricultural Sciences, GKVK, Bangalore-560 065, India

J. Ndunguru
International Laboratory for Tropical Agricultural Biotechnology (ILTAB)/Donald Danforth Plant Science Center, 975 N. Warson Road, St. Louis, MO 63132, USA; and Plant Protection Division, P.O. Box 1484, Mwanza, Tanzania

L.A. Ollenu
Cocoa Research Institute, P.O. Box 8, Tafo, Ghana

D. Pimentel
College of Agriculture & Life Sciences, Cornell University, Ithaca, NY 14853, USA

D. Peters
Laboratory of Virology, Wageningen University, Binnenhaven 11, 6709 PD Wageningen, The Netherlands, dick.peters@wur.nl

G. Radhakrishnan
Advanced Centre for Plant Virology, Indian Agricultural Research Institute, NewDelhi 110012, India

J.W. Randles
Waite Campus, University of Adelaide, Glen Osmond, South Australia

K. Ravi
Division of Molecular Virology, Mahyco Life Sciences Research Center, Jalna, India

D.V.R. Reddy
Donald Danforth Plant Science Center, 975 North Warson Road, St. Louis, MO, USA

Maria Judith B. Rodriguez
Philippine Coconut Authority- Albay Research Center, Banao, Guinobatan, Albay 4503, Philippines.

P. Rott
UMR 385 ENSAM-INRA-CIRAD Biologie et Génétique des Interactions Plante-Parasite, TA 71/09, Avenue Agropolis, 34398 Montpellier Cedex 5, France

Anindya Roy
Department of Biochemistry, Indian Institute of Science, Bangalore 560012, India.

Y. Sachs
Department of Crop Protection, Extension Service, Minstry of Agriculture and Rural Development, Israel

A. Sangaré
Centre National de Recherche Agronomique (CNRA), Laboratoire Central de Biotechnologies, Km 17 Rte Dabou, 01 B.P. 1740, Abidjan 01, Ivory Coast

L. F Salazar
Virology, Crop Protection Department, International Potato Center (CIP), Apartado 1558, Lima 12, Peru

H.S. Savithri
Department of Biochemistry, Indian Institute of Science, Bangalore 560012, India.

G. R Smith
David North Plant Research Centre, Bureau of Sugar Experiment Stations, PO Box 86, Indooroopilly, Q 4068, Australia

K. Thirumala-Devi
Department of Plant pathology, Russell labs, 1630 Linden Drive, University of Wisconsin, Madison, WI, USA

J. E. Thomas
Agency for Food and Fibre Sciences, Department of Primary Industries, Queensland, 80 Meiers Road, Indooroopilly, Qld 4068, Australia

G. Thottappilly
Mahyco Research Foundation, A.G. Heights, Road No. 12, Banjara Hills, Hyderabad- 500034, A.P., India
E-mail: execdirector@satyam.net.in

J.M. Thresh
Natural Resources Institute, University of Greenwich, Chatham Maritime, Kent ME4 4TB, UK
E- mail: kathie.clark@blueyonder.co.uk

J. H. Tsai
Fort Lauderdale Research and Education Center, University of Florida, IFAS, Fort Lauderdale, Florida 33314-7719, USA

R.Usha
Department of Plant Biotechnology, School of Biotechnology, Madurai Kamaraj University, Madurai 625021, India.

A. Varma
Advanced Centre for Plant Virology, Indian Agricultural Research Institute, NewDelhi 110012, India

S. Winter
Plant Virus Division, German Collection of Microorganisms and Cell Cultures, DSMZ, C/o BBA, Messeweg- 11/12, Braunschweig, Germany, E-mail: s.winter@bba.de

Colour Plates

2.1

2.2

2.5

2.3

2.4

Control
Y44
Y85
2.6

Legends to figures

Figure 2.1: Uninfected cassava plants of cultivar CB, Adiopodoumé, coastal region of Ivory Coast

Figure 2.2: CMD-infected cassava plants of cultivar CB at Adiopodoumé, Ivory Coast

Figure 2.3: Intercropping of cassava and maize in Tanzania, both important crops to small-holding farmers

Figure 2.4: Healthy cassava plants derived from virus-free meristem culture and maintained under high phytosanitary conditions by removal of infected plant material during field growth. Cultivar CB is grown on the left, and cultivar Kataoli on the right.

Figure 2.5: Nigerian Cassava landrace TME3, one of the most promising new sources of single gene resistance to CMD.

Figure 2.6: Cassava cultivar TMS 60444 (control) transformed with the replicase gene from ACMV-Kenya to generate lines Y44 and Y85. Both transgenic lines show high levels of CMD resistance when challenged under greenhouse conditions with different species of geminviruses (ACMV, EACMCV, SLCMV).

Fig. 6.1: Symptoms of cassava mosaic disease caused by ACMV (a), EACMV (b) and cassava field (western Kenya) with 100% infections of EACMV-Uganda variant (c).
Fig. 6.2: Symptoms of *Cassava brown streak virus* (CSBV);
Fig. 6.3: CBSV and EACMV mixed infection.
(Photographs: S. Winter)

Legends to figures:

Fig. 7.1 - Yellow blotches and incomplete chevrons induced by PMTV

Fig. 7.2 - Apical leaf curl induced by PALCV

Fig. 7.3 -Bright yellow veins induced by PYVV (courtesy Dr. L.F. Salazar)

Fig. 7.4 -Necrosis induced by TSWV

Legends for Figures:

8.1

8.3

8.5

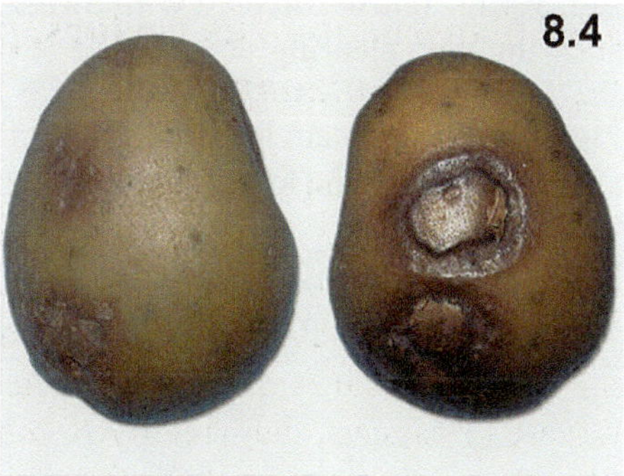
8.4

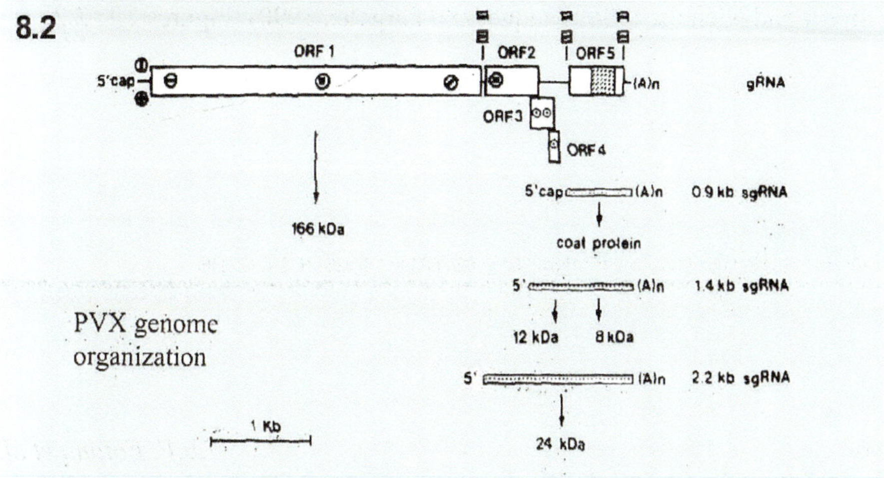
8.2
ORF 1
ORF2
ORF5
5'cap
(A)n
gRNA
ORF3
ORF4
5'cap
(A)n
0.9 kb sgRNA
166 kDa
coat protein
5'
(A)n
1.4 kb sgRNA
12 kDa
8 kDa
PVX genome
organization
5'
(A)n
2.2 kb sgRNA
1 Kb
24 kDa

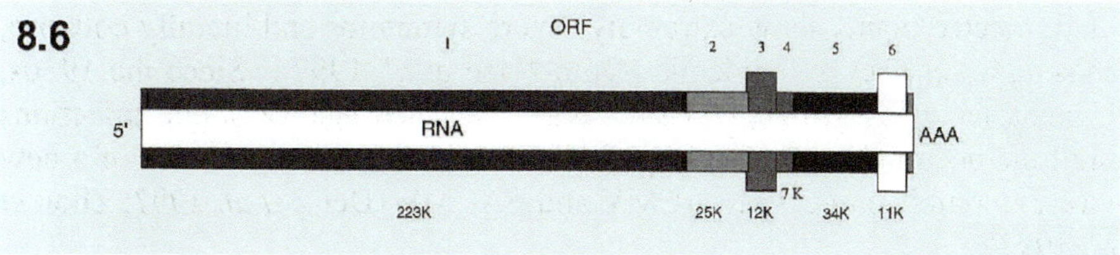
8.6
ORF
1
2
3
4
5
6
5'
RNA
AAA
7K
223K
25K
12K
34K
11K

9.2

9.6

9.5

9.3

9.4

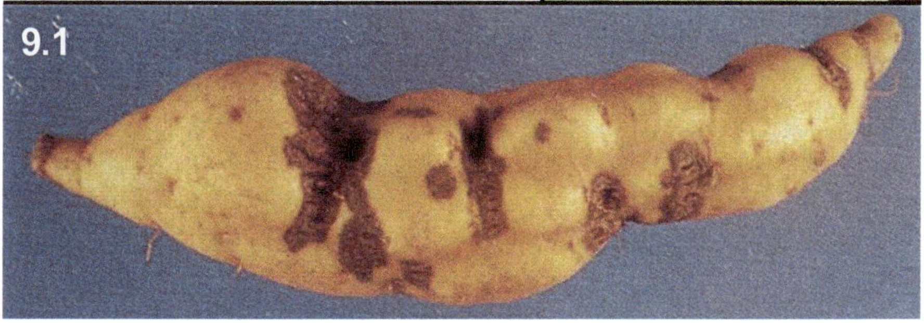
9.1

Fig. 9.7 Sweet potato cv Jewel infected with SPLSV showing whitish speckling.

Fig. 9.8 Symptoms of *Ipomoea crinkle leaf virus* on *I. setosa.* Left- healthy *plant.*

Fig. 9.9 Sweet potato plants (cv. Georgia Jet) severely affected by yellowing and stunting caused by double infection with *Cucumber mosaic virus* and *Sweet potato sunken vein virus.*

Fig. 9.10 *Ipomoea setosa* infected with C-6 virus showing fine chlorotic spots and vein clearing.

Figure 10.1 Symptoms of virus infections in yam, *Dioscorea alata*. Yam mild mosaic virus (left), *Yam mosaic virus* (middle), *Dioscorea bacilliform virus* mixed infection with Yam mosaic virus (right).

Fig. 12.1: A maize field in Nigeria infected with MSV. Please note the severely affected plants in the front and resistant plants in the background.
Fig. 12.2 a and b: Symptoms due to MSV. Symptoms are severe on susceptible lines, while very mild symptoms are observed on tolerant lines.
Fig. 12.3 : MSV vector (*Cicadulina* sp.) on maize plants
Fig. 12.4: Symptoms due to Maize stripe virus: Characterized by chlorotic bands of increasing width.
Source: Courtesy IITA (Nigeria) and H.W. Rossel

Legends to figures:

Fig. 16.1 – Symptoms of chlorotic rosette on peanut plants. Note chlorotic and deformed leaflets especially at the apical positions.

Fig. 16.2 - Typical field symptoms caused by *Tobacco streak virus* on peanut. Note stunting of plants and necrotic blotches on leaflets.

Fig. 16.3 – Field symptoms induced by *Tobacco streak virus* (TSV) (right) and Peanut bud necrosis disease (PBND) (left). TSV caused complete necrosis and death. PBND caused severe stunting and axillary shoot proliferation.

Fig. 16.4 – Typical symptoms induced by Peanut bud necrosis disease. Note severe stunting and axillary shoot proliferation.

Fig. 16.5 – Evaluation of field-resistant cultivars to Peanut bud necrosis disease in North India. Susceptible cultivar is on the left side and a field-resistant one is on the right side.

Legends to figures:
Fig. 17.1 –Symptoms induced Bean common mosaic disease.
Fig. 17.2 – Beans infected with Bean severe mosaic virus.
Fig. 17.3 – Beans infected with Bean golden mosaic virus.
Fig. 17.4 – Beans infected with Bean dwarf mosaic virus.
Fig. 17. 5 – Beans infected with Bean yellow mosaic virus.
Fig. 17.6 – Beans infected with Cucumber mosaic virus.

Preface

The global population size has passed the 6 billion mark and is increasing by roughly 80 million annually. Almost all population growth is in developing countries. Increases in food supply is therefore a necessity, cannot come only from recruiting new land for agriculture, but has to come mainly from increase production of food per hectare. Various diseases and pests may reduce yields significantly. It is estimated that virus and virus-like diseases cause about 20% of the losses caused by the various plant disease agents. Some virus diseases can have a disastrous effect on crop yields and threaten the food-producing potential especially in developing countries. Once infected with a virus, there are few economically or biologically viable means of curing the plant. Development of resistant cultivars, or supply of virus-tested planting material, is the basic requirement for effective management of plant virus diseases. A sound knowledge of each virus, its variants or strains and their distribution is a vital prerequisite for reliable breeding for resistance. Virus indexing is important for providing healthy planting material and for collection, storage and international exchange of plant materials.

Some of the disease phenomena observed in developing countries are still awaiting identification, while for many of the well-known viruses feasible control methods are not yet available. Also the change of cropping systems, intensified farming and movement of germplasm may escalate previously minor problems and allow emergence of new viruses. All these require the need for studying the various virus diseases in their local environment. Proper identification, enabling classification, is an essential first step for later control of the disease. This allows prediction of certain biological properties. So far, many viruses, especially in developing countries, have not been properly identified and the occurrence of a disease was based mainly on symptoms. This often led to confusion regarding the distribution of the disease.

This book is a comprehensive up-to-date treatise on virus and virus-like diseases of the major crops in developing countries and their detection, isolation, biological and molecular characterization, transmission and possible approaches for their control. Also included are chapters on the general impact of these diseases, epidemiology, quarantine and technology transfer. Many of the crop chapters include possibilities for control by cloning resistance genes and later developing transgenic plants. Many scientists see transgenic virus resistance as a major approach to control virus diseases. Some critics see this as a dangerous development.

We hope that this book will contribute towards the knowledge required by scientists in both developing and developed countries. We further hope that the funding agencies will use this book to prioritize their projects.

The contributing authors are internationally recognized specialists who were invited because of their specific knowledge and experience. Each has attempted to include all relevant recently published information; however, with the rapid advances in methodology, especially in diagnosis and production of transgenic resistant genotypes, valuable additional information will undoubtedly continue to be produced. Nevertheless, the principles presented here should long remain valid for the effective control of these diseases.

We thank all the authors for their contributions and for their patience and understanding with the editors' reviewing. Special thanks are due to D. Ronen for preparing graphics and photographic illustrations.

The editors express their sincerest thanks to the Mahyco Research Foundation, Mumbai, India, Technical Centre for Agricultural and Rural Cooperation (CTA), Wageningen, The Netherlands, and the Gatsby Charitable Foundation, UK for supporting this venture.

Gad Loebenstein and George Thottappilly

August 2003

Contents

Fig. 19.1 *Banana bunchy to virus*-infected banana, New Caledonia
Fig. 19.2 *Banana bunchy top virus*-infected banana, New Caledonia
Fig. 19.3 *Banana streak virus* – Chlorotic streaks produced on a leaf of cv. Williams banana
Fig. 19.4 *Banana streak virus* - Necrotic streaks produced on a leaf of cv. Williams banana
Fig. 19.5 *Banana bract mosaic virus* - Mosaic pattern on banana bracts.
Fig. 19.6 *Banana bract mosaic virus* - Mosaic pattern on the pseudostem of banana, after removal of the outer leaf sheaths.

Figure 20.1 Top left, a commercial papaya orchard in the Puna district of Hawaii before PRSV was detected in the island. Top right, devastation caused by PRSV (1994) in the same area where commercial orchards flourished short time ago. Bottom left, after deregulation was granted seeds of transgenic papayas were sown and 13 months later non transgenic papayas (left row) were severely affected by the disease while 'Rainbow' (right row) was already harboring fruits. Bottom right, a commercial orchard 3 years later, with transgenic 'Rainbow' still bearing fruits and showing no symptoms of infection by PRSV.

Fig. 22.1 Sugarcane streak mosaic

Legends to Figures

Figure 23.1.1 (**A**) Typical symptoms of cadang-cadang: nuts become smaller, more spherical and scarified around the equator (left), inflorescences become necrotic (middle) and stipules remain attached to bases of the fronds (right). (**B**) CCCVd- inoculation of a 3-day old coconut sprout using a hand-primed high-pressure injector. (**C**) The unusual brooming syndrome observed in a CCCVd-inoculated coconut palm: severe stunting, much shorter frond and reduced leaf laminae. (**D**) Leaflets from a brooming palm (top) compared with those from cadang-cadang infected (middle) and healthy palms (bottom). (**E**) 20% non-denaturing PAGE assay of extracts from palms showing varying intensities of the brooming symptom (nos. 1-5 and a mixture of their isolates in lane no.6; M denotes CCCVd marker consisting of the commonly detected variants 246, 247, 296 and 297).

Figure 23.1.1

Fig. 23.3.1 Lethal yellowing on an immature palm

23.2.1a

23.2.1b

23.2.2

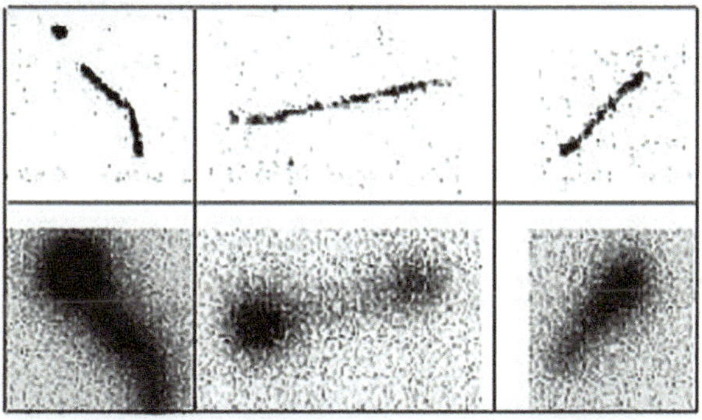

23.2.3

Fig. 23.2.1. Symptoms of CFD on naturally infected MRD hybrids: **a)** late stage with collapse of affected fronds; **b)** necrosis on petioles.

Fig. 23.2.2. Distribution of CFDV DNA in the gut of three adult females of *Myndus taffini.* Insects collected from diseased trees at Saraoutou, Vanuatu, were dissected, their gut removed and stained briefly with aqueous toluidine blue for ease of manipulation. The gut was extended and applied to a nylon membrane (upper panel). After crosslinking, the membrane was hybridized with ^{32}P-labelled CFDV DNA probe and autoradiographed (lower panel). In this trial, 4/17 female and 0/10 male adults showed the distribution of radioactive signal along the isolated gut.

Fig. 23.2.3. Arial view of a variety trial at Saraoutsu Research Station, showing both the gradients of disease incidence from the native forest into the planting, and the variation in susceptibility of different varieties. Note particularly the two rows of tolerant VTT adjacent to the forest – marked by an arrow. The other dense plantings next to the forest are replanted areas.

Figure 24. 1. Symptoms expressed by *Citrus tristeza virus* (CTV), citrus viroids, and citrus blight. A. Hamlin sweet orange tree on sour orange rootstock undergoing decline due to decline strains of CTV. B. Fruit size of Hamlin fruit from the tree illustrated in A, compared to fruit from a non-declining tree, immediately behind the declining tree in A. Depending on the time of year, amount of fruit already set, and length of time of the tree to completely die due to decline strains of CTV, the fruit size may be significantly reduced. C. Stem pitting due to CTV on twigs from a Marsh grapefruit tree showing pits with gumming beneath due to stem pitting strains of CTV. Stem pitting due to CTV can occur on sweet orange also. Fruit size and quality is reduced on trees affected with stem pitting strains of CTV. D. Citrus exocortis induced bark scaling on *Poncirus trifoliata* rootstock, note that the bark scaling occurs on the rootstock only and not on the scion. E. Variation of symptoms due to citrus viroids expressed on the viroid indicator plant Etrog citron clone 861. The plant on the right, labeled "H" is healthy, the rest of the plants were inoculated from field sources containing various mixtures of citrus viroids. F. A Valencia sweet orange tree on rough lemon rootstock showing the symptoms of citrus blight (CB), a disease of unknown etiology which may be graft transmitted using root grafting. The trees affected with CB first develop an off color green appearance, often followed by zinc deficiency-like chlorosis on the southwest quadrant of the tree (Northern hemisphere). Canopy dieback then occurs and the tree declines over a period of several years. Fruit size and yield are reduced, blossom set is delayed 2-3 weeks compared to healthy trees.

24.1A
24.1B
24.1C
24.1D
24.1E
24.1F

Figure 24.2. Symptoms expressed by fastidious prokaryotes causing diseases in citrus. A. Citrus variegated chlorosis (CVC), caused by *Xylella fastidiosa,* causes a general stunting in affected trees, and the tree takes on a yellow or chlorotic appearance. The tree in the foreground is CVC infected while the tree immediately behind is growing normally. B. With CVC, leaf symptoms begin with an interveinal chlorosis which is apparent on both sides of the leaf. Gummy slightly raised lesions are then formed on the underside of the leaf in the areas which are chlorotic. C. CVC reduces fruit size, the fruit have hard rinds, and the fruit ripen early and have a higher than normal sugar content. The Natal sweet orange fruit on the left came from a tree apparently free of CVC, the fruit on the right came from a CVC infected tree. D. Yellow shoot symptom on a Valencia tree infected with an African strain of Huanglongbing (HLB), commonly called citrus greening. The sectored branches showing general chlorosis are common on early symptomatic HLB affected trees. E. Witches' broom of lime (WBDL) from Oman. The witches' brooms start to develop in the inner parts of the tree canopy, then gradually spread and appear throughout the tree's canopy. The WBDL affected trees eventually die. The WBDL in Oman and surrounding countries is caused by *Candidatus* Phytoplasma aurantifolia. F. A 20 year old Swingle citrumelo tree in Jamaica showing witches' broom symptoms. A phytoplasma was identified from samples taken from this tree (R. Lee, unpublished).

24.2A
24.2B
24.2C
24.2D
24.2E
24.2F

Legends to figures:

Fig. 25.1 – Typical symptoms induced by *Tomato yellow leaf curl virus* in tomato cv. Daniela (photo . G. Loebenstein)

Fig. 25.2 – Shoestring leaves induced by *Cucumber mosaic virus* in tomato cv. Fakulta 38

Fig. 25.3 – Typical symptoms induced by *Tobacco mosaic virus* on tomato fruits cv. Hazera 84

Fig. 25.4 - *Tomato apical stunt viroid* induced symptoms on cherry tomatoes.

Figure 26.1. (A) Vein banding, blisters and leaf deformation caused by *Watermelon mosaic virus* in melon. (B) Severe stunting, mosaic and leaf deformation caused by *Zucchini yellow mosaic virus* in zucchini squash. (C) Yellowing of older leaves caused by *Cucurbit aphid borne yellows virus* in melon. (D) Severe yellow mosaic caused by *Watermelon chlorotic stunt virus* in watermelon. (E) Yellowing of older leaves caused by *Cucurbit yellow stunting disorder virus* in melon. (F) Vein clearing caused by *Cucumber vein yellowing virus* in cucumber.

Fig. 27.1 Yellow dwarf mosaic disease of onion. **Fig. 27.2** Mosaic disease of eggplant caused by PVY. **Fig. 27.3** Okra leaf curl disease. **Fig. 27.4** Okra yellow vein mosaic disease. **Fig. 27.5** Mosaic disease of chayote. **Fig. 27.6** Mosaic disease of pea. **Fig. 27.7** Eanation mosaic disease of pea. **Fig. 27. 8** Bobone disease of colocasia. **Fig. 27.9** Symptoms induced by TSWV on sweet pepper. **Fig. 27.10** Leaf curl disease of sweet pepper.

Fig. 28.1

Legends to figures:

Fig. 28.1. Infection by tospoviruses. a) top necrosis induced in tomato; b) infection in sweet pepper; c) infection in tobacco; d) mosaic in early infected groundnut; e) early infection in lettuce (notice the necrosis; f) necrosis in potato tubers; g) local lesions on a petunia leaf disk exposed to viruliferous thrips (left) , feeding damage evoked by non-viruliferous thrips (middle), and an unexposed disk (right).

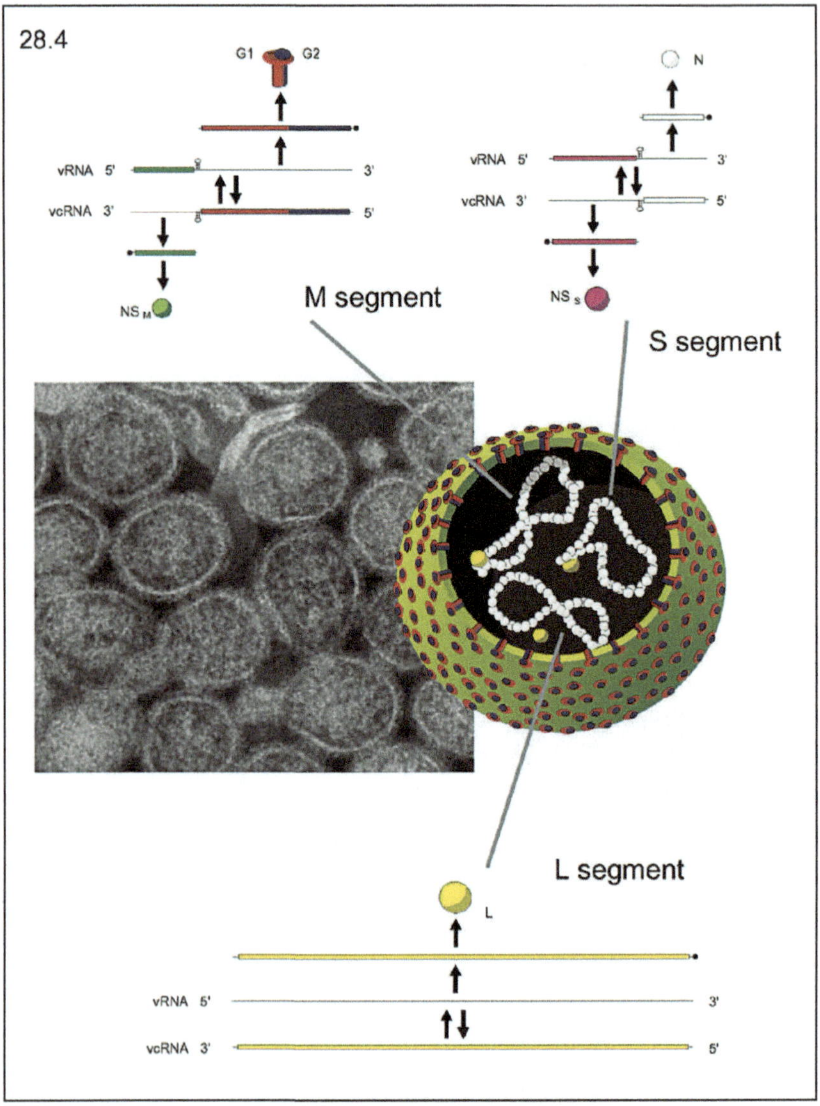

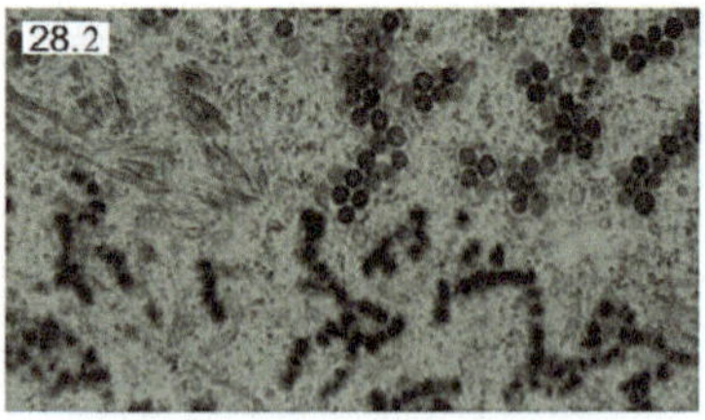

Fig. 28. 4. Electronmicroscopy of *Tomato spotted wilt virus,* its schematic morphology and outline of the genome expression. The virus particles show the lipid membrane with the projections consisting of the G1 and G2 proteins. The core of the schematic particle consists of the three viral nucleocapsids. White: nucleocapsid (N), yellow: RNA-dependent RNA (L) polymerase. The schematic particle is enveloped by two glycoproteins (red =G1, blue is G2. NSs and NSm are the nonstructural proteins. Double arrows indicate replication, where as single arrows indicate transcription and translation. Courtesy of Dr. M. Prins.

Fig. 29.1 Field symptoms of cotton leaf curl disease (CLCuD).
Fig. 29.2 Diagnostic vein thickening symptom of CLCuD. Fig. 29.3 Severe upward leaf curling due to CLCuD. Fig. 29.4 Yellow vein symptom due to CLCuD. Fig. 29.5 Blister mosaic (CLCuD). Fig. 29.6 Cotton anthocyanosis.

Legends to Figures:

Fig.30.1 - Severe symptoms of *Tobacco streak virus* in sunflower from Jalna, Maharasta state, India, infected at early stages of plant development.

Fig. 30.2 - Sunflower plant with severe necrosis symptoms caused by infections of *Tobacco streak virus* in India

Fig. 30.3 - Chlorotic ringspot and mottle symptoms on leaves of sunflower from Zambia, presumably infected with a virus related to sunflower yellow blotch virus.

Fig. 31.1 (b) Cardamom leaf with typical mosaic symptoms caused by *Cardamom mosaic virus*. (a) Healthy cardamom leaf.

1. The Impact of Plant Virus Diseases in Developing Countries

J.M. Thresh
Natural Resources Institute, University of Greenwich, Chatham Maritime,. ME4 4TB, UK

1.1 Introduction

Pests and pathogens affect virtually all crops in tropical, sub-tropical and temperate agro-ecologies, causing losses that are usually substantial and sometimes devastating. Those due to viruses are of particular importance in developing countries that are heavily dependent on agricultural production for food security, employment and export earnings. The overall impact of virus diseases in developing countries is considered here and the chapter is intended as an introduction to subsequent ones on specific crops and diseases.

Several of the diseases considered featured in recent publications (Thresh, 1991; Thurston, 1998; Rybicki and Pietersen, 1999). Earlier, sugarcane mosaic was discussed as an example of a catastrophic virus disease (Klinkowski, 1970) and cassava mosaic disease in Africa was regarded as a serious threat to other regions (Thurston, 1973).

1.2 Developing Countries and the Status of Agriculture

There is no general agreement on the definition of a developing country. For this chapter, FAO statistics on land use, crop production and productivity, human populations and the numbers employed either directly or indirectly in agriculture were used to identify countries which are most heavily dependent on agriculture and where there is relatively little alternative means of employment.

The latest published statistics categorize more than half the total populations of Africa and Asia as agricultural. The comparable proportions in South America and Oceania, and especially in Europe and North/Central America, are much less. Within each region there are considerable differences between countries. For example, the proportion of the economically active employed in agriculture in Africa ranges from 3% in Reunion to 92% in Burkina Faso and it is less than 50% in only 16 of the 52 countries. There is even greater variation in Asia where the proportion ranges from only *c.* 1% in Brunei to 93% in Nepal and is less than 50% in 36 of the 49 countries. The proportion is less than 50% in all other countries except Haiti in Central America and two countries of Oceania (Papua New Guinea and Solomon Islands).

These figures justify an emphasis here on countries of Africa and Asia, where more than 50% of populations depend on agriculture, including large areas of West, Central, Southern and East Africa and the most populous countries of East and South-East Asia, including China and India. Another notable feature is that all the African countries and many in Asia are within the tropics. The important exceptions include China, Bangladesh and parts of India. However, the 50% criterion is arbitrary and somewhat misleading in that it excludes Nigeria (34%), Philippines (40%), Indonesia (49%) and Pakistan (47%). These countries have large urban conurbations and some degree of industrialization, as in Nigeria and Indonesia, which both produce oil. They also have vast rural areas where agriculture is paramount, as in many South and Central American countries.

An important feature of developing countries is the continuing high rate of human population growth compared with developed countries, where overall there has been a marked decline from 1.28% in 1955 to 0.51 in 2000 (Maskey, 2001). In developing countries the rate of growth was 2.00% in 1955, reached a maximum of 2.35 in 1965 and then declined slowly to 2.10 in 1985 and 1.84 in 2000. An important feature of these demographic trends is that the proportion of young people in developing countries is almost double the proportion elsewhere. Consequently, *actual* populations will continue to increase for several years in developing countries even though the *rate* of growth is declining. This creates an increasing demand for food and raw materials at a time when there is concern for the environment, the loss of biodiversity and the need to decrease rates of deforestation, soil erosion and the overall decline in soil fertility.

1.3 Agriculture in Developing Countries

It is difficult to generalize on agriculture in developing countries because of the big differences in the climates and altitudes encountered and in the crops grown. Nevertheless, there are important general features:

- Agriculture is mainly concerned with the production of staple food crops for farming families, and for sale in rural markets or urban centres and sometimes for export.
- Many other crops are grown for export, mainly to industrialized countries. There are powerful incentives for farmers to produce such crops as an important source of income. There is also strong government encouragement to decrease imports of food and raw materials and to enhance export earnings.
- Much of the production of both subsistence and export crops is by small-scale producers who farm mainly rain-fed areas of land, using manual labour or draught animals. There is only limited use of fertilizers, herbicides, pesticides, irrigation and mechanization, especially in Africa. Moreover, there is inadequate access to or use of technical information and advice because of the illiteracy of many farmers and the lack of effective extension services.
- In many developing countries the traditional methods of small-scale subsistence agriculture are being supplemented or replaced by 'industrial' methods of production, as practised on large commercial farms or plantations, usually under strong central management and with access to labour, machinery and other inputs. Sugarcane, banana and cotton are notable examples. A similar approach is now being adopted with the main food crops including maize, rice, groundnut and cassava. There is an expanding market for vegetables, fruits and ornamentals for export to the industrialized countries of Europe, Asia and North America and to affluent oil-producing countries of the Middle East.
- Agricultural productivity in developing countries is generally low, unreliable and at times severely affected by drought or floods leading to food shortages. Initially, low productivity was not usually of great concern, but the situation is changing as populations expand and an increasing proportion live in urban areas and is not involved in agriculture. Furthermore, continued deforestation is leading to soil erosion and loss of soil fertility, while fresh land is less readily available and HIV/AIDS decrease the availability and productivity of the rural workforce.

These trends, and the generally low prices received for many export crops have had serious consequences for developing countries. It is particularly important to increase food production and to do so in a sustainable manner without damaging human health or the environment.

1.4 Plant Virology in Developing Countries

Plant virology began at the end of the 19th century and many of the early studies on virus diseases were by plant pathologists employed by Colonial authorities in Africa or Asia, where considerable attention was given to both export and subsistence crops. After the Second World War (1939-1945) electron microscopy, ultra-centrifugation, serology, electrophoresis, chromatography and other techniques became available in laboratories of developed countries, facilitating the detection, isolation, characterization and control of viruses. The increased number of plant virologists contributed to enhanced food production and also to research on the structure and physico-chemical properties of viruses.

These developments did not occur in less developed regions, and expatriate staff were withdrawn from many of the colonial countries when they became independent. The consequences were even greater for plant virology than for other agricultural sciences because of the need for highly specialized training and expensive equipment, which were seldom available. There were also problems due to the unreliability of water and power supplies and difficulty in procuring and maintaining laboratory equipment.

The International Institute of Tropical Agriculture, Ibadan, Nigeria, and other research centres in Africa and elsewhere have in recent decadces made substantial contributions to crop improvement (Thurston, 1977; Bos, 1992). Laboratories and organizations in industrialized countries, have also trained and advised staff from developing countries and seconded virologists on long- or short-term assignments overseas. Much of the support has been through Government Aid Programmes, but there have also been substantial contributions from International Charitable Organizations.

Ideally, developing countries should become largely self-reliant and play a full part in the global scientific effort, as shown by China, India and some other developing countries. However, many African countries have made only limited progress, due in part to political instability, civil unrest and conflict. The unstable state of national economies can also affect budgets for research and development and necessitate staff reductions. These can undermine attempts to increase food production and earnings from export crops, as emphasised in a recent review of the virus disease situation in Latin America (Morales and Anderson, 2001).

The increasing reluctance of many aid donors to allocate funds for agricultural research is also a constraint. This policy seems to be based on the view that past research funding has not been used effectively and that the main requirement is for more effective means of 'technology transfer'.

Undoubtedly, extension services are generally inadequate and they operate under severe constraints. Nevertheless, this does not justify diverting the research effort to technology transfer at a time when there are numerous unsolved problems, not least those posed by many of the diseases considered in this and subsequent chapters.

1.5 Viruses and Virus Diseases in Developing Countries

Despite the difficulties and the frequent lack of continuity, there has been a substantial research effort on the viruses and virus diseases of crop plants in developing countries. Some of the most important and widely studied of these viruses are listed 'operationally' in Table 1.1. They form a very diverse assemblage from different taxonomic groups and are grouped according to their means of spread by animal or fungal vectors, or by contact. Some of the viruses are also seed-borne or disseminated in vegetative propagules, and hence are readily spread within and between countries.

An additional feature of the viruses listed is that they differ considerably in host range and there are big differences between crops in their longevity, growth habit and means of propagation. Crops that are seriously affected by viruses range from herbaceous annuals to long-lived woody perennials and include crops grown from seed and those usually propagated vegetatively. This emphasises the versatility of viruses and their ability to adapt to a wide range of hosts and circumstances.

Viruses of short-lived herbaceous annuals grown from seed are only likely to cause problems if they spread rapidly to infect a large proportion of the crop within the few weeks or months of the most vulnerable period of growth. In these circumstances initial sources of inoculum due to seed-borne infection, weed/wild hosts or older crops nearby are of particular epidemiological importance in initiating outbreaks. Rapid spread is less crucial for long-lived perennials because spread can occur continuously or discontinuously over a long period and so time is an unimportant epidemiological parameter (Vanderplank, 1963). Crops that are propagated vegetatively are particularly prone to damage by viruses, because infection tends to build up in successive cycles of propagation. This explains why multiple infections with two or more viruses are a feature of many crops, including yams, sweet potato, *Solanum* potato and citrus. As a result of the degeneration farmers tend to discard severely affected stocks and retain those relatively unaffected. This creates a strong selection pressure in favour of host tolerance and relatively avirulent strains of virus.

Mutual adaptation also occurs between viruses and their hosts in indigenous pathosystems that have co-evolved over prolonged periods. Consequently, the damage sustained tends to be relatively slight compared with so-called 'new encounter' diseases (*sensu* Buddenhagen, 1977)). These occur when either the host or the pathogen is introduced to a completely new area where the pathogen or host is not already established. Maize streak, cassava mosaic, groundnut rosette and several other important African diseases have arisen in this way (Thresh, 1991).

Much additional information on specific crops and their viruses and virus diseases is presented in subsequent chapters. This chapter draws on experience with a wide range of viruses to illustrate the various ways in which they influence crop productivity and rural livelihoods, and considers their effects and overall impact on national economies.

1.6 Crop Loss Assessment and Disease Surveys

There are numerous reports of the losses caused by plant virus diseases and their impact on farmers, rural communities, national economies and international trade. However, the information from developing countries is very incomplete and mainly relates to specific countries or diseases and often to particular periods when losses have been so exceptional as to necessitate attention. Moreover, much of the information is qualitative rather than quantitative and it is seldom possible to make valid estimates of the losses sustained and of the overall economic impact in a locality, country or region. These deficiencies are not easily resolved because for each important disease there is need to assess the losses sustained by the main varieties grown and for comprehensive field surveys of disease incidence and severity.

There have been many observations on crop loss in farm plantings subject to natural infection, or in experimental trials in which plants are infected from the outset or at different stages of crop growth. From observations on many different pathosystems, including those listed in Table 1.2, it is evident that many factors influence the yield losses caused by viruses. Some of the most important are:

- *The sensitivity/tolerance of the host variety affected.* Cassava mosaic, maize streak and banana streak diseases, can kill or severely damage some varieties and yet have little or no effect on others (Table1.2);
- *The virulence of the virus strain involved.* Strains can range from being very damaging to almost benign, as demonstrated in early studies on cacao swollen shoot disease (Table 1.2) and recently with cassava mosaic disease (Owor, 2003);
- *Soil fertility and the overall conditions under which crops are grown,* including such factors as spacing, harvest date and any thinning,

pruning, irrigation or fertilizer treatments adopted. For example, the effects of cassava mosaic disease were much less at a fertile site than where cassava had been grown repeatedly for many years (Spittel and van Huis, 2000);

- *The presence or absence of other viruses, or other pathogens and pests.* These can lead to complex interactions (Bos, 1982);
- *The proportion and distribution of diseased plants within the affected stand.* This can influence the extent to which healthy plants compensate for the impaired growth of their diseased neighbours. The principle has been discussed in general terms (Hughes, 1988) and competition and compensation have been demonstrated within mixed stands of healthy and mosaic-affected cassava (Otim-Nape *et al.* 1997).
- *The stage of growth at which plants are infected.* Plants infected from the outset or at an early stage of crop growth are usually more severely affected than those infected later, as demonstrated in studies on maize streak and cassava mosaic diseases (Bosque-Perez *et al.* 1998; Fargette *et al.* 1988).
- *The generally close and negative relationship between crop loss and disease severity.*

Clearly, it is not easy to make detailed assessments of all these and other factors influencing the extent of the yield loss due to each of the viruses occurring in different crops and agro-ecologies. This explains why the information available is totally inadequate and why it is difficult to interpret the data obtained from surveys of disease in farmers' plantings. The problems are apparent from the data presented in Table 1.2 showing that for each pathosystem the losses sustained depend on the sensitivity of the host variety and the virulence of the virus strain considered.

1.6.1 Surveys of disease incidence and severity

Comprehensive surveys of disease incidence and severity are required in any attempt to assess the losses caused by virus diseases. However, they require substantial staff time, funds and resources if they are to be done on a sufficiently large scale to provide results that are truly representative (Barnett, 1986). The full magnitude of the task is apparent from the FAO statistics on the vast areas of the main crops grown.

Such considerations explain why so few data are available for any crop, even in developed countries (Teng and Shane, 1984; Waterworth and Hadidi, 1998). However, there are exceptions and in Ghana, there has been massive expenditure on surveys of the prevalence of cacao swollen shoot disease (Thresh and Owusu, 1986). The commitment was justified due to the importance of cacao to producers, traders and processors, and as a crucial source of export earnings (1.7.2).

In recent years, considerable attention has been given to cassava mosaic disease in Uganda and elsewhere in Africa, following the onset of a particularly damaging epidemic. This was first reported in parts of Uganda in 1988 and it then spread to other areas and into neighbouring countries (Otim-Nape and Thresh, 1998; Otim-Nape *et al.* 2000; Legg, 1999). Reconnaissance surveys of representative plantings have been made in Uganda and elsewhere to monitor the progress of the epidemic and to identify the areas at risk. There have also been surveys in eastern and southern Africa to determine the importance of both cassava mosaic and cassava brown streak diseases and in West Africa as part of more comprehensive assessments of cassava pests and diseases (Thresh *et al.* 1997; Calvert and Thresh, 2002). Although the number of plantings surveyed was small in relation to the vast areas of cassava grown, important evidence was obtained on the differences in the incidence and severity of cassava mosaic disease within and between the thirteen countries assessed and of the sometimes big changes that have occurred in the status of the disease and in the varieties being grown.

Comparable studies on virus diseases of annual crops grown from seed are few, due in part to the need to consider the most appropriate stage of growth when plantings should be assessed and the often big differences between years in the incidence of infection. This is apparent from observations on rice yellow mottle and several other important diseases in Africa and on rice tungro disease in South-East Asia. Some of the most detailed observations have been on tungro disease in Bali and elsewhere in Indonesia as part of regular assessments of rice pests and diseases. Similar pest and disease surveys are operated in Malaysia and India, but seldom elsewhere. However, there are routine scouting operations to monitor the incidence and distribution of pests and diseases in large commercial enterprises growing banana and other export crops (Thresh, 1988; Smith *et al.* 1998). The information is required to determine the need for control measures and to influence cropping practices and planting policy, but much of the information obtained is not readily available.

An important consequence of the few data available from comprehensive surveys is that there is undue reliance on general impressions and on limited observations at sites or in circumstances that are not necessarily representative. Serious outbreaks and severe losses are notable and feature prominently in the plant pathology literature, whereas less extreme damage does not.

The problems that arise from this somewhat biased approach are apparent from experience with rice tungro disease in South-East Asia. There are many reports in the literature of severe outbreaks at different times and places, as discussed later (1.7.1). However, these reports should be considered in the context of the large areas of rice grown. These greatly exceed the tungro-affected area and there is a low overall probability of serious losses occurring at any one time or place.

Another reason for over-estimating the importance of virus diseases is the common and ambiguous use of the term *'area affected'*. This usually refers to the total area in which a disease occurs, irrespective of incidence, whereas it implies that virtually all plants in the area are affected. Another misleading usage is to quantify *'crop loss'* in a country or region based on the results of small-scale comparisons of healthy and diseased plants in experimental plots or farmers' fields. Disease incidence is seldom total and so the trial results usually exaggerate the losses sustained and do not take account of compensatory growth by any remaining healthy plants.

1.7 The Impact of Virus Diseases

Lack of information is a severe impediment to any comprehensive assessment of the impact of virus diseases in developing countries and in considering any possible long-term trends in disease incidence and severity. Nevertheless, attempts are justified so as to assess the relative importance of different diseases, to set priorities for research and extension activities and to evaluate the implications and success or otherwise of the cropping systems and control measures practised. Different approaches have been adopted, but all depend heavily on personal experience, on the limited literature available and on dialogue with virologists and other crop protection specialists. This was the approach adopted by Herdt (1991) in evaluating the losses due to pests and diseases of rice and by Geddes (1990, 1992) and Geddes and Iles (1991) in assessing the relative importance of pests and diseases in Africa, South Asia and Indonesia.

1.7.1 Food Production and Food Security

A general feature of viruses is that they decrease crop growth and yield. The extent to which they do so is very variable and influenced by many factors, as discussed earlier (1.6). The implications for rural communities depend on economic and other factors, including:

- The extent to which the farmers and their dependents rely on the crop as a staple food or source of income;
- The availability of alternative staple food crops and the ability of farming families to grow these or purchase them in local markets;
- The extent of any food reserves from previous crops and the length of the 'hungry period' before new produce can be expected;

- The frequency of damaging epidemics and the time taken for rural communities to recover.

Rice is the staple food for an estimated 2.7 billion people in Asia. The ability to feed such large populations is largely due to the 'green revolution' in South-East Asia that has occurred since the 1960s. The introduction of high-yielding, early-maturing, day-length insensitive, fertilizer-responsive rice varieties, improved methods of irrigation and water management greatly increased production. However, yields have been affected by pests and diseases that first appeared or became increasingly important with the adoption of intensified methods of cultivation (Thresh, 1989; Bos, 1992).

Rice tungro virus disease has caused particular problems in many countries of South-East Asia. It was first reported in 1963 (Rivera and Ou, 1965), although it is known to have had a much longer history in the region. Severe outbreaks occurred in the Philippines, India, Bangladesh, Thailand and Malaysia in the 1960s and 1970s and they were associated with the introduction of high-yielding varieties and large increases in populations of the main leafhopper vector *(Nephotettix virescens)* (Sogawa, 1976). A feature of tungro disease is its 'explosive' spread and early infection leads to almost total crop loss and may necessitate imports of food from elsewhere. Moreover, small-scale farmers are very severely affected, especially those dependent on credit to buy seed, fertilizer or insecticide. This explains why rice tungro disease has such a high political 'profile' and why considerable attention has been given to developing resistant varieties and other approaches to control (Thresh, 1989).

Progress has been made in developing vector-resistant and more recently virus-resistant varieties (Azzam and Chancellor, 2002). There has also been much use of insecticides to control the leafhopper vector, with deleterious effects on natural enemies and human health, as discussed later (1.7.4.). Moreover, vector resistance has not been durable and additional severe outbreaks of tungro occurred in several Asian countries in the 1980s and 1990s (Azzam and Chancellor, 2002). Thus, tungro has remained a problem in endemic 'hot spot' areas from which it has the potential to spread to other locations and develop into the large-scale epidemics that occur sporadically and unpredictably.

Attempts to meet the increasing demand for rice in Africa and to decrease the present heavy dependence on imports have been affected by rice yellow mottle disease, which was first reported and studied in Kenya (Bakker, 1974). It has since been recorded in many other countries of East and West Africa. Some of the worst epidemics have been in the Republic of Niger where large areas were devastated in 1984 and the overall incidence of infection exceeded 25% (Reckhaus and Adomou, 1986; John *et al.* 1984).

Yellow mottle disease occurred in all four agro-ecologies assessed in a survey of thirteen West African countries between 1993 and 1994

(Awoderu, 1991). The overall incidence was 35% at lowland sites, which suggests that losses are considerable in what is a very labour-intensive crop. Moreover, there are costs to human health, as in Madagascar and other areas where insecticides are used to control the beetle vectors (Reckhaus and Andriamasintaheno, 1997). Some of the areas worst affected by yellow mottle are abandoned and new land is cleared of natural forest or other vegetation and cultivated, causing considerable environmental damage.

Viruses of cassava cause very different problems from those of rice because cassava is propagated vegetatively from hardwood stem cuttings. Farmers in Africa frequently plant infected cuttings which explains why cassava mosaic and cassava brown streak diseases have been disseminated widely and may be prevalent even in areas where there is only limited spread by vectors (Calvert and Thresh, 2002). This has serious consequences because cassava is the main staple food in many areas of Africa and supplies an estimated 200 million people. In sensitive varieties cassava brown streak disease causes a necrosis of the tuberous roots that becomes more severe as crops mature. Consequently, many farmers harvest prematurely before much necrosis has occurred and so yields are affected. Moreover, roots are not available when most required to provide food security. The full significance of these effects has become apparent only recently in coastal areas of northern Mozambique, although chronic food insecurity is a long-standing problem in what is a very impoverished region (Hillocks *et al.* 2002).

Cassava mosaic disease is more widespread than brown streak and occurs in all the cassava-growing areas of Africa, and also in India and Sri Lanka. It has long been regarded as the most important disease of cassava in Africa, but there are big differences between regions in the magnitude of the problem (Thresh *et al.* 1997). Effective means of disease control would increase production, or release land for other crops, or allow longer periods of fallow to restore soil fertility. This is apparent from studies in Côte d'Ivoire where it was estimated that mosaic decreased national yields by 500,000 tonnes compared with actual production at the time of 800,000 tonnes (Fargette *et al.* 1988). Similar losses occur in many other parts of Africa, indicating the benefits to be gained from effective disease control.

The most serious losses due to cassava mosaic disease have been during occasional severe epidemics. These have occurred at different times and places in Africa and most recently in Uganda and neighbouring countries. The first of the recent reports was from a localized area of Uganda in 1988. The epidemic then spread to many other areas and it was referred to as a pandemic when it progressed into western Kenya, northwestern Tanzania and Rwanda (Otim-Nape and Thresh, 1998; Legg, 1999; Otim-Nape *et al.* 2000). The pandemic was characterized by unusually large population densities of the whitefly vector (*Bemisia tabaci*) and rapid spread of mosaic in areas where the disease had hitherto been largely benign. The most vulnerable varieties, including some that were being grown almost exclusively over large areas, were so severely damaged that they were abandoned and farmers resorted to sweet potato or other crops. Serious food shortages ensued and there were famine-related deaths and also reports of young village girls being exchanged for cassava or forced into early marriage to provide dowries that were used to buy food.

At the height of the epidemic in Uganda it was estimated that the equivalent of four administrative districts totalling *c.* 60,000 hectares of cassava were out of production. These could have been expected to produce *c.* 600,000 tonnes of cassava worth *c.* US$60 million at a conservative valuation of US$100 per tonne. Comparable losses are now occurring in adjoining countries that were affected later than Uganda (Legg, 1999). There has been a need for massive intervention to alleviate the food crisis through food distribution programmes and the introduction of mosaic-resistant varieties to replaces the susceptible ones grown previously.

Various attempts have been made to evaluate the overall losses caused by cassava mosaic disease in Africa. The most recent was based on the results of surveys of mosaic incidence in eight African countries in the 1990s (Thresh *et al.* 1997). It was assumed that 50-60% of all plants are infected and that they sustain losses of 30-40%. This suggests losses of 15-24 %, which is equivalent to 12-23 million metric tonnes worth US$1200-2300 million; actual production in 1995 was 73 million tonnes. Clearly, this loss in value is unrealistic because markets could not readily absorb such a big increase in production without undermining prices. Nevertheless, the estimates indicate the enormous magnitude of the mosaic problem and are consistent with the view that pests and diseases have a disproportionate impact on poor households (Lenné, 2000). This view is also supported by experience with maize streak, groundnut rosette, rice tungro and other diseases considered elsewhere in this volume.

1.7.2. Export Crops

Crops grown for export are of crucial importance in the economies of many developing countries, as a means of employment, rural income and source of 'hard' currency. The losses caused by virus diseases and the need to adopt resistant varieties or other control measures influence production costs and overall productivity.

This was apparent from early experience with sugarcane Fiji disease in the Pacific Region and with sugarcane mosaic disease in several continents. Fiji disease was first reported on the islands in 1886 and by 1906 the disease was destroying thousands of hectares of susceptible varieties and the whole future of sugar production was threatened until Fiji-resistant varieties were developed and grown widely (Mungomery and Bell, 1933). There have since been other epidemics in Fiji in the 1950s and 1960s and the disease has also caused problems elsewhere in the Pacific Region and in Madagascar.

The impact of sugarcane mosaic disease on sugar production and sugarcane breeding programmes has been even greater than that of Fiji disease. Mosaic was first described in Java, Indonesia and in the early decades of the 20th century it was spread widely to other regions in vegetative material. This led to damaging outbreaks in almost all sugarcane-growing areas of the tropics and sub-tropics, including Argentina, Brazil, Egypt, Hawaii, Louisiana and Puerto Rico (Summers *et al.* 1948). In some areas the devastation led to severe financial losses for commercial companies and their employees. The crisis was only overcome through the development of mosaic-resistant varieties. However, there have been problems due to the occurrence of resistance-breaking strains of *Sugar cane mosaic virus* and a continuing need to develop new varieties resistant to the most prevalent virus strains.

Cacao swollen shoot disease has also had profound effects on export crop production since it was reported in what is now Ghana in the 1930s. Ghana was at that time the most important cacao-producing country in the world and export earnings accounted for its relative prosperity compared with other African countries. No attempt could be made to control swollen shoot disease until after the Second World War (1939-1945). The detailed surveys mentioned earlier (1.6.1.) then showed the disease to be present in all the main cacao-growing areas of West Africa. It was particularly prevalent in the Eastern Region of Ghana which previously had accounted for a large proportion of national production. Millions of trees were killed and millions more were removed in attempts to bring the disease under control (Thresh, 1988; Thresh and Owusu, 1986). In Eastern Region alone production fell from 128,000 tonnes in 1936-37 to only 48,000 in 1951-52 and annual losses at the time were put at 50,000-120,000 tonnes worth £12-30 million. There has been a major diversion of funds and resources for 'cutting out' and replanting operations in attempts to maintain production and yet only limited control has been achieved, mainly in the least-affected areas. Consequently, the main production of cacao in Ghana is now outside the Eastern Region. The competitive position of Ghana and Nigeria was affected and the leading producers are now Brazil, where swollen shoot is absent, and Côte d'Ivoire where it is much less damaging than in Ghana or Nigeria. These developments have had a particularly drastic effect on the Ghanaian economy and rural work force.

Citrus differs from cacao as grown in Africa in that it is propagated vegetatively. This facilitates the dissemination of citrus viruses and explains why *Citrus tristeza virus* occurs in almost all countries where citrus is grown. The virus has had a profound impact on production in many areas of the tropics and sub-tropics. This became apparent from experience in Argentina, Brazil and Gold Coast (now Ghana) in the 1930s when tristeza caused widespread die-back and death of trees. An estimated 30 million trees were destroyed by tristeza in South America and severe losses also occurred elsewhere (Bar-Joseph *et al.* 1981). This necessitated a switch from the tristeza-sensitive sour orange rootstock *(Citrus aurantium)* that had been widely used previously for its resistance to *Phytophthora* root rot.

Tristeza disease has continued to cause problems in Argentina, Colombia, Venezuela and elsewhere in South America. This has been associated with the dispersal of the brown citrus aphid *(Toxoptera citricidus)* to new areas. The aphid is the most efficient vector of tristeza virus and serious losses occurred soon after its first appearance in Argentina (Ducharme *et al.* 1951) and Venezuela (Mendt *et al.* 1984), where 6 million of an estimated 6.5 million trees were killed in the 1980s (FUSAGRI, 1987). *T. citricidus* continues to spread and 180 million trees on sensitive sour orange rootstocks were considered to be at risk from tristeza in the Caribbean basin (Yokomi *et al.* 1994). The countries threatened include Mexico where more than 400,000 hectares of citrus are grown and preparations are being made to combat what is regarded as an inevitable epidemic (Rocha-Peña *et al.* 1995).

Cotton is grown in many developing countries and it is an important source of income for farmers and those engaged in transporting and marketing the crop. There is also a big labour requirement for ginning, textile mills and factories. Leafcurl disease caused by whitefly-borne viruses has long been known as a serious constraint to production, as evident from early experience in Sudan and elsewhere in Africa (Tarr, 1951).

Leafcurl has become increasingly widespread in recent decades and the problems have been particularly severe in Mexico and the Pacific lowlands of Guatemala, El Salvador and Nicaragua (Morales and Anderson, 2001). For example, cotton production was virtually halted in Guatemala in 1964 where 7,000 hectares were destroyed by virus disease in 1975. Epidemics also severely affected production in Nicaragua between 1970 and 1973. Cotton production in Central America then began a downward trend, which was accentuated by the high cost of pest and disease control operations including frequent applications of insecticides. By 1990 the cotton area had declined to only 10% of the total in 1975, with serious consequences for rural incomes and livelihoods.

More recently leafcurl disease has severely affected cotton production in Pakistan. This followed a big increase in cotton plantings in the early 1980s and by the 1990s cotton contributed *c.* 60% of all foreign exchange earnings (Briddon and Markham, 2000). The main area of production was in the Punjab, which alone grew 2.5 million hectares. Leafcurl disease has been known in the region since the late 1960s, but it attracted little attention until 1988. The disease then became increasingly important and ultimately affected the entire cotton-growing area of the Punjab. The estimated losses have been US$5 billion in the five years between 1992 and 1997.

Whitefly-borne viruses have also caused increasing problems in tomato and several other vegetable crops in Asia, the Mediterranean region and the Americas. Losses in tomato have been particularly severe in the Americas, as exemplified by data from the Dominican Republic between 1988-89 and 1994-95 (Polston and Anderson, 2001). Damage of up to 95% was reported, based on loss of crop quality and quantity estimated to be worth US$10 million in 1988 and US$50 million between 1989 and 1995.

Similar problems have occurred elsewhere in the region from Mexico in the north to Sao Paulo, Brazil, in the south. The implications have been profound as crop production in some areas has been halted or curtailed at least for a time in attempts to break the cycle of infection. This occurred whilst governments sought to boost sales of non-traditional export crops to decrease the massive external debts incurred following the 1970s global oil crisis (Morales and Anderson, 2001).

1.7.3 Crop Improvement Programmes

There are many examples of the ways in which plant viruses influence crop improvement programmes. The impact of viruses has been particularly important in attempts to improve the productivity of sweet potato in the many parts of Africa where sweet potato virus disease (SPVD) is a serious threat. The disease is usually contained by the cultivation of resistant landraces, but many of these are inherently lower yielding than susceptible genotypes. Attempts to enhance productivity by introducing improved varieties from outside Africa have been largely unsuccessful because of their vulnerability to SPVD. This has led to the conclusion that one of the main deleterious effects of SPVD is in restricting the use of the improved high-yielding genotypes (Aritua *et al.* 1998).

There is a contrasting situation with cacao in West Africa, where in the early decades of the 20th century, production was based almost entirely on trees of the Amelonado variety introduced from South America. The variety was so severely affected by cacao swollen shoot disease in the 1930s and early 1940s that further introductions were made of a wide range of genotypes from Trinidad in an effort to identify virus-resistant material (Posnette, 1951). The 1944 introductions included some that proved to be partially resistant or tolerant of infection. They also led to the recognition of hybrid vigour that was achieved by inter-crossing individuals of different Amazonian cacao populations being grown in Trinidad.

The Trinidad introductions and their progeny have profoundly affected cacao breeding programmes and cacao production in Ghana and elsewhere in West Africa. The parents have also been transported to other cacao-growing regions and there is now extensive use of vigorous, early-bearing hybrids, even in countries where swollen shoot does not occur. An additional benefit arising from the damage caused by the disease is that it led to a recognition of the need for comprehensive research and development programmes on all aspects of cacao production (Thresh, 1991).

Banana improvement programmes have been adversely affected because of the need to restrict the movement of germplasm within and between regions to decrease the risk of disseminating viruses . Initially *Banana bunchy top virus* was the main concern because it has a limited distribution in Africa and it has not been reported in the Americas. Consequently, there are restrictions and delays on the movement of vegetative material into these regions from elsewhere and all consignments must pass through stringent quarantine procedures (Diekmann and Putter, 1996). *Banana streak virus* has created additional problems because it is extremely variable and difficult to detect (Lockhart and Olszewski, 1993; Geering *et al.* 2000). Moreover, parts of the viral DNA can be incorporated within the host genome (Ndwora *et al.* 1999). In this form the virus does not cause symptoms, but it can be reactivated to do so in tissue culture or under environmental conditions as yet not fully elucidated. This has implications for the movement of germplasm in tissue culture,which hitherto had been regarded as a safe and effective procedure.

Cassava provides another example of the way in which viruses have influenced crop improvement projects. Throughout the 20th century there was only limited interaction and exchange of material between cassava breeding programmes in Africa and those in Asia and the Americas. This is because the most important viruses of cassava are largely restricted to only one region and they are absent or unimportant elsewhere (Calvert and Thresh, 2002). Consequently, there has been a reluctance to obtain germplasm from Africa because of the risk of introducing African cassava mosaic geminiviruses (CMGs) or *Cassava brown streak virus*. Asian CMGs and some of the New World viruses of cassava are also a threat to other regions. Moreover, South American material is extremely vulnerable to CMGs in Africa and cannot be utilized and exploited unless first inter-crossed with African genotypes of cassava to incorporate at least some degree of virus resistance.

Some of these problems can be overcome by adopting stringent quarantine controls and also by introducing seed rather than vegetative material. This is now being done to ensure that African sources of resistance to CMGs are available in South/Central America for use in resistance breeding programmes, should the need arise. There have also been extensive introductions of seed from South America for evaluation in Nigeria (Porto *et al.* 1994). They include hybrids between South American and African genotypes. Nevertheless, progress in all regions has been hampered by the limited range of germplasm available due to the restraints imposed by viruses.

1.7.4 Pesticide Use

Many important viruses of tropical crops are transmitted by arthropod vectors (Brunt *et al.* 1990; Table 1.1). This raises the possibility of effective virus control by using pesticides. In Africa, their use is limited because only few farmers can afford to purchase chemicals and equipment (Abate *et al.* 2000; Thresh, 2003). Nevertheless, insecticides are used widely to control direct pests including whitefly vectors on cotton and tobacco in several African countries. Insecticides are also being used increasingly on rice, on vegetable crops around major African cities and on horticultural or flower crops for export.

The problems that can arise through the use and mis-use of insecticides are exemplified by experience with cotton in the massive Gezira irrigation scheme of the Sudan which accounts for a large proportion of the country's export earnings. Insecticides have been used widely to control direct pests and also the whitefly vector of cotton leafcurl disease *(Bemisia tabaci).* However, insecticide-resistant races of *B. tabaci* developed, populations of natural enemies were destroyed and there was a resurgence of pest populations (Castle, 1999). This led to the adoption of other insecticides and an increase in the rates and frequency of application in attempts to restore control. Production costs increased and there were concerns over the environmental implications and for the health of operators. Similar problems could arise elsewhere with the increased use of insecticides on food crops, creating hazards to the health of operators, consumers and wildlife.

In Central/South America, the Caribbean, India and many other Asian countries, insecticides have been used widely and repeatedly to control the whitefly vector *(B. tabaci)* of geminiviruses causing severe epidemics of cotton leafcurl, tomato leafcurl and other diseases of horticultural crops (Polston and Anderson, 1997; Morales and Anderson, 2001). There are many reports of insecticide-resistance and of resurgences in whitefly populations due to the destruction of natural enemies of the vector.

Highly toxic insecticides have also been used extensively to control the leafhopper vectors of rice tungro disease in South-East Asia. Applications, as granules or sprays, to seed beds or transplanted/direct-seeded crops are made prophylactically, or when symptoms of tungro disease or its leafhopper vectors first appear, or only when they exceed a critical threshold. In some countries, incentives and subsidies or loans were given to farmers by governments in attempts to boost the use of insecticides and help achieve self-sufficiency in rice production. It then became apparent that insecticides posed a serious hazard to farm workers (Loevinsohn, 1987) and destroyed natural enemies, resulting in damaging outbreaks of the rice brown planthopper *(Nilaparvata lugens)* and the viruses it transmits (Kenmore *et al.* 1984). These developments led to a reassessment of the role of insecticides in rice production and to the development and promotion of a more rational Integrated Pest Management (IPM) approach (Matteson, 2000.). Considerable efforts have been made to disseminate the IPM 'message' through mass programmes with farmers. However, a major constraint to adoption has been the potential or actual threat posed by tungro disease which can lead to the widespread application of insecticides where they would not otherwise be used.

1.7.5 Political Considerations

Rural populations form a major proportion of the electorate in developing countries. Such communities are so heavily dependent on crop production for their food and well being that any factors that undermine productivity can have serious political consequences. This became apparent in the British Colonial Territory of Gold Coast (now Ghana) in the late 1940s when severe forms of cacao swollen shoot disease caused serious concern (1.7.2). This became even greater when the colonial authorities enforced a strict eradication policy that required the destruction of all diseased cacao trees and wherever possible the symptomless 'contact' trees alongside (Posnette, 1943; Thresh, 1988). These measures were extremely unpopular with farmers and led to serious opposition, culminating in disputes, riots and fighting. The eradication programme was suspended whilst an international panel of eminent scientists was commissioned to review the overall approach to control and to advise on future policy.

The farmers were supported by one of the main political parties led by Kwame Nkrumah who accused the authorities of attempting to destroy farming livelihoods and divert production elsewhere. This was a major factor contributing to the emergence of Nkrumah and the national independence movement. In 1957 Gold Coast became the first colonial country to gain independence and it was followed later by many other countries in Africa and elsewhere. Swollen shoot disease was not the direct cause of Ghanaian independence, but it undoubtedly contributed to the manner and speed with which this was achieved.

Political issues also became apparent during the pandemic of cassava mosaic disease in Uganda, (1.7.1). Farmers in the most severely affected areas made strong representations to local and national government authorities demanding an immediate remedy. Moreover, there were even allegations that a particularly damaging strain of virus had been introduced deliberately as a form of biological warfare that would undermine political stability and food security! When mosaic-resistant varieties of cassava became available they were initially in very short supply and there was an insatiable demand for stem cuttings (Otim-Nape *et al.* 2000). Those responsible for the multiplication and distribution programme had to withstand considerable pressure from politicians anxious to control distribution and thereby influence the electorate. Other problems arose when there were unsubstantiated claims culminating in press reports that some of the officially released mosaic-resistant varieties had caused poisoning and deaths due to high concentrations of toxic cyanogens in the roots. Similar problems may arise in northern Mozambique where cassava brown streak currently causes particularly serious problems (1.7.1).

There are many other examples of government intervention in attempts to avoid the losses due to virus diseases. They include quarantine controls and restrictions on the movement of planting material into or within countries or regions. There may also be national or local government legislation to enforce the use of resistant varieties and phytosanitation, as in Uganda following the serious losses due to cassava mosaic disease (Jameson, 1964). There are also restrictions on when irrigation water is available and rice can be planted in parts of Malaysia and Indonesia. This is to ensure a break between successive crops and so decrease the carry-over of inoculum and the risk of damaging outbreaks of rice tungro disease (Koganezawa, 1998). Crop sanitation measures have also been enforced in attempts to control cotton leafcurl disease in the Sudan Gezira (Tarr, 1951). Similarly, there may be restrictions on when or where susceptible crops are grown. An example is the break in tobacco production enforced in Zimbabwe to disrupt the cycle of infection of leafcurl disease (Shaw, 1976). More recently there have been temporary bans on vegetable crop production in the Comayagun valley of Honduras and on tomato production in the Dominican Republic following devastating losses due to whitefly-borne viruses (Alvarez and Abud-Antún, 1995; IICA, 1996). A different type of intervention was the action of the President of Indonesia who in 1986 banned certain pesticides and withdrew subsidies. This was crucial in developing the IPM approach to rice pests discussed earlier (1.7.4).

1.7.6 The Cost of Control Measures

The cost of adopting control measures can be substantial and must be considered in assessing the impact of virus diseases. Moreover, the costs are of particular importance because they are incurred to avoid or decrease the losses that would otherwise occur and so do not increase overall yield potential. Consequently, expenditure on control involves at least some diversion of resources, expenditure and personnel that could have been used for crop improvement projects. This is illustrated by experience in Ghana where for years the costs of the 'cutting out' and replanting programme against cacao swollen shoot disease and the associated surveys dominated the budget of the Department of Agriculture at the expense of all other activities with cacao or other crops.

The implications of resistance-breeding programmes are more complex. Substantial costs are incurred, but these are usually within comprehensive crop breeding programmes that seek to meet the requirements of farmers and consumers and overcome all the main biotic and abiotic constraints to yield. Such an approach has been adopted at the International Agriculture Research Centres in breeding programmes on many staple food crops. Host-plant resistance to the most important viruses or their vectors has been a key requirement, although one that may not be met easily. Sources of resistance are not always available, or they are not readily combined with the other attributes required. This is apparent from experience with cassava and cassava mosaic disease at the International Institute of Tropical Agriculture and in earlier cassava breeding programmes in Africa (Jennings, 1994).

There have also been problems in groundnut breeding programmes in identifying sources of resistance to *Peanut clump virus* and in combining resistance to *Groundnut rosette virus* with the short-duration growth habit required in many groundnut-producing areas of sub-Saharan Africa. Short-duration rosette-resistant varieties are now available, but this has required substantial expenditure over many years (Subrahmanyam *et al.* 1998). Even now, seed stocks of rosette-resistant varieties are limited, rates of seed multiplication are low and substantial technology-transfer efforts will be required to ensure that sufficient material is available to supply all the farmers at risk. The difficulty is compounded because many farmers are too poor to purchase seed and there is a lack of effective seed production and distribution systems in many parts of sub-Saharan Africa (DeVries and Toennissen, 2001).

There are even greater problems in producing the large amounts of planting material required of crops propagated vegetatively. This partially explains the inadequate use of virus-free or virus-resistant material, despite the substantial benefits to be gained from doing so. Special agricultural extension projects and substantial funding were required to ensure the up-take of mosaic-resistant cassava varieties in Nigeria and later in Uganda (Nweke *et al.* 1996; Otim-Nape *et al.* 2000). Only limited use was made of micro-propagation in these projects, but this technique is being used increasingly in Africa and many Asian countries for banana and other crops.

1.8 Discussion

Information from diverse sources has been used here to indicate some of the ways in which plant virus diseases influence rural livelihoods and the economic well-being of developing countries. The assessment, though incomplete, highlights the importance of plant virus diseases and the benefits to be gained from effective means of control. Virus diseases present very diverse problems in that some change little from year to year, whereas others occur sporadically. Farmers tend to adjust to 'chronic' diseases of the former category, although they may do so unintentionally by simply selecting varieties that grow and yield satisfactorily, even when infected. Farmers may also anticipate the losses that are likely to occur and increase the area planted to achieve the production required. However, such a deliberate or inadvertent policy of 'living with' disease may be misleading and lead to inefficient production in terms of land, labour and resources. This is because 'cryptic' losses due to infection can be substantial and yet they may be apparent only on comparing the yield of infected and healthy plants. For example, a 37% yield loss due to cassava mosaic disease was detected in a variety of cassava that was widely grown in Côte d'Ivoire, although the variety was almost totally infected and regarded by farmers as satisfactory (Fargette *et al.* 1988). Virologists in developed countries have given considerable attention to losses of this type and to the benefits of adopting virus-free planting material of vegetatively propagated crops (Waterworth and Hadidi, 1998). Little attention has been given to these issues in developing countries where the main effort has been on diseases that cause obvious damage.

Diseases that occur sporadically cause particular difficulties because the sudden impact of occasional devastating epidemics is more damaging than similar losses occurring over several seasons. Farmers have great difficulty in coping with total crop failure and the consequent loss of income and/or food supplies from which they may be slow to recover (Lenné, 2000). There may also be a loss of seed stocks for new plantings, as recorded in Malawi following the 1994-95 epidemic of groundnut rosette disease which led to a 23% decline in the area planted the following season (Naidu *et al.* 1998).

Another general feature of the diseases considered here is that the situation is dynamic and continually changing in response to weather conditions, changes in cropping practices and economic circumstances. Moreover, changes in host, virus and vector populations can have a profound impact on disease incidence and severity. This explains why epidemiology can be regarded as a continuing 'battle' or 'struggle' of the genes, in which neither host nor pathogen gains ultimate supremacy (Gibbs, 1983; Thresh, 1990). There are certainly many examples of the way in which disease problems are closely related to the varieties grown and the cropping systems adopted, and this has led to the view that *"disease patterns are to a great extent a product of our plant breeding and agricultural practices."* (Simmonds, 1962). Consequently, the current labile situation is likely to continue and losses due to viruses may become even greater than at present.

It has been argued that losses due to viruses have increased following an intensification of cropping practices and that they will continue to do so

(Bos, 1992). There is insufficient evidence to confirm or deny this somewhat pessimistic view, although the argument is a plausible one based on experience with diverse crops indicating the various ways in which modern methods of crop production facilitate the spread of pests and pathogens, including viruses and their vectors. Much attention has been given to the big differences between the habitats provided by crops compared with natural vegetation. Indeed, it is suggested that problems seldom occur in undisturbed natural ecosystems and that they are largely an inevitable consequence of cropping practices or other forms of human intervention that disrupt long-standing co-evolutionary equilibria established between hosts and their pests and pathogens (Browning, 1981). Some of the crucial developments have been:-

- The continuing trend away from an original dependence on wild plants and animals to crop and livestock production using a relatively small number of domesticated species farmed on an ever-increasing scale.
- The adoption of small numbers of specially bred varieties grown singly and without intercrops, rather than the use of numerous local landraces, often grown as mixtures and with intercrops. There is only limited information on the implications of these trends for virus spread, but this is likely to be greatly facilitated by a decrease in crop genetic diversity. This is evident from the few published studies in which virus spread was decreased by adopting varietal mixtures or intercrops (Thresh, 2003).
- The large-scale use of varieties that have not always been tested adequately and which prove to be extremely vulnerable to pests or pathogens that previously were absent or unimportant. This is the concept of 'genetic vulnerability' of which several examples have been given in this chapter and previously (Thresh, 1990).
- The movement of crops to entirely new regions where they encounter pathogens not present at the original centres of origin, or where conditions favour pathogens already affecting the crop.
- An increase in the amount and speed of the traffic in plant material between countries in large commercial consignments and also in relatively small amounts for crop improvement or breeding programmes. Viruses that are seed-borne or disseminated in vegetative propagules present particular problems. There have also been serious losses following the inadvertent dissemination of virus vectors.
- The trend towards the repeated utilization of the same sites and away from the use of bush fallow and previously uncultivated land. This facilitates the survival of crop debris as sources of inoculum and leads to a build-up of weeds, soil-borne pests and pathogens.

- The use of irrigation, some form of protected cropping or cold/drought-tolerant varieties to extend natural growing seasons. This increases crop continuity and productivity, but decreases the separation between crops and has undesirable implications in enhancing pest, pathogen and weed populations. It also avoids the disruptions to such populations that would otherwise occur during prolonged periods of cold or drought.
- The increased use of fertilizers and of herbicides for effective weed control, which facilitate the colonization and reproduction of arthropod pests, including virus vectors.
- The increased use of insecticides to control vectors and direct pests of crops has led to the destruction of natural enemies, the development of insecticide-resistant populations and pest resurgence.

The various trends can be used to interpret and explain the current prevalence of many virus diseases, which can be seen as the largely inevitable outcome of changing agricultural and horticultural practices. However, this concept is not new and the renowned UK virologist F.C. Bawden (1955) commented that, *'If we are to reap the full rewards of high farming we must also learn to control virus diseases.'*

The challenge remains and it has become even greater as human populations continue to increase, especially in the least developed countries of Africa and Asia. Moreover, an increasing proportion of the total work force is not concerned with agriculture and the ravages of HIV/AIDS impair the availability and efficiency of rural labour. A further consideration is that there is a continuing decline in the area of unexploited land that could be used for agriculture and in the fertility of much of the land being used.

These developments provide an incentive to develop cropping practices that enhance yields and are less conducive to virus spread than those now used. There is also a need for effective measures to control virus diseases as a convenient and effective means of increasing productivity and thereby alleviating both rural and urban poverty. This would increase production, or free land and resources for other crops, or to permit longer periods of fallow and so help maintain or even restore fertility. The approach is particularly appropriate if the increased yields of largely healthy crops are achieved by using resistant varieties or other measures that require little or no additional labour or expenditure and pose no serious threat to human health or to the environment.

Much research has been concerned with developing means of controlling virus diseases of crops grown in the developing countries of Africa (Thresh, 2003) and elsewhere. However, the overall effort has been inadequate and not commensurate with the magnitude of the problems encountered and the importance of the losses sustained. The main concern has been with a few particularly important diseases, including several of those discussed in this and subsequent chapters. Moreover, totally inadequate attention has been given to disseminating virus-resistant varieties and other technical innovations to the millions of farmers now experiencing disease problems.

Additional research is required on a wider range of viruses and a greater effort is needed to ensure that research findings are passed on to farmers and adopted widely as a means of decreasing or avoiding current losses. These are formidable requirements that are not easily met and the current international effort should continue. However, even more is required and the need is great, whether based on humanitarian or economic considerations.

References

Abate, T., van Huis, A., and Ampofo, J.K.O. 2000. Pest management strategies in traditional agriculture: an African perspective. *Annu. Rev. Entomol.* 45: 631-659.

Ahmed, A.H. 1984. Incidence of peanut mottle virus in the Sudan Gezira and its effect on yield. *Trop. Pest Manag*, 30: 166-169.

Alvarez, P.A. and Abud-Antún, A.J. 1995. *Rep. Rep. Dominicana.* CEIBA 36: 39-47.

Aritua, C., Adipala, E., Carey, E.E. and Gibson, R.W. 1998. The incidence of sweet potato virus disease and virus resistance of sweet potato grown in Uganda. *Ann.Appl. Biol.* 132: 399-422.

Awoderu, V.A. 1991. Rice yellow mottle virus in West Africa. *Trop. Pest Manag.* 37: 356-362.

Azzam, O., and Chancellor, T.C.B. 2002. The biology, epidemiology and management of rice tungro disease in Asia. *Plant Dis.* 86: 88-100.

Bakker, W. 1974. *Characterization and ecological aspects of rice yellow mottle virus in Kenya.* Agricultural Research Reports 829 Pudoc, Wageningen, Netherlands, 152 pp.

Bar-Joseph, M., Roistacher, C.N., Garnsey, S.M. and Gumpf, D.J. 1981. A review on tristeza, an ongoing threat to citriculture. *Proc. Intern. Soc. Citriculture* 1: 419-423.

Barnett, O.W. 1986. Surveying for plant viruses: design and considerations. pp.147-166. In: *Plant Virus Epidemics: Monitoring, Modelling and Predicting Outbreaks.* G.D. McLean, R.G. Garrett and W.G. Ruesink (eds.). Academic Press, Sydney.

Bawden, F.C. 1955. The spread and control of plant virus diseases. *Ann. appl. Biol.* 42: 140-147.

Bos, L. 1982. Crop losses caused by viruses. *Crop Protection* 1: 263-282.

Bos, L. 1992. New plant virus problems in developing countries: a corollary of agricultural modernization. *Adv. Virus Res.* 38: 349-407.

Bosque-Pérez, N.A., Olojede, S.O. and Buddenhagen, I.W. 1998. Effects of maize streak virus disease on the growth and yield of maize as influenced by varietal resistance levels and plant stage at time of challenge. *Euphytica* 101: 307-317.

Briddon, R.W. and Markham, P.G. 2000. Cotton leaf curl virus disease. *Virus Res.* 71: 151-159.

Browning, J.A. 1981. The agro-ecosystem-natural ecosystem dichotomy and its impact on phytopathological concepts. pp.159-172 In: *Pests, Pathogens and Vegetation* (J.M. Thresh (ed.)). Pitman, London. 517 pp.

Brunt, A., Crabtree, K. and Gibbs. A. 1990. *Viruses of Tropical Plants.* CAB International, Wallingford, UK. 707pp.

Buddenhagen, I.W. 1977. Resistance and vulnerability of tropical crops in relation to their evolution and breeding. *Ann. New York Acad. Sci.* 287: 309-326.

Calvert, L.A. and Thresh, J.M. 2002. The viruses and virus diseases of cassava. pp.237-260 In: *Cassava: Biology, Production and Utilization.* R.J. Hillocks, J.M. Thresh and A.C. Bellotti (eds.) CABI, Wallingford, UK.

Castle, S.J. 1999. Agricultural intensification and pest outbreaks: a reappraisal of events in the Sudan Gezirà. *Ann. Entomol. Soc. America* 92: 840-852.

Crowdy, S.H. and Posnette, A.F. 1947. Virus diseases of cacao in West Africa: II Cross-immunity experiments with viruses 1A, 1B and 1C. *Ann. appl. Biol.* 34: 403-411.

Dahal, G., Ortiz, R., Tenkouano, A., Hughes, J. d'A., Thottappilly, G., Vuylsteke, D. and Lockhart, B.E.L. 2000. Relationship between natural occurrence of banana streak badnavirus and symptom expression, relative concentration of viral antigen, and yield characteristics of some micropropagated *Musa* spp. *Plant Pathol.* 49: 68-79.

Delfosse, P., Reddy, A.S., Legrève, A., Devi, P.S., Maraite, H. and Reddy, D.V.R. 1999. Indian peanut clump virus (IPCV) infection on wheat and barley: symptoms, yield loss and transmission through seed. *Plant Pathol.* 48: 273-282.

DeVries, J. and Toenniessen, G. 2001. *Securing the Harvest: Biotechnology, Breeding and Seed Systems for African Crops.* CABI Publishing, Wallingford, UK. 208pp.

Diekmann, M. and Putter, C.A.J. 1996. *Musa* spp. FAO/IPGRI Technical Guidelines for the Safe Movement of Germplasm, No. 15. FAO/IPGRI/INIBAP, Rome. 28pp.

Ducharme, E.P., Knorr, L.C. and Speroni, H.A. 1951. Observation on the spread of tristeza in Argentina. *Florida Citrus Mag.* 13(9): 10-14.

Fargette, D., Fauquet, C. and Thouvenel, J.-C. 1988. Yield losses induced by African cassava mosaic virus in relation to the mode and the date of infection. *Trop. Pest Manag.* 34: 89-91.

FUSAGRI, 1987. Comportamiento de la naranja Valencia sobre diferentia patrones. Fundación Servicio para el Agricultor, Venezuela. Noticias Agricolas 9: No. 22.

Geddes, A.M.W. 1990. *The Relative Importance of Crop Pests in Sub-Saharan Africa.* Bulletin No. 36, Natural Resources Institute, Chatham, UK. 69pp.

Geddes, A.M.W. 1992. *The Relative Importance of Pre-harvest Crop Pests in Indonesia.* Bulletin No. 47. Natural Resources Institute, Chatham, UK. 70pp.

Geddes, A.M.W. and Iles, M. 1991. *The Relative Importance of Crop Pests in South Asia.* Bulletin No. 39. Natural Resources, UK. 102pp.

Geering, A.D.W., McMichael, L.A., Dietzgen, R.G. and Thomas, J.E. 2000. Genetic diversity among *Banana streak virus* isolates from Australia. *Phytopathology* 90: 921-927.

Gibbs, A.J. 1983. Virus ecology – "Struggle" of the genes. pp.537-558 In: *Physiological Plant Ecology: III Response to the Chemical and Biological Environment.* O.L. Lange, P.S. Nobel, C.B. Osmond and H. Ziegler (eds.). Springer-Verlag, Berlin.

Goodman, R.M. and Oard, J.H. 1980. Seed transmission and yield losses in tropical soybeans infected by soybean mosaic virus. *Plant Dis.* 64: 913-914.

Hahn, S.K. 1979. Effects of viruses (SPVD) on growth and yield of sweet potato. *Exper. Agric.*15: 253-256.

Hasanuddin, A. and Hibino, H. 1989. Grain yield reduction, growth retardation, and virus concentration in rice plants infected with tungro-associated viruses. *Trop. Agric. Res.* Series, No. 22, 56-73.

Herdt, R.W. 1991. Research priorities for rice biotechnology. pp.19-43 In: *Rice Biotechnology.* G.S. Khush and G. Toenniessen (eds.). CAB International, Wallingford, UK. 320pp.

Hillocks, R.J., Thresh, J.M., Tomas, J., Botao, M., Macia, R. and Zavier, R. 2002. Cassava brown streak disease in northern Mozambique. *Intern. J. Pest Manag.*48: 179-182.

Hughes, G. 1988. Modelling the effect of spatially heterogeneous pest injury on crop yields. *Crop Res.*28: 137-144.

IICA 1996. Honduras: diagnostico del sector agropecuario. DIPRAT, 172pp.

Jameson, J.D. 1964. cassava mosaic disease in Uganda. *East African Agric. Forestry J.* 29: 208-213.

Jennings, D.L. 1994. Breeding for resistance to African cassava mosaic geminivirus in East Africa. *Trop. Sci.* 34: 110-122.

John, V.T., Thottappilly, G. and Awoderu, V.A. 1984. Occurrence of rice yellow mottle virus in some Sahelian countries in West Africa. *FAO Plant Protect. Bull.* 32: 86-87.

Kenmore, P.E., Cariño, F.O., Perez, C.A., Dyck, V.A. and Gutierrez, A.P. 1984. Population regulation of the rice brown planthopper *(Nilaparvata lugens* Stål) within rice fields in the Philippines. *J. Plant Protect.Tropics* 1: 19-37.

Klinkowski, M. 1970. Catastrophic plant diseases. *Annu. Rev. Phytopath.* 8: 37-60.

Koganezawa, H. 1998. Present status of controlling rive tungro virus. pp.459-469 In: *Plant Virus Disease Control.* A. Hadidi, R.K. Khetarpal and H. Koganezawa (eds.). APS Press, St. Paul, Minnesota.

Legg, J.P. 1999. Emergence, spread and strategies for controlling the pandemic of cassava mosaic virus disease in east and central Africa. *Crop Protect.* 18: 627-637.

Lenné, J. 2000. Pests and poverty : the continuing need for crop protection research. *Outlook Agric.* 29: 235-250.

Lockhart, B.E.L. and Olszewski, N.E. 1993. Serological and genomic heterogeneity of banana streak badnavirus: implications for virus detection in *Musa* germplasm. pp.102-113 In: *Breeding Banana and Plantain for Resistance to Diseases and Pests* (J. Ganry, ed.). CIRAD-INIBAP, Montpellier.

Loevinsohn, M.E. 1987. Insecticide use and increased mortality in rural central Luzon, Philippines. *The Lancet,* 13 June. pp.1359-1362.

Maskey, R. 2001. Population growth and food production : issues, problems and prospects. *Outlook Agric.* 30: 155-160.

Matteson, P.C. 2000. Insect pest management in tropical Asian irrigated rice. *Annu. Rev. Entomol.* 45: 549-574.

Mendt, R., Plaza, G., Boscan, R. Martinez, J. and Lastra, R. 1984. Spread of citrus tristeza virus and evaluation of tolerant rootstocks in Venezuela. pp.95-99 In: *Proc. 9th Conf. IOCV.* S.M. Garnsey, L.W. Timmer and J.A. Dodds (eds.). Riverside, California.

Morales, F.J. and Anderson, P.K. 2001. The emergence and dissemination of whitefly-transmitted geminiviruses in Latin America. *Arch. Virol.* 16: 415-441.

Mungomery, R.W. and Bell, A.F. 1933. Fiji disease of sugarcane and its transmission. *Div. Pathol. Bull. 4. Queensland Bureau of Sugar Experiment Stations*, Brisbane. 28pp.

Naidu, R.A., Bottenberg, H., Subrahmanyam, P., Kimmins, F.M., Robinson, D.J. and Thresh, J.M. 1998. Epidemiology of groundnut rosette virus disease: current status and future research needs. *Ann. appl. Biol.* 132: 525-548.

Ndowora, T., Dahal, G., La Fleur, D., Harper, G., Hull, R., Olszewski, N.E. and Lockhart, B. 1999. Evidence that badnavirus infection in *Musa* can originate from integrated pararetroviral sequences. *Virology* 255: 214-220.

Nweke, F.I., Ugwu, B.O. and Dixon, A.G.O. 1996. *Spread and performance of improved cassava varieties in Nigeria.* Working Paper No. 15. Collaborative Study of Cassava in Africa. IITA, Ibadan, Nigeria. 34pp.

Olorunju, P.E., Kuhn, C.W., Demski, J.W., Misari, S.M. and Ansa, O.A. 1991. Disease reactions and yield performance of peanut genotypes grown under groundnut rosette and rosette-free field environments. *Plant Dis.* 75: 1269-1273.

Omunyin, M.E., Gathuru, E.M. and Mukunya, D.M. 1988. Effect of bean common mosaic virus on growth and yield of beans. *East African Agric. Forestry J.* 54: 7-10.

Otim-Nape, G.W. and Thresh, J.M. 1998. The current pandemic of cassava mosaic virus disease in Uganda. pp.423-443. In: *The Epidemiology of Plant Diseases*. D. Gareth-Jones (ed.). Kluwer Publishers, Netherlands.

Otim-Nape, G.W., Thresh, J.M. and Shaw, M.W. 1997. The effects of cassava mosaic virus disease on yield and compensation in mixed stands of healthy and infected cassava. *Ann. appl. Biol.* 130: 503-521.

Otim-Nape, G.W., Bua, A., Thresh, J.M., Baguma, Y., Ogwal, S., Ssemakula, G.N., Acola, G., Byabakama, B, Colvin, J., Cooter, R.J. and Martin, A. 2000. *The current pandemic of cassava mosaic virus disease in East Africa and its control.* UK Natural Resources Institute. 100pp.

Owor, B. 2003. Effect of cassava mosaic geminiviruses (CMGs) on growth and yield of a cassava mosaic disease (CMD) susceptible cultivar in Uganda and cross protection studies. *MSc Thesis. Makerere University,* Kampala, Uganda. 134pp.

Polston, J.E. and Anderson, P.K. 1997. The emergence of whitefly-transmitted geminiviruses in tomato in the western hemisphere. *Plant Dis.* 81: 1358-1369.

Porto, M.C.M., Asiedu, R., Dixon, A. and Hahn, S.K. 1994. An agroecologically-oriented introduction of cassava germplasm from Latin America into Africa. pp.118-129 In: *Tropical Root Crops in a Developing Economy.* F. Ofori and S.K. Hahn (eds.). *Proc. 9th Symp. Intern. Soc.Tropical Root Crops*, Accra, 1991. IITA.

Posnette, A.F. 1943. Control measures against swollen shoot virus disease of cacao. *Trop. Agric. Trinidad* 20: 116-123.

Posnette, A.F. 1951. Progeny trials with cacao in the Gold Coast. Emp*ire J. Exp. Agric.* 19: 242-252.

Reckhaus, P.M. and Adamou, I. 1986. Rice diseases and their economic importance in Niger. *FAO Plant Prot. Bull.* 34: 77-82.

Reckhaus, P.M. and Andriamasintseheno, H.F. 1997. Rice yellow mottle virus in Madagascar and its epidemiology in the northwest of the island. *J. Plant Dis. Prot.* 104: 289-295.

Rivera, C.T. and Ou, S.H. 1965. Leafhopper transmission of 'tungro' disease of rice. *Plant Dis. Reptr.*49: 127-131.

Rocha-Peña, M.A., Lee, R.F., Lastra, R., Niblett, C.L., Ochoa-Corona, F.M., Garnsey, S.M. and Yokomi, R.K. 1995. Citris tristeza virus and its aphid vector *Toxoptera citricida*: threats to citrus production in the Caribbean and Central and North America. *Plant Dis.* 79: 437-445.

Rybicki, E.P. and Pietersen, G. 1999. Plant virus disease problems in the developing world. *Adv. Virus Res.* 53: 127-175.

Shaw, M.J.P. 1976. Insect-borne diseases of tobacco in Rhodesia and the role of the tobacco-free period. *Rhodesian Agric. J.* 76: 87-90.

Simmonds, N.W. 1962. Variability in crop plants, its use and conservation. *Biol. Rev.* 37: 442-465.

Smith, M.C., Holt, J., Kenyon, L. and Foot, C. 1998. Quantitative epidemiology of banana bunchy top virus disease and its control. *Plant Pathol.* 47: 177-187.

Sõgawa, K. 1976. Rice tungro virus and its vectors in tropical Asia. *Rev. Plant Prot. Res.* 9: 21-46.

Spittel, M.C. and Van Huis, A. 2000. Effect of cassava mosaic disease, soil fertility, plant spacing and their interactions on cassava yields in Zanzibar. *Intern. J. Pest Manag.* 46: 187-193.

Subrahmanyam, P., Hildebrand, G.L., Naidu, R.A., Reddy, J.L. and Singh, A.K. 1998. Sources of resistance to groundnut rosette disease in global groundnut germplasm. *Ann. appl. Biol.* 132: 473-485.

Summers, E.M., Brandes, E.W. and Rands, R.D. 1948. *Mosaic of sugarcane in the United States, with special reference to strains of the virus.* Technical Bulletin No. 955. USDA, Washington. 103pp.

Tarr, S.A.J. 1951. *Leaf Curl Disease of Cotton.* Kew: Comm. Mycological Inst., UK. 55 pp.

Teng, P.S. and Shane, W.W. 1984. Crop losses due to plant pathogens. C.R.C. *Critical Rev. Plant Sci.* 2: 21-47.

Thresh, J.M. 1988. Eradication as a virus disease control measure. pp.155-194. In: *Control of Plant Diseases: Costs and Benefits.* B.C. Clifford and E Lester (eds.). Blackwell Scientific, London.

Thresh, J.M. 1989. Insect-borne viruses of rice and the green revolution. *Tropical Pest Manag.*35: 264-272.

Thresh, J.M. 1990. Plant virus epidemiology : The battle of the genes. In: *Recognition and Response in Plant-Virus Interactions.* R.S.S. Fraser (ed.). NATO ASI Series H. *Cell Biology 41: 93-121.* New York, Springer-Verlag.

Thresh, J.M. 1991. The ecology of tropical plant viruses. *Plant Pathol.* 40: 324-339.

Thresh, J.M. 2003. The control of plant virus diseases in sub-Saharan Africa: the possibility and feasibility of an integrated approach. *Proc. Intern. Conf. Integrated Pest Management in sub-Saharan Africa.* In Press.

Thresh, J.M. and Owusu, G.K. (1986) The control of cocoa swollen shoot disease in Ghana: an evaluation of eradication procedures. *Crop Prot.* 5: 41-52.

Thresh, J.M., Fargette, D. and Otim-Nape, G.W. 1994. Effects of African cassava mosaic geminivirus on the yield of cassava. *Tropical Science* 34: 26-42.

Thresh, J.M., Otim-Nape, G.W., Legg, J.P. and Fargette, D. 1997. African cassava mosaic virus disease: the magnitude of the problem. *Afric. J. Root, Tuber Crops* 2: 13-19.

Thurston, H.D. 1973. Threatening plant diseases. *Annu. Rev. Phytopath.* 11: 27-52.

Thurston, H.D. 1977. International crop development centers: a pathologist's perspective. *Annu. Rev. Phytopath.* 15: 223-247.

Thurston, H.D. 1998. *Tropical Plant Diseases.* Second Edition, APS Press, St. Paul, Minnesota, 200pp.

Vanderplank, J.E. 1963. *Plant Diseases: Epidemics and Control.* Academic Press, New York and London, 349pp.

Waterworth, H.E. and Hadidi, A. 1998. Economic losses due to plant viruses. pp.1-13. In: *Plant Virus Disease Control.* A. Hadidi, R.K. Khetarpal and H. Koganezawa (eds.). APS Press, St. Paul, Minnesota.

Yokomi, R.K, Lastra, R., Stoetzel, M.B., Damsteegt, V.D., Lee, R.F., Garsney, S.M., Gottwald, T.R. Rocha-Peña, M.A. and Niblett, C.L. 1994. Establishment of the brown citrus aphid (Homoptera: Aphididae) in Central America and the Caribbean basin and transmission of citrus tristeza virus. *J. Econ. Entomol.* 87: 1078-1085.

Table 1.1. Representative viruses of crops in developing countries and their means of spread*

Contact-borne	*Cassava common mosaic virus* (V) *Tobacco mosaic virus* *Tomato mosaic virus* (S)
Fungus-borne	*Peanut clump virus* (S)
Mite-borne	Pigeon pea sterility mosaic virus
Aphid-borne: non-persistent	*Bean common mosaic virus* (S *Papaya ringspot virus* *Peanut stunt virus* (S) *Soybean mosaic virus* (S) *Sugarcane mosaic virus* (V) *Sweetpotato feathery mottle virus* (V) *Yam mosaic virus* (V)
Aphid-borne: persistent	*Banana bunchy top virus* (V) *Citrus tristeza virus* (V) Groundnut rosette viruses
Hopper-borne	*Fiji disease virus* (V) *Maize rayado fino virus* *Maize streak virus* *Rice hoja blanca virus* *Rice grassy stunt virus* *Rice ragged stunt virus*
Mealybug-borne	*Banana streak virus* (V) *Cacao swollen shoot virus* (V) Pineapple wilt mealybug-associated viruses (V)
Whitefly-borne	Cassava mosaic geminiviruses (V) Cotton leafcurl viruses Tomato yellow leafcurl viruses
Thrips-borne	*Groundnut bud necrosis virus* *Tomato spotted wilt virus*
Beetle-borne	*Bean rugose mosaic virus* *Cowpea mosaic virus* (S) *Okra mosaic virus* *Rice yellow mottle virus*

S = also seed-borne; V = also disseminated in vegetative propagules
*Names of recognized virus species are printed in italics.

Table 1.2. Yield loss studies on eleven important diseases of diverse crops.

Crop Disease	Yield loss	Reference
Banana streak	0-82%	Dahal *et al.* (2000)
Bean common mosaic	47-86%	Omunyin *et al.* (1988)
Cacao swollen shoot	0-100%	Crowdy and Posnette (1947)
Cassava mosaic	0-82%	Thresh *et al.* (1994)
Groundnut rosette	0-100%	Olorunji *et al.* (1991)
Maize streak	17-71%	Bosque-Perez *et al.* (1998)
Peanut clump	>60%	Delfosse *et al* (1999)
Peanut mottle	24%	Ahmed (1984)
Rice tungro	20-85%	Hasanuddin and Hibino (1989)
Soybean mosaic	45-86%	Goodman and Oard (1980)
Sweet potato virus disease	78%	Hahn (1979)

2. Principles for control of virus diseases in developing countries

I.B.F. Fofana[1;2], A. Sangaré[3], J. Ndunguru[1;4], K. Kahn[1] and C.M. Fauquet[1]
[1]International Laboratory for Tropical Agricultural Biotechnology (ILTAB)/Donald Danforth Plant Science Center, 975 N. Warson Road, St. Louis, MO 63132, USA; [2]Université de Cocody, UFR des Biosciences, Laboratoire de Génétique, 22 BP 582 Abidjan 22, Ivory Coast; [3]Centre National de Recherche Agronomique (CNRA), Laboratoire Central de Biotechnologies, Km 17 Rte Dabou, 01 B.P. 1740, Abidjan 01, Ivory Coast; [4]Plant Protection Division, P.O. Box 1484, Mwanza, Tanzania

2.1 Introduction

By 2050, 90% of the planet's population will reside in the developing countries of the southern hemisphere (Brink *et al.* 1998). Achieving global food security necessitates a doubling of food production in the next 50 years (James, 1997). While farmers from developed countries benefit from massive financial subsidies and investment, aid directed to developing country agriculture is pitifully small.

The development of a sustainable agriculture in developing countries requires strong commitment from both local governments and the international community. The last century has seen a boom in agricultural development worldwide, but the challenge of nourishing the entire world population still remains, particularly for people living in developing countries. Unlike Asia and Latin America, the African Sub-Saharan region did not benefit from the massive yield gains of the green revolution. Today, it is essential that developing countries benefit from current advances in biotechnology, particularly in tissue culture, rapid propagation and genetic engineering. It is also important to better adapt the management of farming systems to the reality of developing countries.

In Africa the situation is particularly complex, and varies from one country to another. There are many differences in factors that directly impact farmers, such as the agro-ecological environment and the degree of drought in some regions. Indirectly impacting factors include disparity in the commitment of the public sector, civil wars and social problems. Nonetheless, industrial cultures such as coffee and cocoa can be a source of considerable income to farmers, even when produced on a relatively small-scale by local stakeholders. Other success stories include banana production in Cote d'Ivoire (Ivory Coast), where multinational companies lead a thriving industry. In contrast, small-scale farmers possessing less than 1 hectare generally cultivate the food crop cassava; cassava productivity lags well behind other crops, and yields over the past 30 years have stagnated.

Achieving food security and well being for the poorest of this world requires strong policies from both public and private sectors, reinforced by non-governmental organizations (such as Rockefeller Foundation, Ford Foundation, McKnight Foundation, Bill and Melinda Gates Foundation and many others). There are many challenges to overcome, resulting from a complex interplay of environmental factors and human activities and policies. From the biological perspective, agriculture in tropical countries is subjected to a greater biotic pressure than in temperate climates. This is the direct consequence of a lack of winter seasons, which tend to break pest and disease cycles, thereby limiting their impact on crops. Any method to control plant viruses will require effective integrated pest management strategies, since insects transmit most plant virus diseases. In the present analysis we will focus on the scientific and biological aspects impacting plant virus diseases in agriculture, rather than the political and societal aspects. We will address the principles for control of virus diseases that can best be applied in developing countries. The major principals for virus control will be discussed, supported by illustrative examples.

2.2 Plant virus diseases in developing countries

Plant viruses constitute a serious threat to developing countries for four reasons: 1- the absence of a winter season to break insect pest cycles responsible for transmission of most of the viruses, 2- a very high ratio between wild species and cultivated species, favoring natural reservoirs of infectious material, 3- a lack of investment in research and development compared to developed countries, 4- the quasi absence of implementation of available technologies. Other phenomena that make it difficult to control viruses include virus recombination and the adaptation of insects to transmit new or different viruses.

Severe epidemics usually appear with out-breaks of vector populations that are generally unpredictable. Vector adaptability to different hosts (crop varieties or wild species) can result in the emergence of new viral diseases. For instance, prevalence of the B biotype of the whitefly vector *Bemisia tabaci* which feeds on a wider range of plants compared to the A biotype (dozens) has resulted in the worldwide spread of viruses that were previously limited to weeds (Brown, 1994; Rybicki and Pietersen, 1999). The contribution of annual and perennial weed hosts in the ecology and epidiomology of alfalfa mosaic virus in green pepper fields has been demonstrated in Zambia (Ndunguru and Kapooria, 2000). Similarly, the spread of cucumber mosaic virus in vegetables is partly attributed to its wide host range, which includes herbs (Ndunguru and Kapooria, 1997). Some viruses, including *Maize streak virus* (MSV), are able to infect a wide range of Graminae species (Mesfin *et al.* 1992) and various indigenous African grasses (Rose, 1978) and can therefore spread widely in crops through adaptation of the vector. In the same way, Vanitharani *et al.* (1996), using *Indian cassava mosaic virus* (ICMV) DNA-A as an hybridization probe, demonstrated that the crop plants *Vigna radiata*, *V. mungo* and the weeds *Acalypha indica*, *Croton sparsiflorus* and *Vernonia cinerea* all harbored whitefly-transmitted geminiviruses. A very informative case is the cotton leaf curl disease resulting from the association of *Cotton leaf curl virus* (CLCuV) with a circular ssDNA (DNA-1) and a DNA satellite DNA β (Mansoor *et al.* 1999; 2003. These authors noticed that DNA-1 shares some homology with a genomic component of members of *Nanoviridae*, a family of DNA viruses that are normally transmitted by aphids or plant-hoppers. Therefore, it seems likely that DNA-1, a nanovirus-like DNA normally transmitted by aphids, has become whitefly transmissible, probably due to *trans*-encapsidation and therefore co-transmission of the DNA-1 and CLCuV.

In some instances, due to their small genomes, viruses may evolve rapidly via recombination or mutation, leading to the possibility that very virulent strains could appear, with a dramatic impact on yield. Changes as little as one or two nucleotides can account for the severity of existing virus strains (Chatterji *et al.* 1999; Li *et al.* 2001). Recombination between virus genomes of different species may cause the appearance of new diseases (Deng *et al.* 1997; Zhou *et al.* 1997; Padidam *et al.* 1999b). Pita *et al.* (2001) identified recombination and pseudo-recombination events between cassava infecting viruses as a key determinant in the emergence of severe cassava mosaic disease in Uganda.

Finally, we must pay attention to the human factor. The spread of many plant virus diseases around the world is firstly due to human “errors”. Modern modes of transportation and international trade involve the transportation of plants, viruses and their vectors, thereby causing the emergence of new diseases. Humans are responsible for the introduction of *Tomato yellow leaf curl virus* (TYLCV) from the Middle East to the Dominican Republic and its subsequent spread over the United States (Polston *et al.* 1999). In East Africa, 60% of farmers were unaware of viral diseases of beans, sweet potato and papaya, leading to their spread and high disease incidences (Gibson *et al.* 2000; Ndunguru, 2002; Ndunguru and Rajabu, 2002). The Ugandan cassava mosaic outbreak in the 1990s has been correlated with a recombination event between two geminiviruses that were separated by the Rift Valley before modern transportation allowed them to be in the same infected plant (Otim-Nape *et al.* 1997). Cotton leaf curl disease (CLCuD), reported in Pakistan in 1960 (Hussain and Mahmood, 1988; Hussain et al. 1991), came to prominence only after the release of the cotton variety (S12) in 1988 (Briddon and Markham, 2000). From 1988 onwards, the disease continued to spread and became a national disaster, causing a loss estimated at 5 billion dollars (US) in the 5 years between 1992 and 1997 (Briddon and Markham, 2000).

Despite these difficulties, a number of achievements have been made in recent years in the attempt to control viruses in developing countries and are worth reporting as examples for other non-controlled diseases. We can consider three types of control: a- preventing transmission of the disease from one generation to the next or from one field to the next, b- controlling viruses via breeding for virus resistance, and c- controlling viruses via genetic engineering.

2.3 Control of virus vectors, phytosanitation and agricultural management

Control of plant virus vectors is a complex problem. A small number of vectors can cause a rapid spread of the virus. The density of vectors colonizing the crop is of limited value in predicting spread (Fereres, 2000). Moreover, for vegetatively propagated crops (e.g. cassava and sweet potato), the plant material is generally infected at planting. Concerning the mosaic disease of cassava, the whitefly *B. tabaci* is responsible for the transmission of viruses to the plant but the disease is mostly perpetuated through cuttings (Storey and Nichols, 1938; Fauquet and Fargette, 1990) (Figures 2.1; 2.2).

Control of virus vectors is mainly by use of pesticides and physical barriers (Hilje *et al.* 2001; Palumbo *et al.* 2001). However, these measures enjoy limited success and are generally outside the means of small-scale farmers in developing countries. The virus population can unexpectedly rise to very high levels leading to an extensive use of pesticide in attempts to eliminate the vector before it transmits the virus (Lapidot and Friedmann, 2002). However, non-persistently aphid-borne viruses like *Potato virus Y* (PVY) and *Soybean mosaic virus* (SMV) are transmitted within seconds or minutes (Satapathy, 1998; Fereres, 2000) rendering pesticides useless.

Physical barriers such as fine-mesh (Cohen and Antignus, 1994) or UV-absorbing plastic sheets and screens (Antignus *et al.* 2001) that were shown to control TYLCV on tomato are costly and not appropriate for field conditions in developing countries. In contrast, crop borders, crop barriers (Simons, 1957) or intercropping (Thresh, 1982) are readily accessible and can serve to some extent to limit vector movement and virus spread within and between fields (Fereres, 2000). These strategies can also serve as a source of natural enemies of the vector and seem to be easy for farmers to handle. The question is whether such systems are effective. In Zambia, it has been reported that the incidence of PVY in green pepper was low when the crop was grown in fields isolated from other solanaceous crops (Ndunguru and Kapooria, 1999). Forest trees surrounding pepper fields were found to limit the movement of the aphid vector *Myzus persicae*. In contrast, in the case of cassava it was clearly demonstrated that barriers were not efficient in reducing the primary infection caused by whiteflies in virus-free fields, and thus were not practical for African farmers (Fargette and Fauquet, 1988). However, a reduction in disease spread was observed when growing cassava as an inter-crop (Fargette and Fauquet, 1988; Ahohuendo and Sarkar, 1995; Fondong *et al.* 2002) and also as a mixture of different varieties (Sserubombwe, 1998; Sserubombwe *et al.* 2001) (Figure 2.3). There is no doubt that barrier crops like maize, sorghum, okra or cotton are themselves at least as important to the farmer as the protected crop. Furthermore, a barrier crop or an intercrop can itself be subjected to infestation from other viruses, generating subsequent problems. This threat is augmented for example by the worldwide spread of the B biotype of the vector *B. tabaci* that has a wider host range compared to the previously predominant A biotype.

An additional cultural practice to control viral disease is high-density planting, as illustrated by the groundnut rosette virus disease spread by aphid vectors. In Uganda, high plant density was shown to significantly reduce aphid infestation and disease spread on the improved groundnut cultivar *"Igola-1"* (Adipala *et al.* 1998). In the coastal forest area of the Ivory Coast,

2.1
2.2
2.5
2.3
2.4
Control
Y44
Y85
2.6

Legends to figures

Figure 2.1: Uninfected cassava plants of cultivar CB, Adiopodoumé, coastal region of Ivory Coast

Figure 2.2: CMD-infected cassava plants of cultivar CB at Adiopodoumé, Ivory Coast

Figure 2.3: Intercropping of cassava and maize in Tanzania, both important crops to small-holding farmers

Figure 2.4: Healthy cassava plants derived from virus-free meristem culture and maintained under high phytosanitary conditions by removal of infected plant material during field growth. Cultivar CB is grown on the left, and cultivar Kataoli on the right.

Figure 2.5: Nigerian Cassava landrace TME3, one of the most promising new sources of single gene resistance to CMD.

Figure 2.6: Cassava cultivar TMS 60444 (control) transformed with the replicase gene from ACMV-Kenya to generate lines Y44 and Y85. Both transgenic lines show high levels of CMD resistance when challenged under greenhouse conditions with different species of geminviruses (ACMV, EACMCV, SLCMV).

ACMV spread was shown to be limited in plots with high (10,000) density planting (Fargette *et al.* 1990).

Small-scale farmers in developing countries need clean planting material or resistant cultivars when available. Meristem culture (Wang and Hu, 1980) is widely used to produce virus-free plants. Heat treatment of planting material was shown to be effective for elimination of many viruses (George, 1993). Heat treatment does not eliminate virus from dormant seeds and is more effective on actively growing vegetative material (George, 1993). Chemotherapy has been suggested for the prevention of virus replication (Cassells and Long, 1980) or for degradation of pre-existing viruses (Simpkins *et al.* 1981). Nevertheless, the routine use of chemotherapy remains questionable and not suitable to developing countries.

In China, diseases caused by potyviruses resulted in drastic reductions of yield and constitute the major constraint to sweet-potato production (http://europa.eu.int/comm/research/conferences/2003/sadc/). By eliminating viral disease from sweet potato planting materials, sweet potato production increased by 30 to 40%, without extra chemicals or genetic improvement (http://www.futureharvest.org/growth/china_sweet.bkgnd.html). In Vietnam, use of *Potato leaf roll virus* (PLRV)-free potato tuber seeds was less successful because the rate of recontamination was high. The level of infection reached 60 to 90 % of the plants in the fields after only one or two seasons (http://lanra.dac.uga.edu/potato/asia/vietnam.htm).

In the case of cassava, it was shown that re-infestation of virus-free cassava plant material greatly depends on local virus pressure. In some places, stakes become infected in less than 4 weeks (as in the southern region of Ivory Coast), rendering the method useless. In other areas such as central Ivory Coast, fields remained virus-free for months and production increased markedly (Fargette *et al.* 1987, Thresh *et al.* 1998) (Figure 2.4). Evidently, the local pressure of inoculation is a combination of vector population, virus source and susceptibility of the plant host. For cassava, spread of viruses from plant to plant is mainly through cuttings (Fargette *et al.* 1987; 1994). Therefore sanitation would be effective if done on a large scale and over many years. This should result in a drastic reduction in infection pressure from old cassava fields to new cassava fields. Naturally the outcome of the system would also depend on the cassava cultivars used in a region, as the higher the level of resistance of the cultivar, the more difficult is the uptake of the virus by the whitefly from infected plants to inoculate new virus-free plants, thereby limiting the virus reservoir and consequently decreasing inoculation pressure.

To conclude, it is clear that phytosanitation for vegetatively propagated crops is a relevant technique that has been proven effective for many viruses in many crops all over the world, but it needs further attention and investment. For phytosanitation to be effective, it has to be implemented on a large scale and should be linked to a strong agricultural policy with a certification system to encourage farmers to use virus-free seed material, but it is truly adaptable for developing countries.

2.4 Breeding for resistance to plant viruses

Efficient management of plant virus diseases can be achieved either by control of the plant virus vectors or phytosanitation methods, but it is clear that both methods are dependent on the level of resistance of the genotypes used. Breeding for resistance is therefore an integral component of the control of viruses. Plant virus resistance is the ultimate tool to stop a virus moving from one generation to the next, and should be used whenever possible. Breeding avoids the use of costly agricultural, phytosanitation and certification methods. However, to become a useful and widespread solution, virus resistance must be combined with other desirable agronomical traits in high yielding, farmer-preferred varieties.

Need for proper virus identification

Developing a breeding program first requires the correct identification of infecting viruses. According to Fauquet *et al.* (2000, 2003), the lack (until recently) of consistency in viral nomenclature constitutes a recurrent problem in breeding for resistance. The concept that we only need to expose the progenies of the plants within a breeding program to local pressure and then deploy the inbred lines for use in many different environments has been successful in some instances but failed in others. Mixed infection is a common situation in fields because individual vectors can transmit more than one virus strain or species. Viruses vary considerably so that cultivars resistant to sweet potato virus diseases (SPVD) in Nigeria were found susceptible in Uganda (Mwanga *et al.* 1991). For this reason, breeders must work in collaboration with virologists and molecular biologists to establish a reliable correlation between a particular virus and the disease displayed by the plant, and finally with resistant plants obtained.

In the case of cassava, for a long time it was thought that there was only one virus in the entire African continent, while today we know that at least 8 species with many strains exist in Africa and India (Fauquet and Stanley, 2003). Fondong *et al.* (2000) and Pita *et al.* (2001) using PCR with primers specific to the coat protein (CP) gene of *African cassava mosaic virus* (ACMV) and *East African cassava mosaic virus* (EACMV), demonstrated that two distinct viruses infected cassava in Cameroon and Uganda. They also demonstrated a synergistic interaction between the two viruses. Such dual infected plants show extremely severe symptoms and literally collapse, causing a very severe epidemic (Otim-Nape *et al.* 1997). Since the 1930s, Uganda has been growing cassava plants selected against single infections until the occurrence of severe epidemics attributed to the presence of a new virus, recombinant between ACMV and EACMV (Deng *et al.* 1997; Zhou *et al.* 1997).

Another important concept for breeding for virus resistance is the evaluation and understanding of the results obtained. This involves two aspects, the use of an appropriate screening system and the evaluation of the type of resistance obtained.

Need for a proper virus resistance type evaluation

The use of field assessment of resistance alone can sometimes be misleading. For a susceptible variety, plants without symptoms can be observed even under high vector pressure as demonstrated on tomato infected with TYLCV (Vidasky *et al.* 1998). In the same way, the observation that the same cultivars can present healthy, mildly or severely infected plants in the same field is not uncommon with cassava (Fauquet, unpublished data).

It is also important to evaluate the type of resistance obtained through breeding. Often, it is not known if the plants are resistant to virus replication, to virus movement in the plants or simply to vector inoculation. Resistance to virus replication is certainly the most desirable but very rare. Often, resistance results from prevention of virus movement within the host plant. Biochemical barriers can prevent virus movement from the inoculated cell to adjoining cells (Ponz and Bruening, 1986; Bol and van Kan, 1988; van Loon, 1989), or from the inoculated leaf to other leaves of the plant (Barker and Harrison, 1986; Lei and Agrio, 1986; Caranta *et al.* 2002) and could act in different types of cells. The most common mechanism of virus resistance, however, relates to insect transmission, when the virus vector cannot inoculate the plants or cannot feed on the host. Consequently the rate of secondary transmission (from one plant to the next) is greatly reduced, translating in the field to a limited overall spread of contamination. Understanding the type of resistance is of primary importance in predicting the durability of the resistance and therefore for the decision making process for its deployment in developing countries. For example, the introduction of high yielding cultivars of rice led to serious epidemics of tungro viral disease in India and the Philippines after a few years of intensive cultivation (Inoue and Ruay-Aree, 1977; Buddenhagen, 1983; Manwan *et al.* 1985). The breakdown of the resistance is attributed to the fact that the tungro-resistant high-yielding cultivars were resistant only to the vector (Hibino *et al.* 1987).

Need for new sources of resistance

Only a few virus resistance genes have been identified to date. These include the Tobacco *N* gene conferring resistance to *Tobacco mosaic virus* (TMV) (Whitham *et al.* 1996), Potato *Rx* resistant gene to *Potato virus X* (PVX), Potato *Ry* (*sto*) gene conferring resistance to *Potato virus Y* (PVY) and *Tobacco etch virus* (Hinrichs *et al.* 1998), Bean *I* resistance gene to *Bean common mosaic virus* (BCMV) (Ali, 1950) and Arabidopsis *HRT* family genes conferring resistance to *Turnip crinkle virus* (Dempsey *et al.* 1997). Nevertheless, breeding programs often focus on finding and using sources of virus resistance in cultivated plants and related wild species. These genes must be transferred to the target crop. There are often barriers to the introgression of desired resistance traits into cultivated crops (e.g. some plants will only self-pollinate, while others demonstrate sporophytic or gametophytic incompatibilities, delays in pollination, etc). The success of a breeding program is highly dependent on the level of resistance of the original resistant parent. For maize, different varieties presenting resistance to *Maize streak virus* (MSV) have been described since the 1930s (Fielding, 1933; Rose, 1938; Gorter, 1959; Storey and Howland, 1967). In the 1970s, an important source of resistance called "La Revolution" was described on Reunion Island (Etienne and Rat, 1973). In South Africa, the national breeding program produced several tolerant or resistant lines in the 1980s (Fourie and Pienaar, 1983). This success was followed by the description of immunity to MSV in inbred lines and in an experimental hybrid (Barrow, 1992). At the *International Institute for Tropical Agriculture* (IITA), breeding programs for resistance to MSV started in 1975 (Dabrowski, 1987; Efron *et al.* 1989). Maize varieties *TZ-Y* (Tropical Zea-Yellow) that combine resistance to MSV with other desirable traits have been developed at IITA in collaboration with the *International Maize and Wheat Improvement Center* (CIMMYT) (Kim *et al.* 1989). *TZ-Y* resistant plants were derived from a cross between *Tuxpeno Planta Baja* from CIMMYT and a source from East Africa. Kim *et al.* (1989) describe a dominant monogenic resistance for "La Revolution" and a polygenic source of 3 major resistance genes for *TZ-Y*. Numerous breeding programs in Africa are using resistance sources developed at IITA to incorporate resistance into their varieties (Efron *et al.* 1989) adapted to local ecological habitats and preferred by farmers and consumers. Despite all these efforts MSV is still a threat to farmers in 2003.

Breeding programs for resistance to TYLCV have been based on the transfer of resistance genes from accessions of wild origin into cultivated tomato (*Licopersicum esculentum*) . Resistant F1 hybrids were obtained by crossing with wild relatives (reviewed in Lapidot and Friedmann, 2002) *L. peruvianum* (Pilowsky and Cohen, 1990; Lapidot *et al.* 1997; Friedmann *et al.* 1998), *L. peruvianum* and *L. pimpinellifolium* (Vidasky *et al.* 1998), *L. chilense* (Zamir *et al.* 1994) and *L. hirsutum* (Vidasky and Czosnek 1998). Different commercial resistant cultivars have been released, *TY20* (Pilowskiy and Cohen, 1990; Rom *et al.* 1993), line *H24* (Kallo and Banerjee, 1990; Hanson *et al.* 2000). In most cases, resistance to TYLCV appears to be controlled by multiple genes (Lapidot and Friedmann, 2002; reviewed in Pico *et al.* 1996, 1999). One very resistant line is *TY172* derived from *L. peruvianum* (Friedmann *et al.* 1998) for which attempts to produce disease symptoms by grafting with a susceptible infected donor were unsuccessful. Resistance in line *TY172* is controlled by at least three interacting genes (Friedmann *et al.* 1998).

In the case of cassava, the first generation of cassava mosaic disease (CMD) resistant plants was obtained from interspecific crosses between different cassava cultivars and the wild relative *Manihot glaziovii* in the 1930s in East Africa (Storey and Nichols, 1938). This program also continued in Western Africa at IITA (Hahn *et al.* 1980). The resistance was described as stable and largely additive in nature, with a heritability of about 60% (Hahn and Howland, 1972). However , later studies revealed that resistance to CMD is polygenic and recessive (Jennings, 1976; Hahn *et al.* 1980), and therefore difficult to transfer with other traits. Improved cassava cultivars for yield show mild and moderate symptoms under high disease pressure (Thresh *et al.* 1994) but are not adapted to farmers' preferences for taste and processing. More recently, several land races with a high level of resistance were identified at IITA (Fregene *et al.* 2000) (Figure 2.5). AFLP analysis of the CMD resistant land races and *M. glaziovii* derived lines showed significant genetic differences, suggesting a low probability that these land races escaped from the back cross derivatives with the first source of resistance (Fregene *et al.* 2000). The availability of a genetic map for cassava (Fregene *et al.* 1997) has made it possible to rapidly identify sources of genes responsible for different types of resistance. Classical genetic analysis and molecular genetic mapping have demonstrated that a major dominant gene controls the new source of resistance (Akano *et al.* 2002). A bulk segregant analysis (BSA) approach was used to identify a simple sequence-repeat (SSR) marker gene linked to the CMD resistance gene: the SSR marker is located on linkage group R of the male parent-derived molecular genetic map. The gene has been designated *CMD2* and is flanked by the SSR and a RFLP marker at 9 and 8cM respectively. Nevertheless, more studies are required to further understand the resistance mechanism of the Nigerian resistant land races.

Breeding for resistance is a difficult enterprise because the overall goal is to achieve resistance in combination with other important agronomic characteristics. This usually requires several back–crosses. For instance, some high yielding tobacco cultivars, resistant to *Tobacco mosaic virus* (TMV) were accompanied by poor leaf quality (Johnson and Main, 1983). For cassava, the major problem with some of the TMS lines derived from *M. glaziovii* is that farmers in many regions do not like them. In Tanzania, farmers from the Mara region (East of the lake Victoria) prefer bitter cultivars while those from Karega region (West of the lake Victoria) prefer sweet cultivars (Ndunguru, unpublished data). Moreover, cassava has many different uses and it is important that the agronomic qualities of the cultivar match the habits and traditional practices of the consumer.

Successful breeding depends on understanding the interaction between four components: the viruses, their vectors, the host plants and the environment. The relationship between the host plant and the insect vectors is influenced by (i) the feeding habits and multiplication of the insect, and (ii) the modes of inoculation of the virus, therefore making it complicated to analyze the impact of insects on virus resistance.

It is obvious that breeding programs need access to new sources of resistance. In the limited cases where they are available, viral resistance genes have been used with great success. However, viruses are subject to molecular variability and rapid evolution, and unexpected situations such as synergism can arise, resulting in explosive epidemics. To achieve durable resistance, it is imperative that modern technologies are used to accessing new sources of resistance, which must be made available to breeders.

In designing a virus resistance-breeding program, it is also important to consider the preferences of the consumers, particularly in developing countries. In many respects, consumers from developing countries are much more discerning about their food than those in developed countries, simply because they themselves generally process the raw material which imposes many constraints. In addition, customs and culture hold a much larger sway than in modern societies. For all these reasons, an increasing number of scientists and biologists believe that genetic engineering will be a key for better virus resistant plants in the near future, and will complement existing breeding programs.

2.5 Engineering resistance to viral diseases

Genetic engineering is commonly considered a non-conventional method for the control of plant diseases, in opposition to breeding or conventional methods. Conventional methods rely on natural resistance, whereas thus far genetic engineering has mostly involved using pathogen derived-resistance (PDR). For viruses, this strategy has been based on the use of virus-derived molecules, full-length viral coding sequences or negative dominant mutated viral genes, defective interfering (DI) molecules and more recently RNA mediated resistance. Initially, resistance to TMV was obtained by constitutive expression of a CP transgene in plants (Powell-Abel *et al.* 1986). Subsequently, coat protein mediated resistance (CP-MR) was used to produce many transgenic plants with a high resistance level. Resistance against diverse strains or isolates was obtained for *Alfalfa mosaic virus* (AIMV) (Loesch-Fries *et al.* 1987), TMV (Nelson *et al.* 1988), PVX (Hemenway *et al.* 1988), *Cucumber mosaic virus* (CMV) (Namba *et al.* 1991; Quemada *et al.* 1991), *Tomato mosaic virus* (ToMV) (Sanders *et al.* 1992), *Tomato spotted wilt virus* (TSWV) (de Haan *et al.* 1992), *Rice stripe virus* (RSV) (Hayakawa *et al.* 1992), PVY (Perlak *et al.* 1994) and *Potato leaf-roll mosaic virus* (PLRV) (Kaniewski *et al.* 1994). In the case of *Rice tungro spherical virus* (RTSV), a virus that contains more than one CP gene, resistance was achieved by expressing the CP genes CP1, CP2 and CP3 individually or together (Sivamani *et al.* 1999). The most dramatic example of CP-MR in tropical countries is the resistance induced against *Papaya ringspot virus* (PRSV) first developed in Hawaii (Gonsalves, 2002) and subsequently used in several countries in Asia and Central America <http://www.isaaa.org/publications/brief/Brief_11.htm> (See chapter 20). However, the application of CP-MR has been fairly limited because the resistance is ineffective for viral strains distantly related to PRSV and because the disease is widespread in many regions. A similar limitation was described for CMV (Quemada *et al.* 1991) and *Tobacco rattle virus* (TRV) (Van Dun and Bol, 1988).

Replicase-mediated resistance (replicase-MR) constitutes the most important PDR strategy after CP-MR. Replicase-MR was also first demonstrated with TMV (Golemboski *et al.* 1990). Later on, this strategy was successfully used for CMV (Anderson *et al.* 1992), PVX (Braun and Hemenway, 1992; Baulcombe, 1994), PVY (Audy *et al.* 1994), and PLRV (Kaniewski *et al.* 1994; Thomas *et al.* 1994). For RTSV, moderate resistance (60%) was achieved for plants producing anti-sense replicase sequence while plants expressing the full-length Rep gene, as well as a truncated Rep gene, in the (+)- sense orientation were 100% resistant (Huet *et al.* 1999).

Concerning cassava and CMD, attempts to trigger resistance through transgenesis are presently restricted to the model plant *Nicotiana benthamiana* (Stanley and Townsend, 1985; Hong and Stanley, 1995; Sangaré *et al.* 1999; Hong *et al.* 1996). Resistance was not achieved using CP-MR (Fauquet, unpublished data) but was obtained using defective interfering (DI) DNA (Stanley and Townsend, 1985), or using a full-length coding sequence of the replication-associated-protein (Rep) gene (Hong and Stanley, 1995), and finally using a mutant of the putative NTP binding domain of the Rep gene (Sangaré *et al.* 1999). Work is now underway in cassava, with field-testing planned in Africa for 2003/2004. Recently, cassava plants expressing the complete Rep gene coding sequence of ACMV were generated and tested at the *International Laboratory for Tropical Agricultural Biotechnology* (ILTAB at the Danforth Plant Science Center, St Louis). Extensive challenges in greenhouse conditions showed these plants to be highly resistant or immune to several cassava geminivirus species (Chellappan, *unpublished data*; Figure 2.6). Southern blot analysis revealed very low accumulation of replicase gene transcripts in the highly resistant lines of cassava while the moderately resistant lines displayed significant levels of accumulation of replicase RNA. Similarly to CMD resistance, in the case of RTSV, 100% resistance was obtained for plants accumulating low levels of viral RNA, so that the resistance was attributed to co-suppression (Huet *et al.* 1999). For the same RTSV, a protein mediated strategy (CP-mediated) led only to a moderate level of resistance ranging from 17% to 73% of seedlings that escaped infection. These results suggest a promising future for RNA mediated strategies particularly involving gene silencing based mechanisms.

One major issue related to PDR is the specificity of the resistance mechanism. For instance, protection generated by defective interfering DNA (Frischmuth and Stanley, 1991) is confined to closely related virus strains that express compatible *trans*-acting proteins. Thus, DI DNA originating from a Kenyan isolate of ACMV conferred similar levels of protection against Kenyan and Nigerian isolates that can form viable pseudo-recombinants (Stanley *et al.* 1985), but provided no protection against the more distantly related viruses *Tomato golden mosaic virus* (TGMV) and *Beet curly top virus* (BCTV) (Stanley *et al.* 1990). Another interesting example is the apparent immunity, conferred by a pathogen-derived transgene encoding a hairpin RNA, to *Barley yellow dwarf virus* (BYDV) (Wang *et al.* 2000). In this example the immunity is restricted to BYDV and the transgene is ineffective against *Cereal yellow dwarf virus* (CYDV), a related virus also infecting barley. Similarly, resistance to *Rice yellow mottle virus* (RYMV) was conferred to cultivated African rice varieties through an RYMV transgene, coding for an RNA-dependent RNA polymerase (Pinto *et al.* 1999). However, the durability of this resistance in the African environment is questionable because of the existence of a wide range of strains in Africa (Fargette *et al.* 2002) and the possibility that some may break resistance.

Despite some successes, CP-MR or replicase-MR strategies have only enjoyed limited deployment in the field due to a lack of investment and the complexity of biosafety regulations. The newly emerging technology of gene silencing, also called small interfering RNA (siRNA) or RNA interfering (RNAi) technology (Pooggin *et al.* 2003), seems much more attractive as it bypasses many of these concerns for practical use and therefore deployment. Gene silencing is very specific, requiring a high degree of homology between the transgene and the invading virus. This usually results in a narrow range of protection, limited to very closely related viruses. However, as more becomes known about the mechanisms involved, we will better learn to exploit this natural defense mechanism, perhaps leading to broad range protection.

An emerging concern with virus-derived transgenes is their capacity to recombine with the virus genome (Gal *et al.* 1992; Schoelz and Wintermantel, 1993) leading to the fear that *de novo* viruses could emerge from using such technologies. The use of mutated proteins also raises biosafety concerns and it is usually necessary to demonstrate that these mutations do not cause any kind of toxicity or allergic reactions in consumers.

A further strategy involving the G5 protein (Padidam *et al.* 1999a), a single stranded DNA binding protein isolated from an inovirus phage, was shown to trigger resistance in tobacco and tomato against different geminivirus species infecting cassava and tomato from different parts of the world (Yadav, *unpublished data*). This strategy is particularly promising because it could offer wide spectrum resistance in any plant for any geminivirus.

At present, testing of genetically engineered resistant plants in developing countries is limited, largely due to the need to establish and maintain biosafety regulatory frameworks. Kenya and South Africa are presently the most advanced countries in sub-Saharan Africa in this regard. In Kenya, the *Kenyan Agricultural Research Institute* (KARI) has already conducted field trials of sweet potato expressing a *Sweet potato feathery mottle virus* (SPFMV) transgene. Realizing the importance of transgenic research in combating plant diseases, several countries are now formulating agricultural biosafety policies that will allow introduction and evaluation of transgenic plants including virus resistant plants. In East Africa, for example, through the *Association for Strengthening Agricultural Research in Eastern and Central Africa* (ASARECA), many countries have developed and reviewed several biosafety regulations and policies for introduction, evaluation and development of transgenic crops.

2.6 General discussion

Recent history does not offer any single miracle solution to control plant viruses in developing countries. There are, however, many encouraging examples demonstrating that there are ways to efficiently block viruses, including MSV in maize, CMD in cassava, Tungro disease in rice, TYLCV in tomato and others. Although some of these approaches were discovered almost 100 years ago, there are only few examples, if any, where these solutions have been successfully implemented at a regional or continental level for long periods of time. The difficulties are compounded by poor knowledge about the viruses, their vectors, and resistance mechanisms in plants, and above all the difficulty of combining known plant resistance genes with other desirable traits. Moreover, there is a chronic lack of funding to address these issues and no attempt to deploy combined strategies at a national or international level. Cassava is perhaps the most illustrative example, and can be considered in analogy with potatoes in developed countries. When European nations decided after World War II to improve the viral situation of potatoes in Europe, instead of adopting a single strategy, a combination of possible solutions was preferred. These included: breeding for elevated resistance, sanitation and implementation of certification systems with the involvement of scientists, farmers' associations and a strong and sustained international coordinated policy. Therefore a single solution based strategy will probably not work for cassava in developing countries either.

One major component is the human factor. Farmers need to be considered in breeding and biotechnological projects so that their preferences are taken into account. The concept of the 1970s that breeding for resistance was the panacea, and farmers would take whatever was provided for them, is definitively dead and lessons should be learned from that failure. This task is particularly daunting in developing countries, where a multitude of different people groups and cultures are found in many countries. With no winter and extensive biodiversity, the number and variety of insects and viruses are incredible and better control measures have to be developed.

Modern technologies offer a range of new possibilities to overcome the many difficulties of controlling plant virus diseases. It is already becoming easier and cheaper to propagate virus free material using tissue culture techniques. PCR technology will drive the cost of certification down, thereby boosting plant sanitation. The number of virus sequences in publicly available databases is increasing exponentially; this will directly support better diagnosis and genetic engineering and will improve our understanding of virus resistance. Marker assisted breeding will shorten the work of breeders, offering a precise tool to move natural or *de novo* genes by clusters of genes to interesting cultivars for better adaptability to the environment and to farmer preferences. Finally, genetic engineering will offer the unprecedented possibility to return to original sources of plant resistance in wild species where no crosses are possible and transfer these sources to improved varieties in a short period of time. The recently discovered plant defense mechanism of gene silencing offers an increased chance of success in controlling viruses, with the possibility of simplified biosafety regulations for an RNA-based approach.

Better control of plant viruses will dramatically improve the lives of millions of people in developing countries. To achieve this, a substantial increase in research funding is needed, from developing countries themselves as well as the international community. Finally, developing nations need to accept these new technologies to improve the productivity of their food crops.

References

Adipala, E., Warren, H.L., Epieru, G., Takan, J.P., Khyamanywa, S. and Willson, H. 1998. Comparative performance of cv *Igola-1* and other local groundnut cultivars for the control of rosette disease. <www.ag.vt.edu/ipmcrs/meetings/symp98/abstracts1.html>

Ahohuendo, A.C. and Sarkar, S. 1995. Partial control of the spread of African cassava mosaic virus in Benin by intercropping. *Zeit. Pfanzen. Pfanzen.* 102: 249-256.

Akano, A.O., Dixon, A.G.O., Mba, C., Barrera, E. and Fregene, M. 2002. Genetic mapping of a dominant gene conferring resistance to cassava mosaic disease. *Theor. appl. Gen.* Published online. DOI10.1007/s00122-002-0891-7.

Anderson, J.M., Palukaitis, P. and Zaitlin, M. 1992. A defective replicase gene induces resistance to cucumber mosaic virus in transgenic tobacco plants. *Proc. Nat. Acad. Sci. USA* 89: 8759-8763.

Antignus, Y., Nestle, D., Cohen, S. and Lapidot, M. 2001. Ultraviolet-deficient greenhouse environment affects whitefly attraction and flight behavior. *Env. Entomol.* 30: 394-399.

Audy, P., Palukaitis, P., Slack, S. and Zaitlin, M. 1994. Replicase-mediated resistance to potato virus Y in transgenic tobacco plants. *Mol. Plant-Microbe Interact.* 7: 528-530.

Barker, H. and Harrison, B.D. 1986. Restricted distribution of potato leafroll virus antigen in resistant potato genotypes and its effect on transmission of the virus by aphids. *Ann. appl. Biol.* 109: 595-604.

Barrow, M.R. 1992. Development of maize hybrids resistant to maize streak virus. *Crop Prot.* 11: 267-271.

Baulcombe, D. 1994. Replicase-mediated resistance: A novel type of virus resistance in transgenic plants. *Trends Micro.* 2: 60-63.

Bol, J.K. and van Kan, J.A.L. 1988. The synthesis and possible functions of virus-induced proteins in plants. *Microbiol. Sci.* 5: 47-52.

Braun, C.J. and Hemenway, C.L. 1992. Expression of amino-terminal portion of full-length viral replicase genes in transgenic plants confers resistance to potato virus X infection. *Plant Cell* 4: 735-744.

Brink, J.A., Woodward, B.R. and DaSilva, E.J. 1998. Plant biotechnology: a tool for development in Africa. *Electronic J. Biotechnol.* ISSN: 0717-3458, 1: 1-10.

Brown, J.K. 1994. Current status of Bemisia tabaci as a plant pest and virus vector in agro-ecosystems worldwide. *FAO Plant Prot. Bull.* 42: 3-32.

Buddenhagen, I.W. 1983. Crop improvement in relation to virus diseases and their epidemiology. In: *Plant* virus *epidemiology*. R.T. Plumb and J.M. Thresh (eds). Blackwell, Oxford. pp. 25-37.

Caranta, C., Pflieger, S., Lefebvre, V., Daubèze, A.M., Thabuis, A. and Palloix, A. 2002. QTLs involved in the restriction of cucumber mosaic virus (CMV) long-distance movement in pepper. *Theor. appl. Gen.* online. DOI 10.1007/s001220100753, 104: 586-591.

Cassells, A.C. and Long, R.D. 1980. The regeneration of virus-free plants from cucumber mosaic virus and potato virus Y infected tobacco explants cultured in the presence of virazole. *Z. Naturforschung* 35c: 350-351.

Chatterji A., Padidam, M., Beachy R.N. and Fauquet C.M. 1999. Identification of replication specificity determinants in two strains of tomato leaf curl virus from New Dehli. *J. Virol.* 73: 5481-5489.

Cohen, S. and Antignus, Y. 1994. Tomato yellow leaf curl virus, a whitefly-borne geminivirus of tomatoes. *Adv. Dis. Vector Res.* 10: 259-288.

Dabrowski, Z.T. 1987. *Cicadulina ghaurii* (Hem., Euscelidae): Distribution, biology and maize streak virus (MSV) transmission. *J. appl. Entomol.* 103: 489-496.

De Haan, P., Gielsen, J.JL., Prins, M., Wijkamp, I.G., van Schepen, A., Peters, D., Van Grinsven, M.Q.J.M. and Goldbach, R. 1992. Characterization of RNA-mediated resistance to tomato spotted wilt virus in transgenic tobacco plants. *Bio/Technology* 10: 1133-1137.

Deng, D., Otim-Nape, G.W., Sangaré, A., Ogwal, S., Beachy, R.N. and Fauquet, C.M. 1997. Presence of a new virus closely related to East African cassava mosaic geminivirus, associated with cassava mosaic outbreak in Uganda. *Afr. J. Root Tuber Crops* 2: 23-28.

Efron, Y., Kim, S.K., Fajesmisin, J.M., Mareck, J.H., Tang, C.Y., Dabrowski, Z.T., Rossel, H. Thottappilly, G. and Buddenhagen, I.W. 1989. Breeding for resistance to maize streak virus: a multidisciplinary team approach. *Plant Breeding* 103: 1-36.

Etienne, J. and Rat, B. 1973. Le stripe: Une maladie importante du mais a la Reunion. *Agro. Trop.* 28: 11-17.

Fargette, D. and Fauquet, C.M. 1988. A preliminary study on the influence of intercropping maize and cassava on the spread of African cassava mosaic by whiteflies. *Ann. appl. Biol.* 17: 195-202.

Fargette, D., Fauquet, C.M., and Thresh, J.M. 1994. Analysis and modeling of the temporal spread of African cassava mosaic virus and implications for disease control. *Afr. Crop Sci. J.* 2: 449-458.

Fargette, D., Fauquet, C., Grenier, E. and Thresh, J.M. 1990. The spread of African cassava mosaic virus into and within cassava fields. *J. Phytopath.* 130: 289-302.

Fargette, D., Thouvenel, J.-C., and Fauquet, C. 1987. Virus content and symptom expression of cassava leaves infected by African cassava mosaic virus. *Ann. appl. Biol.* 110: 65-73.

Fargette, D., Pinel, A., Halimi, H., Brugidou, C., Fauquet, C. and Van Regenmortel, M. 2002. Comparison of molecular and immunological typing of isolates of *Rice yellow mottle virus. Arch. Virol.* 147: 583-596.

Fauquet, C.M. and Stanley, J. 2003. Geminivirus classification and nomenclature; progress, achievements and problems. *Ann. appl. Biol.* 142: 165-189.

Fauquet, C.M., Maxwell, D.P., Gronenborn, B. and Stanley, J. 2000. Revised proposal for naming geminiviruses. *Arch. Virol.* 145: 1743-1761.

Fereres, A. 2000. Barrier crops as a cultural control measure of non-persistently transmitted aphid-borne viruses. *Virus Res.* 71: 221-231.

Fielding, J. 1933. Field experimental work on rotation crops. In: *Empire Cotton Growing Assoc. Progr. Rep.* 1931-1932. pp. 10-14.

Fondong,, V., Pita, J.S., Rey, M.E.C., de Kochko, A., Beachy, R.N., and Fauquet, C.M. 2000. Evidence of synergism between African cassava mosaic virus and the new double recombinant geminivirus infecting cassava in Cameroon. *J. Gen. Virol.* 81: 287-297.

Fondong, V.N., Thresh, J.M. and Zok, S. 2002. Spatial and temporal spread of cassava mosaic virus disease in cassava grown alone and when intercropped with maize and/or cowpea. *J. Phytopathol.* 150: 365-374.

Fourie, A.P. and Pieenar, J.H. 1983. Breeding for resistance to maize streak virus: A report on the vaal Harts breeding programme. Tech. Commun.-S. Afr. Dept. Agric. Pretoria. *The Department* 182: 41-50.

Fregene, M., Angel, F., Gomez, R., Rodriguez F., Chavariaga, P., Roca, W., Tohme, J. and Borniabale, M. 1997. A molecular genetic map of cassava (*Manihot esculenta* Crantz). *Theor. appl. Gen.* 95: 431-441.

Fregene, M., Bernal, A., Duque, M., Dixon, A. and Tohme, J. 2000. AFLP analysis of African cassava (*Manihot esculenta* Crantz) germplasm resistant to the cassava mosaic disease (CMD). *Theor. appl.*Gen 100: 678-685.

Friedmann, M., Lapidot, M., Cohen, S. and Pilowsky, M. 1998. A novel source of resistance to tomato yellow leaf curl virus exhibiting a symptomless reaction to viral infection. *J. Am. Society Hort. Sci.* 123: 1004-1007.

Frischmuth, T. and Stanley, J. 1991. African cassava mosaic virus DI DNA interferes with the replication of both genomic components. *Virology* 183: 539-544.

Gal, S., Pisan, B., Hohn, T., Grimsley, N. and Hohn, B. 1992. Agroinfection of transgenic plants leads to viable cauliflower mosaic virus by intermolecular recombination. *Virology* 197: 525-533.

George, E. 1993. Plant progagation by tissue culture, Part 1, *The Technology*. (2nd ed.). pp574.

Gibson, R.W., Jeremiah, S.C., Aritua, V., Msabaha R.P., Mpembe, I. and Ndunguru, J. 2000. Sweet potato virus disease in Sub-Sahara Africa: evidence that neglect of seedlings in the traditional farming system hinders the development of superior resistance landraces. *J. Phytopathol.* 148: 441-447.

Golemboski, D.B., Lomonossoff, G.P. and Zaitlin, M. 1990. Plants transformed with a tobacco mosaic virus nonstructural gene sequence are resistant to the virus. *Proc. Nat. Acad. Sci. USA* 87: 6311-6315.

Gonsalves, D. 2002. Coat protein transgenic papaya: "acquired" immunity for controlling papaya ringspot virus. *Curr. Top. Microbiol. Immunol.* 266: 73-83.

Gorter, G.J.M.A. 1959. Breeding maize for resistance to streak. *Euphytica* 8: 234-240.

Hahn, S.K, and Howland, A.K. 1972. Breeding for resistance to cassava mosaic. In: *IITA Proceedings of the Cassava Mosaic Workshop*. S.K. Hahn (ed), Ibadan, Nigeria. pp. 4-7.

Hahn, S.K., Terry E.R. and Leuschner, K. 1980. Cassava breeding for resistance to cassava mosaic disease. *Euphytica* 29: 673-683.

Hanson, P.M., Bernacchi, D., Green, S., Tanksley, S.D., Venkataramappa, M., Padmaja, A.S., Chen, H., Kuo, G., Fang, D. and Chen, J. 2000. Mapping a wild tomato introgression associated with tomato yellow leaf curl virus resistance in a cultivated tomato line. *J. Am. Society Horti. Sci.* 125: 15-20.

Hayakawa, T., Zhu, Y., Itoh, K. Kimura, Y., Izawa, T., Shimamoto, K. and Toriyama S. 1992. Genetically engineered rice resistance to rice stripe virus, an insect-transmitted virus. *Proc. Nat. Acad. Sci.. USA* 89: 9865-9869.

Hemenway, C. Fang, R-X., Kaniewski, W., Chua, N-H. and Tumer, N. 1988. Analysis of the mechanism of protection in transgenic plants expressing the potato virus X coat protein or its antisense RNA. *EMBO J.* 7: 1273-1280.

Hibino, H., Tiongco, E.R., Cabunagan, R.C. and Flores, Z.M. 1987. Resistance to rice tungro-associated viruses in rice under experimental and natural conditions. *Phytopathology* 77: 871-875.

Hilje, L., Costa, H.S. and Stansly, P.A. 2001. Cultural practices for managing *Bemisia tabaci* and associated viral diseases. *Crop Protect.* 20: 801-812.

Hong, Y. and Stanley, J. 1995. Regulation of African cassava mosaic virus complementary-sense gene expression by N-terminal sequences of the replication-associated protein AC1. *J. Gen. Virol.* 76: 2415-2422.

Huet, H., Mahendra, S., Wang, J., Sivamani, E., Ong, C.A., Chen, L., de Kochko, A., Beachy, R.N. and Fauquet, C. 1999. Near immunity to Rice tungro spherical virus achieved in rice by a replicase-mediated resistance strategy. *Phytopathology* 89: 1022-1027.

Inoue, H. and Ruay-Aree, S. 1977. Bionomics of green rice leafhopper and epidemics of yellow orange leaf virus diseases in Thailand. *Trop. Agri. Res. Ser.* 10: 117-121.

James, C. 1997. Progressing public-private sector partnership in International Agriculture Research and Development. In: *ISAAA* Briefs 4: 1-32.

Jennings, D.L. 1976. Cassava *Manihot esculenta* (Euphorbiacea). In: *Cultivated crops*. J. Harlan (ed). Longman, London. pp 81-84.

Johnson, C.S. and Main, C.E. 1983. Yield/quality tradeoffs of tobacco mosaic virus-resistant tobacco cultivars in relation to disease management. *Plant Dis.* 67: 886-890.

Kallo, G. and Banerjee, M.K. 1990. Transfer of Tomato leaf curl virus resistance from *Lycopersicon hirsutum* f. *glabratum* to *L. esculentum*. *Plant Breeding* 105: 156-159.

Kaniewski, W., Lawson, C., Loveless, J., Thomas, P. Mowry, T., Reed, G., Mistky, T., Zalewski, J. and Muskopf, Y. 1994. Expression of potato leafroll virus (PLRV) replicase genes in Russet Burbank potatoes provide immunity to PLRV. In: *Proceed. 3rd EFPP Conference*, M. Manka, (ed), *J. Phytopathol.* pp. 289-292.

Kim, S.K., Efron, Y., Fajesmisin, J. M. and Buddenhagen, I.W. 1989. Mode of gene action for resistance in maize to Maize streak virus. *Crop Sci.* 29: 890-894.

Lapidot, M., Friedmann, M., Lachman, O., Yehezkel, A., Nahon, S., Cohen, S. and Pilowsky, M. 1997. Comparison of resistance level to Tomato yellow leaf curl virus among commercial cultivars and breeding lines. *Plant Dis.* 81: 1425-1428.

Lapidot, M. and Friedmann, M. 2002. Breeding for resistance to whitefly-transmitted geminiviruses. *Ann. appl. Biol.* 140: 109-127.

Lei, J.D. and Agrio, G.N. 1986. Mechanisms of resistance in corn to maize dwarf mosaic virus. *Phytopathology* 76: 1034-1040.

Li, Q., Ryu, K.H. and Palukaitis, P. 2001. Cucumber mosaic virus-plant interactions: Identification of 3a protein sequences affecting infectivity, cell-to-cell movement, and long-distance movement. *Mol. Plant-Microbe Interact.* 14: 378-385.

Loesch-Fries, L.S., Merlo, D., Sinnen, T., Burshop, L., Hill, K., Krahn, K., Jarvis, N., Nelson, S., and Halk, E. 1987. Expression of alfalfa mosaic virus RNA 4 in transgenic plants confers virus resistance. *EMBO J.* 6: 1845-1851.

Mansoor, S., Briddon, R. W., Zafar, Y. and Stanley, J. 2003. Geminivirus disease complexes: an emerging threat. *Trends Plant Sci.* 8: 128-134.

Mansoor, S., Khan S.H., Bashir, A., Saeed M., Zafar Y., Malik K.A., Briddon R., Stanley J. and Markham P. G. 1999. Identification of a novel circular single-stranded DNA associated with cotton leaf curl disease in Pakistan. *Virology* 259: 190-199.

Manwan, I., Sama, S. and Rizvi, S.A. 1985. Use of varietal rotation in the management of tungro disease in Indonesia. *Indonesian Agric. Res. Develop. J.* 7: 43-48.

Mesfin, T., Bosque-Perez, N.A., Buddenhagen, I.W., Thottappilly, G., and Olojede, S. O. 1992. Studies of maize streak virus isolates from grass and cereal hosts in Nigeria. *Plant Dis.* 76: 789-795.

Mwanga, R.O.M., p'Obwoya, C.N.O., Otim-Nape, G.W., and Odongo, B. 1991. Sweetpotato improvement in Uganda. In: The Role of Root Crops in Regional Food Security and

Sustainable Agriculture. *Proc. East. South. Afr. Regional Root Crops Workshop 4th*. N. Alvarez and R. Asiedu, (Eds). IITA, Ibadan, Nigeria. pp. 59-67.

Namba, S., Ling, K., Gonsalves, C., Gonsalves, D. and Slightom, J.L. 1991. Expression of the gene encoding the coat protein of cucumber mosaic virus (CMV) strain WL appears to provide protection to tobacco plants against infection by several different CMV strains. *Gene* 107: 181-188.

Ndunguru, J. 2002. Disease hampers bean production. *Agriforum Newsletter* 19: 11.

Ndunguru, J. and Kapooria, R. G. 1997. Detection of cucumber mosaic virus in vegetables and herbs in Zambia. *Afr. Pl. Protect.* 3: 57-58.

Ndunguru, J. and Kapooria, R.G. 1999. Identification and incidence of virus diseases of capsicum annum in the Lusaka Province of Zambia. *EPPO Bull.* 29: 183-189.

Ndunguru, J. and Kapooria, R.G. 2000. The Rrole of non-capsicum hosts in the ecology of Alfalfa mosaic alfamovirus in pepper growing areas of Lusaka Province, Zambia. *J. Myco. Pl. Pathol.* 30: 11-14.

Ndunguru, J. and Rajabu, C.A. 2002. Papaya ring spot virus disease in the Lake Victoria basin. *Trop. Sci.* 42: 11-16.

Nelson, R.S., McCormick, S.M., Delanney, X., Dube, P., Layton, J., Anderson, E.J., Kaniewska, M., Proksch, R.K., Horsch, R.B., Rogers, S.G., Fraley, R.T. and Beachy, R.N. 1988. Virus tolerance, plant growth, and field performance of transgenic tomato plants expressing coat protein from tobacco mosaic virus. *Bio/Technology* 6: 403-409.

Otim-Nape, G.W., Bua, A., Baguma, Y. and Thresh, J.M. 1997. Epidemics of the severe cassava mosaic virus in Uganda and efforts to control it. *Afr. J. Root Tuber Crops* 2: 42-43.

Padidam, M., Beachy R.N. and Fauquet, C.M. 1999a. A phage single-stranded DNA (ssDNA) binding protein complements ssDNA accumulation of a geminivirus and interferes with viral movement. *J. Virol.* 73: 1609-1616.

Padidam, M., Sawyer, S. and Fauquet C.M. 1999b. Possible emergence of new geminiviruses by frequent recombination. *Virology* 265: 218-225.

Palumbo, J.C., Horowitz, A.R. and Prabhaker, N. 2001. Insecticidal control and resistance management for *Bemisia tabaci*. *Crop Protect.* 20: 739-766.

Perlak, F. Kaniewski, W., Lawson, C., Vincent, M. and Feldman, J. 1994. Genetically improved potatoes: their potential role in integrated pest management. In: *Proc. 3rd EFPP Conference*, M. Manka (ed). *J. Phytopathol.* pp 451-454.

Pico , B., Diez, M.J. and Nuez, F. 1996. Viral diseases causing the greatest economic losses to the tomato crop. II. The tomato yellow leaf curl virus - a review. *Scienta Hortic* 67: 151-196.

Pico, B., Ferriol, M., Diez, M.J. and Nuez, F. 1999. Developing tomato breeding lines resistant to tomato yellow leaf curl virus. *Plant Breeding* 118: 537-542.

Pilowsky, M. and Cohen, S. 1990. Tolerance to tomato yellow leaf curl virus derived from *Lycopersicon peruvianum*. *Plant Dis*. 74: 248-250.

Pinto, Y.M., Kok, R. A. and Baulcombe, D.C. 1999. Resistance to rice yellow mottle virus (RYMV) in cultivated African rice varieties containing RYMV transgenes. *Nature Biotechnology* 17: 702-707.

Pita, J.S., Fondong, V.N., Sangaré, A., Otim-Nape, G.W., Ogwal, S. and Fauquet, C.M. 2001. Recombination, pseudorecombination and synergism of geminiviruses are determinant keys to the epidemic of severe cassava mosaic disease in Uganda. *J. Gen. Virol.* 82: 655-665.

Polston, J.E., McGovern, R.J. and Brown, L.G. 1999. Introduction of Tomato yellow leaf curl virus in Florida and implications for the spread of this and other geminiviruses of tomato. *Plant Dis*. 83: 984-988.

Ponz, F. and Bruening, G. 1986. Mechanisms of resistance to plants viruses. *Annu. Rev. Phytopathol.* 24: 355-381.

Pooggin, M. Shivaprasad, P.V., Veluthambi, K. and Hohn, T. 2003. RNAi targeting of DNA virus in plants. *Nature Biotechnology* 21: 131-132.

Powell-Abel, P., Nelson, R.S., De, B., Hoffmann, N., Rogers, S.G., Fraley, R.T. and Beachy, R.N. 1986. Delay of disease development in transgenic plants that express the tobacco virus coat protein gene. *Science* 232: 738-743.

Quedema, H.D., Gonsalves, D. and Slightom, J.L. 1991. Expression of coat protein gene from cucumber mosaic virus strain C in tobacco: Protection against infections by CMV strains transmitted mechanically or by aphids. *Phytopathology* 81: 794-802.

Rom, M., Antignus, Y., Gidoni, D., Pilowsky, M. and Cohen, S. 1993. Accumulation of tomato yellow leaf curl virus DNA in tolerant and susceptible tomato lines. *Plant Dis.* 77: 253-257.

Rose, D.J.W. 1978. Epidemiology of maize streak disease. *Annu. Rev. Entomol.* 23: 259-282.

Rose, F.M. 1938. Rotation crops. In: *Empire Cotton Growing Assoc. Progr. Rep.* 1939-1940. pp. 21-25.

Rybicki, E.P. and Pietersen G. 1999. Plant virus disease problems in the developing world. *Adv. Virus Res.* 53: 127-175.

Sanders, P.R., Sammons, B., Kaniewski, W., Haley, L., Layton, J., Lavallee, B.J., Delannay, X., and Tumer, N.E. 1992. Field resistance of transgenic tomatoes expressing the tobacco mosaic virus or tomato mosaic virus coat protein genes. *Phytopathology* 82: 683-690.

Sangaré, A., Deng, D., Fauquet, C.M. and Beachy R.N. 1999. Resistance to African cassava mosaic virus conferred by a mutant of the putative NTP-binding domain of the Rep gene (AC1) in *Nicotiana benthamiana*. *Mol. Biol. Rep.* 5: 95-102.

Satapathy, M.K. 1998. Chemical control of insect and nematode vectors of plant viruses. In: *Plant Virus Disease Control.* (eds) A. Hadidi, R.K. Khetarpal and H. Koganezawa, APS Press, St. Paul, MN, USA. pp. 188-195.

Schoelz, J.E. and Wintermantel, W.M. 1993. Expansion of viral hosts range through complementation and recombination in transgenic plants. *Plant Cell* 5: 1669-1679.

Simon, J. 1957. Effects of insecticides and physical barriers on field spread of pepper veinbanding mosaic virus. *Phytopathology* 47: 139-145.

Simpkins, I., Walkey, D.G.A. and Neely, H.A. 1981. Chemical suppression of virus in cultured plant tissues. *Ann. appl. Biol.* 99: 161-169.

Sivamani, E., Huet, H., Shen, P., Ong, C.A., de Kochko, A., Fauquet, C. and Beachy, R.N. 1999. Rice plant (*Oryza sativa* L.) containing Rice tungro spherical virus (RTSV) coat protein transgenes are resistant to virus infection. *Mol. Breeding* 5: 177-185.

Sserubombwe, W. 1998. Progress of cassava mosaic virus disease (CMD) and its effects on growth and yield. M Sc. Thesis, Makerere University, Kampala, Uganda.

Sserubombwe, W.S., Thresh, J.M., Otim-Nape, G.W. and Osiru, D.O.S. 2001. Progress of cassava mosaic virus disease and whitefly vector populations in single and mixed stands of four cassava varieties grown under epidemic conditions in Uganda. *Ann. appl. Biol.* 135: 161-170.

Stanley, J. and Townsend, R. 1985. Characterization of DNA forms associated with cassava latent virus infection. *Nucl. Acids Res.* 13: 2189-2330.

Stanley, J., Frischmuth, T. and Ellwood, S. 1990. Defective viral DNA ameliorates symptoms of geminivirus infection in transgenic plants. *Proc. Nat. Acad. Sci. USA* 87: 6291-6295.

Stanley, J., Townsend, R. and Curson, S.J. 1985. Pseudorecombinant between cloned DNAs of two isolates of cassava latent virus. *J. Gen. Virol.* 66: 1055-1061.

Storey, H.H. and Howland, A.K. 1967. Inheritance of resistance in maize to the virus of streak disease in East Africa. *Ann. appl. Biol.* 59: 429-436.

Storey, H.H. and Nichols, R.F.W. 1938. Studies on the mosaic disease of cassava. *Ann. appl. Biol.* 25: 790-806.

Swanson, M.M. and Harrison, B.D. 1994. Properties, relationships and distribution of cassava mosaic geminiviruses. *Trop. Sci.* 34: 15-25.

Thomas, P., Kaniewski, W., Reed, G. and Lawson, C. 1994. Transgenic resistance to potato leafroll virus in Russet Burbank potatoes. In: *Proc. 3rd EFPP Conference*, M. Manka, (ed). *J. Phytopathol.* pp. 551-554.

Thresh, J.M. 1982. Cropping practices and virus spread. *Annu. Rev. Phytopath.* 20: 139-218.

Thresh, J.M., Fargette, D, and Otim-nape, G.W. 1994. Effects of African cassava mosaic geminivirus on the yield of cassava. *Trop. Sci.* 34: 26-42.

Thresh, J.M., Otim-Nape, G.W. and Fargette, D. 1998. The control of African cassava mosaic virus disease: phytosanitation and/or resistance. In: *Plant Virus Disease Control.* (eds) A. Hadidi, R.K. Khetarpal and H. Koganezawa, APS Press, St. Paul, MN, USA. pp. 670-677.

Van Dun, C.M.P. and Bol, J.P. 1988. Transgenic tobacco plants accumulating Tobacco rattle virus coat protein resist infection with tobacco rattle virus and pea early browning virus. *Virology* 167: 649-652.

Van Loon, L.C. 1989. Stress proteins in infected plants. In: *Plant-Microbe Interaction.* T. Kosuge and E.W. Nester (eds). Mc Graw-Hill, New York, USA. pp: 198-237.

Vanitharani R., Karthikeyan, A.S., Anuradha, S. and Veluthambi K. 1996. Genome homologies among geminiviruses infecting *Vigna*, cassava, *Acalypha*, *Croton* and *Vernonia*. *Curr. Sci.* 70: 63-69.

Vidasky, F., Leviatov, S., Milo, J., Rabinowitch, H.D., Kedar, N., and Czosnek, H. 1998. Response of tolerant breeding lines of tomato, *Lycopersicum esculentum*, originating from three different sources (*L. peruvianum*, *L. pimpinellifolium* and *L. chilense*) to early controlled inoculation by tomato yellow leaf curl virus (TYLCV). *Plant Breeding* 117: 165-169.

Vidasky, F. and Czosnek, H. 1998. Tomato breeding lines resistant and tolerant to tomato yellow leaf curl virus issued from *Lycopersicon hirsutum*. *Phytopathology* 88: 910-914.

Wang, M.B., Abbott, D.C. and Waterhouse, P.M. 2000. A single copy of a virus-derived transgene encoding hairpin RNA gives immunity to barley yellow dwarf virus. *Mol. Plant Pathol.* 1: 347-356.

Wang, P.J. and Hu, C.Y. 1980. Regeneration of virus-free plants through *in vitro* culture. In: *A. Fietchter* (ed.). pp. 61-99.

Zamir, D., Ekstein-Michelson, I., Zakay, Y., Navot, N., Zeidan, M., Sarfatti, M. Eshed, Y., Harel, E., Pleban, T., Van Oss, H., Kedar, N., Rabinowitch, H.D. and Czosnek, H. 1994. Mapping and introgression of a tomato yellow leaf curl virus tolerance gene, TY-1. *Theor. appl.Genet.* 88: 141-146.

Zhou, X., Liu, Y., Calvert, L., Munoz, C., Otim-nape, G.W., Robinson, D.J., Harrison, B.D. 1997. Evidence that DNA-A of a geminivirus associated with severe cassava mosaic disease in Uganda has risen by interspecific recombination. *J. Gen. Virol.* 78: 2101-2111.

3. Epidemiology of tropical plant viruses

J.M. Thresh[1], D. Fargette[2] and M.J. Jeger[3]
[1]*Natural Resources Institute, University of Greenwich, Chatham Maritime, Kent ME4 4TB, UK;* [2]*IRD BP 64501 Montepellier, Cedex 5, 34394, France;* [3]*Department of Agricultural Sciences, Imperial College London, Wye Campus, Wye, Ashford, Kent, TN25 5AH, UK*

3.1 Introduction

Epidemiology is broadly defined as the study of disease in host populations. Basic epidemiological concepts and principles in relation to some of the most important areas of study are reviewed in this chapter, with particular reference to tropical virus diseases. One of the main aims is to show the role of epidemiology in developing and evaluating control measures.

3.2 Modes of spread

Viruses must have one or more effective means of spread from infected to uninfected plants if they are to persist in host populations and not be at risk of extinction (Thresh, 1985). A distinction can be made between virus spread above and below ground and also between autonomous spread and by means of humans or vectors. These distinctions are apparent in Table 1 of Chapter 1 which lists the means of spread of some of the most important viruses of tropical crops. Some of these viruses are seed-borne, or disseminated in vegetative propagules. Others have fungal or insect vectors and there is marked specificity in that each virus is transmitted by only one or more species of a single taxonomic group. For example, aphid-borne viruses are not transmitted by other types of insect or by mites, nematodes or fungi.

In considering the effectiveness of the different means of dissemination it is important to distinguish between virus spread over short distances from foci of infection within crops and that occurring over greater distances into or between crops. Local spread within crops is effective in enabling viruses to *exploit* habitats already colonized, whereas distant spread is more difficult to achieve, but leads to the *colonization* of entirely new habitats (Thresh, 1974). It is notable that each of the different means of dissemination can lead to local spread, but only some can achieve effective colonization over longer distances, notably by active winged vectors that can fly or be blown far and also by virus dissemination in infected seed or vegetative propagules.

Insect vectors have very diverse life cycles and characteristics, although a crucial feature of all species is that they are active at one or more stages of their life cycle (Thresh, 1985). For example, the first-instar nymphal stages of mealybug vectors are small and well-adapted to dispersal by wind, whereas the later stages and wingless female adults are relatively immobile and make little contribution to distant spread. Although adult male mealybugs have wings they do not have functional mouthparts and so cannot transmit viruses. With other types of insect vector including leafhoppers, planthoppers, thrips, whiteflies, beetles and aphids, at least some of the adults are winged and they are well adapted to disperse far. This is exemplified by experience with *Nilaparvata lugens* the planthopper vector of *Rice grassy stunt virus* and *Rice ragged stunt virus* that is dispersed annually over long distances from tropical to temperate regions of Asia (Thresh, 1983a).

Knowledge of the means of virus spread is of crucial importance in attempts to develop effective methods of control. There are obvious benefits to be gained from selecting only virus-free propagules to minimize the losses caused by seed-borne viruses and those of crops propagated vegetatively. It is also important to know which viruses are transmitted by soil-inhabiting fungi so that infested sites can be avoided and appropriate crop rotations can be adopted. Identification of an arthropod vector can lead to effective control by pesticides, or by non-chemical means such as inter-cropping, barrier crops and use of natural enemies.

3.3 Epidemiological cycles and cycles of infection

When healthy plants of a susceptible variety are exposed to virus inoculum, the virus may or may not become established in the plants and replicate. If infection does occur, and the virus becomes systemic, the host plant can act as a source of inoculum for further spread. Infectiousness may occur before observable symptoms. Moreover, symptoms may or may not develop at the initial site of infection.

The duration of the *latent period* (time from infection to infectiousness) is important epidemiologically because it influences the potential number of *infection cycles* that can occur during the life-span of the host, although in reality plants often become resistant to infection as they age and approach maturity. This so-called *'mature plant resistance'* is known to be an important feature of many crops including *Solanum* potato (Beemster, 1957) and cassava (Fargette *et al.*, 1994). Moreover, vectors may not be abundant throughout the duration of the crop and their activity may be curtailed during periods of unfavourable weather. Thus spread may be restricted to brief 'windows of opportunity' when vectors are present, active and infectious whilst plants are still at a susceptible stage of growth (Thresh, 1983a).

The concept of infection cycles is simplistic as temperature and other factors influence crop growth and virus multiplication. Nevertheless, it is useful in indicating the potential for rapid spread of viruses of many herbaceous annual crops that are grown from seed, especially during the early active phase of crop growth. By contrast, cassava is propagated vegetatively and with cassava mosaic geminiviruses the number of infection cycles is restricted by the onset of mature plant resistance. With tree crops such as cocoa and citrus, vegetative growth is sporadic and virus is slow to become systemic throughout the canopy of branches. Repeated cycles of infection are possible only because of the longevity of the crop.

Such differences in epidemiological behaviour can be considered in ecological and evolutionary terms and related to the type, continuity and stability of the habitats that crops provide for viruses (Thresh, 1980). Those infecting short-lived herbaceous annuals are unlikely to persist unless they have the ecological competence to invade and multiply rapidly whenever and wherever suitable habitats occur. They must also have effective means of survival between growing seasons. For example, *Bean common mosaic virus* is seed-borne, *Peanut clump virus* persists in the long-lived resting spores of a soil-inhabiting fungus vector (*Polymyxa graminis*) and *Maize dwarf mosaic virus* infects *Sorghum halepense* and other perennial weeds in maize fields. Viruses of long-lived perennials are not subject to such limitations.

3.4 Spatial patterns of spread

Information on the distribution of diseased plants within crop stands may facilitate identification of the initial source(s) of inoculum from which spread occurs. It may also be possible to infer the means of spread and the most effective control strategy to adopt. Spread can occur from infected sources within crops and from sources that are nearby or remote. The distinctions between local and distant and between crop and non-crop hosts can be somewhat arbitrary, as spread may occur from diverse sources. Initial foci that occur within crops from the outset present the greatest risk and can lead to rapid infection, even if spread occurs only over short distances by contact, or by vectors of limited mobility. Spread into crops from remote sources is less likely to occur, especially if spread is entirely dependent on vectors that are able to fly, or that are blown far by wind currents.

Such considerations explain why there is such variation in the patterns and rates of spread of virus diseases. Patches of disease often occur within crops following local spread around initial foci of infection due to soil-borne inoculum, the debris of previous crops, the use of infected seed, or the occurrence of weed hosts within the stand. Patches can also occur as a consequence of spread around primary foci of infection that are initiated by incoming arthropod vectors, such as the aphid vector (*Aphis craccivora*) of groundnut rosette disease (Naidu *et al.*, 1998).

New outbreaks of *Cacao swollen shoot virus* are initiated by *'jump spread'* of viruliferous wind-borne first-instar mealybug nymphs. Subsequent *'radial spread'* is by mealybugs walking mainly short distances between the canopy branches of adjacent trees to form obvious patches of disease within plantations (Cornwell, 1958). The patches continue to expand and give rise to additional *'satellite'* outbreaks that ultimately coalesce and form continuous *'areas of mass infection'*. Such patterns of spread have been reported with a wide range of virus diseases, but with differences in their rate of progress and in the relative importance of 'jump' and 'radial' spread (Jeger and Thresh, 1993).

When spread is caused mainly by incoming viruliferous arthropod vectors from outside sources there may be *'edge effects'* around the margins of crop stands. Such information can be helpful in devising management practices to decrease the incidence of infection by planting away from and upwind of known sources of inoculum and in large compact blocks to decrease the proportion of plants in the vulnerable peripheral areas. There may also be advantages in planting a barrier crop or wind-break around the field margin, or border rows of a resistant variety. However, there is limited scope for such approaches in the tropics where fields are typically small and farmers have little latitude in choice of site.

A feature of many crop diseases is that there are clear trends in the incidence of disease with increasing distance along one or more transects across the planting being assessed. Such changes are referred to as *disease gradients* (Gregory, 1968). Those associated with changes in soil fertility or micro-climate or with physical features of the environment are termed *environmental gradients*. In contrast, *infection gradients* are associated with spread from initial foci of infection and incidence tends to decrease with increasing distance from the source of inoculum (Rose, 1973; Thresh and Lister, 1960; Fargette *et al*., 1993). Gradients are visualised by plotting disease incidence against distance using appropriate scales. The gradients tend to be curvilinear and concave as disease incidence decreases most rapidly near the source and then declines to a low *'horizon of infection'* (Fig. 3.1), which is the distance beyond which spread is so infrequent that infection becomes insignificant.

For any particular disease, gradients tend to be shallower and over longer distances when conditions favour spread, compared with situations or seasons when condition are less favourable (Fig. 3.1). For viruses with wind-borne vectors the disease gradients tend to be steeper and more circumscribed up-wind than down-wind from the source. Consequently it is seldom possible to give a simple unqualified advice on the most appropriate isolation distance to adopt to avoid serious risk of infection. The 'safe' distance depends not only on site and season but also on the size of the source of inoculum and the extent of the losses that are considered acceptable. This explains why substantial isolation may be justified and appropriate for producing 'elite' specially selected virus free propagules but not for routine crop production (Thresh, 1976).

The mode of spread and type of vector are other important considerations. Shallow gradients of spread over great distances are due solely to the most mobile stages of the vector and their occurrence may be restricted to clearly defined migration periods. Spread at other times and by less active individuals or development stages of the vector species results in relatively steep gradients around the source. Thus, the initial patterns of infection by migrants moving considerable distances may be quite different from those due to subsequent local movement by their less mobile progeny. Where different forms of the vector coexist the observed gradients of infection represent the superimposition of two or more separate distributions. However, there is seldom any discontinuity or inflexion in the disease gradients observed. There is considerable uncertainty as to the relative importance in local spread of the short- and long-flying forms of certain leafhopper vectors, including the *Cicadulina* spp. that transmit *Maize streak virus* (Rose, 1972). Conflicting claims have also been made on the performance of apterous and alate aphids as vectors and their relative importance seems to differ between viruses and between regions. Nevertheless, there is general agreement that in some crops the relatively

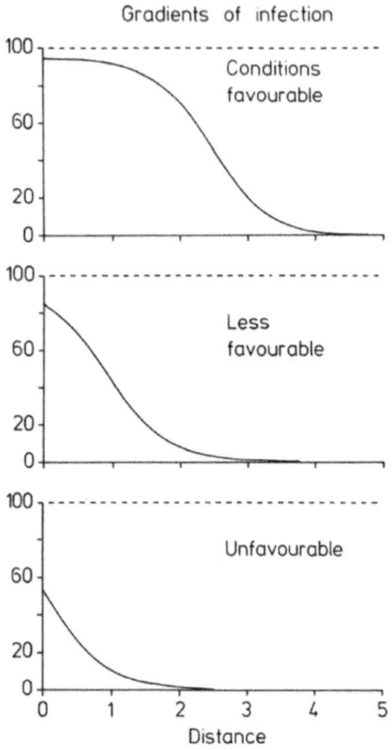

Fig. 3.1 Generalized gradients of disease obtained by plotting incidence (%) on the vertical scale against distance on the horizontal scale (arbitrary units) for spread under favourable (top), less favourable (middle) and unfavourable conditions (bottom).

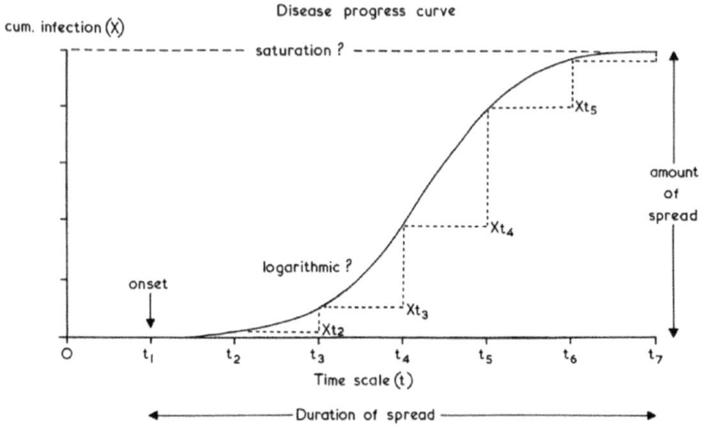

Fig. 3.2 Generalized curve of disease progress obtained by plotting cumulative incidence (X) against time (t) (arbitrary units). The sequential observations X_t can be used to estimate absolute and relative rates of spread, maximum disease incidence and area under the disease progress curve.

immobile forms of the vector often reach plants that have been infected already by the more active members of the population.

The general 'blurring' of disease gradients with time makes it increasingly difficult to assess whether new infections are due to local spread from secondary foci, or to continuing spread from primary foci. The spread of some diseases appears to increase dramatically once a critical level of infection has been exceeded and control then becomes increasingly difficult to achieve, as is apparent from experience with cocoa swollen shoot disease in West Africa (Chapter 20).

Disease gradients are difficult to handle mathematically in seeking to compare gradients between sites, seasons and diseases. Transformations have been used widely to convert the usual curvilinear relationships between disease incidence and distance from the source to straight lines and so enable the use of conventional regression techniques. One approach is to transform disease incidence by using logarithms of incidence values after allowing for multiple infection, or by double logarithmic transformations of both incidence and distance (Gregory, 1968). Other more complex mathematical approaches are also possible (Jeger, 1999), and have been applied to cocoa swollen shoot (Jeger and Thresh, 1993) and rice tungro (Holt and Chancellor, 1999).

3.5 Temporal patterns of increase

Temporal patterns of disease increase are concerned with changes in disease incidence with time. The amount of disease and the rate of increase within the life-span of a crop are important in determining the losses caused and in evaluating the effectiveness of host plant resistance and other approaches to control.

Temporal patterns of spread are visualised from disease progress curves produced by plotting cumulative incidence, usually as a percentage or proportion, against time (Fig. 3.2). The most appropriate time-scale depends on the nature and longevity of the crop being studied and ranges from days to years. It is often necessary to compromise between the need for frequent observations to follow the details of disease progress and the additional time and expense incurred (Nutter, 1997). Disease progress curves are often sigmoid in shape as the incidence of many diseases tends to increase slowly at the outset and also in the later stages of crop growth as crops mature, or when almost all plants become diseased. Increase is usually most rapid at the intermediate stages of crop growth.

Disease progress curves for plant viruses merit detailed scrutiny to obtain the maximum possible information from the observations made (Thresh, 1983; Nutter, 1997). Various means of quantifying disease increase can be derived including:- *time of onset* of disease; *duration* and *rate* of increase; *final* or *maximum disease incidence*; and other derived variates such as *area under the disease progress curve* (AUDPC), many of which are highly intercorrelated. These criteria have been used widely in studying the temporal dynamics of many tropical virus diseases of diverse crops. More sophisticated mathematical procedures for analysis have been developed (Chan and Jeger, 1994; Fargette *et al.,* 1994; Holt *et al.*, 1997; Jeger *et al.,* 1998; Madden *et al.*, 2000), but these largely remain in the strategic domain and have yet to find practical application.

The greatest disease threat occurs when there is rapid increase over a long period and from many initial foci. However, numerous initial foci, or a prolonged period of increase, can lead to high incidences of disease despite low rates of disease progress. Initial disease incidence at the outset of crop growth and the rate and duration of increase can be manipulated to influence final disease incidence. In evaluating control measures it is helpful to distinguish phytosanitation or other means of decreasing the initial amount of disease and *'rate reducing'* measures such as host plant resistance, or the use of pesticides to control vectors (Thresh, 1983a). The effectiveness of phytosanitation is inversely related to values of the rate of spread that determines the 'explosiveness' of an epidemic (Putter, 1980). This is consistent with the finding that the removal of infected plants ('roguing') is an effective means of controlling slow-spreading diseases such as cocoa swollen shoot (Thresh, 1988) but not for rapidly spreading diseases such as rice tungro (Holt and Chancellor, 1996).

Monocyclic and *polycyclic* diseases were distinguished by Vanderplank (1963). The former spread mainly or exclusively from initial sources of inoculum and the newly infected plants do not contribute to further spread. In contrast, newly affected plants contribute to the spread of polycyclic diseases and there are repeated cycles of infection depending on the duration of the crop and the length of the latent period.

Plant virus diseases are usually polycyclic, but several monocyclic diseases have also been reported in the tropics. For example, 'kromnek' disease of tobacco in southern Africa is spread into crops by thrips vectors dispersing from infected weed hosts and there is no subsequent spread within crops. The incidence of kromnek disease is decreased if thinning the stand is delayed until the main influx of infective vectors has occurred and so diseased or asymptomatic plants are removed (Vanderplank and Anderssen, 1945). A ringspot disease of sunflower in eastern and southern Africa also behaves monocyclically as inoculum is carried into crops by aphids and there is no spread between sunflower plants (Theuri e*t al.,* 1987). A tomato leaf curl virus in India that is transmitted by whiteflies, seems to share these characteristics (Holt *et al.,* 1999). Maize streak differs in that it behaves as a monocyclic disease in some situations and as a polycyclic disease in others (Rose, 1978). The behaviour of *Bean common mosaic virus* is also complex as it usually causes a polycyclic disease, but is monocyclic in stands of hypersensitive bean varieties that become poor sources of inoculum when infected.

3.6 Disease incidence and crop loss

Information on disease incidence and crop loss is required by researchers, policy-makers and donors in order to determine priorities for research and extension projects, to make effective use of the personnel and other resources available and to evaluate the cost-effectiveness of control measures. However, reliable and representative data are not readily obtained because of the huge scale of crop production in the tropics, the great diversity of crops and varieties grown, the wide range of agro-ecologies utilized and the differences in disease incidence that can occur between countries and seasons (Chapter 1).

Crop loss information has been obtained by comparing the growth and yield of healthy and diseased plants in specially planted trials involving controlled inoculations, or by sampling healthy and naturally infected plants within farmers' fields. Important results have been obtained with diverse pathosystems (eg Fargette *et al.* 1988; Thresh *et al.* 1994; Bosque-Perez *et al.* 1998), but it is difficult to assess their overall significance given the variation in virulence of the virus strain(s) present, the sensitivity of the crop varieties grown, and by the soil fertility, cropping practices and environmental conditions encountered.

Another complication is that healthy plants within a stand may partially compensate for the impaired growth of their diseased neighbours (Otim-Nape *et al.* 1997). The extent to which compensation occurs will depend on spacing and other agronomic factors and on the distribution of diseased plants and will be less when they occur in patches (Hughes, 1988). There may also be a critical incidence of disease in a crop stand below which there is little or no effect on overall productivity. Consequently, the results of yield trials cannot readily be extrapolated to estimate the losses occurring in a country or region as a whole, although they can be useful in providing a broad indication, especially if the results of field surveys of disease incidence are also available.

Such surveys of disease incidence are expensive and demanding of time, resources and expertise. Ideally they should be done throughout the region being considered and over a sequence of several years. These are exacting requirements that are seldom met and this largely explains the paucity of the information available. However, very detailed and repeated surveys of the distribution of cocoa swollen shoot disease were made throughout the cocoa growing areas of Ghana and elsewhere in West Africa in attempts to eliminate or at least contain the disease (Thresh and Owusu, 1986). Recently there have been surveys of the incidence and severity of cassava mosaic disease in many of the most important cassava-growing countries of Africa, and in India (Thresh *et al.*, 1997, 1998b).

Annual crops grown from seed present particular difficulties as the results of surveys will be greatly influenced by the stage of growth and the season. Moreover, there is great variation in disease incidence between years and it is difficult to draw meaningful conclusions from the results of a single survey. Accordingly caution is required when interpreting some of the data and statements that appear in the literature.

In evaluating the losses caused by virus disease it is important to consider their effects in restricting the areas that are considered suitable for crop production, or the range of varieties that can be grown successfully. This is apparent from experience with banana where the movement of genotypes and breeding lines has been prevented or severely curtailed by the need for stringent quarantine regulations. Special quarantine procedures were developed to overcome the problem posed by *Banana bunchy top virus* and *Banana streak virus* and so facilitate the movement of *Musa* germplasm (Diekmann and Putter, 1996). Quarantine problems have also arisen with cassava and partly explain why so little use has been made of African genotypes by breeding programmes in South America and India. Moreover, South American genotypes tend to be so severely affected by cassava mosaic disease if grown in Africa that their traits can only be exploited after inter-crossing with mosaic-resistant material.

The possibility of forecasting the incidence of virus diseases and the losses they cause has received considerable attention from epidemiologists (Thresh, 1986). Early studies in South Africa by Storey and Bottomley (1928) established a positive relationship between the final incidence of groundnut rosette disease and the total rainfall during the preceding dry season. This influenced the survival of 'volunteer' self-grown groundnut seedlings and crop debris and hence the abundance of initial sources of inoculum. Other positive relationships between disease incidence and pre-season rainfall that influences the prevalence of vectors and inoculum sources have been established with cotton leafcurl disease and maize streak disease. Data on the probability of cocoa swollen shoot disease spreading to neighbouring trees has also been used to evaluate and modify eradication measures (Thresh and Lister, 1960; Thresh and Owusu, 1986). Observations on the progress of the 1990s epidemic of the severe form of cassava mosaic disease in Uganda have been used to anticipate the threat to neighbouring countries and the need to propagate stocks of virus-resistant varieties for release to farmers (Legg, 1999).

3.7 Cropping practices and virus spread

Cropping practices are important in influencing the incidence and effects of virus diseases and the prospects for control (Thresh, 1982). The situation is particularly complex in the tropics because of the very diverse crops grown and the wide range of agro-ecologies utilized. Moreover, rain-fed agriculture is increasingly being supplemented by irrigation and so cropping practices that have been little-changed for centuries are being used together with modern 'industrial' techniques (Thresh, 1991). This creates difficulties in achieving effective control because measures shown to be suitable for some groups of farmers may be inappropriate for others.

In assessing the impact of cropping practices on virus spread it is convenient to distinguish between those adopted at or before planting and those deployed later. It has long been recognised that the incidence of groundnut rosette disease is increased by weeding early (Hayes, 1932). Sowing late and at wide spacing were also shown to enhance spread and led to the recommendation that farmers should sow early and at close spacing (A'Brook, 1964; Davies, 1976). Planting date is also an important factor influencing the spread of cassava mosaic (Fargette *et al.*, 1994) and many other diseases.

The increased use of irrigation in the tropics has permitted crops to be grown at times or in places when or where it would otherwise be difficult. This has the effect of shortening or eliminating any natural break between growing seasons. There is enhanced survival of viruses, vectors, weed hosts, crops and crop debris and virus spread is greatly facilitated, as reported with *Rice yellow mottle virus* (Bakker, 1974), rice tungro viruses (Azzam and Chancellor, 2002) and *Maize streak virus* (Rose, 1978). The increased use of irrigation and intensive cropping practices are also likely to accentuate the problems caused by virus diseases in the vegetable crops now being grown increasingly around large urban centres in the tropics.

Jones (1981) discussed some of the contrasting features of 'traditional' and 'modern' agriculture based on experience with viruses of *Solanum* potato in South America, Europe and North America. Similar considerations apply to many other crops that are cultivated in different ways in the tropics. There is little information on the extent to which the different practices influence the prevalence and control of virus diseases. However, there has been considerable debate on the implications of the trend towards decreased crop diversity as complex multiple cropping systems are replaced by single crops grown over large contiguous areas and the numerous landraces of crops are displaced by small numbers of specially bred cultivars.

Diversity in the crops and varieties grown is considered to be an important and robust feature of traditional agriculture that provides a substantial degree of resilience and enables farmers to sustain production (Smithson and Lenné, 1996). Particular crop species or varieties in the mixture may withstand pests and diseases or adverse weather conditions and yield satisfactorily, even though others succumb.

There is little evidence to support or deny the merits of intercropping as a means of avoiding virus disease problems because epidemiology trials are usually done with single crops and varieties. Moreover, there have been few theoretical analyses of the effects of plant genetic diversity on virus spread (Power, 1991). In SE Asia (Way and Heong, 1994) and central America in particular (Castro *et al.,* 1992) the effects of diversity and mixed cropping on the dynamics and behaviour of insect pests and vectors have received attention. Several of the few experiments on intercropping have provided evidence of beneficial effects in decreasing virus spread; with *Bean common mosaic virus* in bean grown together with maize (van Rheenan, *et al.* 1981) and with cassava mosaic viruses when cassava was intercropped with maize or cowpea (Fargette and Fauquet, 1988; Fondong *et al.* 2002). However, the consequences of intercropping are not always consistent or substantial, as noted with *Maize streak virus* when maize was interplanted with sorghum or cowpea (Page *et al.*, 1999). In other studies field beans suppressed the growth of groundnut intercrops, even though there was a lower incidence of rosette disease compared with groundnut grown alone (Farrell, 1976). These results emphasise the complex effects of intercropping and the need for much additional multi-disciplinary research if clear and acceptable recommendations are to be made to farmers.

There is even less information on the implications of the trend away from the traditional use of varietal mixtures and the adoption of a relatively small number of improved genotypes. In Uganda cassava mosaic disease was more damaging in areas where only a single variety predominated than where many varieties were being grown (Otim-Nape *et al.* 2001). The disease also spread less rapidly to a susceptible variety grown with resistant ones as a mixture than when the susceptible variety was grown alone (Sserubombwe *et al.*, 2001).

3.8 The ecological approach

Plant pathologists have made considerable use of the so-called *'epidemiological triangle'* to facilitate an understanding of the complex inter-relationships between *pathogens*, *hosts* and their *environment*. With plant virus diseases this led to the concept of the *'ecological trinity'* of *viruses*, *hosts* and *vectors* within the overall context of the *environment,* as developed by the American entomologist Walter Carter. The interactions of viruses and vectors with crops, weeds and wild hosts and the overall influence of cropping practices and other environmental factors create the conditions for epidemics to occur. The need for an ecological approach became apparent in studies on cocoa swollen shoot disease (Tinsley, 1964) following the detection of wild indigenous tree hosts of the causal virus and its mealybug vectors and by the finding that there are several vector species and that these occur in association with more than 120 other insect species.

Crop hosts can be viewed as *apparent* ('easily found') or *non-apparent* ('cryptic') species, depending on their longevity, size and other growth characteristics that influence the ease with which plants are located and colonized by arthropod vectors. *Apparent species* include trees and shrubs, especially those that occur widely and in large continuous stands. *Non-apparent species* are typified by herbaceous hosts that are often ephemeral and restricted to particular sites or seasons. Crops can be regarded as providing either transient or stable habitats that are invaded readily or with difficulty, depending on their size, stability and degree of isolation and on the mobility of the vector. Concepts derived from biogeography are relevant in plant virus epidemiology (Thresh, 1980). It is also advantageous to consider the evolutionary history of viruses, vectors and hosts. Many important crops that have been from their center of origin to other continents include maize, groundnut, cocoa, sweetpotato, cassava and Asian rice (*Oryza sativa*). Consequently, crops may have had long or short periods of co-existence and co-evolution with their viruses (Buddenhagen, 1977).

The plant virus literature contains numerous references to severe epidemics and serious losses. This is understandable but can give a somewhat exaggerated, biased and misleading impression of the magnitude of the problems caused by viruses. Severe epidemics undoubtedly occur, but usually they can be regarded as infrequent and often restricted to certain crops, varieties, areas and seasons. Otherwise, virus diseases are but one of many constraints that must be addressed if productivity is to be increased.

If damaging outbreaks are indeed exceptional events, then this implies that there are usually effective factors that impede their occurrence. It is important to identify any such factors so that they can be sustained, exploited and even enhanced to improve crop health and increase productivity. Three attributes or mechanisms that operate singly, or in combination limit the vulnerability of crops to viruses - *evasion, resistance* to infection or to vectors and *tolerance* of infection. Evasion operates if plants escape infection because they grow at times or in places when/where viruses or their vectors seldom occur. This is more likely to occur with transient, sparsely distributed species growing seasonally in mixed stands together with non-host plants and weeds, than it is with long-lived species growing widely and throughout the year in sole stands subject to rigorous weed control. Moreover, the situation is dynamic in that a period of increased disease pressure may lead to a decrease in host prevalence, or in the proportion of susceptible genotypes and so to an eventual relaxation of disease pressure. Host prevalence or the proportion of susceptibles is then likely to increase, so facilitating a resurgence of disease and continuing cycles of increasing and decreasing prevalence (Buddenhagen, 1977). Although this type of cyclical behaviour is arguably less likely to occur with modern methods of crop production, the implications for virus spread and the losses caused have yet to be evaluated in detail.

Resistance to or tolerance of infection is an alternative or complement to evasion. Many types of resistance have been described that are effective against viruses or their vectors. Mature plant resistance and the apparent immunity of any one plant species to all but a few viruses are distinct and epidemiologically important characteristics. Other types of resistance include those associated with gene-for-gene relationships between host and pathogen, as reported with *Bean common mosaic virus* (Drijfhout, 1978). There is also quantitative 'rate-reducing' resistance, as described with *Cacao swollen shoot virus* (Thresh *et* al., 1988) and cassava mosaic geminiviruses (Thresh *et al.* 1998a). Tolerance of infection is quite different and in the strict sense of the term is the ability of plants to grow and yield satisfactorily despite sustaining a virus content that causes serious damage in sensitive varieties.

A limitation of both resistance and tolerance to viruses is that they may be overcome by new or more aggressive pathogenic strains; 'breakdown' of resistance to virus vectors has also been reported. Moreover, varieties that are resistant in one region or country may succumb to the same disease elsewhere. Decreases in virus incidence due to the introduction of resistant varieties have been followed some years later by a resurgence as new virus strains or vector biotypes have appeared and become prevalent. Examples include *Sugarcane mosaic virus* and *Tomato mosaic virus* and the breakdown of resistance to the main leafhopper vector of rice tungro viruses (Thresh, 1989). These examples portray a dynamic equilibrium in which neither host nor virus gains permanent ascendancy.

The concept of a dynamic equilibrium provides a convenient and appropriate basis from which to evaluate the adoption of intensified cropping practices. The ability of plants to avoid or withstand diseases when growing in natural habitats is undermined when they are cultivated in dense stands of genetically uniform monocrops grown at the same sites in successive years. The situation is further exacerbated by the use of irrigation or other means of extending the natural growing season to facilitate year-round production and also by selecting a small number of high-yielding genotypes for use on a large scale (Thresh, 1982). Crops and varieties have been transferred to entirely new areas where they have been affected by novel viruses or more virulent strains than those encountered previously. The increased traffic in plant material has also facilitated the dissemination of viruses and their vectors.

These developments have contributed to the view that disease epidemics seldom occur in undisturbed natural ecosystems, that they are largely the outcome of agricultural practices and that they are increasing as traditional methods are abandoned, cropping systems become increasingly specialized and there is increased traffic in plant material. There is insufficient evidence to confirm or deny these suppositions. Nevertheless, many epidemics can be interpreted as major perturbations of previously stable situations as a consequence of changes in the crops grown and the method of cultivation adopted. Cocoa swollen shoot, cassava mosaic, groundnut rosette and maize streak are all *'new encounter'* diseases (*sensu* Buddenhagen, 1977) that followed the transfer of non-indigenous hosts from the Americas to Africa, where they have been severely affected by viruses spreading from indigenous hosts.

A more recent perturbation has been the 1990s pandemic of cassava mosaic disease. This is associated with a particularly virulent strain of the causal virus that is considered to be a novel recombinant of two different cassava mosaic geminiviruses (Chapter 12). Other examples of this type are known and it seems inevitable that there will be problems with other whitefly-borne viruses due to the continued expansion in range of the B biotype of *Bemisia tabaci*. This has a wider host range than the biotypes already present and is associated with recent virus epidemics in tomato, beans and other crops in the Americas (Brown, 1994; Morales and Anderson, 2001) and India (Banks *et al.*, 2001).

3.9 Molecular epidemiology

Epidemiological data are usually obtained by visual assessment of the incidence of diseased plants. It is seldom feasible to test for the occurrence of the virus or viruses responsible, because suitable methods of virus detection that are robust, reliable and able to achieve the large throughput required are seldom available. Similarly it is seldom possible to determine the strains of virus that are encountered in epidemiology trials and field surveys. These are important constraints, especially where there is a general lack of laboratory facilities and trained personnel.

The epidemiological importance of virus strains was apparent in early studies on tropical plant viruses. Distinct virus strains were recognized from the type and severity of the symptoms expressed by the crop host and some were shown to be more damaging than others. For example, avirulent forms of the 1A strain of *Cacao swollen shoot virus* that occur naturally in Ghana will protect against many of the most damaging virulent strains (Posnette and Todd, 1955). There have been similar findings with *Citrus tristeza virus* (van Vuuren *et al,* 1993). Moreover, certain strains of *Sugarcane mosaic virus* and *Bean common mosaic virus* can infect and damage varieties selected as being resistant to other virus strains (Spence and Walkey, 1994). Strain variation is a crucial feature of the epidemiological competence of viruses that enables them to adapt to changing physical and biotic environments.

Initially virus strains were distinguished by the symptoms they cause in crop plants or when inoculated to indicator hosts. This led to the selection of a range of differential varieties to distinguish strains. Serology was then developed as a convenient means of testing many samples and it has provided important information on the occurrence and variability of many viruses, including those causing rice yellow mottle and cassava mosaic diseases (N'Guessan et al., 2000; Swanson and Harrison, 1994). Originally, cassava mosaic disease was assumed to be caused by a single whitefly-borne geminivirus, although different strains were distinguished using polyclonal antisera and later by PCR and DNA sequencing (Hong *et al.* 1993). DNA sequence data have been used in the most recent studies to distinguish eight cassava mosaic geminiviruses of which six occur in Africa and the other two in Asia (Fauquet and Stanley, 2003).

Banana streak disease, first reported in Côte d'Ivoire and Morocco is now known to occur in many countries of sub-Saharan Africa and elsewhere in the tropics. The disease is caused by DNA-containing viruses that are so variable it has been difficult to develop reliable methods of detection based on serology or PCR (Lockhart and Olszewski, 1993). Complexity of a different type has become apparent in molecular studies on groundnut rosette disease. The aetiology of rosette is complex and a distinct assistor virus is required for the virus that causes rosette symptoms to be transmitted by the aphid vectors. More recent studies have established the role of a satellite RNA in causing rosette symptoms and in facilitating the transmission of *Groundnut rosette virus* by vectors. Techniques have been developed to assay each of the three components and are now being used in epidemiology studies and to assess the behaviour of rosette-resistant varieties of groundnut (Naidu et al., 1998).

These examples indicate how the molecular analysis of virus isolates can facilitate an understanding of epidemiology and have led to the term *molecular epidemiology* (Garcia-Arenal et al., 2000). Molecular techniques are being used increasingly in plant virus epidemiology to characterise virus populations, to elucidate virus/vector relationships and to study the interactions between virus strains so as to determine their origins and epidemiological effects.

Mutations accumulate with time in the genomes of viruses, because many lack fidelity during replication. Some mutations confer phenotypic differences, enabling the virus to infect different host species, or to be transmitted by different means. It is now possible to make inferences on the origin of epidemics by collating and analysing nucleotide sequences in order to reconstruct the history of virus spread through host populations. Phylogenic studies clarify the *relationships* between strains, whereas such information was not accessible, or could only be interpreted broadly and indirectly, using earlier serological and biological data.

The biochemical information now being obtained broadens the scope of 'classical' epidemiology. Analysis of phylogenetic structure provides insights into epidemiological processes and the history of epidemics (Harvey *et al.* 1996). The epidemiological and aetiological characteristics of several of the animal viruses groups have been mapped onto their phylogeny to reveal striking correlations between the topological arrangements of the viruses and their epidemiological characteristics. It is inevitable that plant virus epidemiology will follow a similar trend (Harrison and Robinson, 1988; Gibbs *et al.* 1999). For instance, the quasi-species structure of a virus population, its spatial and temporal variation, recombination phenomenon and virus interactions are critical for unravelling the complex evolutionary relationships between viruses, vectors and plant hosts and the ecological niches they exploit (Gaunt *et al.* 2001).

3.10 Discussion

The epidemiology of plant viruses in the tropics features prominently in the plant pathology literature (Thresh, 1991) and several tropical viruses are considered in the highly influential text *'Plant Diseases: Epidemics and Control'* (Vanderplank, 1963). However, it should be emphasised that there is totally inadequate information on the prevalence, distribution and spread of many tropical plant viruses and on the losses they cause. There are likely to be many as yet undiscovered viruses and there is detailed epidemiological information on only a few of the viruses known to occur. Moreover, most epidemiological studies have been of limited duration and restricted to specific agro-ecologies that are not always truly representative.

These are important limitations on the ability to mount and sustain effective control measures. The problems that arise are apparent from experience with cassava mosaic disease in Africa. The disease has been known since 1894 and arguably it has received more attention than any other virus disease of an African food crop (Thresh, 1991). Nevertheless, the information available in the early 1990s was not sufficient to explain the very damaging regional pandemic that was first reported in Uganda in 1988, or to provide an effective means of control. This has necessitated much additional research and led to important new findings on the whitefly vector and on the nature and distribution of the geminiviruses responsible and on the interactions between them. Nevertheless, many uncertainties remain on the causes of the current pandemic in East Africa, and on the distribution and implications of the different cassava mosaic geminiviruses and races/biotypes of *Bemisia tabaci* now known to occur (Chapter 12). There is also a need to resolve the long-standing uncertainty concerning the need for phytosanitation if mosaic-resistant varieties are adopted.

There is similar uncertainty over the origins and indigenous hosts of the viruses responsible for groundnut rosette disease and no information on the source(s) of inoculum from which spread occurs to groundnut crops in areas where there is such a prolonged dry season that sources of inoculum do not persist between growing seasons (Naidu *et al.*, 1998). There is a need to assess the epidemiological significance of the early observation that infection of groundnut with rosette disease enhances the production of winged (alate) forms of the aphid vector (Rèal, 1955). Additional studies are also required on *Rice yellow mottle virus* because it is still unclear whether the main means of spread is mechanical or by beetle vectors and there are conflicting claims on the relative importance of the two processes (Abo *et al.*, 2000). Even less is known on the means of spread of *Cassava brown streak virus* that is now causing such severe losses in eastern and southern Africa (Chapter 12).

These examples demonstrate the deficiencies in current knowledge of tropical plant viruses. These will not be overcome easily because epidemiology is an exacting science and fully-trained practitioners are few, especially in the tropics where national virologists seldom have the opportunity to specialise on particular crops or diseases. Furthermore, experience has shown the merits of a multi-disciplinary approach involving virologists and those concerned with vectors, biometricians, plant breeders and other specialists. Such teams have operated at various times and places but for restricted periods in studies on only a few particularly important diseases, including rice tungro, cocoa swollen shoot, maize streak, groundnut rosette and cassava mosaic. Multi-disciplinary teams are not easily assembled or sustained, because they require a substantial commitment of funds over prolonged periods to overcome the inevitable vagaries of field experimentation in different agro-ecologies and in contrasting seasons.

National programmes seldom have the resources required to undertake epidemiological studies and the main emphasis in plant virology at the International Agricultural Research Centres has been on diagnosis and to support resistance breeding projects, rather than on epidemiology. Moreover, with few exceptions the emphasis of collaborators in advanced laboratories in Europe and North America has been on virus characterisation and aetiology and more recently on biochemical and biotechnological aspects. A major change of emphasis will be required if progress is to be made in solving the many plant virus problems in the tropics. This is particularly so for Africa if it is to develop effective control measures that are comparable in effectiveness to those that have made such a big contribution to food production in temperate regions and to some extent in Asia and South/Central America. The challenge is to obtain a basic understanding of virus epidemiology so that effective and sustainable control measures can be developed to exploit fully the advances being made in other aspects of crop technology and to do so despite decreasing soil fertility and a decline in the rural work-force and without damaging the environment.

References

Abo, M.E., Alegbejo, M.D., Sy, A.A. and Misari, S.M. (2000) An overview of the mode of transmission, host plants and methods of detection of rice yellow mottle viruses. *J. Sustainable Agric.* 17: 19-36.

A'Brook, J. (1964) The effect of planting date and spacing on the incidence of groundnut rosette disease and of the vector, *Aphis craccivora* Koch, at Mokwa, northern Nigeria. *Ann. Appl, Biol.* 54: 199-208.

Azzam, O. and Chancellor T.C.B., (2002) The biology, epidemiology and management of rice tungro disease in Asia. *Plant Dis.* 86: 88-100.

Bakker, W. (1974) Characterisation and ecological aspects of rice yellow mottle virus in Kenya. *Agric. Res. Rept.* No. 829. Wageningen: Pudoc.

Banks, G.K., Colvin, J., Chowda Reddy, R.V., Maruthi, M.N., Muniyappa, V., Venkatesh, H.M., Kiran Kumar, M., Padmaja, A.S., Beitia, F.J. and Seal, S.E. (2001) First report of the *Bemisia tabaci* B biotype in India and an associated *Tomato leaf curl virus* disease epidemic. *Plant Dis.* 85: 231.

Beemster, A.B.R. (1957) Some aspects of mature plant resistance to viruses in the potato. In: *Proc. Third Conf. Potato Virus Dis.*. Lisse-Wageningen, 24-28 June 1957. Pp. 212-217.

Bosque-Pérez, N.A., Olojede, S.O. and Buddenhagen, I.W. (1998) Effect of maize streak virus disease on the growth and yield of maize as influenced by varietal resistance levels and plant stage at time of challenge. *Euphytica* 101: 307-317

Brown, J.K. (1994) The status of *Bemisia tabaci* Genn. as a plant pest and virus vector in agro-ecosystems worldwide. *FAO Plant Prot. Bull.* 42: 3-32.

Buddenhagen, I.W. (1977) Resistance and vulnerability of tropical crops in relation to their evolution and breeding. *Ann. N. Y. Acad. Sc.* 287: 309-326.

Castro, V., Rivera, C., Isard, S.A., Gámez, R., Fletcher, J. and Irwin, M.E. (1992) The influence of weather and microclimate on *Dalbulbus maidis* (Homoptera, Cicadellidae) flight activity and the incidence of diseases within maize and bean monocultures and bicultures in Tropical America. *Ann. Appl. Biol.* 121: 469-482.

Chan, M.S. and Jeger M.J. (1994) An analytical model of plant virus disease dynamics with roguing and replanting. *J. Appl. Ecol.* 31: 413-427.

Cornwell, P.B. (1958) Movements of the vectors of virus diseases of cacao in Ghana I. Canopy movement in and between trees. *Bull. Ent. Res.* 49: 613-630.

Davies, J.C. (1976) The incidence of rosette disease in groundnut in relation to plant density and its effect on yield. *Ann. Appl. Biol.* 82: 489-501.

Diekmann, M. and Putter, C.A.J. (1996). FAO/IPGRI Technical Guidelines for the Safe Movement of Germplasm No. 15 *Musa* spp. 2nd Edition. FAO/IPGRI, Rome.

Drijfhout, E. (1978) Genetic interaction between *Phaseolus vulgaris* and bean common mosaic virus with implications for strain identification and breeding for resistance. *Agricultural Research Report 872.* Centre for Agricultural Publishing and Documentation, Wageningen. 90pp.

Fargette, D. and Fauquet, C. (1988) A preliminary study on the influence of intercropping maize and cassava on the spread of African cassava mosaic virus by whiteflies. *Asp. Appl. Biol.* 17: 195-202.

Fargette, D., Fauquet, C. and Thouvenel, J.-C. (1988) Yield losses induced by African cassava mosaic virus in relation to the mode and the date of infection. *Trop. Pest Man.* 34: 89-91.

Fargette, D., Jeger, M., Fauquet, C. and Fishpool, L.D.C. (1994) Analysis of temporal disease progress of African cassava mosaic virus. *Phytopathology* 84: 91-98.

Fargette, D., Muniyappa, V., Fauquet, C.M., N'Guessan, P. and Thouvenel, J.-C. (1993) Comparative epidemiology of three tropical whitefly-transmitted geminiviruses. *Biochimie* 75: 547-554.

Farrell, J.A.K. (1976) Effects of intersowing with beans on the spread of groundnut rosette virus by *Aphid craccivora* Koch (Hemiptera, Aphididae) in Malawi. *Bull. Entomol. Res.* 66: 331-333.

Fauquet, C.M. and Stanley, J. (2003) Geminivirus classification and nomenclature: progress and problems. *Ann. Appl. Biol.* 102: 165-189.

Fondong, V.N., Thresh, J.M. and Zok, S. (2002) Spatial and temporal spread of cassava mosaic virus disease in cassava grown alone and when intercropped with maize and/or cowpea. *J. Phytopathol.* 150: 365-374.

Garcia-Arenal, F., Escru, F., Aranda, M.A., Alonso-Prados, J.L., Malpica, J.M. and Fraile, A. (2000) Molecular epidemiology of *Cucumber mosaic virus* and its satellite RNA. *Virus Res.* 71: 1-8.

Gaunt, M., Sall, A., Lamballerie, X., Falconar, A., Dzhivanian, I. and Gould, E. (2001) Phylognetic relations of flaviviruses correlate with the epidemiology, disease association and biogeography. *J. Gen. Virol.* 82: 1867-1876.

Gibbs, A.J., Keese, P.L., Gibbs, M.J. and Garcia-Arenal, F. (1999) Plant virus evolution : past, present and future, In: E. Domingo, R.W. Webster and J. Holland (eds.) *'Origin and evolution of viruses'*, pp.263-285. Academic Press.

Gregory, P.H. (1968) Interpreting plant disease dispersal gradients. *Ann. Rev. Phytopathol.* 6: 189-212.

Harrison, B.D. and Robinson, D.J. (1988) Molecular variation in vector-borne plant viruses: epidemiological significance. *Phil. Trans. Royal Soc. B London* 321: 447-462.

Harvey, P.H., Leigh Brown, A.J., Maynard Smith, J. and Nee, S. (1996) *New uses for new phylogenies.* Oxford University Press, 349 pp.

Hayes, T.R. (1932) Groundnut rosette disease in the Gambia. *Trop. Agric., Trin.* 19: 211-217.

Holt, J. and Chancellor, T.C.B. (1996) Simulation modelling of the spread of rice tungro virus disease. The potential for management by roguing. *J. Appl. Ecol.* 33: 927-936.

Holt, J. and Chancellor, T.C.B. (1999) Modelling the spatio-temporal deployment of resistant varieties to reduce the incidence of rice tungro disease in a dynamic cropping system. *Plant Pathol.* 48: 453-461.

Holt, J., Colvin, J. and Muniyappa, V. (1999) Identifying control strategies for tomato leaf curl virus using an epidemiological model. *J. Appl. Ecol.* 36: 625-633.

Holt, J., Jeger, M.J, Thresh, J.M. and Otim-Nape, G.W. (1997) An epidemiological model incorporating vector population dynamics applied to African cassava mosaic virus disease. *J. Appl. Ecol.* 34: 793-806.

Hong, Y.G., Robinson, D.J. and Harrison, B.D. (1993) Nucleotide sequence evidence for the occurrence of three distinct whitefly-transmitted geminiviruses in cassava. *J. Gen. Virol.* 74: 2437-2443.

Hughes, G. (1988) Modelling the effect of spatially heterogeneous pest injury on crop yields. *Crop Res.* 28: 137-144.

Jeger, M.J. (1999) Improved understanding of dispersal in crop pest and disease management: current status and future directions. *Agric and Forest Meteorol.* 97: 331-349.

Jeger, M.J. and Thresh, J.M. (1993) Modelling reinfection of replanted cocoa by swollen shoot virus in pandemically disease areas. *J. Appl. Ecol.* 30: 187-196.

Jeger, M.J., VandenBosch, F., Madden L.V. and Holt, J. (1998) A model for analysing plant virus transmission characteristics and epidemic development. *IMA J. Math Appl. Mod* 15: 1-18.

Jones, R.A.C. (1981) The ecology of viruses infecting wild and cultivated potatoes in the Andean region of South America. In: J.M. Thresh (ed.) *Pests, Pathogens and Vegetation*, pp.89-107. Pitman, London.

Legg, J.P. (1999) Emergence, spread and strategies for controlling the pandemic of cassava mosaic virus disease in east and central Africa. *Crop Prot.* 18: 627-637.

Legg, J.P., French, R., Rogan, D., Okao-Okuja, G. and Brown, J.K. (2002) A distinct *Bemisia tabaci* (Gennadius) (Hemiptera: Sternorrhyncha: Aleyrodidae) genotype cluster is associated with the epidemic of severe cassava mosaic virus disease in Uganda. *Mol. Ecol.* 11: 1219-1229.

Lockhart, B.E.L. and Olszewski, N.E. (1993) Serological and genomic heterogeneity of banana streak badnavirus: implications for virus detection in *Musa* germplasm. In: J. Ganry (ed.) *Breeding Banana and Plantain for Resistance to Diseases and Pests*, pp.105-113. CIRAD-INIBAP, Montpellier.

Madden, L.V., Jeger, M.J. and Vandenbosch, F. (2000) A theoretical assessment of the effects of vector-virus transmission mechanism on plant virus disease epidemics. *Phytopathology* 90: 576-594.

Maruthi, M.N., Colvin, J., Seal, S. and Thresh, J.M. (2002) First report of a distinct begomovirus infecting cassava from Zanzibar. *Plant Dis.* 86: 187.

Morales, F.J. and Anderson, P.K. (2001) The emergence and dissemination of whitefly-transmitted geminiviruses in Latin America. *Arch. Virol.* 146: 415-441.

Naidu, R.A., Bottenberg, H.H., Subrahmanyam, P., Kimmins, F.M., Robinson, D.J. and Thresh, J.M. (1998) Epidemiology of groundnut rosette virus disease: current status and future research needs. *Ann. Appl. Biol.* 132: 525-548.

Ndowora, T., Dahal, G., LaFleur, D., Harper, G., Hull R., Olszewski, N.E. and Lockhart, B. (1999) Evidence that badnavirus infection in *Musa* can originate from integrated pararetroviral sequences. *Virology* 255: 214-220.

N'Guessan, P., Pinel, A., Caruana, M.L., Frutos, R., Sy, A., Ghesquière, A. and Fargette, D. (2000) Evidence of the presence of two serotypes of rice yellow mottle sobemovirus in Côte d'Ivoire. *Eur. J. Plant Pathol.* 106: 167-178.

Nutter, F.W. (1997) Quantifying the temporal dynamics of plant virus epidemics : a review. *Crop Prot.* 16: 603-618.

Otim-Nape, G.W., Alicai, T. and Thresh, J.M. (2001) Changes in the incidence and severity of cassava mosaic virus disease, varietal diversity and cassava production in Uganda. *Ann. Appl. Biol.* 138: 313-327.

Otim-Nape, G.W., Thresh, J.M. and Shaw, M.W. (1997) The effects of cassava mosaic virus disease on yield and compensation in mixed stands of healthy and infected cassava. *Ann. Appl. Biol.* 130: 503-521.

Page, W.W., Smith, M.C., Holt, J. and Kyetere, D. (1999) Intercrops, *Cicadulina* spp. and maize streak virus disease. *Ann. Appl. Biol.* 135: 385-393.

Posnette, A.F. and Todd, J. McA. (1955) Virus diseases of cacao in West Africa IX. Strain variation and interference in virus 1A. *Ann. Appl. Biol.* 43: 433-453.

Power, A.G. (1991) Virus spread and vector dynamics in genetically diverse plant populations. *Ecology* 72: 232-241.

Power, A.G. (1992) Patterns of virulence and benevolence in insect-borne pathogens of plants. *Crit. Rev. Plant Sci.* 11: 351-372.

Putter, C.A.J. (1980) The management of epidemic levels of endemic diseases under tropical subsistence farming conditions. In: J. Palti and J. Kranz: (eds.) *Comparative Epidemiology,* pp.93-103. Pudoc, Wageningen, Netherlands. 122pp.

Rèal, P. (1955) Le cycle annuel du puceron de l'arachide (*Arachis leguminosae* Theob.) en Afrique noire Francaise et son déterminisme. *Review Pathologie Végétale* 34: 3-122.

Rose, D.J.W. (1972) Dispersal and quality in populations of *Cicadulina* species (Cicadellidae). *J. Animal Ecol.* 41: 589-609.

Rose, D.J.W. (1973) Distances flown by *Cicadulina* spp. (Hem., Cicadellidae) in relation to distribution of maize streak disease in Rhodesia. *Bull. Entomol. Res.* 62: 497-505.

Rose, D.J. (1978) Epidemiology of maize streak disease. *Ann. Rev. Entomol.* 23: 259-282.

Smithson, J.B. and Lenné, J.M. (1996) Varietal mixtures: a viable strategy for sustainable productivity in subsistence agriculture. *Ann. Appl. Biol.* 128: 127-158.

Spence, N.J. and Walkey, D.G.A. (1994) *Bean common mosaic virus and related viruses in Africa.* NRI Bulletin 63, Chatham, U.K. Natural Resources Institute.

Sserubombwe, W.S., Thresh, J.M., Otim-Nape, G.W. and Osiru, D.O.S. (2001) Progress of cassava mosaic virus disease and whitefly vector populations in single and mixed stands of four cassava varieties grown under epidemic conditions in Uganda. *Ann. Appl. Biol.* 138: 161-170.

Storey, H.H. and Bottomley, A.M. (1928) The rosette disease of peanuts (*Arachis hypogea* L.). *Ann. Appl. Biol.* 15: 26-45.

Swanson, M.M. and Harrison, B.D. (1994) Properties, relationships and distribution of cassava mosaic geminiviruses. *Trop. Sci.* 34: 15-25.

Theuri, J.M., Bock, K.R. and Woods, R.D. (1987). Distribution, host range and some properties of a virus disease of sunflower. *Trop. Pest Man.* 33: 202-206.

Thresh, J.M. (1974) Vector relationships and the development of epidemics: the epidemiology of plant viruses. *Phytopathology* 64: 1050-1056.

Thresh, J.M. (1976) Gradients of plant virus diseases. *Ann. Appl. Biol.* 82: 381-406.

Thresh, J.M. (1980) An ecological approach to the epidemiology of plant virus diseases. In: J. Palti and J. Kranz (eds.) *Comparative Epidemiology,* pp.57-70. Pudoc, Netherlands.

Thresh, J.M. (1982) Cropping practices and virus spread. *Ann. Rev. Phytopathol.* 20: 193-218.

Thresh, J.M. (1983a) Progress curves of plant virus disease. *Adv.Appl. Biol.* 8: 1-8.

Thresh, J.M. (1983b) The long-range dispersal of plant viruses by arthropod vectors. *Phil. Trans. Roy Soc.* London B. 302: 497-528.

Thresh, J.M. (1985) Plant virus dispersal. In: D.R. MacKenzie, C.S. Barfield, G.G. Kennedy, R.D. Berger, and D.J. Taranto (eds.) *The Movement and Dispersal of Agriculturally Important Biotic Agents,* pp.51-106. Claistors Publishing, Baton Rouge, Louisana.

Thresh, J.M. (1986) Plant virus disease forecasting. In: G.D. McLean, R.G. Garrett and W.G. Ruesink (eds.) *Plant Virus Epidemics: Monitoring, Modelling and Predicting Outbreaks,* pp.359-386. Academic Press, Sydney.

Thresh, J.M. (1988) Eradication as a virus disease control measure. In: B.C. Clifford and E. Lester (eds.) *Control of Plant Diseases: Costs and Benefits.* pp.155-194. Blackwell, Oxford.

Thresh, J.M. (1989) Plant virus epidemiology: the battle of the genes. In: R.S.S. Fraser (ed.) *Recognition and Response in Plant-Virus Interactions.* NATO ASI Series. Series H: *Cell Biol.* 41: 93-121.

Thresh, J.M. (1991) The ecology of tropical plant viruses. *Plant Pathol.* 40: 324-339.

Thresh, J.M. and Lister, R.M. (1960) Coppicing experiments on the spread and control of cacao swollen-shoot disease in Nigeria. *Ann. Appl. Biol.* 48: 65-74.

Thresh, J.M. and Owusu, G.K. (1986) The control of cocoa swollen shoot disease in Ghana: an evaluation of eradication procedures. *Crop Prot.* 5: 41-52.

Thresh, J.M., Fargette, D. and Otim-Nape, G.W. (1994) Effects of African cassava mosaic geminivirus on the yield of cassava. *Trop. Sci.* 34: 26-42.

Thresh, J.M., Otim-Nape, G.W. and Fargette, D. (1998a) The components and deployment of resistance to cassava mosaic virus disease. *Int. Pest Man. Rev.* 3: 209-224.

Thresh, J.M., Owusu, G.K., Boamah, A. and Lockwood, G. (1988) Ghanaian cocoa varieties and swollen shoot virus. *Crop Prot.* 7: 219-231.

Thresh, J.M., Otim-Nape, G.W., Legg, J.P. and Fargette, D. (1997) African cassava mosaic virus disease: the magnitude of the problem. *African J. Root and Tuber Crops* 2: 13-19.

Thresh, J.M., Otim-Nape, G.W., Thankappan, M. and Muniyappa, V. (1998b) The mosaic diseases of cassava in Africa and India caused by whitefly-borne geminiviruses. *Rev. Plant Pathol.* 77: 935-945.

Tinsley, T.W. (1964) The ecological approach to pest and disease problems of cocoa in West Africa. *Trop. Sci.* 6: 38-46.

Vanderplank, J.P. (1963) *Plant Diseases: Epidemics and Control.* Academic Press, New York and London.

Vanderplank, J.E. and Anderssen, E.E. (1944) Kromnek disease of tobacco; a mathematical solution to a problem of disease. Science Bulletin, Department of Agriculture, South Africa. No. 240. 6pp.

Van Rheenan, H.A., Hasselbach, O.E. and Muigai, S.G.S. (1981) The effect of growing beans together with maize on the incidence of bean diseases and pests. *Neth. J. Plant Pathol.* 87: 193-199.

Van Vuuren, S.P., Collins, R.P. and Da Graça, J.V. (1993) Evaluation of citrus tristeza virus isolates for cross protection of grapefruit in South Africa. *Plant Dis.* 77: 24-28.

Way, M.J. and Heong, K.L. (1994) The role of biodiversity in the dynamics and management of insect pests of tropical irrigated rice – a review. *Bull. Entomol. Res.* 84: 567-587.

4. Plant pathogens and quarantines and the implication for global trade and world food

David Pimentel
College of Agriculture & Life Sciences, Cornell University, Ithaca, NY 14853, USA

4.1 Introduction

In developing countries plant pathogens cause pre-harvest food losses ranging from 13% to 16%, despite all the pesticides applied plus other control measures employed (Pimentel *et al.* 2001). This significant food loss is taking place when many people, especially people in developing countries are starving and/or are malnourished. The World Health Organization (WHO, 1996, 2000) reports that more than 3 billion people suffer from malnourishment (insufficient calories, protein, vitamins, iron, and iodine). This is the largest number and proportion of malnourished people ever in history.

The malnourishment problem is growing worse daily because of the rapidly growing world population (Figure 4.1). The world population doubled in the last 40 years to the current level of more than 6 billion and is projected to double again to 12 billion in just 50 years (PRB, 2001). This desperate over-population situation cannot be reduced, no matter what actions can be taken immediately. The problem is population growth momentum. For example, if the world's people decided tomorrow to limit the number of children born per female to only 2 children, it would take more than 70 years for the world population to stabilize (Population International, 2000). In this 70-year period, the world population would have more than doubled to 12 billion (Pimentel *et al.* 1999). The population momentum is due to the young structure of most population in most nations (PRB, 2001), who are 15 to 30 year of age.

The rapidly increasing world population is stimulating global trade. The growing world trade has benefits and costs. The benefits are to many world economies, however, some nations do not benefit from the global trade (GEO 2000). In addition, some countries use quarantines for pests and pesticides to restrict trade with some nations (DePolo, 1999; McDonell, 2002). A major problem with increased trade is the spread of invasive plant pathogens and other pests to nations of the world (Pimentel *et al.* 2000; 2001).

In this article, I examine plant pathogens and quarantines against pests and the implications on global trade. In addition, an assessment is made on the relationship of plant pathogens and other pests to the world food problem.

4.2 Losses of World Food to Pests

Worldwide, approximately 70,000 different pest species damage agricultural crops. Included in these species are about 50,000 species of plant pathogens, 9,000 species of insects and mites, and 8,000 species of weeds (Pimentel, 1997). In general, about 10% of these species are considered major pests. For plant pathogens and weeds, most (60% to 85%) species are invasive species that were introduced into each nation (Pimentel *et al.* 2000; Pimentel *et al.* 2001). For insect and mite pests, most (60% to 80%) are native species that moved from feeding on native vegetation to feeding on crops, which were introduced into the region (Pimentel, 1988; Hokkanen and Pimentel, 1989). For example, the Colorado beetle moved from feeding on a weed to feeding on the potato that was introduced into Colorado from Bolivia and Peru.

Despite the yearly investment of about $35 billion per year worldwide for the application of 3 million metric tons of pesticides (Table 4.1) plus the use of various biological and other non-chemical controls, more than 40% of potential world food production is lost to pests (Oerke *et al.*, 1994; Pimentel, 1997). Worldwide, plant pathogens cause about 12% of crop losses, weeds about 13%, and insect pests about 15%. The value of crop losses to pests is estimated to be more than $\$1 \times 10^{12}$ per year. Probably developing countries suffer the most from crop attacks by plant pathogens. It is reported that plant pathogens cause several billions of dollars in crop losses each year (APS, 2001).

In the United States, yearly crop losses caused by pests are estimated to be about 37% (12% to plant pathogens, 12% to weeds, and 13% to insects) (Pimentel, 1997). In total, pests in the United States are destroying an

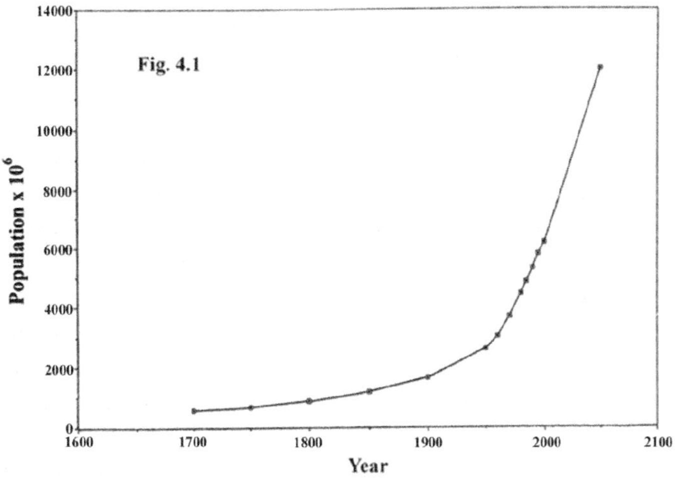

Fig. 4.1.Increase of world population from 1700 onwards

Table 4.1. Estimated annual pesticide use (Pimentel, unpublished data).

Country/Region	Pesticide Use (million tons)
United States	0.5
Canada	0.2
Europe	1.0
Other developed	0.5
China	0.2
Asia, other developing	0.3
Latin America	0.2
Africa	0.1
Total	3.0

estimated $100 billion per year in crops despite the application of 0.5 million metric tons of pesticide plus non-chemical controls (Pimentel *et al.* 2000). Currently, the U.S. invests about $8 billion in pesticides, which saves about $32 billion per year in crops, and non-chemical pest controls also save approximately $32 billion per year. The dollar return per dollar invested in pesticides is about $4 while the return for biological controls ranges from $50 to $150 per dollar invested. Without pesticides and non-chemical controls, the crop losses inflected by pests would range from 60% to 70%.

4.3 Global Movement of Pests

From 95% to 99% of the crops in all nations are introduced species (Pimentel *et al.*, 2001). Most plant pathogens, from 60% to 85%,are introduced species , many of them with the crops themselves.. (Pimentel *et al.* 2000; Pimentel *et al.* 2001). In the United States 65% of the plant pathogens are introduced, in India about 74%, in Brazil about 85%, in Australia about 82%, and in South Africa about 85% are introduced (Pimentel *et al.*, 2001). It is assumed that percentage of virus plant pathogens introduced into developing nations is similar to plant pathogens in general and also range from 60% to 85%. Again, most of the virus plant pathogens were introduced when the alien crop species were introduced into each nation.

There are an estimated 51,300 species of parasitic and non-parasitic plant pathogens in the United States (Pimentel *et al.* 2000). Most of these plant pathogens are fungi while only approximately 1,300 species of viruses, or 2.5%, are plant pests. In India, also a small percentage (2%) of the plant pathogens are viruses or about 650 species (Patel and Patel, 1985; Pandotra, 1997). Although many fungal plant pathogens can be controlled using fungicides, there are no known chemical controls for virus plant pathogens. The control methods for virus diseases of crops include: controlling the insect vectors of the virus, using virus free plant stocks, using host-plant resistant stocks, employing crop rotations, employing biological controls, and preventing the introduction of any new virus disease organisms into the nation through quarantines (Kiritani *et al.* 1991; MacGregor, 1986).

Virus plant pathogens associated with international trade can be placed into three ecological categories (Hollings, 2002): (1) Opportunistic viruses with wide plant-host ranges with efficient means of spread. Most of these viruses exist worldwide now. (2) Virus pathogens with narrow host ranges with highly efficient means of spread, but are localized geographically. (3) Highly-host specific virus pathogens, with relatively inefficient methods of spread.

4.4 Quarantines for the Prevention of Plant Pathogens

All nations have animal and plant health inspection services to prevent the introduction of alien plant and animal species invasions, including plant pathogens. At best these quarantines or inspection services are marginally effective, however, they do play an important role in preventing at least some plant pathogens from invading the valuable crops of all nations. For example, the USDA Animal and Plant Health Inspection Service during 1997 intercepted 860 plant pathogen disease species that are potential pests of commercially valuable crop at U.S. ports of entry (USDA, 2001). The USDA also intercepted 3,180 species of insects and related arthropods and 10 weed seed species.

The Food and Agricultural Organization (FAO, 2002) of the United Nations recommended measures that would stimulate trade without spreading plant pests. The new measures are intended to improve consistency in the ways that nations conduct their surveillance for potential pests and to certify that plant products for export are pest-free. If the measures were implemented, the new standards would encourage continued growth in agricultural trade without increasing the risks of spreading plant pathogens and insect pests (FAO, 2002).

Some nations have prepared detailed procedures and in some cases long lists of species of plant pathogens that are risks to the crops grown in the nation (Singh, 1983; Kahn, 1989; Smith *et al.* 1997). Implementing the quarantine rules and regulations are left to custom and agricultural people at the port of entry of each nation but these regulations are not fully effective. I recall in 1980 when I was attempting to bring a virus that was found in Russia in the wax moth into the United States, the Canadian officials told me that it was all right to transport the virus in Canada. In addition, the U.S. Department of Agriculture quarantine official indicated that I could bring the Russian virus into the United States, as long as I did not bring the wax moth. Interestingly, the wax moth species has a worldwide distribution and is widespread in the United States.

Regulations for plant quarantines require that plants that are being shipped to the United States port of entry have an appropriate certificate tag, label, or stamp issued by the appropriate U.S. regulatory agency (PPRegs, 2001). Also, the state in the United States to receive the plant material has to be clearly identified. The certificate must clearly specify that the plant material is pest free.

Detecting virus plant pathogens in plants is extremely difficult. Plants that are infected may not show any outward symptoms of the virus disease when being introduced. The new plant pathogen testing technologies using "Enzyme-linked Immunosobent Assay" or ELISA can be an effective bioassay for detecting virus infections in plants (PacWest, 2001). However, the plant pathologist must know which plant pathogen he/she is searching for and use the appropriate technology, otherwise the virus plant pathogen will not be detected.

One means of removing virus plant pathogens from plant stocks is employing "heat therapy" where the plants are grown for several weeks at temperatures around 38° C (PacWest, 2001). Then the shoots or samples of the plants are grown *in vitro* and eventually regenerated into plants and retested to determine, if the virus plant pathogen has been successfully eliminated. Establishing virus disease-free germplasm of pome, stone, and other fruit crops requires a long period of time and is costly (Hurtt and Waterworth, 1997).

All known crops are infected by virus plant pathogens, thus great care must be taken to make sure that the quarantine procedures are sound to prevent the introduction of pests. Once the virus plant pathogen gains entrance into the crop system, it is difficult or impossible to control and or exterminate the plant pathogen (Comstock, 1996).

4.5 Movement of Germplasm

The movement of germplasm involves a risk of accidentally introducing plant pests along with the host plant. In particular, pathogens that maybe initially symptomless, such as viruses, pose a special risk. In order to manage this risk, effective testing (indexing) procedures are required to ensure that distributed material is free of pests that are of quarantine concern.

This has prompted FAO and International Plant Genetic Resources Institute (IPGRI) to launch a collaborative technically and scientifically focused programme for the safe and expeditious movement of germplasm. This programme has generated a series of technical guidelines for various crops that provide information on disease indexing and other procedures that will help researchers ensure phytosanitary safety and to complement phytosanitary regulations and mechanisms existing in countries (IPGRI, 2002).

In addition seed companies are iseeking new germplasm to develop new seed types to farmers. At the same time, the companies need ways to protect their investments. Protection of intellectual property right related to new genetic crops in biotechnology is essential to guarantee a return on their investments. Therefore, the seed companies with any kind of new plant variety that enhances the quality and yield of the material seek protection. As a result for the need of protection of intellectual property, the Food and Agriculture Organization (FAO) and the Consultative Group on Agricultural Research (CGIAR) have been involved in setting guidelines for document agreements related to the exchange of genetic material. These documents deal with the material transfer agreement (MTA) for the exchange of germplasm (Ortiz, 1999).

FAO and CGIAR are centers recognized worldwide and are involved with development investors for international public goods. Material transfer agreements are utilized in the use and transfer of germplasm accessions to other nations. The material held by CGIAR is based on an agreement with the FAO of the United Nations (Ortiz, 1999; (FAO, 2002). The CGIAR has committed itself to the integrity of science and promoting social benefits through research and partnerships for the use of international goods, such as germplasm. CGIAR is committed to fairness and giving agricultural helps to the rural poor and disadvantaged members of society (Ortiz, 1999). The prime objective of the material transfer agreement is to protect the germplasm or breeding lines used by various nations, especially developing countries (Ortiz, 1999; (FAO, 2002). Thus, material transfer agreements guarantee the continued and free availability of genetic materials from the CGIAR centers. As many developing countries do not have the facilities and technical expertise to deal with the direct transfer of germplasm from seed companies, the developing countries have, therefore, to rely on FAO and CGIAR.

4.6 Barriers to Global Trade

The major concern for many nations is introducing new plant pathogen pests and this is having negative impacts on global trade. A growing concern is the apparent "excessive" caution in recent years in some nations that are resulting in perhaps unjustified barriers to global trade (McDonell, 2002). When the regulations are too restrictive to prevent pest invasions then smuggling may result and/or people may find legal ways to find a way around the regulations (Liebhold *et al.* 1995).

In some cases phytosanitary measures are being used as non-tariff or technical barriers to trade (Ivess *et al.* 1994; Graaf, 1994; Calvin and Krissoff, 1998; Blank, 2000). In addition to phytosanitary measures being used as non-tariff or technical barriers to trade, pesticides residues on foods are also being used as barriers. Some nations, like India, still heavily depend on DDT and other chlorinated insecticides for insect control. In the case of India, it was recently reported that 70% of the insecticides in use were DDT and BHC (Pimentel 1997). DDT and BHC are both long-lived, persisting for 30 years or more in the soil. When crops are grown in the soil, DDT and other pesticides may be splashed on the crop. Most developed countries have banned DDT and BHC, therefore if the food exports are contaminated with these pesticides they are not accepted.

Even when acceptable pesticides are used, the tolerance levels for the pesticides on the food crops may be above the acceptable level for the nation. These apparent food safety concerns can be translated into barriers to international trade (Nemajovskky and Centner, 1991; Vvas, 1993/1994; DePolo, 1999).

4.7 Quarantine Regulations Used to Protect Markets

The World Trade Organization forbids the use quarantine regulations to protect markets. However, under some circumstances some developed nations can escape detection using quarantine to protect markets. Although it was not a case of using quarantine regulations to protect the U.S. beef market, in the 1970s when the United States banned DDT and established a tolerance level of 5 ppm of DDT in foods, the regulation did have negative impacts on developing countries. For example, in the early 1970s Guatemala was shipping beef to the U.S. and two boatloads of beef had to be dumped because the beef contained DDT above the 5 ppm limit (ICAITI, 1977). After this event, Guatemala established laboratories to test the beef for DDT before it was shipped. It should be noted that any beef that was contaminated with DDT above the 5 ppm tolerance level was sold on the Guatemalan market.

A great many countries, especially developed countries, have sophisticated regulations for both plant and animal quarantines. Developing countries are generally familiar with the standard procedures for the export crops such as cocoa, coffee, rubber, tea, timber, and others (FAO, 2002a). However, new quarantine regulations may exclude some crops from some markets because the developing countries lack the technical and management knowledge to fulfill the new requirements (FAO, 2002a,b). For example, a developed country may require a "certification assurance". This requirement includes a quality assurance system to be implemented by the exporting industry and audited by the quarantine authorities in the developing country (FAO, 2002a,b). The new system may be too complicated and too expensive to be implemented by the developing country.

He reduction of trade barriers may result in higher food prices for developing countries (FAO, 2002b. These higher prices might at the same time provide incentives for domestic producers to increase production because of higher market prices for their products.

4.8 Conclusion

A major danger exists for all nations for the invasion and introduction of virus plant pathogens. The threat of plant pathogen invasions is growing because of the rapid growth in the world population and increasing global trade. Although viral plant pathogens are not as important as fungal plant pathogens in agriculture, the virus plant pathogens still do cause significant crop losses worldwide. What makes virus plant pathogens such a major problem is that they are exceedingly difficult to detect in plants and equally difficult to control once the crop is infected.

Plant pathogens despite all control measures are causing between 12% and 16% crop losses worldwide and are thus contributing to the world malnourishment problem. The World Health Organization reports that more than 3 billion people are currently malnourished and the number increase daily.

References

APS (2001). American Phytopathological Society. http://www.apsnet.org/members/ppb/ps.APS%20Biotech%20Statement.pdf. December 29, 2001.

Blank, K.G. (2000). The TBT agreement – content and difference to the SPS agreement. *Bull. Int. Dairy Fed.*.349.: 11-13.

Calvin, L. and Krissof, B. (1998). Technical barriers to trade: a case study of phytosanitary barriers and the U.S. – Japanese apple trade. *J, Agri. a Resource Economics*. 23 : 351-366.

Comstock, J.C. (1996). Major diseases affecting sugarcane production in the United States. Sugarcane Germplasm Conservation and Exchange: Rep. International Workshop Held in Brisbane, Queensland, Australia. June 28-30, 1995. 24-26.

DePolo, J. (1999). Food safety and international trade. *Futures (East Lansing)*. 16 : 17-21.

FAO (2002). Agricultural Committee Recommends Measures to Stimulation Trade without Spreading Plant Pests. http://www.fao.org/NEWS/1997/970506-e.html (January 1, 2002). Food and Agricultural Organization of the United Nations.

FAO. (2002a).Lessons Learnt. Quarantine Regulations Used to Protect Markets. http://fao.org/inpho/vlirary/move_rep/x0296e/x0296e03.htm

FAO (2002b). World Livestock Trade. Agric.Dep., Food and Agric. Org. United Nations. http://www.fao.org/ag/magazine/0204sp1.htm.

FAO (2002). Material Transfer Agreement for FAO-Designated Germplasm. Inter. Network Improvement of Banana and Plantain. http://www.inibap.org/research/mta1_eng.htm

Graaf, J. de (1994). Plant health and international trade is seed tomatoes. *Can. J. Plant Path.* 16 : 146-149.

Hokkanen, H.M. T. and Pimentel, D. (1989). New associations in biolgical control: theory and practice. *Can. Entomologist*. 121 :828-840.

Hollings, M. (2002). International trade in ornamental plants: phyosanitary and plant virus quarantine problems. *Acta Hort.*. http:www.actahort.org./books/110/110 17.htm. (January 1, 2002).

Hurtt, S.S. and Waterworth, H.E. (1996). Indexing and establishing disease-free germplasm of pme and stone fruit crops. Suagrcane Germplasm Conservation and Exchange: Rep. International Workshop Held in Brisbane, Queensland, Australia. June 28-30, 1995. 90-94.

Ivess, R.J., Kanawi, D., Bannick, K. and Kuila, G. 1994. Quarantine – a client oriented approach: the process of quarantine in PNG and its present status. Papua New Guinea *J.Agriculture, Forestry and Fisheries*. 37 : 139-146; 147-151.

ICAITI (1977). An Environmental and Economic Study of the Consequences of Pesticide Use in Central American Cotton Production. Guatemala, Guatemala, C.A.: Instituto Centroamericano de Investigacion y Technlogia Industrial.

IPGRI (2002). http://www.ipgri.cgiar.org/publications/guideline.htm

Kahn, R.P. (1989). *Plant Proection and Quarantine*. Three Volumes. CRC Press: Baca Raton, FL.

Kiritani, K., Su, H.J., Chu, Y.I. (1991). *Integrated Control of Plant Virus Diseases*. Republic of China on Taiwan: food and Fertilizer technology Center for the Asian and Pacific Region.

Liebold, A.M., MacDonald, W.L., Bergdahl, D. and Mastro, V.C. (1995*). Forest Science Monograph*. (no.30). 49 pp.

McDonell, I. (2002). Bring on the barriers to trade. http://exoticpests.apsnet.org/Discussion/disc8/0000003b.mtm. (January 1, 2002).

Nemajovsky, A.R. and Centner, T.J. (1991). Pesticide residues in food: the Delaney Clause and global harmonization of Pesticide standards. *Agribuiness*. 7 : 187-196.

Oerke, E.C., Dehne, H.W., Schonbeck, F., and Weber, A. (1994*). Crop Production and Crop Protection: Estimated Losses in Major Food and Cash Crops*. Elsevier: Amsterdam.

Ortiz, R. (1999). Material Transfer Agreement and Germplasm Exchange. Intern. Crops Res. Inst. for the Semi-Arid Tropics. http://.icrisat.org/text/research.grep/homepage/grephomepage/mta.htm. FAO (2002). Material Transfer Agreement for FAO-Designated Germplasm. International Network for the Improvement of Banana and Plantain. http://www.inibap.org/research/mta1_eng.htm.

PacWest (2001). Pathogen Testing. http://www.ars-grin.gov/ars/PacWest/Corvalis/ncgr.path.html. (December 28, 2001).

Pandotra, V.R. (1997). *Illustrated Fungi of North India with Special Reference to J & K State*. International Book Distributors Dehra Dunn India.

Patel, B.N. and Patel, G.J. (1985). Virus diseases of tobacco and their management. In Jayarj, S. (ed.). *Integrated Pest and Diseases Management. Proceedings of the National Seminar*, Tamil Nadu Agricultural University, Coimbatore, India. 261-266 pp.

Pimentel, D. (1988). Herbivore population feedings pressure on plant host: feedback evolution and host conservation. *Oikos*, 53: 289-302.

Pimentel, D. (1997*). Techniques for Reducing Pesticides: Environmental and Economic Benefits*. John Wiley: Chichester, UK.

Pimentel, D., Bailey, O., Kim, P., Mullaney, E., Calabrese, J., Walman, F., Nelson, F., and Yao, X. (1999). Will the limits of the Earth's resources control human populations? *Envir., Devel. Sustainability*, 1: 19-39.

Pimentel, D., Lach, L., Zuniga, R., and Morrison, D. (2000). Environmental and economic costs of nonindigenous species in the United States. *BioScience*, 50 , 53-65.

Pimentel, D., McNair, S., Janecka, J., Wightman, J., Simmons, C., O'Connell, C., Wong, E., Russel, L., Zern, J., Aquino, T., and Tsomondo, T. (2001). Economic and environmental threat of alien plant, animal and microbe invasions. *Agriculture, Ecosystems and the Environment*, 84: 1-20.

PPRegs. (2001). Summary of Plant Protection Regulations. http://www.agi.state.al.us/PPRegs.htm. December 29, 2001.

PRB (2001). World Population Data Sheet. Washington, DC: Population Reference Bureau.

Singh, K.G. (1983). Plant quarantine development in the ASEAN and threats by pests. In, Exotic Plant Quarantine Pests and Procedures for Introduction of Plant Materials. Singh,

K.G. (Ed.). ASEAN Plant Quarantine Centre and Training Institute, Serdang, Selangor, Malaysia.
Smith, I.M., McNaara, D.G., Scott, P.R., Holderness, M. and Burger, B. (1997). Quarantine Pests for Europe. CAB International in Association with European and Mediterranean Plant Protection Organization, Cambridge, UK.
UNEP. (2000). Geo 2000: Global perspectives, social and economic background. (http://www.unep.org/geo2000/english/0026.htm) (March 7, 2002). UNEP.
USDA. (2001). Animal and Plant Health Inspection Service. U.S. Dept. of Agriculture. http://www.us-ecosystems.org/croplands/biological invasions/ (December 29, 2001).
Vvas, V.S. (1993/1994). Changes in agrarian structure and agricultural development in poor countries in Asia. *APO Productivity Journal*. Winter: 133-153.
WHO (1996). Micronutrient Malnutrition – Half of the World's Population Affected (pp 1-4 no. Press release WHO No. 78). World Health Organization.
WHO (2000). Malnutrition Worldwide. http://www.who.int/nut/malnutrition_worldwide.htm, July 27, 2000.

5. Transfer of crop protection technologies in developing countries

Reuben Ausher and Yair Sachs
Department of Crop Protection, Extension Service, Minstry of Agriculture and Rural Development, Israel

5.1 Introduction

An appropriate way of evaluating agricultural extension systems is to study their end product or message. One would expect the ideal product to be professional, objective, practical, reliable, up-to-date, market-driven, environment-friendly, integrated and accessible. It should improve farmers' managerial abilities and stimulate their use of knowledge (Ausher, 1996). Obviously such a message could be generated when all three components, namely growers, research and extension are involved in its formulation, diffusion and adoption. Extension's particular niche in this context would be the identification of knowledge gaps, initiating ideas for technology development, responsibility for applied research, and disseminating technologies to producers. During the last two decades the performance and outcome of agricultural extension met with an unfortunate crisis in the generation and transfer of technologies in both industrialized and developing countries. By and large, agricultural extension systems are slow in adjusting to the changing environment, characterized by a proliferation of factors within the once much simpler farmer-extension-research relationship. Nowadays, agricultural extension has to cope with the challenges to intensify and diversify production and promote at the same time market-oriented, environment-friendly and sustainable production systems. Ausher, (1997) indicated that the technology development capability of extension is being depleted of its assets: the problem-oriented, empirical, pragmatic and "cheap and nasty" approach. Followed by the field-level expertise, a knack for integrative field diagnosis, field-centered interdisciplinary activity and accountability.

5.2 Agricultural Extension Systems

According to a survey of the Food and Agriculture Organization of the United Nations (FAO, 1990), worldwide, agricultural extension employs at least 800,000 extension workers and hundreds of thousands more farmer technicians or leading farmers, reaching about 1,200 million people. Currently, about 80 percent of the world's extension services are publicly funded and delivered by civil servants (World Bank, 1997). Universities and nongovernmental organizations deliver about 12 percent of services, and the private sector another 5 percent. Government involvement tends to be at many levels and in many forms. It may fund, staff, or facilitate extension by establishing conducive regulations and policies for other providers – and it may pursue a range of purposes. Some of the generic problems of extension are caused in part by this complexity. Most governments regard agricultural extension as a technology development instrument responsible for the generation, diffusion and adoption of agricultural technologies and consider this function as producing "public good" and carrying exclusive responsibility for the technology diffusion component (Birkhaeuser, *et al.* 1991; Umali-Deininger, 1996). Public funding has been also justified for extension programs targeting social goals such as poverty alleviation, population distribution, and supporting rural women. These days government intervention in agricultural production, providing agricultural extension services could be regarded as a major indirect support channel complying with the requirements of the World Trade Organization (WTO). FAO surveyed the status of agricultural extension in 113 countries contacting 207 agricultural extension organizations that were considered to be generally representative of agricultural extension systems throughout the world (Swanson *et al.* 1990). Eighty-six percent of these agencies were funded by ministry of agriculture or similar government agency. Based on this survey, FAO estimated that in 1988, $6,000 million were spent on public agricultural agencies worldwide, representing an average expenditure per extension worker of $8,522. The annual per farmer spending on extension services ranged from $2 to $3 in low-income countries to $65 in high-income countries. The main features of the 1970s in the developing world included integrated rural development and the rise of the training and visit (T&V) extension system. The first attempted to provide remedies on a broad, comprehensive basis subsequently diluting the efforts. Training and Visit focused exclusively on agricultural extension, concentrating efforts to enhance agricultural production without changing other production-related inputs. Extension workers in the developing world who enjoyed their transfer from non-technical to technical activities and receiving regular training accepted Training and Visit with much enthusiasm. In the period 1975-1995, the T&V extension approach was adopted in scores of countries, with the active promotion and support of the World Bank. In most countries where the system was adopted, after several years of operation, it was scaled back due to fiscal difficulties, and the scheme was simplified or abandoned. In many countries where T&V was significantly scaled down or abandoned, no rigorous extension effectiveness or economic impact studies were done, but positive indicators of performance were provided by Monitoring and Evaluation units, extension administrators and other commentators, based on non-rigorous studies and subjective field observations. An interesting illustration of the point being made is the case of Kenya. An economic study conducted in the early nineties by Bindlish and Evenson, (1997) suggested very high economic rates of return to T&V extension investments in Kenya. Gautam and Anderson (1999) challenged these findings and suggested that the system had faced fiscal strains since 1996 and its operations were adversely affected. These experiences show according to Feder *et al.* (1999), that funding difficulties can adversely affect national extension operations, even in the face of positive indications on performance. The 1980s brought an increasing emphasis on participatory approaches. Concerns on increased participation of women and preserving ecosystems were added, along with attempted cost recovery, privatization and decentralization schemes. In the 1990s, alternatives were tested out in an environment of fiscal stringency. In this context, countries are in a process of veering away from the traditional centrally administered and publicly delivered extension. Chile adopted a centrally administered but privately delivered system. Colombia opted for locally administered and publicly delivered, while Venezuela adopted a locally administered and privately delivered system with a small support unit located in the Ministry of Agriculture. In the latter attempts were also made to establish farmer's councils to supervise and gradually fund private entities: universities, firms and other non-governmental organizations, which were contracted to deliver extension (Nielson, D., unpublished observations). World Bank project experiences indicated, (Purcell, 1997), that there is no single extension method which could be universally used to improve public sector extension and strategies need to be developed which are suited to the circumstances of each country. The extension principle that advice must be relevant to the conditions of the targeted farmer clients needs to be paramount. The ultimate expression of this principle would be in the clients "owning" the service and being actively involved in determining its direction and shape. Any extension strategy must consider sustainability of its service components. Special emphasis has to be given to the fiscal capacity and willingness to support staff-intensive services (Feder *et al.* 1999).

5.3 Extension work in crop protection

Ausher (1986) indicated the existence of five organizational forms providing crop protection advice: (1) Crop advisors at the local level are generalists (often including even animal husbandry in their scope). Crop protection specialists serve only at the district, regional, and state levels as in Kenya, India, and Peru. (2) Crop advisors and crop protection specialists serve at the local level (some specializing in crops or in disciplines) and are supported by experts at regional or national centers and research institutions. This is the case in most of Western Europe, Israel and much of the United States. (3) Crop protection advisors operate outside the general farm advisory structure, either in separate units of the Ministry of Agriculture, as in Thailand, Togo and Germany, or in research institutions, as in Portugal. (4) Each farm employs its own crop protection advisor or consulting firm with the Ministry of Agriculture providing overall guidance and issuing warnings. This system was prevalent in Eastern Europe on state and cooperative farms. It could be found in developing countries in the form of single commodity-focused extension where crops are produced on large farms, plantations or as special projects. (5) Either no public advisory service system exists or the advisory organization has no crop protection specialists and relies on representatives of chemical firms, commercial dealers and consulting firms for advice. This is the case in some regions of southern and western Europe and in developing countries such as Chile and Brazil. France, Norway and Taiwan provide examples of the prominent role farmers' organizations can play in funding and organizing agricultural extension (Ameur, 1994; Umali and Schwartz, 1994; Haug, 1991; Nagel, 1994). The approach that separates extension from research prevails in Europe, the UK and Israel and most developing countries. The approach that unites extension and research is exemplified by the American land-grant system. In the developing countries, state extension plays more of a coordinating role while the actual fieldwork addressing major current and mainly new problem areas is carried out in the form of special projects. The latter is based on the participation of universities, regional research and development, farmers associations and processing plants. This is the case of *ad-hoc* organizations e.g. in the Dominican Republic when faced with severe problems of viral infestation on tomato and melon. The overall project coordination being done under the leadership of a government-related non-governmental organization, Junta Agroempresarial Dominicana (JAD). The role of regional interstate technical assistance bodies such as the Organismo Internacional Regional de Sanidad Agropecuaria (OIRSA) active in Latin America and the Caribbean is on the rise sharing crop protection expertise over the whole continent. It is hard to evaluate which system is the better one. In our view, crop protection advisors who specialize in crops, covering all major plant protection disciplines, namely, plant pathology, virology, entomology and weed control, have easier contact with both crop specialists and farmers than research-oriented advisors who specialize in various disciplines. The crop and farmer oriented extension structure supports the advisor in his decision- making in the field about cropping techniques and their impact on crop health, in the area of all pests' diagnosis, and pest management. Furthermore, this integrated structure enhances the capabilities of crop protection advisors to cope with Integrated Pest Management (IPM). This discipline looks at the overall situation of pests in the field, the losses caused by them, and their rational control with the help of a wide array of chemical and non-chemical methods. The crop-specialized advisor can master the overall problems of pest control ensuing of the single or combined use of insecticides, fungicides and herbicides, application techniques and equipment, waiting periods and market requirements. The field advisor specializing in just one of the three major plant protection disciplines is less equipped to make these decisions. In the particular case of plant viruses in vegetatively propagated crops, control practices focus heavily on crop-related factors like: nursery management, selection of promising propagation stocks, grafting and use of virus-free rootstocks. The need for good knowledge of plant propagation, varieties and agro techniques turn crop husbandry extension specialists, collaborating with plant virologists into leaders of the domain.

5.4 Promotion of Integrated Pest Management schemes in the developing world

In general, promotion of Integrated Pest Management (IPM) schemes in the developing world falls under two main categories: single commodity-focused schemes and area wide pest management. The first approach is the most common one and suitable mainly for the conditions of large commercial farms specialized in the cultivation of processing and export crops such as cotton, rubber, oil palms, citrus, etc. As far as smallholders are concerned, the application of pest monitoring and IPM is size-dependent. Surrendered by adjacent pest-infested and often heavily sprayed plots, it is almost impossible to sustain a reasonable IPM regime in small plots. Two new IPM platforms address the problem area of promoting IPM on smallholders' plots in the developing world. The first, the Farmer Field School approach (FFS) widely applied in Asia and more recently introduced in Africa, and the Area Wide Pest Management (APM) concept developed in the US, Australia and Israel. The first attempt to apply the latter under a developing world setting has been done in the framework of a regional cooperation effort on pest management bringing together Egypt, Jordan, the Palestinian Authority, Israel and the US.

The Field School approach. The IPM FFS emerged out of a decade of experimentation in implementing participatory farmer training activities in the Philippines, beginning in the late 1970s. Refinements in the Philippine program and the launching of a major new effort in Indonesia in the late 1980s led to the birth of the FFS movement that has since spread across the region and around the world. Conceptually, the FFS approach weaves together reinforcing elements of adult education, agro ecology and local organizational development. Simpson and Owens (2002) indicated that the educational focus of the approach is perhaps its most distinct feature. According to Loevinsohn *et al.* (1998) who have evaluated its performance in Africa, the FFS enhanced capacity among small farmers to practice IPM. Considerable interest is being shown in it in other parts of the world, and it is being intensively promoted by several organizations, notably FAO and the Word Bank, through its Global IPM Facility. Quizon *et al.* (2001), insisted that critical evaluation is essential to ensure that the approach is not oversold or asked to do more than it can reasonably be expected to. The FFS aims to go beyond the passing of messages to enlarge the concepts and principles that farmers have at their disposal as they manage their agro ecosystems. In Asian rice environments, this is done through discovery-based learning, in groups generally of 15-20 that meet in the farm environment. Among the key principles that farmers discover and grasp through regular observation of their fields, joint experimentation and discussion, are:

Food webs: there are insects that attack crop plants, and insects that attack insects. Farmers see this for example by placing an herbivorous insect in a muslin-covered cage, together with a rice plant and a spider.

Crop compensation: Plants are able to recover from even severe damage at an early stage of growth. Farmers grasp this through experiments in which they cut different numbers of leaves or tillers and observe the effect on later growth and harvest.

Life cycles: Farmers learn that what they might have thought was different organisms are in fact stages of the same species. They do this, for example, by collecting egg masses, and observing them in cages until the larvae emerge and eventually pupate.
Crop nutrition and tolerance: Farmers see that a well-nourished and watered crop is able to tolerate substantial insect or disease attack.

The group discusses management options, for example in soil and pest management, before being tested in a common "IPM field". The importance of trying out options, whether they are proposed by extension, or by the farmers, and continuing such experimentation after the FFS, is repeatedly stressed. The FFS is led generally by a front line extension worker and continues for an entire crop season, usually meeting once a week. The trainers themselves typically go through an intensive residential Training of Trainers (TOT) course of several months' duration. Loevinsohn *et al.* (1998) surveying FFS systems applied in Kenya concluded that the FFS model is a method essentially to permit "rediscovery" by farmers of principles which are known to scientists and other farmers. The principles are new and important to the FFS farmers, make possible important changes in their management and may stimulate continuing experimentation in different farm enterprises and directions. It is essentially a better form of extension-based research. But the FFS is not well suited to "discovery" situations where farmers don't have answers to problems and no one else does either. Some of the problems farmers confronted in Kenya, included the variable performance of botanical pesticides on vegetables, and went beyond simple adaptive questions to applied or strategic issues. It is essential that farmers speak with scientists, and scientists have to speak with them. More intensive participatory research, with a greater concentration of experimentation, evaluation and follow-up experimentation would be required in these cases to develop suitable solutions. This does not happen in an FFS where there are only a few, unreplicated trials and where the links to research and the most current knowledge available are very indirect. Simpson and Owens (2002) in a qualitative data collection reviewed the introduction of the FFS approach to Ghana and Mali following the data of Ketelaar (1995), who indicated its cost saving potential. However, none of the FFS farmers in Mali had adopted the new IPM practices at the whole-field level. The most consistent and striking examples occurred among participants in the FFS on vegetables, where certain management practices, such as the use of neem infusions and mulching, covered in the FFS on tomatoes and cabbages, were transferred to other vegetable crops. As it would be expected of any experimental effort, many of these initial adaptations were later rejected. The approach, however, has made significant contribution in providing the opportunity for farmers to understand important systems-concepts and relationships to local social development. Quizon *et al.* (2001) examined the issue of fiscal sustainability of FFS extension approach to reaching farmers over a large geographical area. The Philippine and Indonesian experiences with field schools cast doubts on the fiscal viability of the FFS approach for disseminating knowledge-intensive technologies, such as IPM, to large farmer populations. While several village-level studies and limited-size pilot experiences may attest to the viability of field schools in specific local circumstances, overall national-level experiences indicate that fiscal sustainability can be difficult to surmount at "scaled-up" levels. Some as a way out of the fiscal dilemma views farmer-led field schools if part of the cost is shifted to the community. However, farm survey data from Indonesia indicate that the extent of the takeover of training responsibilities by farmers was minor.

Area wide pest management. A second IPM platform that was tried out in the industrialized world and is making its first steps to serve the developing world is the Area Wide Pest Management (APM). Kogan, (1998) envisaged a concept covering with an IPM approach all major crops of a given area. This provides the opportunity of tackling the problem of migratory pests, and of pests moving from one host crop to another. Control efforts could focus on the multiplication sources of pests, reducing pest population's generation after generation. This could enable the introduction of more environment-friendly and selective pest control methods such as sex pheromones to disrupt the mating behavior of insect pests. The APM approach developed along three major stages: a "one crop-one pest" situation applied for the area wide control of the codling moth on apples in the western US, a "one crop-multiple pest" situation prevalent in the cotton industry in Australia, and a "multiple crop-multiple pest" situation applied in Israel since 1996. The latter was exposed to the circumstances of small holders in the framework of a US-funded regional IPM promotion effort encompassing vegetable growers in Egypt, Jordan, the Palestinian Authority and Israel. Ausher (1999) and Hameiri (1999) outlined that the project is focused on the reduction in the use of pesticides, and the promotion of cultivation practices complying with pesticide tolerances. This educated farmers to rationalize pest control, protecting the environment and public health, extending the lifetime of essential insecticides and reducing pest control costs. A state steering committee, regional steering committees and day-to-day management core groups guide the program in Israel. Local pest scouts monitor units of 150 ha and perform two weekly field visits at peak season. Two regional pest scouts supervise the local scouts' work, visiting fields on a weekly basis and responding to calls. Pest control policy is annually configured. Weekly recommendations are issued under the leadership of the regional crop protection extension specialist. The pest resistance laboratory carries out routine checks of main insect pests, reporting regularly to scouts and extension on resistance curves and trends. Pest scouts, extension and pest resistance laboratory staff, growers and regional field staff of chemical industry meet on a weekly basis to report on pest and resistance situation, formulating control recommendations. A quantitative project performance evaluation shows thatt in general in the transition season from conventional control to IPM, a reduction of about 30% is attained in the use of pesticides. More specifically, data on seasonal pest control applications and their costs on processing tomatoes, cotton, corn and watermelon (year 2000) were collected in the Bet-Shean/Gilboa area and compared to the neighboring Yezreel Valley in which APM had not been adopted. No significant differences were detected between the two areas, with the exception of corn. In the latter, according to Hameiri (personal communication) Bet-Shean/Gilboa reduced significantly pest control treatments and subsequent costs by growing in the early season and evading higher pest populations. Spillover of information between extension and growers in two close and homogenous regions could explain the similarity of their pest control performance. The pest resistance monitoring program addresses five major insect pests: cotton pink bollworm (*Pectinophora gossypiella*), cotton budworm (*Helicoverpa armigera),* silverleaf whitefly (*Bemisia tabaci*), Egyptian cotton leafworm (*Spodoptera litoralis*), and the potoato tuber moth (*Phtorymaea operculella*). Over the last three years the program covered also indoor vegetables grown by private growers in the Ismailia area in Egypt, along the Jordan Valley in both the Palestinian Authority and in Jordan, and family farms in Israel. According to Ausher (2001) indoor vegetables and fresh herbs grown as export crops in Israel, require growers to be trained on Quality Management to comply with requirements of the European marketplace. Growers have to be acquainted with standards set by the International Standardization Organization (ISO), the umbrella organization of the European retailer chains-EUREPGAP-2000, and with local extension-formulated crop production and protection protocols. The scouts and growers are invited to attend the weekly meetings of the area wide project's technical team and to receive specific written information and recommendations on a weekly basis.

5.5 Virus disease control problems in vegetatively propagated perennial crops in the developing world

Crops intensively cultivated in the developing countries include citrus, tropical and subtropical fruit crops, grapes and deciduous orchards.

Citrus. Viral infection of citrus is spread mainly through vegetative propagation and vectors. Viral diseases belong to two main categories. First, mechanically transmitted diseases in which grafting and cuttings act as the main source of infection and second, vector-transmitted viral diseases such as tristeza. Resistant cultivars and rootstocks, preferably from local species, are a possible answer. Strategically, the main focus of control efforts should prevent import of diseases with plant material by the adoption of a rigorous registration procedure, monitoring the movement of plant and propagation material (Sadowsky, personal communication). Scion brought in from other countries or regions has to be checked for being virus-free by the best available technological means. Since only a small part of viral diseases affecting citrus is seed-borne, appropriate rootstocks could be selected from seed-propagated germplasm. However, several major viral diseases of citrus such as psorosis and xyloporosis are latent and their symptoms appear only after 10-15 years. Virus-free germplasm has to be maintained under conditions that prevent infection: in insect-proof screen houses, under meticulous sanitation practices. Stocks from screen houses are planted in foundation blocks for propagation purposes and subsequently released to growers. The stocks have to be monitored for both health ("pathogen-free") and for horticultural traits ("true-to-type"). The indexing effort is an interdisciplinary one necessitating the collaboration of research plant virologists, plant pathologists, and extension crop husbandry and crop protection specialists. The mission of agricultural extension in the generation and diffusion of virus control technologies in the developing world consists of the identification of problems at field level, teaching the main problem areas and promoting the concept of proper handling of propagation material. One of the more delicate and responsible technical roles of extension in this context is the clear definition, of the accepted impurity levels of propagation material. In many countries tolerance to local strains of pathogens as in impure local material, is preferable to absolutely clean material. that is usually more susceptible. For example, grafting nucellar "Jaffa" on Palestine sweet lime caused severe losses inflicted by *Phytophthora*. Inoculating citrus with dwarfing viroids reduces juvenile growth and susceptibility to growth-related pathogens. Hence, in the developing world it is often recommended to utilize local propagation material that may carry a certain amount of viral infection rather than use imported virus-free material. For example, Tristeza-infected lime trees in Guyana produce acceptable commercial yields, while in Brazil orange trees free of tristeza often produce much lower yields. On the other hand, Uruguay imports virus-free grape seedlings from European nurseries and they perform well for many years. (Sigler, personal communication). These considerations require the presence of a reliable extension system to assess the cost-effectiveness and impact of deviations from a textbook approach. In view of its role within an integrative professional system, agricultural extension should examine local propagation material that passed through natural selection cycles. To this end, extension subject matter specialists should be well aware of the state-of-the-art in both their country and overseas. Proper recording of all propagation-related procedures and the adoption of Quality Management principles is the only way through which errors, lack of attention or bad judgment could be detected and remedied in time. Collaboration of extension subject matter specialists from the developing world with researchers and extension workers from overseas is a must in order to keep staff abreast of new technological developments. Due to this type of cooperation, the citrus industry in the developing world enjoys a much higher technical level than other commodities such as tropical fruits, avocado, grapes and deciduous fruit crops (Sigler, personal communication). No certification systems are in place to service these crops and, although known in theory, indexing is not being applied. The first steps for improving the quality of all non-citrus fruit crops would be the earmarking of healthy commercial trees of outstanding productivity, identification of reliable growers, and monitoring the performance of the selected trees. Such trees could provide propagation material (scions, cuttings, buds, etc.) for nurseries. In many countries legislation is still in its infancy and there is an uncontrolled movement of propagation material across the borders. Import of clean material from reliable sources could be authorized for establishing centers for nuclear budwood. By and large, viral diseases don't cause major economic losses in deciduous fruit crops but they are of major significance for grapes and tropical fruit crops. In this respect, the situation of the large and private farms is much better than that of the smallholders.

5.6 Virus disease control problems in annual crops in the developing world

Several virus diseases cause losses to annual crops in the developing world affecting food supply and growers' income. Among the more important viral diseases, with severe outbreaks recorded during the last decade is the *Tomato yellow leaf curl virus* (TYLCV) in tomato. It has spread over the Mediterranean basin, the Caribbean and parts of the U.S. The same is true for bean yellow on beans in Central America, and various non-persistent viral diseases in cucurbits in many countries. Various strains of the *Tobacco mosaic virus* (TMV) affect tomato and pepper. More recently, soil-borne viral diseases such as the *Melon necrotic leaf spot virus* (MNSV) affecting melon and watermelon have gained importance.

Tomato yellow leaf curl virus (TYLCV) originated in the Middle East. It is spread persistently by the whitefly (*Bemisia tabaci)*. It heavily affects table tomato and to a somewhat smaller degree processing tomato and Cherry varieties. Its other cultivated hosts are beans and Lisianthus. The losses on tomato plants may reach 100%. In addition to the Middle East, the disease is present in the western Mediterranean and northern Africa, namely Spain and Morocco. In the 1980s the disease reached the Dominican Republic and in a short period it devastated the tomato industry of the country. Later tomato crops in Cuba and Jamaica became infected. In the late 1990s it was found in Florida and in Honduras. The pathogen was identified as being identical to the Middle Eastern type. In the late 1980s two different biotypes of the vector were identified. The one found in Florida was called B or Poinsettia biotype, present for years in the Middle East. Most probably the B type reached the American continent in the 1980s and in a short while TYCLV reached the Caribbean. It is not known whether the disease was carried on infected plant parts or on plants carrying infective whitefly. Extension and research workers in the newly affected areas formulated a systematic series of control measures. They rely on close collaboration established with other countries in which the disease was already established. Among the main approaches are the identification of the viral pathogen and vector, and screening for additional host plants, including symptomless carriers. Thus, new hosts were found (Brown *et al.* 1991, Jorn. Cient. 1992). Due to favorable climatic conditions, and the continuous presence of both vector and host plants, it was not possible to break the infection cycle. The Dominican Republic (DR) adopted a mandatory annual no-growth period of 2-3 months for tomato. Due to the presence of wild alternate hosts, this was only partially successful. Furthermore, due to shortages of tomato the moratorium in growing was implemented only partially on a regional timetable, and small holders continued growing tomatoes illicitly. Due to the low effectiveness of this approach present efforts focus on: (i) breeding of resistant cultivars. The first cultivars (Ty varieties) were tolerant to TYCLV but of poor horticultural qualities. Improved tomato varieties are presently grown successfully in the Dominican Republic and in Egypt. Use of TYLCV-resistant varieties will also reduce the frequent chemical applications, which are presently recommended for the control of TYCLV. (ii) A small part of the tomato-cultivated area in the DR was put under screen houses based on an Israeli design, providing high quality fruit. (iii) Production of seedlings under insect-proof screens. Thus, the transplanted tomato plants got infected at a much later stage.(iv) Yellow plastic mulch applied in the fields attracts whitefly, which desiccate on the sun-heated mulch. This practice prevents early infection of transplanted tomato plants. The implementation of these practices in the DR was the outcome of a crash program bringing together the public sector and non-governmental organizations that provided agricultural extension-advisory services to their growers (Melamed, personal communication). As described by Griver (2002) these organizations adapted expertise from other countries. Studies of whitefly populations led to the identification of additional strains. The majority of the whitefly population of Spain was identified as biotype Q, which is more resistant to neonicotinoid insecticides. Massive use of Imidacloprid caused a reduction of the B biotype and enhanced the proliferation of the Q biotype. The same trend was found in Israel where the Q biotype became predominant. Both biotypes are effective TYLCV vectors.

Non-persistent viruses in melon. Melon production is a major export industry in the Caribbean area. Viral diseases such as the *Cucumber mosaic virus* (CMV) and the *Zucchini yellow mosaic virus* (ZYMV) affect a large part of the fields. Progressive producers in the Caribbean followed extension recommendations, and were successful in reducing losses by spraying with JMS type mineral oils to prevent viral infection.

Cross protection. Infecting watermelons with avirulent strains of ZYMV is becoming an accepted practice in the last decade. Mainly private firms, with limited involvement of extension, promote the practice.

Tobacco mosaic virus (TMV). TMV affects tomato and pepper amongst other hosts. Over the last years, many tomato cultivars had been bred for resistance to TMV strains. However, new susceptible varieties were released and new TMV biotypes developed. It is a challenge for extension to determine the losses caused by TMV to tolerant or resistant varieties. Seed dressing and disinfecting equipment are helpful in the reduction of inoculum sources.

Melon necrotic spot virus (MNSV). Cucurbits grown in warm countries are severely affected by soil-borne MNSV. The latter is carried by the fungus *Olpidium spp.* Soil treatments that control the fungal vector is effectively reducing MNSV infection.

5.7 A final word

Advisory work focused on plant viral diseases is part of the crop protection extension system. In turn, crop protection extension, addressing losses caused to crops by both parasitic and non-parasitic, environmental factors, is a major component of the overall extension-advisory work in both the industrialized and the developing world. Under intensive cropping regimes, often more than 50% of the crop advisors' work is devoted to crop protection. Despite the crisis affecting agricultural extension, a wide array of successful extension systems and modules are found in various countries. The role of both policy makers and extension specialists is to identify the weak points of their country's structure and to select and apply in a pragmatic way successful extension mechanisms and modules that could improve their systems' needs.

References

Ameur, C.H. 1994. Agricultural Extension: A Step Beyond the Next Step. *World Bank Technical Paper* 247. Washington D.C.

Ausher, R. 1986. Extension Around the World. *Plant Disease*. 70 : 95.

Ausher, R. 1996. Crisis in the Generation and Transfer of Agricultural Technologies-Possible Remedies, 541-554. *Food Security and Innovations, Successes and Lessons learned.* Int'l Symposium. (F. Heidhues andA. Fadani, eds.) Peter Lang.

Ausher, R. 1999. Promotion of Areawide Pest Management (APM) in Israel. *Phytoparasitica* 27: 83-84.

Ausher, R. 2001. Quality Management in Israel's Agriculture. *Phytoparasitica* 29 : 279-283.

Bindlish, V. and Evenson, R. E. 1997. The Impact of T&V Extension in Africa: The Experience of Kenya and Burkina Faso. *The World Bank Research Observer* 12 : 183-201.

Birkhaeuser, D., Evenson, R.E. and Feder, G. 1991. The Economic Impact of Agricultural Extension: A Review. *Economic Development and Cultural Change* 39 : 607-650.

Brown J.K., Lastra, R .and Bird, J. 1991. First documentation of whitefly-transmitted geminiviruses causing widespread disease in cotton, tobacco and tomato in Dominican Republic and in tomato in Puerto Rico. *Fitopatologia.* 26: 47.

FAO 1990. *Global Consultation on Agricultural Extension.* Rome.

Feder, G., Willett, A. and Zijp, W. 1999. Agricultural Extension-Generic Challenges and Some Ingredients for Solutions. *Policy Res. Worki. Pap.* No. 2129. Washington, D.C.

Gautam, M. and Anderson, J. 1999. Reconsidering the Evidence on Returns to T&V Extension in Kenya. *Policy Research Working Paper* 2098, World Bank, Washington D.C.

Griver, S. 2002. What's Damaging the Crops? Area-Wide Integrated Pest Management in the Dominican Republic. *Shalom, Magazine Alumni Mashav Training Courses.* 2. 29-30.

Hameiri, Y. 1999. Areawide Intercrop Integrated Pest Management ("Hamabia") in the Gilboa, Bet-Shean and Jordan Valley Regions. *Phytoparasitica* 27: 84 (abstr.).

Haug, R. 1991. Public-Private Cooperation: Farmer-Led Research/Extension Circles in Norway. In *Agricultural Extension*: Worldwide Institutional Innovation and Forces for Change. Rivera, W.M. and Gustafson D.J. eds. Amsterdam, Elsevier.

Jornada Cientifica-Tecnica Sobre el Cultivo de Tomate. 1992. Grupo Tomate/Mosca Blanca. Proyecto CATIE/MAG-MIP (NORAD-ASDI). 23 pgs.

Ketelaar, J.W.H., Millomeda, G. and Pulmano A. 1995. Report on Training of Trainers and Farmers Field Schools for Rice Integrated Pest Management in Ghana from May 31-October 6, 1995. FAO, Manila.

Kogan, M. 1998. Integrated Pest Management: Historical Perspectives and Contemporary Development. *Annu. Rev. Entomol.* 43: 243-270.

Loevinsohn, M., Meijerink, G., and Salasya., B. 1998. Enhancing capacity to manage resources: assessing the farmer field school approach. *Sec. Meeting Integrated Pest Manag. Caribbean,* sponsored by CARDI and CTA. Kingston, Jamaica, February 4-6.

Nagel, U.J. 1997. Alternative Approaches to Organizing Extension. *In Improving Agricultural Extension: A Reference Manual.* FAO. Rome.

Purcell, D. 1996. Agricultural Extension and the World Bank. 555-565. *Food Security and Innovations, Successes and Lessons Learned. Int. Symp.* F. Heidhues and A. Fadani, eds. Peter Lang.

Quizon, J., Feder, G. and Murgai, R. 2001. Fiscal Sustainability of Agricultural Extension: The Case of the Farmer Field School Approach. *J. Internat. Agric. Extension Education* 8: 13-23.

Simpson, B.M. and M. Owens. 2002. Farmer Field Schools and the Future of Agricultural Extension in Africa. *J. Internat. Agric. Extension Education* 9: 29-36.

Swanson, B. E., Farner, B. J. and Bahal, R. 1990. The Current status of Agricultural Extension Worldwide. In *Global Consultation on Agricultural Extension.* FAO, Rome.

Umali-Deininger, D. 1996. New Approaches to an Old Problem: The Public and Private Sector in Extension. *Extension Workshop, Alternative Mechanisms for Funding and Delivering Extension.* World Bank. Washington D.C.

Umali, D. and Schwartz, L.. 1994. Public and Private Agricultural Extension: Beyond Traditional Frontiers. *World Bank Discussion Paper* 236. Washington D.C.

World Bank. 1997. Rural Development: From Vision to Action. Sector Strategy. *Environmentally and Socially Sustainable Development Studies and Monographs.* Series 12. Washington D.C.

6. Cassava

G. Thottappilly[1], J.M. Thresh[2], L.A. Calvert[3] and S. Winter[4]
[1] Mahyco Research Foundation, A.G. Heights, Road No. 12, Banjara Hills, Hyderabad-500034, A.P., India, E-mail: execdirector@satyam.net.in ; [2]Natural Resources Institute, University of Greenwich, Chatham Maritime, Kent ME4 4TB, UK; E- mail: kathie.clark@blueyonder.co.uk ; [3]Centro Internacional de Agricultura Tropical (CIAT), AA 6713, Cali, Colombia, E-mail: l.calvert@cgiar.org ; [4]Plant Virus Division, German Collection of Microorganisms and Cell Cultures, DSMZ, C/o BBA, Messeweg- 11/12, Braunschweig, Germany, E-mail: s.winter@bba.de

6.1 Introduction

Cassava, *Manihot esculenta* (Euphorbiaceae), also known as tapioca, yuca, mandioca or manioc, was domesticated during 4000-2000 B.C. in South America and reportedly originated in Brazil (Abraham, 1956; Karakacha, 2001). The Portuguese took cassava to the Gulf of Guinea in Africa in the 16th century (Fauquet and Fargette, 1990). By the 18th century, the crop had reached the East Coast of Africa, Madagascar, Reunion and Zanzibar. After the mid-18th century, it reached India and Ceylon (now Sri Lanka). Another theory is that cassava was directly introduced from South America into India in 1840 (Abraham, 1956). By the 20th century, cassava was grown throughout much of sub-Saharan Africa. However it remained as a subsistence crop until the 1920s/1930s when it became the major staple food crop of Africa (Carter *et al.* 1992). Currently the crop is grown in more than 39 African and 56 other countries.

Cassava is the staple or subsistence food for about one fifth of the world's population (Edison, 2000) and it is grown and consumed by many of the world's poorest and most food-insecure households. Moreover, it has a wide range of uses: food security crop (consumed fresh or processed), cash crop, animal feed, and raw material for industrial uses such as starch and alcohol production (Kisgeci, 1989; Wright, 1996). The leaves are used widely as a vegetable and contain proteins, minerals and vitamins A and C (Smith, 1988; Edison, 2002). Thus cassava constitutes an important source of income in rural and often resource-poor farmers, many of whom are women. The total world production of cassava in 2001 was 179 million tonnes, of which *c.* 53% were produced in Africa while Asia and South America contributed about 28 and 17%, respectively (source: FAOSTAT, 2001; www.fao.org). Nigeria, Democratic Republic of Congo, Thailand, Indonesia, and Brazil are the biggest cassava producers. Production and consumption of cassava is expected to rise in the next 20 years due to human population growth, low and stagnant *per capita* incomes and rapid urbanisation, all of which generate demand for a cheap starchy staple to feed the rural poor and urban consumers (Scott *et al.* 2000).

Since cassava is mainly propagated vegetatively, it is particularly prone to viral infection, which tends to build up in successive cycles of propagation (Calvert and Thresh, 2002). The long growth cycle of the crop (6-24 months), increases the exposure to viruses and/or vectors.

Eighteen different viruses have been reported from cassava (Table 6.1) (Frison and Feliu, 1991; Thresh *et al.* 1994a; Brunt *et al.*1996; Rybicki and Pietersen, 1999; Calvert and Thresh, 2002). Those of known economic importance are marked by an asterisk in Table 6.1 and discussed in the text. However, minor diseases could become major constraints in future, so a knowledge of 'all' viruses of cassava, their characterization, and diagnostic tools for their detection permit early responses to outbreaks of new diseases.

The distribution of each cassava virus is generally restricted to only one of the continents in which the crop is grown, or to an even more localized geographic area (Calvert and Thresh, 2002). Therefore, the virus diseases of Africa, Asia and the Pacific region, South and Central America are discussed separately.

This chapter complements a recent detailed review by Calvert and Thresh (2002). Here, we consider the identification of cassava diseases, their etiology, available information about the viruses responsible and their strains, present and future control measures, and finally the possible directions of future research.

Table 6. 1: The viruses of cassava

	VIRUS	DISTRIBUTION	Transmission			KEY REFERENCES
A	VIRUSES REPORTED FROM AFRICA**		SAP	VECTOR	SEED	
1	*African cassava mosaic virus* (ACMV)* *Geminiviridae: Begomovirus*	Occurs in all cassava growing areas of Africa and the adjacent islands	Yes	Whitefly	No	Bock and Harrison, 1985 Calvert and Thresh, 2002
2	*East African cassava mosaic virus (EACMV)** *Geminiviridae: Begomovirus*	Kenya, Tanzania, Malawi, Zimbabwe, Madagascar, Nigeria, Cameroon, Ghana, Guinea, Togo	Yes	Whitefly	No	Zhou *et al* .1997; Fondong *et al.* 1998; Ogbe *et al.* 1999,2003c; Offei *et al.* 1999; Winter, 1998
2A	Uganda variant of EACMV (EACMV-Uv)* *Geminiviridae: Begomovirus*	Uganda, Kenya, Tanzania, Rwanda, Sudan, Democratic Republic of Congo and Congo Republic	Yes	Whitefly	NS	Otim-Nape *et al.* 1997a; Zhou *et al.*1997, 1998; Karakacha, 2001; Legg *et al.* 2001a; Neuenschwander *et al.* 2002;
3	*South African cassava mosaic virus (SACMV)* *Geminiviridae: Begomovirus*	South Africa, Madagascar	Yes	Whitefly	NS	Berrie *et al.* 1997, 1998, 2001; Ranomenjanahary *et al.* 2002
4	*Cassava brown streak virus(CBSV)** *Potyviridae: Ipomovirus*	Tanzania, Kenya, Uganda Mozambique,Malawi, Zimbabwe and Zambia	Yes	Whitefly: *B. afer?*	NS	Storey, 1936; Nichols, 1950;; Monger *et al.* 2001a,b; Calvert and Thresh, 2002; Hillocks *et al.* 2002
5	*Cassava Ivorian bacilliform virus (CIBV)* *Ourmiavirus*	Côte d'Ivoire	Yes	NS	NS	Fargette *et al.* 1991
6	Kumi virus 'A' (bacilliform particles of 50 x 18 nm)	Kumi district of eastern Uganda	Yes	NS	NS	Harrison *et al.* 1995; Thresh *et al.* 1994a
7	Kumi virus 'B' (Isometric particles of 30 nm diameter)	Kumi district of eastern Uganda	Yes	NS	NS	Harrison *et al.* 1995; Thresh *et al.* 1994a
8	*Cassava 'Q' virus (Previously code named Cassava 'C' virus)*	Côte d'Ivoire, Malawi, Zimbabwe and Cameroon	Yes	NS	NS	Aiton *et al.* 1988a; Frison and Feliu,1991; Harrison *et al.* 1995; Thresh *et al.* 1994a
	Cassava common mosaic virus(CsCMV) *Potexvirus* (see 12)	Côte d'Ivoire	Yes	NS	No	Aiton et al. 1988b; Brunt *et al.* 1996

B	**ASIA / PACIFIC Region**					
9	*Indian cassava mosaic virus* (ICMV)* *Geminiviridae: Begomovirus*	Spreads in India and Sri Lanka	Yes	Whitefly	No	Abraham, 1956; Malathi *et al.* 1987; Mathew and Muniyappa, 1992,1993
10	*Sri Lankan cassava mosaic virus* (SLCMV) *Geminiviridae: Begomovirus*	Sri Lanka	Yes	Whitefly	NT	Saunders *et al.* 2002
11	*Cassava green mottle virus (CGMV) Comoviridae: Nepovirus*	The Solomon Islands	Yes	Possible trans-mission by nematodes	No	Lennon *et al.* 1987
	Cassava common mosaic virus (CsCMV) *Potexvirus* (see 12)	Taiwan	Yes	NS	No	Chen *et al* .1981
C	**SOUTH AND CENTRAL AMERICA**					
12	*Cassava common mosaic virus(*CsCMV)* *Potexvirus*	Tropical America from Mexico to Paraguay	Yes	NS	No	Costa and Kitajima, 1972a; Calvert and Thresh, 2002
13	Cassava frogkin 'virus'*(FSV)	Colombia, Brazil, Venezuela, Peru and Costa Rica	No	The whitefly *B. tuberculata* is suspected	No	Lozano and Nolt, 1989; Calvert and Thresh, 2002
14	*Cassava vein mosaic virus* (CVMV)* *Caulimoviridae*	Brazil	Yes	NS	NT	Lin and Kitajima, 1980; Calvert and Thresh, 2002
15	*Cassava Colombian symptomless virus* (CCSV) *Potexvirus*	Colombia	Yes	NS	NT	Aiton and Harrison, 1988
16	*Cassava virus X* (CsVX) *Potexvirus*	Colombia, Venezuela	Yes	NS	NT	Lennon *et al.* 1986b; Chaparo-Martinez and Trujillo-Pinto, 2001
17	*Cassava American Latent virus (CALV) Comoviridae: Nepovirus*	Brazil and Guyana	Yes	NS	NT	Walter *et al.* 1989
18	Cassava symptomless Rhabdovirus	Brazil	No	NS	NS	Brunt *et al.* 1996

NS= not studied; NT= not tested in cassava

** Recently, Fauquet *et al.* (2003) updated list of *Begomovirus* species affecting cassava in Africa and proposed three new species: *East African cassava mosaic Cameroon virus*, *East African cassava mosaic Malawi virus* and *East African cassava mosaic Zanzibar virus.*

6.2 Africa

Nine viruses have been reported from cassava in Africa (Table 6.1). Eight of these occur only in Africa, although, a disease resembling cassava mosaic disease, that occurs in India and Sri Lanka, is caused by virus species similar to those in Africa. Like several other viruses including *Maize streak virus* and *Rice yellow mottle virus*, the principal viruses of cassava in Africa presumably originate from indigenous hosts, infecting cassava after its introduction (Calvert and Thresh, 2002). The limited exchange of breeding and other cassava material between continents largely explains the non-overlapping zones of virus occurrence. The report of *Cassava common mosaic virus* (CsCMV) in Côte d'Ivoire (Aiton *et al.* 1988b), is attributed to inadvertent introduction of infected cassava material from South America, where the virus is prevalent. However, there is no confirmation of the establishment of this virus in West Africa. There is virtually no information on three viruses (Kumi viruses 'A' & 'B' and 'Q' virus). Only those causing cassava mosaic and cassava brown streak diseases are described here in detail.

6.2.1 Cassava mosaic disease (CMD) was first reported in what is now Tanzania by Warburg (1894) and referred to as 'Kräuselkrankheit'. CMD has since been observed throughout all cassava growing areas of Africa and its adjacent islands (Fauquet and Fargette, 1990; Thottappilly, 1992; Thresh *et al.* 1998c; Thresh and Fargette, 2001). The pioneering work on CMD confirmed that the causal pathogen is transmitted by whiteflies (Storey and Nichols, 1938) and it was later shown to be a geminivirus (Bock *et al.* 1978, 1981; Bock and Woods, 1983). Originally one geminivirus was regarded as the causal agent of CMD. However, subsequent studies (Hong *et al.* 1993; Zhou *et al.* 1997; Rey and Thompson, 1998) have shown that several similar, but distinct whitefly-transmitted geminiviruses cause CMD in Africa and they can occur singly or in combinations. The most important of these viruses are *African cassava mosaic virus* (ACMV) and *East African cassava mosaic virus* (EACMV), which are now, ascribed to the genus *Begomovirus* of the family *Geminiviridae*. Consequently, if not otherwise specified, the viruses causing CMD in Africa are referred to here as cassava mosaic begomoviruses (CMBs). From 1988 to the present, a major pandemic of an unusually severe form of CMD has been spreading throughout East and Central Africa, causing massive losses and affecting the region's food security (Otim-Nape *et al.* 1997a; Legg *et al.* 2001a,b). A distinct strain of EACMV, the Uganda variant referred to as EACMV-UG or UgV (Deng *et al.* 1997; Zhou *et al.* 1997), is associated with the problem. This unique strain has had serious consequences for cassava cultivation and it is the single most significant threat to cassava production in Africa.

Thresh *et al.* (1997) distinguished *'epidemic'*, *'endemic'* and *'benign'* situations. In *epidemic* situations, whitefly vectors spread CMD rapidly and symptoms are severe. Use of CMD-resistant varieties of cassava is essential if production is to be restored. The epidemic situation now prevails in large areas of East Africa and in western regions of Central Africa (Otim-Nape *et al.* 2000; Legg and Thresh, 2000; Thresh and Fargette, 2003). In *endemic* situations, there is usually a high incidence of CMD with moderate severity, probably because farmers have selected somewhat tolerant varieties over the years (Thresh and Fargette, 2003). In *benign* situations, the incidence of CMD is low and seldom exceeds 20%. However, a disease situation can change rapidly and drastically, as exemplified by an early experience in Madagascar and elsewhere (Cours *et al.* 1997) and more recently in Uganda where it switched from benign to epidemic and is now reverting to an endemic phase (Otim-Nape *et al.* 2000).

6.2.1.1 Crop losses: Crop loss data have been obtained for CMD in many African countries (Fauquet and Fargette, 1990; Thresh *et al.* 1994b). It is difficult, however, to extrapolate from them to estimate the overall yield losses for the African continent because the losses are influenced by virus strain(s) present, sensitivity of the varieties grown and environmental conditions (Thresh and Fargette, 2003). Also, healthy plants within a stand may compensate for the impaired growth of their diseased neighbours (Otim-Nape *et al.* 1997b).

Yield losses with individual cassava cultivars have been reported from different countries to range from 20 to 95% and Fauquet and Fargette (1990) estimated the average yield loss due to CMD to be 50%. The virus can infect virtually all cassava cultivars grown in Africa and on plausible assumptions the continent-wide losses have been estimated at 12-23 million tonnes of fresh tuberous roots per year, worth about US $1200-2300 million (Thresh *et al.* 1997).

The mode and time of infection affect the magnitude of yield reduction, and plants grown from infected cuttings sustain greater yield losses than those infected subsequently by vectors (Fargette *et al.* 1988). Yield losses were not significant when symptoms appeared later than four months after planting. Symptom severity correlated negatively with shoot weight, storage root yield and harvest index (Fauquet and Fargette, 1990; Otim-Nape *et al.* 1994b). However, some resistant genotypes sustain storage root yield losses even when expressing mild symptoms or when latently infected (Seif, 1982).

Yield losses in Uganda at the peak of the recent EACMV-UG pandemic (Otim-Nape *et al.* 2000) were immense and farmers had to abandon cultivation of cassava in the worst affected areas. The consequent food shortages led to many famine-related deaths (Otim-Nape *et al.* 1998a). It was assumed that each year an area of *c.* 60,000 ha of cassava yielding 600,000 metric tonnes worth US$ 60 million was lost due to CMD. Similarly, the losses due to the epidemic in western Kenya were estimated to exceed US $ 14 million in 1998 alone (Legg, 1999).

6.2.1.2 Host range: CMBs have been reported from seven *Manihot* species and from other closely related Euphorbiaceous species including *Manihot glaziovii* and *Jatropha multifida* (Fauquet and Fargette, 1990). *Hewittia sublobata* (Convolvulaceae) is suspected to be a natural host in Kenya (Bock *et al.* 1981) and *Laportea (Flueryа) aestuans* (Urticaceae) in Nigeria (Rossel *et al.* 1987), although the virus detected has not been transmitted from these species back to cassava. Shoyinka *et al.* (2001) reported the occurrence of a CMB from naturally infected castor (*Ricinus communis)* in Nigeria. Moreover, ACMV was suspected from chayote (*Sechium edulis*) with mosaic symptoms in Nigeria because on sap inoculation of *Nicotiana benthamiana*, the symptoms produced were similar to those caused by ACMV (Rossel *et al.* 1987). However recently, the DNA A-component of a geminivirus isolated from chayote was sequenced (Gene Bank, accession No. AJ-223191, G.Thottappilly, S. Winter, D. Maxwell and S. Kamal, unpublished) revealing that this virus is more closely related to a *Tobacco leaf curl Zimbabwe virus* (TbLCZV), than to ACMV (Paximadis *et al.* 1999).

6.2.1.3 Symptoms: CMD causes leaf symptoms in cassava of variable type and severity (Fig. 6.1a,b,c). The symptoms are expressed either as a 'green mosaic' or a 'yellow mosaic'. Leaves affected by 'green mosaic' have contrasting sectors of normal and light green tissue. Leaves affected by 'yellow mosaic' are more obvious, with contrasting areas of normal green and yellow tissues. Severe chlorosis is often associated with premature leaf abscission, and an obvious decrease in growth and yield.

Cassava varieties differ in the type, extent and severity of the symptoms caused by CMD (Gibson and Otim-Nape, 1997). Varieties with some degree of tolerance or resistance show mild to moderate symptoms, especially during the late stages of crop growth when recovery often occurs. The severity of symptoms is influenced by environmental factors (Gibson and Otim-Nape, 1997). Leaves produced during hot weather tend to be less affected than those produced at other times. Symptoms of CMD are sometimes confused with injury due to pests or nutrient deficiency. The cassava green mite (*Mononychellus tanajoa*) and zinc deficiency cause particular problems in diagnosis.

Simple numerical scoring systems based on the extent and severity of the leaf symptoms have been used in screening for resistance to CMD. Scales of 0 (no symptoms) to 5 (severe mosaic) or 1 to 5 have been used to quantify differences in varietal reaction and in yield loss studies. The concentration of ACMV was assessed by ELISA in relation to symptom severity (Fargette *et al.* 1996; Ogbe *et al.* 2003 a) and symptom expression was not necessarily a reflection of virus concentration in some of the genotypes. Consequently, in assessing the resistance of cassava to CMBs, virus concentration should also be considered in addition to symptom severity scores.

6.2.1.4 Transmission: CMBs are not transmitted through cassava seeds (Storey and Nichols, 1938), but can be transmitted mechanically to *Nicotiana clevelandii*, *N.glutinosa*, *N. benthamiana, N. rustica*, *N. tabacum* '*xanthi*', *Datura stramonium* and *Nicandra physaloides* (Bock *et al.* 1978; Rossel and Thottappilly, 1984, 1985; Rossel *et al.* 1987). The viruses are transmitted by grafting either by using a wedge (Seif, 1982) or bud (Adejare and Coutts 1982).

The only known insect vector of CMBs is the whitefly, *Bemisia tabaci* (Hemiptera: Aleyrodidae). *B. tabaci* is thought to be of Indian origin and was probably introduced into Africa, whereas *B. afer* which also occurs on cassava is suspected to be of African origin (Fishpool and Burban, 1994). There is no information on the role of *B. afer* as a vector of cassava viruses in Africa.

Virus dissemination between fields and over long distances is primarily through the use of infected stem cuttings as planting material (Fauquet *et al.* 1988), and secondarily by the whitefly vector, *B. tabaci*. Two biotypes of *B. tabaci* have been identified in Côte d'Ivoire (Burban *et al.* 1992). One colonised only cassava and eggplant, while the other was polyphagous on okra, eggplant and other plants, but not on cassava. Four biotypes were later distinguished in Uganda, two on cassava and one each on sweet potato and cotton (Legg *et al.* 1994). A molecular analysis to reveal the population structure of *B.tabaci* found on cassava showed that the cassava-associated whiteflies in Africa are unique populations that are distinct from all others (Abdullahi *et al.* 1998). The cassava whitefly populations from West and Central Africa showed less than 70% genetic identity to the East African group (Abdullahi *et al.* 2003b), suggesting a separate status of the two populations. There is recent evidence for two genotype clusters, Ug1 and Ug2, in *B. tabaci* populations from Uganda (Legg *et al.* 2002) with Ug2 being associated with the severe CMD epidemics in that region. However, no evidence of differences between them in virus transmission efficiency was obtained when using whitefly populations from epidemic and non-epidemic areas (Maruthi *et al.* 2001, 2002b).

Transmission experiments with whiteflies were done in Nigeria (Chant, 1958), Côte d'Ivoire (Dubern, 1994) and Kenya (Seif, 1981). Based on current knowledge it is likely that the transmission studies in coastal East Africa (Storey and Nichols, 1938; Seif, 1981) were with EACMV and those in Congo and West Africa with ACMV.

Single whiteflies can transmit CMBs, but optimum rates of transmission were achieved with 10 whiteflies per cassava plant (Chant, 1958; Dubern, 1994); Seif (1981) observed a low level of virus transmission to resistant genotypes with 10-15 whiteflies per plant. Unstarved insects acquired virus after 5 h on infected cassava compared with 3.5 h for those starved. The latent period is *c.* 6-8 h in the insect and a period of 10-30 minutes is needed for inoculation into healthy cassava test plants. Virus is retained by an infective whitefly for *c.* 9 days (Dubern, 1994). Each of the four nymphal instars of the insect can transmit virus (Chant, 1958; Dubern, 1994), which is retained through the moult; however, there is no transovarial transmission (Dubern, 1994). Up to 1.7% of individual adult whiteflies were shown to be infective when collected in heavily infected cassava fields in Côte d'Ivoire and transferred to young test seedlings of cassava (Fargette *et al.* 1990).

6.2.1.5 Etiology: For many years CMD was assumed to be a virus disease because it was transmissible by grafts and whiteflies, although the causal agent was not identified. The situation changed in the 1970s when geminate particles were observed regularly in sap preparations of diseased cassava (Harrison, 1985; Harrison *et al.* 1977) and were transmitted by sap inoculation to *Nicotiana clevelandii* (Bock *et al.* 1978). The status of the virus so isolated was, however, at first unclear because it could not be transferred to *N. clevelandii* from all the CMD-affected plants tested and Koch's postulate was not proven. Hence, the virus was initially referred to as 'cassava latent virus' (CLV) and this name continues to appear occasionally in the literature. However, the name became inappropriate when an additional herbaceous host was used (*N. benthamiana*), and sap inoculations from herbaceous hosts to healthy, susceptible cassava from South America produced typical CMD symptoms (Bock and Woods, 1983; Rossel and Thottappilly, 1984,1985; Rossel *et al.* 1987). The etiology of CMD was thus established and the disease was shown to be caused by a geminivirus referred to as 'African cassava mosaic virus' (ACMV). Later, Briddon *et al.* (1998) successfully proved the Koch's postulates inducing characteristic symptoms of CMD in cassava following the introduction of cloned ACMV (Nigeria) genomic components and the biolistic gene gun method of inoculation.

Geminiviruses are characterised by their bisegmental (geminate) shape and size (30 x 20 nm particles) and they have a capsid protein of *c.* 30 kDa. Each geminate particle encapsidates a small (2.5-3.0 kb) circular single-stranded (ss) DNA genome that replicates in the nuclei of host cells via a double-stranded (ds) DNA intermediate (Saunders *et al.* 1991). The whitefly-transmitted viruses that infect dicotyledonous plants form the genus *Begomovirus,* family *Geminiviridae.* Their genomes are mostly bipartite, with the components DNA-A and DNA-B, each *c.* 2750-2800 nt in size. All cassava begomoviruses (CMBs) are typical of the genus with two circular single-stranded DNA components (Stanley, 1983; Stanley and Gay, 1983). DNA-A contains the coat protein gene (CP, AV1) in the virus-sense direction and four genes on the complementary strand (Hong *et al.* 1993; Haley *et al.* 1995) that code for a replication-associated protein (Rep, AC1), a transcriptional activator protein (TrAP, AC2) and a replication enhancer protein (REn, AC3). Only Old World begomoviruses carry a Pre-CP on the virion-sense strand (AV2). An ORF on the complementary strand, AC4 that is contained within AC1 is present in some begomoviruses. However, the function of the gene products of the respective ORFs has not been elucidated unequivocally. DNA-B encodes the nuclear shuttle protein (NSP, BV1) on the viral strand controlling the transport between nucleus and cytoplasma (Pascal *et al.* 1994; Sanderfoot *et al.* 1996), and the MP (BC1), implicated in virus cell-to-cell movement on complementary strand (Noueiry *et al.* 1994; Ward *et al.* 1997).

The coat protein (CP) gene (AV1) plays a crucial role in vector transmissibility, whereas both DNA components are required for efficient transmission by whitefly (Liu *et al.* 1997). CP determines virus-vector interactions, while the DNA-B is responsible for virus movement, which places the virus in the appropriate plant tissues for efficient acquisition by vectors (Stanley *et al.* 1985; Liu *et al.* 1997).

Zhou *et al.* (1997) reported that EACMV-UG DNA-A was 16% and 84% similar to those of ACMV and EACMV, respectively. The coat protein gene sequence of a new CMB, *South African cassava mosaic virus* (SACMV) was 80% similar to that of ACMV and 89% to that of EACMV (Berrie *et al.* 1997). In a more detailed study, the nucleotide sequence of the coat protein gene of SACMV was found to be more related to *Tomato yellow leaf curl virus* (TYLCV), a monopartite begomovirus, than to EACMV (Berrie *et al.* 1998). However, SACMV is bipartite like EACMV and in serological reactions, it is 95% and 75% identical to EACMV and TYLCV, respectively. Also, SACMV (Berrie *et al.* 1998, 2001) was proven to be the result of a recombination event, but although there are striking genome similarities between SACMV and EACMV isolates, SACMV is considered to be a distinct virus species (Fauquet *et al.* 2003). It is also reported from Madagascar (Ranomenjanahary *et al.* 2002).

A further recombinant virus, an EACMV strain (EACMV-CM) first reported from Cameroon (Fondong *et al.* 2000a), appears to have a widespread distribution in West Africa. It was reported from Côte d'Ivoire (Pita *et al.* 2001a) and also detected in all cassava growing regions of Nigeria (S. Winter, unpublished data).

6.2.1.6 Strains: Originally, CMD was assumed to be caused by a single whitefly-borne geminivirus, although four types of 'isolate' were recognized: West Africa, Kenya coastal, India and Angola (Bock and Harrison, 1985). These different isolates were grouped into three clusters: group A included those from Angola, Côte d'Ivoire, Nigeria, and western Kenya; Group B from coastal Kenya, Madagascar and Malawi; group C from India and Sri Lanka (Harrison *et al.* 1991). Hong *et al.* (1993) analyzed nucleotide sequences from these strains and regarded the virus isolates from different geographic origins as distinct geminiviruses. Thus, groups A and B were renamed *African cassava mosaic virus* (ACMV) and *East African* *cassava mosaic virus* (EACMV), respectively, while the Indian isolate was named *Indian cassava mosaic virus* (ICMV). Swanson and Harrison (1994) confirmed the distinctiveness of the viruses using monoclonal antibodies (MAbs). Improved diagnostic techniques later resulted in identification of other CMBs. In Uganda and neighbouring countries, a new virus variant was detected and identified. This varient resembles EACMV from which most of its genome is derived, except for the core region of the coat protein gene, which is identical to that of ACMV (Harrison *et al.* 1997b). This recombinant virus has been referred to as a distinctive strain of EACMV (EACMV-UG) (Deng *et al.* 1997) or, as the Uganda variant cassava mosaic virus (UgV) (Zhou *et al.* 1997). Recently Fauquet *et al.* (2003) revised the taxonomic criteria for species demarcation and proposed six CMB species in Africa, of which ACMV, EACMV and SACMV are well characterized, while *East African cassava mosaic Cameroon virus* (EACMCV), *East African cassava mosaic Malawi virus* (EACMMV) and *East African cassava mosaic Zanzibar virus* (EACMZV) are all regarded as distinct from typical EACMV and hence await further confirmation as species. The recombinant virus referred to as a Uganda variant is a distinctive strain of EACMV (EACMV-UG2).

Swanson and Harrison (1994) demonstrated that ACMV occurs in West and Central African countries up to the west of the Great Rift Valley and also in South Africa; EACMV in countries east of the Rift Valley including coastal Kenya, coastal Tanzania, Malawi, Zimbabwe and Madagascar; and ICMV in India and Sri Lanka. It was, therefore, considered that EACMV and ACMV had distinct, largely non-overlapping geographical distributions and that cassava grown in diverse ecologies had acquired each virus from different indigenous host plant species. Later studies showed that EACMV and ACMV occur over a much wider area.

In a recent virus survey in Côte d'Ivoire, Pita *et al.* (2001a) revealed mixed infections with EACMV-CM and ACMV in 40% of all samples tested and they reviewed the current knowledge on the molecular and biological diversity of CMBs. ACMV and EACMV occur across geographical zones in sub-Saharan Africa (Ogbe *et al.* 1997a,b,1999; Fondong *et al.* 1998, 2000a; Winter, 1998; Offei *et al.* 1999; Berry and Rey, 2001). Both ACMV and EACMV viruses occur in single and mixed infections in cassava causing mild and severe symptoms (Fondong *et al.* 2000a; Obge et al. 2003b). An epidemiological significance, however, can only be attributed to the recombinant EACMV-UG virus. The potential threat presented by this virus moving further west to the major areas of cassava production in West Africa (Karakacha *et al.* 2001), is considerable and enormous efforts will be required to reduce the impact. The recombinant virus has been detected in samples from Uganda, Tanzania, Rwanda, Kenya and southern Sudan (Harrison *et al.* 1997b; Zhou *et al.* 1998; Legg *et al.*

2001a). It was also found in samples from Kisangani, Federal Republic of the Congo, in 1998 (S.Winter, unpublished). In 2000, EACMV-UG infections were prevalent in Bandudu and the Bas-Congo (Boma) area of the Democratic Republic of Congo marking the most westwards expansion of this particularly severe strain of EACMV (Karakacha *et al.* 2001; Neuenschwander *et al.* 2002).

Further strains of EACMV, which presumably arose through recombination were reported from Cameroon (Fondong *et al.* 2000a) and from South Africa (Berrie *et al.* 1998, 2001), providing evidence for the diversification and evolution of CMBs. This, however, is especially true for EACMV genomes. While ACMV isolates, irrespective of their geographical origin, have highly conserved genome sequences, those of EACMV are genetically diverse also due to frequent occurrence of recombinations within the components DNA A and DNA B (Pita *et al.* 2001a,b). The genetic variability of EACMV is shown by the quite diverse geographic isolates obtained from Kenya, Tanzania and Malawi. The recent description of another distinctive virus species from Zanzibar, EACMV-Z (Maruthi *et al.* 2002a), which is also prevalent in coastal Kenya (Bull *et al.* 2003), provides further evidence for the high molecular variability within EACMV, rendering species demarcation of CMBs difficult.

Ogbe (2001) used a panel of 12 MAbs to characterize CMBs in Nigeria. Nine serotypes of ACMV were identified using ELISA. EACMV was found in the humid forest, derived/coastal, and southern Guinea savannas, while ACMV occurred in all agroecologies. These MAbs were raised against a Kenyan isolate (Thomas *et al.* 1986) which shared extensive homologies with ACMV from Nigeria and similar host range and symptoms (Bock *et al.* 1981). Some Nigerian isolates differ extensively from those of Kenya, which probably explains why 13% of the 290 isolates failed to react with the MAbs. Biological assay and polymerase chain reaction (PCR) results further confirmed the occurrence of variants of EACMV in Nigeria (Ogbe *et al.* 2003b,c).

6.2.1.7 Epidemiology: The first epidemiological evidence on CMD was provided by Storey (1938). Although alternative hosts of CMD have since been reported, they seem to be of little or no current epidemiological importance for spread to cassava (Fauquet and Fargette, 1990). Harrison (1987) noted that the whitefly population in heavily infested plots was not proportional to the amount of virus spread that was occurring, and concluded that virus-infected cassava represents the determining factor for the spread of CMD. Thus the dissemination of CMBs through infected cassava cuttings, together with transmission by whiteflies, accounts for virus incidence. All secondary spread can be attributed to viruliferous whiteflies moving between or within cassava plantings. This is consistent with the findings of epidemiological studies in Côte d'Ivoire, Kenya and Uganda, that spread of

CMD into and within experimental plantings is related to the number of adult whiteflies recorded and also to the overall incidence of infection in the area (Legg *et al.* 1997; Otim-Nape, *et al.* 1998b).

There are usually periods of high and low whitefly populations in the different agroecologies where cassava is grown. In the rainforest and transition forest zones of Nigeria, for example, the major peak in whitefly populations is usually between April and June (Leuschner, 1977). Planting cassava during the period of low whitefly population is expected to reduce CMD incidence during the early growth stage of the crop. Storey and Nichols (1938) found that the spread of CMD at a site near Tanga in northern Tanzania was fastest from February to May, and then decreased rapidly and remained at a low level until December. This sequence corresponded with the increase and decrease of populations of the whitefly vector in Côte d'Ivoire (Fauquet and Fargette, 1990; Fargette *et al.* 1994a). Up to 80% of the cassava planted in March was infected within 2 months, but the comparable incidence in August plantings did not exceed 4%. Disease incidence largely reflected fluctuations in whitefly populations and was related to variations in climatic factors, such as temperature, rainfall, and wind. Little spread occurred in a savannah region, which was outside the main cassava production area. Whether this is because conditions were less favourable for the whitefly vector and/or virus was not determined.

The distribution of immigrant whiteflies and of plants newly affected by CMD is influenced by the direction of the prevailing wind and by the effects of wind turbulence around and within stands. The incidence of CMD tends to be greatest at the margins of plantings, especially those facing the direction of the prevailing wind (Fargette *et al.* 1985; Fauquet and Fargette, 1990). This is associated with the tendency of incoming whitefly vectors to alight preferentially and become most numerous on plants in the outer rows (Fargette *et al.* 1985; Fishpool *et al.* 1995; Colvin *et al.* 1998). Progress of CMD with time is influenced by cultural practices (Fauquet and Fargette, 1990). Primary spread from outside sources accounted for *c.* 70% of transmission by vectors, field infection was mainly from outside sources, which led to internal foci that contributed to limited, secondary spread. Within a healthy cassava field, the dispersal gradient from an infected source, although occurring in all directions, did not exceed several meters and was probably related to the relatively limited flight of whiteflies within fields.

Adult whiteflies are found more frequently on the top five leaves of the shoot where feeding and egg laying occur (Fauquet and Fargette, 1990; Fishpool and Burban, 1994). Populations increase to a maximum 4-5 months after planting and then decline. There is also a decrease in virus concentration with plant age (Fargette *et al.*1987, 1994b). Cassava genotype, climate and their interactions have significant effects on populations of *B.*

tabaci and the development of CMD (Abdullahi *et al.* 2003a). A positive correlation between the incidence and severity of CMD was observed, but this did not correlate with the whitefly population density.

Temperatures of 30°-33°C favour the development of *B. tabaci* (Fishpool and Burban, 1994), increase its fecundity and reduce its developmental periods (Atiri *et al.* 2000), leading to a build up of the insect population. High temperatures of 25°-30°C also favour the replication of CMBs (Sequeira and Harrison, 1982).

There are reports on the effect of rain on whitefly populations. Leuschner (1977) observed that heavy rains did not affect the nymphal development of the vector, but caused a decline in adult whitefly populations (Fishpool and Burban, 1994). Vector population was related to crop growth, which depends on rainfall, but heavy rain affects adult whiteflies. However, vector population and the rate of virus spread were high during the rainy season in Kenya (Bock, 1994a), Côte d'Ivoire (Fauquet and Fargette, 1990; Fargette *et al.* 1994a) and also in Nigeria (Leuschner, 1977). This may be because the rate of leaf production is high during the rainy season, which is likely to favour whitefly population development and spread of CMD (Bock, 1994a; Fargette *et al.* 1994a; Legg, 1994). The cassava genotypes, however, dictate the rate of spread of virus and vectors, being more rapid for susceptible genotypes than for resistant ones (Hahn *et al.* 1980; Thresh *et al.* 1994c; Otim-Nape *et al.* 1998a,b).

There is more spread of CMD into new planting than within fields (Fargette *et al.* 1985; 1990), depending on the layout of the fields. It was observed that viruliferous whiteflies could be transported far by wind, but new plantings are likely to be at particular risk where sources of infection are nearby and upwind in relation to the direction of the prevailing wind (Thresh and Otim-Nape, 1994).

6.2.1.8 Identification and differentiation of CMBs: Genomes of CMBs have a high plasticity and naturally occurring variations due to inter- and intra-species recombination occur at a higher frequency than in genomes of single stranded-RNA viruses (Harrison and Robinson, 1999). Geminiviruses can recombine (Etessami *et al.* 1989) and exchange sequences between genomic components (Roberts and Stanley, 1994). Intra- and interspecies recombination events have probably contributed to begomovirus diversity (Padidam *et al.* 1999). This is regarded as one of the forces driving begomovirus evolution (Padidam *et al.* 1999). The description of recombinant and pesudorecombinant viruses (Fondong *et al.* 2000a; Pita *et al.* 2001b) and of particular virus strains (Maruthi *et al.* 2002b) occurring in single or mixed virus infections (Harrison *et al.* 1997b; Pita *et al.* 2001a; Ogbe *et al.* 2003c) emphasizes the high variability within the genomes of CMBs. Moreover, the limited sampling of infected cassava and the description of probably only a few extreme variants currently render any statement on the epidemiological significance of a particular virus type impossible.

The study of CMBs is impeded by the difficulty of clearly identifying the viruses present in a particular plant sample. The often-striking symptoms of 'mosaic' in diseased plants do not have any informative character and symptom type – severe/mild – rather reflect host plant genotype, virus strain virulence and environmental conditions. Cassava plants respond to attack by various viruses in only a limited number of ways, which can make virus diagnosis difficult. ACMV, EACMV, SACMV, ICMV, *Sri Lankan cassava mosaic virus*, *Cassava common mosaic virus* and possibly others all cause similar leaf mosaic symptoms in cassava and hence cannot be distinguished by symptoms. Furthermore, even though suspicious symptoms may occur, serological methods frequently fail to detect any CMB due to low virus concentrations or to inhibitors in plant tissue samples. Development of an effective, rapid, sensitive and reliable tool to differentiate between viral isolates is therefore required.

Serological assays: After the first successful attempts to purify virus particles (Bock *et al.* 1977) and to produce an antiserum, CMBs were detected by using gel-diffusion tests (Bock *et al.*1981), by DAS-ELISA (Sequeira and Harrison, 1982) and also by immunosorbent electron microscopy (ISEM) (Roberts *et al.*1984). In gel diffusion tests, Bock *et al.* (1981) observed spur formation between two CMB isolates, demonstrating that isolates were related, but serologically distinct, and so providing early evidence for strain variation. Moreover, polyclonal antibodies did not clearly resolve serological variations and strain differences among CMBs. This only became possible when monoclonal antibodies (MAbs) were produced against ACMV (Thomas *et al.* 1986). Extensive studies on serological relationships among CMBs were conducted adopting the commonly used triple-antibody sandwich ELISA format (TAS-ELISA), using a polyclonal antibody, the MAbs as detecting antibody and an anti-mouse antibody conjugate as reporting antibody (Thomas *et al.* 1986; Swanson and Harrison, 1994).

A characteristic serological reaction pattern was obtained from the differential reactivities of the monoclonal antibodies resulting in an epitope profile for each virus (Harrison *et al.* 2002). Three reaction clusters were obtained representing ACMV, EACMV, and ICMV. For practical purposes, in routine virus indexing, it is not feasible to use a comprehensive panel of 11 MAbs to construct an epitope profile for a particular virus sample; although, a selection of the most critical MAbs can be useful for identification of CMBs.

Serological assays provide the basis for a robust virus diagnosis, performed under field conditions and enabling large scale testing. Currently, serology does not distinguish mixed infection of EACMV and ACMV or between ACMV and UgV. Despite their limitations, serological assays provide the basis for management decisions, i.e., in the production of virus-free planting material, and for surveys to obtain general information on the occurrence and spread of geminiviruses in a given area. It is important to have appropriate serological reagents for routine indexing and it should cover all known CMBs that cause CMD.

Monoclonal antibodies: A panel of MAbs is recommended for detection and differentiation of CMBs. The antibodies produced at the Scottish Crop Research Institute (SCRI), SCR 17 and 20 detect ACMV, EACMV and ICMV, whereas SCR 58 and 60 detect only ICMV (Table 6.2). While a specific antibody exists for detection of ACMV (SCR 33), none exists for the detection of EACMV. Presence of this virus can only be concluded from the negative reaction with SCR 33 and a positive reaction with the CMB general antibody SCR 23. Consequently, it is not possible to detect EACMV in mixed infections with ACMV.

Table 6.2: Reactions of MAbs SCR (Scottish Crop Research Institute) in TAS-ELISA using a rabbit polyclonal antiserum against ACMV for virus trapping *

MAb	Begomoviruses		
	ACMV	EACMV	ICMV
SCR 17	+	+	+
SCR 20	+	+	+
SCR 23	+	+	-
SCR 33	+	-	-
SCR 58	-	-	+
SCR 60	-	-	+

Adapted from Ogbe (2001). + positive reaction, - no reaction

A similar reagent kit for serological diagnosis of CMBs is available at the DSMZ Plant Virus Division, Germany. Here a polyclonal antibody raised against ACMV is used to trap virus antigen. A panel of MAbs raised against *Tomato yellow leaf curl virus* (TYLCV), purified ACMV and bacterially-expressed EACMV coat protein, is used for differential detection of CMBs. Table 6.3 shows that several monoclonal antibodies react or fail to react with epitopes present in a particular virus strain or species. When young freshly expanding leaves of cassava are assessed, CMBs can be detected reliably in TAS-ELISA and a preliminary evaluation of the presence of a particular virus type can be made. However, the presence of EACMV-UG and/or

EACMV can only be evaluated by inference from differential reactions obtained with the panel of MAbs (Table 6.3).

Table 6.3: Differential reactions of 5 monoclonal antibodies in TAS-ELISA with specific CMB species and strains in cassava.

MAb 1C1	MAb 6E9	MAb 4F10	MAb 1H2	MAb 4G7	Virus defined
+	+	+	+	-	ACMV-NG
+	+	+	-	-	ACMV-KE
+	-	+	+	-	UgV
+	-	-	+	-	EACMV
-	-	-	-	+	ICMV

An ELISA system using MAbs for coating and trapping CMBs has been reported (Ogbe *et al.* 1997b; Ogbe, 2001). Although TAS-ELISA is used in several laboratories to test cassava samples from different geographic origins, the assay does not resolve multiple virus infections, e.g., mixed infections with UgV and ACMV. Moreover, since its detecting capacity is confined to young symptomatic cassava leaves, subliminal virus infections are not diagnosed reliably. A further limitation to resolving CMB viruses and strains, is that the recombinant virus strain EACMV-UG and the SACMV from South Africa and Madagascar are not reliably detected and/or differentiated by serology. While in serological assays, EACMV-UG largely behaves like an ACMV, SACMV is serologically an EACMV strain. The only information for definition of EACMV-UG is obtained from a reaction panel (Table 6.3) with lack of reaction of MAb 6E9 discriminating EACMV-UG from ACMV isolates. To identify SACMV, a panel of SCR MAbs was used in TAS-ELISA (Berrie *et al.* 1998). Here, the lack of reaction with MAb SCR16 distinguished SACMV from the otherwise typical EACMV-type reaction pattern revealed with the 19 SCR MAbs used.

Clearly the resolution capacity of serological assays for discrimination of all CMB viruses, strains and variants is limited and its informative value reduces with every new virus described. Consequently, for identification of a specific virus strain, or a specific virus feature, i.e. a recombination event, only molecular approaches will suffice.

Molecular methods: Serological assays are best used for virus surveys, for which high sensitivity is not essential. However, when a particular virus isolate reflects a variation due to a recombination in genome regions outside the coat protein, as with SACMV, serological tests will not distinguish the diverging virus. Genome information, especially of the DNA A genomic components of several CMBs, provided the basis for the development of PCR-based assays (Harrison *et al.* 1997a) for virus discrimination. Such molecular tests, however, have a limited field applicability and are often neither robust nor sufficiently reliabile for routine use. Specific nucleotide sequence primers have been designed to detect ACMV, EACMV, ICMV and EACMV-UG in polymerase chain reaction (PCR). PCR is a sensitive and widely used technique for detection and identification of geminiviruses (Rojas *et al.* 1993; Deng *et al.* 1994). With specific primers available for each virus type, PCR can be used successfully to identify a particular virus strain and to resolve mixed infections. This is one of the important applications of the technique since mixed infections are common in the field (Harrison *et al.* 1997b; Fondong *et al.* 2000a; Ogbe *et al.* 2003b,c).

A PCR test was used successfully in many laboratories to detect CMBs (Deng *et al.* 1994). The degenerate oligonucleotide primers (Table 6.4) flanking a characteristic segment of begomovirus DNA A, including the TAATATTACC motif that is conserved in all whitefly-transmitted begomoviruses and the 5' amino terminus of the begomovirus coat protein gene are used for a genus-specific amplification of begomovirus sequences. An informative PCR fragment of *c.* 550 nt is amplified by this genus-specific PCR test. Further identification of the virus being considered can be based subsequently on sequence analysis of the genomic fragment, providing nucleotide information sufficient for evaluation of its species affiliation.

Table 6.4. Primers recommended for amplification and discrimination of begomovirus species and strains in cassava samples from Africa

Primer	Sequence (5′→ 3′)	Strand	Position	Specificity
ACMV AL1/F*	gcggaatccctaacattatc	Sense	AC1	ACMV
ACMV ARO/R*	gctcgtatgtatcctctaaggcctg	Complement	AV2	
EACMVA2469s	tggagatgaggcaccccatc	Sense	AC1	All EACMV including recombinant EACMV-Ug
EACMVA391c	tcctccgcaccttggatacg	Complement	CP	
EACMV588Us	cactggtatggtccgatgtg	Sense	CP	EACMV
EACMV L CPc	tctttattaatttgtcactgcat	Complement	CP	
UV AL1 /F2*	gtaattgggaaagggcctct	Sense	AC1	EACMV-Ug
CAGUCO772	caaagtcttggggcgcattg	Complement	CP	
SACMV2543s	agttgtcggagttgatct	Sense	AC1	SACMV
SACMV252c	gcctgcaaatatttaatagc	Complement	AV2	
Begomo 146 ¥	taatattacckgwkgvccsc	Sense	CR	All CMBs
Begomo 672 ¥	tggacyttrcawggbccttcaca	Complement	CP	

*Primers designed by Zhou *et al.*1997; ¥ Primers designed by Deng *et al.* 1994

A series of specific PCR primers (Table 6.4) will discriminate the different CMBs described so far. The primers located in specific motifs present in only a particular species or variant are now preferred for discrimination of CMBs and suitable for resolving mixed virus infections. In a multiplex PCR reaction, several primer combinations can also be used to reveal the virus present in a particular cassava sample in a single PCR reaction. Despite the limitations of PCR-type assays for field evaluation, these tests can be designed according to the resolution required and provide a basis for genetic evaluation of CMBs.

Recently, Berry and Rey (2001) described heteroduplex mobility assays (HMA) to differentiate CMBs. The assay involves amplifying the highly conserved core region of the coat protein gene of field samples followed by denaturing and annealing with a number of reference strains. The HMA profiles differentiated four different viral species and 11 different virus strains. This technique is sensitive and rapid and has the added advantage of being able to detect mixtures of viruses in field-grown cassava (Berry and Rey, 2001).

6.2.1.9 Control: For control of CMD, the main emphasis has been on the development and deployment of resistant varieties (Hahn *et al.* 1980; Fargette *et al.* 1996; Thresh *et al.* 1998a,b). There has also been some use of phytosanitation, involving the selection of cuttings from healthy plants and subsequent removal (roguing) of diseased plants (Thresh *et al.* 1998a), and the production and use of virus-free tissue-cultured planting material (Adejare and Coutts, 1981; Ng *et al.* 1990,1992). Transgenic technologies for CMD control are being developed (Taylor *et al.* 2002).

6.2.1.9.1 Cultural Control and Phytosanitation

Adjusting planting date: Adjusting cropping practices can decrease the losses caused by CMD, as for example by adopting planting dates so as to avoid exposing young plants to infection at times when the largest numbers of viruliferous whiteflies are likely to occur (Adipala *et al.* 1998). Planting away from and upwind of existing sources of infection and also in large compact blocks to minimize 'edge' effects are likely to be advantageous (Thresh and Otim-Nape, 1994; Fargette *et al.* 1985).

Intercropping: Some experiments in Africa have shown the beneficial effects of intercropping in decreasing virus spread when cassava was intercropped with maize and/or cowpea (Fondong *et al.* 1997, 2002). However, the earlier results with a maize intercrop were not always consistent (Fargette and Fauquet, 1988).

Varietal Mixtures: There is less information on the implications of varietal mixtures, although it has recently been observed in Uganda that the effects of CMD have been much worse in areas where only a single variety predominated than where several varieties were being grown (Otim-Nape *et al.* 2001). Moreover, CMD spread less rapidly to a susceptible variety grown together with resistant varieties than when the variety was grown alone (Sserubombwe *et al.* 2001). These preliminary results indicate the potential importance of intercropping and varietal mixture, but there is a need for further research on the whole range of agricultural practices before definitive advice can be given to farmers.

Crop Spacing: Infection progressed fastest at low plant density and final incidence was lowest with high density. Therefore, close spacing is recommended (Fargette *et al.* 1990). CMD incidence is also increased around breaks or discontinuities in the crop canopy, which facilitate the alighting and establishment of viruliferous vectors (Fargette *et al.* 1985).

Selecting healthy plants: Since infected stem cuttings are the primary source of CMBs, the use of virus-free cuttings is critical in the control of CMD (Bock, 1994b). Plants grown from virus-free cuttings establish more readily, grow faster and outyield those from infected cuttings (Thresh and Otim-Nape, 1994). Although healthy plants may become infected during growth, such plants sustain less yield loss than plants that were infected at planting (Fargette *et al.* 1988).

Cuttings from healthy plants should be used as planting material in the next season. Despite the advantage of selecting virus-free cuttings, farmers seldom do this in Africa. This may be due among many other factors to the scarcity of virus-free cuttings, the inability or unwillingness of farmers to select cuttings from symptomless plants, and the lack of awareness by farmers of the damage caused by CMD (Thresh *et al.* 1998a).

Roguing: Roguing of infected plants is recommended at an early stage of growth before foci for further virus spread are established (Thresh and Otim-Nape, 1994). The practice is not effective in reducing CMD under high disease pressure, especially for genotypes that do not have substantial resistance to virus infection (Fargette *et al.* 1990; Thresh *et al.* 1998a). Farmers know that infected plants will yield something and hence are reluctant to rogue such plants. Although roguing has been recommended widely, some researchers have questioned the approach. Otim-Nape *et al.* (1997b) reported that healthy cassava plants compensated for loss in yield of their diseased neighbours and that there was no significant yield difference between plots with 100% and 50% healthy plants. Fargette *et al.* (1990) observed that spread of CMD between fields was more important than within fields and so roguing had little or no benefit.

6.2.1.9.2 Production of virus-free planting materials: There are no technical difficulties in producing basic stocks of cassava cuttings free of CMBs. This has been done simply by careful selection amongst the stocks already available and also by exploiting the reversion phenomenon (Fargette and Vie, 1995; Thresh *et al.*1998a, b), or by heat therapy and/or meristem-tip culture (Ng *et al.* 1992).

Exploiting reversion: CMBs are not always fully systemic and uninfected cuttings can be obtained from some branches of infected plants, especially those of resistant varieties not expressing symptoms. At least some branches of infected plants of resistant varieties are free from virus through 'reversion' under natural conditions (see Fargette *et al.* 1994b). The extent of reversion depends on the genotype, and is regarded as a component of the resistance of cassava to CMBs (Fargette *et al.* 1996; Thresh *et al.*1998 a, b; Fondong *et al.* 2000b). Reversion has been exploited to select and produce healthy cuttings for CMD epidemiological studies in Côte d'lvoire (Fargette *et al.* 1985, 1988).

In resistant genotypes, cuttings obtained from the lower portions of the main stem are more likely to grow into virus-infected plants than cuttings from the upper portion of the main stem and stem branches (Cours, 1951; Njock *et al.* 1994). Recent studies showed that restriction of virus movement into axillary buds is an important aspect of resistance in CMD (Ogbe *et al.* 2002). This probably explains why reversion of resistant genotypes is such a feature of resistant gene. The benefits of using healthy planting material can be considerably enhanced when planted in an area of low infection pressure (Akano *et al.* 1997). Plants established under conditions of low vector population could serve as particularly suitable sources of cuttings in subsequent season.

Heat therapy: Keeping infected cassava plants at 37-39°C for 28-105 days in growth chambers eliminated CMBs (Chant, 1959; Kaiser and Teemba, 1979; Kaiser and Louie, 1982) from growing shoots.

Meristem culture: Virus-free cassava plants have been produced through meristem-tip culture (Kaiser and Teemba, 1979; Adejare and Coutts, 1981; Ng *et al.* 1992). Since the meristematic regions in some genotypes are not always invaded by CMBs, it was possible to obtain a small proportion of virus-free plants from 0.2-0.5 mm meristem-tips when cultured in *vitro.* Heat therapy before meristem-tip culture led to a greater proportion of virus-free plants than meristem culture alone. Studies conducted in Nigeria using cuttings derived from virus-free *in vitro* plantlets showed that under high CMD infection pressure susceptible varieties were readily re-infected within two months of planting. However, resistant varieties remained symptomless for at least one complete growing cycle (IITA, 1986; Akano *et al.* 1995, 1997). The situation was totally different when virus-free cuttings were planted in a low CMD infection pressure environment, where plants of both susceptible and resistant varieties remained symptomless over a 2-3 year period. It was also observed that virus-free cuttings planted in the main growing season were reinfected earlier than those planted later. This suggested that changing cropping practices might contribute to a reduction of disease pressure and spread.

Planting of cuttings derived from *in vitro* virus-free plantlets of CMD-resistant varieties in a high CMD infection pressure area gave higher tuberous root yield and number of stem cuttings as compared to plants derived from standard cuttings (Ng, 1986). This yield increase was sustained for 2 years in such an environment. Yield increases of 50 to 100% were also reported by planting cuttings derived from *in vitro* plantlets in Congo compared to conventional field cuttings (Mabanza *et al.* 1995).

Meristem culture will not automatically guarantee total freedom from viruses (Thottappilly and Rossel, 1988; Thottappilly *et al.* 1999). Before the inter-continental transfer of planting materials, therefore, the tissue culture-derived plantlets should first be tested for the presence of virus(es).

6.2.1.9.3 Vector Control:

Cultural control of whiteflies: There is a dearth of information on the effect of cultural control of *B. tabaci* on the incidence of CMD in Africa. Apart from being a vector of CMD, the insect seldom causes economic damage on cassava.

Host plant resistance: According to Bellotti *et al.* (2001), host plant resistance to whiteflies is rare in cultivated crops. On screening 5000 cassava clones against the whitefly, *Aleurotrachelus socialis*, several cultivars with moderate to high levels of resistance were identified (Bellotti *et al.* 2001). When *A.socialis* fed on resistant genotypes, including M Ecu-72 and M Ecu-64, lower oviposition, longer development periods, reduced size, and higher mortality were observed than when feeding on susceptible genotypes. This finding should be pursued in Africa or Asia using *B.tabaci* and whitefly-resistant genotypes from S. America.

Insecticidal applications: Farmers occasionally use insecticides in attempts to restrict the spread of CMD by controlling the whitefly vector. However, the use of insecticides is unlikely to be effective and is not recommended. It also is costly and with risks to farmers, consumers, natural enemies of arthropods and the environment.

6.2.1.9.4 Use of resistant genotypes of cassava

Given the difficulty in controlling plant viruses using chemical or cultural methods, the use of resistant cultivars is the ideal method of CMD control and one that has been used most widely and successfully (Nichols, 1947; Cours, 1951; Jameson, 1964; Otim-Nape *et al.* 1994a; 1997a; Thresh *et al.* 1994c, 1998c).

The search for CMD resistance started in the 1920s, but the most influential programmes began at Amani in Tanzania in East Africa in the late 1930s (Jennings, 1994) and later in Madagascar. In Madagascar, for example, all local varieties and many cassava accessions from elsewhere were screened (Cours-Darne, 1968). Varieties such as "Bouquet de la Reunion", "Java 12/28", and "Criolina" were identified and released to farmers. However, more effective resistance was needed, and this was sought by crossing *M. esculenta* with several wild species including the tree cassava species, *M. glaziovii* (Nichols, 1947; Cours, 1951; Jennings, 1957). The hybrids had non-tuberous roots but some plants, though infected by CMD, showed only mild and transient symptoms of CMD. At Amani, three backcrosses to cassava were used to restore root quality and maintain resistance to CMD (Nichols, 1947). Intercrosses between third backcross selections produced hybrids such as 5312/12, 53122/17, and 5318/34. These produced good quality roots and effective virus resistance (Jennings, 1994). Open-pollinated seeds from these hybrids were sent to many African countries including Nigeria where clone 58308 was selected and utilized (Hahn *et al.* 1980, 1989; Jennings, 1994). This clone had poor root yield but high resistance to CMD. It was used extensively at the International Institute of Tropical Agriculture (IITA), Ibadan, Nigeria (Hahn *et al.* 1989) to develop resistant genotypes including TMS 30001, TMS 30395, TMS 30555, TMS 30572, and TMS 4(2) 1425. Some of IITA-improved varieties have been widely tested in diverse environments and shown to out-yield the local varieties. Seeds of CMD-resistant genotypes and *in-vitro* virus-free clones were sent by IITA to over 30 national programmes in Africa for evaluation and selection under specific agroecologies (Ng *et al.* 1992; Mahungu *et al.* 1994). At the National Root Crops Research Institute (NRCRI), Nigeria, such seeds and breeder's stock materials were used to develop additional resistant genotypes such as NR 8082 and NR 8083 (Nnodu and Ogbe, 1994).

Although highly resistant varieties of this type are available in many countries, they are not always widely grown due to the lack of adequate quantities of planting material and in many countries farmers continue to grow local varieties including, some that may have little or no resistance to CMD. This explains why the disease is so prevalent in many areas and why such serious losses are occurring during the current pandemic in East Africa.

Epidemiological studies and field evaluation of the resistant genotypes reveal that they are not readily infected (eg. Hahn *et al.* 1980), and when infected, show mild symptoms, which may be restricted to some shoots (Jennings, 1960; Fargette *et al.* 1996). They tend to recover from symptoms with age (Njock *et al.* 1996) and have been shown to contain low virus concentrations (Fargette *et al.* 1996). However, the severity of symptoms expressed was not necessarily a reflection of virus concentration in some of the genotypes (Obge *et al.* 2003a). Another important feature of the resistant genotypes is the ability to provide uninfected cuttings even when infected ('reversion').

Resistance to virus infection differs from resistance to the whitefly vector. Fargette *et al.* (1996) found that cassava genotypes differed widely in whitefly infestation. In contrast, Hahn *et al.* (1980) observed similar number of whiteflies on CMD-resistant and susceptible genotypes and thus inferred that resistance to the vector was unlikely to explain the disease resistance. Nevertheless, variation in the suitability of cassava as a host for *B. tabaci* has been noted (Legg, 1994; Fargette *et al.* 1996). This variation could be exploited in breeding for resistance to the vector, since several studies have shown a correlation between whitefly population and disease incidence (Leuschner, 1977; Otim-Nape *et al.* 1998b).

Resistance in cassava to CMD derived from *M. glaziovii* is polygenic, and this involves recessive genes that are inherited additively, with a heritability of exceeding 60% (Hahn *et al.* 1989; Jennings, 1994; Mahungu *et al.* 1994; Mba and Dixon, 1997). More recently, classical genetic analysis and molecular mapping confirmed the polygenic nature of the *M. glaziovii* source of resistance to CMD (Akano *et al.* 2002). The resistance has been effective for more than 40 years in East and West Africa and no breakdown has occurred due to the emergence of resistance-breaking strains of CMBs. A notable example of the successful use of resistant varieties was their contribution to overcoming the effect of the recent pandemic of EACMV-UG in Uganda, where IITA genotypes have been widely used (Otim-Nape *et al.* 1994a). Following the outbreak of EACMV-UG, the involvement of farmers in multilocational trials for the selection of resistant genotypes led initially to the adoption of TMS 30572, TMS 60142, and TMS 30337, which had been introduced from IITA to Uganda as virus-indexed *in vitro* plantlets (Otim-Nape *et al.* 1994a) and they were adopted widely (Otim-Nape *et al.* 2000). Other TMS varieties and Ugandan selections have been released subsequently.

Whilst IITA was deploying TMS-type resistance, the need to diversify resistance and expand the gene pool was recognised and pursued (Dixon *et al.* 2001). A novel source of resistance was identified recently in a Nigerian cassava landrace (TME-3) that confers virtual immunity to CMD (Dixon *et al.* 2001; Fregene *et al.* 2001a). Classical genetic analysis and molecular genetic mapping showed that a major dominant gene designated CMD2 confers this resistance (Akano *et al.* 2002) and a simple sequence repeat marker (SSRY28) that is linked to this resistance gene was identified. This provides an opportunity to apply marker-assisted breeding for efficient selection of this trait. CMD2 is being cloned (Fregene and Puonti-Kaerlas, 2002) opening the possibility of utilizing it for genetic engineering to confer resistance in cassava.

It is evident that improving cassava production requires a change from traditional production practices towards a type of crop management that considers all factors sustaining high and stable harvest. With virus diseases in cassava, the progress to be made will depend on the 'seed' developed. Breeding for resistance against cassava viruses will undoubtedly provide the most durable and promising approach towards improving cassava production in Africa. For CMB resistance, a number of promising cassava breeding materials have been developed recently (Fokunang *et al.*, 2000; Ariyo, *et al.*, 2002). Some varieties have already been released to help restore cassava production in areas most severely affected by CMD. Several of the breeding lines, currently under development and investigation, do not permit virus replication and virus movement. After an initial virus invasion and symptom development, newly formed leaves appear symptomless, virus is not detected and hence virus infection is aborted. As shown by inoculation experiments with defined virus species and strains, resistance in these cassava lines appears to act against a range of CMBs and in both single and mixed infections. Furthermore, the resistance is effective against ICMV, which is not present in Africa; hence a broad range of CMBs seems to be covered by this resistance (S.Winter, unpublished data). With increased knowledge of the viruses involved in CMD and new methods available for artificial inoculation by gene gun or agroinoculation, the previous reliance on natural infection pressure and on a phenotypic screening of symptoms, can be avoided. The standardisation of inoculation protocols permits the accurate assessment of the resistance level of the material to be screened. Furthermore, the recovery and reversion phenomena, which are well-utilized resistance mechanisms in cassava, can be elucidated.

Understanding interactions between cassava, CMBs and the respective biotype of whiteflies may suggest new methods of CMD control. Biotechnology is likely to be of importance in facilitating control. The reintroduction into cassava of cloned ACMV DNA (Nigeria) by biolistic inoculation (Bridon *et al.* 1998) and cloned SACMV DNA by agroinoculation (Berrie *et al.* 2001) has boosted basic research. This led to site-directed mutagenesis of the viral DNA and facilitated a study of gene functions. Biotechnological techniques such as PCR, AFLP, RAPD, etc., are providing new tools to identify genes influencing virus resistance.

A great stimulus was provided by the development of molecular markers linked to resistance genes. Thus resistance screening, independent from virus infections can be conducted and screening at an early stage in resistance development can be enhanced significantly. Molecular markers, linked to different sources of resistance improve the understanding of the genetic components that influence virus resistance and mapping of resistance genes will define genes unique or common to all sources and thus facilitate pyramiding resistance in cassava. Considerable progress has been made in developing a comprehensive molecular genetic map (Fregene *et al.* 2001b) and a clustering of cassava accessions into groups having differential resistance has been achieved (Fregene *et al.*, 2000). Furthermore, progress has been reported in localizing resistance genes (Akano *et al.,* 2002).

6.2.1.9.5 Transgenic resistance: Since 1986, attempts have been made to introduce virus-derived sequences into plants to protect against infection of the homologous virus or related strains (Powell-Abel *et al.* 1986; Hull and Davies, 1992; Dasgupta *et al.* 2003). Obviously, protocols for the production of transgenic cassava plants must first be developed for routine use (Taylor *et al.* 2002). Progress has been made in the development of cassava transformation and regeneration systems (Fregene and Puonti-Kaerlas,

2002). Increasing numbers of cassava genotypes have been regenerated and transgenic plants obtained.

Cloning of natural virus-resistance genes may be another area of promise. All resistant crops developed by conventional methods are based on genes that occur naturally in plants. However, it is difficult to transfer such genes into unrelated crop species. With genetic engineering it may eventually be possible to transfer naturally occurring resistance genes to crops of interest. This requires a prior identification and isolation of the resistance gene from one species and then transformation of cassava.

Several attempts have been made to produce transformed cassava using ACMV coat protein genes and a replication-associated protein gene (AC1) (Fauquet *et al.* 1992; Hong and Stanley, 1996). A particle bombardment technique using a gene gun (Zhang *et al.* 2000) was used successfully to transform cassava. Despite the successes reported, research has yet to begin on the evaluation and production of transgenic cassava resistant to CMBs in sub-Saharan Africa (For details refer to Fofana *et al.*, Chapter 2, in this volume). Meanwhile, there is considerable public opposition to any such trials.

6.2.2 Cassava Brown streak disease (CBSD), recognized since the early 1930s in East Africa, causes symptoms distinct from those of CMD. CBSD was more important than CMD in some coastal areas of Tanzania (Storey, 1936). However, subsequent research on CBSD has been sporadic and it has been inferred only recently that the causal virus is of the genus *Ipomovirus* (*Potyviridae*) (Monger *et al.* 2001a,b). Many uncertainties still remain, especially concerning the existence of a second CBSD-associated virus as suggested earlier (Lennon *et al.* 1986a; Brunt, 1990). The effects of CBSD on crop yield and the natural means of spread should also be elucidated.

6.2.2.1 Geographical distribution: In early studies on CBSD, the disease was reported in coastal areas of Kenya, Tanzania, and Malawi and assumed to be present in adjacent areas of coastal Mozambique (Nichols, 1950). Moreover, it was also reported in Tanzanian material that had been introduced to Uganda (Jameson, 1964).

In the 1990s, the disease was found in only one planting in Uganda (Thresh *et al.* 1994a) and in 62 of the 325 plantings examined in Tanzania (19%), where the mean incidence in the country as a whole was only 6% (Legg and Raya, 1998). The incidence was much higher in the lowland coastal areas of Kenya and Tanzania and on Oguja Island of Zanzibar (Thresh and Mbwana, 1998; Hillocks *et al.* 1999; Calvert and Thresh, 2002).

The occurrence of CBSD in Nampula and Zambézia provinces of northern Mozambique, the two most important cassava-growing areas of the country was confirmed in 1999 (Hillocks *et al.* 2002). The overall incidence, based on assessments of leaf and stem symptoms, was 49% in Zambézia and 28% in Nampula, and the incidence was much higher than that of CMD in some districts, especially in lowland coastal areas (Calvert and Thresh, 2002). Moreover, root symptoms were not assessed and the leaf symptoms were sometimes inconspicuous and not readily distinguished, which suggested that the survey results underestimate the true incidence of infection.

There may have been movement of infected planting material across the border from Mozambique into Zimbabwe and Zambia, where CBSD is known to occur. The disease has not been reported in South Africa or Angola, or in any of the countries of West and Central Africa, although symptoms closely resembling those of CBSD were seen on a recent visit to Democratic Republic of Congo (N.M. Mahungu, personal communication).

6.2.2.2 Crop losses: There is only limited information on the effects of CBSD on growth and yield. In studies on a local Kenyan variety, the main effect was on the quality of the tuberous roots and not on root weight or number (Bock, 1994c). Necrosis decreases the value of the roots produced, which become unusable and unsaleable if the damage is extensive. This may force farmers to harvest prematurely before much deterioration of the roots occurs, but this incurs a yield penalty and food security is impaired because roots cannot be left in the ground until required at times of food shortage (Calvert and Thresh, 2002).

6.2.2.3 Symptoms: The name 'brown streak' was given to CBSD because of the brown, elongate necrotic lesions that develop on the young green stem tissue of affected plants. The symptoms may be confused with the superficial circular necrotic spots of unknown cause that develop on the stems of some varieties (Nichols, 1950). The leaf symptoms of CBSD are variable and distinct from those of CMD, as they affect only mature leaves (Fig. 6.2). Symptoms are severe when mixed virus infections with EACMV occur (Fig. 6.3). The most easily recognizable leaf symptoms occur as a characteristic 'feathery' chlorosis that is closely orientated along the secondary and tertiary veins of affected laminae.

CBSD causes necrosis of the tuberous roots, which also develop characteristic constrictions. Some varieties do not express root necrosis, or do so only at a late stage of crop growth. These varieties are damaged much less severely than those that develop extensive symptoms at an early stage.

6.2.2.4 Transmission: Natural spread of CBSD occurs between plants and it is attributed to an arthropod vector or vectors as yet unidentified. However, only a few transmission experiments have been done, mainly involving the aphid *Myzus persicae* and the whitefly species *Bemisia tabaci* and to a lesser extent *B. afer* (= *Bemisia hancockii*). The two whitefly species have been considered because these two sap-feeding insects have had a long association with cassava in Africa. However, transmission experiments using *B. tabaci* and CBSV-infected herbaceous hosts failed to verify *B. tabaci* as the vector of CBSV (I. Abdullahi, personal communication). It is also notable that *B. afer* seems to be particularly common in coastal areas of eastern and southern Africa where CBSD is most prevalent. This emphasises the need for additional studies with *B. afer* and also with insect species that visit but do not colonize cassava, as many potyviruses, are transmitted by transient vectors in a nonpersistent manner. At least some of the spread of CBSD to cassava may be from hosts other than cassava, as CBSV has been detected only in eastern and southern Africa and it is assumed to have indigenous hosts.

6.2.2.5 Viral characteristics: From the outset, CBSD was assumed to be caused by a virus because it was graft-transmissible in the absence of a visible pathogen. The first evidence of a virus was obtained by sap inoculation from cassava to herbaceous hosts and back to cassava (Lister, 1959) and also by electron microscopy (Kitajima and Costa, 1964). Virus isolates in herbaceous hosts were later shown to have elongate particles 650-690 nm long (Lennon *et al.* 1986a). They resembled viruses of the genus *Carlavirus*, but no serological relationship was demonstrated at the time with any definitive virus of this type.

There was evidence that two different elongate viruses occur in CBSD-affected plants (Lennon *et al.* 1986a; Brunt, 1990) and isolates in herbaceous hosts were shown to induce 'pin-wheel' inclusions of the type produced by potyviruses. Hence a cassava brown streak associated-carlavirus and a cassava brown streak potyvirus were distinguished (Brunt *et al.* 1996). Recent studies indicate that CBSD is caused by a virus of the genus *Ipomovirus*, family *Potyviridae*, but this has not been confirmed as Koch's postulates have not been fulfilled (Monger *et al.* 2001a,b).

Bacterial-expressed coat protein of the virus reacted with an antiserum against a putative 'carlavirus' preparation from CBSD. However carlavirus sequences have never been confirmed in CSBD-affected plants and as an ipomovirus is constantly associated with CBSD, it is likely to be the sole cause of CBSD.

Amino acid sequence comparison of the C-terminal core region of the CBSV coat protein gene revealed a close relationship of CBSV with only two other known ipomoviruses, *Cucumber vein yellowing virus*, CVYV, (>70% sequence identity; S.Winter, unpublished data) and *Sweet potato mild mottle virus*, SPMMV, (> 42% sequence identity; Monger *et al.* 2001a). There is no relationship with sequences of carlaviruses.

Further studies by Monger *et al.* (2001b) provided some insights into the molecular variability of the virus causing CBSD and on the two virus hypothesis for the etiology of CSBD. Mechanical transmission of CBSV from different cassava cultivars collected in one region of Tanzania resulted in a differential symptom expression in *N.benthamiana* but only ipomovirus sequences were amplified from infected plants. These sequences differed considerably in the region of analysis and showed a nucleotide divergence of up to 8%. This probably also explains symptom variation in the indicator plants. Nevertheless, it remains to be elucidated, whether these sequence variants are separate virus isolates or, a reflection of the plasticity of the CBSV genome. The inherent variation in CBSV poses a problem to virus detection and identification, which needs to be addressed if reliable diagnostic procedures are to be developed for CBSV indexing.

6.2.2.6 Diagnosis: Until recently, all reports on the distribution of CBSD in cassava-growing regions of Africa were based on assessment of the characteristic leaf symptoms in cassava. However, symptom evaluation, although a good indication of virus incidence in the field, is not sufficient for virus screening or quarantine indexing. It is also unreliable for CBSD in cassava because of latent infections and environmental and varietal influences on symptom development. Furthermore the expression of symptoms in mature leaves only, i.e. relatively late in crop growth, provides information on the health status of the plant that is of limited value to support crop management decisions.

CBSV is transmissible mechanically to several herbaceous species, of which *N. benthamiana, N. glutinosa, N. clevelandii, N. debneyi, Datura metel* and *D. stramonium* (Karakacha, 2001) developed the most pronounced symptoms. Mechanical transmissions to a set of differential hosts provide good indications of the presence of CBSV, especially in mixed CMB/CBSV infections where the symptoms of CBSV are often masked by the more conspicuous symptoms of CMBs. However, these bio-assays are not amenable to large scale application.

Antisera raised against purified particle preparations were produced (Lennon *et al.* 1986a; DSMZ, Germany) and sufficed to detect CBSV in herbaceous hosts, but virus detection in cassava by serology was unreliable. This is probably because CBSV concentration in younger plant parts and immature cassava leaves is generally lower than in older parts (in contrast to CMB concentrations).

From the molecular analysis of the 3`-terminal RNA sequences of CBSV and sequence resolution of isolate variations, it was possible to define oligonucleotides within conserved regions at the 3'- end of the virus coat protein and to develop a PCR assay for virus detection. Using RT-PCR, it was possible to amplify all CBSV isolates described so far (Monger *et al.* 2001b). With this approach, it is possible to test for CMBs and CBSV concurrently in cassava and to test for virus at an early stage. Although routine methods have yet to be developed, great progress has been made recently in diagnosis of CBSV.

6.2.2.7 Epidemiology: CBSD has been neglected compared with CMD, and there is inadequate information on its distribution, effects, and mode of spread. Moreover, CBSD has been recognized only recently in Mozambique, even though it is prevalent in large areas of the country and undermines food security (Hillocks *et al.* 2001). There is no adequate explanation as to why the disease seems to be prevalent only in southern and eastern Africa, or why it seldom occurs at altitudes exceeding 700 m above sea level (Hillocks *et al.* 1999). The original indigenous host from which CBSD is assumed to have spread to cassava has not been determined, and the putative insect vector is not known. There is little information on the epidemiology and there are many uncertainties, which impede the development of effective management strategies. One of the problems has been the lack of assured virus-free stocks of planting material for epidemiological experiments and for use by farmers. Another has been the failure to identify the natural means of spread between plants. However, there is evidence of such spread as clones introduced from West Africa or other areas that are free from CBSD have become infected when grown at sites in Mozambique, Malawi, Kenya, and Tanzania. Plants raised from seed introduced from West Africa have also become infected at these sites. There is little evidence on temporal or spatial patterns of spread, but this is known to have been slow in an experiment at a site in coastal Kenya (Bock, 1994c) and rapid in some years at sites in southern Tanzania and in coastal areas of northern Mozambique (R.J. Hillocks and J. M. Thresh, unpublished data).

6.2.2.8 Control: Virus-free stocks can be produced by rigorous selection (Mtunda *et al.* 1999) and in future the sensitive methods of virus-detection now being developed may facilitate this. It is also possible to use meristem-tip and/or heat therapy to eliminate CBSV from clones that seem to be totally infected (Kaiser and Teemba, 1979).

Currently, farmers try to use cassava varieties that do not develop severe root necrosis, or do so only at a late stage of crop growth. There is only limited information available on CBSD-resistant varieties, but resistance to CBSD and CMD was identified in early breeding programmes in Tanzania (Jennings, 1957). More recently field surveys on the incidence of CBSD in Tanzania and Mozambique (Hillocks *et al.*, 2002) have indicated some local cultivars with promising resistance characters that are now being evaluated. With the diagnostic tools available, and the information from field surveys, which have distinguished regions with high or low disease incidence, virus-free zones can be defined where propagation of cassava cuttings is done best. Until good resistance is developed in cassava, which ideally should be against both CMBs and CBSV, the use of virus-free planting material along with other phytosanitary measures is the best means for control for CSBD.

The use of virus-resistant varieties and effective methods of phytosanitation may help to bring CBSD under control. Future progress is likely to be more rapid than in the past following the recognition of the prevalence of CBSD and the stimulus to further research provided by the losses now being encountered in Mozambique and elsewehere.

6.3 Asia and the Pacific region

Cassava is grown in many countries of South East Asia and the Pacific. In Asia, a cassava mosaic disease (CMD) was reported in India and Sri Lanka. Two distinct begomoviruses, *viz.*, *Indian cassava mosaic virus* (Hong *et al.* 1993) and the recently reported *Sri Lankan cassava mosaic virus* (Saunders *et al.* 2002) cause CMD in Asia. An early report of CMD in Indonesia (Muller, 1931) has not been confirmed and the symptoms were later attributed to a mineral deficiency (Bolhuis, 1949). *Cassava green mottle virus* has been detected in cassava originating from the Pacific region (Table 6.1). The prevalence and importance of *Cassava common mosaic virus* reported in Taiwan (Chen *et al.* 1981) is not known and the virus is not considered here.

6.3.1 Cassava mosaic disease was first noticed in India in 1942 (Abraham, 1956), and it was reported authentically in 1966 (Alagianagalingam and Ramakrishnan, 1966). Within India, the overall incidence of CMD was higher in the two main cassava-growing states of Kerala (23%) and Tamil Nadu (30%) than in Andhra Pradesh (<1%) and Karnataka (5%), which are outside the main cassava-growing areas (Mathew, 1989). There is a need for more comprehensive surveys, especially as CMD seems to have become more prevalent in recent years.

6.3.2 Crop losses: In India, losses in tuberous root yield due to primary infection of CMD vary from negligibile to 84%, depending upon the cultivar and time of infection (Narasimhan and Arjunan, 1976). Losses were only 19-26% in the hybrids tested and in the widely grown M4 from Malaysia (Thankappan and Chacko, 1976). In other experiments, losses were 42% in the popular variety Kalikalan, and ranged from 17 to 36% in nine selected hybrids (Malathi *et al.* 1985). Losses were even less in a later trial with M4 (7-10%) and four hybrid varieties (9-21%) and there was a positive correlation between yield loss and symptom severity scores (Nair and Malathi, 1987). Jeeva (1997) studied the effect of secondary infection of ICMV by whiteflies on growth and tuberous root yield of five cassava varieties and meristem-derived and clonally multiplied disease-free planting material. Root yield was reduced by 13-31%.

6.3.3 Host range: The natural hosts of ICMV are cassava, and *Manihot glaziovii.* In Kerala, *Jatropha curcas*, a hedge plant of the Euphorbiaceae shows mosaic symptoms. The infected leaf samples reacted in ELISA with ACMV antiserum (G. Thottappilly, unpublished), but an ICMV probe for

DNA B gave no signal (V.G. Malathi, personal communication), indicating that *J.curcas* is not infected with ICMV. Recently, Rajinimala (2002) and R. Rabindran (personal communication) identified ICMV as causing a new virus disease in bitter gourd using ELISA and PCR with ICMV-specific primers.

6.3.4 Symptoms: The symptoms of CMD in India resemble those reported in Africa and the name CMD is adopted in some publications and Indian cassava mosaic disease (ICMD) in others (Thresh *et al.* 1998c). The apparent similarity of the mosaic diseases of cassava in India and in Africa justifies the use of the same name in the two continents, but further work may show that there are important differences (as for example in the effectiveness of the different types of host plant resistance, or in alternative hosts).

6.3.5 Transmission: ICMV is not transmitted through cassava seeds (Malathi *et al.* 1987). Sap inoculation to *Nicotiana benthamiana* produced typical symptoms similar to those of African CMBs.

Although ten species of whitefly breed on cassava in different parts of the world, only *B. tabaci* and *Aleurodicus dispersus* colonize cassava in India (Palaniswami *et al.* 1994, 1996, 2001). The role of *A. dispersus* in ICMV transmission has not been investigated. Populations of *B. tabaci* are active throughout the year, and its population dynamics and distribution have been studied (Lal and Pillai, 1977,1982; Pillai and Daniel, 1979; Palaniswami and Pillai, 1991). In India, B. *tabaci* is the main insect vector of CMD. Successful transmissions have been reported using whiteflies transferred from infected cassava to healthy cassava and herbaceous hosts, and also between herbaceous hosts. High rates of transmission were achieved in some experiments, as between cassava plants (19%) and from cassava to *Nicotiana tabacum* cv. Jayasri (100%), *N. rosulata* (67%), and 11 other *Nicotiana* spp. (20-25%) using 50 whiteflies per test plant (Mathew and Muniyappa, 1993). In recent growth chamber studies, high rates of transmission were also reported (Maruthi *et al.* 2002b). However, such successful transmission seems to be exceptional and not readily reproducible. Other studies reported much lower rates of transmission (Nair, 1975), or were completely unsuccessful (Malathi *et al.* 1985; Palaniswami *et al.* 1996). Nevertheless there is no doubt that *B. tabaci* is the vector of ICMV and studies on epidemiology, control, and whitefly population dynamics support this assumption (Mahto and Sinha, 1978). Paranidharan (1994) reported that though the whitefly population was high from December to February at Coimbatore, there was no spread of CMD. Nair (1988) noted that maximum field spread of ICMV occurred in Trivandrum (Kerala) in November. According to Palaniswami *et al.* (1996), the whitefly population in cassava fields was high during February-March and was significantly and positively correlated with maximum temperature. Whitefly spread CMD to 4-30% of the plants in the stand by harvest.

Recently, Palaniswami *et al.* (2000) showed the presence of two host-specific biotypes of *B.tabaci* in India - cassava and sweet potato. The cassava biotype develops successfully on cassava, eggplant, and tobacco, but not on sweet potato, cotton or tomato. In contrast, the sweet potato biotype infested sweet potato, cotton, eggplant, tomato and tobacco, but not cassava. Eggplant and tobacco were common hosts for both strains. Lisha *et al.* (2003) reported that only the cassava biotype transmitted ICMV from infected to healthy cassava and not the sweet potato biotype. The cassava biotype achieved 40-100% transmission of ICMV under different acquisition access and inoculation access periods. This may explain why a lower rate of transmission was reported earlier when whiteflies from the field were collected and tested without distinguishing the cassava biotype. Rajinimala (2002) used eggplant to rear whiteflies, and transmission experiments with them were successful. Recent studies on transmission of ICMV and different CMBs from Africa by *B. tabaci* collected from India and different locations in Africa suggest a strong tendency of vectors to transmit the local viruses with significantly higher efficiencies than the non-local ones under glasshouse conditions (Maruthi *et al.* 2002b). The above observation can probably be explained by differences in amino acid sequences of the viral coat protein.

6.3.6 Etiology: A geminivirus was isolated from CMD-affected plants in India (Malathi and Sreenivasan, 1983; Mathew and Muniyappa, 1992). When testing four CMD isolates with polyclonal antisera prepared against African and Indian isolates (Malathi *et al.* 1985, 1987), three of the Indian isolates reacted positively with antisera to an African CMB, but were serologically distinct from African isolates and so were regarded as being a separate strain of an African CMB. In subsequent tests using a panel of monoclonal antibodies, Indian isolates were distinguished from those from East and West Africa and later distinguished as a separate virus. Furthermore, begomoviruses from different host species in the same geographical region tend to be antigenically more closely related to one another than to ACMV (Swanson *et al.* 1992; Harrison *et al.* 1997a). The epitope profile of ICMV is more similar to that of *Cotton leaf curl virus* from Pakistan than to ACMV. Recently, a distinct geminivirus was identified from cassava in Sri Lanka showing mosaic symptoms and named '*Sri Lankan cassava mosaic virus*' (SLCMV) (Saunders *et al.* 2002).

Using DNA A of ICMV as a general probe and DNA B as a specific probe, Rani *et al.* (1996) tested geminiviruses infecting *Vigna*, cassava, *Acalypha, Croton,* and *Vernonia* in India. ICMV DNA A hybridized intensely to DNA from infected *Vigna radiata, V.mungo, Acalypha indica, Croton sparsiflorus,* and *Vernonia cinerea*, indicating that all these plants contained whitefly-transmitted geminiviruses. DNA B of ICMV, used as a specific probe, hybridized only to DNA from infected cassava and *Acalypha,* suggesting that these geminiviruses are related.

6.3.7 Diagnosis: Malathi *et al.* (1989) detected ICMV in meristem-derived, clonally multiplied symptom-free planting material of cassava cultivars by using ELISA, so demonstrating that this is sufficiently sensitive to detect ICMV, even in symptomless plants. Specific monoclonal antibodies to identify the viruses causing CMD in India are available and have been used in several studies (Swanson and Harrison, 1994; Rajinimala, 2002). Ogbe (2001) used the following primer combination to detect ICMV by PCR:

ICMV-F1: 5'- ttc tct ctc ctc aat cgg ta- 3'

ICMV-R1: 5'-act cag gga act cgt tta gt-3'

PCR is highly sensitive for detecting ICMV.

6.3.8 Epidemiology: Cassava in India is grown in diverse environments, including the lowland humid forest areas of coastal Kerala, the upland foothills of the Western Ghats, and the irrigated areas of Tamil Nadu, where there is a prolonged dry season. The primary spread of ICMV is through infected cuttings. Epidemiological studies have used virus-free stocks of selected planting material or clones derived from meristem-tip culture. Several cultivars were included in experiments done in three successive seasons (Nair, 1985). The final incidence of CMD did not exceed 1.3% in plots containing initial disease foci, and it was even less in plots without such sources. In monthly plantings at a site near Bangalore in Karnataka State, which is outside the main cassava-growing area, there was little or no spread of the disease (Mathew, 1989). In another trial near Trivandrum, six cultivars were established in plots, which contained initial sources of inoculum and where CMD was also prevalent in the surrounding plantings. There was substantial spread to the susceptible cv. Kalikalan (50% incidence), but little disease in the five more resistant cultivars (1-10%) (Nair, 1988). In a more comprehensive study, there was greater spread to plots, which contained initial sources of inoculum (overall incidence 5.7%) than to those without such sources (2.8%). However, the source effect was not consistent at each of the four sites or in the five cultivars tested, and it was largely due to the big difference in incidence in cv. Kalikalan at the site where most spread occurred (Nair and Thankappan, 1990).

In hybrids H43 and H226, the incidence was 30 and 15%, respectively, while in H97, H165, and M4 it was less than 5 % at the same sites. However the incidence was as high as 52% in the susceptible cv. Kalikalan (Palaniswami *et al.* 1996).

6.3.9 Control: In India, cassava is grown under conditions very different from those in much of Africa. The relatively high productivity of cassava in India is associated with generally good cropping practices. Moreover, Indian crops are unaffected by either the cassava green mite or the cassava mealybug, which had damaging effects in many parts of Africa.

As in Africa, the main approaches to control of CMD in India are through phytosanitation and use of resistant varieties. Some attention has also been paid in India to the use of insecticides to control the whitefly vector in attempts to reduce spread (Nair, 1977, 1981; Mahto and Sinha, 1978). However, the results have been unsatisfactory and the use of insecticides is not recommended (Malathi *et al.* 1985).

Use of carefully selected disease-free planting material is the only effective method of control of cassava viruses in the absence of resistant varieties. Farmers were advised to paint-mark the diseased stems during the active growth period of the crop to differentiate them from healthy ones at harvest, when many of the symptom-bearing leaves were likely to have abscissed (Malathi *et al.* 1987). Secondary spread of the disease in the field through vectors can also be minimised by periodic roguing of newly infected plants and volunteer cassava from previous crops. However, these measures are seldom used by farmers and the incidence, severity, and losses due to CMD are increasing progressively in many parts of India, rendering much of the crop unfit for use as planting material (Thankappan *et al.* 1996).

Virus-free stocks have been obtained by rigorous selection and through the use of meristem-tip therapy leading to substantial increases in yield compared with infected material. Such stocks have been used in experiments and shown to remain largely free of CMD in areas where there is limited spread by whiteflies (Nair, 1990; Nair and Thankappan, 1990). However, only limited attempts have been made to encourage the widespread adoption of such material, as the cost of producing tissue culture planting material is high.

Spread of ICMV by the vector is reported to be low in improved cultivars (Hrishi *et al.* 1977). Nair and Daniel (1983) further reported varietal preference by the vector, but it is not related to ICMV resistance or virus susceptibility. Varieties with green petioles were preferred by whitefly to those with red or red-green petioles and soft leaves were preferred to coarse ones. Those varieties with erect leaf orientation had double the number of adult whiteflies of those with horizontal or downward orientation. The finding could be exploited in breeding.

Breeding for CMD resistance has been carried out by the Central Tuber Crops Research Institute, Thiruvananthapuram, Kerala State (CTCRI, 2000). The two hybrids (H-97 and H-165) released in 1971 were field-tolerant. Other hybrids such as H-226 and Sree Visakham and other varieties such as Sree Sahya, Sree Prakash, Sree Rekha (TCH-1) and Sree Prabha (TCH-2) are susceptible. Sree Jaya and Sree Vijaya are moderately susceptible, while Sree Harsha (2-14), a triploid is susceptible. It is notable that most of the recent releases are susceptible to CMD. However, two high-yielding cassava accessions, CE 185 and CI 800, were identified from the CMD-free accessions (CTCRI, 2002a). Additionally, two high-yielding good quality accessions with high tolerance to CMD (CE-138 and CE 142), were identified. The CMD tolerance of interspecific hybrids from cassava x *M. caerulescence* has also been reported (CTCRI, 2002a).

A new project has began at CTCRI and aims to clone the replicase gene (AC1) of ICMV, transform cassava, and finally test the transgenic plants for their resistance to ICMV infection (CTCRI, 2002b).

6.4 South and Central America

Although seven viruses were reported from South and Central America (Table 6.1), only *Cassava common mosaic virus*, *Cassava vein mosaic virus*, and cassava frog skin disease are described in detail here. The casual agent of frogskin disease is not known, but a virus is suspected because the disease is graft-transmissible. *Cassava virus X* (CsVX) and *Cassava Colombian symptomless virus* (CCSpV) are species of the genus *Potexvirus* that infect cassava (Lennon *et al.* 1986b). They have been detected only in Colombia. There is only one report of *Cassava American latent virus* from Brazil and Guyana, and little is known of its distribution (Walter *et al.* 1989). None of these three viruses cause symptoms; hence it is difficult to determine their distribution and to evaluate their importance. Brunt *et al.* (1996) listed Caribbean mosaic virus from Colombia, but recent studies indicate that this pathogen is similar or identical to that causing cassava frog skin disease (L. Calvert, unpublished data).

Two diseases of procaryotic etiology are notable. 'Cassava antholysis' occurs in Brazil, the Caribbean, Central America, Colombia and Venezuela (Frison and Feliu, 1991). The causal agent is a phytoplasma and dissemination is mainly by vegetative propagation (Jayasinghe *et al.* 1984). The first observed symptoms are virescence followed by phyllody. Cassava witches' broom is also caused by a phytoplasma (Costa and Kitajima, 1972b). The characteristic symptoms are stunting and an abnormal proliferation of branches. The disease is economically important in Brazil and southern Mexico.

6.4.1 Cassava Common Mosaic Disease (CsCMD) was first reported in southern Brazil (Silberschmid, 1938; Costa, 1940) and has since been recorded elsewhere in South America, but rarely in Colombia (Nolt *et al.* 1992). The disease is most prevalent in southern Brazil and Paraguay. The affected plants are sometimes stunted and yield loss can be up to 60% (Costa and Kitajima, 1972a). In general, the disease is unimportant, although there are areas where control efforts are needed (Calvert and Thresh, 2002). Leaves of cassava plants affected by CsCMD develop mosaic and chlorotic symptoms. Severe symptoms are prevalent during relatively cool periods; hence cassava grown in the semi-tropical areas of South America is most affected by the disease.

CsCMD is caused by *Cassava common mosaic virus* (CsCMV), Genus *Potexvirus*. There are no known vectors of CsCMV and the primary source of inoculum is infected planting material. CsCMV is very stable and can be spread mechanically by contaminated machetes and other tools used to prepare cuttings. Although mechanical transmission of CsCMV through sap of infected plantings is inefficient, it is the only known means of plant-to-plant spread. On sap inoculation, CsCMV can infect species of several families of dicotyledonous plants (Silva *et al.* 1963; Kitajima *et al.* 1965), including *Euphorbia* spp., *Cnidoscolus aconitifolius* (chaya), *N. benthamiana* (systemic host) and *Chenopodium amaranticolor* (local lesion host) (Costa and Kitajima, 1972a).

Particle characteristics: The CsCMV virion is a semi-flexuous rod shaped particle, *c.* 15 x 495 nm (Kitajima *et al.* 1965) long containing RNA (Silva *et al.* 1963). Nuclear inclusions typical of potexviruses occur in cassava and *N. benthamiana*. The viral particles of CsCMV contain a single coat protein with a molecular weight of 26,000 daltons (Nolt *et al.* 1991). The complete nucleotide sequence of the single-stranded RNA genome is known (Calvert *et al.* 1996). The genome organization of CsCMV resembles that of a typical potexvirus.

Control: Eliminating (roguing) plants with CsCMV symptoms provides adequate disease control in the field. The symptoms are usually obvious on the first leaves developing on infected stem cuttings. This is the most appropriate time to identify and remove diseased plants. If the plants are not rogued early they should be marked and the stems burned later after harvesting the tuberous roots. Only healthy plants should be selected as a source of vegetative propagules. To minimize the risk of mechanical transmission, cutting tools should be disinfected at regular intervals (Lozano and Nolt, 1989). With care in selecting healthy planting material, CsCMD can be eradicated or its economic impact minimized (Calvert and Thresh, 2002).

Schöpke *et al.* (1998) reported that tobacco plants transformed with the coat protein gene of CsCMV were highly resistant to infection with the virus and cassava cultivars with transgenic resistance may soon be available.

6.4.2 Cassava Vein Mosaic Disease (CVMD) was first reported by Costa (1940). It is most prevalent in remote semi-arid areas inhabited by poor rural communities. Although large areas are affected by CVMD, the disease has received little attention, probably because the symptoms are sporadic and not readily apparent at the end of the cassava-growing period (Calvert and

Thresh, 2002). CVMD is common in the semi-arid zone of northeastern Brazil, although there are a few reports from the southern part of the country. The disease is common in the Brazilian States of Ceará, Pernambuco, Alagoas, Piaui and Bahia (Calvert *et al.* 1995).

The leaf symptoms of CVMD occur in flushes. After an infected stem cutting sprouts, the first four to six leaves show vein chlorosis. Some leaves develop mosaic patterns affecting the entire lamina. Leaf deformation and epinasty are common in severe infections. Plants then recover from infection by producing several apparently symptomless leaves, followed by another series of leaves with symptoms. The expression of symptoms is influenced by the prevailing climatic conditions and symptoms are more pronounced in semi-arid areas than in coastal regions of northeastern Brazil. Except during the period just after sprouting, CVMD does not seem to affect plant vigour, however the premature senescence and abscission of affected leaves reduces the total leaf area. As infected cassava matures, it is often difficult to find leaves with mosaic symptoms (Calvert and Thresh, 2002).

Virus particle: CVMD is caused by *Cassava vein mosaic virus* (CVMV), which has isometric particles, *c.* 50 nm in diameter (Kitajima and Costa, 1966). The genome consists of dsDNA, *c.* 8,200 nucleotides long. Based on particle structure and the dsDNA genome (Lin and Kitajima, 1980), CVMV was regarded as a tentative member of the caulimovirus group. However, the analysis of the complete sequence of CVMV revealed a genome organization different from that of either the genus *Caulimovirus* or *Badnavirus* (Calvert *et al.* 1995). Consequently, the virus is ascribed to a unique genus, "Cassava vein mosaic-like viruses" in the family *Caulimoviridae*.

Control: Little is known about the epidemiology or control of CVMV. The only known host of the virus is cassava and an important means of dissemination is through infected propagules. Disease spread in the field suggests the presence of a vector, but so far none has been identified (Calvert and Thresh, 2002). For virus control, 'roguing' may be effective if diseased plants are identified and removed soon after sprouting. Although the full economic importance of CVMD is not well quantified, considerable crop losses can occur, especially when drought occurs during early stages of crop development (Calvert and Thresh, 2002).

6.4.3 Cassava Frogskin Disease (CFSD) is a virus-like disease that affects cassava and it was first reported from southern Colombia (Pineda *et al.* 1983). A similar disorder in northern Colombia was called Caribbean mosaic disease because mosaic symptoms were expressed by the cassava landrace 'Secundina'. In the Amazon regions of Brazil and Colombia, CFSD is called 'jacare' (caiman) because of the distinctive ridges on affected roots. All three disorders are manifestations of the same disease since very similar symptoms are expressed in standard indicator varieties of cassava grown under uniform conditions (Calvert and Thresh, 2002).

The disease affects cassava grown in isolated, indigenous Amazonian Indian communities. They regarded it as a physiological disorder of certain varieties rather than as a disease. CFSD was 'first discovered' in 1971, when an apparently new disease that caused severe losses occurred in the mid-altitude Andean mountains of southern Colombia. This was later named cassava frogskin disease (Hernández *et al.* 1975). By the 1980s, it was prevalent throughout most cassava-growing regions of Colombia and it also spread to Venezuela and Costa Rica. Recently, CFSD was reported in Panama affecting plants grown from stem cuttings imported from Costa Rica (Calvert and Thresh, 2002). Preliminary evidence suggests that the disease was eradicated in Panama by eliminating all such imports. CFSD also occurs in the Amazon region of Peru.

In most genotypes, the leaves of infected plants are symptomless and appear normal (Calvert and Thresh, 2002). The stems of these plants may be slightly enlarged, especially in its basal sections, due to a lack of starch in the roots. Because of their apparent vigour, farmers preferentially select these infected stems. Root symptoms range from very mild to very severe, and the severity of the symptoms varies with age of the roots, climatic conditions, and variety. Hot and dry conditions inhibit symptom development, whereas cooler temperatures enhance symptoms. Severe stunting is the most characteristic symptom in a few cassava genotypes. Severely affected roots do not accumulate starch and often show zones of constrictions. Newly infected plants usually have mild or no symptoms unless infection had occurred early in plant development (Calvert and Thresh, 2002). A few landraces develop mosaic symptoms on the leaves and appear stunted.

Although some studies indicate that the whitefly *Bemisia tuberculata* is a vector of CFSD (Angel *et al.* 1990), it has been difficult to verify this by vector transmissions performed under controlled conditions. In the field, the disease spreads very slowly, but progressively through both stem cuttings and plant-to-plant transmission. In a controlled trial, which excluded use of infected stem cuttings, disease incidence exceeded 10% after three crop cycles. The rate of virus spread increased with the incidence of infected plants in the field (Calvert and Thresh, 2002). If cassava is planted sequentially, the rate of spread between plantings can be rapid. This is apparently due to the vector preference for younger plants, with each planting equivalent to a new cassava cycle.

CFSD is readily graft-transmitted and currently the best diagnostic test available is grafting to the indicator cassava variety 'Secundina' (CIAT accession M Col 2063) (Calvert, 1994). The buds of the rootstock stem should be removed to increase the likelihood of successful union. Inoculated plants should be grown in areas where temperatures are normally below 28°C, to ensure optimum symptom expression. After 3 or 4 weeks, plants can be evaluated for mosaic symptom expression on the newly developing leaves, which indicates the presence of CFSD.

Virus particles: The causal agent of CFSD has not been proven definitively, although isometric virus-like particles 70-80 nm in diameter occur consistently in thin sections of the leaves, petioles, stems and roots of affected plants (Calvert and Thresh, 2002). At least nine species of dsRNA occur in infected plants, and the RNA genome has been partially cloned. The predicted proteins encoded by two cDNA clones reveal homology with genes encoded by RNA 1 and RNA 5 of *Rice ragged stunt virus* (L. Calvert, unpublished data). Viroplasma-like bodies also occur in leaves of infected plants. The symptoms of hyperplasia in the root cortex are similar to tumours caused by plant reoviruses. While root symptoms, particle morphology, dsRNA pattern, and sequence homology are consistent with the causal agent being a reovirus, additional experiments are needed to prove that a reovirus causes CFSD.

Control: Farmers can reduce the incidence of CFSD by rigorous selection of healthy planting material. Roots should be inspected for symptoms at harvest and only cuttings from apparently healthy plants that bear normal roots should be selected. If this is done, disease incidence can be maintained at low levels. Where CFSD incidence is substantial, it is advisable to collect propagation material from disease-free areas. The disease can be eliminated from infected plants by thermotherapy and meristem culture *in vitro* (Maffla *et al.* 1984). This is applicable, if *in vitro* propagation systems for producing cassava are in place. The use of CFSD-free material is an effective control strategy if the disease pressure in neighbouring fields is not too excessive.

More than 500 cassava lines have been screened for tolerance to CFSD. In *c.* 20 percent of lines, a considerable level of resistance was maintained after several years of exposure to the disease. Resistance was determined by rating disease symptoms during six successive crop cycles. Since such resistant plants were still carrying CFSD, which was transmitted by graft inoculation, the cassava lines express tolerance to the disease. The starch and yield of these varieties are normal compared to susceptible lines. Some of the cassava lines with resistance to CFSD are commercial varieties and widely grown. Although environmental conditions influence the expression of CFSD symptoms, the resistance or tolerance appears to be stable. The systematic development of resistant varieties should effectively reduce the economic losses associated with CFSD.

6. 5 General discussion and outlook

Although there are some large-scale growers, in general, cassava is a low-cost carbohydrate food predominantly grown by resource-poor farmers in small areas of land. There is great scope for expanding the cultivation and utilization of cassava in developing countries as a cheap source of carbohydrate. Cassava is grown in many different ecological situations and the crop is highly adaptable to poor soil conditions, is drought-resistant and thus provides a stable but often low yield. Virus diseases seriously impede cassava production, especially in Africa. There the crop is grown in many diverse ecologies, from coastal regions to semi-arid zones of the Sahel and mosaic diseases caused by begomoviruses are regarded as the prime constraint to cassava production.

Since cassava has reached the status of a crop able to ameliorate the pressing problems of food shortage and for income generation, research efforts on CMBs have been boosted substantially. In contrast, many of the other cassava viruses reported in the literature have received only little attention and their taxonomic and epidemiological status is still unclear. The causal agent of the cassava brown streak disease has been identified only recently but knowledge on its epidemiology and control remains inadequate. The etiology of cassava frogskin disease reported from South America is still unresolved.

It is notable that, other than two unrepeated observations, cassava viruses from South and Central America have not been reported from Africa or Asia. This implies that the cassava crop was probably free from viruses at the time of vegetative introduction to Africa and Asia, or that only seeds were introduced. The viruses reported from Africa have not been reported from Asia, and the viruses reported from India and Sri Lanka have not been reported elsewhere. It appears that the indigenous viruses already present in each region subsequently infected the crop.

Of the eighteen different viruses of cassava (Table 6.1), only two have a full description in *'Descriptions of Plant Viruses'* (Costa and Kitajima, 1972a; Bock and Harrison, 1985). This shows the dearth of knowledge on other cassava viruses, and emphasizes the urgent need to determine yield losses, characterise the viruses and to develop diagnostic assays.

It is important to have a thorough knowledge of the viruses associated with each cassava-producing region. High priority should be given to disease surveys to record the virus situation and to monitor newly emerging virus problems. It is important to develop diagnostic methods for all viruses of cassava, including those currently causing only minor diseases. Diagnostic tools for routine detection of SACMV, EACMV-UG, and SLCMV are needed.

The CMBs that differ from each other in sequence identity are considered to be of independent origin. Rapid progress has been made recently in advancing the knowledge on CMBs as detailed in this chapter. It is obvious that whitefly-transmitted geminiviruses are becoming increasingly important with novel geminiviruses and new whitefly biotypes. For CMBs, this has been shown by many reports on the emergence of new virus types through recombination between and among virus species contributing to a high genetic diversity of begomoviruses (Padidam *et al.* 1999).

Two distinct CMBs are reported from Asia and it is necessary to carry out surveys to monitor the emergence of new virus types so that early counter measures can be recommended and adopted. There is also uncertainity concerning the epidemiology and mode of spread of cassava brown streak, cassava frogskin, and other diseases. A need exists to confirm the role of whitefly and other potential vectors in the spread of these diseases. In future studies, molecular techniques should be utilized in virus transmission studies.

Experience with CMD in Africa over many years, and more recently with cassava frogskin disease in South America, show that epidemiological situations are labile and prone to rapid changes. The recent invasion of the *B. tabaci,* 'B' biotype into Latin America, which seems to be adapting to cassava (Calvert and Thresh, 2002), might pose the risk of establishing CMBs in South America. This is the first time the vector of CMBs has infested cassava in Latin America. The 'B' biotype has recently been reported from India and associated with tomato leaf curl virus disease epidemics (Banks *et al.* 2001).

Other insects, i.e. *B. afer* in Africa and *Aleurodicus dispersus* in Asia have received little or no attention with regards to virus transmission capabilities. These whitefly species and others (*Aleurotrachelus socialis, Trialeurodes variabilis*), might have a role to play in the transmission of cassava virus diseases and continued research on vectors and viruses of cassava is needed.. Thus it will still take a considerable time before many of the basic findings translate into pragmatic solutions to support pest and disease management and crop improvement programmes.

Use of resistant cultivars, production and distribution of healthy planting material, improved cultural practices and eventually strategic use of transgenic crops could provide more sustainable solutions to cassava virus problems. Host plant resistance to viruses and vectors, eventually remain one of the most important means of disease control.

Strict statutory planting material certification schemes may not be practical or justifiable in many developing countries. Nevertheless, some developing countries are using tissue culture to 'clean up' and guarantee virus-free planting material. For producing planting material within a particular country, it is easier to adopt rapid multiplication of techniques and material selected for virus-freedom by vigorous selection.

Yield losses due to CMBs may increase if virulent strain(s) reach new areas by natural spread or through the movement of infected planting material. The occurrence of different viruses or virus combinations in different regions undermines the effectiveness of resistance breeding programmes. Until these issues are resolved, it is important not to move infected cassava between different countries or regions to avoid transfer of CMBs from Africa to the Indian sub-continent and *vice versa*, or to the Neotropics. Germplasm exchange is important for breeding purposes. However, it is essential to follow strict quarantine regulations when disseminating vegetative material. So far, there is no report of any virus of cassava being transmitted through seeds. Hence, currently this may be the safest way to exchange germplasm.

There is now considerable interest in transgenic research, and novel techniques for cassava virus control are being developed. Until recently, the progress in this field was impeded by the lack of a reliable cassava regeneration and transformation protocols. Now, considerable progress has been made in these areas (Fregene and Puonti-Kaerlas, 2002)..

While resistance identified in germplasm resources provide the backbone in breeding for resistance to CMD, pathogen-derived genes present an additional significant source of resistance. It is also important to ensure, that screening efforts consider other viral pathogens, e.g. CBSV, and target the appropriate host genotypes with the most desired traits. It is expected that these approaches together with conventional methods will ultimately yield cassava cultivars resistant to CMD.

A co-ordinated regional effort is required in order to make rapid and sustained progress (Legg *et al.* 1997). The resources in many developing countries are limited, especially for transgenic research, implementation of biosafety guidelines and molecular diagnostics. These efforts should be done on a regional basis and as a collaborative undertaking as for cowpea viruses in Africa (Thottappilly *et al.* 1993). The ultimate aim must be development of safe, economic and durable control strategies.

Acknowledgements

The authors are obliged to the following for their constructive criticisms of portions of this chapter in draft form: I. Dasgupta, V.T. John, G. Loebenstein, T. Makeshkumar, V.G. Malathi, A.V. Mathew, N.G. Nair, Jacob Mignouna, N.Y.C. Ng, M.S. Palaniswami, D.R. Mohan Raj, F. Ogbe, R. Rabindran and A. Schilder.

References

Abdullahi, I., Atiri, G., Winter, S. and Thottappilly, G. 1998. Differentiation of whitefly (*Bemisia tabaci*) biotypes from different regions in Africa and their relation to the occurrence of cassava mosaic geminiviruses (in German). *Mittl. Biolog. Bundesan.* Braunschweig 357: 284.

Abdullahi,I., Atiri,G.I., Dixon, A.G.O., Winter, S. and Thottappilly, G. 2003a. Effects of cassava genotype, climate and the *Bemisia tabaci* vector population on the development of African cassava mosaic geminivirus (ACMV). *Acta Agronom. Hungarica* 51(1): 37-46.

Abdullahi, I., Winter, S., Atiri, G. and Thottappilly, G. 2003b. Molecular characterization of whitefly, *Bemisia tabaci* (Hemiptera: *Aleyrodidae*) populations infesting cassava. *Bull. Entomol. Research* 93: 97-106.

Abraham, A. 1956. Tapioca cultivation in India. *Farm Bulletin No.* 17. Indian Council of Agricultural Research, New Delhi. 20 pp.

Adejare, G.O. and Coutts, R.H.A. 1981. Eradication of cassava mosaic disease from Nigeria cassava clones by meristem-tip culture. *Plant Cell Tissue Organ Culture* 1: 25-32.

Adejare, G.P. and Coutts, R.H.A.1982. Isolation and characterization of a virus from Nigeria Cassava plants affected by the cassava mosaic disease, and attempted transmission of the disease. *Phytopathol. Z.* 103: 198-210.

Adipala, A., Byabakama, B.A., Ogenga-Latigo, M.W. and Otim-Nape, G.W. 1998. Effect of planting date and varietal resistance on the development of cassava mosaic virus disease in Uganda. *African Plant Protection* 4: 71-79.

Aiton, M.M. and Harrison, B.D. 1988. Cassava Colombian symptomless virus (CCSV), In: *Annual Report 1987*, p. 192, Scottish Crops Research Institute, Dundee.

Aiton, M.M., Lennon, A.M., Roberts, I.M. and Harrison, B.D. 1988a. Two new cassava viruses from Africa. In: *Abstracts 5th Inter. Congr. Plant Pathol.*, Kyoto, p. 43.

Aiton, M.M., Roberts, I.M. and Harrison, B.D. 1988b. Cassava common mosaic potexvirus from mosaic-affected cassava in the Ivory Coast. *Ann. Rept.* 1987, p. 191, Scottish Crop Research Institute, Dundee.

Akano, A., Ng, S.Y.C., Asiedu, R. and Atiri, G. 1995. Performance in three agroecologies of virus-tested cassava genotypes derived from meristem culture. In: *Proc. sixth Triennial Symp. Intern. Soc. Tropical Root Crops*-Africa Branch, M.O. Akoroda and I.J. Ekanayake (eds.), pp. 194-198.

Akano, A.O., Atiri, G.I., Ng, S.Y.C. and Asiedu, R. 1997. Effect of African cassava mosaic disease on growth and yield components of virus-tested cassava genotypes derived from meristem culture in early and late planting periods in three agroecologies of Nigeria. *African J. Root and Tuber Crops* 2: 44-48.

Akano, A.O., Dixon, A.G.O., Mba, C., Barrera, E. and Fregene. M. 2002. Genetic mapping of a dominant gene conferring resistance to cassava mosaic disease. *Theor. Appl. Genetics* 105:521-525.

Alagianagalingam, M.N. and Ramakrishnan, K. 1966. Cassava mosaic in India. *South Indian Horticulture* 14: 71-72.

Angel, J.C., Pineda, B.L., Nolt, B. and Velasco, A.C. 1990. Mosca blancas (Homoptera: Aleyrodidae) asociadas a transmision de virus en yuca. *Fitopatol. Colomb.* 13: 65-71.

Ariyo, O., Dixon, A.G.O. and Atiri, G.I. 2002. The relative resistance of cassava cultivars to African cassava mosaic disease (ACMD) as determined by two methods: Rank-sum and the area under diseased progress curve. *Arch. Phytopath. Pflanz.* 35: 23-20.

Atiri, G.I., Njukeng, A.P. and Ekpo, E.J.A. 2000. Climate in relation to plant virus epidemiology and sustainable disease management in West Africa. *J. Sust. Agric.*16: 17-30.

Banks, G.K., Colvin, J., Chowda Reddy, R.V., Maruthi, M.N., Muniyappa, V., Venkatesh, H.J., Kiran Kumar, M., Padmaja, A.S., Beitia, F.J. and Seal, S.E. 2001. First report of the

Bemisia tabaci B biotype in India and an associated tomato leaf curl virus disease epidemic. *Plant Dis.* 85: 231.

Bellotti, A.C., Bohorquez, A., Arias, B., Mba, C. and Tohme, J. 2001. Host plant resistance for management of cassava pest with emphasis on whiteflies. In: *Cassava, An Ancient Crop For Modern Times: Food, health, culture*, N.J. Taylor, F. Ogbe and C.M. Fauquet (eds.), pp. S8-02, Donald Danforth Plant Sciences Center, St. Louis, USA.

Berrie, L.C., Palmer, K.E, Rybicki, E.P., Hiyadat, S.H., Maxwell, D.P. and Rey, M.E.C. 1997. A new isolate of African cassava mosaic virus in South Africa. *African J. Roots and Tuber Crops* 2: 49-52.

Berrie L.C., Palmer K.E., Rybicki E.P. and Rey M.E.C. 1998. Molecular characterisation of a distinct South African cassava infecting geminivirus. *Arch. Virol.* 143: 2253-2260.

Berrie, L.C., Rybicki, E.P. and Rey, M.E.C. 2001. Complete nucleotide sequence and host range of *South African cassava mosaic virus*: further evidence for recombination amongst begomoviruses. *J. Gen. Virol.* 82:53-58.

Berry, S. and Rey, M.E.C. 2001. Differentiation of cassava-infecting begomoviruses using heteroduplex mobility assays. *J. Virol. Methods* 92: 151-163.

Bock, K.R., 1994a. The spread of African cassava mosaic geminiviruses in coastal and western Kenya. *Trop. Science* 34: 92-101.

Bock, K.R. 1994b. Control of African cassava mosaic geminivirus by using virus-free planting material. *Trop. Science* 34: 102-109.

Bock, K.R. 1994c. Studies on cassava brown streak disease in Kenya. *Trop. Science* 34: 134-145.

Bock, K.R. and Harrison, B.D. 1985. African cassava mosaic virus. In: *Descriptions of Plant viruses.* No. 297. Assoc. Applied Biologists, Wellesborne, Warwick, UK, 6pp.

Bock, K.R. and Woods, R.D. 1983. Etiology of African cassava mosaic disease. *Plant Dis.* 67: 994-995.

Bock, K.R., Guthrie, E.J., Meredith, G. and Barker, H. 1977. RNA and protein components of maize streak and cassava latent viruses. *Ann. appl. Biol.* 85: 305-308.

Bock, K.R., Guthrie, E.J. and Meredith, G. 1978. Distribution, host range, properties and purification of cassava latent virus, a geminivirus. *Ann. appl. Biol.* 90: 361-367

Bock, K.R., Guthrie, E. J. and Figueriedo, G. 1981. A strain of cassava latent virus occurring in coastal districts of Kenya. *Ann. appl. Biol.* 99: 151-159.

Bolhuis, G.C. 1949. Waarnemingen ouer de zg mosaiekziekte big cassava op Java. *Buitenzoig, Java General Agricultural Research Communication* No.92, 9pp.

Briddon, R.W., Liu, S. M.S. Pinner, S.M.S. and Markham, P.G. 1998. Infectivity of *African cassava mosaic virus* clones to cassava by biolistic inoculation. *Arch. Virol.* 143: 2487-249.

Brunt, A.A. 1990. Cassava brown streak carlavirus. Cassava brown streak potyvirus. In *Viruses of Tropical Plants.* A. Brunt, K. Crabtree and A. Gibbs (eds.). pp. 157-158, CAB International, Wallingford, UK.,707 pp.

Brunt A.A., Crabtree, K., Dallwitz, M.J., Gibbs, A.J. and Watson, L. 1996. *Viruses of plants.* Descriptions and lists from the vide database. CAB International, Oxon, 1484pp.

Bull, S.E., Karakacha, H.W., Briddon, R.W., Nzioki, S., Maruthi, M.N., Stanley, J. and Winter, S. 2003. Occurrence of *East African cassava mosaic Zanzibar virus* (EACMVZV) in coastal Kenya. New Disease Reports, British Society of Plant Pathology, Vol. 7. http://www.bspp.org.uk/ndr/July

Burban, C., Fishpool, L.D.C., Fauquet, D., Fargette, D. and Thouvenel, J.-C. 1992. Host-associated biotypes within West African populations of the whitefly *Bemisia tabaci* (Genn.) (Hom., Aleyrodidae). *J. appl. Entomol.* 113: 416-423.

Calvert, L.A. 1994. The safe movement of cassava germplasm. In: *International network for cassava genetic resources.* Report of the first meeting of the International Network for Cassava Genetic Resources, held at Cali, Colombia, 18-23 August, 1992. International

Plant Genetic Resources Institute, Rome. (*International crop network series 10*). pp.163-165.

Calvert, L. and Thresh, J.M. (2002) The viruses and virus diseases of cassava. In: *Cassava: Biology, Production and Utilization*. R.J. Hillocks, J.M. Thresh and A. Bellotti (eds.), pp. 237-260, *CAB International*, Wallingford, UK.

Calvert, L.A., Ospino, M.D. and Shepherd, R.J. 1995. Characterization of cassava vein mosaic virus: a distinct plant pararetrovirus. *J. Gen. Virol.* 76, 1271-1278.

Calvert, L.A., Cuervo, M., Ospina, M.D., Fauquet, C. and Ramirez, B.C. 1996. Characterization of cassava common mosaic virus and a defective RNA species. *J. Gen. Virol.* 77, 525-530.

Carter, S.E., Fresco, L.O. and Jones, P.G. 1992. *An atlas of cassava in Africa: historical, agroecological and demographic aspects of crop distribution:* Centro Internacional de Agricultura e Tropical (CIAT), Cali, Colombia, 86pp.

Chant, S.R. 1958. Studies on the transmission of cassava mosaic virus by *Bemisia* spp (Aleyrodidae). *Ann. appl. Biol.* 46: 210-215.

Chant, S.R. 1959. A note on the inactivation of mosaic virus in *cassava (Manihot utilissima Pohl.)* by heat treatment. *Emp. J. Exp. Agric.* 27: 55-58.

Chaparro-Martinez, E.I. and Trujillo-Pinto, G. 2001. First report on cassava virus x in cassava in Venezula. *Plant Dis.* 85:1119.

Chen, C.T., Ko, N.J. and Chen, M.J. 1981. Electron microscopy of cassava common mosaic in Taiwan. *Rep. Taiwan Sugar Res. Insp.* 93: 20-27.

Colvin, J., Fishpool, L.D.C., Fargette, D., Sherington, J. and Fauquet, C. 1998. *Bemisia tabaci* (Hemiptera: Aleyrodidae) trap catches in a cassava field in Côte d'Ivoire in relation to environmental factors and the distribution of the African cassava mosaic disease. *Bull. Entomol. Research* 88: 369-378.

Costa, A.S. 1940. Observacoes sobre o mosaico comum e o mosaico das nervuras da mandioca (*Manihot utilissima* Pohl.). *J. Agronomia (Piracicaba)* 3: 239-248.

Costa, A.S. and Kitajima, E.W. 1972a. Cassava common mosaic virus. In: *CMI/AAB Descriptions of Plant Viruses* No. 90., Commonwealth Agricultural Bureaux, Slough,4pp.

Costa, A.S. and Kitajima, E.W. 1972b. Studies on virus and mycoplasma diseases of the cassava plant in Brasil. Page 18, In: *Proc. Cassava Mosaic Workshop,* IITA, Ibadan, Nigeria.

Cours, G. 1951. Le Manioc a Madagascar. Memoires de l'Institute Scientifique de Madagascar. *Serie B, Biologie Vegetale* 3: 203-400.

Cours, G., Fargette, D., Otim-Nape, G.W. and Thresh, J.M. 1997. The epidemic of cassava mosaic virus disease in Madagascar in the 1930s-1960s: lessons for the current situation in Uganda. *Tropical Science* 37: 238-248.

Cours-Darne, G. 1968. Improving cassava in Africa. The Abidjan Conference. *Agricultural Research Priorities for Economic Development in Africa* 2: 330-339.

CTCRI, 2000. Tuber crops varieties released by CTCRI. *Technical Bulletin Series:24.* Central Tubers Crop Research Institute, Trivandrum, Kerala. 40 pp.

CTCRI, 2002a. Annual Report 2000-2001. Central Tubers Crop Research Institute, Trivandrum, Kerala. 167 pp.

CTCRI, 2002b. Research Highlights 2001-2002. Central Tuber Crops Research Institute, Trivandrum, Kerala, 26pp.

Dasgupta,I., Malathi, V.G. and Mukherjee, S.K. 2003. Genetic engineering for virus resistance. *Current Science,* 84: 341-354.

Deng, D., McGrath, D.F., Robinson, D.J. and Harrison, B.D. 1994. Detection and differentiation of whitefly-transmitted geminiviruses in plants and vector insects by the polymerase chain reaction with degenerate primers. *Ann. appl. Biol.* 125: 327-336.

Deng, D., Otim-Nape, G.W., Sangare, A., Ogwal, S., Beachy, R.N. and Fauquet, C.M. 1997. Presence of a new virus closely related to East African cassava mosaic geminivirus associated with cassava mosaic outbreak in Uganda. *Afric. J. Root Tuber Crops* 2: 23-28.

Dixon, A.G.O., Whyte, J.B.A., Mahungu, N.M. and Ng, S.Y.C. 2001. Tackling the cassava mosaic disease (CMD) challenge in sub-saharan Africa: The role of host-plant resistance and germplasm deployment. In: *Cassava, an ancient crop for modern times: Food, health, culture*, N.J. Taylor, F. Ogbe and C.M. Fauquet (eds.). pp. S8-05, Donald Danforth Plant Sciences Center, St. Louis, USA.

Dubern, J. 1994. Transmission of African cassava mosaic geminivirus by the whitefly *(Bemisia tabaci). Trop. Science* 34: 82-91.

Edison, S. 2000. Tuber crops–scope for value addition. In: *The Hindu Survey of Indian agriculture* 2000, pp. 77-82, Hindu, Chennai, India.

Edison, S. 2002. Tuber crops: Energy and nutritional storehouses. In: *The Hindu Survey of Indian agriculture* 2002, pp. 81-86, Hindu, Chennai, India.

Etessami, P., Watts, J.and Stanley, J. 1989. Size reversion of African cassava mosaic virus coat protein gene deletion mutants during infection of *Nicotiana benthamiana. J.Gen.Virol.* 70: 277-289.

FAOSTAT, 2001; www.fao.org

Fargette, D. and Fauquet, C. 1988. A preliminary study on the influence of intercropping maize and cassava on the spread of African cassava mosaic virus by whiteflies. *Aspects Appl. Biol.* 17: 195-202.

Fargette, D. and Vie, K. 1995. Simulation of the effects of host resistance, reversion and cutting selection on incidence of cassava mosaic virus and yield losses in casava. *Phytopathology* 85: 370-375.

Fargette, D., Fauquet, C. and Thouvenel, J.C. 1985. Field studies on the spread of African cassava mosaic. *Ann. appl. Biol.* 106: 285-294.

Fargette, D., Thouvenel, J.C. and Fauquet, C. 1987. Virus content of leaves of cassava infected *by* African cassava mosaic virus. *Ann. appl. Biol.* 110: 65-73.

Fargette, D., Fauquet, C. and Thouvenel, J.C. 1988. Yield losses induced by African cassava mosaic virus in relation to the mode and the date of infection. *Trop. Pest Management* 34: 89-91.

Fargette, D., Fauquet, C., Grenier, E. and Thresh, J.M. 1990. The spread of African cassava mosaic virus into and within cassava fields. *J. Phytopathol.* 130: 289-302.

Fargette, D., Roberts, I.M. and Harrison, B.D. 1991. Particle purification and properties of cassava Ivorian bacilliform virus. *Ann. appl. Biol.* 119: 303-312.

Fargette, D., Jeger, M., Fauquet, C. and Fishpool, L.D.C. 1994a. Analysis of temporal disease progress of cassava mosaic virus. *Phytopathology* 84: 91-98.

Fargette, D., Thresh, J.M. and Otim-Nape, G.W. 1994b. The epidemiology of African cassava mosaic geminivirus: reversion and the concept of equilibrium. *Trop. Science* 34: 123-133.

Fargette, D., Colon, L.T., Bouveau, R. and Fauquet, C. 1996. Components of resistance of cassava to African casava mosaic virus. *European J. Plant Pathol.* 102: 645-654.

Fauquet, C. and Fargette, D. 1990. African cassava mosaic virus; etiology, epidemiology and control. *Plant Dis.* 74: 404-411.

Fauquet, C., Fargette, D. and Thouvenel, J.C. 1988. Some aspects of the epidemiology of African cassava mosaic virus in Ivory Coast. *Trop. Pest Management* 34: 92-96.

Fauquet, C., Bogusz, D., Chavarriaga, P., Franche, C., Schopke, C. and Beachy, R.N. 1992. Cassava viruses and genetic engineering. In: *Biotechnology: Enhancing Research on Tropical Crops in Africa*, G. Thottappilly, L.M. Monti, D.R. Mohan Raj and A.W. Moore (eds.). pp. 287-296 , CTA/I.I.T.A. co-publication, IITA, Ibadan, Nigeria.

Fauquet, C.M., Bisaro, D.M., Briddon, R.W., Brown, J.K., Harrison, B.D., Rybicki, E.P., Stenger, D.C. and Stanley, J. 2003. Revision of taxonomic criteria for species demarcation in the family Geminiviridae, and an updated list of begomovirus species. *Arch Virol.* 148: 405-421.

Fishpool, L.D.C. and Burban, C. 1994. *Bemisia tabaci*: the whitefly vector of African cassava mosaic geminivirus. *Trop. Science* 34: 55-72.

Fishpool, L.D.C., Fauquet, C., Fargette, D., Thouvenel, J.C., Burban, C. and Colvin, J. 1995. The phenology of *Bemisia tabaci* (Homoptera: Aleyodidae) populations on cassava in southern Côte d'Ivoire. *Bull. Ent. Research* 85: 197-207.

Fokunang, C. N.., Akem, C., Dixon, A.G.O. and Ikotun, T. 2000. Evaluation of a cassava germplasm collection for reaction to three major diseases and the effect on yield. *Genetic Resour. Crop Evolution* 47: 63-71.

Fondong, V.N., Adipala, E. and Thresh, J.M. 1997. Spread of African cassava mosaic geminivirus at four agroecological locations of Cameroon. *African Crop Science Conf. Proc.* 3: 1035-1049.

Fondong, V.N., Pita, F.S., Rey, C., Bechy, R.N. and Fauquet, C.M. 1998. First report of the presence of East African cassava mosaic virus in Cameroon. *Plant Dis.* 82: 1172.

Fondong, V.N., Pita, J.S., Rey, M.E.C., de Kochko, A., Beachy, R.N. and Fauquet, C.M. 2000a. Evidence of synergism between African cassava mosaic virus and a new double-recombinant geminivirus infecting cassava in Cameroon. *J. Gen. Virol.* 81: 287-297.

Fondong, V.N. Thresh, J.M. and Fauquet, C. 2000b. Field experiments in Cameroon on cassava mosaic virus disease and the reversion phenomenon in susceptible and resistant cassava cultivars. *Internat. J. Pest Management* 4: 211-217.

Fondong, V.N., Thresh, J.M. and Zok, S. 2002. Spatial and temporal spread of cassava mosaic virus disease in cassava grown alone and intercropped with maize and/or cowpea. *J. Phytopath.* 150: 365-374.

Fregene, M. and J. Puonti-Kaerlas. 2002. Cassava Biotechnology. In: *Cassava: Biology, Production and Utilization.* R.J. Hillocks,J.M. Thresh and A.C. Bellotto (eds.), Chapter 10, pp. 179-207. CABI Publishing, Wallingford, UK.

Fregene, M., Bernal, A., Duque, M., Dixon, A. and Tohme, J. 2000. AFLP analysis of African cassava (*Manihot esculenta* Crantz) germplasm resistant to cassava mosaic disease (CMD). *Theo. Appl. Genetics* 100: 678-685.

Fregene, M., Matsumura, H., Akano, A., Dixon, A. and Terauchi, R. 2001a. Progress towards cloning the single dominant gene conferrring immunity to cassava mosaic disease (CMD). In: *Cassava, an ancient crop for modern times: Food, health, culture,* N.J. Taylor, F. Ogbe and C.M. Fauquet (eds), pp. S5-09, Donald Danforth Plant Sciences Center, St. Louis, USA.

Fregene, M., Okogbenin, E., Mba, C., Angel, F., Suarez, M.C., Janneth, G., Chavarriaga, P., Roca, W., Bonierbale, M. and Tohme J. 2001b. Genome mapping in cassava improvement: Challenges, achievements and opportunities. *Euphytica* 120: 159-165.

Frison, E.A. and Feliu, E. (eds.). 1991. *FAO/IBPGR Technical guidelines for the safe movement of cassava germplasm.* FAO, Rome. 48pp.

Gibson, R.W. and Otim-Nape, G.W. 1997. Factors determining recovery and reversion in mosaic-affected African cassava mosaic virus resistant cassava. *Ann. appl. Biol.* 131:259-271.

Hahn, S.K., Terry, E.R. and Leuschner, K. 1980. Breeding cassava for resistance to cassava mosaic disease. *Euphytica* 29: 673-683.

Hahn, S.K., John, C., Isoba, C. and Ikotun, T. 1989. Resistance breeding in root crops at the International Institute of Tropical Agriculture (IITA), Ibadan, Nigeria. *Crop Protection* 8: 147-168.

Haley, A., Richardson, K., Zhan, X. and Morris, B. 1995. Mutagenesis of the BC1 and BV1 genes of African cassava mosaic virus identifies conserved amino acids that are essential for spread. *J. Gen. Virol.* 76: 1291-1298.

Harrison, B.D. 1985. Advances in geminivirus research. *Ann. Rev. Phytopathology* 23: 55-82.

Harrison, B.D. 1987. Properties and geographical variation of geminivirus isolates from mosaic affected cassava. In: *International Seminar on African Casava Mosaic Disease and its control*. pp 36-42, CTA, FAO, ORSTOM, IITA, IAPC publication, Yamoussoukro, Côte d'lvoire.

Harrison, B.D. and D.J. Robinson, 1999. Natural genomic and antigenic variation in whitefly-transmitted geminiviruses (begomoviruses). *Annu. Rev. Phytopathol.* 37: 369-389.

Harrison, B.D., Barker, H., Bock, K.R., Guthrie, E.J. and Meredith, G., and Atkinson, M. 1977. Plant viruses with circular single-stranded DNA. *Nature* 270: 760-762.

Harrison, B.D., Swanson, M.M., McGrath; P.F and Fargette, D. 1991. Patterns of antigenic variation in whitefly-transmitted geminiviruses. Pages 88-90, *Rept Scottish Crop Res Insititute 1990,* Dundee.

Harrison, B.D., Swanson, M.M., and Robinson, D.J. 1995. Cassava viruses in the Old World. In: *Proc. Sec. Intern. Scientific Meeting Casssava Biotechnology Network*, Bogor, Indonesia, 22-26 August 1994. pp. 463-472, Working Document No. 150. CBN/CRIFC/ AARD/CIBT.

Harrison, B.D., Liu, Y.L., Zhou, X., Robinson, D.J., Calvert, L., Munoz, C. and Otim-Nape, G.W. 1997a. Properties, differentiation and geographical distribution of geminiviruses that cause cassava mosaic disease. *Afric. J. Root and Tuber Crops* 2: 19-22.

Harrison, B.D., Zhou, X., Otim-Nape, G.W., Liu, Y. and Robinson, D.J. 1997b. Role of a novel type of double infection in the geminivirus-induced epidemic of severe cassava mosaic in Uganda. *Ann. appl. Biol.* 131: 137- 148.

Harrison, B.D., Swanson, M.M. and Fargette, D. 2002. Begomovirus coat protein: serology, variations and functions. *Physiol. Molec. Plant Pathol.* 60: 257-271.

Hernández, A., Calderón, H., Zárate, R.D. and Lozano, J.C. 1975. El cuero de sapo de la yuca (*Manihot esculenta* Crantz). *Noticias Fitopatológicas* 4: 117-118.

Hillocks, R.J., Raya, M.D. and Thresh, J.M. 1999. Distribution and symptom expression of cassava brown streak disease in southern Tanzania. *Afric. J. Root and Tuber Crops* 3: 57-62.

Hillocks, R.J., Raya, M.D., Mtunda, K. and Kiozia, H. 2001. Effects of brown streak virus disease on yield and quality of cassava in Tanzania. *J. Phytopathol.* 149: 389 391.

Hillocks, R.J., Thresh, J.M., Tomas, J., Botao, M., Macia, R. and Zavier, R. 2002. Cassava brown streak disease in northern Mozambique. *Intern. J. Pest Management* 48: 178-181.

Hong, Y. and Stanley, J. 1996. Virus resistance in *Nicotiana benthamiana* conferred by African cassava mosaic virus replication-associated protein (AC1) transgene. *Molec. Plant-Mocrobe Interact.* 9: 219-225.

Hong, Y.G., Robinson, D.J. and Harrison, B.D. 1993. Nucleotide sequence evidence for the occurrence of three distinct whitefly-transmitted geminiviruses in cassava. *J. Gen. Virol.* 74: 2437-244.

Hrishi, N., Rajendran, P.G., Kamalam, P., Maini, S.B., Nair, N.G., and Lal, S.S. 1977. Genetic improvement of cassava for yield and quality: Screening hybrids/varieties against cassava mosaic disease. *Annu. Report,* pp. 15-16, Central Tuber Crops Res. Inst., Trivandrum, India.

Hull, R. and Davies, J.W. 1992. Approaches to nonconventional control of plant virus diseases. *Critic. Rev. Plant Science.* 11: 17-33.

IITA, 1986. Infection rate of selected cassava varieties by cassava mosaic virus. In: *IITA Annual Report and Research Highlights 1985*. P. 111, IITA, Ibadan, Nigeria.

Jameson, J.D. 1964. Cassava mosaic disease in Uganda. *East Afric. Agric. Forestry J.* 29: 208-213.

Jayasinghe, U., Pineda, B. and Lozano, J.C. 1984. Antholysis in cassava (*Manihot esculenta* Crantz.,) possibly caused by mycoplasma like organisms. *Phytopathol. Z.* 109: 295-300.

Jeeva, M.L. 1997. Effect of secondary infection of cassava mosaic disease on growth and tuber yield of cassava. *J. Mycol. Pl. Pathol.* 27: 78-80.

Jennings, D.L. 1957. Further studies in breeding cassava for virus resistance. *East Afr. Agric. J.* 22: 213-219.

Jennings, D.L. 1960. Observations on virus diseases of cassava in resistant and susceptible varieties: I mosaic disease. *Emp. J. Experimental Agric.* 28: 23-34.
Jennings, D.L. 1994. Breeding for resistance to African cassava mosaic geminivirus in East Africa. *Trop. Science* 34: 110-122.
Kaiser, W.J. and Louie, R. 1982. Heat therapy of cassava infected with African cassava mosaic disease. *Plant Dis.* 66: 475-477.
Kaiser, W.J. and Teemba, L.R. 1979. Use of tissue culture and thermotherapy to free East African cassava cultivars of African cassava mosaic and cassava brown streak diseases. *Plant Dis. Reptr.* 63: 780-784.
Karakacha, H.W. 2001. Serological and molecular characterization of begomoviruses infecting cassava (*Manihot esculenta* Crantz) in Africa. Ph.D thesis, Univ. Hannover, 166pp.
Karakacha, H.W., Koerbler, M., Epampuka, P.B., Ayecho, P., and Winter, S. 2001. Characterization and distribution of begomoviruses causing - Uganda variant - cassava mosaic virus disease in East and Central Africa. *Proc. of the 3rd Geminivirus Symposium,* Norwich, UK, 24-28.July 2001, p. 115.
Kisgeci, J. 1989. Agricultural biomass for energy, CNRE study No. 4, University of Novi Sad. FAO 21: 58-59, Novi Sad.
Kitajima, E.W. and Costa, A.S. 1964. Elongated particles found associated with cassava brown streak. *East Afr. Agric. Farming J.* 29: 28-30.
Kitajima, E.W. and Costa, A.S. 1966. Particulas esferoidais associadas ao virus do mosaico das nervuras da mandioca. *Bragantia* 25: 211-221.
Kitajima, E.W., Wetter, C., Oliveira, A.R., Silva, D.M. and Costa, A.S. 1965. Morfologia do virus do mosaico comum da mandioca. *Bragantia* 24: 247-260.
Lal, S.S. and Pillai, K.S. 1977. *Annual Report.* Cent. Tuber Crops Res. Inst. Trivandrum. pp. 67-82.
Lal, S.S., and Pillai, K.S., 1982. Ecological studies on whitefly *Bemisia tabaci* (Genn.) infesting cassava in Kerala. *Entomon* 7: 101-102.
Legg, J.P. 1994. Bemisia tabaci: the whitefly vector of cassava mosaic geminiviruses in Africa: An ecological perspective. *African Crop Science J.* 2: 437-448.
Legg, J.P. 1999. Emergence, spread and strategies for controlling the pandemic of cassava mosaic virus disease and east and central Africa. *Crop Protect.* 18: 627-637.
Legg, J.P. and Raya, M.D, 1998. Survey of cassava virus diseases in Tanzania. *Int. J. Pest Management* 44: 17-23.
Legg, J.P. and Thresh, J.M. 2000. Cassava mosaic disease in East Africa: a dynamic disease in a changing environment. *Virus Res.* 71: 135-149.
Legg, J.P., Gibson, R.W. and Otim-Nape, G.W. 1994. Genetic polymorphism amongst Ugandan populations of *Bemisia tabaci* (Gennadius) (Homoptera: Aleyrodidae), vector of African cassava mosaic geminivirus. *Trop. Science* 34: 73-81.
Legg, J.P., James, B., Cudjoe, A., Saizonou, S., Gbaguidi, B., Ogbe, F., Ntonifor, N., Ogwal, S., Thresh, J.M. and Hughes, J. 1997. A regional collaborative approach to the study of ACMD epidemiology in sub-Saharan Africa. *Afri. Crop Sci. Conf. Proc.* 3: 1021-1033.
Legg, J.P., Okao-Okuja, G., Mayala, R. and Muhinyuza, J.B. 2001a. Spread into Rwanda of the severe cassava mosaic virus disease pandemic and associated variant of East African cassava mosaic virus (EACMV-UG). *Plant Pathol.* 50: 796.
Legg, J.P., Owor, B. and Okao-Okuja, G. 2001b. PCR-based diagnostics characterize the effects of cassava mosaic geminiviruses on yield and cassava mosaic disease epidemiology in Uganda. In: *Cassava, an ancient crop for modern times: Food, health, culture,* N.J. Taylor, Ogbe, F. and Fauquet, C.M. (eds), p. S8-09, Donald Danforth Plant Sciences Center, St. Louis, USA.
Legg, J.P., French, R., Rogan, D., Okao-Okuja, G., and Brown, J. 2002. A distinct *Bemisia tabaci* (Gennadius) (Hemiptera: Sternorrhyncha: Aleyrodidae) genotype cluster is

associated with the epidemic of severe cassava mosaic virus disease in Uganda. *Molec. Ecol.* 11: 1219-1229.
Lennon A.M., Aiton, M.M. and Harrison, B.D. 1986a. Cassava viruses from Africa: Third country quarantine of cassava. In: *Annual Report 1985*, p 168. Scottish Crop Res. Inst., Dundee, Scotland.
Lennon, A.M., Aiton, M.M. and Harrison, B.D. 1986b. Cassava viruses from South America. In: *Annual Report 1985*. p. 167. Scottish Crop Res. Inst., Dundee, Scotland.
Lennon, A.M., Aiton, M.M. and Harrison, B.D. 1987. Purification and properties of cassava green mottle, a previously undescribed virus from the Solomon Islands. *Ann. appl. Biol.* 110: 545-555.
Leuschner, K. 1977. Whiteflies: Biology and transmission of African cassava mosaic disease. In: *Proceedings Cassava Protection Workshop*, T. Brekelbaum, A. Belloti and J.C. Lozano (eds.). pp.51-58, CIAT, Cali, Colombia.
Lin, M.T. and Kitajima, E.W. 1980. Purificão e serologia do virus de mosaico das nervaduras da mandioca. *Fitopatol. Bras.* 5: 419. (Abstr.)
Lisha, S.V., Antony, B., Palanisawmi, M.S. and Henneberry, T.J. 2003. *Bemisia tabaci* (Homoptera: Aleyrodidae) biotypes in India. *J.Econ. Entomol. 96(2): 322-327.*
Lister, R.M. 1959. Mechanical transmission of cassava brown streak virus. *Nature* (London) 183: 1588-1589.
Liu, S., Bedford, I.D., Briddon, R.W. and Markham, P.G. 1997. Efficient whitefly transmission of African Casava mosaic geminivirus requires sequences from both genomic components. *J. Gen. Viorol.* 78: 1791-1794.
Lozano, J.C. and Nolt, B.L. 1989. Pest and pathogens of cassava. In: *Plant Protection and Quarantine: Selected Pests and Pathogens of Quarantine Significance* Vol. 2 R.P. Kahn, ed., pp. 174-175, CRC Press, Boca Raton, Florida.
Mabanza, J., A.V. Rodriguez-Andriyamasi, J. Mahouka and B. Boumba. 1995. Evaluation of cleaned cassava varieties in Congo. In: *Proc. Second Intern. Scientific Meeting Cassava Biotechnology Network*, Vol. 1, pp.194-201, 22-26 August 1994, Bogor, Indonesia,CBN/CRIFC/AARD/CIAT.
Maffla, G., Roa, J.C. and Roca, W.M. 1984. Erradicacion de la enfermedad cuero de sapo de la yuca Manihot esculenta, pro medio del dultive de meristemos. Efecto de la termoterapia, y del tamano del explante sobre la tasa de saneamiento. In: *Congreso Nacional de Cultivo de Tejidos Vegetales.* D. Perea and Z.A Angarita (eds.). pp. 171-175, Bogota, Colombia, 1984. Memorias. Bogoata, U. Nacional de Colombia.
Mahto, D.N. and Sinha, D.C. 1978. Evaluation of insecticides for the control of whitefly, *Bemisia tabaci* Genn. in relation to the incidence of mosaic of cassava. *Indian J. Entomol.* 40: 316-319.
Mahungu, N.M., Dixon, A.G.O. and Kumbira, 1994. Breeding cassava for multiple pest resistance in Africa. *Afric. Crop Science J.* 2: 539-552.
Malathi, V.G. and Sreenivasan, M.A. 1983. Association of gemini particles with cassava mosaic disease in India. *J. Root Crops* 9: 69-73.
Malathi, V.G., Nair, N.G. and Shantha, P. 1985. Cassava mosaic disease. *Technical Bulletin Series 5.* Central Tuber Crops Research Institute, Trivandrum. 18 pp.
Malathi, V.G., Thankappan, M., Nair, N.G., Nambisan, B. and Ghosh, S.P. 1987. Cassava mosaic disease in India. In: *Proc. Intern. Seminar African mosaic disease and its control,* C. Fauquet and D. Fargette (eds.), pp. 189-198, Yamoussoukro, Cote d'lvoire, 4-8 May 1987. CTA, FAO, ORSTOM, IITA, IAPC publication.
Malathi, V.G., Varma, A., and Nambisan, B. 1989. Detection of Indian cassava mosaic virus by ELISA. *Curr. Science* 58: 149-150.
Maruthi, M.N., Colvin, J. and Seal, S. 2001. Mating compatibility, life history traits and RAPD-PCR variation in *Bemisia tabaci* associated with the cassava mosaic disease pandemic in East Africa. *Ent. Exp. Applicata* 99:13-23.

Maruthi, M.N., Colvin, J., Seal, S. and Thresh, J.M. 2002a. First report of a distinct begomovirus infecting cassava from Zanzibar. *Plant Dis.* 86: 187.

Maruthi, M.N., Colvin, J., Seal, S., Gibson, G. and Cooper, J. 2002b. Co-adapation between cassava mosaic geminiviruses and their local vector populations. *Virus Res.* 86: 71-85.

Mathew, A.V. 1989. Studies on Indian cassava mosaic virus disease. Ph.D Thesis, *Univ. Agric. Sciences,* Bangalore, 237 pp.

Mathew, A.V. and Muniyappa, V. 1992. Purification and characterisation of Indian cassava mosaic virus. *Phytopathology* 135:299-308.

Mathew, A.V. and Muniyappa, V. 1993. Host range of Indian cassava mosaic virus. *Indian Phytopathology* 46: 16-23.

Mba, R.E.C. and Dixon, A.G.O. 1997. Heritability estimates for some newly developed cassava clones in Nigeria. *Afric. J. Root Tuber Crops* 2: 56-58.

Monger, W.A., Seal, S., Isaac, A.M. and Foster, G.D. 2001a. Molecular characterization of *Cassava brown streak virus* coat protein. *Plant Pathol.* 50: 527-534.

Monger, W.A., Seal, S., Cotton, S. and Foster, G.D. (2001b). Identification of different isolates of *Cassava brown streak virus* and development of a diagnostic test. *Plant Pathol.* 50, 768-775.

Mtunda, K., Mahungu, N.M., Thresh, J.M., Kilima, M.S.and Kiozya, H.C. 1999. Cassava planting material sanitation for the control of cassava brown streak disease. In: *Food security and crop diversification in SADC countries: the role of cassava and sweet potato. Proceedings of the Scientific Workshop of Southern African Root Crops Network,* M.O. Akoroda and J.M.Teri (eds.), pp. 300-304 , SADC/IITA/CIP, 476pp.

Muller, H.R.A. 1931. Mozaiekziekte bij cassava. *Institut voor Plantenziekten Bull.* 24: 17.

Nair, N.G. 1975. Transmission studies on CMD using whitefly *Bemisia tabaci* Genn. *Annual Report,* Central Tuber Crops Research Institute, Trivandrum. pp. 70

Nair, N.G. 1977. Transmission of cassava mosaic virus. *Annual Report.* Central Tuber Crops Research Institute, Trivandrum, pp 61.

Nair, N.G. 1981. Relationship between cassava mosaic disease spread and whitefly (*Bemisia tabaci* Gen.) population under different insecticide treatments. *J. Root Crops* 7: 15-19.

Nair, N.G. 1985. Nature and extent of spread of cassava mosaic disease. In: *Proceedings of the National Symposium on Production and Utlization of Tuber Crops,* held on Nov. 27-29 1985, CTCRI, Trivandrum and published by Indian Society of Root Crops, pp. 175-178, CTCRI, Trivandrum, India.

Nair, N.G. 1988. Influence of inoculum source and varietal susceptibility on the field spread of Indian cassava mosaic disease. *J. Root Crops* 14: 5-9.

Nair, N.G. 1990. Performance of virus-free cassava *(Manihot esculenta* Crantz) developed through meristem tip culture. *J. Root Crops* 16: 123-131.

Nair, N.G. and Daniel, R.S. 1983. Preference of *Bemisia tabaci* (Genn.) to cassava varieties and their reaction to cassava mosaic disease. *J. Root Crops.* 9 (1-2): 45-49.

Nair, N.G. and Malathi, V.G. 1987. Disease severity and yield loss in cassava (*Manihot esculenta* Crantz.) due to Indian cassava mosaic disease. *J. Root Crops* 13: 91-94.

Nair, N.G. and Thankappan, M. 1990. Spread of Indian cassava mosaic disease under different agroclimatic locations. *J. Root Crops*. ISRC Nat. Sym. Special Volume, 216-219.

Narasimhan, V. and Arjunan, G. 1976. Mosaic disease of cassava – loss in yield and tuber splitting. *Indian Phytopathol.* 29: 428-429.

Neuenschwander, P., Hughes, J.d'A., Ogbe, F., Ngatse, J.M. and Legg, J.P. 2002. The occurrence of the Uganda variant of East African cassava mosaic virus (EACMV-UG) in western Democratic Repulic of Congo and the Congo Republic defines the westernmost extent of the CMD pandemic in East/Central Africa. *Plant Pathol.* 51(3): 385.

Ng, S.Y.C. 1986. Reinfection of selected cassava clones with the cassava mosaic virus and the tuber yield. In: *IITA Root and Tuber Improvement Program Annual Report 1986.* pp. 65-67 International Institute of Tropical Agriculture(IITA), Ibadan, Nigeria.

Ng, S.Y.C., Rossel, H.W., and Thottappilly, G. 1990. The role of tissue culture in the establishment of disease-free germplasm for international distribution, In: *Integrated Pest Management for Tropical Root and Tuber Crops*, S.K. Hahn and F.E. Caveness (eds), pp. 73-77, IITA, Ibadan, Nigeria.

Ng, S.Y.C., Thottappilly, G. and Rossel, H.W. 1992. Tissue Culture in disease elimination and micropropagation, In: *Biotechnology: Enhancing Research on Tropical Crops in Africa.* G. Thottappilly, L.M. Monti, D.R. Mohan Raj, A.W. Moore (eds.), pp. 171-182, CTA/I.I.T.A. co-publication. IITA, Ibadan, Nigeria.

Nichols, R.F.W. 1947. Breeding cassava for virus resistance. *East Afr. Agric. J.* 15: 154-160.

Nichols, R.F.W. 1950. The brown streak disease of cassava: distribution, climatic effects and diagnostic symptoms. *East Afr. Agric. J.* 15: 154-160.

Njock, T.E., Atiri, G.I., Thottappilly, G. and Hahn, S.K. 1994. Distribution and disease recovery of African cassava mosaic virus (ACMV) in stems of resistant and susceptible cassava. In: *Proc. Fifth Triennial Symp. Intern. Soc. Tropical Root Crop-African Branch (ISTRC-AB)*, pp.149-152. Fifth Triennial ISTRC-AB Symposium, Kampala, Uganda, 22-28 November, 1992.

Njock, T.E., Atiri, G.I., Thottappilly, G. and Thresh, J.M. 1996. Disease development and recovery in resistant and susceptible cassava genotypes infected by African cassava mosaic geminivirus. *African Plant Protection* 2: 97-102.

Nnodu, E.C. and Ogbe, F.O. 1994. Re-evaluation of elite cassava varieties for sustained yield and resistance to major cassava diseases. *Nigerian J. Plant Protection* 15: 38-43.

Nolt, B.L., Velasco, A.C. and Pineda, B. 1991. Improved purification procedure and some serological and physical properties of cassava common mosaic virus from South America. *Ann. appl. Biol.* 118: 105-113.

Nolt, B.L., Pineda, L.B. and Velasco, A.C. 1992. Surveys of cassava plantations in Colombia for virus and virus-like diseases. *Plant Pathol.* 41: 348-354.

Noueiry, A.O., Lukas, W.J. and Gilbertson, R.L. 1994. Two proteins of a plant DNA virus coordinate nuclear and plasmodesmal transport. *Cell* 76: 925-932.

Offei, S.K., Owuna-Kwakye, M. and Thottappilly, G. 1999. First report of East African cassava mosaic begomovirus in Ghana. *Plant Dis.* 83: 877.

Ogbe, F.M. 2001. Survey of cassava begomoviruses in Nigeria and the response of resistant cassava genotypes to African cassava mosaic begomovirus infection. Ph.D. thesis, University of Ibadan, Nigeria, 197 pp.

Ogbe, F.O., Legg, J., Raya, M.D., Muimba-kankolongo, A., Theu, M.P., Kaitisha, G., Phiri, N.A. and Chalwe, A. 1997a. Diagnostic survey of cassava mosaic viruses in Tanzania, Malawi and Zambia. *Roots* 4: 12-15.

Ogbe, F.O., Thottappilly, G. and Quin, F.M. 1997b. Implementation in Africa of serological diagnostic test for cassava mosaic geminiviruses. *African J. Root Tuber Crops* 2: 33-36.

Ogbe, F.O., Atiri, G.I., Robinson, D., Winter, S., Dixon, A.G.O., Quin, F.M. and Thottappilly, G. 1999. First report of East African cassava mosaic begomovirus in Nigeria. *Plant Dis.* 83: 398.

Ogbe, F.O., Atiri, G.I., Thottappilly, G., Dixon, A.G.O., Mignouna, H.D. and Quin, F.M. 2001. Evidence of double infection and random occurrence of cassava begomoviruses in sub-Saharan Africa. In: *Root Crops in the 21st Century.* M.O.Akoroda and J.M.Ngeve (eds.). pp.524-529, Proc. Seventh Triennial Symp. Intern/ Soc. Tropical Root Crop African Branch (ISTRC-AB), Cotonou, Benin Republic..

Ogbe F.O., Dixon, A.G.O., G.I. Atiri, A.G.O. and Thottappilly, G.. 2002. Restriction of virus movement into axillary buds in an important aspect of resistance in cassava to *African cassava mosaic virus*. *J. Phytopath.* 150: 546-552.

Ogbe, F.O., Atiri, G.I., Dixon, A.G.O. and Thottappilly, G. 2003a. Symptom severity of cassava mosaic disease in relation to concentration on *African cassava mosaic virus* in different cassava genotypes. *Plant Pathol.* 52, 84-91.

Ogbe, F.O., Atiri, G. I., Dixon, A.G.O. and Thottappilly, G. 2003b. Serological and biological

variations of *African cassava mosaic virus* in Nigeria . *Ann. appl. Biol.* 143: 203-213.
Ogbe, F.O., Thottappilly, G., Dixon, A.G.O. and Atiri, G.I. 2003c. Variants of *East African cassava mosaic virus* and their distribution in double infections with *African cassava mosaic virus* in Nigeria. *Plant Dis.* 87: 229-232.
Otim-Nape, G.W., Bua, A. and Baguma, Y. 1994a. Accelerating the transfer of improved production technologies: Controlling African cassava mosaic virus disease epidemic in Uganda. *Afric. Crop Science J.* 2: 479-495.
Otim-Nape, G.W., Shaw, M.W. and Thresh, J.M. 1994b. The effects of African cassava mosaic geminivirus on the growth and yield of cassava in Uganda. *Trop. Science* 34: 43-54.
Otim-Nape, G.W., Bua, A., Thresh, J.M., Baguma, Y., Ogwal, S., Ssemakula, G.N., Acola, G., Byabakama, B. and Martin, A. 1997a. Cassava mosaic virus disease in Uganda: The current pandemic and approaches to control. NARO, NRI, ODA publication, 65pp.
Otim-Nape, G.W., Thresh, J.M. and Shaw, M.W. 1997b. The effects of cassava mosaic virus disease on yield and compensation in mixed stands of healthy and infected cassava. *Ann. appl. Biol.* 130: 503-521.
Otim-Nape, G.W., Thresh, J.M. and Shaw, M. 1998a. The incidence and severity of cassava mosaic virus disease in Uganda: 1990-1992. *Trop. Science* 38: 25-37.
Otim-Nape, G.W., Thresh, J.M., Bua, A., Baguma, Y. and Shaw, M.W. 1998b. Temporal spread of cassava mosaic virus disease in a range of cassava cultivars in different agro-ecological regions of Uganda. *Ann. appl. Biol.* 133: 415-430.
Otim-Nape, G.W., Bua, A., Thresh, J.M., Baguma, Y., Ogwal, S., Ssemakula, G.N., Acola, G., Byabakama, B., Colvin, J., Cooter, R.J. and Martin, A. 2000. The current pandemic of cassava mosaic virus disease in East Africa and its control. *Natu. Resour. Inst.* Chatham, UK, 100pp.
Otim-Nape, G.W., Alicai, T. and Thresh, J.M. 2001. Changes in the incidence and severity of cassava mosaic virus disease, varietal diversity and cassava production in Uganda. *Ann. appl. Biol.* 138: 313-327.
Padidam, E., Sawyer, S. and Fauquet, C.M. 1999. Possible emergence of new geminiviruses by frequent recombination. *Virology* 265: 218-225.
Palaniswami, M.S. and Pillai, K.S. 1991. Bioecology of *Bemisia tabaci* G. on cassava. *J. Root Crops.* 17(special): 169-173.
Palaniswami, M.S., Pillai, K.S., Nair, R.R. and Mohandas, C. 1994. New cassava pest in India. *Cassava Newsletter (CIAT)* 19(1): 6-7.
Palaniswami, M.S., Nair, R.R., Pillai, K.S. and Thankappan, M. 1996. Whiteflies on cassava and its role as vector of cassava mosaic disease in India. *J. Root Crops.* 22: 1-8.
Palaniswami, M.S., Vijayan, S.L., and Antony, B. 2000. Biotypes in *Bemisia tabaci* Genn.-A first report in India. *CTCRI News,* 17: 5, Central Tuber Crops Research Institute, Thiruvananthapuram, India
Palaniswami, M.S., Antony, B., and Vijayan, S.L. 2001. Indian bibliography of sweet potato whitefly *Bemisia tabaci* (Genn.) 1979-1998. Central Tuber Crops Research Institute, Thiruvananthapuram, India, 144 pp.
Paranidharan, V. 1994. Use of ELISA to detect Indian cassava mosaic virus (ICMV) from meristem derived cassava (*Manihot esculenta* Crantz) and studies on secondary spread by the vector *Bemisia tabaci* Genn. M.Sc thesis, Dept. Plant Pathol, Tamil Nadu Agric. University, Coimbatore.
Pascal, E., Sanderfoot, A.A., Ward. B.M., Medville,R., Turgeon, R. and Lazarowitz, S.G. 1994. The geminivirus BR1 movement protein binds single stranded DNA and localises to the cell nucleus. *Plant Cell* 6: 995-1006.
Paximadis, M., Idris, A.M., Torrez-Jerez, I., Villareal, A., Rey, M.E.C. and Brown, J.K. 1999. Characterization of tobacco geminiviruses in the old and new world. *Arch. Virol.* 144: 703-717.

Pillai, K.S. and Daniel, R.S. 1979. Monthly variations in the population of whitefly *Bemisia tabaci,* the vector of cassava mosaic disease. *J. Root Crops* 5: 8-10.

Pineda, B., Jayasinghe, U. and Lozano, J.C. 1983. La enfermedad "Cuero de Sapo" en yuca (*Manihot esculenta,* Crantz). *ASIAVA* 4:10-12.

Pita, J.S., Fondong, V.N., Sangare, A., Kokora, R.N.N. and Fauquet, C.M. 2001a. Genetic and biological diversity of the African cassava geminiviruses. *Euphytica* 120: 115-125.

Pita, J.S., Fodong, V.N., Sangare, A., Otim-Nape, G.W., Ogwal, S. and Fauquet, C.M. 2001b. Recombination, pseudorecombination and synergism of geminiviruses are determinant keys to the epidemic of severe cassava mosaic disease in Uganda. *J. Gen. Virol.* 82: 655-665.

Powell-Abel, P., Nelson, R.S., De, B., Noffman, N., Rogers, S.G., Fraley, R.T. and Beachy, R.N. 1986. Delay of disease development in transgenic plants that express the tobacco mosaic virus coat protein gene. *Science* 232: 738-743.

Rajinimala, N. 2002. Biological, serological and polymerase chain reaction studies of whitefly transmitted geminiviruses with special reference to bitter gourd yellow mosaic virus disease. M.Sc thesis Dept. Plant Pathology, Tamil Nadu Agricultural University, Coimbatore, 123pp.

Rani, R.V., Karthikeyan, A.S., Anuradha, S., and Veluthambi, K., 1996. Genome homologies among geminiviruses infecting *Vigna*, cassava, *Acalypha, Croton* and *Vernonia. Current Science* 70(1): 63-69.

Ranomenjanahary, S., Rabindran, R. and Robinson, D.J. 2002. Occurrence of three distinct begomoviruses in cassava in Madagascar. *Ann. appl. Biol.* 140: 315-318.

Rey, M.E.C. and Thompson, G. 1998. Cassava mosaic disease in South Africa. *Roots* 5(1): 3-5.

Roberts, S. and Stanley, J. 1994. Lethal mutations within the conserved stem-loop of *African cassava mosaic virus* DNA are rapidly corrected by genomic recombination. *J. Gen. Virol.* 75: 3203-3209.

Roberts, I.M., Robinson, D.J. and Harrison, B.D. 1984. Serological relationships and genome homologies among geminiviruses. *J. Gen. Virol.* 65: 1723-1730.

Rojas, M.R., Gilbertson, R.L., Russel, D.R. and Maxwell, D.P. 1993. Use of degenerate primers in the polymerase chain reaction to detect whitefly-transmitted geminiviruses. *Plant Dis.* 77: 340-347.

Rossel, H.W. and Thottappilly, G. 1984. The etiology of African cassava mosaic finally elucidated? *IITA Research Briefs* 5(3): 2-3.

Rossel, H.W. and Thottappilly, G. 1985. *Virus diseases of important food crops in Tropical Africa.* Intern. Inst. Tropical Agric. (IITA) Ibadan, Nigeria, 67 pp.

Rossel, H.W., Thottappilly, G., Van Lent, J.M.W. and Huttinga, H. 1987. The etiology of cassava mosaic in Nigeria. In: *Proc. Intern. Seminar African Cassava Mosaic Disease and its Control,* C.Fauquet and D. Fargette(eds), pp-43-56, 4-8 May 1987, Yamoussoukro, Côte d'lvoire. CTA, FAO, ORSTOM, IITA, IAPC publication.

Rybicki, E.P. and Pietersen, G. 1999. Plant virus disease problems in the developing world. *Adv. Virus Research* 53: 125-175, Academic Press.

Sanderfoot, A.A., Ingham, D.J. and Lazarowitz, S.G. 1996. A viral movement protein as a nuclear shuttle – the geminivirus BR1 contains domains essential for interaction with BL1 and nuclear localization. *Plant Physiol.* 110: 23-33.

Saunders, K., Lucy, A. and Stanley, J. 1991. DNA forms of the geminivirus *African cassava mosaic virus* consistent with a rolling circle mechanism of replication. *Nucleic Acids Res.* 19: 2325-2330.

Saunders, K., Salim, N., Mali, V.R., Malathi, V.G., Briddon, S.R., Markham, P.G. and Stanley, J. 2002. Characterisation of *Sri Lankan cassava mosaic virus* and *Indian cassava mosaic virus*: Evidence for acquisition of a DNA B component by a monopartite begomovirus. *Virology* 293: 63-74.

Schöpke, C., Taylor, N.J., Masona, M.V., Carcamo, R., Ho, T., Beachy, R.N. and Fauquet, C.M. 1998. Production and characterization of transgenic cassava plants containing the coat protein gene of cassava common mosaic virus. In: *Fourth Intern. Scientific Meet. Cassava Biotechnology Network* (CBN), Carvalho, L.J.C.B, Thro A-M., and A.D. Vilarinhos (Eds). pp. 236-243, Salvador, Bahia, Brazil, November 3-7, 1998. EMBRAPA, Brasilia.

Scott, G.J., Best, R. Rosengrant, M.W. and Bokanga, M. 2000. Roots and tubers in the global food system: A vision statement to the year 2020. *International Potato Center*, Lima, Peru, 111 pp.

Seif, A.A. 1981. Transmission of cassava mosaic virus by *Bemisia tabaci*. *Plant Dis.* 65: 606-607.

Seif, A.A. 1982. Effect of cassava mosaic virus on yield of cassava. *Plant Dis.* 66: 661-662.

Sequeira, J.C. and Harrison, B.D. 1982. Serological studies on cassava latent virus. *Ann. appl. Biol.* 101: 33-42.

Shoyinka, S.A., Thottappilly, G., McGrath, F.F. and Harrison, B.D., 2001. Detection, relationships and properties of cassava mosaic geminivirus in naturally infected castor oil plant, *Ricinus communis* L. in Nigeria, In: *Cassava, an ancient crop for modern times: Food, health, culture,* N.J. Taylor, F. Ogbe and C.M. Fauquet (eds), pp. S8-20, Donald Danforth Plant Sciences Center, St. Louis, USA.

Silberschmid, K. 1938. O mosaico da mandioca. *O Biologico* 4: 177-182.

Silva, D.M., Kitajima, E.W. and Oliveira, A.R. 1963. Obtencao do virus do mosaico comun da mandioca purificado. *Ciencia e Cultura* 15: 304.

Smith, S.S. 1988. Root and tuber processing and utilization. A case study of Sierra Leone. *A paper presented at the IITA-UNICEF Inter-regional Experts Group Meeting.* April 13-19, Ibadan, Nigeria, pp. 306-323.

Sserubombwe, W.S., Thresh, J.M., Otim-Nape, G.W. and Osiru, D.O.S. 2001. Progress of cassava mosaic disease and whitefly vector populations in single and mixed stands of four cassava varieties grown under epidemic conditions in Uganda. *Ann. appl. Biol.* 138: 161-170.

Stanley, J. 1983. Infectivity of the cloned geminivirus genome requires sequences from both DNAs. *Nature* 305: 643-645.

Stanley, J. and Gay, M.R. 1983. Nucleotide sequence of cassava latent virus DNA. *Nature* 301: 260-262.

Stanley, J., Townsend, R. and Curson, S.J. 1985. Pseudorecombinants between cloned DNAs of two isolates of cassava latent virus. *J. Gen. Virol.* 66: 1055-1061.

Storey, H.H. 1936. Virus diseases of East African plants: VI. A progress report on studies of the diseases of cassava. *East Afr. Agric. J.* 2: 34-39.

Storey, H.H. 1938. Virus diseases of East African plants, VII. A field experiment in the transmission of cassava mosaic virus. *East Afr. Agric. J.* 3: 446-449.

Storey, H.H. and Nichols, R.F.W. 1938. Studies of the mosaic disease of cassava. *Ann. appl. Biol.* 25: 790-806.

Swanson, M.M. and Harrison, B.D. 1994. Properties, relationships and distribution of cassava mosaic geminiviruses. *Trop. Science* 34: 15-25.

Swanson, M.M., Brown, J.K., Poulos, B.T. and Harrison, B.D. 1992. Genomic affinities and epitope profiles of whitefly transmitted geminiviruses from the Americas. *Ann. appl. Biol.* 121: 285-296.

Taylor, N.J., Masona, M.V., and Fauquet, C.M. 2002. Transgenic cassava for food security and economic development. In: *Transgenic Plants and Crops*, G.K. Khachatourians, A. McHughen, R. Scorza, Wai-Kit Nip, and Y.H. Hui (eds), pp. 523-546, Marcel Dekker, Inc., Basel, Switzerland.

Thankappan, M. and Chacko, C.I. 1976. Effect of cassava mosaic on the different plant parts and tuber yield in cassava. *J. Root Crops* 2: 45-47.

Thankappan, M., Nair, N.G. and Nair, R.R. 1996. Degeneration of cassava planting materials due to CMD. In: *Tropical Tuber Crops*. G.T. Kurup, M.S. Palaniswami, V.P. Potty, G. Padmaja, S. Kabeerathumma and Santha V. Pillai (eds), pp. 370-374, Oxford & IBH, New Delhi, 597 pp.

Thomas, J.E., Massalski, P.R. and Harrison, B.D. 1986. Production of monoclonal antibodies to African cassava mosaic virus and differences in their reactivity with other whitefly-transmitted geminiviruses. *J. Gen. Virol.* 67: 2739-2748.

Thottappilly, G. 1992. Plant virus diseases of importance to African agriculture. *J. Phytopathol.* 134: 264-288.

Thottappilly, G. and Rossel, H.W. 1988. Application of new virus detection techniques at IITA. In: *IITA meeting reports series 2, The use of biotechnology for the improvement of cassava, yams and plantain in Africa.* pp. 51-54. International Institute of Tropical Agriculture, Ibadan, Nigeria. 73pp.

Thottappilly, G., Ng, S.Y.C., and Winter, S. 1999. Supply of pest free stock. In: *Abstr. XIVth Intern. Plant Protect. Congr. (IPPC),* Jerusalem, Israel, 25-30 July, 1999, p. 86,

Thottappilly, G., Hamilton, R.I., Huguenot, C., Rossel, H.W., Furneaux, M.T., Gumedzoe, M.Y., Shoyinka, S.A., Naik, D.M., Konate, G., Atchem-Agneroh, T., Haciwa, H.C., Anno-Nyako, F.O., Saifodine, N. I., Wangai, A., Lamptey, P., Gubba, A., Mbwaga, D.M., Neya, J. and Offei, S.K. 1993. Identification of cowpea viruses and their strains in tropical Africa in a pilot project involving an advanced laboratory, an international center, and other national programs. *FAO Plant Protection* 41: 65-71.

Thresh, J.M., and Fargette, D. 2001. Virus diseases of tropical crops. Encyclopedia of Life Sciences, pages 1-9. *Nature Publishing Group* (www.els.net).

Thresh, J.M., and Fargette, D. 2003. The epidemiology of African plant viruses: Basic principles and concepts. *IITA Publication* (in press)

Thresh J.M. and Mbwana, M.W. 1998. Cassava mosaic and cassava brown streak virus diseases in Zanzibar. *Roots* 6: 6-9.

Thresh J.M. and Otim-Nape, G.W. 1994. Strategies for controlling African cassava mosaic geminivirus. *Advances in Disease Vector Res.* 10: 215-236.

Thresh J.M., Fargette, D. and Otim-Nape, G.W. 1994a. The viruses and virus diseases of cassava in Africa. *Afric. Crop Science J.* 2: 459-478.

Thresh, J.M., Fargette, D. and Otim-Nape, G.W. 1994b. Effects of African cassava mosaic geminivirus on the yield of cassava. *Trop. Science* 34: 26-42.

Thresh, J.M., Otim-Nape, G.W. and Jennings, D.L. 1994c. Exploiting resistance to African cassava mosaic virus. *Aspects Appl. Biol.* 39: 51-60.

Thresh, J.M., Otim-Nape, G.W., Legg, J.P. and Fargette, D. 1997. African cassava mosaic virus disease: the magnitude of the problem. *Afric. J. Root and Tuber Crops* 2: 13-19.

Thresh, J.M., Otim-Nape, G.W. and Fargette, D. 1998a. The control of African cassava mosaic virus disease: phytosanitation and/or resistance? In: *Plant Virus Disease Control,* A. Hadidi, R.K. Khetarpal and H. Koganezawa (eds.). pp. 670-677, American Phytopathological Society, St. Paul, Minnesota..

Thresh, J.M., Otim-Nape, G.W. and Fargette, D. 1998b. The components and deployment of resistance to cassava mosaic virus disease. *Integrated Pest Manag. Rev.* 3: 209-224.

Thresh, J.M., Otim-Nape, G.W., Thankappan, M. and Muniyappa, V. 1998c. The mosaic diseases of cassava in Africa and India caused by whitefly-borne geminiviruses. *Rev. Plant Pathol.* 77: 935-945.

Walter, B., Ladeveze, I., Etienne, L., and Fuchs, M., 1989. Some properties of a previously undescribed virus from cassava: cassava American latent virus. *Ann. appl. Biol.* 115: 279-289.

Warburg, O. 1894. Die Kulturpflantzen Usambaras. *Mitteilungen Deutschen Schutzgebieten* 7: 131.

Ward, B.M., Medville, R., Lazarowitz, S.G. and Turgeon, R. 1997. The geminivirus BL1 movement protein is associated with endoplasmic reticulum derived tubules in developing phloem cells. *J. Virol.* 71: 3726-3733.

Winter, S. 1998. Analysis of cassava samples from Guinea to study presence of viruses contributing to cassava mosaic disease (CMD). *West African Seed and Planting Material Newsletter* (WASEED, English edition), 2: 7-8.

Wright, R.M. 1996. Jamaica's energy: Old prospects, new resources. *Twin Guinep publishers/ Stephenson' Litto Press.*

Zhang, P., Legris, G., Coulin, P. and Puonti-Kaerlas, 2000. Production of stably transformed cassava plant via particle bombardment. *Plant Cell Reports* 19: 939-945.

Zhou, X., Liu, Y., Calvert, L., Munoz, C., Otim-Nape, G.W., Robinson, D.J. and Harrison, B.D. 1997. Evidence that DNA-A of a geminivirus associated with severe cassava mosaic disease in Uganda has arisen by interspecific recombination. *J. Gen. Virol.* 78: 2101-2111.

Zhou, X., Liu, Y. Robinson, D.J. and Harrison, B.D. 1998. Types of variation in DNA-A among isolates of East African Cassava mosaic virus from Kenya, Malawi and Tanzania. *J. Gen. Virol.* 79: 2835-2840.

Fig. 6.1: Symptoms of cassava mosaic disease caused by ACMV (a), EACMV (b) and cassava field (western Kenya) with 100% infections of EACMV-Uganda variant (c).
Fig. 6.2: Symptoms of *Cassava brown streak virus* (CSBV);
Fig. 6.3: CBSV and EACMV mixed infection.
(Photographs: S. Winter)

7. Potatoes in Warm Climates

S.M. Paul Khurana and I.D. Garg
Central Potato Research Institute, Shimla, India

7.1 Introduction

Most of the warm areas in the world lie between 0 to 34°N and S. Major potato production occurs between 23°N and 34°N and represents the sub-tropical lowlands, where potato is a winter crop (Hijmans, 2001). During the last three decades, potato areas increased in the developing world and North America while it decreased in Europe (FAO 2000, 2001). The marked growth of potato in developing countries affirms its increasing importance as a source of food for growing population, rural employment and income. In the warm climates, potato is grown round the year in the tropical hills, and during winter under short days in the sub-tropical plains. A major part of this area (ca.80%) is in Asia while the remainder 20% is about equally divided between North, South and Central Americas and Africa.

7.2 Virus Surveys

Tropical and sub-tropical climates provide a suitable environment for growth of vegetation and insects round the year. Many of the plants serve as reservoirs for viruses, and the insects as their vectors, including the viruses that infect potatoes. The most important ones are *Potato virus X* (PVX; Genus *Potexvirus*), *Potato virus Y* (PVY; Genus *Potyvirus*), *Potato virus S* (PVS; Genus *Carlavirus*), *Potato leafroll virus* (PLRV; Genus *Polerovirus*), *Potato virus V* (PVV; Genus *Potyvirus*), *Andean potato latent virus* (APLV; Genus *Tymovirus*), *Andean potato mottle virus* (APMV; Genus *Comovirus*), *Potato virus T* (PVT; Genus *Trichovirus*), *Potato yellow mosaic virus*

(PYMV; Genus *Begomovirus*), *Potato yellow vein virus* (PYVV; Genus *Crinivirus*), *Potato mop-top virus* (PMTV; Genus *Pomovirus*), *Tobacco rattle-virus* (TRV; Genus *Tobravirus*), *Tomato spotted wilt virus* (TSWV; Genus *Tospovirus*) and *Potato apical leaf curl virus* (PALCV; Genus *Begomovirus*). Distribution of different viruses and virus resistant cvs/clones, except PVX, PVS, PVY and PLRV which are ubiquitous, are given in Table 7.1.

Some viruses as PVA, PVM and the viroid PSTVd were introduced to the warm climate countries from Europe and North America probably through seed tubers. The important ones responsible for degeneration of the potatoes are PVX, PVY and PLRV.

7.3 The Viruses

The viruses PVV, PYV, PYMV, PYVV, PMTV, TRV and TSWV will be described in this chapter. Information on other viruses of potato is presented in Chapter 8

7.3.1 ***Potato mop top virus*** Genus ***Pomovirus*** (PMTV) causes symptoms known as 'spraing,' consisting of brown arcs and circles in the flesh of potato tubers, sometimes visible also on the tuber surface. Secondarily infected tubers may be cracked and distorted and have blotchy surface markings or reticulate surface cracking. The three most common shoot symptoms in secondarily infected plants are yellow blotching or mottling (particularly on lower leaves), chlorotic V-shaped markings (chevrons) in leaflets, and extreme stunting in shoots (known as mop top) (Fig.7.1). The infection causes some yield losses, but the qualitative losses are more important (Jeffries, 1998).

The virus is transmitted by the motile zoospores of the fungus *Spongospora subterranea* f. sp. *subterranea*. The fungus spores may retain infectious virus for many years. Only up to 50% of the progeny tuber get infection from the infected mother plant.

7.3.1.1 Diagnosis. ELISA can detect the virus in indicator plants but may miss the detection in potato, as virus concentration is low or erratically distributed. Sap inoculations induce on *Chenopodium amaranticolor* or *C.quinoa* concentric brown local lesions and in addition necrosis bordering veins in *C.quinoa*. Nucleic acid probes have been produced (Mills, 1987).

7.3.1.2 Virion Morphology and Genome. Fragile rod-shaped particles, with two predominant lengths, 100-150 nm and 250-300 nm, and a diameter of 18-20. The genome consists of three single-stranded positive-sense RNA

Table 7.1 List of potato viruses and virus resistant varieties in warm climates

Country	Viruses reported	Virus resistant varieties/clones
China	PVA, PVM, AlMV, PMTV, TRV, PAMV, TNV, PSTVd	Kexin 1, Hutou and Yuejin (virus resistant), Kexin 2 (PLRV resistant), Fengshoubai and Changshu 4 (mosaic resistant) (Gitomer, 1996)
India	PVA, PVM, Geminivirus, Tospovirus,	Kufri Lalima (resistant to PVX, PVY), Kufri Badshah (resistant to PVX), Kufri Kuber (resistant to PVY and PLRV), Kufri Sindhuri (tolerant to PLRV) (Shekhawat and Naik, 1999)
Egypt	PVA, PVM, AlMV, PSTVd	Russet Nile (resistant to mosaics; Taghian *et al.* 1997)
Cuba	PVM, TRV, PYMV	Lizette and Ibis (Estevez *et al.* 1994)
Maxico	PVA,PVM,AlMV,TRV	--
Bolivia	PVV, PVT, PYV, APLV	CIP Nos. 702311, 702318 (resistant to PVX), 702317, 703243, 703346 (resistant to PLRV) (Huaman, 1989)
Brazil	APLV,TRV	Omega and Palma (resistant to PVY) (Camara *et al.* 1986); Cristal and 2 CR 11149-1-79 (resistant to PVY; Daniels, 2000), Embrapa/CIP-PP063 and Embrapa/CIP-PP084 (immune to PVX and PVY; Brune *et al.* 1999)
Peru	PVV, APLV, APMV, PVT, PYVV, TRV, SALCV, PMTV, PYV	CIP Nos.700455, 700775; 701172, 701225, 701299 and 701535 (resistant to PVX and PVY), 700523, 701171 (resistant to PVX and PLRV) and 700754, 702867 and 703294 (resistant to PVY and PLRV) (Huaman, 1989)
Colombia	APLV, APMV, PYVV	CIP Nos.703292, 703295, 703506, 703531 704220 (resistant to PLRV, Huaman, 1989)
Chile	PVM, APMV, APLV, PSTVd, PYV	SerranaINTA, Line 76, Line 536 (ME) and Line 21 (resistant to PLRV) (Santos-Rojas et al., 1990); clones CON 08900979, CON 08900921, CON 08900928 and CON 08900993 (resistant to mosaic; Theoduloz *et al.* 1992)
Venezuela	PYMV, PYVV	-

molecules; RNA 1 encodes the viral replicase (Savenkov *et al.* 1999), RNA-2 contains four ORFs, three of which correspond to the triple gene block (TGB) involved in cell-to-cell movement of the virus (Lauber *et al.* 1998). RNA-3 encodes coat protein and another read-through protein (Sandgren *et al.* 2001).

7.3.2 ***Potato virus V*** Genus ***Potyvirus*** (PVV). The disease is virtually symptomless in the majority of potato cultivars. A few develop mosaic and necrotic spotting on the lower leaves, others develop severe systemic necrosis and leaf dropping. The natural host range narrow, restricted to potato and tomato (*Lycopersicon esculentum*). PVV is transmitted non-persistently by the aphids *Macrosiphum euphorbiae* and *Myzus persicae,* and mechanically by sap.

7.3.2.1. Diagnosis. PVV causes systemic veins clearing in tomato and systemic necrosis in *S.demissum* A. PVV can be detected by ELISA. However, commercially available PVV polyclonal antibodies produced against European isolates may not detect all isolates of PVV (Alvarez and Fernandez-Northcote, 1996).

7.3.2.2 Particle Morphology. Filamentous particles 700 - 720 x 12-13 nm.

7.3.3 *Potato yellow mosaic virus* Genus ***Begomovirus*** (PYMV) causes bright yellow mosaic, leaf distortion and dwarfing. Depending on incidence and intensity, tuber yield loss may be appreciable. The natural host range is narrow; limited to potato and *Solanum* spp. and tomato (*Lycopersicon esculentum*). PYMV has been reported from Venezuela.The virus is transmitted by the whitefly*Bemisia tabaci.*

7.3.3.1 Diagnosis. PYMV can be detected by ELISA and sap inoculation to *Nicotiana benthamiana, N. tabacum* cv. *Xanthi nc* or *Petunia hybrida.* The virus induces chlorotic spots on inoculated leaves, most clearly seen in *P. hybrida*, and systemic golden-yellow mottling on newly developing leaves and deformation and dwarfing of plants.

7.3.3.2 Particle Morphology. Bigeminate (bisegmented), 18-20 nm in diameter. Biological properties are very similar to *Tomato yellow mosaic virus* (ToYMV). Though PYMV has been reported to be molecularly distinct form ToYMV (Roberts *et al.* 1988), Morales *et al.* (2001) showed PYMV to be a strain or variant of ToYMV

Other Begomoviruses infecting potatoes are *Potato defomring mosaic virus* in Brazil (Daniels and Castro, 1985; Costa *et al.* 1988*), Potato apical leaf curl virus* in India (Fig. 7.2) (Garg *et al.* 2001) and *Solanum apical leaf curling virus* in Peru (Hooker and Salazar, 1983).

7.3.4 *Potato yellow vein virus* Genus ***Crinivirus*** (PYVV) initially causes leaf veins to become bright yellow, and later the entire leaf lamina turns yellow (Fig. 7.3). Under favourable conditions, the entire plant may becomes bright yellow. However, infected plants not always produce symptoms. Yield reduction may be about 50% (Saldarriaga *et al.* 1988) and infected plants produce smaller numbers of tubers. The natural host range is narrow limited to potato, wild *Lycoperisicon* spp., *Polygonum mepalense* and *Solanum nigrum*. The virus is transmitted by the whitefly *Trialeurodes vaporariorum* (Saldarriaga *et al.* 1988) but not mechanically through sap.

7.3.4.1 Diagnosis. The virus can be detected by graft transmission to *Solanum tuberosum* or clone A-6 causing bright yellow veins, but infected plants may not always produce symptoms. For reliable detection DNA probes prepared from cloned PYVV HSP 70 cDNAs can be used (Salazar *et al.* 2000).

7.3.4.2 Particle Morphology. Flexuous particles. Their dimensions have still to be determined, due to instability of virions and their tendency to aggregate (Salazar *et al.* 2000).

7.3.5 ***Potato yellowing virus*** Genus ***Alfamovirus*** (PYV) causes mosaic, yellowing and premature senescence of the plant . Symptomless infections may occur (Jeffries, 1998). The natural host range is confined to potato. However, the experimental host range is wide and the virus can infect 44 of 51 species in seven plant families (Fuentes and Jaysinghe 1993).
The virus is widespread in Peru, Chile (Valkonen *et al.* 1992) and Bolivia.

PYV is transmitted semipersistently by *Myzus persicae,* mechanically and through true potato seeds (TPS) (ca 20%).

7.3.5.1 Diagnosis. Sap inoculation to *Capsicum annuum*, e.g. cvs. Gold Spike, Golden Calwonder, Jalapeno, causes systemic leaf distortion, yellow mosaic, and vein clearing. Inoculation to *Nicotiana tabacum* cv. Samsun induces systemic line patterns, mosaic, and chlorotic spots. The virus can be diagnosed by ELISA.

7.3.5.2 Particle Morphology bacilliform with 4-5 sizes, between 21-60 nm x 18 nm. PYV is serologically not related to AlMV (Fuentes, 1991).

7.3.6 ***Tobacco rattle virus*** Genus ***Tobravirus*** (TRV). Tubers from infected plants may show spraing and non-virus particle-producing (NM) type of the virus can be isolated from these tubers. Plants grown from these tubers exhibit 'stem mottle', distortion, stunting and mottling confined to one or a few shoots, or a few brown flecks in the tubers from which normal virus particle producing (M) type of the virus can be isolated. Transient yellow chevrons (Fig.7.6) are sometimes detected on a few leaflets of some stems. Virus can be detected in all leaves.

Some cultivars are resistant to infection and will not show symptoms. There may be appreciable loss of saleable yield because of spraing in the tubers (Jeffries, 1998).

The natural host range is wide and includes, besides potato, *Gladiolus* spp., *Hyacinthus* spp., *Lactuca sativa, Narcissus* spp., *Spinacia oleracea, Beta vulgaris,* tobacco, *Tulipa* spp. and many weeds which act as virus reservoirs.

The virus is transmitted by the trichodorid nematodes *Paratrichodorus* spp. and *Trichodorus* spp. The nematodes retain the virus for many months but lose it through moults. Transmission through TPS occurs at very low levels (Horvath *et al.* 1996).

7.3.6.1 Diagnosis. Serological detection is unreliable due to considerable diversity in the protein of different isolates and due to the fact that NM isolates do not have coat protein. Sap inoculation to *Chenopodium amaranticolor* induces necrotic local lesions and on *Phaseolus vulgaris* cv. Prince local pin-point lesions. Labelled cDNA probes based on RNA-1 (Robinson, 1989) and RT-PCR (Robinson, 1992; Crosslin and Thomas, 1995; Weidemann, 1995) enable reliable detection.

7.3.6.2 Particle Morphology and Nucleic Acids. Virion population consists of rigid tubular particles of two predominant lengths, 190 nm and 50-115 nm x 22 nm. Normal particle producing isolates (M-type) have two species of genomic RNA (RNA-1 and RNA-2) and are readily transmitted to test plants by sap inoculation and by nematodes. Other isolates (NM-type) have only one RNA species (RNA-1), do not produce particles, are transmitted with difficulty by sap inoculation and are not transmitted by nematodes.

7.3.7 ***Tomato spotted wilt virus*** Genus ***Tospovirus*** (TSWV). Primary infection by TSWV results in top and/or stem necrosis, with the formation of concentric rings or spots on leaves and stems Shoots may have a stunted rosette appearance One or a few stems on a plant and only some progeny tubers may become infected. Infected tubers may be small and malformed with necrotic damage and may show reduced sprouting or no sprouting. The disease is important in areas where thrips activity is high.

The natural host range is very wide and includes many cultivated crop plants (Peters, 1998; and this book).

TSWV is transmitted by many species of thrips namely *Frankliniella occidentalis* , *F. fusa, F. intonsa, F. bispinosa* , *F. schultzei, Thrips palmi, T. setosus* and *T. tabaci* in a persistent propagative manner. *T.palmi* is a pest on potato in Brazil.

7.3.7.1 Diagnosis: Sap inoculation to *Nicotiana glutinosa, N. tabacum* cvs. Samsun and White Burley induces large slightly sunken necrotic local lesions and systemic necrosis and *on Petunia hybrida* causes specific local lesions. For detection by ELISA, antisera to the nucleocapsid (N) protein have been used (Resende *et al.* 1991; De Avila *et al.* 1993). Nucleic acid probes (Huguenot *et al.* 1990) and RT-PCR are also reliable (Adam *et al.* 1996).

7.3.7.2. Particle Morphology. Particles are enveloped, roughly spherical and ca 70 to 110 nm in diameter, and with knob-like surface projections of glycoproteins. Three RNA segments forming the genome are encapsidated by the nucleocapsid protein into three pseudo-spherical particles.

7.3.7.3. Other Tospoviruses Infecting Potatoes: are *Groundnut ringspot virus* in Argentina (Granval de Millan *et al.* 1998), *Tomato chlorotic spot virus* in Argentina and Brazil (Nagata *et al.* 1995), and a tospovirus serologically related to *Groundnut bud necrosis virus* in India (Fig.7.4) (Khurana *et al.* 2001; Garg *et al.* 1999).

7.4. Seed Production

Due to the quick decline of potatoes in warm climates, availability of healthy or certified tuber seed is a major constraint in increasing the potato productivity in these areas. International Potato Center (CIP) has bred varieties resistant to PLRV, PVY, late blight and *Pseudomonas solanacearum* and supplied promising materials to national agricultural research programmes, which released locally suited varieties. True potato seeds (TPS) may also be of potential for reducing production costs. CIP has put in efforts in producing TPS lines suitable to many developing countries.

Regarding seed production, countries in the tropics and sub-tropics can be grouped into two categories namely (i) those which import healthy seed from other countries and then multiply them for several generations. Countries in this category generally lack their own variety breeding programme and or do not have vector free periods of sufficient duration favourable for potato cultivation; (ii) those which possess a well established breeding programme and also a vector-free period of 60 to 90 days duration favourable for potato cultivation.

To maintain the health and purity of the selected or released varieties, some countries have started *in vitro* rapid multiplication schemes (genrally supported by governments) to obtain pre-basic seed. The pre-basic seed is further multiplied by either government funded agencies, private companies, farmers' cooperatives or registered growers.

A general scheme, practised with some variations, for production of pre-basic seed through *in vitro* rapid multiplication is as follows:

Popular variety→ "Clean up" through apical meristem culture and thermotherapy →Pathogen tested virus-free (clean) in vitro material

↓ ↓

Disease free *in vitro* plantlet production — *In vitro* microtuber production

↘ ↙

Aphid proof screen /green house

↓

Minituber production → Pre-basic seed

7.3 Potato Producing Countries

7.5.1. China. Depending on altitude, potato is a highland crop grown in all seasons in South and Central China, while winter potatoes are grown only in South China. Potato is a summer crop in North China. Out of about 2.8 million ha total area under potato in China, 1.1 million ha are in South West China, which produces 13 million tonnes of potatoes (Jun *et al.* 1994).

Major potato cultivars are Kexin 1, Kexin 4, Shuangji 1 (or Pushu 1), Anemone, Youyi 3, Red Warba, Fengshoubai, Kexin 2, Fengshou, Zishanyao, Hutou, Yuejin, Changshu-4 and Mira (Zhang *et al.* 1990; Gitomer, 1996).

Though potato viruses have been studied in China since the 1950s Use of sensitive detection techniques (ELISA and NASH) started in the 1970s, after establishment of CIP's Region VIII office in Beijing.

7.5.2 South Korea. Five types of potato crops, depending on altitude and season, are grown in Korea, with spring upland crop occupying more than 60%. Since 1970, many foreign cvs namely Shimabara, Dejima, Superior and Irish Cobbler were introduced. A serious disease problem caused by PLRV and PVY was overcome by a combination of tissue culture, virus eradication and a basic seed program. A seed program, parallel to the national program, enabled farmers to plant virus-free or slightly infected seed on most fields. PLRV and PVY are the two major viruses present in

S.Korea. They reduce yields, particularly at low-elevations, where farmers replant small tubers from their previous crop (Kim *et al.* 1986).

7.5.3 Viet Nam. About 98% of the country's potato area is located in the Red River Delta, where potatoes are grown from November to February. Harvested tuber seeds are stored for about 9 months under hot, humid conditions, resulting in substantial losses and deterioration in seed quality. Seed usually accounts for about 50% of the total production cost (Ho *et al.* 1987).

More than 90% of the potato crop is grown using tuber seeds of local var. "Thuongtin". These are of low quality and lack of good quality tuber seeds is the main constraint in improving potato production.

7.5.4 Indonesia. Three crops a year are obtained in West Sumatra. Indonesia imports seed tubers because of continuous infection with viruses. Potato is grown in the highlands around the year depending on altitude. Tuber seeds are imported from Europe; the main variety being Granola. The Indonesian government launched a project to produce quality seed potatoes for West Java. The scheme produces seed in five stages with Basic seed "A" (Go) obtained from virus-free stem cuttings planted in a screen house. During field multiplication, the crop is inspected for health standard (Tanaka, 1994).

7.5.5 Nepal is located between ca. 26 N to 30°N. Potato is grown during winter in Terai (lower elevation), during spring and autumn in the mid-hills and during summer in high-altitude regions.

Potato is included in different crop rotation schemes and is also inter-cropped with maize and vegetables. It is widely grown in the high hills as a monocrop, which have a suitable climate for quality seed production. A National Potato Development Programme (NPDP) was initiated in 1972 dealing with variety evaluation, multiplication and distribution of clean seed and dissemination of knowledge through extension activities. Low productivity in Nepal is attributed mainly to the use of poor-quality seed since good quality seed is costly and in short supply.

Apical meristem culture and thermotherapy are used to 'clean' the stocks, which are tested with ELISA for viruses. The 'cleaned' stocks are then used to produce pre-basic seed (Ranjit *et al.* 1994).

Cultivars grown in Nepal include Kufri Sindhuri, Kufri Jyoti, Desiree and Achirana Inta, Kathmandu Local (KL), NPI-106, Theru Local (TL).

7.5.6 Bangladesh. Potato is the 3rd largest food crop occupying 0.86%

of total cropped area. It is grown as a winter crop. The main constraints in obtaining higher yields include non-availability of quality tuber seeds, low inputs, sub-optimal standards of cultivation by the farmers, storage and marketing and heavy post harvest losses. Tuber Crops Research Center at the Bangladesh Agricultural Research Institute (BARI) produces 500 to 700 tonnes of pre-foundation seed annually and supplies it to Bangladesh Agriculture Development Corporation (BADC). The later produces certified seeds after multiplication as F-1 and F-2 (Salahuddin, 2000). Imported seed tubers are used for the production of pre-foundation seed in the first step. BADC is able to supply only 6000 t of certified seed each year, which is ca. 7.3% of the requirement of modern varieties and 4.8% of the total requirement of seeds (Rabbani, 1994).

The main potato cultivars grown in Bangladesh are Diamant, Heera, Chamak, Dheera, Patronas, Multa, Kronia, Baraka, Cardinal, Morena, Origo, Mondial, Alisa, Granola, Cleopatra, Binalla and Kufri Sindhuri.

7.5.7 India. Potato is a short duration crop in the Indian tropics and sub-tropics, grown in a wide range of climatic conditions and soils, with flexibility in planting and harvesting time. It is a crop suitable for inclusion in intensive cropping systems and inter-cropping in different regions of the country.

Potato research on developing indigenous varieties and agro-techniques at the Central Potato Research Institute (CPRI), Shimla, resulted in phenomenal increases in production and yields during last 50 years.

Of the 34 varieties developed by CPRI, 13 possessed multiple resistance to different biotic and abiotic stresses. The popular varieties Kufri Jyoti, Kufri Bahar and Kufri Sindhuri showed low rates of degeneration (Vashisth *et al.* 1981).

Tissue culture is being routinely used to eliminate viruses by meristem culture, and micropropagation for preparation of potato tuber seeds. Indian potato varieties developed by CPRI have been characterized employing DNA fingerprinting techniques (Shekhawat and Naik, 1999).

CPRI surveyed a number of locations for the prevalence of aphids. On the basis of these data it was found that seed potatoes could be grown in many parts of the Indian plains with certain minimum precautions (Nagaich *et al.* 1969). This led to the "Seed Plot Technique", which grows seed potatoes during low aphid periods in north Indian plains, during October to early January (Fig.7.5). Presently, about 50,000 virus-free microtubers of Indian varieties produced from meristem derived *in vitro* plants are given annually to the breeders' seed production program of CPRI. Presently virus-tested minitubers are produced only by CPRI. Use of virus detection techniques as ELISA (Khurana and Garg, 1993), nucleic acid spot hybridization (NASH), and immunosorbent electron microscopy (ISEM) (Garg and Khurana, 1992; Garg *et al.* 2000) improved the quality of potato tuber seeds. With the average price of imported potato seed in Pakistan and Sri Lanka (US $ 1120 per ton), the country could save Rs.17424 million annually (about 484 million US $) on import of potato seeds (Shekhawat and Naik, 1999).

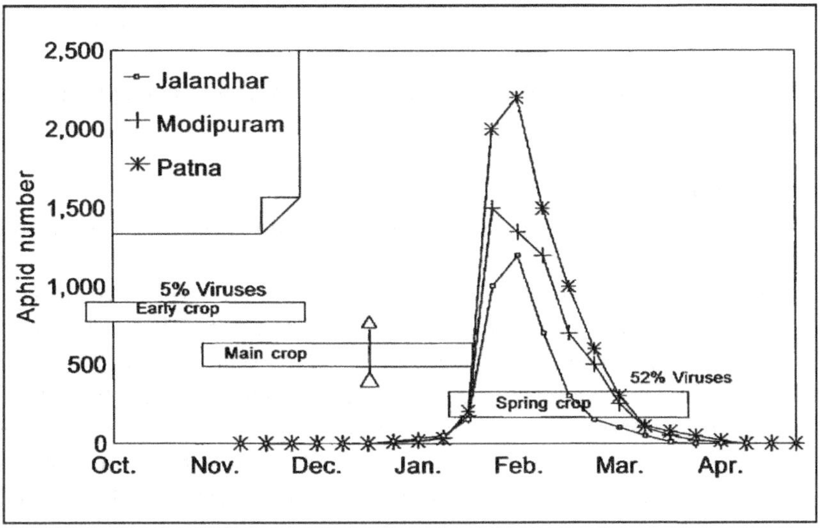

Fig 7.5 Aphid population curve in northern plains of India (average of 10 year data) (Khurana, 1999)

Indian seed stocks are free from PVV, PVY^{N}, PVY^{NTN}, TRV, PMTV, APLV, APMV, PVT, TRSV, PYDV, PYMV, SALCV, PYVV and PYV (Khurana, 1999).

Seed stocks multiplied during autumn (under low aphid activity) in the north west plain show only negligible rates of degeneration (Vashisth *et al.* 1981). *Myzus persicae* and *Aphis gossypii* are important vectors (Verma and Chandla, 1999).

7.5.8 Pakistan. Potato is cultivated during summer in the hills (1500 to 3100 m) and during winter in the plains (lowlands). Main production is in the North West Frontier Province, Punjab and Sindh. The important viruses, which cause heavy yield losses, are PVY, PLRV, PVX and PVS. Losses by PLRV vary from 49 to 60% (during spring) and 7-16% (during autumn/winter). PVY yield losses range from 40 to 70%, PVX from 10 to 25% and PVS from 10 to 20%. PVA, PVM and PMTV are of minor importance. Mainly Dutch cultivars as Desiree, Ultimus, Cardinal, Patrones,

Diamant and Multa are grown. Seeds are partly imported or produced in high hills. About 2000 tonnes of seed are imported annually from the Netherlands and multiplied in autumn-to-autumn cycle (Malik, 1995).

Average yields of fields planted with imported seed grown during autumn declined from 22.8 t/ha (in the first generation) to 17.9 t/ha (in the fifth generation) (Kokab, 1986). Yields in the spring crop after two seasons were 54% (in cv. Multa) and 52.5% (in cv. Patrones). The major potato crop is grown in the plains during autumn.

7.5.9 Iran. The climate in Iran varies from sub-tropical in the south to temperate in the north. In the north (Karaj) Potato is grown during March to November depending on altitude. The varieties grown include Deraga, Aula, Cosmos, Pashandi, Mondial, Volkhano, Diamond, Agria, Marfona, Moren., Red Gold, Yukan Gold, Nethod Gem, Kennebec, Rose Gold, Shepody, Onaway, Norchip and Fricia. In Isfahan region, *M. persicae, A. gossypii* and *Acyrthosiphon pisum* are important vectors of PVY (Sayedoleslaami *et al.* 1995), and the main viruses are PVY (65%), PVX (59%), PLRV (16%) and PVS (2%) in ELISA of leaves. *Egg plant mottled dwarf rhabdovirus* has been reported in potato in Isfahan (Danesh and Lockhart, 1989), as well as two strains of PVY (0 and N) (Danesh *et al.* 1992).

7. 5.10. Syria The General Organisation for Seed Multiplication (GOSM) is responsible for potato seed production. Elite ('E') seed potato (ca. 7500 ton) are imported from Europe and multiplied to obtain locally produced class 'A' seed potatoes. GOSM chooses appropriate farmers from specific areas to multiply the seed potatoes, under their supervision and guidance.The main varieties are Draga, Spunta, Diamant, Nicola and Marfona. Sahel, Famosa, Ilona, Escort and Planta are also multiplied and distributed, and Lira and Monalisa were recently certified for export. Seed production is during spring. GOSM started to produce seed potatoes of high quality that are nearly free of virus diseases, by employing in vitro rapid multiplication techniques and testing stocks by ELISA at different stages of multiplication (Abdellatif, 1994).

7.5.11 Cyprus. Potatoes are an important source of income for many Cypriot farmers. Three crops are grown per year. Cost of potato cultivation is high (ca. Pound sterling 3200 per ha). Potatoes are exported to Europe, mainly to the UK. The recommended varieties are Agria, Atica, Charlotte, Diamant, Esparante, Linzer Delikatess, Liseta, Lola, Monalisa, Ovatio, Sieglinde, Arran Banner, Cara, Nicola and Spunta (Vakis, 1990).

Seed potatoes are produced in Cyprus by single multiplication of imported class A stocks during February-May. The seed produced are kept in cold storage for production of ware potato in autumn (September-October) or in spring of the following year. PVY and PLRV are the important viruses and *Myzus persicae* populations are high from February to April (Ioannou, 1989). Seed potatoes are imported from northern European countries, mainly the Netherlands, the UK and the Irish Republic.

7.6 Africa -General

North African countries use mainly European potato cultivars, as Desiree and Spunta. In tropical Africa potatoes are grown primarily for fresh consumption, using different varieties. Some of these were developed by national programs, often in cooperation with the International Potato Centre, and incorporate multiple disease resistance traits, like Molinera and Dhamana resistant, to PLRV and PLRV and PVY, respectively (Corsini and Brown, 2001).

7.6.1 Morocco. Potato is cultivated in the Sous Valley at two locations, Ait Melloul and Ouled Abdellah. PLRV is the main virus and *M.persicae* was prevalent all the year around in Ait Melloul. This area is therefore not suitable for seed potato production. On the other hand, the dry hot summer at Ouled Abdellah largely eliminated the vector and volunteer potatoes, which enabled production of seed potato during the autumn (Hanafi *et al.* 1995).

7.6.2 Tunisia. Potatoes are mainly cultivated along the Mediterranean coast and only in limited areas inland. Potatoes are grown in spring, autumn and winter. Spring crop is the major one cultivated over 45% of the total potato area. Spring crop is planted with imported seed, while winter crop is planted partly with imported seeds and partly with the produce from the spring crop. Autumn crop uses seed from the spring or winter crops. Cold storage is limited and 50% of the area is planted with imported seed every year (Struik and Wiersema, 1999). When imported elite seed tubers were sown in Haouaria, Kerkouane, Soliman and Menzel Bou Zelfa and plots were rogued three times during the growing season, virus disease incidence after four annual cycles of multiplication was only 1.87% (Hattab *et al.* 1994). The most popular cultivar is Spunta, imported from Europe.

7.6.3 Egypt is the largest potato producer in Africa. About 10% of the produce is exported to Europe and the Middle East. Potatoes in Egypt are traditionally grown in three seasons - spring, autumn and winter. The winter crop is harvested during February for exports. Producers import more than 40,000 t tuber seeds from Europe for planting in the spring (El-Sadany and

Salam, 1991). They also save seed from the spring crop and store it for use in autumn or winter. Potato cultivars include local and introduced cultivars as: Arenda, Ajiba, Liseta, Picaso, Obelix, Spunta, Gigant, Mondial, Alpha, King Edward, Cara, Jearla, Claudia, Domina, Eambo, Grata, Prelina, Arran Banner, Van Goch, Sante. Infection by PVY and AlMV significantly reduced tuber yield. Their occurrence was monitored in relation to populations of *M .persicae, Aphis gossypii*, and *Macrosiphon euphorbiae* during December, 1989 to May, 1992. The average yield per plant infected with AlMV and PVY was 257.8 and 229.5 g compared to 763.2 g in healthy plants (Elnagar *et al.* 1996). PLRV, PVX and PSTVd were also found.

Private companies (Pico, Egaseed, High Tec and Pioneer) produce tuber seed from two TPS registered progenies and distribute them as high quality seed to potato growers (Rizk *et al.* 1991).

7.6.4 Ethiopia. The main potato producing areas are Awassa, Alamaya and Holetta. There is no organisation to produce true to type variety seeds for distribution to farmers. Only local varieties are grown viz. Rogie, Talian, Nech Hararghie, Wolamo, Saburie, Adama, Duramie (Saka), Yeferengi Dinich, Mongor and Sabune. Rogie is the most common and popular var. in Awassa (Borgel *et al.* 1980).

The major limitation to potato production in Ethiopia, as in many other African countries, is the shortage and high cost of quality seed tubers (Brown, 1982). TPS technology holds promise as an alternative to conventional technology. Recently, collaboration with CIP has enabled the selection of late blight resistant and high yielding clones viz. CIP-381376.15; CIP-382121.5, CIP-382173.12 and CIP-387315.2.

7.6.5 Cameroon: Institute of Agricultural Research (IRA) of Cameroon in collaboration with CIP developed three potato cultivars CIPIRA, TUBIRA and BAMIRA adapted to highlands, with resistance to late blight, capable of growing under low fertilizer application. They have a potential tuber yield of 30 to 40 t/ha, with a growing period of 90 to 125 days. Basic seed is produced by employing mass production of rooted stem cuttings developed by IRA/CIP project in Cameroon. (Martin *et al.* 1995).

7.6.6 Burundi. The main diseases are bacterial wilt, late blight and root knot nematode. Other constraints are the lack of appropriate agronomic technology and high quality seeds for farmers. TPS is a good alternative for producing planting material free of diseases and could provide cheaper seed in a short time and in the off-season. It might also be adopted in crop rotations (Berrios, 1995).

7.6.7 Kenya. Potato is grown in high lands in Kenya. The main varieties cultivated include Kerr's Pink, Maritta, B-53, Kinongo, Ngabati, Ngorobu, Annet, Thiba, B-59, Desiree, Tako, Dutch Robjin, Tanganyika, Pimpernel, Kenya Baraka, Roslin Eburu, Roslin Tana, Asante, Tigoni and Nyayo, which is the most popular and widely cultivated one. Collaboration with CIP resulted in the release of cvs. Kenya Dhamana, Kenya Chaguo (Crissman *et al.* 1993). Seed potatoes are produced through a mix of a government directed and supported, and a farmer led system. 'Seed plot technique' is employed for tuber seed production (Crissman, 1989).

Primary infection by PVX and PLRV in cvs. Kenya Baraka and Roslin Eburu decreased tuber yields significantly (21 and 68% in Kenya Baraka and 10 and 35% in Roslin Eburu). Four commercial stocks of Roslin Eburu showed infection of PRLV (2 to 93%), PVY (2 to 17%), PVX (20 to 68%) and PVS (28 to 100%) and tuber yield reductions from nil to 48% as compared to healthy certified stock (Robertson, 1978).

7.6.8 Rwanda is in the highlands of Central Africa. Most of the growers are small and grow potatoes as a subsistence crop. Potatoes are grown twice a year and farmers generally do not use chemical fertilizer and pesticides. The popular varieties are Sangema, Condea and Utila. Seed supply is mostly from informal system. Major constraints are late blight and bacterial wilt. Research on potatoes is carried by the Institute des Sciences Agronomiques du Rwanda (ISAR) which produced basic seed for further multiplication through other institutes (Durr, 1983).

7.6.9 Uganda. Potato is an important food crop and source of income. Most potato production is on a subsistence level in small farms in the southwest and eastern highlands. Following the release of new, widely adapted potato cultivars in 1991, potato growing expanded rapidly in the mid- and low-elevation areas of the country. Lack of low-cost clean seed is still the main constraint for increasing production and yields (Alacho *et al.* 1994). The improved new varieties are Victoria, Kisoro, Kabale, Malirahinda, Rutuku and Sangema, in addition to the local cultivar 'Mali-Kabera'. Six TPS progenies tried in three districts of southern and central Uganda gave good yield (average 26 t/ha) (Alacho *et al.* 1994).

7.6.10 Yemen AR. Three potato crops are grown in a year. Ratio of area in the highland to lowland is 10:4. A Potato Seed Production Centre (SPPC) was established in 1977 in the highlands to multiply imported seed potatoes. It produces 2500 to 3000 t of seed tubers in two multiplications from 100 t of basic seed imported annually. Seed produced by SPPC meets only ca. 12%

of the seed requirement while 88% is supplied by the informal seed system. Degeneration due to viruses is low in the highlands and farmers store their seed locally due to low ambient temperature. Lowland farmers produce only ware potatoes and have to obtain new seed every year from the highlands. Popular cvs. are Diamant, Baraka and Desiree (Struik and Wiersema, 1999).

7. 7 The Americas

7.7.1 Mexico: Both local bred/selected and introduced potato cultivars are grown. They include Alpha, Jwanita, Michoacan, Nortena, Rosita, Tollocan, Atzimba, Lopez, Marciana, Murca, Puebla, Gigant, Mondial, Snowden, Atlantic, Russet Burbank, Montsama ATM, Granola, BGPM-6, Timate, Sangema, Diamante and Greta. The main virus vectors are *Aphis fabae, Macrosiphum euphorbiae* and *Myzus ornatus* (Pinto and Cardenas-Alonso, 1990). Yellow spot disease caused by AlMV was reported (Diaz-Valasis *et al.* 1992). An isolate of PVX designated as 'mosaic isolate of PVX' induced severe mosaic in many cultivars (Diaz-Valasis *et al.* 1995).
Tuber seeds produced from virus-free potato cultivars were practically free from viruses, in field experiments conducted during 1993-97(Cadena-Hinojosa *et al.* 1999).

7.7.2 Cuba. The main cultivars are Desiree, Baraka, Red Pontiac, Kandor, Atlantic, CIP-110, CIP-115, CIP-23, CIP-114, Lizette, Ibis, Claudia, 6-370 (Cubanita), 6-5 (Jorinca), Alkon, Maradona, Spunta, Origo and Aninka. Lizette and Ibis show good field resistance to *Alternaria solani* and viruses and yielded 48 and 35 t/ha, respectively (Estevez *et al.* 1998). Aninka (6-453), Marinka (6-126), Cubanita and Jorinca were the first varieties bred (Estevez *et al.* 1994).

Rate of degeneration due to viruses and their economic significance are given in tables 3 and 4 (Lago and Perez, 1990)

In Cuba exist 14 well-equipped provincial laboratories using ELISA to detect viruses at commercial scale, and NASH for specialized uses (Lago and Perez, 1990

Table 7. 2 Rate of degeneration due to viruses in Cuba.

Province	% of virus disease	
	Ist reproduction (PVX+PVYY+PLRV)	2nd reproduction (PVX+PVY+PLRV)
Pinar del Rio	0.35-5.4	6-29
Habana	0.35-5.4	15-30
Villa Clara	0.38-3.05	44
Matanzas	0.60-2.8	4.6

Table 7.3 Economic importance of potato virus disease in different var. in Cuba

Var.	Disease	Yield (t/ha)	
		Healthy	Diseased
Red Soda	PLRV	25.2	10.0
Viking	PLRV	13.5	7.1
Cariboo	PLRV	18.2	12.4
Red Soda	PVY	25.2	10.0

7.7.3 Latin America - General . The sub-species *andigena* is prevalent at high altitudes of the Andes. Outside the Andes the sub-species *tuberosum* is grown, which is adapted to long day photoperiod and require comparatively higher temperature for foliage and tuber growth than the *andigena.*

About half of the potato production in Latin America is in the Andes. In the Andes potatoes have been cultivated for centuries, with good seed as multiplication of the potatoes is at high altitudes. The farmers at lower altitudes periodically renew their seed stocks from higher altitudes. Opening of new production areas in the lower altitudes of Andes during the recent past, required an improved seed production system to reduce degeneration rates, usually found in the tropical and sub-tropical regions. (Ezeta, 1987).

7.7.3.1 Colombia. Potato is grown on about 162,000 ha with an annual production of 2705,000 tonnes, and an average yield of 16 t/ha. In addition to the common potato viruses, other important viruses and virus resistant varieties/clones are given in Table 7.1. *M. persicae*, a vector of PLRV and PVY was predominant at the 6 sites studied (Meneses, 1990).

7.7.3.2 Ecuador. Potato production is primarily an activity of the small highlands farmer. Seed production from the official program is at a low level and utilized mainly by a few large farmers within a small radius of the government seed production farm. Four potato varieties have been released by the government-funded agencies and account for a significant portion of the seeded area. The development of a rapid multiplication system for pre-basic seed production substantially improved the health quality of official seed. The variety INIAP Santa Catalina released in 1966, was the result of initial screening from Colombian materials. Several promising clones were developed and released as INIAP Maria, INIAP Gabriela and INIAP Espernanza in the 1980s. Most of the breeding work used ssp. *tuberosum* as a female parent and ssp. *andigena* as male parent (Crissman and Uquillas, 1989).

7.7.3.3 Brazil. The National Research Centre for Vegetable Crops (CNPH), EMBRAPA indexes the pre-basic seed, advanced clones and potato germplasm for viruses and viroids. Brazil has reduced its seed imports from 16,000 t to 3500 t (Ezeta, 1987).

In South-Central Brazil, vectors of PVY and PLRV are *M. persicae, Macrosiphum euphorbiae* and *Aulacorthum solani.* Early planting and harvesting avoided widespread infection of the crop (Bertels *et al.* 1971). In the states of Parana, Santa Caterina and Rio de Janeiro *Myzus nicotianae* is an efficient vector of PLRV (Cupertino *et al.*1995). *Solanum aculeatissimum, S. lycocarpum, S. paniculatum, S. variabile* and *S. erianthum* are reservoirs of PLRV in Brazil (Souza-Dias *et al.* 1993).

Cultivars commonly grown in Brazil are Dutch Bintje, Swedish Bintje, Jatte Bintje, Aracy, Elvira, Itarare, Radosa, Tabere, Delta S, Baraka, Achat, Itaiquara, Piraquara and Apua.

Incidence of PLRV, PVY and PVX was determined by field inspections and biological tests on 3rd, 4th and 9th generation plants from originally certified seed lots of 10 European cultivars. Cumulative incidence of PLRV and PVY after 9 generations was very high in most cultivars in Goias (Camara *et al*. 1986).

Andean potato mottle virus strain-B (APMV-B) from Brazil caused severe mosaic and crinkle on potato plants in Santa Catarina (Avila *et al.* 1984). *M. persicae* activity and PLRV infection were found round the year at Itarara (Souza Diaz *et al.* 1984). The 'Bouquet' strains of tomato black ring virus and TRV have also been reported from Brazil.

7.7.3.4 Bolivia. Potato production is in three zones: (i) Valleys at an altitude of less than 2000m. Here disease pressure is high and potato is grown as a cash crop. (ii) Higher valleys located between 2000 to 3000 m, and potato is a cash crop, and both *tuberosum* and *andigena* varieties are grown. (iii) Andean region above 3000 m. Disease pressure is low, and farmers use their own seed for 4 to 5 generations without appreciable degeneration. Potato is a subsistence crop in these areas. Only *andigena* potato varieties are grown in this zone. Seed production for *andigena* potatoes is through the traditional system followed in the Peruvian Andes. Clonal and positive selection techniques are employed by the National Programme on a small scale to produce better quality seed (Struik and Wiersema, 1999).

7.7.3.5 Chile: Viruses and sources of virus resistance are given in Table 7.1. Incidence of PVX, PVY, PVS and APLV ranged from 40, 17-37, 53-88 and 13-33%; respectively. In the central zone of Chile virus incidence was higher than in the southern part (Apablaza, 1980). Important cvs. are Serrana INTA, Line 76, Line 536 (ME), Line 21, Desiree, Spartan, Urgenta, Arka, Ultimus, Corahila, Grata.

Potato yellowing virus (SB-22) (alfamovirus) was also detected in Chile (Fuentes and Jayasinghe, 1993). Results of a survey (1981-82) in Chile (fifth region) showed that PVY occurred in 59% of potato samples, PVS in 38%, PVX in 24% and APLV in 14% of the total samples (Saldias and Apablaza, 1984).

Apical meristem culture, chemo- and thermotherapy is used to eliminate viruses and ELISA to detect viruses. Chile's certified seed production program has operated uninterruptedly since 1959. The contribution of private companies, as Segenta Ltda., Semillas S/Z, Anasac, Porosecor and individual seed producers has gained wide recognition. The Chilean programme also benefits from its location in one of the most favourable world regions for the production of seed potato. The region is characterized by its cool moderate climate, its deep and permeable soils, and the absence of plagues and diseases. Over the last 20 years, an average of 6852 t of seed tubers were produced annually ((Santos Rojas, 1990).

7.7.3.6 Peru: Agriculture is practised in the lower lying river basins, which are best regarded as gaps between sections of the highland plateau. Agricultural land is extremely scarce in the highlands. Potatoes in the Montaro valley are grown between 3000 to 4000 m. Viruses and sources of resistance are given in Table 7.1

The important cultivars are Arran Piolot (group *tuberosum*), Coompis, Chata Blanca, Mantaro, Renacimiento, Sipena, Tarmenia (all *andigena*), Antarki and Marvia.

7.7.3.7 Argentina: Seed tuber production involves cooperation between the government-funded agency National Seed Service Agency (NSSA) and the private company Diagnosticos Vegetables R.L. (DV). NSSA undertakes certification by field inspections, and sends samples to DV for post control diagnosis of virus diseases with ELISA. The growers of non-certified seed also use the post-harvest control test to determine the sanitary conditions of their seed crops. Other services of DV to seed potato growers include: (1) Virus testing during the growing season; (2) Nematode diagnosis and (3) Monitoring aphid population to determine haulm killing date; (4) and advice on seed borne disease epidemiology.

In 1985, DV started a new potato project, Atlantida Semillas, to produce pre-basic seed from *in vitro* cultivated plantlets, grown under controlled green house conditions. CIP has helped DV by providing *in vitro* pathogen-tested cultivars and training their personnel. Atlantida Semillas has commercialized pre-basic seed in the Argentina seed potato market with excellent results (Escarra, 1990).

7.8 Future Perspectives

There is a great diversity in temperatures and day length during potato crop growth in countries with warm climates. Thus, a wide range of potato cultivars suited to these diverse agro-ecological situations is required. At the same time, the majority of these countries do not have their own variety-breeding program. This problem has been solved to some extent through collaboration with CIP, which has helped the National Potato Programmes to select suitable clones, many of which have been released as varieties in some of these countries. With the help of CIP, many countries also established facilities for 'cleaning' the important potato varieties through meristem culture and thermotherapy, and initiated production of pre-basic seed through *in vitro* rapid multiplication. However organised seed production does so far not meet the requirements of the growers. The only exceptions are India (in Asia), Argentina, Brazil, Chile, Mexico, Cuba, Peru and Ecuador (in North, Central and Latin America). Countries where vectors are active around the year depend on import of healthy seed every year, which are then multiplied for a limited time.

Some African countries like Burundi, Rwanda, Kenya, Ethiopia, Uganda and a few Asian countries like Viet Nam, a large proportion of potato is grown as a subsistence crop. Since seed constitutes more than 50% of the cost of production for growers in these countries, alternative technology of using TPS for potato production is being evaluated. The TPS technology holds promise for smallholding growers in these countries thereby enhancing potato production.

In most of the Asian countries, there is a need to increase potato production by providing healthy seed in sufficient quantities and at cheaper prices. Favourable conditions for the production of healthy seed exist in many of these countries and promoting participation of the private sector can increase certified seed production. Some private sector investment has already taken place in some of these countries and is expected to increase.

Availability of GM potato with desirable attributes may boost potato production in these countries thereby ensuring their food and nutritional security.

References

Abdellatif, M. 1995. Seed multiplication in Syria. In: *True Potato Seed in the Middle East and Africa.* B.Hardy, P.Malagamba and C.Martin (eds.). International Potato Center (CIP). Lima, Peru pp. 53-55.

Adam, G., Peters, D. and Goldbach, R.W. 1996. Serological comparison of tospovirus isolates using polyclonal and monoclonal antibodies. *Acta Hortic* 431: 135-158.

Alacho, F.O., Kajuhanzire, R.M., Kanzikwera, R.C. and Sikka, L.C. 1995. Experiences in TPS research and development in Uganda. In: *True Potato Seed in the Middle East and Africa.* B.Hardy, P.Malagamba and C.Martin (eds.). International Potato Center (CIP), Lima, Peru. pp.83-93.

Apablaza, G. 1980. Important virus diseases of potato in Chile (1980). In: *Strategy for Virus Management in Potatoes.* Report of a Planning Conference. April 21-25, 1980. International Potato Center (CIP), Lima, Peru. pp.140-142.

Avila, A.C., Salazar, L.F., Ortega, M. and Daniels, J. 1984. A new strain of Andean potato mottle virus from Brazil. *Plant Dis.* 68: 997-98.

Berrios, D.E. 1995. True potato seed : an alternative to improve potato production in Burundi. In: *True Potato Seed in the Middle East and Africa.* B. Hardy, P.Malagamba and C.Martin (eds.). International Potato Center (CIP), Lima, Peru. pp.17-24.

Bertels, A., Ferreira, E. and Casagrande, W. 1971. Problems of vectors of potato viruses and their control under the conditions of Rio Grande do Sul. *Pes-Agro Bras Agronomia* 6: 291-306.

Borgel, H., Arend, B., Jacob, C. et al. 1980. Production, marketing and consumption of potatoes in the Ethiopian Highlands (Holetta, Awassa, Alemaya). Research Co-operation Addis Ababa University and Technical University Berlin-West. 217p.

Brown, K.J. 1982. Current developments in the research program of CIP in relation to seed potato production. In: *Potato Seed Production for Tropical Africa.* S.Nganga and F.Shideler (eds.). CIP, Lima, Peru. pp.2-4.

Brune, S., Melo, P.E.and Avila, A.C.1999. New potato progenitors immune to PVY and PVX, and resistant to early blight. *Hortic. Bras.*17: 173-174.

Cadena Hinojosa, M.A., Diaz-Valasis, M., Lopez Delgado, H., Zavala-Quintana; T.E., Flores Lopez, R. et al. 1999. Evaluation of potato varieties for tuber seed production in the states of Mexico and Tlaxcala. *Agricultura Tecnica Mexico* 25: 97-105.

Camara, F.L.A., Cupertino, F.P. and Filgueira, F.A.R. 1986.Incidence of viruses in consecutive plantings of potato cultivars in Goias. *Fitopatol Brasil.* 11: 711-716.

Contreras, M.A. and Banse, H.J. 1982. Virus determination in Chilean potato germplasm (Solanum sp.). *Agro-Sur.* 10: 84-89.

Corsini, D.L. and Brown, C.R. 2001. Important potato cultivars. In: *Viruses and Virus-Like Diseases of Potatoes and Production of Seed Potatoes.* G.Loebenstein, P.H. Berger, A.A. Brunt and R.H. Lawson (eds.). Kluwer Academic Publishers, Printed in the Netherlands. pp.39-52.

Costa, A.S., Miranda, H., Granja,, N.P. and Vega, J. 1988. Mosaics amarelo de mosca branca, virose de batata introduzida em Sao Paulo/A type of whitefly transmitted potato yellow mosaic introduced in the state of Sao Paulo. *Summa Phytopathol.* 14: 35 (abstr.)

Crissman, C.C. and Uquillas, J.E. 1989. *Seed Potato Systems in Ecuador. A Case Study.* International Potato Centre (CIP), Lima, Peru. 70p.

Crissman, C.C., Crissman, L.McA. and Carli, C. 1993. *Seed Production Systems in Kenya : A Case Study.* International Potato Center. (CIP), Lima, Peru. 44p.

Crissman, L.M., 1989. Evaluation and choice of potato varieties in Kenya. *Working Paper of Social Sciences Department, Series No.1989*-1. International Potato Center. (CIP), Lima, Peru. 44p.

Crosslin, J.M. and Thomas, P.E. 1995. Detection of tobacco rattle virus in tubers exhibiting symptoms of corky ringspot by polymerase chain reaction. *Am. Potato J.* 72: 605-609.

Cupertino, F.P., Silva, A.M.R., Dusi, A.N., Avila, A.C., Costa, C.L. and Buso, J.A. 1995. Transmission of potato leafroll virus by the red tobacco aphid, Myzus nicotianae. *Fitopatol Brasil* 20: 440-43.

Danesh, D. and Lockhart, B.E.L., 1989. Eggplant mottled dwarf virus in potato in Iran. *Plant Dis.* 73: 856-858.

Danesh, D., Soleimanian, S, Filsouf, F. and Dehghan, M. 1992. Incidence of four potato viruses in the Esfahan region of Iran. *Iranian J Plant Pathol* 28: 1-9.

Daniels, J.2000. Evaluation of potato genotypes for resistance to potato virus Y. *Hortic. Bras.* 18: 145-147.

Daniels, J. and Castro L.A.S. 1985. Occurrencia do virus do mosaico deformante da batata no Rio Grande do Sul. *Fitopatol Bras.* 10: 306 (abstr.)

De Avila, A.C., de Haan, P., Smeets, M.M.L., Resende, R.de. O., Goldback, R. and Peters, D. 1993. Distinct relationships between tospovirus isolates. *Arch. Virol.* 128: 211-227.

Diaz-Valasis, M., Cadena Hinojosa, M.A., Torres Vazquez, E and Cardenas, S.E. 1992. Aetiology of yellow spot of potatoes grown on the central plateau of Mexico. *Revista Maxicana Fitopatologia* 10: 60-63.

Diaz-Valasis, M., Cadena-Hinojosa, M.A., Garcia Espinoza, R. and Estanol Botello, E. 1995. Effect of two isolates of potato X potexvirus in the main potato (Solanum tuberosum) cultivars grown in Mexico. *Rivista Mexicana Fitopatologia* 13: 10-17.

Durr, G. 1983. *Potato Produciton and Utilization in Rwanda.* International Potato Center (CIP) Social Science Department Working Paper 1983-1, CIP, Lima, Peru. 85p.

Elnagar, S., El-Sheikh, M.A.K., Makland, F.M. and Kabeil, S.S.A. 1996. Substantial loss in potato yield due to the early population of vectors of two aphid borne viruses in Egypt. *Bulletin Facult. Agr. Univ. Cairo* 47: 677-690.

El-Sadany, M.M. and Salam, F.E. 1991. *Potato situation and outlook.* USAID Project Report 1991, Cairo, Egypt. 53p.

Escarra, A.M. 1990. Diagnosticos Vegetales S.R.L. – Success in private-public interaction. In: *Control of Virus and Virus-Like Diseases of Potato and Sweet Potato.* International Potato Center (CIP). Report of the 3rd Planning Conference, 20-22 November, 1989. CIP, Lima, Peru. pp.83-85.

Estevez, A., Gonzalez, M.E., Castillo, J., Ortiz, U. and Cordero, M. 1998. Lizette and Ibis, two new Cuban potato varieties. *Cultivos Tropicales* 19: 57.

Estevez,A., Gonzalez, M.E. and Cordero, M. 1994. First Cuban varieties of potato. *Cultivos Tropicales* 15: 74-79.

Ezeta, F.N. 1987. Recent developments in potato seed production in Latin America. In: *Workshop on Recent and Future Developments of the Potato in the World*, April 7-8, 1987, Wageningen. pp.91-94.

FAO 2000. Database at http://apps.fao.org.

FAO 2001. FAO Production Yearbook 1999 Vol.53.. FAO, Rome, pp.92-94.

Fribourg, C.E. 1980. History and distribution of potato viruses in Latin America. *Fitopatologia* 15: 13-24.

Fuentes, S. 1991. Identificacion, caracteristicas, y distribucion de un virus baciliforme aislado de papa (*Solanum tuberosum*). MSc. Thesis, Universidad Nacional Agraria La Molina. Lima, Peru.

Fuentes, S. and Jayasinghe, U. 1993. Potato yellowing, caused by a new bacilliform virus. *Fitopatol* 28: 22-37.

Garg,I.D. and Khurana,S.M.Paul . 1992. Factors influencing immune electron microscopy of flexuous potato viruses. *Acta Virol* 36: 435-442.

Garg, I.D., Hegde,V. , Khurana,S.M.Paul and Singh,R.A. 1999. Ultrastructural studies on different host-tospovirus combinations. *Indian Phytopathol* 52: 215-220.

Garg, I.D., Hedge, V. and Khurana, S.M. Paul. 2000. Effect of pH of antisera, bovine serum albumin and ions on the stability of immunosorbed flexuous potato viruses. *Acta Virol.*44: 67-72.

Garg, I.D., Khurana, SM Paul, Kumar, Shiv and Lakra, B.S. 2001. Association of geminivirus with potato apical leaf curl in India and its immuno-electron microscopic detection. *J. Indian Potato Assoc* 28: 227-232.

Gitomer, C.S. 1996. *Potato and Sweet Potato in China: Systems, Constraints and Potential.* International Potato Center (CIP), Lima, Peru. 191p.

Granval de Millan, N., Piccolo, R. and Garcia, O. 1998. Potato (*Solanum tuberosum* L.) tuber transmission of tospoviruses (TSWV, GRSV and TCSV) in Mendoza, Argentina. In: *Recent Progress in Tospovirus and Thrips Research.* Abstracts of papers and poster presentation at the Fourth International Symposium on Tospoviruses and Thrips in Floral and Vegetable Crops, 2-6 May, 1998. Wageningen, The Netherlands, pp. 46-47.

Hanafi, A., Redcliffe, E.B. and Ragsdale, D.W. 1995. Spread and control of potato leafroll virus in the Souss Valley of Morocco. *Crop Prot.* 14: 145-153.

Hattab, M.M., Charif, C., Hattab, H. and Mnari-Hattab, M. 1994. Study on the multiplicaiton of seed potatoes in Tunisia: virus diseases. *Annales Institut National Recherche Agronomique Tunisie* 67: 31-43.

Hijmans, R.J. 2001. Global distribution of the potato crop. *Am. J. Potato Res.* 78: 403-412.

Ho, T.V., Tuyet, L.T., Tung, P.X. and Vander Zaag, P. 1987. *Potato research and development in Vietnam in recent years.* CIP Circular No.15(3): 1-15.

Hooker, W.J. and Salazar, L.F. 1983. A new plant virus from the high jungle of the Eastern Andes : Solanum apical leaf curling virus (SALCV). *Ann.Appl Biol.* 103; 449-454.

Horvath, J., Wolf, I., Bosze, Z., Pocsai, E. and Kazinczi, G. 1996. Is tobacco rattle tobravirus transmissible by true potato seeds. In: Proc. *9th EAPR Virology Section Meeting*. 18-22 June, 1995. Bled. pp. 71-72.

Huaman, Z. 1989. Source of resistance to viruses in Andean potato cultivars maintained at CIP. In: *Control of Virus and Virus-like Diseases of Potato and Sweet Potato.* Report of the 3rd Planning Conference, pp.117-123. International Potato Center, Lima, Peru.

Huguenot, C., van den Dobbelsteen, G., de Haan, P., Wagemakers, C.A.M., Drost, G.A., Osterhaus, A.D.M.E. and Peters, D. 1990. Detection of tomato spotted wilt virus using monoclonal antibodies and riboprobes. *Arch. Virol.* 110: 47-62.

Ioannou, N. 1989. The infection pressure of potato leafroll virus and potato virus Y in relation to aphid populations in Cyprus. *Potato Res* 32: 33-47.

Jeffries, C.J. 1998. FAO/*IPGRI Technical Guidelines for the Safe Movement of Germplasm No.19. Potato.* Food and Agriculture Organization of the United Nations Rome/International Plant Genetic Resources Institute, Rome. 177p.

Jun, W., Bofu, S. and Hunbao, H. 1994. Improvement of seed potato production technique in Southwest China. *Proc. 4th Asian Potato Association Triennial Conf. Vol.I*, July 5-7, 1994. ET Rasco, Jr. and FB Aromin (eds.), Daekwanreong, Korea. pp.172-176.

Khurana, S.M. Paul. 1999. *Potato Viruses and Viral Diseases.* Technical Bulletin No.35 (revised). Central Potato Research Institute, Shimla. 94.p

Khurana, S.M. Paul and Garg, I.D., 1993. New techniques for detection of viruses and viroids. In: *Advances in Horticulture* Vol.7. K.L. Chadha and JS Grewal (eds.) Malhotra Publishing House, New Delhi, India. pp. 529-566.

Khurana, S.M. Paul, Pandey, S.K., Singh, R.B. and Bhale, U. 1997. Spread and control of the potato stem necrosis. *Indian J Virol* 13: 23-28.

Khurana, SM Paul, Bhale, U. and Garg, ID. 2001. *Stem Necrosis Disease of Potato.* Technical Bulletin No.54, Central Potato Research Institute, Shimla. 5p.

Kim, Y.C., Kim, K.K. and Hahn, B.H. 1986. Impact of the South Korean seed potato program. International Potato Center (CIP), Circular 14(4): 1-5.

Kokab, A.A. 1986. Survey on seed supply, multiplication and degeneration of potato in the plains of Punjab 1985-86. *Pakistan-Swiss Potato Development Programme.* Pakistan Agricultural Research Council, Islamabad, Pakistan. 43p.

Lago, L. and Perez S. 1990. Virus and viroid detection: strategies in Cuba. In: *Control of Virus and Virus-like Diseases of Potato and Sweet Potato.* International Potato Centre (CIP). Report of the 3rd Planning Conference, 20-22 Novembner,1989. CIP, Lima, Peru. pp.89-96.

Lauber, E., Bleykasten-Grosshans, C., Erhardt, M., Bouzoubaa, S., Jonard, G, Richards, K.E. and Guilley, H. 1998. Cell-to-cell movement of beet necrotic yellow vein virus. 1. Heterolog-complementation experiments provide evidence for specific interactions among the triple gene block proteins. *Mole. Pl. Microbe Interact.* 618-625.

Malik N.J. 1995. *Potatoes in Pakistan – A Handbook.* Pakistan-Swiss Potato Development Project. Pakistan Agricultural Research Council, Islamabad, Pakistan. 274p.

Martin C., Demo, P., Njualem, D., Gass, T. and Ntonifor, C. 1995. Late blight resistant varieties for Cameroon. *Am. Potato J.* 72: 513-522.

Mills, P.R. 1987. Comparison of cDNA hybridization and other tests for the detection of potato mop top virus. *Potato Res* 37: 373-381.

Morales, F.J., Lastra, R, de Uzcategui, R.C. and Calvert, L. 2001. Potato yellow mosaic virus: A synonym of tomato yellow mosaic virus. *Arch Virol* 146: 2249-2253..

Nagaich, B.B., Bhardwaj, V.P., Giri, B.K., Anand, S.R. and Upreti, G.C. 1969. Production of disease free seed potatoes in the Indo-Gangetic plains. *Indian J. Agric. Sci.* 39: 238-243.

Nagata, T., De Avila, A.C., Tavares, P.C.T., Barbosa, C. de J., Juliatti, F.C. and Kitajima, E.W. 1995. Occurrence of different tospoviruses in six states of Brazil. *Fitopatol. Bras.* 20: 90-95.

Peters, D. 1998. Plant hosts of tospoviruses. http://www.spg.wau.nl/viro/research/hostlist.html.

Pinto, V.M. and Cardenas Alonso, M. 1990. Detection of potato (*Solanum tuberosum* L.) viruses and evaluation of their incidence in the municipality of Nauzontla, Puebla. *Revista Chapingo* 15: 67-68, 87-89.

Rabbani, M.G. 1994. Prospects of using plant tissue culture techniques for seed potato production in Bangladesh. Proc. *Asian Potato Assoc. 4th Triennial Conf. Vol.I.* ET Rasco, Jr. and FB Aromin (eds.), Daekwanreong, Korea, pp. 191-195.

Ranjit, M., Rai, G.P., Manandhar, A. and Pandey, V. 1994. Virus testing and elimination of viruses from some local cultivars of potato (Solanum tuberosum L.) in Nepal. Proc. *Asian Potato Association 4th Triennial Conference Vol.II.* E.T. Rasco, Jr., F.B. Aromin and C.H. Balataro (eds.). Daekwanreong, Korea, pp. 95-99.

Resende, R. de O., de Avila, A.C., Goldbach, R.W. and Peters, D. 1991. Comparison of polyclonal antisera in the detection of tomato spotted wilt virus using the double antibody sandwich and cocktail ELISA. *J.Phytopathol* 132: 46-56.

Rizk, A., Sabaa, M.F. and El-Bedewy, R. 1991. On-farm agronomic evaluation of different types of potato planting material in Egypt, 1990. *Social Sciences Department Working Paper Series No.1991-5*. International Potato Center (CIP), Lima, Peru. 9p.

Roberts, E.J.F., Buck, K.W. and Coutts, R.H.A. 1988. Characterization of potato yellow mosaic virus as a geminivirus with a bipartite genome. *Intervirology* 29: 162-169.

Robertson, D.G. 1978. The effect of virus diseases on the yield of potatoes. In: *Pathogens and Pests of the Potato in The Tropics*. Proc. 2nd Regional Symp. Potato Production in South East Asia and The Pacific, Feb. 5-16, 1978. Los Banos, Laguna, Philippines.

Robinson, D.J. 1989. Tobacco rattle tobravirus : variation among strains and detection by cDNA probes. *EPPO Bull*. 19: 619-623.

Robinson, D.J. 1992. Detection of tobacco rattle virus by reverse transcription and polymerase chain reaction. *J.Virol. Methods* 40: 57-66.

Salahuddin, A.B.M. 2000. The progress of research in potato and its status in Bangladesh. In: *Potato Global Research and Development Vol.I*. SM Paul Khurana, GS Shekhawat, BP Singh and SK Pandey (eds.). Indian Potato Association, Shimla, India. pp.156-162.

Salazar, L.F. 1996. *Potato Viruses and Their Control*. International Potato Center (CIP), Lima, Peru. 214p.

Salazar, L.F., Muller, G., Querci, M, Zapata, T.L. and Owens, R.A. 2000. Potato yellow vein virus : its host range, distribution in South America and identification as a crinivirus transmitted by Trialeurodes vaporariorum. *Ann Appl Biol*. 137: 7-19.

Saldarriaga, A., Alvarez, A.M. and Jaramillo, J.E. 1988. Effecto del emarillamiento de venas transmitido por Trialeurodes vaporariorum (Westwood) en papa. *Revista Colombiana Entomol* 14: 3-8.

Saldias, G. and Apablaza, G. 1984. Survey of 5 viral diseases on potato and tobacco of the 5th region in Chile. *Ciencia Investigacion Agraria* 11: 129-134.

Sandgren, M., Savenkov, E.I. and Valkonen, JPT. 2001. The read through region of Potato mop top virus (PMTV) coat protein encoding RNA, the second largest RNA of PMTV genome, undergoes structural changes in naturally infected and experimentally inoculated plants. *Arch Virol* 146: 467-477.

Santos Rojas, J. 1990. Production of seed potato in Chile. In: International Potato Center (CIP). *Control of Virus and Virus-like Diseases of Potato and Sweet Potato*. Report of the 3rd Planning Conference, 20-22 November, 1989. CIP, Lima, Peru, pp.179-183.

Santos-Rojas, J., Kalazich, J., Hidalgo, O.A. (ed.) and Rincon, R.H. (ed.) 1990. Evaluation of resistance to field infection with potato leafroll virus (PLRV). Avences-en-el-mejoramiento-genetico-de-la-papa-en-los-paises-del-cono-sur. 101-110.

Savenkov, E.I., Sandgren, M. and Valkonen, JPT 1999. Complete sequence of RNA 1 and presence of t-RNA-like structures in all RNAs of Potato mop-top pomovirus. *J. Gen.Virol.* 80: 2779-2784.

Sayedoleslaami, H., Danesh, D., Naderi, A and Eslami, A. 1995. Alate aphid monitoring for selecting of potential seed potato production districts in Isfahan and Chahr Mahal Bakhtiari provinces of Iran. *Iranian J. Agr Sci* 26: 19-25.

Shekhawat, G.S. and Naik, P.S. 1999. *Potato in India*. Technical Bulletin No.1 (Revised), Central Potato Research Institute, Shimla, India. 99p.

Souza-Dias, J.A.C.-de, Costa, A.S. and Nardin, A.M. 1993. Potato leafroll virus in solanaceous weeds in Brazil explains severe outbreaks of the disease in the absence of known potato donor sources. *Summa Phytopathol* 19: 80-85.

Souza-Dias, J.A.C.-de, Yuki, V.A., Miranda, H.S. and Costa, A.S. 1984. Dissemination of potato leafroll virus and population fluctuation of the vector at the Experiment Station of Itarare, Sao Paulo. *Summa Phytopathol.* 10: 74-76.

Struik, P.C. and Wiersema, S.G. 1999. *Seed Potato Technology Wageningen Pers*, Wageningen, The Netherlands. 383p.

Taghian, A.S., Abd El Moneem and FGM Fahmy. 1997. Establishment of a superior potato somaclone and the differential gene expression during *Alternaria solani* and potato mosaic viruses infection. *Assiut J. Agric. Sci.* 28: 45-46.

Tanaka, S. 1994. Development of propagation schemes and techniques of seed potato production in West Java. Proc. *APA 4th Triennial Conf. Vol.I.* ET Rasco, Jr. and FB Aromin (eds.). Daekwanreong, Korea, pp.143-149.

Theoduloz, C., Saez-Vasquez, J., Poblete, F., Contreras, A., Hubert, E. and Meza-Basso, L.1992. The incidence of potato viruses X, Y and S in the Chilota potato collection. *Am. Potato J.* 69: 827-830.

Vakis, N.J. 1990. Evaluation of potato varieties and clones in Cyprus, 1983-89. *Miscellaneous Report of Cyprus Agricultural Research Institute 1990*, No.38, 22p.

Valkonen, J.P.T., Contreras, M., Pehu, E. and Salazar, L.F. 1992. Naturally occurring viral infections in Solanum brevidens and S.fernandezianum. *Potato Res.* 35: 411-417.

Vashisth, K.S., Verma, A.K. , Chaubey, I.P. and Nagaich, B.B. 1981. Comparative degenerative effects of viral and MLO – diseases during autumn and spring seasons on Indian potato varieties in Punjab. *Seeds & Farms* 7: 125-28.

Verma, K.D. and Chandla, V.K. 1999. *Potato Aphids and Their Management.* Technical Bulletin No.26 (revised). Central Potato Research Institute, Shimla, India. 34p.

Weidemann, H.L. 1995. Detection of tobacco rattle virus in potato tubers and roots by polymerase chain reaction (PCR). *J.Phytopathol.* 143: 455-458.

Zhang, H., Salazar, L.F. and Bo Fu, S. 1990. The advances in virus testing in China. In: *Control of Virus and Virus-Like Diseases of Potato and Sweet Potato.* Report of the 3rd Planning Conference, 20-22 November, 1989. CIP, Lima, Peru. pp.73-81.

F

Legends to figures:

Fig. 7.1 - Yellow blotches and incomplete chevrons induced by PMTV

Fig. 7.2 - Apical leaf curl induced by PALCV

Fig. 7.3 -Bright yellow veins induced by PYVV (courtesy Dr. L.F. Salazar)

Fig. 7.4 -Necrosis induced by TSWV

8. Potatoes in the Central Asian Republics

G. Loebenstein and A. Manadilova
Dept. of Virology, Agricultural Research Organisation, Bet Dagan, Israel; and Institute of Molecular Biology and Biochemistry, Almaty, Kazakhstan

8.1 Introduction

Potatoes are an important crop in the Central Asian Republics. They are the second most important crop in Kazakhstan, after wheat, grown on about 160.000 ha. In Kyrgistan more than 65.000 ha are planted, and smaller acreages in the other Central Asian Republics (Table 8.1). Average yields, however, are low ranging from 5 tons/ha in Turkmenistan to 15 tons/ha in Kyrgistan, compared with yields of 40 tons/ha and more in Western Europe, USA and Israel. In Kazakhstan, the largest potato grower in the Central Asian Republics, yields average 10 tons/ha or less. Potatoes in Kazakhstan and Kyrgistan are planted in April-May, depending on the region and weather and harvested in September- October. In Uzbekistan potatoes are planted in early spring (end of February- beginning of March) and in the summer (June- July). In the northern regions and foothills of Uzbekistan potatoes are planted from end of May- till June. In Tadjikistan potatoes are planted from October till March. "Seed" tubers are at present taken from the same fields grown for market potatoes, and not from special "Certified

Seed" production farms. The low yields are due to adverse weather conditions, lack of a continuous water supply and low quality seed potatoes, often highly infected by viruses. Reliable information is scarce and most information comes from Kazakhstan (Loebenstein, and Sadvakasova, 1997). The main potato cultivars in Kazkhstan are Nevsky , Sante, Polet, Ogonek, Tamasha, Nerly, Tochtar (G. Ligai, personal communication), and in Uzbekistan Belorusskiy Ranniy (early), Sedov, Detskolselskiy, Lorh, Akrab, Tuyimly, Umid, Alvara, Agave, , Ogonek,, Sante (Sh. Nurmatov, personal communication) and Picasso, Romano, Cardinal, Alvara, Orli and others (A.M. Abassov and A.I. Rasulov, personal communication). In Kyrgistan the important cultivars are Ogonek, Shershnevskii Early, Dalma, and Nevsky (Interregional Conference on Potatoes, 1991. In Tadjikistan the main cultivars are Lorch, Cardinal and Nevsky; less spread in some regions are Polet, Narymka and Ranjia Rosa (Aliev *et al.* 1997).

8.2 Virus surveys

Some early surveys in Kazakhstan, between 1975-1986, based on symptoms, with some verification using test plants and precipitin serology in droplets, showed high incidence – 70-80% of streak and crinkle mosaic. Most of these symptoms were associated with *Potato virus X* (PVX; Genus *Potexvirus*), *Potato virus Y* (PVY; Genus *Potyvirus*) and *Potato virus S* (PVS; Genus *Carlavirus*) (Eliseeva *et al.* 1978; Tulegenov *et al.,* 1986). *Potato leafroll virus* (PLRV; Genus *Polerovirus*) symptoms were also observed. Between 1994 and 1996 virus incidence was surveyed in various regions of Kazakhstan, using DAS-ELISA with a commercial (Boehringer) kit for PLRV and high quality antisera for PVX, PVY, PVS and *Potato virus M* (PVM; Genus *Carlavirus*). The later were prepared in Israel and Kazakhstan. Samples were taken from 12 locations in Kazakhstan, at different elevations above sea level, generally two times during the growing season. In the surveys three Kazakh (Russian) – Nevsky, Chortandinsk and Ogenek and 12 European cultivars as Desiree, Diamant, Jaerle, Escort, Impala and others were tested. The origin of the European cultivars was mainly from The Netherlands after being propagated for 2-5 generations in Kazakhstan.

As presented in Table 8.2 virus incidence was generally high often reaching more than 50%. Even in "elite" tubers of cv. Reserve from Zhanater PVX and PVM reached 57% and 25%, respectively. Potatoes are generally highly infected with PVM, mainly the local (Russian) cultivars Nevsky and Chortandinsk, but also cultivars originating from The Netherlands after having been grown for several years in Kazakhstan. Differences in infection rates were observed between locations. In the southern parts of Kazakhstan like the Almaty region virus incidence, including PVY and PLRV, was high, whereas in the Astana region infection rates by PVY were low. In the eastern and northern areas, as Narinkol and Bestobe, respectively, virus infection appeared to be lower. These areas might be considered for future growing of tuber seed potato.

In a preliminary survey of virus incidence in weeds done during 1998 in the Astana and Bestobe regions no infection, as tested by ELISA, was found in: *Amaranthus retroflexus, Chenopodium album, Sonchus arvensis, Convolvulus arvensis and Euphorbia falcate* sampled far from potato fields. However, in potato fields in the Astana area PVS and PVM were found in *Chenopodium album,* while in the Bestobe area no such infection was detected (V. Shvidchenko and Ljuba Sozinova, personal communication).

In several preliminary surveys of viruses in Uzbekistan, using serological tests and test plants, infection rates between 30-60% for PVX, 20-40% for PVM and 5-20% for PVY were observed in the Northern and Central regions. In the plain regions infection rates ranged between 30-80% for PVX, 20-60% for PVS and 5-30% for PVM. In the mountain areas only a few cases of PLRV were found (Anonymous, 1991). It should be mentioned that these data were based on drop precipitation serology and not on ELISA.

In Tajikistan PVX, PVS, PVM PVY and *Potato virus A (*PVA) have been reported to be widespread in the valley zones, compared with their incidence in mountain areas (Aliev *et al.* 1997).

Virus surveys were conducted in the northern part of Kyrgistan during 1969-1974 in the mountain (1500-1800) and hilly areas (800m). In two cultivars- Berlichingen and Shamrock infection rates of PVX were more than 90% in all areas, about 60% for PVS, between 0-45 % for PVY and 0-98% for PLRV. Infection rates of PLRV, PVY and PVM were significantly lower in fields gown above 1500m than in fields in the hilly area (Gaynullina and Gerasimova, 1974).

8.3 Aphid surveys

A few surveys for potato aphid vectors have come to our attention. Thus, P. Leimann in his Ph.D thesis at the Vavilov Institute, St. Petersburg - 1988 (personal communication) reported that in the Astana area the main aphid species, during 1988-1900, were (in decreasing order of importance): *Aphis frangolae, A. nasturtii, A. fabea, Macrosiphum solanii, Myzus persicae* and *Aulocorthum solanii*; *M. persicae* mainly at the end of the season and in low numbers. Alatae start to appear generally in June. Total numbers of aphids trapped in a Moericke trap during the whole growing period were 182, 256 and 102 in 1988, 1989 and 1990, respectively. Highest numbers of aphids appear during the 3rd week of July. Non-alatae also were also sampled from 100 leaves of potato. In 1988 30 aphids were counted and 225 and 105 aphids in 1989 and 1990, respectively. High numbers of ladybirds (*Adonia sp.*) were observed, with a maximum in August. The number of winged aphids – during the whole growing season, is very low. This area seems therefore suitable for growing seed potatoes, provided that virus free material is planted.

During 1994 and 1995 aphid trappings were done in 3 locations (Dr. R. Kh. Kadyrbekov, Institute of Zoology, Almaty, personal communication). Results are presented in Table 8.3. It is evident that in these areas in 1994 or 1995 the number of *Myzus persicae* did not reach the threshold for major virus dissemination, which in Central Russia is 50 aphids in one trap after emergence of the potato, or 10-20 aphids on 100 leaves. These data strengthen the suggestion that the Narinkol area may be suitable for seed potato production.

8.4 The Viruses

As mentioned above in 8.2 the main viruses located during the virus surveys were PVX, PVY, PLRV, PVS and PVM. Here a brief description of these viruses, their mode of transmission and methods for their detection are outlined

8.4.1 Potato X virus Genus *Potexvirus* (PVX)

Most isolates cause only mild leaf mosaic or symptomless infections, especially at higher temperatures. Yield reductions in such plants are generally less than 10%, even though they may contain high virus titers. These plants remain carriers of the virus and are important sources of infection. Other strains induce necrotic streaks, severe mosaic, crinkling and rugosity of leaves and may cause significant yield losses. In mixed infections with PVY, yield reductions of 50% may occur.

This is limited mainly to the *Solanaceae* and includes *Solanum tuberosum, Datura stramonium, Nicotiana* sp., *Lycopersicon esculentum, Petunia* sp., *Solanum nigrum* and *Cyphomandra betacea.* PVX also infects *Brassica campestris ssp. rapa* plants causing stunting, mild mosaic and leaf distortion.

PVX is transmitted by mechanical inoculation. It spreads naturally by contact between infected and healthy plants, on contaminated farm equipment and workers or by animals that have been in contact with diseased plants. Healthy "seed" potato tubers may become infected during storage by contacts of their sprouts with those of infected tubers. Contact between roots may allow infection in the field, but does not seem to be of importance.

Although PVX concentration in the plant is very high, aphids do not transmit the virus. Planthoppers and other chewing insects can transmit the virus due to mechanical contact. Transmission by zoospores of the fungus

Synchytrium endobioticum has also been reported (Nienhaus and Stille, 1965), and infection of *N. debneyi* and *N. benthamiana* by PVX was frequently obtained when they were planted in soil infested by *S. endobioticum* (Salazar, 1966). PVX can be transmitted experimentally by *Cuscuta campestris* (Ladeburg *et al.*1950).

PVX is not transmitted through true seed or pollen.

Strains of PVX have been divided into three groups on the basis of their thermal inactivation point (Köhler, 1962), into 4 groups according to their serological properties (Matthews, 1949) and into 4 groups based on their ability to overcome resistance conferred by two dominant resistance genes Nx and Nb in *Solanum tuberosum* (Cockerham, 1955). For PVX strains able to overcome resistance conferred by both Nx and Rx1 genes, the viral coat protein is the principal determinant (Kavanagh *et al.* 1992). For additional information on strains see Loebenstein, (2001).

8.4.1.1 Diagnostic and Propagation Species

Datura stramonium – systemic chlorotic rings, mosaic and mottling. Symptoms differ according to strain. As *D. stramonium* is immune to PVY it may be used to separate PVX when in complex with PVY.

Gomphrena globosa – all strains of PVX except PVXHB induce local lesions. Four to five days after inoculation small grey spots appear , which later increase in size and develop a reddish halo (Fig.8.1). The middle leaves of *G. globosa* plants with 8-10 leaves on the main shoot are the most susceptible.

Nicotiana tabacum cv. Samsun, Samsun NN, White Burley, etc. – systemic veinclearing, mottling and ringspots.

For propagation *Nicotiana tabacum, N. glutinosa* (to avoid contamination with TMV), and other *Nicotiana* species are best.

8.4.1.2 Purification

Before starting the purification procedure it is imperative to ensure that the virus culture is pure. For PVX it is recommended to inoculate *Gomphrena globosa,* and from a single lesion to inoculate *Datura stramonium,* which is immune to PVY, and then to propagate the isolate in *N. glutinosa.*

PVX can be purified according to Francki and McLean (1968) or by a simplification of this method (Sadvakasova *et al.* 1996). Thus, 200g of fresh or frozen *N. glutiosa* leaves were triturated in the cold in 400ml of 0.05 M borate buffer pH 8.0, containing 0.6 ml thioglycollic acid, 100 ml chloroform and 100 ml carbon tetrachloride. After clarification, by centrifuging for 10 min at 8000 *g*, the supernatant was mixed with activated charcoal and filtered through a celite pad. The water clear virus-containing fraction was then ultracentrifuged for 75 min at 70000 *g* on a 3 ml 33% sucrose cushion in borate buffer. The resulting pellet is resuspended in 2-3 ml 0.05 M borate buffer ph 8.0, containing 0.005 M EDTA and left overnight at 4C with stirring. Finally the virus suspension is centrifuged for 10 min at 3000 *g* and the supernatant collected. See also Thomas and Kaniewski (2001).

For preparation of antisera it is important to ensure that the virus preparation does not contain plant proteins, as these may cause later nonspecific reactions. Thus, it is recommended to check the virus preparation by polyacrylamide gel electrophoresis. Virus concentration can be determined by measuring the adsorbance at OD 260, and calculating according to $E^{0.1\%} = 3$.

8.4.1.3 Particle Morphology and Properties

Virions not enveloped. Nucleopcapsids filamentous, usually flexuous, 470-580 nm long, with a modal length of 515 nm and 13 nm in diameter. Helical symmetry. Axial canal 3.4 nm in diameter and 8 7/8 subunits per turn (Tollin *et al.* 1980). Purified preparations contain one sedimenting component.

A260 / A280: 1.20 . Specific absorbance at 260= 2.97.

Mol. wt: 3.5×10^6 , particles stable.

The virus contains 6% RNA and 94% protein.

Mol. wt. of the coat protein: c. 25×10^3, as deduced from the amino acid sequence obtained from the nt sequence of the coat protein gene. In SDS PAGE intact protein appears to have an apparent mol. wt. of c. 30kDa, but for preparations, which have been kept at room temperature, faster moving bands are observed, with calculated mol. wts. of c. 22-24 kDa (Koenig *et al.* 1970). Infectivity is retained when PVX is deproteinized with proteases, phenol or detergent.

8.4.1.4 Nucleic Acid Component

One molecule of linear positive-sense ssRNA with a mol. wt. of 2.1×10^6 (6,435 nt) and 5 ORFs (Fig. 8.2). The 5'terminus has a methylated nucleotide cap m7 G5 pppGpA. Poly (A) at the 3'terminus. Nucleotide sequence deposited at EMBL/GenBank under the following accession numbers M63141; M28049.Additional accessions can be found at: http://life.anu.edu.au/viruses/CTVdB/56010001.htm.

The genomic RNA of PVX contains 5 ORFs (Fig. 8.2). ORF 1 codes for a 166 K polypeptide that presumably possesses an RNA-dependent RNA polymerase activity (Morozov *et al.* 1990). Products of ORFs 2, 3 and 4 (the "triple block") are involved in cell-to-cell transport of PVX (Lough *et al.* 1998), together with a functional coat protein (Santa Cruz *et al.* 1998). ORF 5 encodes the coat protein.

8.4.1.5 Detection Methods

PVX can be detected easily in potato leaves and dormant and sprouted tubers by serology, using commercial enzyme-linked immunosorbent assay (ELISA) kits. Antisera can also be obtained from the Institute of Molecular Biology and Biochemistry, Almaty, Kazakhstan. Diffusion tests in agar can be used both for detection and determining strain relationships, but PVX virions have to be dissociated into soluble antigens, which migrate more readily into the agar medium. Dissociation of PVX can be achieved by incubating the plant extract with pyrrolidine at a final concentration of 2.5% (Shepard, 1970).

8.4.1.6 Therapy of Individual Plants

PVX–free plants can be obtained by culturing 0.1 mm meristems with or without a leaf primordium (Kassanis and Varma, 1967), although only a small percentage of plantlets was regenerated from the meristems. Combining heat-treatment before the excision of the meristem tip enhanced the proportion of virus-free plants obtained (Stace-Smith and Mellor, 1968). Incorporation of Virazole (ribavirin) into the culture medium yielded a higher percentage of virus-free progeny then comparable cultures without Virazole (Faccioli and Marani, 1998). (See also Faccioli, 2001). It should be emphasised that following therapy, reliable testing for viruses, preferably over two growing periods, and evaluation of trueness-to-type of the cured lines are mandatory.

8.4.1.7 *Potato Virus Y* Genus *Potyvirus* (PVY)*

In many cultivars, the primary symptoms of PVY^{O} and PVY^{C} are leaf mottling and necrosis, and secondary symptoms are leaf mottling and crinkling, plant dwarfing or leaf drop (Fig. 8.3). PVY^{N} causes mosaic primary symptoms, and leaf mottling and mosaic secondary symptoms. Infected plants are more severely affected if PVY occurs in complex with other

*Adapted from Brunt (2001a)

Viruses, especially PVX. Yield losses of 10-80% have been reported (de Bokx and Huttinga, 1981).

PVY^{NTN} causes a damaging disease, in which tubers develop superficial rings that initially are raised but later are sunken and necrotic (Fig. 8.4), and often become more conspicuous during storage.

In ultathin leaf sections cylindrical cytoplasmic inclusions can be observed.

8.4.2.2 Natural Hosts and Experimental Hosts

PVY has a wide natural host range; it occurs naturally in at least 41 species of four families. Various strains occur in potato worldwide especially in solanaceous species.

Strains of PVY are mechanically transmissible to more than 400 plant species (Edwardson and Christie, 1997). Of these, the following are important test species:

Nicotiana glutinosa - mild to severe systemic mottling.

N.tabacum - systemic vein clearing followed by mottling by all strains except PVY^{N}, which induces veinal necrosis (Fig. 8.5).

Solanum demissum × *S. tuberosum* `A6' on detached leaves necrotic lesions.

Physalis floridana- local lesions.

Datura stramonium is immune to PVY strains.

8.4.2.3 Transmission

More than 40 aphid species are known to transmit PVY (Kennedy *et al.* 1962; Sigvald, 1984), but they vary in efficiency. The more important vectors are *Macrosiphon euphorbiae, Aphis fabae,* Myzus *persicae and Rhopalosiphoninus latysiphon*. Aphid transmissibility is dependent on virus-encoded helper component and a DAG triplet in the coat protein.

8.4.2.4 Strains

Three strains, the common or ordinary strain (PVY^{O}), the stipple streak strain (PVY^{C}) and the tobacco veinal necrosis strain (PVY^{N}) have long been differentiated by their differential serological reactions and symptoms in a range of test plant species including *Nicotiana tabacum*, *Physalis floridana* and potato cultivars such as Duke of York; PVY^{o} and PVY^{c} are also differentiated, respectively, by the strain-specific interaction with the genes

Ny and Nc for hypersensitive resistance in potato, and PVY^{N} by the vein necrosis it causes in tobacco. During the last decade PVY^{NTN} has been described, which causes potato tuber necrotic ringspot disease (Beczner *et al.* 1984), and occurs in most European potato-growing countries (Le Romancer and Kerlan, 1991). This isolate will probably reach also the Central Asian Republics, through import of seed potatoes from Europe.

8.4.2.5 Particle Morphology and Properties

The virus can be purified according to Moghal and Francki (1976).The filamentous particles are mostly 730 x 11 nm, which sediment as a single component with an sedimentation coefficient of c.150 S. They consist of a coat protein of c.33 kDa and ss-RNA of c.10 kb. The coat protein consists of 267 amino acids, which corresponds to a mol.wt of 30 kDa.
A_{260}/A_{280}= 1.21-1.22; Specific absorbance at 260nm= 2.3.
Purification according to Moghal and Francki (1976).
The sequences of about 30 PVY isolates have been reported, (van der Vlugt et al. 1993).

8.4.2.6 Nucleic Acid Component

The genomic RNAs of all three strains have been fully sequenced andshown to contain 9698-9704 nt. The arrangement of the genomic RNA and gene products are shown in Fig. 8.6.

See also: http://life.anu.edu.au/viruses/CTVdB/56010001.htm

8.4.2.7 Detection Methods

An indication of the strain can be obtained by the reactions induced in a range of test plant species that include *Solanum demissum x S.tuberosum* A6, *S.demissum* Y and *S.chacoense* "TEL". Presently serological techniques are used for rapid detection and identification.

8.4.3 *Potato Leafroll Virus* Genus *Polerovirus* (PLRV)

Symptoms of primary infection by PLRV consist of pallor and upward rolling of young leaves, especially at the base, with an upright habit. Secondary symptoms, in plants grown from infected tubers, are stunting of the shoots and upward rolling of leaflets, especially of basal leaves (Fig. 8.7), which become rigid and leathery.

Necrosis may develop in the phloem tissue of stems and petioles, and excessive callose occurs in the sieve tubes of stems and tubers. Callose staining with resorcin blue, the Igel-Lange test, (de Bokx, 1967) was often used for detecting PLRV in tubers before the introduction of ELISA. Carbohydrates accumulate in the leaves- sometimes 2-3 times more than in healthy leaves, with a corresponding reduction in the tubers - due to impaired phloem transport. This is probably due not only to necrosis in the phloem cells, but perhaps also because of blockage of photoassimilate movement from the chloroplast into the cytosol by the triose-phosphate translocator (Riesmeier *et al.* 1993). This could also be due to blockage of sucrose loading into the phloem of PLRV-infected plants by the sucrose transporter protein. Tuber symptoms are rare, but some cultivars such as Green Mountain and Russet Burbank have internal net necrosis, which can be seen when tubers are cut. Sometimes tubers from infected plants may develop thin sprouts - "spindling sprouts". This is an inconsistent symptom and depends on cultivar, virus strain and environmental conditions.

In plants grown from infected tubers (secondary infection), yields may be reduced by 33-50%. Tubers of infected plants are small to medium. Even greater losses are observed when PLRV occurs in complex with PVX or PVY.

The virus is mainly restricted to species of the Solanaceae, including: *Datura stramonium* – interveinal necrosis; *Solanum villosum* , *Physalis floridana* and *Datura tatula.* Non-solanaceous host species include *Amaranthus caudatus, Capsella bursa-pastoris, Celosia argentea, Gomphrena globosa* (Tamada *et al.* 1984; Fox *et al.* 1993; Souza-Dias *etal.* 1993).

The virus occurs worldwide in potato growing areas, especially in climatic regions that favor high vector populations and in agro-systems where use of virus-tested "seed" potatoes is not common.

8.4.3.1 Transmission

Several aphid species have been reported to transmit PLRV (Kennedy *et al.* 1962). *Myzus persicae* seems to be the most efficient and important vector. *Macrosiphum euphorbiae* transmits potato strains less efficiently, but is an efficient vector of Australian tomato yellow top isolates.

PLRV is transmitted in a persistent manner, characterised by minimum access times for acquisition and inoculation of c. 1 h each. There is a latency period between acquisition and transmission of the virus, and the minimum time for transmission is c. 12 h. Transmission frequency increases with an increase in the access-feeding period of up to 2 days or more. Both larvae and adults can transmit the virus. Aphids remain infective after moulting and remain viriferous for life. PLRV is circulative in the vector aphids and can be found in the haemolymph of *M. persicae.*

Recently it was shown that endosymbiontic bacteria (*Buchnera* sp.) are involved in the persistent transmission of PLRV by *M. persicae* (Hogenhout *et al.* 1996). PLRV is not transmitted by mechanical inoculation, by seed or pollen; it is transmitted experimentally by grafting.

8.4.3.2 Strains

Isolates have been detected that, although they are serologically indistinguishable, induce symptoms of different severity on potato cultivars or *Physalis pubescens (syn. P.floridana)*, or by their ease of transmission by *M. persicae* (Tamada *et al.* 1984

Tomato yellow top strains (Thomas, 1984) that cause yellow edge symptoms in tomato leave, cause inconspicuous symptoms or symptomless infection in potato, and are readily transmitted by *Macrosiphum euphorbiae.*

8.4.3.3 Diagnostic and Propagation Species

Datura stramonium - systemic interveinal yellowing

Physalis pubescens (syn P.floridana) – severe stunting when young seedlings become infected (Fig. 8.8); systemically infected leaves develop mild interveinal necrosis, older leaves become slightly rolled.

Solanum tuberosum ssp. tuberosum (potato) – stunting and leafroll (Fig. 8.7).

For propagation plants the following are used: *Physalis pubescens, Datura stramonium and Solanum tuberosum (*especially for maintenance by storing dormant infected tubers at 4° C).

8.4.3.4 Purification

From *P. pubescens* or *D. stramonium* as described by Takanami and Kubo (1979).

8.4.3.5 Particle Morphology and Properties

Virions are isometric *c.* 24 nm in diameter, and not enveloped. Purified preparations contain one sedimenting component of 115 S (Takanami and Kubo, 1979). A260/ A280: 1.78;

Specific absorbance at 260= 8.6. The virus contains 30% nucleic acid and 70% protein. Mol.wt of the coat protein is 26.3 kDa (Rowhani and Stace-Smith, 1979).

The genome consists of a single-stranded, linear messenger sense RNA molecule of 5.88-5.99 kb (mol. wt. 2.0 x10^6), which is covalently linked to a small protein (VPg) of 32 amino acids at it's N-terminal (Rowhani and Stace-Smith, 1979). The RNA does not contain a Poly A sequence at it's 3'terminus, and is infectious when used to inoculate tobacco mesophyll protoplasts (Mayo *et al.* 1982). Sequence database accession codes include: D00530 Em(40)_vi:PLVGR Gb(84)_vi:PLVGR 5,987bp.

The genomic sequence of PLRV consists of 8 open reading frames (ORFs) (Fig. 8.9) (Ashoub *et al.* 1998). A full-length cDNA copy of the virus has been expressed in protoplasts of *Chenopodium quinoa,* and in transgenic potato plants that accumulated the 17 kDa movement protein and had symptoms typical of PLRV (Prüfer *et al.* 1997).

8.4.3.6 Detection Methods

PLRV can be detected in potato leaves by serology, using commercial ELISA kits However, the concentration of PLRV varies and, in plants grown at temperatures of c. 30°C or in older plants, ELISA may not always detect infection. It is also difficult to detect PLRV by ELISA in unsprouted tubers (Hill and Jackson, 1984).

Various methods based on PCR or nucleic acid probes are now being developed and evaluated. Thus, PLRV could be detected in tubers within one day by immunocapture and a fluorogenic 5'nuclease RT-PCR assay (Schoen *et al.*1996; Russo *et al.* 1999). Using a digoxigenin-labeled cRNA probe, PLRV was easily detected in dormant tubers (Loebenstein *et al.* 1997). The limit of PLRV detection with this probe was 1 pg/ml compared with 2 ng/ml by ELISA. The method has been tested experimentally in Kazakhstan (Manadilova, unpublished). Such methods, if adopted by testing laboratories, may increase in importance in inspection schemes by eliminating the present necessity of sprouting dormant tubers.

8.4.3.7 Therapy of Individual Plants

PLRV can be eliminated by keeping tubers for c. 25 days at 37.5°C in a humid atmosphere (Lizárraga *et al.* 1991) Heat treatment causes tubers to deteriorate and is therefore not used for commercial stocks, but is mainly employed to obtain virus-free plants of small special potato lines which can then be propagated. In certain regions in India, an increase in PLRV was observed when cold storage replaced open-air storage; apparently the high ambient temperatures eliminated PLRV from the tubers (Quak, 1987). Tissue culture, with or without thermotherapy, is the most common practice to retrieve virus-free potatoes in general and PLRV-free plants in particular (Lizárraga *et al.* 1991). It should, however, be emphasized that following the therapy step reliable testing for viruses, preferably over two growing periods and evaluation of trueness-to-type of the cured lines are essential.

8.4.4 *Potato Virus M* Genus *Carlavirus* (PVM)*

Most isolates of the virus alone infect symptomlessly many potato cultivars but some may induce very mild leaf chlorosis and leaf distortion in other cultivars or, in a few cultivars (e.g., Arran Victory), severe leaf mottling, chlorosis, crinkling and rolling ("paracrinkle") and stunting. However, even isolates that induce only mild symptoms can reduce tuber yield by 10-18%. Infection of intolerant cultivars by more severe isolates, which are especially common in some eastern European countries, can result in tuber yield losses of 40-75%. PVM frequently occurs in complex with other viruses (especially PVS), and symptoms are usually more severe.

Infected cells often contain irregularly-shaped inclusions consisting of endoplasmic reticulum, ribosomes, virus particles and vesicles (Edwardson and Christie, 1997). PVM particles also occur in infected plants either scattered throughout cytoplasm of immature parenchyma cells and sieve tubes, or as aggregates forming paracrystals or banded bodies.

Potato is the major natural host of PVM and is especially common in Eastern Europe and the former-USSR.

The virus has been detected occasionally in weed species including *Datura metel* in Egypt, *Fumaria officinalis, Galium aparine, Solanum dulcamara* in Europe and *Solanum fernandezianum* in S. America (Valkonen *et al.* 1992).

PVM is experimentally transmissible to potato and to another 122 species in 9 genera of the *Solanaceae* and to 47 species in the *Amaranthaceae, Caryophyllaceae, Chenopodiaceae, Compositae, Cucurbitaceae, Fabaceae, and Rubiaceae* . Of these, the following are useful indicator plant species:

Chenopodium amaranticolor, C. quinoa, Phaseolus vulgaris, Vigna unguiculata - chlorotic or necrotic lesions in inoculated leaves, but no systemic infection.

Chenopodium album, Nicotiana glutinosa, N. tabacum, Physalis pubescens (syn. P.floridana) – no infection

Lycopersicon esculentum, Nicotiana occidentalis, N. rosulata - symptomless systemic infection. Several cultivars of *Lycopersicon esculentum* (e.g., cv. Marmande) are immune to potato virus S (PVS) but can be infected systemically with PVM; such cultivars can be used to separate PVM from PVS with which it often occurs in complex in potatoes (Bagnall *et al.* 1959).

Nicotiana debneyi - brown ring-like local lesions and, rarely, symptomless systemic infection

8.4.4.1 Transmission and Strains

The virus is transmitted readily in the non-persistent manner by aphid species including *Myzus persicae* and *Aphis nasturtii* and, less efficiently, by *Aulacorthum solani, Macrosiphum euphorbiae.*

The virus is not seed-borne. Although it is experimentally transmissible by mechanical inoculation to potato and test plants it is not transmitted at a significant level by plant contact.

---

* Adapted from Brunt (2001b)

Numerous isolates differing in virulence occur in potato crops worldwide; a serologically distinct strain (PVM-ID) has been described recently (Cavileer *et al.*1998).

8.4.4.2 Particle Morphology and Properties

The virus has particles mostly measuring 650 x 12 nm. Purified virus preparations have a u.v. absorbance spectrum with maximum and minimum absorbance at c.260 and c.244 nm, respectively, an A260/280 ratio of 1.2, an extinction coefficient of 2.88, a sedimentation coefficient of 157 S and a buoyant density in caesium chloride of 1.32 g/cm^3 (Tavantzis, 1984).

The particles contain c. 6% positive-sense single-stranded RNA and c. 94% protein. The genomic RNA is encapsidated in a coat protein composed of c.1,800 identical polypeptides each of 33 kDa. The polypeptide subunits are arranged in a primary helix having a pitch of c.3.4 nm and with c.10 subunits per turn of the helix (Koenig, 1982).

PVM is serologically related to other *Carlaviruses* as *Carnation latent, and Potato S virus* (Koenig, 1982)

Each particle contains a single molecule of RNA of 8,534 nucleotides (M_r 2.5 x 10^6), the full sequence of which has been reported (Zavriev *et al.* 1991). The genomic RNA (Genbank Accession number X53062) contains six large open reading frames (ORFs) (Fig. 8.10). Infected plants contain a double- stranded form of the genomic RNA (M_r 5.0 x 10^6) (Tavantzis, 1984).

8.4.4.3 Detection Methods

Symptoms on mechanically inoculated test plants species can indicate the occurrence of PVM. However, as the virus induces no conspicuous symptoms in many potato cultivars, and often occurs in complex with other viruses, it is best to detect it by ELISA (Dedic, 1995). The double antibody sandwich type of ELISA can detect as little as 5 ng virus /ml of plant sap; this is equivalent to the detection of virus in infective potato sap diluted to c.1/30,000 (Faccioli and Colombarini, 1996). ELISA on nitrocellulose membranes (dot-ELISA) is also now often used (Dedic, 1995) and allows virus to be detected in potato sap diluted to 1/64,000, the equivalent to a concentration of 0.005 ng of virus/ml of sap. Diagnosis by direct tissue blotting is reported to be more sensitive, faster and less expensive than ELISA (Samson *et al.* 1993). Time-resolved fluoroimmunoassay, especially using monoclonal antibodies, is 10-20 times more sensitive than DAS-ELISA (Jarvekulg *et al.* 1989).

The virus can be readily detected by the polymerase chain reaction (Badge *et al.* 1996).

8.4.4.4 Therapy of Individual Plants

The virus is best controlled by the production, propagation, and planting of elite virus-tested nuclear stocks. Such stocks are obtainable by meristem tip culture. However, because the virus occurs in potato meristem, meristem tip culture is more effective if preceded by or combined with thermotherapy (35^{0}C) and/or chemotherapy using Ribavirin and 2,4-dioxo-hexahydro-1,3,5-triazine (Faccioli and Colombarini, 1996). Production schemes utilizing elite virus-tested stocks, now used in many countries, are fully reviewed by Salazar (1996).

8.4.5 *Potato Virus S* Genus *Carlavirus* (PVS)*

Many isolates of PVS^{O} alone induce no conspicuous symptoms in many potato cultivars. However, in susceptible cultivars some isolates cause undulation of leaf margins and some rugosity of leaf surfaces (Fig 8.12). The importance of PVS is mainly dependent upon its local incidence, the virulance of the isolate, tolerance of the potato cultivar and environmental conditions. However, most isolates reduce tuber yield by 3-20%). The virus, however, often occurs in complex with other viruses and may then exacerbate the severity of infection.

Infected plant cells contain aggregated virus particles, seen as paracrystalline inclusions (Edwardson and Christie, 1977).

Potato is the major host of PVS^{O} and PVS^{A}, although the latter also occurs in pepino (*Solanum muricatum*) in New Zealand and the Netherlands. $PVSA^{O}$ has also been reported to occur occasionally in some weed species including *S. brevidens, S.chacoense, S. dulcamara,* *S. spegazzini* and *Geranium dissectum* (Valkonen *et al.*1992).

PVS^{O} occurs worldwide in potato cultivars. PVS^{A} has been reported in potatoes in some S.American countries.

8.4.5.1 Transmission

Some isolates are not aphid-transmissible but others are readily transmitted by several aphid species including *Aphis fabae, A. nasturtii, Myzus persicae* and *Rhopalosiphum padi* . PVS is not seed-transmitted but is sufficiently infectious to be transmitted mechanically from infected to healthy field-grown potato plants.

---

*Adapted from Brunt (2001c)

9.4.5.2 Experimental Hosts

Although the virus has a very restricted natural host range, it is transmissible by mechanical inoculation to at least 56 other solanaceous species and to 33 species in 12 other families (Valkonen *et al.* 1992). Useful indicator plant species are:

Chenopodium album, C.amaranticolor, C.quinoa - chlorotic local lesions, often with a green halo on older leaves. PSV^{A}, but not PVS^{O}, infect *C. quinoa* systemically and induce chlorotic spotting.

Lycopersicon esculentum - symptomless systemic infection by PVS^{A} but immune to PVS^{O}.

Nicotiana clevelandii - conspicuous chlorosis of systemically infected leaves. Useful host for propagation and maintenance of virus cultures.

Nicotiana debneyi - vein-clearing, mottling and necrosis of systemically infected leaves. This species is immune to PVM, and allows PVS to be separated from complexes of the two viruses.

8.4.5.3 Strains

For long it was considered that only minor variants of the virus occurred in potato cultivars. Later it was shown that some variants from several S. American countries induce systemic leaf mottling in *Chenopodium amaranticolor* and *C. quinoa*, and are more readily transmissible by aphids and plant contact; and are now designated as the Andean strain (PVS^{A}) to distinguish it from the commonly occurring ordinary strain (PVS^{O}).

8.4.5.4 Particle Morphology and Properties

PVS has particles mostly measuring c. 650 x 12 nm, which contain c. 5% single-stranded RNA and 95% protein. The protein subunits have a mol. wt of 33-34 kDa.

The genomic RNA (Genbank Accession number D00461) contains c.7.500 nucleotides, the sequence of which at the 3' terminus has been sequenced (Foster and Mills, 1992). The RNA has six ORFs, which encode proteins of 10, 33, 7, 11, 25 and 41 kDa. The arrangement and functions of the ORFs are similar to those of PV M.

8.4.5.5 Detection Methods

Best detected by immunosorbent electron microscopy (Milne and Luisoni, 1977) and ELISA in tubers and leaves (Banttari and Franc, 1982). Direct tissue blotting is reported to be more sensitive, cheaper and faster than conventional ELISA (Samson *et al.* 1993). Monoclonal antibodies have been produced for the specific identification of PVS^{A} (Cerovska and Filigarova, 1995).

Nucleic hybridization is a sensitive and reliable procedure for the detection of PVS using radioactive (Weidemann and Koenig, 1990) or non-radioactive probes (Audy *et al.*1991). More recently, PCR has been shown to be c. 1,000-fold more sensitive than ELISA as a diagnostic procedure (Badge *et al.* 1996).

8.4.6 *Potato Virus A* Genus *Potyvirus* (PVA)*

The severity of symptoms depends on virus isolate, potato cultivar and environmental conditions, but infection can reduce tuber yield by up to 40%. Leaves of potatoes infected with PVA alone may remain symptomless or develop a mild mosaic, rugosity and marginal distortion; hypersensitive cultivars develop tip necrosis. In complex with PVX or PVY, PVA induces leaf crinkling and significant yield losses.

Its only known natural hosts are potato and *Solanum betacea* (tamarillo).

PVA is widely distributed in potato growing areas in Europe and N.America, but does not occur in the Andean region of S.America. It occurs less commonly than PVY in tropical countries.

At least ten aphid species are its vectors, but the more important are *Aphis frangulae, Aulocorthum solani, Macrosiphum euphorbiae* and *Myzus persicae.* Like other potyviruses, the transmission of PVA is dependent on the virus-encoded helper component proteinase and the presence of an aspartic acid-alanine- glycine (DAG) sequence in the coat protein. PVA is not seed-transmitted.

PVA has a restricted experimental host range that is limited to the Solanaceae. The most useful of these are:

Nicotiana tabacum cvs Samsun and White Burley - vein clearing and dark-green vein-banding of systemically infected leaves.

Nicandra physalodes - systemic leaf mottling, severe necrosis, rugosity and palnt stunting.

Solanum demissum x S.tuberosum cv. Aquila (=A6) and *S.demissum* "SdA" – systemic vein necrosis..

---

* Adapted from Brunt (2001d)

8.4.6.1 Strains

Of the five strains groups generally recognized by the hypersensitive response of potato cultivars (Oruetxebarria *et al.* 2000), three from potato are recognized by their differential interactions with potato cv. King Edward. Another strain consists of potato isolates, which, after repeated experimental propagation in *Nicotiana spp.*, fail to infect potato systemically or to be aphid transmissible (Rajamaki *et al.*1998). A strain from *Solanum betacea* (tamarillo) is distinguishable by experimental host reactions; like the strains from potato and *Nicotiana* spp., its genomic sequence has been determined (Kekarainen *et al.* 1999).

8.4.6.2 Particle Morphology and Properties

The filamentous particles, mostly measure 730 x 11 nm.
The genomic RNA of 9565 nucleotides has been completely sequenced and has at its 3' terminus an untranslated region of 227 or 229 nucleotides (Kekarainen *et al.* 1999). The arrangement of the genomic RNA is generally similar to that of PVY.

8.5.6.3 Detection Methods

The test plant species 'A6' is a very sensitive indicator host for PVA
PVA is conveniently detected in leaves and tubers by serological tests, of which DAS-ELISA (Vetten *et al.*1983), dot-ELISA (Weidemann, 1988) and direct tissue blotting (Samson *et al.* 1993) are now commonly used; the use of MAbs in ELISA has also proved reliable for mass testing (Browning *et al.*1995).

8.5 Therapy

Virus-induced yield losses are best minimized by the production, large-scale propagation and distribution of virus-free stocks. These are obtainable by meristem tip culture especially if preceded by thermotherapy or if antiviral agents such as Virazole or Ribavirin are included in the culture media (Faccioli, 2001). Although the use of elite virus-tested stocks in production schemes is well established in many countries, virus-resistant cultivars are also utilized. The use of transgenic resistant plants has great potential (Berger and German, 2001), but disputes over genetically modified (GM) plants hamper their widespread use.

8.6 Quality of Planting Material, Yields and Certification Schemes

As stated in the introduction there are practically no national certification schemes to provide farmers with high quality tuber seeds. In most cases farmers use tuber seeds from their own fields, from other farms with a reputation for 'good" seeds or from a few special state seed producing farms. In general virus incidence in the "tuber seeds" is high, resulting subsequently in low potato yields. For example, when imported certified seeds from Holland were planted in Kazakhstan during 1993-1994 yields reached more than 40 tons/ha, compared to local seeds that yielded about 10 tons/ha. FAO also provided to Tajikistan in 1998 high quality seed potato, with a subsequent increase in yields. However, these high yields did not continue in Kazakhstan or Tajikistan when the next and third generations from these imports were used for plantings. It should, however, be mentioned that this fourfold increase is probably also due to better cultivation technologies and irrigation.

Production of local certified tuber seed is now in progress in several countries. Thus, in Tajikistan attempts to produce elite tubers from microtubers obtained from meristems have been reported (Aliev *et al,* 1997), and in Kazakhstan several seed growing farms use 3 year schemes to produce elite tubers, starting from meristems (Ligai, personal communication). However, these schemes, many of them in the planning stage, don't test their source material carefully for presence of viruses often using low quality antisera. No data on yield performance of the elite seed have come to our attention, except from Tadjikistan where yields of 30-40 tons/hectare were reported (Aliev *et al.* 1997), and no data of virus incidence in the seed potatoes were reported.

A pilot scheme conducted by the Agricultural Institute of Astana, Kazakhstan and the Institute of Molecular Biology and Biochemistry in Almaty, within a US-AID project (Levy *et al.* 2001) may serve as a prototype for production of elite potato seed. By using microtubers, obtained from meristems of virus-tested potatoes, grown under insect protected-conditions, and open field propagation of foundation and elite tubers, resulted in only a low infection rate of 2.5% by PVY. These elite seed gave an increase in yields of about 90% when compared with yields of commercial fields in the vicinity, i.e. 25.3 t/ha compared with 13.3 t/ha, respectively. It might be mentioned that these fields in the northern region of Kazakhstan were not fertilized and with irregular irrigation.

8.7 Future Perspectives

The use of low quality seed potato, heavily infected by viruses seems to be a major, though not the only, reason for the low yields of potatoes in the

Central Asian Republics. Virus infection aggravates the effects on yields due to lack of fertilizers, irregular irrigation and adverse weather conditions. Use of higher quality seed in those few instances tested in the Central Asian Republics at least doubled the yields. It seems therefore most advisable to build up the production of certified seed. The present European and US clonal propagation schemes from "base" plants (Watad *et al.* 2001), where 6-7 increase generations are done in the open field before releasing the seed to the ware potato grower, seem not to be suitable for the local conditions. In these long cycle schemes plants have a good chance to become reinfected, especially if grown in areas with virus sources and vectors. These schemes also require a long term investment before seed potato can be sold to farmers. Therefore, it seems that rapid propagation schemes are preferable. In such a scheme single node cuttings are prepared *in vitro* from carefully virus-tested base plants, and grown in a greenhouse to produce sterile plantlets or microtubers. The sterile material is planted in a virus-free greenhouse in soilless medium to produce (gen0) minitubers. Generation 1 (gen1) is planted in a screenhouse in fumigated soil to prevent seed borne diseases. Generation 2 (gen2) and 3 (gen3) are produced in fields remote from other potatoes and in areas with low aphid infestation. If carried out precisely and efficiently, production costs of gen3 should be less then US$ 300 per ton, compared with US$ 800 for imported "certified seed". These short cycle schemes will probably become the norm in those countries interested to build up a certified seed program from scratch.

It should also be mentioned that genetically modified (GM) potatoes, with multivirus resistance genes may be the ultimate answer, once they prove themselves under local conditions, and consumers will accept them. Such potatoes are already available but require public acceptance of GM products in general (Berger and German, 2001).

References

Aliev, K.A., Karimov, B.K. and Karimov, B.B. 1997. Cultivation of the improved potatoes in Tadjikistan, Institute of Physiology of Plants and Genetics, Academy of Science. Dushanbe.

Audy, P., Parent, J.G. and Asselin, A. 1991. A note on four non-radioactive labeling systems for dot hybridization of potato viruses. *Phytoprotection* 72:81-86.

Ashoub, A., Rohde, W. and Prüfer, D. 1998. *In planta* transcription of a second subgenomic RNA increases the complexity of the subgroup 2 luteovirus genome. *Nucleic Acids Res.* 26: 420-426.

Badge, J.,Brunt, A.,Carson, R.,Dagless, E.,Karamagioli, M.,Phillips, S., Seal, S.,Turner, R. and Foster, G.D.1996. A carlavirus-specific PCR primer and partial nucleotide sequence provides further evidence for the recognition of cowpea mild mottle virus as a whitefly-transmitted carlavirus. *Eur.. J. Plant Pathology* 102:305-310.

Bagnall, R.H.,Larson, R.H. and Walker, J.C.1956.Potato viruses M, S and X in relation to interveinal mosaic of the Irish Cobbler variety. *Bull. Wisconsin Agric. Res. Stn 198*, 45 pp.

Banttari, E.E. and Franc, G.D. 1982. ELISA with single or combined antisera for viruses S and X in potato tubers and plants. Amer.Pot. J. 59: 375-387.

Berger, P. and German, T. 2001. Biotechnology and resistance. In: Virus and Virus-like Diseases of Potatoes and Production of Seed-potatoes. (eds.) G. Loebenstein, P.H. Berger, A.A. Brunt and R. Lawson. Kluwer Academic Publishers, Dordrecht, pp. 341-363.

Beczner, L., Horvath, J., Romhanyi, I. and Forster, H.1984. Studies on the aetiology of tuber necrotic ringspot disease in potato. *Potato Res.* 27:339-352.

Browning,I.A., Burns,R., George,E.L. and Darling,M.1995. Development and evaluation of ELISA assays incorporating monoclonal antibodies for the detection of potato virus A. *EPPO Bull.* 25:259-268.

Brunt, A.A. 2001a. Potato virus Y. In: *Virus and Virus-like Diseases* of *Potatoes and Production of Seed-potatoes.* (eds.) G. Loebenstein, P.H. Berger, A.A. Brunt and R. Lawson. Kluwer Academic Publishers, Dordrecht, pp. 81-86.

Brunt, A.A. 2001b. Potato virus M. In: *ibid.* pp. 101-106.

Brunt, A.A. 2001c. Potato virus S. In: *ibid.* pp. 109- 112.

Brunt, A.A. 2001d. Potato virus A. In: *ibid.* pp. 77- 81.

Brunt, A.A., Crabtree, K., Dallwitz, M.J., Gibbs, A.J. and Watson, L. 1996. *Viruses of Plants.* CAB International, Wallingford, UK.

Cavileer,T.D., Clark,R.C., Corsini,D.L. and Berger,P.H. 1998. A new strain of potato carlavirus M. *Plant Dis.*82:98-102.

Cerovska,N. and Filigarova,M.1995. Specific detection of the Andean strain of potato virus S by monoclonal antibodies. *Ann.appl.Biol.*127: 87-93.

Cockerham, G. 1955. Strains of potato virus X. *Proc. 2nd Conf. Potato Virus Diseases.* Lisse-Wageningen 1954, pp. 89 - 92.

De Bokx, J.A. 1967. The callose test for the detection of potato leafroll virus in potato tubers. *Eur. Potato J.* 10: 221-234.

De Bokx, J.A. and Huttinga, H. 1981. Potato virus Y. C.M.I./A.A.B. Descr. Pl. Viruses 242, 6 pp.

Dedic,P. 1995. The reliability of potato virus M detection in the progeny of primarily infected potato plants. *Ochr. Rostl.* 31: 277-285.

Edwardson, J.R. and Christie, R.G.1997. Viruses infecting peppers and other solanaceous crops; Monogr. Agric.Exp.Stn, Univ. Florida 18, Vol.1: 106-123.

Eliseeva, Z., Tulegenov, T. and Katin, I. 1978. Potato Y virus identification. *Vestnic Agric. Scci.* pp. 40-44 (in Russian).

Faccioli, G. 2001. Control of potato viruses using meristem and stem-cutting cultures, thermotherapy and chemotherapy. In: *Virus and Virus-like Diseases* of *Potatoes and Production of Seed-potatoes.* (eds.) G. Loebenstein, P.H. Berger, A.A. Brunt and R. Lawson. Kluwer Academic Publishers, Dordrecht, pp. 365-390.

Faccioli, G. and Colombarini, A.1996. Correlation of potato virus S and potato virus M contents of potato meristem tips with the percentage of virus-free plantlets produced in vitro. *Potato Res.* 39:129-140.

Faccioli, G. and Marani, F. 1998. Virus elimination by meristem tip culture and tip micrografting. In: *Plant Virus Disease Control.* (eds) A. Hadidi, R.K. Khetarpal and H. Koganezawa, pp. 346- 380. APS Press, St. Paul, Minnesota.

Food and Agriculture Organization of the United Nations, FAO, Statistical Databases 2002, FAOSTAT http://apps.fao.org/

Foster,G.D. and Mills,PR.,1990.Detection of strains of potato virus S by nucleic acid spot hybridisation. *Potato Res.*33:487-495.

Francki, R.I.B. and McLean, G.D. 1968. Purification of potato virus X and preparation of infectious ribonucleic acid by degradation with lithium chloride. *Austr. J. biol. Sci.* 21: 1311-1318.

Gaynullina, R.G. and Gerasimova, Z. P. 1974. Virus surveys of potatoes in Kyrgistan. In: *Virus and Virus Diseases of Plants.* Naukova Dumka, Kiev (in Russian).

Hill, S.A. and Jackson, E.A. 1984. An investigation on the reliability of ELISA as a practical test for detecting potato leafroll virus and potato virus Y in tubers. *Plant Pathology* 33: 21-26.

Hogenhout, S. A., Verbeek, M., Hans, F., Houterman, P.M., Fortass, M., Wilk, F. van der, Huttinga, H. and Heuvel, J. F. J. M. van den. 1996. Molecular bases of the interactions between luteoviruses and aphids. *Agronomie* 16: 167- 173.

Jarvekulg, L., Sober, J., Sinijaru, R., Toots, I. and Saarma, M.1989.Time-resolved fluoroimmunoassay of potato virus M with monoclonal antibodies. *Ann. appl. Biol.* 114: 279-291.

Kassanis, B. and Varma, A. 1967. The production of virus-free clones of some British potato varieties. *Ann. appl. Biol.* 59: 447-450.

Kavanagh, T., Goulden, M., Santa Cruz, S., Chapman, S., Barker, I. and Baulcombe, D. 1992. Molecular analysis of a resistance-breaking strain of potato virus X. *Virology* 189: 609-617.

Kekarainen,T., Metits,A., Oruetxebarria,I., Rajamaki,M.-L. and Valkonen,J.P.T. 1999. Comparison of the complete sequences of five different isolates of potato virus A (PVA). *Arch.Virol.*144: 1-12.

Kennedy, J.S., Day, M.F. and Eastop, V.F. 1962. *A Conspectus of Aphids as Vectors of Plant Viruses. Commonwealth Institute of Entomology, London,* 114 pp. 50: 263 -266.

Koenig, R.1982. Carlavirus Group. *CMI/AAB Descr.Plant Viruses* No.259, 4 pp.

Koenig, R., Stegemann, H., Francksen, H. and Paul, H. 1970. Protein subunits in the potato virus X group. Determination of the molecular weights by polyacrylamide electrophoresis. *Biochim. Biophys. Acta* 207: 184 – 189.

Köhler, E.1942. Untersuchungen uber das "K-Virus" der Kartoffel.1. *Mitt. Angew.Botanik* 24:118-130.

Ladeburg, R.C., Larson, R.H. and Walker, J.C. 1950. Origin, interrelation and properties of ringspot strains of virus X in American potato varieties. *Res. Bull. Wisc. Agr. Exp. Stn,* 165, 47 pp.

Le Romancer, M. and Kerlan,C. 1991. La maladie des nécroses annulaires superficielles des tubercules; une infection de la pomme de terre, due á virus Y. *Agronomie* 11:889-900.

Levy, D., Loebenstein, G., Shvidchemko, V. and Manadiliva, A. 2001. Rapid propagation of virus-tested Kazakh potato cultivars. *Final Rept. to the U.S. Agenc. Intl. Devel.* 17p.

Lizárraga, R., Panta, A., Jayasinghe, U. and Dodds, J. 1991. Tissue culture for elimination of pathogens. CIP Research Guide 3. International Potato Center (CIP), Lima, Peru, 21pp.

Loebenstein, G. and Sadvakasova, G. 1977. Advanced Detection Methods for Viruses in Potatoes. Final Report to the CDR Program , Agency for International Development, USA.

Loebenstein, G., Akad, F., Filatov, V., Sadvakasova, G., Manadilova, A., Bakelman, H., Teverovsky, E., Lachman,O. and David, A. 1997. Improved detection of potato leafroll luteovirus in leaves and tubers with a digoxigenin-labeled cRNA probe. *Plant Dis.* 81:489-491.

Loebenstein, G. 2001. Potato X virus. In :*Virus and Virus-like Diseases* of *Potatoes and Production of Seed-potatoes.* (eds.) G. Loebenstein, P.H. Berger, A.A. Brunt and R. Lawson. Kluwer Academic Publishers, Dordrecht, pp. 87-94.

Matthews, R.E.F. 1949. Studies on potato virus X. *Ann. appl. Biol.* 36: 449-474.

Mayo, M.A., Torrance, L., Cowan, G., Jolly, C.A., Macintosh, S.M., Orrega, R., Barrera, C. and Salazar, L.F. 1996. Conservation of coat protein sequence among isolates of potato mop-top virus from Scotland and Peru. *Arch. Virol.* 141: 1115-1121.

Milne, R. and Luisoni, E. 1977. Rapid immune electron microscopy of virus preparations. In: *Methods in Virology.* (eds.) K. Maramorosch and H. Koprowsky. Vol. 6, pp.264-281, Academic Press , New York.

Moghal, S.M. and Francki, R.I.B. 1976. Towards a system for the identification and classification of potyviruses: I. Serology and amino acid composition of six distinct viruses. *Virology* 73: 350-362.

Morozov, S. Y., Kanyuka, K.V., Levay, K.E. and Zavriev, S.K. 1990. The putative RNA replicase of potato virus M : obvious sequence similarities with potex- and tymoviruses. *Virology* 179: 911-914.

Morozov, S.Y., Miroshnichenko, N.A., Solovyev, A.G., Fedorkin, O.N., Zelenina, D.A., Lukasheva, L.I., Karasev, A.V., Dolja, V.V. and Atabekov, J.G. 1991. Expression strategy of the potato virus X triple gene block. *J. gen. Virol.* 72: 2039-2042.

Nienhaus, F. and Stille, B. 1965. Übertragung des Kartoffel-X-Virus durch Zoosporen von *Synchytrium endobioticum. Phytopahol. Z.* 54: 335 -337.

Oruetxebarria,I., Kekarainen,T., Spetz,C. and Valkonen,J.P.T.,2000. Molecular characterisation of Potato virus V genomes from Europe indicates limited spatio-temporal strain differentiation. *Phytopathology* (in press).

Prüfer, D., Schmitz, L., Tacke, E., Kull, B. and Rohde,W. 1997. In vivo expression of a full-length cDNA copy of potato leafroll virus (PLRV) in protoplasts and in transgenic plants. *Mol. Gen. Genetics* 253: 609-614.

Quak, F. 1987. Therapy of individual plants. In: *Viruses of Potatoes and Seed-Potato Production.* (eds) J.A. de Bokx and J.P.H. van der Want, pp. 151-161. Pudoc, Wageningen, The Netherlands..

Rajamaki,M., Merits,A., Rabenstein,F,Andrejeva,J., Paulin,L., Kekarainen,T., Kreuze,J.F., Forster,R.L.S. and Valkonen,J.P.T. 1998. Biological, serological and molecular differences among isolates of potato A potyvirus. *Phytopathology* 88:311-321.

Riesmeier, J.W., Flügge, U.I., Schulz, B., Heineke, D., Heldt, H.W., Willmitzer, L. and Frommer, W.B. 1993. Antisense repression of the chloroplast triose phosphate translocator affects carbon partitioning in transgenic potato plants. *Proc. Natl Acad. Sci.* USA 90: 6160-6164.

Rowhani, A. and Stace-Smith, R. 1979. Purification and characterization of potato leafroll virus. *Virology* 98: 45-54.

Russo,P., Miller,L., Singh,R.P. and Slack, S.A. 1999. Comparison of potato leafroll virus and potato virus Y detection in seed potato samples tested by Florida winter field inspection and RT-PCR . *Amer. Potato J.* 76: 313-316.

Sadvaksova, G., Tam, Y. and Loebenstein, G. 1996. A simple method for purification of potato virus X. *Appl. Biochem. Microbiol.* 32: 393-396 (in Russian).

Salazar, L.F. 1996. *Potato Viruses and their Control, 214 pp.* Intern. Potato Center, Lima, Peru.

Samson, R.G., Allen, T.C. and Whitworth, J.L.1993. Evaluation of direct tissue blotting to detect potato viruses. *Amer. Potato J.* 70:257-265.

Schoen C.D., Knorr,D. and Leone,G. 1996. Detection of potato leafroll virus in dormant potato tubers by immunocapture and a fluorogenic 5' nuclease RT-PCR assay. *Phytopathology* 86: 993-999.

Shepard, J.F. 1970. A radial-immunodiffusion test for the simultaneous diagnosis of potato viruses S and X. *Phytopathology* 60: 1669-1671.

Sigvald,R. 1984. The relative efficiency of some aphid species as vectors of potato virus Y. *Potato Res.* 27:285-290.

Singh,R.P. and Somerville,T.H. 1992. Evaluation of the enzyme-amplified ELISA for the detection of potato virus A,M, S, X, Y and leafroll . *Amer. Potato J.* 69:21-30.

Stace-Smith, R. and Mellor, F.C. 1968. Eradication of potato viruses X and S by thermotherapy and axillary bud culture. *Phytopathology* 58:199-203.

Takanami, Y. and Kubo, S. 1979. Enzyme-assisted purification of two phloem-limited plant viruses: tobacco necrotic dwarf and potato leafroll. *J. gen. Virol.* 44:153-159.

Tamada,, T., Harrison, B.D. and Roberts, J..M. 1984. Variation among British isolates of potato leafroll virus. *Ann. appl. Biol.* 104: 107-116.

Tavantzis, S.M.1984. Physicochemical properties of potato virus M. *Virology* 133:427-430.

Thomas, J.E. 1984. Characterisation of an Australian isolate of tomato yellow top virus. *Ann. appl. Biol.* 104: 79-86.

Thomas, P.E. and Kaniewski, W.K. 2001. Isolation and purification. In: *Virus and Virus-like Diseases* of *Potatoes and Production of Seed-potatoes.* (eds.) G. Loebenstein, P.H. Berger, A.A. Brunt and R. Lawson. Kluwer Academic Publishers, Dordrecht, pp. 285-310.

Tollin, P., Wilson, H.R. and Bancroft, J.B. 1980. Further observations on the structure of potato virus X. *J. gen. Virol.* 49: 407-410.

Tulegenov, T., Eliseeva, Z. amd Akhatova, F. 1986. Spreading and detriment of potato Y virus in Alma-Ata region. *Trans. Microb. Virol.* Inst. Acad. Sci. Kazakhstan, pp. 145-151, (in Russian).

Valkonen,J.P.T., Contreras,A., Pehu,E. and Salazar,L.F. (1992). Naturally occurring viral infections in Solanum brevidens and S.fernandezianum. *Potato Res.* 35:411-417.

Van der Vlugt, R.A.A., Leunissen,J. and Goldbach,R.1993. Taxonomic relationships between distinct potato virus Y isolates based on detailed comparison of the viral coat proteins and 3'-nontranslated regions. *Arch.Virol.*131: 361-375

Weidemann,H,L.1988. Rapid detection of potato viruses by dot-ELISA. *Potato Res.*31:485-492.

Weidemann, H.L. and Koenig, R. 1990. Differentiation of isolates of potato virus S which infect Chenopodium quinoa systemically by means of quantitative cDNA hybridization tests. Plant Dis. & Prot. 97: 323-327.

Zavriev, S.K.,Kanyuka, K.V. and Leavy, K.E.1991. The genomic organisation of potato virus M RNA. *J. gen. Virol.* 72: 9-14.

Table 8.1
Areas and Production of Potatoes in the Central Asian Republics

Country	Area (ha)		Production (ton)	
	2000	2001	2000	2001
Kazakhstan	159.100	164,000	1,692,600	1,600,000
Kyrgistan	69,000	67,000	1,046,000	1,056,000
Tajikistan	23,000	23,000	303,000	304,799
Turkmenistan	6,000	6,000	29,000	30,000
Uzbekistan	51,300	55,000	729,800	800,000

From FAOST

Table 8.2

Location and Elevation	Cultivar and Generation **	Sampling Date	PVX	Percentage Infection PVY	PLRV	PVS	PVM
Almaty, 1050m	Escort 3rd	8.95	4.0	64.0	20.0	8.0	25.0
	Escort *	7.96	0	96.4	85.7	46.4	53.5
	Desiree*	8.96	0.	100	7.1	82.1	42.9
Chemolgan, 70km west of Almaty 900m	Jaerle 2nd	9.94	12.5	7.7	1.7	NT	6.0
	Nevsky*	6.94	71.0	28.5	14.2	NT	NT
	Chortandinsk *	6.94	48	25	14.8	NT	NT
	" " 2nd	9.95	0	3.5	0	NT	0
1200m	Jaerle 2nd	6.95	18	10.7	0	NT	28
	Rikea 3rd	9.94	25.0	50.0		NT	
Kerbukak 90 km north-east of Akmaty 1500m	Impala*	9.94	37.0	29.0	0	NT	0
	Mixture of cultivars *	6.94	18.0	4.5	13.6	NT	NT
Kaskelen 20km west of Almaty 400m	Nevsky*	8.96	35.7	3.6	7.1	14.2	78.5
	Diamant*	8.96	0	100	7.1	7.1	3.6
	Romano*	8.96	0	0	3.6	52.2	0
	Sante*	8.96	0	3.6	12.5	52.2	0
Zhanater 10km east of Djambul 1200m	Reserve elite	7.95	57	3.5	3.5	NT	25
	Reserve 1st	9.95	96	0	39.3	92.8	NT
	Desiree 2nd	8.95	0	0	3.5	NT	0
	Desiree 3rd	9.95	39.1	35.0	4.3	4.3	NT
	Nevsky *	6.96	45	5.0	0	40	85
Kegen 250km east of Almaty 1500m	Chortandinsk *	9.94	39	0	0	NT	NT
Narinkol 350km east of Almaty 1200m	Escort *	8.94	21.4	0	0	NT	NT
	Escort 1st	9.95	0	0	0	0	NT
	Nevsky *	6.94	12.5	3	0	NT	NT
	Frisea 3rd	8.96	3.5	3.5	3.5	NT	14
Bestobe 250 km north-east of Astana 300m	Nevsky *	9.94	26	3.0	0	NT	0
Akmola 400m	Nevsky *	7.97	19.7	1.5	28.9	5.4	0
	Nevsky *	9.2000	2.8	0	0	4.2	8.5

** Generation after certified seed * Generation not known **Determined by ELISA
NT- not tested

Table 8.3
Aphid Trappings in 3 Locations in Kazakhstan

Location and Elevation	Date	Total No. of Aphids	*Myzus persicae*	*Aphis fabae*
Narinkol 1550m	1.8.94	11		
	23.9.94	16		
Chemolgan 1550	6.7. 94	25	3	2
	15.7. 94	93	4	2
	21.7. 94	214	3	3
	28.7. 94	648	4	2
	4.8. 94	815	2	12
	9.8. 94	1	0	0
	1.9. 94	1	0	
Zhanaters 1200	12.7. 95	0	0	
	28.7. 95	26	3	
	23.8. 95	7	0	
Myrbulak 1200	12.7. 95	0	0	
	28.7. 95	54	6	
	9.8. 95	15	0	
	23.8. 95	7	0	

Weekly trappings in Moericke type traps, by Dr. Kh. Kadyrbekov

Legends for Figures:

Fig. 8.1 Local lesions induced by *Potato virus X* on *Gomphrena globosa.*

Fig. 8.2 Genome organization and expression strategy of *Potato virus* ***X.***

Fig. 8.3 Leaf drop symptoms induced by PVY^{o} in cv. Pinto. Courtesy of Dr. J. Valkonen.

Fig. 8.4 Necrotic rings on potato tubers caused by PVY^{NTN}. Courtesy of Dr. C. Kerlan , INRA, Le Rheu, France.

Fig. 8.5 Vein necrosis symptoms induced by PVY^{N} in *Nicotiana tabacum* cv. Samaun nn. Courtesy of Dr. J. Valkonen.

Fig. 8.6 Genome organization of PVY. Courtesy of Dr. A. Gal-On , ARO.

Fig. 8.7 Potato cv. Desirée showing leafroll symptoms.

Fig. 8.8 Stunting of *Physalis pubescens (sym. P. floridana);* left - after aphid inoculation with PLRV; right – healthy.

Fig. 6.10 Genome organization of *Potato virus M* (PVM) (modified from Zavriev *et al.* 1991). Courtesy of Association of Applied Biologists.

8.1

8.3

8.5

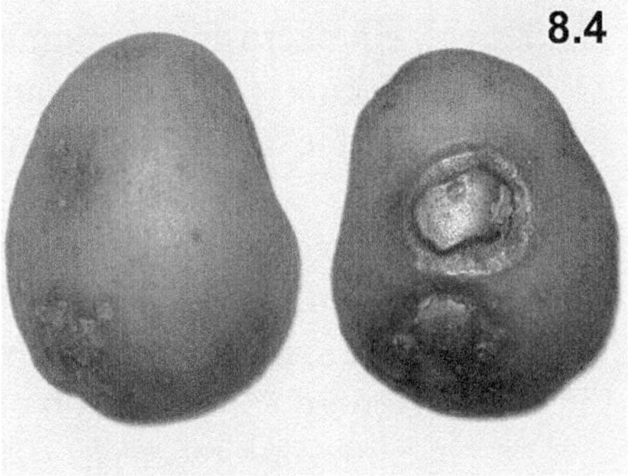
8.4

8.2

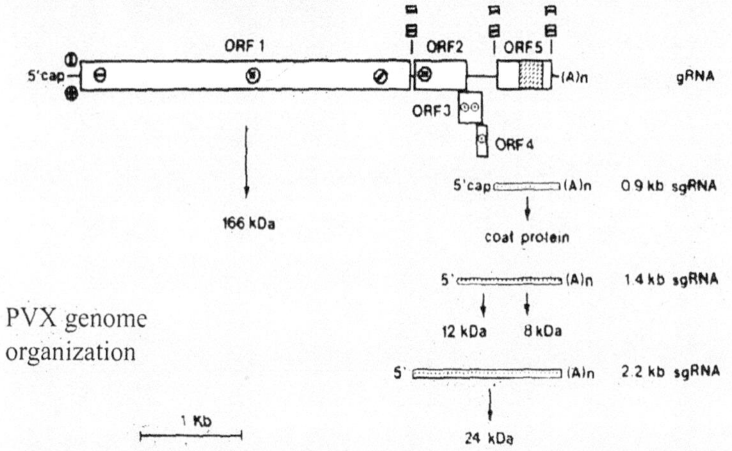
ORF 1
ORF2
ORF5
5'cap
(A)n
gRNA
ORF3
ORF4
5'cap
(A)n
0.9 kb sgRNA
166 kDa
coat protein
5'
(A)n
1.4 kb sgRNA
12 kDa
8 kDa
PVX genome
organization
5'
(A)n
2.2 kb sgRNA
1 Kb
24 kDa

8.6

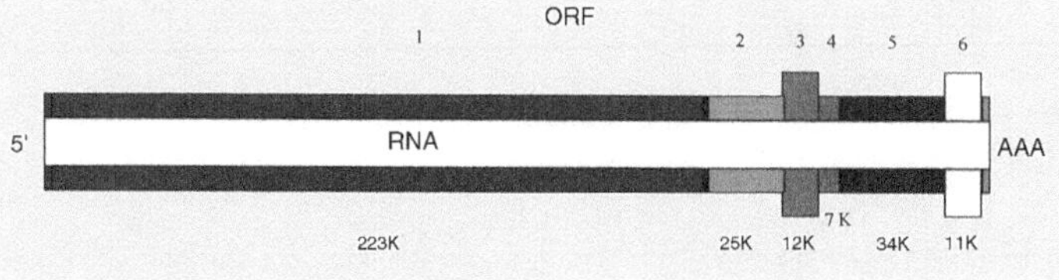
ORF
1
2
3
4
5
6
5'
RNA
AAA
7 K
223K
25K
12K
34K
11K

9. Sweet Potato

G. Loebenstein[1], S. Fuentes[2], J. Cohen[1] and L. F Salazar[2]

[1]*Dept. of Virology, Agricultural Research Organization, Volcani Center, Bet Dagan, 50-250, Israel. E-mail: gadtalma@inter.gov.il; and* [2]*Virology, Crop Protection Department, International Potato Center (CIP), Apartado 1558, Lima 12, Peru.*

9.1. Introduction

Sweet potato, *Ipomoea batatas*, is the seventh most important food crop in the world in terms of production. They are grown on about 9 million hectares, yielding c~ 140 million tons, with an average yield of about 15 ton/ha (FAOSTAT, 2001). They are mainly grown in developing countries, which account for over 95% of world output. The cultivated area of sweet potato in China, about 6.6 million ha, accounted for 70% of the total area of sweet potato cultivation in the world. China produces about 100 million tons, circa 80% of the total world production. Vietnam is the second largest producer. Sweet potato is a 'poor man's crop', with most of the production done on a small or subsistence level. Sweet potato produces more biomass and nutrients per hectare than any other food crop in the world. It is well suited to survive in fertile tropical soils and to produce tubers without fertilizers and irrigation and is one of the crops with a unique role in relief famine. Thus, for example, across East Africa's semiarid, densely populated plains, thousands of villages depend on sweet potato for food security and the Japanese used it when typhoons demolished their rice fields. Sweet potato is grown for both the leaves, which are used as greens, and the tubers, for a high carbohydrate and beta-carotene source. Yields differ greatly in different areas or even fields in the same location. Thus, the average yield in African countries is about 4.7 tons/ha, with yields of 8.9, 4.3, 2.6 and 6.5 ton/ha in Kenya, Uganda, Sierra Leone and Nigeria, respectively. The yields in Asia are significantly higher, averaging 18.5 tons/ha. China, Japan, Korea, Thailand and Israel have the highest yields with about 20, 24.7, 20.9, 12 and 33.3 tons/ha, respectively. In South America the average yield is 12.2 tons/ha, with Argentina, Peru and Uruguay in the lead with 18, 11 and 10 tons/ha, respectively. For comparison, the average yield in the US is 16.3 tons/ha (all data are averages for 2000 and 2001 from the FAOSTAT 2002).

These differences in yields are mainly due to variation in quality of the propagation material. Sweet potatoes are vegetatively propagated from vines, root slips (sprouts) or tubers, and farmers in African and other countries often take vines for propagation from their own fields year after year. Thus, if virus diseases are present in the field they will inevitable be transmitted with the propagation material to the newly planted field, resulting in a decreased yield. Often these fields are infected with several viruses, thereby compounding the effect on yields. In China, on average, losses of over 20% due to sweet potato virus diseases are observed (Gao *et al.* 2000), mainly due to *Sweet potato feathery mottle virus* (SPFMV) and *Sweet potato latent virus* (SwPLV). The infection rate in the Shandong province reaches 5-41% (Shang *et al.* 1999). In countries were care is taken to provide virus-tested planting material as, amongst others in the US and Israel, yields increase markedly, up to 7 times and more

9.2. The Main Viruses

Sweet potato feathery mottle virus Genus *Potyvirus* (SPFMV); is the most common sweet potato virus worldwide. Certain isolates in the US cause much economic damage by inducing cracking or internal corkiness in some cultivars (Fig 9.1). In Africa, SPFMV causes a severe sweet potato virus disease (SPVD) in a complex infection with the whitefly-transmitted *Sweet potato sunken vein virus* Genus *Crinivirus* (SPSVV) [Syn. *Sweet potato chlorotic stunt* Genus *Crinivirus* (SPCSV)].

Most sweet potato cultivars infected by SPFMV alone show only mild circular spots on their leaves (Fig. 9.2) or light green patterns along veins. However, when infected together with the whitefly-transmitted SPSVV stunting of the plants, feathery vein clearing and yellowing of the plants are observed (Fig. 9.3). In controlled experiments SPFMV-infection alone did not reduce yields compared to virus-free controls, while the complex infection with SPCSV reduced yields by 50% or more (Milgram *et al.* 1996). In Brazil yield increases of 118% were observed when virus-tested cuttings derived from meristems of heat-treated plants were compared to yields from cuttings taken from fields (Pozzer *et al.* 1995). This is a unique report where SPFMV alone markedly decreased yields, and it might be that another virus could have been present.

SPFMV is transmitted in a nonpersistent manner by aphids, including *Aphis gossypii, Myzus persicae, A. craccivora* and *Lipaphis erysimi.* Aphid transmissibility is dependent on virus encoded helper proteinase and a DAG triplet in the protein coat. The virus can be transmitted mechanically to various *Ipomoea sp.*, as *I. batatas, I. setosa, I. nil, I. incarnata* and *I. purpurea,* and *Nicotiana benthamiana* and *Chenopudium amaranticolor* (for some strains). The virus is transmitted by grafting but not by seed or pollen or by contact between plants.

The virus can best be diagnosed by grafting on *I. setosa*, causing vein clearing (Fig. 9.4) followed by remission, or on *I. incarnata* and *I. nil* inducing systemic vein clearing, vein banding and ringspots. SPFMV can be diagnosed by ELISA, and antisera are commercially available. However, ELISA reliably detects SPFMV only in leaves with symptoms. It is best to sample several leaves from a plant, as the virus seems to be unevenly distributed, especially for meristem-derived plantlets. This has to be followed up by indexing on a susceptible indicator (Green *et al.* 1988). SPFMV can also be detected by membrane immunobinding (MIBA, also termed NCM-ELISA) and by using a riboprobe (Abad and Moyer, 1992). The latter although being very sensitive, detecting 0.128 pg of RNA, compared with 179 pg of capsid protein, was still somewhat less sensitive then grafting on a sweet potato indicator (cv. Jewel). The use of a riboprobe also requires a well-equipped laboratory and radioactive materials or digoxigenin for labeling the probe. SPFMV has also been identified by PCR, utilizing degenerate genus-specific primers, designed to amplify the variable 5' terminal region of the potyvirus coat protein gene (Colinet *et al.* 1998).

SPFMV can be purified from infected *I. nil* leave material (Cali and Moyer, 1981), by rather complex purification procedures, and directly from sweet potatoes (infected also with SPSVV), by a relatively simple method (Cohen *et al* 1988). This method yielded 50-100 mg/kg virus, 5-10 times more when compared with the previous methods. It should be mentioned that virus yields from plants infected by SPFMV alone, without SPSVV, are markedly lower.

Virions are filamentous; not enveloped; usually flexuous; with a modal length; of 830-850 nm. The genome consists of single stranded linearRNA, with a Poly(A) region. Total genome size 11.6 kb. Genome is unipartite; largest (or only) genome part 11.6 kb (Moyer and Cali, 1985). The complete nucleotide sequence of a sweet potato feathery mottle potyvirus severe strain (SPFMV-S) genomic RNA was determined (Sakai *et al.* 1997). The viral RNA genome (S strain) is 10,820 nucleotides long, excluding the poly(A) tail, and contained one open reading frame, potentially encoding 3493 amino acids. Except in the regions of P1 and P3, the polyprotein has a high level of amino acid identity with those of other potyviruses (Sakai *et al.* 1997). The helper component-proteinase of SPFMV facilitates systemic infection spread of *Potato virus X* in *I. nil* (Sonoda *et al.* 2000).

Many strains of SPFMV (Moyer, 1986), isolates, variants and serotypes of SPFMV have been reported, as the russet crack (RC), C, S, C1 (from Peru), strain 835 (from Guatemala) and others (Colinet and Kummert, 1993; Karyeija *et al.* 2001). By comparing coat protein gene sequences of isolates it was shown that isolates from East Africa form a separate cluster (Kreuze *et al.* 2000).

Though SPFMV alone generally causes only minor damage, its control is imperative as in combination with other viruses its effect on plant growth and yields may become substantial. The proven approach is to prepare virus-tested plants from meristems. The plants are then grown under insect-proof conditions, and propagation is continued by cuttings. The farmer plants a certain number of vines in the open field and continues to propagate until the stand of the field is complete. The following year he will again start from virus-tested tested plants. This scheme has been in operation for more than 15 years in Israel, and presently it is difficult to find SPFMV in the country.

Breeding of SPFMV resistant plants was initiated by CIP (Mihovilovich *et al.* 2000). Several clones that were resistant to SPFMV in CIP's tests were found to be susceptible, when exposed to Israeli (unpublished) and Ugandan isolates (Karyeija *et al.* 1998). Apparently, strain diversity requires that breeding and selection have to be done in various locations. On the other hand a substantial number of African sweet potato landraces have resistance to this virus (Carey *et al.* 1997).

A promising approach, which still has to show its effectiveness in the field, is the development of transgenic sweet potatoes with coat- mediated (CP) resistance to SPFMV (Okada *et al.* 2001). Also, CP- mediated resistance introduced into several African varieties and cultivar CPT-560 is presently evaluated in Kenya in a cooperative project between Monsanto Co., USA and Kenya Agriculture Research Institute (KARI). One of the main questions to await answer will be to see if these transgenic lines will be resistant to the 'complex' infection with SPSVV, causing the SPVD.

An interesting idea for controlling SPFMV involves the introduction of a rice cysteine- inhibitor gene. This gene inhibits the proteolysis of the viral polyprotein, thereby interfering with viral replication. Improved resistance to SPFMV was observed in 18 of the 25 transgenic lines after challenge-inoculation with the russet crack strain of SPFMV (Cipriani *et al.* 2001). However, when stem cuttings were prepared from the tolerant transgenic plants and grafted with healthy *I. setosa* scions, virus symptoms appeared on the scions. Apparently, the transgenic sweet potato plants still contained some virus.

Sweet potato sunken vein virus Genus *Crinivirus* (SPSVV). {Possible synonym: *Sweet potato chlorotic stunt virus* (SPCSV)}. Infection of sweet potato by SPSVV alone produced on cv. Georgia Jet mild symptoms consisting of slight yellowing of veins, with some sunken secondary veins on the upper sides of the leaves and swollen veins on their lower sides (Fig. 9.5) (at 28-29°C). Upward rolling of the 3-5 distal leaves was also observed. Similar symptoms were also obtained on graft-inoculated plants of cv. Papel and Camote Negro (Cohen *et al.* 1992). The disease could easily be detected in the field where plants had erect branches, with some upward rolling leaves. Effects on yields by SPSVV alone are minor (Milgram *et al.* 1996), but in complex infection with SPFMV yield losses of 50% and more are observed.

SPSVV is transmitted by the whitefly *Bemisia tabaci* in a semi-persistent manner, requiring at least one hour for acquisition and infection feeding and reaching a maximum after 24 h for both of them. The virus is graft transmissible, but not by mechanical inoculation. The virus was transmitted by whiteflies to *I. setosa, N. clevelandii, N. benthamiana* and *Amaranthus palmeri* from *I. setosa* and by grafting to other Ipomoeas.

The virus is best being diagnosed on a pair of sweet potato plants- one healthy, the other infected by SPFMV. On the healthy plants hardly any symptoms will become apparent, while (if carrying SPFMV) severe symptoms of SPVD will appear. The virus can also be diagnosed by immunosorbent electron microscopy (ISEM). So far ELISA does not detect the virus reliably.

SPSVV can be purified from infected *I. nil* or *N. clevelandii,* maintained at 28°C (Cohen *et al.* 1992). Normal length of virus particles from purified preparations was 850 nm, the diameter ~ 12 nm, with an open helical structure typical for closteroviruses In leaf dip preparations from *I. setosa* infected whith an isolate from Nigeria a modal length of 950 nm was obtained (Winter *et al.* 1992). The coat protein of the Israeli and Nigerian isolates had a mol. wt of ~ 34kd and ~ 29 kd, respectively (Cohen *et al.* 1992; Hoyer *et al.* 1996). Double-stranded RNA from the Israeli isolate consisted of one major band of M_r 10.5 kbp, and 2 minor bands of 9.0 and 5.0 kbp. Northern blot analysis of the Nigerian isolate revealed the presence of a large dsRNA with an estimated size of ~9.0 kbp and several smaller ones. Several Kenyan cDNA clones revealed an open reading frame (ORF) of 774 nucleotides coding for the coat protein (Hoyer *et al.* 1996). The isolates from Israel, Nigeria, Kenya and the US are serologically closely related (Hoyer *et al.* 1996).

For control of SPSVV it is best to start from meristems and test the plants on a pair of sweet potato plants as described above

Sweet potato virus disease (SPVD) is caused by the interaction of SPFMV and SPSVV. Characteristic symptoms of the disease include veinclearing (Fig.9.3), chlorosis and stunting. The disease was described by Schaefers and Terry (1976) in Nigeria and is the most important virus (complex) disease in East Africa, where sweet potato is often the main food staple (Karyeija *et al.* 1998). The disease was described in Israel by Loebenstein and Harpaz (1960). It can cause losses of fifty to eighty percent, especially in Uganda and Kenya, though in another study from Uganda, losses were much smaller, probably due to relatively high levels of virus resistance in their landraces (Gibson *et al* 1997). In a 3-year field study in Cameroon SPVD reduced root yields by 56-90% in susceptible varieties (Ngeve and Bouwkamp, 1991); and yield reductions of 78% due to SPVD were reported from field trials in Nigeria (Hahn, 1979). In Israel in a 2-year field experiment yield reductions of ~50% were observed in plots planted with SPVD- infected cuttings (Milgram *et al.* 1996).

As stated above the optimal approach for controlling SPVD is to supply virus-tested planting material. In practice, however, in Uganda and other African countries this has sofar not been achieved. In Uganda, where sweet potatoes are grown on a large scale the most important control practice, as perceived by farmers, is to plant cuttings from symptomless parents, and destroying diseased plants (Gibson *et al.* 2000). Although SPVD- resistant sweet potato landraces are available, these were perceived by farmers to have poor and late yields. Cultivars Tanzania and New Kawogo were relatively resistant to SPVD. SPFMV in these cultivars was difficult to detect, because of low virus titer, and was seldom acquired by aphids. However, resistance to SPFMV was not apparent once infected also with SPSVV (Aritua *et al.* 1998). In SPVD- infected plants SPFMV titers were much higher than in sweet potato plants infected with SPFMV alone (Cohen *et al.* 1988; Karyeija *et al.* 2000). This increase is not associated with enhanced virus movement but is due to a synergistic interaction with the crinivirus (Karyeija *et al.* 2000).

Sweet potato mild mottle virus Genus *Ipomo*, (SPMMV). Synonyms – sweet potato B virus (Sheffield, 1957). SPMMV has so far been reported from West Africa (Kenya, Uganda, Burundi and Tanzania), Indonesia, Papua New Guinea, India, and Egypt.

SPMMV can cause leaf mottling, stunting and loss of yields. Cultivars differ greatly in their reaction to the virus, some being symptomlessly infected, others apparently immune. The virus is transmitted persistently by *B. tabaci,* by grafting and by mechanical inoculation. It is not transmitted by seed nor by contact between plants. The virus was transmitted to plants in 14 families (Hollings *et al.* 1976). Diagnostic species: *N. tabacum, N. glutinosa*- vein clearing, leaf puckering, mottling and distortion; *C. quinoa* – local lesions, not systemic; *I. setosa* - conspicous systemic vein chlorosis (Fig 9.6). Diagnosis can be confirmed by serological tests of *I. setosa.* Commercial antisera are available. After 3-4 weeks, new growth is almost symptomless. Sap transmission from sweet potato to test plants is often difficult. The virus is best maintained in *N. glutiniosa, N. clevelandii and N. tabacum,* and can be assayed quantitatively on *C. quinoa,* where SPMMV induces local lesions.

The virus can be purified from systemically infected *N. tabacun* (Hollings *et al.* 1976). Virions are flexous rod shaped particles, 800-950 nm in length, containing 5% RNA and 95% protein. The genome consists of single stranded RNA. The viral RNA was cloned and the assembled genomic sequence was 10,818 nucleotides in length with a polyadenylated tract at the 3' terminus. The sequence accession code is Z48058 Em (44)n. Almost all known potyvirus motifs are present in the polyprotein of SPMMV, except some motifs in the putative helper-component and coat protein, which are incomplete or missing. This may account for its vector relations (Colinet *et al.* 1998). The coat protein has a molecular weight of 37,700.

Sweet potato latent virus Genus *Potyvirus* (SwPLV) is widespread in China and has been reported also from Taiwan, Japan, Indonesia, Philippines, Uganda, kenya, and India.

SwPLV may cause mild chlorosis but in most cultivars the infection is symptomless. Symptoms often disappear after infection, but the plants remain infected. Crystal inclusions are observed in the nucleus and pinwheels in the cytoplasm.

There is no known vector, and the virus can be transmitted by mechanical inoculation and by grafting. It is not transmitted by seed.

Diagnostic species: *N. benthamiana* – systemic mosaic and stunting; *N. clevelandii*- systemic pin-prick chlorotic lesions; *C. quinoa, C. amaranticolor* – brown necrotic local lesions, no systemic infection; *I. setosa* – systemic mottle. Diagnosis can be confirmed by serological tests of *I. setosa.*

The virus is best maintained in *N. benthamiana* or *N. clevelandii* and can be assayed on *C. quinoa* or *C. amaranticolor.*

The virus can be purified according to Liao *et al.* (1979). Virus particles are flexous rods, 750-790 nm in length. The capsid protein has a molecular weight of 36,000. By using poly- and monoclonal antibodies (MAbs) some epitopes common to SwPLV and SPFMV are found. These can easily be differentiated when potyvirus cross-reactive MAbs are used, indicating a distant relationship (Hammond *et al.* 1992).

Combining reverse transcription and polynerase chain reaction (PCR) and using degenerate oligonucleotide primers derived from the conserved regions of potyviruses, it was possible to identify SwPLV, as well as SPFMV and *Sweet potato G virus* (Colinet *et al.* 1998), and to differentiate between 2 strains of SwPLV (Colinet *et al.*1997).

Sequence data accession codes are: X84011 (89) SwPLV (Chinese) mRNA for coat protein . X84012 (89) SwPLV (Taiwan) mRNA for coat protein.

The best way to control this virus, as well as other viruses infecting sweet potato is by establishing propagation nurseries derived from virus-tested mother plants.

Sweet potato virus G Genus *Potyvirus* (SPVG) is widespread in China, (Colinet *et al.* 1994, 1998). The virus is transmitted by aphids. A partial sequence of SPVG (X76944) has been obtained after RT-PCR, showing an identity of around 70% and 80% in the amino acid sequence between the complete and conserved core of the coat protein of SPFMV, respectively (Colinet *et al.* 1998). The SPVG coat protein has 355 amino acids while that of SPFMV has 316 amino acids (Colinet *et al* 1994). Comparison with coat protein sequences of known potyviruses indicates that SPVG is a member of the genus *Potyvirus*. Strains of SPVG seem to occur in China (Colinet *et al.* 1998). SPVG-CH2 shared 89.2% and 90.6% amino acid sequence identities with SPVG-CH in the N-CP and the N-terminal region of the coat protein core (N-CP core), respectively. Sequence identity was much lower with SPFMV and even more so with SwPLV and other potyviruses. No biological properties (including host range and symptomatology) or viral characteristics have come to our attention.

Sweet potato mild speckling virus Genus *Potyvirus* (SPMSV) was isolated from plants cv. Morada INTA from Argentina showing symptoms of "sweet potato chlorotic dwarf disease" (Di Feo *et al.* 2000). These plants show chlorosis, dwarfing, vein clearing, and leaf distortion (Alvarez *et al.* 1997; Di Feo *et al.* 2000). The severity of the disease depends on the presence of SPMSV in the complex. Symptoms in plants infected with SPFMV and SPSVV (Syn. SPCSV) were milder than those in plants infected with the three viruses (Di Feo *et al.* 2000).

SPMSV has been found in Peru, Argentina, Philippines, Egypt and Indonesia. SPMSV is transmitted mechanically and by *M. persicae* in a nonpersistent manner. However, it does not react with the MAb PTY-1 that recognizes a cryptotope conserved in most aphid-transmitted potyviruses (Di Feo *et al.* 2000; Jordan and Hammond, 1991). Its host range is restricted to *Convolvulaceae, Chenopodiaceae*, and *Solanaceae*. *I. setosa* and *I. nil* react with vein clearing, blistering, leaf deformation and mosaic; *N. benthamiana* with vein clearing, and reduction, deformation and down rolling of leaves, and *C. quinoa*, and *N. tabacum* "Samsun" with local infections (Fuentes *et al.* 1997). SPMSV can easily be detected by ELISA.

This virus has flexuous particles of c.800 nm. The coat protein sequences of SPMSV showed 63% identity with SPFMV, 68 to 70% with SwPLV, 57% with SPVG, and 73% with *Potato virus Y* (Alvarez et al, 1997). This high resemblance favors inclusion of SPMSV within the *Potyvirus* genus.

Sweet potato leaf speckling virus Possible Genus *Polerovirus* (SPLSV) was isolated from a sweet potato accession DLP 1541 from CIP's germplasm collection, showing slight leaf curl and whitish speckling symptoms (Fuentes *et al* 1996) (Fig. 9.7). It is found occasionally in farmers' materials. The virus was also found in Cuba. Infected plants of cv. "Jewel" yielded c. 20% less under greenhouse conditions. Virus causes dwarfing and leaf curling in addition to chlorotic and necrotic spotting in *I. nil* and *I. setosa.* SPLSV is transmitted by grafting and by *Macrosiphon euphorbiae* in a persistent manner, but not by *M. persicae, A. gossypi*i or mechanically. Diagnosis can be confirmed by nucleic acid spot hybridization (NASH) from *I. setosa* or *I. nil* (Fuentes *et al.* 1977). Concentration of SPLSV increases towards the bottom of the plant. On leaves, the virus was detected in the basal part of the leaves and petioles, but not in the apical halves.

SPLSV can be purified from leaves or roots of infected plants (Fuentes *et al.* 1996). Virions are isometric c.30 nm in diameter. Reverse transcription and PCR amplification of RNA from SPLSV-infected plants, using degenerate primers designed to amplify sequences in Luteovirus- or PLRV-RNA, yielded c. 500 bp and c. 600 bp DNA fragments, respectively. The nucleotide sequence of these fragments encoded two polypeptides (coat protein and 17K) characteristic of luteo- and poleroviruses sequences. The coat- and the 17K protein of SPLSV have similarities to of 70% and 63% those of PLRV, respectively. NASH using a PLRV probe can therefore detect SPLSV in sap. However, the heterologous probe gives a lower signal than the homologous reaction with PLRV. SPLSV was not detected in ISEM or ELISA tests using PLRV antibodies.

Sweet potato chlorotic flecks virus (SPCFV) was isolated from sweet potato accession DLP 942 from the CIP′s germplasm collection, showing fine chlorotic spots (flecks) (Fuentes and Salazar, 1992). SPCFV was detected in samples from Peru, Japan, China, Cuba, Panama, Bolivia, Colombia, Brazil, Uganda, Philippines, India, Indonesia, and Austria (Fuentes and Salazar, 1992; Gibson *et al.* 1997; Usugi *et al.* 1991). This virus is transmitted mechanically to *Convolvulaceae* and *Chenopodiaceae* species. SPCFV induces in *I. nil* fine chlorotic spots and vein clearing on the first and second true leaf. The virus can be recovered and detected from symptomatic leaves, but not from symptomless ones. No clear symptoms are observed on infected *I. setosa* (only transient vein clearing on 1-2 leaves). In *C. murale* and *C. quinoa* react with local necrotic lesions. Graft inoculated cvs. Nemañete, Jewel, Luby 3074, IITA TIS 2498 did not show symptoms. SPCFV was not transmitted by the aphid *M. persicae* or through seed from infected *I. nil*. Infectivity of SPCFV is lost at dilution between 10^{-2} and 10^{-3}, and after storage between 1h and 1 day. The thermal inactivation point is between 50 and 60°C. Some cytological alterations (hypertrophy of chloroplasts) were observed in *I. nil* but no inclusion bodies ("pinwheels").

SPCFV seems to be serologically related to an uncharacterized Japanese isolate (sweet potato symptomless virus – SPSV). Although SPCFV and SPSV are serologically related, they differ in some characteristics. SPSV has flexuous short and long particles c. 710 to 760 nm and 1,430 to 1,510 nm in length, and it infects only *Ipomoea* spp., while SPCFV particles are flexuous rod c. 750-800 in length. Also, *I. nil* infected with SPSV showed stronger symptoms (vein clearing, curling and vein necrosis) than those caused by SPCFV. The thermal inactivation point of SPSV (70-80°C) is higher than that of SPCFV.

Virus can be purified from cotyledons and leaves of infected *I. nil* (12 day after inoculation). The virus particles are flexuous rod c. 750-800 in length with a capsid polypeptide of M_r 34.5 KD (Fuentes and Salazar, 1992)

Sweet potato caulimo-like virus (SPCV) has been found together with SPFMV in widely scattered geographical locations – Madeira, New Zealand, Papua New Guinea and Solomon Islands (Atkey and Brunts, 1987), China (Gao *et al.* 2000) and Uganda (WISARD Project Information, 1999). Also reported from Kingdom of Tonga, Austria, Kenya, Egypt, and Puerto Rico.

Sweet potato plants infected by SPCV are usually without symptoms, though a few leaves occasionally have chlorotic or purple spots, probably induced by SPFMV (Atkey and Brunts, 1987). Virus is not transmitted by mechanical inoculation, not by contact between plants or by seed.

Graft inoculation of *I. setosa* seedlings with SPCV resulted in appearance of chlorotic flecks along minor veins or circular interveinal spots on several leaves. Such leaves sometimes become completely chlorotic, wilt and die. Fifty nm caulimo-like particles can be detected in negatively stained sap of *I. setosa* (but not from *I. batatas*), and c. 10 times more by ISEM. In the cytoplasm of infected *I. setosa* leaf cells numerous isometric particles and intracellular inclusions are readily detected. The inclusions are spherical or ovoid , up to 4µm in diameter with a large central lacuna and usually several smaller peripheral lacunae; they do not contain virions (Atkey and Brunts, 1987).

The viral genome consists of double-stranded DNA and the coat of one M_r 42000-44000 protein (Brunt *et al.*1996).

Sweet potato leaf curl virus Genus *Begomovirus* (SPLCV) was first reported by Shinkai (1979) and Liao *et al.* (1979) from Japan and Taiwan. The virus has also been reported from USA and samples from Brazil, China, Korea, Puerto Rico (Lotrakul *et al.* 2002), and Peru. In infected young sweet potato plants upward curling of leaves and vein swelling on young plants are observed, though later only few symptoms remain and plants become symptomless. SPLCV can cause up to 30% reductions in yield. The virus is transmitted by *B. tabaci* in a persistent manner and by grafting, but not mechanically or by seeds. Various *Ipomoea* species were susceptible to SPLCV, as I. purpurea causing leaf curl and stunt (Osaki and Inouye, 1991), *I. aquatica* - yellow vein symptoms, *I. nil, I. setosa* and *N. benthamiana* - leaf curl symptoms (Lotrakul et *al.* 1998). The disease is caused by a gemini virus belonging to the *Begomovirus* group (Onuki *et al.* 2000). Cytopathic changes in the nucleoplasm, as fibrillar rings and crystalline arrays of virus-like particles, typical of gemini-infected tissue were also observed (Osaki and Inouye, 1991). The virus was partially purified yielding typical geminate particles with a size of *c.* 18 x 30 nm, and Western blot analysis revealed serological relationships between SPLCV and *Bean golden mosaic virus* (Onuki *et al.* 2000). A complete sequence of SPLCV (AF104036, 2828 nt) has been determined by Lotrakul and Valverde (1999). Its genomic DNA and organization is similar to that of monopartite begomoviruses. The partial nucleotide sequence identity (from 87% to nearly 100%) of the AC1 gene of different SPLCV isolates and the phylogenetic analysis of them suggest that there may be more than one species of the SPLCV (Lotrakul *et al.* 2002).

Ipomeoa crinkle leaf curl virus (ICLCV). In 1992, Cohen *et al.* (1997) observed on *I. setosa,* grafted with scions from sweet potato cv. Georgia Jet plants, introduced from an unknown source in North America, atypical symptoms of little leaf and crinkle symptoms (Fig. 9.8). Geminate particles were observed in crude sap preparations. The virus was transmitted by *B. tabaci* in a persistent manner and by grafting, but not mechanically. The virus was transmitted to several *Ipomoea* species, including *I. hederacea, I. trifida, I. nil, I. littoralis* and *I. setosa* and induced symptoms on them, including on some cultivars of *I. batatas,* but not on cv. Georgia Jet. It might be mentioned that SPLCV did not infect *I. hederacea* and symptoms induced by ICLCV on *I. setosa* and *I. nil* differed from those described for SPLCV (Chung *et al.* 1985). Based on the host range ICLCV is considered to be distinct from sweet potato crinkle leaf curl virus (SPLCV).

Cucumber mosaic virus Genus *Cucumovirus* (CMV) is one of the most widespread plant viruses, recorded in more than 190 species, belonging to more than 40 families (Francki *et al.*1979). CMV has been isolated from *I.*

setifera (Migliori *et al.* 1978) and Martin (1962) succeeded in transmitting CMV by mechanical inoculation to *I. nil, I. purpurea, I. lacunosa,* and *I. trichocarpa* but not to *I. batatas* cv. Puerto Rico. However, sweet potatoes can easily be infected by CMV if the plant is already carrying the whitefly-transmitted SPSVV (Cohen and Loebenstein, 1991). Aphid, mechanical and graft inoculations by CMV were obtained when the acceptor plants were already infected by SPSVV. Transmission of CMV to healthy sweet potato plants failed. In some fields in Israel during the 80es heavy infections together with SPFMV anf SPSVV caused severe yellowing and stunting (Fig. 9.9). Later, when farmers used certified planting material such symptoms were hardly found. Apparently, the presence of another virus (SPSVV) is required for replication or translocation of CMV in sweet potato. CMV occurs worldwide,. but in sweet potato has been reported so far only from Israel, Uganda, Kenya, Japan, and Egypt.

Sweet potato yellow dwarf virus Genus *Ipomovirus,* Family *Potyviridae* *(*SPYDV) in *I. batatas* was first reported from Taiwan by Liao *et al.* (1979). The virus in sweet potato causes severe systemic leaf chlorosis and stunting. The virus was transmitted by *B. tabaci* in a persistent manner, as well as by grafting and by mechanical inoculation. Susceptible hosts include in addition to sweet potato, *Chenopodium* spp., *Gomphrena globosa, Datura stramonium, Cassia occidentalis, Sesamum indicum* (Chung *et al.* 1986). SPYDV induces in *I. setosa* stunting, general leaf chlorosis, small distinct chlorotic spot, and vein chlorosis. Virions are filamentous, not enveloped, flexuous, with a clear modal length of 750 nm. The viral coat protein has a molecular weight of 33,000. Pinwheel inclusions are present in the cytoplasm of infected cells.

Sweet potato vein mosaic virus Possible Genus *Potyvirus* (SPVMV) was first reported from Argentina by Nome (1973). In sweet potato, the virus causes a disease called "batata crespa" and infected plants shows vein clearing, mosaic, twisting of leaves, stunting, and noticeable reduction in number and size of the fleshy roots. The virus can be transmitted in a nonpersistent manner by *M. persicae*, as well as by grafting and by mechanical inoculation (Nome *et al.* 1974). The virus was transmitted to several *Ipomoea* species, including *I. alba, I. nil, I. setosa, I. lacunosa, I. hederacea, I. tricolor, I. trichocarpa, I. kurtziana, I. fistulosa,* and *I. angulata,* generally inducing systemic vein clearing and mosaic. Twisting, chlorosis and size reduction of leaves were marked in *I. setosa, I. nil, I. hederacea,* and *I. batatas*. Virus particles are flexuous rod, 761 nm in length. Virions are found in the cytoplasm of all parts of the host plant. Pinwheels and scrolls and laminated aggregates inclusions are present in the cytoplasm of infected cells (Brunt *et al.* 1996). Unfortunately the original culture and antiserum are not available.

Sweet potato C-6 virus. The virus was isolated from sweet potato Sosa 29, from the Dominican Republic, showing chlorotic spots (Fuentes, 1994). Its host range is restricted to *Convolvulaceae*. The virus induces on *I. nil* and *I. setosa* fine chlorotic spots and vein clearing (Fig. 9.10) but on sweet potato cv Paramonguino first chlorotic spots and later mottle symptoms. This is followed by yellowing and leaf drop. The virus is transmitted by grafting and by mechanical inoculation (with a low efficiency), using sap from *I. nil* roots but not from leaves. Attempts to transmit the virus with *M. persicae* were unsuccessful. The virus was detected in samples from Peru, Uganda, Cuba, USA, Dominican Republic, Philippines, Indonesia, Egypt, and Puerto Rico. C-6 virus gave a mild reaction with Potato virus S (PVS) antiserum.

C-6 virus was purified from infected *I. setosa* plants (Fuentes, 1994). Virus particles are flexuous rods, 750-800 nm in length. No "pinwheels" were observed in infected *I. nil*, but chloroplasts and mitochondria showed hypertrophy.

Sweet potato C-3 virus, a suspected flexuous virus, was isolated from a sweet potato co-infected with SPFMV and showing leaf deformation, vein clearing and mosaic, from Brazil (Fuentes and Salazar, 1989). The virus is transmitted by grafting but neither mechanically nor by aphid *M. persicae*. It induces mosaic, leaf deformation and vein clearing in *I. setosa*. Infected sweet potato cv. Paramonguino showed mottle symptoms and cv Georgia Red interveinal mottling. *N. benthamiana* graft-inoculated with C-3 showed yellowing vein, mosaic and leaf deformation, but *I. nil* did not show symptoms.

Sweet potato chlorotic stunt virus Probably Genus *Potyvirus* (SPCSV). This name was tentatively given to a *Potyvirus* isolated from sweet potatoes from the Caribbean area showing stunting and chlorotic leaves (Brown *et al.* 1988). This virus should not be confused with the *Crinirus,* termed here *Sweet potato sunken vein virus,* which was termed by others *Sweet potato chlorotic stunt virus*. The virus was also detected in Kenya, Uganda (Wambugu, 1991), and Zimbabwe (Chavi *et al.* 1997). This potyvirus is not transmitted by aphids, but is readily sap-transmissible to a wide range of herbaceous species of which *N. benthamiana* is a good propagation host and *C. amaranticolor* and *C. quinoa* are convenient local lesion assay hosts. Virus particles are flexuous rod, 850-950 nm in length, which contain ss-RNA and a single capsid polypeptide of M_r 43 KD. The virus induces “pinwheels" characteristic of potyvirus This potyvirus can be detected by ISEM and ELISA.

Sweet potato ringspot virus Genus *Nepovirus* (SPRSV) was observed in *I. batatas,* imported from Papua New Guinea into the UK. (Brown *et al.* 1988). Also reported in Kenya (Wambugu, 1991). The virus on sweet potato occasionally causes chlorotic ringspots, which disappear soon after infection. The virus can be transmitted by grafting and mechanical inoculation. The virus infects a wide range of experimental hosts, including *I. setosa* – faint systemic chlorotic leaf mottling; *N. megalosiphon* – necrotic rings and systemic mottling; *N. benthamiana* – systemic leaf deformations; *C. quinoa, C. murale* – transient systemic chlorosis (www.ncbi.nlm.nih.gov/ICTVdb/). Virions are not enveloped, 28 nm in diameter, which sediment as three components (T, M and B) with sedimentation coefficients of 47, 81 and 130 S and buoyant densities in cesium chloride of 1.32, 1.45 and 1.52 g cm^{-3}, respectively. All components contain a single polypeptide of M_r 56.6 x 10^3. T component particles contain only protein but those of M and B components contain ss-RNA 8,848 nt (M_r 2.93 x 10^6) and 6,670 nt (M_r 2.31 x 10^6), respectively. Virions can be found in leaves and roots. Although the virus has physico-chemical properties resembling those of nepoviruses, it showed no serological relationship to any of 13 recognized nepoviruses (including *Arabis mosaic virus, Cassava green mottle virus, Cherry leaf roll virus, Raspberry ringspot virus, Strawberry latent ringspot virus* and *Tobacco ringspot virus*) (Brunt and Brown, 1990).

New Potyviruses

Four virus isolates, LSU-1, LSU-2, LSU-3, and LSU-5, were obtained from sweet potato plants in Louisiana, USA (Sim, 2001) and Peru (unpublished date). The LSU viruses react with MAb PTY-1 (Jordan and Hammond, 1991) and weakly or not with polyclonal antiserum to SPFMV. LSU-2 and 5 can be transmitted to *N. benthamiana* inducing vein clearing, and to *C. quinoa* inducing chlorotic local lesions. LSU-1 induces chlorotic spotting on *I. tricolor,* while LSU-2 and 5 induce vein mosaic. On *I. setosa* – LSU-1 and 3 induce mild, transient chlorotic spotting and LSU-2 and 5 vein mosaic. All LSU viruses are transmitted mechanically. LSU-1, but not LSU-2 and 5, was transmitted by aphid *A. gossypii* and *M. persicae* in a nonpersistent manner. Inclusion bodies consisting of bundles, pinwheels, and scrolls were observed in epidermal, mesophyll, and vascular tissues of *I. nil* and *I. setosa.* Antisera for LSU viruses were produced after purification (Di Feo *et al.*2000). The nucleotide sequences of the 3' end region of the NIb through the 5' end of the coat protein gene (CP) of LSU-1 and 3 were identical, and showed 98% identity in nucleotide and 100% identity in amino acid sequences to the corresponding region of the CP gene of SPVG. LSU-2 and 5 had 95% identity in nucleotide and 100% identity in amino acid sequences with each other, but only 66% and 62% identity in the nucleotide sequence and amino acid sequence of the CP gene with LSU-1 and 3. However, phylogenetic analysis indicated that LSU-2 and 5 were more closely related to LSU-1, LSU-3 and SPVG than to the group of SPFMV strains. The potyvirus nature of the LSU viruses was confirmed by biological assay (including host range and symptomatology), cytopathology, ELISA, and RT-PCR. Apparently, LSU-1 and LSU-3 seem to be strains of *Sweet potato virus G*. LSU-2 and LSU-5 seem to be isolates of a new member of genus Potyvirus (C. Clark and R. Valverde, personal communication).

Two previously undescribed viruses affecting sweet potato in North Carolina, USA have been reported by Moyer *et al.* (2002). These two isolates gave no reaction with antiserum for SPFMV, SPMMV, SwPLV, SPCFV, SPMSV, C-6, SPCV, and SPSVV, but showed novel dsRNA profiles. The first is a 9-10 kbp single species and may correspond to a potyvirus, the second is a 7.5 kbp dsRNA with two putative sub-genomic fragments. Both isolates tested negative in ELISA with a universal antiserum for potyviruses. From the first profile one isolate, GW Beauregard, was selected for dsRNA purification, cloning and sequencing. This isolate when graft inoculated on cv. Tanzania induced severe mosaic, vein banding and leaf deformation The CP gene sequences were used for the phylogenetic analysis. Sequence of the putative potyvirus was more closely related to SPMMV than SwPLV, SPVG, and SPFMV.

Viruses infecting wild *Ipomoeas* or sporadically sweet potato

Viruses infecting *I. aquatica* and *I. indica* were reported by Brunt *et al.* (1996) and Banks *et al.* (1999), respectively.

Tobacco streak virus and Tobacco mosaic virus (Clark and Moyer, 1988), and Sweet potato virus II (Moyer *et al.* 1989) have also been reported infecting sweet potato.

Little-leaf Phytoplasma Disease also referred to as witches' broom, was first described by Van Velsen (1967) from Papua and New Guinea. It has been reported also from the Ryuku islands (Yang, 1969), Tonga (Kahn and Monroe, 1969), Solomon Islands (Dabek and Sagar, (1978), Taiwan (Yang, 1969) and Northern Australia (Gibb *et al.* 1995). First symptoms are veinclearing of otherwise normal leaves. Subsequent symptoms include large reductions in the size of leaves and internodes and an erect habit of the tips of the vines. Plants which become infected at an early stage yield poorly, producing only one or two pencil-thin tubers (Pearson *et al.* 1984). Typical phytoplasma bodies were observed in phloem sieve cells. Meristem tip culture was found to be effective in eliminating the phtytoplasma from the infected sweet potato tissue (Green *et al.* 1989). The phytoplasma could be diagnosed by polymerase chain reaction (PCR), and the full-length chromosome was determined as 600 kb, which is one of the smallest phytoplasma genome sizes (Gibb *et al.* 1995).

9.3. Control

At present the best way to control virus diseases in sweet potato is to supply the grower with virus-indexed propagation material. Such programs are operating in Israel and in the Shandong province of China (Gao *et al.* 2000). In Israel , as a result of planting virus-tested material, yields increased at least by 100%, while in China increases ranged between 22-92%. The payoff to the farmer has been high and in Israel use of certified material is common practice, while in China the use of pathogen-free material is being extended. In African countries such programs are operating only on a limited scale, because sweet potatoes are grown mainly as a food security crop, and not as a commercial one.

Breeding programs might be a future answer and such programs are in operation in Uganda, combining SPVD resistance with desirable agronomic traits such as yield, earliness and acceptable culinary quality. Progress has been made and several cultivars were released (Turyamureeba *et al.* 1998). It will have to be seen if these cultivars will retain their resistance in other places where different strains of SPVD may be present. Thus, several clones that were resistant to SPFMV in CIP's tests were found to be susceptible, when Israeli (unpublished) and Ugandan isolates were tested (Karyeija *et al.* 1998).

As *Ipomoea* spp. are susceptible to most sweet potato viruses it is of importance to survey weeds as hosts and potential reservoirs of viruses.

Transforming sweet potato with viral (CP) or other genes (rice cysteine-inhibitor gene) to obtain resistance is an interesting approach. It will however await answers if these transformed plants will be resistant also to 'complex' infections.

The best way at present is the orthodox approach of supplying virus-indexed planting material.

9.4. Concluding remarks

The demand for root and tuber crops in developing countries to ensure a sustainable food supply is predicted to grow steadily over the next decades. Sweet potato is one of the major crops, and much of it is grown in developing countries, for supplying carbohydrates and beta-carotene. In Sub-Saharan Africa it serves as a food security crop, and is important also in relief famine. As sweet potato is propagated vegetativley control of virus diseases are an important factor in obtaining high yields. Over 20 viruses have so far been reported to infect sweet potato (Table 8.1), but several of them are still not characterized. Antisera are available only for a limited number of viruses. Broad-spectrum PCR for detecting new uncharacterised viruses has been demonstrated especially with potyviruses and geminiviruses. It is, therefore, not possible at present to determine weather the new isolated viruses are strains of previously reported viruses or represent a new virus species. Their identification, economic impact and effect in mixed infections, especially with Poty- and Geminiviruses are an important task for the future.

Research on sweet potato viruses has not received the attention it deserves and the international community and it's organizations should increase their efforts to strengthen sweet potato research in general and sweet potato virus focussed research in particular.

Acknowledgement

We thank Dr. J. Moyer who kindly provided illustrations.

References

Abad, J.A. and Moyer, J.W. 1992. Detection and distribution of sweet potato feathery mottle virus in sweet potato by in vitro-transcribed RNA probes (riboprobes), membrane immunobinding assay, and direct blotting. *Phytopathology* 82: 300-305.

Alvarez, V., Ducasse, D.A., Biderbost, E., and Nome, S.F. 1997. Sequencing and characterization of the coat protein and 3' non-coding region of a new swet potato potyvirus. *Arch. Virolology* 142: 1635-1644.

Aritua, V., Alicai, T., Adipala, E., Carey, E.E. and Gibson, R.W. 1998. Aspects of resistance to sweet potato virus disease in sweet potato. *Ann. appl. Biol.* 132: 387-398.

Atkey, P.T. and Brunt, A.A. 1987. Electron microscopy of an isometric caulimo-like virus from sweet potato (*Ipomoea batatas*). *J. Phytopath.* 118: 370-376.

Banks, G.K., Bedford, I.d., Beitia, F.J., Rodrigues-Cerezo, E., and Markham, P.G. 1999. A Novel Geminivirus of *Ipomoea indica* (*Convolvulaceae*) from Southern Spain. Plant Dis. 83: 486.

Brown, J.D., Brunt, A.A. and Hugo, S.A. 1988. Studies on viruses isolated from sweet potato (*Ipomoea batatas) Rep. Glasshouse Crops Res. Inst. for 1986-1987,* pp. 104-108..

Brunt, A.A. and Brown, J. D. 1990. Sweet potato ringspot, a newly recognised Nepo-like virus from Ipomoea batatas. (Abstract). *Int. Virol. Congress, Berlin.*

Brunt, A.A., Crabtree, K., Dallwitz, M.J., Gibbs, A.J., and Watson, L. (eds.). 1996. Viruses of Plants: Descriptions and Lists from the VIDE Database. CAB INTERNATIONAL, Cambridge, UK. 1484 pp.

Cali, B.B. and Moyer, J.W. 1981. Purification, serology, and particle morphology of two russet crack strains of sweet potato feathery mottle virus. *Phytopathology* 71: 302-305.

Carey, E.E., Mwanga, R.O.M., Fuentes, S., Kasule, S., Macharia, C., Gichuki, S.T. and Gibson, R.W. 1997. Sweet potato viruses in Uganda and Kenya: results of a survey. *Proc. Sixth Triennial Symp. Inter. Soc. Tropical Crops – Africa Branch* p. 22-28.

Chavi, F., Robertson, A.I., and Verduin, B.J.M. 1997. Survey and characterization of viruses in sweetpotato from Zimbabwe. *Plant Dis.* 81: 1115-1122.

Chung, M.L., Hsu, Y.H., Chen, M.J. and Chiu, R.J. 1986. Virus diseases of sweet potato in Taiwan. In: *Pl. Virus Diseases of Hort. Crops in the Tropics and Subtropics*, No. 33, p. 84-90. FFTC Book Series.

Chung, M.L., Liao, C.H., Chen, M.J., and Chiu, R.J. 1985. The isolation, transmission and host range of sweetpotato leaf curl disease agent in Taiwan. *Plant Prot. Bull* (Taiwan) 27: 333-342.

Cipriani, G., Fuentes, S., Bello, V., Salazar, L.F., Ghislain, M. and Zhang, D.P. 2001. Transgene expression of rice cysteine proteinase inhibitors for the development of resistance against sweetpotato feathery mottle virus. *Scientists and farmer: Partners in research for the 21 st Century*. P.267-271.

Clark, C.A. and Moyer, J.W. 1988. Compendium of Sweetpotato Diseases. The American Phytopathological Society. Minnesota, USA. 74 pp.

Cohen, J. and Loebenstein, G. 1991. Role of a whitefly-transmitted agent in infection of sweet potato by cucumber mosaic virus. *Plant. Dis.* 75: 291-292.

Cohen, J., Franck, A., Vetten, H.J., Lesemann, D.E. and Loebenstein, G. 1992. Purification and properties of closterovirus-like particles associated with a whitefly-transmitted disease of sweet potato. *Ann. appl. Biol.* 121: 257-268.

Cohen, J., Milgram, M., Antignus, Y., Pearlsman, M., Lachman, O. and Loebenstein, G. 1997. Ipomoea crinkle leaf curl caused by a whitefly-transmitted gemini-like virus. *Ann. appl. Biol.* 131: 273-282.

Cohen, J., Salomon, R. and Loebenstein, G. 1988. An improved method for purification of sweet potato feathery mottle virus directly from sweet potato. *Phutopathology* 78: 809-811.

Colinet, D. and Kummert, J. 1993. Identification of a sweetpotato feathery mottle virus isolate from China (SPFMV-CH) by the polymerase chain reaction with degenerate primers. *J. Virol. Methods* 45: 149-159.

Colinet, D., Kummert, J. and Lepoivre, P. 1997. Evidence for the assignment of two strains of SPLV to the genus Potyvirus based on coat protein and 3' non-coding region sequence data. *Virus Res.* 49: 91-100.

Colinet, D., Kummert, J., and Lepoivre, P. 1994. The complete nucleotide sequences of the coat protein cistron and 3' non-coding region of a newly-identified potyvirus infecting sweetpotato, as compared to those of sweetpotato feathery mottle virus. *Arch. Virolology* 139: 327-336.

Colinet, D., Kummert, J., Lepoivre, P., and Semall, J. 1994. Identification of distintict potyvirus in mixedly-infected sweetpotato by polymerase chain reaction with degenerate primers. *Phytopathology* 84: 65-69.

Colinet, D., Nguyen, M., Kummert, J., Lepoivre, P., and Xia, F.Z. 1998. Differentiation among potyviruses infecting sweet potato based on genus- and virus-specific reverse transcription polymerase chain reaction. *Plant Dis.* 82: 223-229.

Dabek, A.J. and Sagar, C. 1978. Witches' broom chlorotic little leaf of sweet potato in Guadalcanal, Solomon Islands, possibly caused by mycoplasma-like organisms.*Phytopath. Z.* 92: 1-11.

Di Feo, L., Nome, S.F., Biderbost, E., Fuentes, S., and Salazar, L.F. 2000. Etiology of sweet potato chlorotic dwarf disease in Argentina. *Plant Dis.* 84: 35-39.

FAOSTAT 2001 and 2002. FAO Statistical Databases. URL http://apps.fao.org/

Francki, R.I.B., Mossop, D.W. and Hatta, T. 1979. Cucumber mosaic virus. No. 213. *Descriptions of Plant Viruses.* Commonwealth Mycol. Inst. Assoc. Appl. Biol. Kew, Surrey, UK.

Fuentes S. and Salazar, L.F. 1989. Identification of sweetpotato [*Ipomoea batatas* (L.) Lam.] viruses. *Fitopatologia* 24: 43 (Abstr. in Spanish).

Fuentes, S. 1994. Preliminary identification of a sweet potato virus (C-6). *Fitopatologia* 29: 38 (Abstr., in Spanish).

Fuentes, S. and Salazar, L.F. 1992. Identification of a new sweetpotato virus. *Fitopatologia* 27: 50 (Abstr., in Spanish).

Fuentes, S., Arellano, J., and Meza, M.A. 1997. Preliminary studies of a new virus, C-8, affecting sweetpotato. *Fitopatologia* 32: 9-10 (Abstr. in Spanish).

Fuentes, S., Mayo, M., Jolly, C.A., Nakano, M., Querci, M., and Salazar, L.F.. 1996. A novel luteovirus from sweet potato, sweet potato leaf speckling virus. *Ann. appl. Biol.* 128: 491-504.

Fuentes, S., Querci, M., Salazar, L.F., and Mayo, M. 1997. Detection of SPLSV by Nucleic Acid Spot Hybridization (NASH) Test. *Fitopatología* 32: 204 (Abstr. in Spanish).
Gao, F., Gong, Y.F., Zhang, P.B., Gao, F.,Gong, Y.F. and Zhang, P.B. 2000. Production and development of virus-free sweet potato in China. *Crop Protect.* 19: 105-111.
Gibb, K.S., Padovan, A.C. and Mogen, B.D. 1995. Studies on sweet potato little-leaf phytoplasma detected in sweet potato and other plant species growing in Northern Australia. *Phytopathology* 85: 169-174.
Gibson, R.W., Jeremiah, S.C., Aritua, V., Msabaha, R.P., Mpembe, I. and Ndunguru, L. 2000. Sweet potato virus disease in sub-Saharan Africa: evidence that neglect of seedlings in the traditional farming system hinders the development of superior resistant landraces. *J. Phytopath.* 148: 441-447.
Gibson, R.W., Mawanga, R.O.M., Kasule, S., Mpembe, I. and Carey, E.E. 1997. Apparent absence of viruses in most symptomless field-grown sweet potato in Uganda. *Ann. appl. Biol.* 130: 481-490.
Green, S.K., Kuo, Y.J. and Lee, D.R. 1988. Uneven distribution of two potyviruses (feathery mottle virus and sweet potato latent virus) in sweet potato plants and ins implication on virus indexing of meristem derived plants. *Tropical Pest Management* 34: 298-302.
Green, S.K., Luo, C.Y. and Lee, D.R. 1989. Elimination of mycoplasma-like organisms from witches' broom infected sweet potato. *J. Phytopathol.* 126: 204-212.
Hahn, S.K. 1979. Effects of viruses (SPVD) on growth and yield of sweet potato. *Exp. Agric.* 15: 253-256.
Hammond, J., Jordan, R.L., Larsen, R.C. and Moyer, J.W. 1992. Use of monoclonal antisera and monoclonal antibodies to examine serological relationships among three filamentous viruses of sweetpotato. *Phytopathology* 82: 713-717.
Hollings, M., Stone, O.M. and Bock, K.R. 1976. Purification and properties of sweet potato mild mottle virus, a whitefly-borne virus from sweet potato *(Ipomoea batatas)* in East Africa. *Ann. appl. Biol. 82: 511- 528.*
Hoyer, U., Maiss, E., Jelkmann, W., Lesemann, D.E. and Vetten, H.J. 1996. Identification of the coat protein gene of a sweet potato sunken vein closterovirus isolate from Kenya and evidence for a serological relationship among geographically diverse closterovirus isolates from sweet potato. *Phytopathology* 86: 744-750.
Jordan, R. and Hammond, J. 1991. Comparison and differentiation of potyvirus isolates and identification of strain-, virus-, subgroup-specific and potyvirus group-common epitopes using monoclonal antibodies. *J. Gen. Virol.* 72: 25-36.
Kahn, R.P. and Monroe, R.L. 1969. Detection of sweet potato little leaf agent (witches' broom) associated with a mycoplasma-like organism. *Pl.Prot. Bull.* FAO 17: 104-106.
Karyeija, R.F., Gibson, R.W. and Valkonen, J.P.T. 1998. Resistance in sweet potato virus disease (SPVD) in wild East African Ipomoea. *Ann. appl. Biol.* 133: 39-44.
Karyeija, R.F., Kreuze, J.F., Gibson, R.W. and Valkonen, J.P.T. 2000. Synergistic interactions of a potyvirus and a phloem-limited crinivirus in sweet potato plants. *Virology* 269: 26-36.
Karyeija, R.F., Kreuze, J.F., Gibson, R.W. and Valkonen, J.P.T. 2001. Variability of sweetpotato feathery mottle virus in Africa. *Afr. Crop Science J.* 9: 293-299.
Kreuze, J.F., Karyeija, R.F., Gibson, R.W. and Valkonen, J.P.T. 2000. Comparisons of coat protein gene sequences show that East African isolates of sweet potato feathery mottle virus form a genetically distinct group. *Arch. Virology* 145: 567-574.
Liao, C.H., Chien, K., Chung, M.L., Chiu, R.J. and Han, Y.H. 1979. A study of a sweet potato virus disease in Taiwan. I. Sweet potato yellow spot virus disease. *J. agric. Res. China* 28: 127-137.
Loebenstein, G. and Harpaz, I. 1960. Virus diseases of sweet potatoes in Israel. *Phytopathology* 50: 100-104.

Lotrakul, P. and Valverde, R.A. 1999. Cloning of a DNA-A-like genomic component of sweet potato leaf curl virus: nucleotide sequence and phylogenetic relationships. Mol. Plant Pathol. On-line publication/1999/0206lotrakul.

Lotrakul, P., Valverde, R.A., Clark, C.A., Hurt, S., and Hoy, M.W. 2002. Sweetpotato Leaf Curl Virus and Related Geminiviruses in Sweetpotato. Proc. 1st IS on Sweet potato. Ed. T. Ames, ISHS. *Acta Hort.* 583:135-141.

Martin, W.J. 1962. Susceptibility of certain Convolvulaceae to internal cork, tobacco ringspot and cucumber mosaic viruses. *Phytopathology* 52: 607-611.

Migliori, A., Marchoux , G. and Quiot, J.B. 1978. Dynamique des populations du virus de la mosaique du concombre en Guadelupe. *Ann. Phytopathol.* 10: 455-466.

Mihovilovich,R., Mendoza, H.A. and Salazar, L.F. 2000. Combining ability for resistance to sweet potato feathery mottle virus. *HortScience* 35: 1319-1320.

Milgram, M., Cohen, J. and Loebenstein, G. 1996. Effects of sweet potato feathery mottle virus and sweet potato sunken vein virus on sweet potato yields and rate of reinfection on virus-free planting material in Israel. *Phytoparasitica* 24: 189-193.

Moyer, J.F. 1986. Variability among strains of sweet potato feathery mottle virus (Abstr.) *Phytopathology* 76: 1126.

Moyer, J.W. and Cali, B.B. 1985. Properties of sweet potato feathery mottle virus RNA and capsid protein. *J. Gen. Virol.* 65: 1185-1189.

Moyer, J.W., Abad, J.A.,New, J, and Bell, J. 2002. Isolation, Identification and Detection of Undescribed RNA Sweepotato Viruses. Proc. 1st IS on Sweet potato. Ed. T. Ames, ISHS. *Acta Hort.* 583:121-127.

Moyer, J.W., Jackson, G.V.H., and Frison, E.A. (eds). 1989. *FAO/IBPGR Technical Guidelines for the Safe Movement of Sweet Potato Germplasm.* Food and Agriculture Organization of the United Nations, Rome/ International Board For Plant Genetic Resources, Rome.

Ngeve, J.M. and Bouwkamp, J.C. 1991. Effects of sweet potato virus disease (SPVD) on the yield of sweet potato genotypes in Cameroon. *Exp. Agric.* 27: 221-225.

Nome, S.F. 1973. Sweetpotato vein mosaic in Argentina. *Phytopatho Z.* 77: 44-45

Nome, S.F., Shalla, T.A., and Peterson, L.J. 1974. Comparison of virus particles and intracellular inclusions associated with vein mosaic, feathery mottle, and russet crack diseases od sweetpotato. *Phytopathol. Z.* 79: 169-178.

Okada, Y., Saito, A., Nishiguchi, M., Kimura, T., Mori, M., Hanada, K., Sakai, J., Miyazaki, C., Matsuda, Y. and Murata, T. 2001. Virus resistance in transgenic sweet potato [Ipomoea batatas L. (Lam)] expressing the coat protein gene of sweet potato feathery mottle virus. *Theor. Appl. Genetics* 103: 743-751.

Onuki, M., Honda, Y. and Hanada, K. 2000. Geminate particle morphology of sweet potato leaf curl virus in partially purified preparation and its serological relationship to two Begomoviruses by western blotting. *J. Gen. Plant Pathol.* 2: 182-184.

Osaki, T. and Inouye, T. 1991. Transmission characteristics and cytopathology of a whitefly-transmitted virus isolated from sweet potato leaf curl disease. *Bull. Univ. Osaka Prefecture, Series B, Agric. Biol.* 43: 11-19.

Pearson, M.N., Keane, P.J. and Thagalingham, K. 1984. Little leaf: A disease of sweet potato in Papua New Guinea probabky caused bt mycoplasma-like organisms. *Phytopath. Z.* 109: 269-276.

Pozzer, l., Dusi, A.N., Lima, M.I. and Kitajima, E.W. 1995. Characterization of a Brazilian isolate of sweet potato feathery mottle virus infecting sweet potato. *Fitopat. Brasil.* 20: 65-71.

Sakai, J., Mori, M., Morishita, T., Tanaka, M., Hanada, K., Usugi, T. and Nishigushi, M. 1977. Complete nucleotide sequence and genome organization of sweet potato feathery mottle virus (S strain) genomic RNA: the large coding region of the P1 gene. *Arch. Virology* 142: 1553-1562.

Salazar, L.F. 1996. Potato Viruses and their Control. International Potato Center, Lima, Peru.

Schaefers, G.A. and Terry, E.R. 1976. Insect transmission of sweet potato agents in Nigeria. *Phytopathology* 66: 642-645.
Shang, Y.F., Yang, C.L., Zhao, J.H., Li, C.H., Lu, X.B., Wang, S.J., Xin, X.Q., Shang, Y.F., Yang, C.L., Zhao, J.H., Li, C.S., Lu, X.B. Wang, S.J. and Xin, X.Q. 1999. Progress in the researches and application of virus-free sweet potato in Shandong province. *Res. Progress Plant Protect. Plant Nutrit.* 1999, p. 104-108. China Agric. Press, Beijing, China.
Sheffield, F.M.L. 1957. Virus diseases of sweet potato in East Africa. I. Identification of the viruses and their insect vectors. *Phytopathology* 47: 582-590.
Shinkai, A. 1979. Sweet potato leaf curl disease. *Agric. and Hortic.* 54: (1) the Commencement.
Sim, J. 2001. Cytopathology, detection, and identification of viruses infecting sweet potato. PhD. Thesis. The Departemnet of the Plant Pathology and Crop Physiology. Louisiana State University, U.S.A. 137 pp
Sonoda, S., Koiwa, H., Kanda, K., Kato, H., Shimono, M. and Nishiguchi, M. 2000. The helper component-proteinase of sweet potato feathery mottle virus facilitates systemic spread of potato virus X in Ipomoea nil. *Phytopathology* 90: 944-950.
Takanami, Y and Kubo, S. 1979. Enzyme-assisted purification of two phloem-limited viruses: tobacco necrotic dwarf and potato leafroll. *J. Gen. Virol.* 44: 153-159.
Turyamureeba, G., Mwanga, R.O.M., Odongo, B., Ocitti p'Obwoya, C. and Carey, E.E. 1998. Application for inclusion of a crop/ variety in the National Cultivar List. Submission to the Variety Release Committee for release of sweet potato varieties. The Ugandan Ministry of Agriculture, Animal Industries & Fisheries, Uganda.
Usugi, T., Nakano, M., Shinkai, A., and Hayashi, T. 1991. Three filamentous viruses isolated from sweetpotato in Japan. *Ann. Phytopathol. Soc. Japan* 57: 512-521.
Van Velsen, R.J. 1967. "Little leaf", a virus disease of *Ipomoea batatas* in Papua and New Guinea. *Papua New Guinea Agric. J.* 18:126-128.
Wambugu, F.M. 1991. In vitro and epidemiological studies of sweetpotato (*Ipomea batatas*)(L.) Lam. virus diseases in Kenya. PhD Thesis. University of Bath. 271 pp.
Winter, S., Purac, A., Leggett, F., Frison, E.A., Rossell, H.W. and Hamilton, R.I. 1992. Partial purification and molecular cloning of a closterovirus from sweet potato infected with the sweet potato virus disease complex from Nigeria. *Phytopathology* 82: 869-875.
Yang, I.L. 1969. Studies on witches broom in sweet potato in Taiwan. *J. Taiwan Agric. Res.* 18: 50-60.

Fig. 9.1. Internal cork disease caused by *Sweet potato feathery mottle virus* (Courtesy J.W. Moyer).

Fig. 9.2. Sweet potato plant infected with *Sweet potato feathery mottle virus* showing circular spots.

Fig. 9.3. Sweet potato cv. Georgia Jet infected by both *Sweet potato feathery mottle virus* and *Sweet potato sunken vein virus* causing stunting of the plants, feathery vein clearing and yellowing of the plants

Fig. 9.4 *I. setosa* infected with *Sweet potato feathery mottle virus*, showing vein clearing

Fig. 9.5. Sweet potato leaf (right) infected with *Sweet potato sunken vein virus*-swollen veins on the lower side. (left – healthy leaf).

Fig. 9.6. Symptoms of *Sweet potato mild mottle virus* on *I. setosa* leaves. Healthy leaf is at top left (Courtesy Dr. J. Moyer).

Fig. 9.7. Sweet potato cv Jewel infected with SPLSV showing whitish speckling.

Fig. 9.8. Symptoms of *Ipomoea crinkle leaf virus* on *I. setosa.* Left- healthy plant.

Fig. 9.9. Sweet potato plants (cv. Georgia Jet) severely affected by yellowing and stunting caused by double infection with *Cucumber mosaic virus* and *Sweet potato sunken vein virus.*

Fig. 9.10. *Ipomoea setosa* infected with C-6 virus showing fine chlorotic spots and vein clearing.

9.2

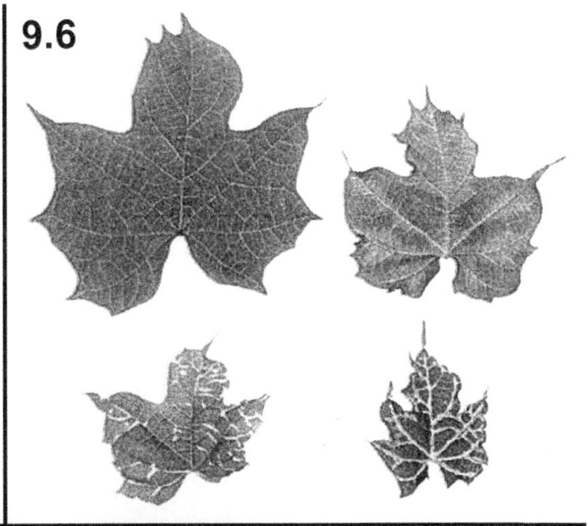
9.6

9.5

9.3

9.4

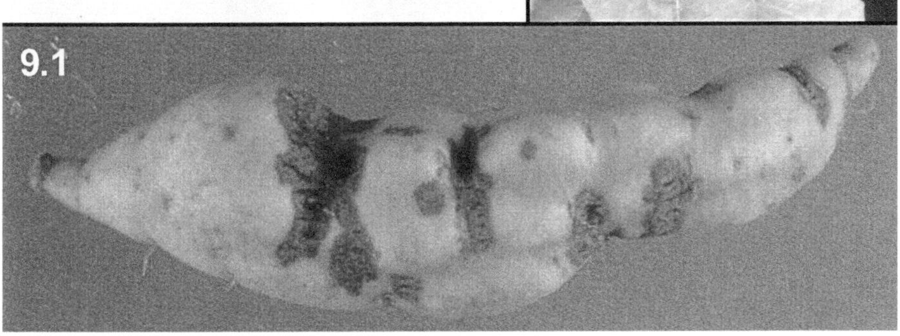
9.1

9.11
9.7
9.9
9.10
A
B
9.8

Table 8.1. A list of recognized viruses known to infect sweet potato

Virus	Shape and size	Vector	Distribution	References
SwPLV-***Potyvirus***	Flexuous, 750-790 nm	Aphid	Uganda, Kenya, Taiwan, Japan, China, Indonesia, Philippines, India	10, 11, 25
SPMSV-*Potyvirus*	Flexuous, 800 nm	Aphid	Argentina, Peru, Indonesia, Philippines, Egypt	1, 12, 18
SPLSV-*Polerovirus*?	Isometric, 30 nm	Aphid	Peru, Cuba	17
CMV- *Cucumovirus*	Isometric, 30 nm	Aphid	Israel, Uganda, Kenya, Japan, Egypt.	7, 13
SPVMV-Potyvirus?	Flexuous, 761 nm	Aphid	Argentina	29, 30
SPSVV-Crinivirus	Flexuous, 850-950 nm	Whitefly	Africa (Nigeria, Uganda, Kenya, Zaire, Zambia, Madagascar), Asia (Taiwan, China, Indonesia, Philippines), America (USA, Brazil, Argentina, Peru), Israel, Egypt	6
SPMMV-Ipomovirus	Flexuous, 800-950 nm	Whitefly	West Africa (kenya, Uganda, Burundi, Tanzania), Indonesia, Papua New Guinea, India, Egypt	11, 21, 32
SPYDV-Ipomovirus	Flexuous, 750 nm	Whitefly	Taiwan	25
SPLCV-Begomovirus	Geminate	Whitefly	Taiwan, Japan, Korea, China, USA, Puerto Rico, Peru	25, 26, 33
ICLCV-Geminivirus?	Geminate	Whitefly	USA?	8

Table 8.1 (continued)

Virus	Shape and size	Vector	Distribution	References
SPVG-Potyvirus	Flexuous	Unknown	China	9, 11
SPCFV	Flexuous, 750-800 nm	Unknown	Peru, Japan, Brazil, China, Cuba, Panama, Colombia, Bolivia, Indonesia, Philippines, Uganda, India, Austria	16, 20, 35
C-6	Flexuous 750-800 nm	Unknown	Peru, Uganda, Cuba, Dominican Republic, Indonesia, Philippines, USA, Puerto Rico, Egypt	14
SPCV-Caulimo-like	Isometric, 50 nm	Unknown	South Pacific region (Kingdom of Tonga, Papua New Guinea, New Zeland, Salomon Islands, Austria), Madeira, kenya, Uganda, China, Egypt, Puerto Rico,	2, 19
SPCSV-Potyvirus	Flexuous, 850-950 nm	Unknown	Caribbean Region, Kenya, Uganda, Zimbabwe	3
SPRSV-Nepovirus	Isometric, 28 nm	Unknown	Papua New Guinea, Kenya	3, 4
TSV-Ilar-like	Isometric, 30 nm	Unknown	Guatemala	5
C-3	Flexuous?	Unknown	Brazil. Unknown in other countries	15
TMV-Tobamovirus	Rod, 300 nm	None	USA	5

1) Alvarez *et al.* 1997; **2)** Atkey and Brunt, 1987; **3)** Brown *et al.* 1988; **4)** Brunt *et al.* 1996; **5)** Clark and Moyer, 1988; **6)** Cohen *et al.* 1992; **7)** Cohen and Loebentein, 1991; **8)** Cohen *et al.* 1997; **9)** Colinet *et al.* 1994; **10)** Colinet *et al.* 1997; **11)** Colinet *et al.* 1998; **12)** Di Feo *et al.* 2000; **13)** Francki *et al.* 1979; **14)** Fuentes, 1994; **15)** Fuentes and Salazar, 1989; **16)** Fuentes and Salazar, 1992; **17)** Fuentes *et al.* 1996; **18)** Fuentes *et al.* 1997; **19)** Gao *et al.* 2000; **20)** Gibson *et al.* 1997; **21)** Hollings *et al.* 1976; **22)** Karyeija *et al.* 1998; **23)** Karyeija *et al.* 2001; **24)** Kreuze *et al.* 2000; **25)** Liao *et al.* 1979; **26)** Lotrakul *et al.* 2002; **27)** Moyer, 1986; **28)** Moyer *et al.* 1989; **29)** Nome, 1973; **30)** Nome *et al.* 1974; **31)** Pozzer *et al.* 1995; **32)** Sheffield, 1957; **33)** Shinkai, 1979; **34)** Sim, 2001; **35)** Usugi *et al.* 1991.

10. Yam

G. I. Atiri[1], S. Winter[2] and O. J. Alabi[3]

[1]Department of Crop Protection & Environmental Biology, University of Ibadan, Ibadan, Nigeria.; [2] Plant Virus Division, German Collection of Cell Cultures and Microorganisms, c/o BBA Messeweg 11/12, D-38104, Braunschweig, Germany; [3]Research Fellow, International Institute of Tropical Agriculture, PMB 5320, Ibadan, Nigeria.

10.1 Introduction

Food yams are economically important staple foods in tropical and subtropical regions of the world. They belong to the genus *Dioscorea*, family *Dioscoreaceae* and order *Dioscoreales*. More than 600 species are known, with the most important cultivated edible yams being *D. alata, D. rotundata, D. cayenensis, D. dumetorum, D. esculenta* and *D. bulbifera*. The highest levels of yam production world-wide is attributed to white yam, *D. rotundata, while D. alata*, the greater Asiatic or water yam, is considered to be the most widely distributed species of this crop (Onwueme, 1978).

Over 90 % of the world's yam production of about 40 million tons (FAOSTAT, 2002) is produced in the region extending from Côte d'Ivoire to Cameroon commonly referred to as the `yam zone´ of West Africa. Nigeria, with 27.5 million tons of yam tubers produced on 27.5 million ha is the world's leading yam-producing country. Substantial production also takes place in South America (Brazil, Venezuela and Colombia), Papua New Guinea, the West Indies and Japan (Onwueme and Sinha, 1991).

Yam is a vegetatively propagated crop, grown for the high caloric value of its tubers. Vegetative propagation, insect transmission and uncontrolled introductions of infected yam germplasm have contributed to the widespread distribution of viruses in this crop. Hence virus diseases are known for long in yam, not only as the cause of production constraints, but also presenting serious impediments to international trade of the commodity. In 1965, 50 % of exported yams from Barbados and neighboring islands were rated below acceptable market standards due to Internal brown spotting (IBS) disease (Jeffers and Headley, 1967). The international exchange of yam germplasm is restricted due to the occurrence of several yam viruses subject to quarantine regulations (Ng, 1992). Extensive indexing programs had to be developed to address this phytosanitary problem (Njukeng, 1998).

The first reports of mosaic symptoms on yam date back to 1936/1937, from Sierra Leone (Deighton, 1936) and Puerto Rico (Cook, 1937). A more extensive survey for viruses in yams in West Africa was conducted in 1964 by J. Craig. At that time, yam mosaic virus (YMV) was found at 11 sites in Western Nigeria, causing moderate to severe symptoms in white yam. Although this virus survey was not comprehensive since it did not cover many of the major yam-growing areas, and viruses were not unequivocally identified, it showed that mosaic diseases could contribute to significant yield reductions. The virus survey throughout West Africa conducted by Njukeng (1998) confirmed that the potyvirus YMV was the most prevalent virus found in seven countries. In Nigeria, YMV was found in all yam growing regions with a particular high prevalence in the humid forest and coastal savannah regions of the country, reaching disease incidences of >80% (Njukeng *et al.* 2002).

In the Caribbean, two common diseases of *Dioscorea* spp. were attributed to virus infections: the internal brown spotting disease of yam tubers and the mosaic, mottle and greenbanding symptoms on leaves.. Bacilliform particles in addition to flexuous rod shaped particles were found associated with internal brown spot disease of *D. alata* cv. White Lisbon in Barbados (Harrison and Roberts, 1973). A bacilliform virus was also found in *D. cayenensis* (Mohamed and Mantell, 1976) and in other regions of yam cultivation in the West Indies, later recognized as the causal agent of the internal brown spot disease (Phillips *et al.* 1999). Several other viruses were reported from different cultivated edible yams but also from *D. floribunda*, *D. composita* and *D. opposita* grown for medicinal purposes, as sources of steroid precursors.

Many of the virus reports reflect on the description of putative members of the *Potyviridae* and in fact, potyviruses are the viruses most frequently found in all varieties grown across the yam-growing world (deAvila, *et al.* 1982; Porth *et al.* 1987; Hughes *et al.* 1997; Bousalem and Dallot 2000; Kondo 2001). However, many of the earlier descriptions do not provide precise information on the causal viruses, and virus identity and relationships with other (poty)viruses mostly remained elusive. Since molecular data as a basis for the definition of a taxonomic status were not available to support earlier studies, identifications as distinct viruses remained tentative.

In this chapter, we evaluated the available information about viruses associated with yam diseases in the world. Table 18.1 gives a summary of the viruses reported from yam, providing a biased reflection on the putative strain relationships among viruses with distinct names.

10.2 The Main Viruses

10.2.1 *Yam mosaic virus* Genus *Potyvirus* (YMV)

First reported from *D. cayenensis* in Côte d'Ivoire (Thouvenel and Fauquet, 1979) the virus is now known to infect most edible yams and occurs in all yam-growing regions of West Africa and in the West Indies (Brunt *et al.* 1996). The virus causes significant economical losses (Thouvenel and Dumont, 1990) and presents a serious impediment in the international movement of germplasm.

The host range of YMV is narrow and confined mainly to the *Solanaceae* and *Dioscoreaceae*. It induces systemic mosaic and veinbanding in *D. alata* (Figure 10.1 middle), *D. rotundata, D. cayenensis, D. dumetorum, D. praehensilis,*and *D. preusii.* In *Nicotiana benthamiana,* often used as the maintenance and propagation host, it induces systemic mosaic and veinbanding symptoms. Symptoms are variable, differ considerable even in leaves of the same plant, and range from diffuse mottling, vein banding and mosaic to severe chlorosis, leaf deformation and shoestringing.

The virus is transmitted mechanically and by several aphid species in a nonpersistent manner. These include *Aphis gossypii, A. craccivora, Rhopalosiphum maidis* and *Toxoptera citricidus* (COPR, 1978). Seed transmission has not been reported.

Mucilaginous substances in yam leaves make purification difficult, and therefore *N. benthamiana* is the preferred host for virus purification. Purification is generally done by chloroform clarification, ammonium sulphate precipitation and differential centrifugation followed by sucrose density gradient centrifugation (Fauquet and Thouvenel, 1987).

Serological tests have been developed and are widely used for virus detection in routine indexing schemes (Njukeng, 1998; Dongo, 2001). Using monoclonal antibodies for detection and discrimination of YMV isolates, Goudou-Urbino and associates (1996) defined two serogroups. In this study, a considerable biological variability among the virus isolates was shown which was not correlated with geographic origin of the viruses collected.

For sensitive detection of two potyviruses that infect yam including YMV isolates, an immunocapture RT-PCR test was described (Mumford and Seal, 1997). This test was subsequently developed further into a reliable and robust molecular typing test for world isolates of YMV, taking into account the existing high genetic diversity of this potyvirus (Bousalem *et al.* 2000a,b). This test permits the sensitive detection of YMV directly in yam tubers and provides information on the type of YMV present in a particular sample.

Aleman-Verdaguer and associates (1997) studied the genetic variability of 6 YMV isolates, by comparing genetic variation in different cistrons of the potyvirus genome. For all virus genes, a similar phylogeny was found. Comparing the 3'- terminal region of the RNA genome, the coat protein of YMV isolates was found most variable relative to the divergence found with other potyvirus coat protein sequences (Moury *et al.* 2002). Upon phylogenetic analysis, this high genetic variability was further structured into 9 distant molecular groups with the most diverse and divergent groups formed by isolates collected from Africa and infecting the complex *D. rotundata-D. cayenensis*.

For YMV in Africa, 6 distinct molecular groups were determined as suggested by phylogenetic analysis, YMV originated from this continent. The ancestral yam host for this virus is the complex *D. rotundata-D. cayenensis* from which subsequent independent transfers to *D. alata* and to *D. trifida* occurred during virus evolution. Furthermore, a geographic distinction among virus isolates from the Caribbean, South America and Africa can be drawn with isolates forming a clade with well-separated branches.

The large diversity in YMV is attributed to the differential accumulation of mutations and the contribution of recombination events (Bousalem *et al.* 2000a). The phylogenetic data obtained for isolates of YMV showed nine of distinct virus populations. Several of these virus groups were directly implicated in severe epidemics. The accurate identification of severe strains or recombinant viruses within virus populations is required to analyze YMV epidemics, support resistance breeding and to asses the risks of introducing new virus strains into the environment.

***Yam mosaic virus* strains and isolates.** Several potyviruses with a cytopathology and host range similar to YMV were described and Goudou-Urbino *et al* (1996) discussed the status of some of the viruses described previously. Porth *et al.* (1987) already showed that Dioscorea greenbanding mosaic virus, (Reckhaus and Nienhaus, 1981) and YV-N (Terry, 1976) and

Dioscorea alata virus, DaV, from Togo (Porth *et al.* 1987) are serologically closely related to YMV. The Dioscorea trifida virus, which was first reported from *D. trifida* in Guadeloupe (Migliori and Cadilhac, 1976), and later found in French Guyana (Boeglin *et al.* 1993), can be considered to be related to YMV (Goudou-Urbino *et al.* 1996). In *D. trifida* the virus often occurred in mixed infections with *Cucumber mosaic virus* (Migliori and Cadilhac, 1976). It has a very narrow host range and although it can be transmitted to *D. alata* and *D. cayenensis,* natural infections occur only in *D. trifida.* In this plant, symptoms of virus infections are mosaic, chlorotic flecking, veinbanding and leaf malformation.

The Dioscorea trifida virus presents an interesting example of the diversity and evolution of this potyvirus. While a relationship of the phylogeny of YMV isolates and yam species is not obvious for most YMV isolates, a correlation of YMV isolates with *D. trifida* was demonstrated (Bousalem *et al.* 2000a). Isolates of YMV collected in the Caribbean and in French Guiana were restricted to this host, suggesting the existence of a particular pathotype. A further isolate of YMV from *D. trifida* collected in Guadeloupe (Bousalem *et al.* 2000) was found to be a recombinant virus from a YMV group infecting *D. alata/cayenensis/rotundata* and the *D. trifida* virus. This virus was implicated in a severe epidemic of *D. trifida* in Guadeloupe providing further evidence for continuous evolutionary processes and diversification of YMV.

10.2.2 **Yam mild mosaic virus** Genus ***Potyvirus*** (YMMV)

A virus with potyvirus-like particle morphology was reported from *D. alata* in Togo (Reckhaus, 1980). Infected plants showed a diffuse mottling on young leaves (Figure 10.1 left), , which was masked in older plant parts. Potyvirus-like particles were found to be associated with the disease and in thin sections of infected *D. alata* plants, cytoplasmatic inclusions were found in root and leaf parenchyma cells. The virus was not mechanically transmissible and distinct in cytopathology and biological characters (Reckhaus, 1980), and serologically related to YMV (Porth *et al.* 1987).

A similar virus was recently described from *D. alata* in Nigeria (Odu *et al.* 1999). In this study, a potyvirus was transmitted from infected *D. alata* to cowpea, *Vigna unguiculata,* using *Aphis craccivora* as vector. Back transmission was achieved by mechanical inoculation from infected cowpea to healthy *D. alata*. . Cowpea was the only alternative host for this particular virus. An antiserum against Dioscorea alata virus (DAV) from Nigeria reacted with a virus previously described as yam virus I (Hughes, 1986), but a serological relationship with YMV was not found.

DAV and Yam virus I can be considered as isolates of the same virus and although detailed studies are pending, it can be assumed that the diseases referred to as *D. alata* vein yellowing (Rossel, 1982; Thottappilly, 1983) or Dioscorea alata chlorosis virus (Rossel and Thottappilly, 1985) are caused by isolates and/or strains of the same virus species.

This virus often causes mild symptoms in yam and hence the species name *Yam mild mosaic virus*, YMMV. Mumford and Seal (1997) provided first molecular evidence for a classification of YMMV as a distinct potyvirus that infects yam. While the virus is mainly infecting *D. alata*, it was also detected in *D. rotundata* (Rick Mumford pers. commun.). Recently the 3'-terminal sequence of YMMV RNA comprising the complete coat protein was analyzed (Fuji *et al.* 1999). This sequence and the partial sequences of 20 further YMMV isolates, available in the Genebank, provides a comprehensive picture of the YMMV from diverse locations. For better clarity, we will henceforth use the virus acronym YMMV instead of its synonyms, DAV or YMV1 or else.

The distribution of YMMV coincides with cultivation of its main host *D. alata.* The virus was described from Papua New Guinea, Barbados and Indonesia and the Solomon Islands (Hughes, 1986) and it was recently detected in field gene banks from Fiji and Vanuatu in the South Pacific (Lebas *et al.* 2002). The partial nucleotide sequences (core- and carboxy-terminus of the coat protein and 3'-non translated region) for several virus isolates were retrieved and are available from Genebank. YMMV was also found in Colombia, Mauritius, Japan, Papua New Guinea, Sri Lanka and Martinique (Bousalem and Dallot, 2000). The most comprehensive molecular data were provided by Mumford and associates (unpublished, sequences available in Genebank), which confirmed the widespread presence of YMMV in all yam-growing areas of Nigeria and throughout West Africa (Thottappilly, 1992; Hughes *et al.* 1998), providing evidence that YMMV is the second most prevalent virus in yam after YMV.

The 3'- terminal sequence comprising the complete (266 amino acids) coat protein (Fuji *et al* 1999) reveals that YMMV is only distantly related to YMV (57% amino acid identity) and to Japanese yam mosaic virus, JYMV, (61% identical amino acids).

Upon a phylogenetic analysis of the 3'-NTR sequences from all virus isolates available (Winter, unpublished) it becomes evident, that isolates of YMMV obtained from different parts of the world are quite diverse, as is the case for YMV genomes (Bousalem *et al* 2000a). However, although a comprehensive genetic analysis of YMMV isolates is pending, a relationship between the phylogeny of the YMMV isolates and the geographic origin can be drawn. YMMV isolates from Africa, comprising the most diverse sequences form a clade well separated from virus isolates found in Asia and South America (Winter, unpublished).

The viruses can be detected by, ELISA and ISEM (Odu *et al.* 2001). The robust and convenient IC-RT-PCR method prescribed by Mumford and Seal (1997) permitted the detection and discrimination of YMMV and YMV and was recommended for detection of YMMV especially in tubers. ELISA is a robust means of virus detection in infected yam leaves and provides the basis for epidemiological studies of YMMV, while the structure of virus population can be resolved by analysis of 3'-terminal sequence of YMMV isolates.

10.2.3 *Dioscorea bacilliform virus* Genus *Badnavirus* (DBV)

A virus disease complex from D. *alata* cv. White Lisbon from Barbados was described in 1967 by Coursey (1967) and Jeffers and Headly (1967). A virus with bacilliform particles in mixed infection with a potyvirus (Harrison and Roberts, 1973) was found associated with internal brown spot symptoms of the cultivar White Lisbon. The name yam internal brown spot was subsequently used for a virus with bacilliform particles causing symptoms in tubers of *D. alata* that was not mechanical transmissible.

Similar bacilliform virus particles were found in *D. cayenensis* leaves in Barbados (Mohamed and Mantell, 1976) and in other Caribbean countries. Phillips and associates (1999) found a bacilliform virus either alone or in mixed infections with DAV, YMV or an unidentified filamentous virus in *Dioscorea* species including *D. alata, D. rotundata* and *D. trifida.* These viruses were found in samples from West Africa, Barbados, Guyana and Brazil, This and further studies in West Africa (Dongo, 2001; Odu *et al.* 2001) provided evidence for the worldwide distribution of badnavirus(es) in different yam genotypes.

The bacilliform virus from Barbados was subsequently designated as *Dioscorea alata badnavirus*, (DaBV) (Phillips *et al.* 1999). A further reference to Yam internal brown spot virus was not made.

DaBV causes leaf crinkling, distortion, vein chlorosis and vein necrosis in *D. alata,* with severe symptoms in cv Sagbe from Nigeria, in which the virus reaches high concentrations (Figure 10.1 right). DaBV infections appeared not to be associated with a discoloring of yam tubers as described for the yam internal brown spot disease and tuber symptoms were never reported by any other author. DaBV is transmissible by mechanical inoculation of partially purified preparations to several other *Dioscorea* species (Phillips *et al.* 1999), inducing chlorotic spotting, vein-banding and mosaic symptoms. Mechanical transmission to herbaceous hosts outside the *Dioscoreaceae* failed, but Odu and associates (2001) reported graft transmission of DaBV from *D. alata* to cowpea TVu 22. The mealybug, *Planococcus citri,* readily transmitted DaBV.

Virus purifications from infected *D. alata* cv. Sagbe plants resulted in low virus yields, but the antisera produced react well in ISEM and ELISA. Using serological detection as a basis for virus surveys (Hughes, 1986; Phillips *et al.* 1999; Odu, 2001) different bacilliform viruses were detected in several species of *Dioscorea* in Africa, Asia, South and Central America and Oceania, indicating a close serological relationship. Serological reactions with antisera against *Banana streak virus* (BSV) and *Sugarcane bacilliform virus* (ScBV) provided further evidence forassigning this virus to the genus *Badnavirus*.

The complete nucleotide sequence of the DaBV genome has been determined by Briddon and associates (1999). The size and organization of the ORFs I-III and the coding capacity was similar to those ofother badnaviruses. DaBV is a distinct pararetrovirus species of the genus *Badnavirus,* named *Dioscorea bacillform virus* (DBV). It is most similar to other mealybug-transmitted *Badnaviruses,* sharing the highest sequence identity (50%) over the complete virus genome with a recently sequenced bacilliform virus from Taro, *Taro bacilliform virus*, TaBV (Yang *et al.* 2003). From the reconstruction of the full-length sequence of the DaBV genome, it can be deduced that the genetic variability of the virus may be quite high (Briddon *et al.* 1999). This virus is distributed throughout the yam- growing world probably through exchange of yam germplasm. With the indexing methods now available, further distribution of DBV should be prevented.

10.2.4 *Dioscorea latent virus* Genus *Potexvirus* (DLV)

This virus was first reported from Puerto Rico (Lawson *et al.* 1973; Waterworth *et al.* 1974), but probably occurs in other areas in Central America where its two natural hosts, *D. composita* and *D. floribunda*, are cultivated. Lebas *et al.* (2002) detected DLV in about 4.4 % of leaf samples collected in Fiji and Vanuatu in the South Pacific.

This virus was first found in *D. floribunda* showing greenbanding and mosaic symptoms, in a mixed infection with the potyvirus, *Dioscorea greenbanding virus* (Lawson *et al.* 1973; Hearon *et al.* 1978). The virus is readily sap transmitted to *N. megalosiphon* and *N. benthamiana,* where it induces symptomless systemic infections. It did not infect *D. alata* and DLV infections were symptomless when the virus was singly infecting *D. composita* and *D. floribunda.* The virus is not transmitted by contact between plants, vector or pollen transmission.

Virus particles of DLV are flexuous rods 450-500 nm long. Infected *D. floribunda* plants contain amorphous, vacuolated, cytoplasmic inclusions, and banded inclusions consisting of closely packed virus particles (Hearon *et al.* 1978). The virus has a single stranded RNA genome, approximately 7.5 kb in size and a coat protein with a molecular weight of approximately 25kDa (Phillips *et al.* 1986). This assigns DLV to the potexviruses although the flexuous filamentous particles are shorter than those of most other members of this genus.

An antiserum against purified DLV reacted weakly with *Lily virus X and Commelina virus X,* indicating that the virus is a member of the genus *Potexvirus*.

10.3 Viruses with limited distribution

10.3.1 *Chinese yam necrotic mosaic virus* Genus *Macluravirus* (ChYNMV)

The virus was first observed in Japan from Chinese yam, *D. opposita cv* Nagaimo causing chlorotic and necrotic spots, mosaic and netting (Fukumoto and Tochihara, 1978). Primary infections, through seed tubers can result in crop yield losses up to 45%.

ChYNMV has flexous filamentous particles 660 nm in length with a coat protein of 38 kDa. The virus is restricted to Chinese yam, it does not infect any other host and was assigned as a tentative member of the Genus *Carlavirus*.

However, based on the recently available partial sequence of the 3'-terminal region of the ChYNMV genome (Kondo, 2001) this virus is now considered to be a member of the *Potyviridae*, since its nucleotide sequence had a sequence corresponding to the C-terminal regions of a potyviral polyprotein. Alignment of the amino acid sequences of the putative coat protein cistron, which contained several key motifs of the *Potyviridae,* further inducated a close relationship of ChYNMV to species of the genus *Macluravirus*, *Narcissus latent virus* and *Maclura mosaic virus*. This was further substantiated by the presence of cylindrical cytoplasmic inclusions and straight laminated aggregates found in ChYNMV infected cells.

ChYNMV is a distinct virus species of the genus *Macluravirus*. Despite its regional importance, it is only of limited significance due to its restriction to a single host. Once appropriate sanitation schemes will be in operation, there is a good chance to eradicate this virus from *D. opposita*.

10.3.2 *Japanese yam mosaic virus* Genus *Potyvirus* (JYMV)

The virus was first described in 1978 (Okuyama and Saka) as a potyvirus related to YMV. Infected plants of the Japanese yam, *D. japonica,* showed mosaic and green-banding of leaves resembling symptoms reported for YMV. Virus infections often resulted in significant yield reductions (Okuyama and Saka, 1978). However, while YMV is transmissible to various herbaceous hosts, only *D. japonica,* is susceptible to this virus. Plants raised from virus free stocks planted in the field became readily infected with JYMV (Kameya *et al.* 1999) indicating the presence of an alternative host and efficient aphid vectors.

The complete sequence of a JYMV was determined (Fuji and Nakamae, 1999). It was shown that the virus infecting *D. japonica* is distantly related to YMV, but sharing a close relationship with potyviruses infecting *Dasheen mosaic virus, Zantedeschia mosaic virus* and *Konjac mosaic virus* (Lesemann and Winter, 2000).

As with ChYNMV, the restricted host range of JYMV limits its significance to production of *D. japonica* in Japan. Control is difficult, since healthy plantlets became readily infected in the field. A cross protection strategy was suggested, using attenuated strains, and mild strains of JYMV have been selected and characterized (Fuji and Nakamae, 1999).

10.3.3 Viruses reported from yam

Several viruses were reported from *Dioscorea* spp., occurring either singly or in mixed infections together with one or several potyviruses. However, with the exception of ***Beet mosaic virus*** (BtMV) previously described as Dioscorea ring mottle virus (Porth and Nienhaus, 1983), none of the viruses were attributed to a distinct disease or regarded as a significant pathogen in yam.

Porth and associates (1987) identified BtMV in *Dioscorea alata* from Togo on the basis of host range, serology and cytopathology. They were unable to demonstrate BtMV as the pathogen causing ring mottle disease in yam since back transmission tests failed. Although it is evident that *D. alata* is a host for BtMV, it remains to be proven that BtMV is an indigenous West African virus.

The ubiquitous ***Cucumber mosaic virus*** (CMV) was found during several virus surveys (Migliori and Cadilhac 1976; Fauquet and Thouvenel, 1987; Hughes *et al.* 1997) in West Africa and in Guadeloupe. Although the virus was found in *D. alata* and in *D. trifida* its role in a disease was not proven. It can be speculated that CMV as an opportunistic virus is involved in synergistic infections resulting in more severe symptoms.

An isometric virus was found during surveys for viruses infecting yam in Nigeria. It was tentatively called ***Dioscorea* mottle virus**, syn. Dioscorea mild chlorosis virus (Hughes *et al.* 1997). Strains of this virus were isolated from *D. alata* and *D. rotundata* (Dongo, 2001) and all isolates reacted with an antiserum against *Cowpea mottle virus.* While the full taxonomic characterization of this virus as a member of the *Tombusviridae* is still pending, its causative association with a disease and an epidemiological significance has still to be proven.

Dioscorea dumetorum virus was reported from yam, *D. dumetorum*, in Nigeria (Hughes *et al.* 1998; Odu *et al.* 2001) and partial sequences of virus isolates from Bangladesh and Sri Lanka are available (Mumford unpublished data). From these partial sequences it is evident that the potyvirus from *Dioscorea dumetorum* qualifies as a distinct species, only distantly related to other potyviruses that infect yam.

10.4 Epidemiology and control

Although the etiology of the major yam virus diseases is known quite well, only limited knowledge is available on their epidemiology and control. However, from the characteristics of yam and its cultivation, and from the biology of the viruses infecting yam, a few generalizations may be drawn.

In vegetative propagated crops virus is disseminated through infected stem (vine) cuttings and tuber pieces. The incidence of the diseases resulting from the use of infected tubers is therefore increasing from season to season (Craig, 1964). Yam potyviruses are disseminated mainly by tubers than by insect vectors (Seal, 2000). Selection of propagation materials from healthy yam plants and seed certification for virus freedom is therefore the best practical method of virus disease control.

Indexing and certification schemes, such as those used for potato seed stocks in Europe are still absent, or at best, are at a developmental stage. At the International Institute of Tropical Agriculture (IITA), which has the worldwide mandate for research on yams, a combination of heat treatment of the mother plants for one month, followed by meristem culture, have been used to eliminate YMV from white yam (Ng, 1992). Chemotherapy, using vidarabine, ribavirin and 2-thiouracil was sometimes combined with heat treatment (Malaurie *et al.* 1998). These virus-tested plants were then micropropagated and distributed as plantlets and microtubers to national programs in India, Fiji and countries in Africa. Similar work conducted in the Caribbean using apical meristems led to a system for producing yam setts that were free of viruses (Mantell and Haque, 1979; Degras, 1989; 1993).

Several yam potyviruses are transmitted by aphids. Thus, after initial infection through use of infected seed pieces, aphids may spread the viruses to other plants. In some areas such as the rainforest zone of West Africa, where infection pressures and vector population are high, the rate of re-infection of healthy plants may be high. Re-infection may be reduced by the use of appropriate insecticides to suppress vector populations, although this adds to the cost of yam production. In contrast, the absence of vectors, e.g. in Barbados facilitates the use of virus-free meristem-derived materials (Degras, 1993).

10.5 Outlook and conclusions

Most of the diseases of edible yam with putative virus etiology are caused by potyviruses, termed with different virus names. Most of the past reports provided only fragmentary information on the taxonomic status of the virus(es). It is the merit of molecular analysis that more comprehensive information on these viruses can now be provided

The most significant viruses in yam are *Yam mosaic virus* and *Yam mild mosaic virus*. These distinct potyviruses are widely distributed, present in all yam growing regions of the world and occur as diverse populations. The badnavirus *Dioscorea bacilliform virus* probably consists also of diverse strains occurring in different locations.. This diversity resulted in the description of more than 15 different potyvirus diseases occurring in *Dioscoreaceae*. The molecular data presently available provide a basis for the assumption that strains and isolates of only 3 viruses are responsible for the most significant diseases in yam, although some additional viruses, e.g. Japanese yam mosaic virus, have been described.

While YMV, YMMV and DBV remain the main targets for resistance breeding, knowledge of all viruses infecting yam at various locations and their epidemiological significance is needed. More research is required on the variability of bacilliform viruses occurring in yam while the clarification of the epidemiological significances of the *Dioscorea dumetorum virus* and Dioscorea *mottle virus* provides scope for further studies.

While in the past virus indexing in yam was not easy, serological assays and molecular tests are now available for accurate virus testing in leaves and tubers of yam (Mumford and Seal, 1997, Hughes *et al.* 1998; Njukeng, 1998; Bousalem *et al.* 2000b; Njukeng *et al.* 2002).

Yam is a vegetatively propagated crop and dissemination of viruses is mainly by using infected planting material. Hence certification schemes providing planting material free of viruses are essential for improving yam productivity and yield. However, the common practices, to use own "farmer's seeds" for seed production hampers these efforts to reduce virus incidence. On farm and without appropriate testing methods and facilities, it is almost impossible to select yam seed materials free of viruses. Therefore, the most significant contribution to increased yam production will come from resistance breeding. There are perhaps over 4000 yam accessions from wild and cultivated species in germplasm collections in Africa and the Caribbean. Some of these are being evaluated in multilocational trials for virus resistance. For instance, moderate levels of resistance to YMV have been identified in 32 accessions of *D. rotundata* (IITA, 1993).

Recent data suggest that some landraces of *D. alata* and *D. rotundata* exhibit moderate to high levels of resistance to YMV and other viruses in Nigeria (Odu *et al.* 2001; Alabi, 2002). The genetic basis of resistance in white yam to a Nigerian isolate of YMV was investigated in three tetraploid *D. rotundata* genotypes. Major dominant genes were found to control resistance and these may facilitate the transfer of this resistance into the genetic background of elite *D. rotundata* (Mignouna *et al.* 2001).

Although DNA markers have been widely used to assess the genetic diversity of cultivated yams such as *D. rotundata* and *D. cayenensis* in West Africa (Mignouna, 1999), another possible application is to determine markers that are linked to genes conferring resistance to diseases (Mignouna *et al.* 2001). The advantage of molecular markers over conventional genetic markers include the fact that they can be used for genetic assessment *in vitro* without relying on plant development and are independent of the environment influencing their expression (Bosland and Votava, 2000).

Acknowledgement

The authors are indebted to Dr. Rick Mumford, CSL, UK, for providing helpful information on yam virus isolates.

References

Alabi, O. J. 2002. *Identification of Resistance in Yams (Dioscorea spp.) to Dioscorea Mild Mottle Virus* genus *Comovirus*. M.Sc. Project Report, University of Ibadan. 58 pp.

Aleman-Verdaguer, M. E., Goudou-Urbino, C., Dubern, J., Beachy, R. N. and Fauquet, C. 1997. Analysis of the sequence diversity of the P1, HC, P3, NIb and CP genomic regions of several yam mosaic potyvirus isolates: implications for the intraspecies molecular diversity of potyviruses. *J. Gen Virol.* 78: 1253-1264.

Asiedu, R. 1999. Potential for molecular breeding of yams at IITA. In: *DNA Marker-assisted Improvement of the Staple Crops of Sub-Saharan Africa.* J. H. Crouch and A. Tenkouano (eds.). IITA, Ibadan and CTA, Wageningen. Pp 203-206.

Boeglin, M., Labonne, G., Degras, L., Quiot-Douine, L. and Quiot, J. B. 1993. Mise en évidence de la présence du virus de la mosaique de l'igname sur igname (*Dioscorea trifida*) en Guyane. *Comm., Caribbean Food Crop Soc. Martinique, July 1993.*

Bosland, P. W. and Votava, E. J. 2000. *Peppers: Vegetable and Spice Capsicums.* CABI Publishing, UK. 204 pp.

Bousalem, M. and Dallot, S. 2000. First report and molecular characterization of *Yam mild mosaic virus* in *Dioscorea alata* on the island of Martinique. *Plant Dis.* 84: 200

Bousalem, M., Dallot, S. and Guyader, S. 2000b. The use of phylogenetic data to develop molecular tools for the detection and genotyping of Yam mosaic virus. Potential application in molecular epidemiology. *J. Virol Meth. 90: 25-36.*

Bousalem, M., Douzery, E. J. and Fargette, D. 2000a. High genetic diversity, distant phylogenetic relationships and intraspecies recombination events among natural populations of Yam mosaic virus: a contribution to understanding potyvirus evolution. *J. Gen Virol.* 81: 243-255.

Briddon, R. W., Phillips, S., Brunt, A. and Hull, R. 1999. Analysis of the sequence of *Dioscorea alata bacilliform virus;* comparison to other members of the *Badnavirus. Virus Genes* 18: 277-283.

Brunt, A. A., Crabtree, K., Dallwitz, M. J., Gibbs, A. J., Watson, L. and Zurcher, E. J. (eds.). 1996 onwards. *Plant Viruses Online: Descriptions and Lists from the VIDE Database. Version*: 20th August 1996. URL http:biology.anu.edu.au/Groups/MES/vide.

Cobley, L. S. and Steele, W. M. 1976. *The Botany of Tropical Crops.* Longman, London.

COPR. 1978. *Pest Control in Tropical Root Crops. PANS Manual No. 4.* Centre for Overseas Pest Research, London.

Coursey, D. G. 1967. Internal brown spot, a condition in yams in Barbados. *J. Agric Soc. Trinidad Tobago* 67: 473-482.

Craig, J. 1964. Studies of a mosaic disease of white yam (*Dioscorea rotundata*) in the Western Region of Nigeria. *Rep. USAID Plant Path. Advisor.* 4 pp.

deAvila, A. C. DE, Gama, M. DE I.C.S., Kitajima, E. W., 1982. Deteccao de um potyvirus em inhame (Dioscorea spp.) no Brasil. Fitopatologia Brasileira 7: 447-452.

Degras, L. 1989. Biotechnologies for yam. In: *Plant Biotechnologies for Developing Countries.* A. Sasson and V. Costarini (eds.). CTA, Wageningen & FAO. pp 229-233.

Degras, L. 1993. *The Yam: A Tropical Root Crop.* Macmillan Press, London.

Dongo, L. N. 2001. *Survey of Viruses Infecting Yams and Partial Characterization of Three Isometric Virus Isolates in Major Yam-growing Areas of Nigeria.* Ph.D. Thesis, Univ. Ibadan, Ibadan. 187 pp.

Dongo, L. N., Hughes, J. d'A. and Atiri, G. I. 2001. Three new isometric viruses infecting yams in Nigeria. *Proc. 7th Triennial Symp. Intern. Soc. For Tropical Root Crops, Africa-Branch, Cotonou, Benin, 1998.*

Emehute, J. K. U., Ikotun, T., Nwauzor, E. C. and Nwokocha, H. N. 1998. Crop Protection of Yams. In: *Food Yams: Adv. in Research.* G. C. Orkwor, R. Asiedu and I. J. Ekanayake (eds). IITA and NRCRI, Nigeria.

Fauquet, C. and Thouvenel, J. C. 1980. *Viral Diseases of Crop Plants in Ivory Coast.* ORSTOM, Paris. 128 pp.

Fauquet, C. and Thouvenel, J. C. 1987. *Plant Viral Diseases in the Ivory Coast.* Documentation Techniques, ORSTOM, Paris. 243 pp.

Francki, R. I. B., Mossop, D. W. and Hatta T. 1979. Cucumber mosaic virus. *CMI/AAB Descriptions of Plant Viruses* 213, 1979.

Fuji, S. and Nakamae, H. 1999. Complete nucleotide sequence of the genomic RNA of a Japanese yam mosaic virus, a new potyvirus in Japan. *Arch. Virol.* 144: 231-240.

Fuji, S., Mitobe, I., Nakamae, H. and Natsuaki, K. T. 1999. Nucleotide sequence of coat protein gene of yam mild mosaic virus, isolated from Papua New Guinea. *Arch. Virol.* 144: 1415-1419

Fukumoto, F. and Tochihara, H. 1978. Chinese yam necrotic mosaic virus. *Ann. Phytopath. Soc. Japan* 44: 1-5.

Geering, A. D. W., McMichael, L. A., Dietzgen, R. G. and Thomas, J. E. 2000. Genetic diversity among banana streak virus isolates from Australia. *Phytopathology* 90 : 921-927.

Goudou-Urbino, C., Givord, L., Konate, G., Boeglin, M., Quiot, J. B. and Dubern, J. 1996. Differentiation of yam virus isolates by using symptomatology, western-blot and monoclonal antibodies. *J. Phytopath.* 144, 235-240.

Goudou-Urbino, C. 1995. *La Mosaic de l'igname. Aspects Epidemiologiques au Burkina Faso et Variabilite du Virus.* Thése Doct. Univ. Montpellier 11, France. 147 pp.

Haque, S. Q. and Mantell, S. H. 1980. Status of the virus diseases of yams (*Dioscorea* spp.) in the Commonwealth Caribbean. *INRA Symp., Intern. Sem. Yam 1981*, pp 85 – 92.

Harrison, B. D. and Roberts, I. M. 1973. Association of virus-like particles with internal brown spot of yam (*Dioscorea alata). Trop. Agric. (Trin.)* 50: 335-340.

Hearon, S. S., Corbett M. K., Lawson, R. H., Gillaspie Jr., A.G. and Waterworth, H. E., 1978. Two flexuous-rod viruses in Dioscorea-floribunda: symptoms, identification, and Ultrastructure. *Phytopathology* 68: 1137-1146 1978

Hughes, J. D'A. 1986. *Viruses of the Araceae and Dioscoreaceae: Their Isolation Characterization and Detection.* Ph.D. Thesis, University of Reading. 160 pp.

Hughes, J. d'A., Dongo, L. N. and Atiri, G. I. 1997. Viruses infecting cultivated yams (*Dioscorea alata* & *D. rotundata*) in Nigeria. *Phytopathology* (Supplement) 87: S 45.

Hughes, J. d'A., Dongo, L. N. and Ng, S.Y.G. 1998. Diagnosis of yam viruses. *Trop. Agric.* 75: 45-48.

Hull, R. and Lane, L. C. 1973. Unusual nature of components of a strain of pea enation mosaic-virus. *Virology* 55: 1-13.

ICTV. 2000. V*irus taxonomy.* Classification and Nomenclature of Viruses. 7th Report of the International Committee on Taxonomy of Viruses. Academic Press

IITA, 1993. *Root and Tuber Improvement Programme: Archival Report (1989-1993). Part III. Yams.* Crop Improvement Div. IITA, Ibadan

IITA. 2002. *Yams.* International Institute of Tropical Agriculture Publ., Ibadan. http://www.iita.org/crop/yam.htm.

Jeffers, W. de C., and Headley, W. 1967. The problem of internal spotting of yams (*Dioscorea* sp.) in Barbados. *Min Agric Bull* 46: 7. Kondo T. 2001.

Kameya, I. M., Yamaguchi, K., Hara, T., Ito, S., Fuji, S., Kajihara, H. and Tanaka, S. 1999. Detection of Japanese yam mosaic virus from Chinese yam (*Dioscorea opposita* Thumb. Cv. Ichoimo) by immunocapture PCR and differentiation of virulent and attenuated isolates by PCR-RFLP analysis. *Ann. Phytopathol. Soc. Jpn.* 65: 494-497.

Kondo, T. 2001. The 3'-terminal sequence of chinese yam necrotic mosaic virus genomic RNA: a close relationship with macluravirus. *Arch Virol.* 146: 1527-1535.

Lawson, R. H., Hearon, S. S., Smith, F. F. and Kahn, R. P. 1973. Electron microscopy and separation of viruses in *Dioscorea floribunda. Phytopathology* 63: 1435 (Abst).

Lebas, B., Canning, E., Kenyon, L. and Seal, S. 2002. *Yam Viruses of the South Pacific Islands.* Pest Management Dept., Univ, Greenwich, Natural Resources Institute, UK.

Lesemann, D.- E. and Winter, S. 2000. *Konjac Mosaic Virus, Dasheen Mosaic Virus* and Unknown Potyviruses Infecting *Zantedeschia* Spp. and other Cultivated Araceae. *Acta Hortic.* 568: 135-141.

Lisa, V., Boccardo, G. D'Agostino, G., Dellavalle, G. and D'Aquilio, M. 1981. Characterization of a potyvirus that causes zucchini yellow mosaic. *Phytopathology* 71: 667-672.

Malaurie, B., Trouslot, M., Berthaud, J., Bousalem, M., Pinel, A. and Dubern, J. 1998. Medium-term and long-term *in vitro* conservation and safe international exchange of yam (*Dioscorea* spp.) germplasm. *EJB Electr. J. Biotechnology 1: 1-15.*

Mantell, S. H. and Haque, S. Q. 1979. A rapid propagation system for yams. *Yams Virus Project Bull.* (No.1).

Mayo, M. A. and Pringle, C. R. 1998. Virus taxonomy-1997. *J. Gen. Virol.* 79: 649-657.

Migliori, A. and Cadilhac, B. 1976. Contribution à l'étude de la maladie du virus de l'igname *Dioscorea trifida* en Guadeloupe. *Ann. Phytopath* 8: 73-78.

Mignouna, H. D. 1999. Application of DNA markers in yam. In: *DNA Marker-assisted Improvement of the Staple Crops of Sub-Saharan Africa.* J. H. Crouch and A. Tenkouano (eds.). IITA, Ibadan and CTA, Wageningen. Pp 200-202.

Mignouna, H. D., Njukeng, P., Abang, M. M. and Asiedu, R. 2001. Inheritance of resistance to *Yam mosaic virus* genus *Potyvirus,* in white yam (*Dioscorea rotundata*). *Theor. Appl. Genet.* 103: 1196 – 2000.

Mohamed, N. A. and Mantell, S. H. 1976. Incidence of virus symptoms in yam (Dioscorea sp.) foliage in the Commonwealth Carribbean. *Trop.Agric.* (Trinidad). 53, 255-261.

Moury, B., Morel, C., Johansen, E. and Jacquemond, M. 2002. Evidence for diversifying selection in Potato virus Y and in the coat protein of other potyviruses. *J. Gen Virol.* 83: 2563-2573.

Mumford R.A. and Seal S.E. 1997. Rapid single-tube immunocapture RT-PCR for the detection of two yam potyviruses. *.J, Virol Meth.* 69: 73-79.

Ng, S. Y. C. 1992. Micropropagation of white yam (*Dioscorea rotundata* Poir). In: *Biotechnology in Agriculture and Forestry.* Y. P. S. Bajah (ed.) pp 135-159. Springer-Verlag.

Ng, S. Y. C. 1992. Tissue culture of root and tuber crops. In: *Biotechnology: Enhancing Research on Tropical Crops in Africa.*

Njukeng, A. P., Atiri, G. I., Hughes, J.d'A., Agindotan, B. O., Mignouna, H. D. and Thottappilly, G. 2002. A sensitive TAS-ELISA for the detection of some West African isolates of *Yam Mosaic Virus* (YMV) in *Dioscorea* spp. *Trop. Science* 42: 65-74.

Njukeng, A.P., Hughes, J. d'A., Atiri, G.I. and Ekpo E.J.A. 2002 Distribution of *Yam Mosaic Virus* in Nigeria. *VIIIth Intern. Plant Virus Epidemiology Symp.,* Aschersleben, 12.-17.5.2002.

Njukeng, P. N. 1998. *Development of Sensitive Indexing Techniques for Diagnosing Yam Mosaic Potyvirus in Yams, Dioscorea species.* Ph.D Thesis, Univ. Ibadan, Ibadan. 209 pp.

Noordam, D. 1973. *Identification of Plant Viruses: Methods and Experiments.* Centre for Agricultural Publishing and Documentation, Wageningen.

Odu, B. O., Hughes, J. d'A., Shoyinka, S. A., and Dongo, L. N. 1999. Isolation, charcterisation and identification of a potyvirus from *Dioscorea alata* L. (water yam) in Nigeria. *Ann. appl. Biol.* 134: 65-71.

Odu, B. O., Shoyinka, S. A., Hughes, J. d'A., Asiedu, R. and Oladiran, A. O. 2001. Yam viruses in Nigeria. *Proc. 7th Triennial Symp. Intern. Soc. for Tropical Root Crops, Africa Branch, Cotonou, Benin 1998.*

Okuyama, S. and Saka, H., 1978. Yam mosaic virus. *Sci. Rep. Fac. Agric. Ibaraki Univ.* 26, 29-34 .

Olatunde, O. J. 1999. *Viruses of Yams in Ghana.* M.Sc. Diss., Univ. Greenwich, U. K.

Onwueme, I. C. 1978. The *Tropical Tuber Crops.* John Wiley & Sons, Chichester, UK. 234pp.

Onwueme, I. C. and Sinha, T. D. 1991. *Field Crop Production in Tropical Africa.* CTA, Wageningen.

Okuyama, S., Saka, H. (1978). Yam mosaic virus. *Science Reprt. Fac. Agric. Ibaraki Univ.* 26: 29-34.

Orkwor, G. C. 1998. The importance of yams. In: *Food Yams: Advances in Research.* G. C. Orkwor, R. Asiedu and I. J. Ekanayake (eds.). IITA and NRCRI, Nigeria.

Phillips, S. and Brunt, A. A. 1988. Dioscorea latent virus. *CMI/AAB Descriptions of Plant Viruses No. 335.*

Phillips, S., Briddon, R. W., Brunt, A. A. and Hull, R. 1999. The partial characterization of a *Badnavirus* infecting the greater Asiatic or water yam (*Dioscorea alata*). *J. Phytopath.* 147: 265 – 269.

Phillips, S., Piggott, J. and Brunt, A. A. 1986. Further evidence that *Dioscorea latent virus* is a potexvirus *Ann. appl. Biol.* 109: 137-145.

Porth, A. and Nienhaus, F., 1983. Dioscorea alata ring mottle virus, a new potyvirus of yam in Togo. *Z.PflKrankh. Pfl.Schutz* 90: 352-362.

Porth, A., Lesemann, D. E., Vetten, H. J. 1987. Characterisation of potyvirus isolates from west African yams. *J. Phytopath.* 120: 166-183.

Reckhaus, P. 1980. *Untersuchungen zur Ätiologie Virus Verdächtiger Erkrankungen an Yam (Dioscorea spp) in West Afrika (Togo).* Ph.D. Thesis, University of Bonn, Bonn. 103 pp.

Reckhaus, P. and Nienhaus, F. 1981. Etiology of a virus disease of white yam *Dioscorea rotundata* in Togo. *Z. PflKrankh. PflPath. PflSchutz* 88: 492-509.

Regenmortel, M.H.V., Fauquet, C.M., Bishop, D.H.L., Carstens, E.B., Estes, M.K., Lemon, S.M., Maniloff, J., Mayo, M.A., McGeoch, D.J., Pringle, C.R., and Wickner, R.B. 2000. *Virus Taxonomy: Classification and Nomenclature of Viruses*: Seventh Rep. Intern. Comm. Taxonomy of Viruses. Academic Press, New York, 1162 p.

Rehm, S. and Espig, G. 1991. *The Cultivated Plants of the Tropics and Subtropics.* CTA, Wageningen and Margraf, Weikersheim.

Rossel, H.W., 1982a. Dioscorea alata vein yellowing. *Ann. Rep. 1982, IITA*, Ibadan, Nigeria, 109.

Rossel, H. W. and Thottappilly, G., 1984. A possible new potyvirus in sweet potatoes. *Ann. Rep. 1984, IITA,* Ibadan, Nigeria, 147.

Rossel, H. W., Thottappilly, G., 1985. *Virus Diseases of Important Food Crops in Tropical Africa.* IITA, Ibadan, Nigeria.

Seal, S. 2000. *Development of rapid tests for identification and differentiation of yam virus variability.* WISARD Information, Univ. Greenwich and Natural Resources Inst., UK.

SPC. 1992. *South Pacific Foods Leaflets.* South Pacific Commission, New Caledonia.

Terry E. R. (1976). Incidence, symptomatology and transmission of a yam virus in Nigeria. In : *Proc. Fourth Symp. Intern. Soc. Tropical Crop Crops*, pp170-176. CIAT, Cali Columbia, 1976.

Théberge, R. L. 1985. *Common African Pests and Diseases of Cassava, Yam, Sweet Potato and Cocoyam.* IITA, Ibadan.

Thottappilly, G. 1992. Plant virus diseases of importance to African Agriculture *J. Phytopath.* 134: 265-288.

Thottappilly, G., 1983, Dioscorea alata vein yellowing. *Ann. Rep. 1983 IITA*, Ibadan, Nigeria, 131.

Thottappilly, G., Crouch, J. H. and Ortiz, R. 1999. DNA markers for plant health management. In: *DNA marker-assisted Improvement of the Staple Crops of Sub-Saharan Africa.* J. H. Crouch and A. Tenkouano (eds.). IITA, Ibadan and CTA, Wageningen. Pp 101-106.

Thottappilly, G., Mignouna, H. D. and Omitogun, O. G. 2000. The use of DNA markers for rapid improvement of crops in Africa. *African Crop Sci. J.* 8: 99 – 108.

Thottappilly, G., Monti, L., Mohan Raj, D.R. and Moore, A.W. (eds.) 1992. *Biotechnology: Enhancing Research on Tropical Crops in Africa.* CTA/IITA co-publication. IITA, Ibadan, Nigeria. Pp 135-141.

Thouvenel, J. C. and Dumont, R. 1990. Perte de rendement de l'igname infectée par le virus de la mosaique en Côte d'Ivoire. *L'agronomie Tropicale* 45: 125-129.

Thouvenel, J. C. and Fauquet, C. 1977. Une mosaique de l'igname cause par un virus filamenteux en Côte d'Ivoire. *Compt. Rendues Hebdomadaires Scéan. l'Acad.* 284: 1947-1949.

Thouvenel, J. C. and Fauquet, C. 1979. Yam mosaic, a new potyvirus infecting Dioscorea cayenensis in the Ivory Coast. An*n. appl. Biol.* 93: 279-283.

Thouvenel, J. C. and Fauquet, C. 1986. Yam Mosaic Virus. *CMI/AAB Descriptions of Plant Viruses* No. 314.

Waterworth, H. E., Lawson, R. H. and Kahn, R. P. 1974. Purification, electron microscopy and serology of *Dioscorea latent virus*. *J. Agric. Univ. Puerto Rico* 58: 351-357.

Yang, I. C., Hafner, G. J., Dale, J. L. and Harding, R. M. 2003. Genomic characterisation of taro bacilliform virus. *Arch. Virol. 148 (5): 937-949.*

Table 10.1 Viruses reported from natural infections of *Dioscorea* spp.

Virus	Distribution	Synonyms/strains	References
Yam mosaic virus *Potyviridae* Genus *Potyvirus*	World wide	YMV	Thouvenel and Fauquet, 1986
		Dioscorea greenbanding mosaic virus	Reckhaus and Nienhaus, 1981
		YV-N	Terry, 1976
		Dioscorea alata virus	Porth *et al.* 1987
		Discorea trifida virus	Migliori and Cadilhac, 1976
Yam mild mosaic virus *Potyviridae* Genus *Potyvirus*	World wide	YMMV for YMV 1	suggested by A. Brunt
		YMMV	Fuji *et al.* 1999
		Yam virus I	Hughes, 1986
		Dioscorea alata virus	Odu *et al.* 1999
		D. alata vein yellowing virus	Thottappilly, 1983
		D. alata chlorosis virus	Rossell and Thottappilly, 1985
		Dioscorea greenbanding virus	Hearon *et al.*1978
Dioscorea bacilliform virus *Caulimoviridae* Genus *Badnavirus*	World Wide	DBV	Phillips *et al.* 1999
		Dioscorea alata badnavirus	Briddon *et al.* 1999
		Yam internal brown spot disease	Harrison & Roberts (1973)
		Dioscorea bulbifera badnavirus	Phillips *et al.* 1999
Dioscorea dumetorum virus *Potyviridae* Genus *Potyvirus*	Nigeria, Sri Lanka, Bangla Desh		Hughes *et al.* 1997; Odu *et al.* 2001
Dioscorea latent virus Genus *Potexvirus*	Central America, South Pacific	DLV	Lawson *et al.* 1973
Japanese yam mosaic virus *Potyviridae* Genus *Potyvirus*	Japan	JYMV	Okuyama and Saka, 1978
Chinese yam necrotic mosaic virus *Potyviridae* Genus *Macluravirus*	Nagaimo in Japan	ChYNMV	Fukumoto and Tochihara, 1978
Cucumber mosaic virus *Bromoviridae* Genus *Cucumovirus*	Guadeloupe, West Africa	CMV	Fauquet and Thouvenel, 1987 Migliori and Cadilhac 1976
Beet mosaic virus	Togo	Dioscorea alata ring mottle virus	Porth and Nienhaus, 1983
Dioscorea mild chlorosis mosaic *Tombusiridae* Genus *Carmovirus*	Nigeria		Dongo *et al.* 2001

Virus names in *italics* are species names acknowledged by ICTV.
? - speculative for sufficient information not available.

Figure 10.1 Symptoms of virus infections in yam, *Dioscorea alata*. Yam mild mosaic virus (left), *Yam mosaic virus* (middle), *Dioscorea bacilliform virus* mixed infection with Yam mosaic virus (right).

11. Rice

Lee A. Calvert[1], Hiroki Koganezawa[2], Denis Fargette[3], and G. Konate[4]
[1]Centro Internacional de Agricultura Tropical Cali, Colombia: [2]National Agricultural Research Center for Western Region, Nishifukatsu, Hiroshima 721-8514, Japan: [3]IRD, BP 64501, 34394. Montpellier cedex 5, France [4]INERA, Laboratoire de Virologie, Station de Kamboinse, 01 BP 476 Ouagadougou, Burkina-Faso

11.1 Introduction

Rice, *Oryza sativa* (Graminae), was domesticated earlier than 5000 B.C. in northeast of India. Its cultivation then spread throughout Southeast Asia and southern China. In Africa, *Oryza glaberrima* was domesticated along the Niger River probably around 1500 B.C. It is principally grown in West Africa.

Rice is the most important staple or subsistence food in the world. The total world production of rice in 2002 was 526 million tonnes, of which *ca.* 90% were produced in Asia while Latin America and Africa each contributed about 4% (source: FAOSTAT, 2002; www.fao.org). China, India, Indonesia, Bangladesh, Vietnam, Thailand and Myanmar are the biggest rice producers and account for over 75% of the production. Both production and consumption of rice are rising most rapidly in areas outside of Asia especially in the developing countries of Latin America and Africa.

Twelve different viruses have been reported from rice (Table 11.1). The distribution of each rice virus is generally restricted to only one of the continents in which the crop is grown. An exception is rice stripe necrotic virus, for which strains have been found in both Africa and Latin America.

The most important virus from each continent was selected for a more in depth review. Several other viruses are also important and are briefly discussed. In this review, we consider the identification of rice diseases, available information about these viruses and their strains, and control measures.

11.2 Tungro: *Rice tungro bacilliform virus* Genus *Rice tungro bacilliform-like viruses* (RTBV) and *Rice tungro spherical virus* Genus *Waikavirus* (RTSV)

Tungro, which means 'degenerated growth' in the Ilocano language of the Philippines, has become more prominent since the introduction of the high-yielding varieties by the International Rice Research Institute (Rivera and Ou, 1965). Soon after the release of modern semi-dwarf type varieties such as IR8, outbreaks of the disease were observed in many South and Southeast Asian countries. Before the name tungro became popular, diseases showing tungro-like symptoms have been known under different names including "mentek" and "penyakit habang" in Indonesia, "penyakit merah" in Malaysia, "yellow orange leaf" in Thailand and "leaf yellowing" in India (Ou, 1985). In addition, local farmers in many areas of the Philippines, Malaysia and Indonesia have various local names for the disease. In each region of the Philippines, the rice farmers have indigenous names including accep na pula (Tagalog), po-ol/po-or (Pangasinan), cadang-cadang (Bicol), los-ong (Negros), sebukaw (Bohol) and tagostos (Iloilo). This indicates that tungro has been widespread in South and Southeast Asia and farmers have been aware of the disease for a long time.

The causal agent of Tungro was reported to be a spherical virus, which was named rice tungro virus (Galvez, 1968a). Later, Saito *et al.* (1975) reported the association of another virus with the disease, and then it was established that tungro is caused by two viruses, rice tungro bacilliform virus (RTBV) and rice tungro spherical virus (RTSV) (Hibino *et al.* 1979).

Tungro disease is widespread in most countries of South and Southeast Asia, from India to West Papua (Sogawa, 1976; Davis *et al.* 2000). It has not been reported from Laos and Cambodia. It also occurs in southern China (Lin *et al.* 1994). The major endemic areas are West Bengal of India, Bali and Java of Indonesia, and Mindanao of the Philippines where rice is grown all the year round. A disease caused by RTSV alone was reported in Japan (Shinkai, 1977).

Tungro causes periodic epidemics and has caused severe crop losses in South and Southeast Asia for the past 40 years (Koganezawa, 1998; Satapathy and Anjaneyulu, 2001)). Annual crop losses were estimated between 680 million (Herdt, 1991) and 1.5 billion dollars (Satapathy and Anjaneyulu, 2001). Before the recognition of tungro, Serrano (1957) reported that the accep an pula (tungro) caused 30% yield losses in 1940s in the Philippines. Satapathy and Anjaneyulu (2001) speculated that the great Bengal famine in 1942-43, which claimed two million lives, was caused by tungro. Although the famine has been attributed to rice brown spot, it has not been so widespread or devastating as compared with Tungro in India and other countries after the famine. In pot-experiments, grain yield reduction is as high as 100% in doubly infected plants, and 40% in highly susceptible varieties infected with RTSV alone (Hasanuddin and Hibino, 1989).

The important natural host of tungro viruses is rice. In artificial inoculation tests, tungro viruses can infect several Graminaceous weeds such as *Eleusine indica*, *Echinochloa colonum* and *E. crus-galli*. The transmission rate from weeds to rice is extremely low, and from an epidemiological viewpoint, infected weeds do not play an important role as disease sources. Natural infection of tungro in wild rice species was observed in the Thailand and Philippines (Hino et al. 1974; Tiongco et al. 1993a).

In susceptible varieties such as Taichung Native 1, plants infected with both RTBV and RSTV, are stunted and the leaves become discolored turning yellow or yellow-orange. Plants infected with RTBV alone develop similar, albeit milder symptoms than those caused by double infection. RTSV alone usually does not cause clear symptoms under tropical conditions. The African rice species *O. glaberrima* is an exception because an infection of RTSV causes stunting and the leaves become pale yellow (Cabauatan *et al.* 1993). Dual infection of RTBV and RTSV causes systemic necrosis in *O. glaberrima* and *O. barthii* (Kobayashi and Ikeda, 1992).

For purification the same procedure was used for both RTBV and RTSV. The infected plants were homogenized in 0.01 M EDTA, pH 8.0. Since the viruses are restricted to the vascular bundles, high yield of the viruses can be obtained by treating homogenates with cell wall degrading enzymes (Cabauatan and Hibino, 1988). However, care must be taken in the analysis of coat proteins, because coat proteins are partially degraded during this treatment (Miranda et al. 1995). Homogenate was clarified by heating it at 40 °C for 1 hour. Further purification was done by differential centrifugation, followed by sucrose density gradient centrifugation.

11.2.1 Virus characteristics

Both RTSV and RTBV multiply independently in rice plants. In rice plants, RTSV particles are restricted to the phloem tissues, while RTBV particles are found in both phloem and xylem cells (Santa Cruz *et al.* 1993). RTSV particles are isometric and 30 nm in diameter. RTSV consists of a genome of a 12 kb polyadenylated single-stranded RNA and is encapsidated by three coat proteins. It belongs to the genus *Waikavirus* within the family *Sequiviridae* (Mayo *et al.* 2000). 'Waika' means stunting in Japanese, and susceptible japonica varieties infected with RTSV are stunted. RTBV particles are bacilliform, 100-300 nm in length, and 30-35 nm in width. RTBV contains a circular double-stranded DNA of 8 kb with site-specific discontinuities. It is classified in the group of "rice tungro-bacilliform-like viruses" within the family *Caulimoviridae*. RTBV genome has four major ORFs, and these are predicted to encode proteins of 24, 12, 194, 46 kDa (Hull, 1996). The 194 kDa protein is predicted to be encoded as a polyprotein, which consists of the putative movement, coat, aspartic protease, reverse transcriptase and ribonuclease H proteins (Hull, 1996; Marmey *et al.* 1999). Recent advances in the Rice Genome Research Program revealed that the japonica rice genome contains a number of integration of 194K-protein sequences of RTBV, indicating that RTBV existed in very ancient times (Harper *et al.* 2002).

Several efforts have been made to characterize the strains of RTBV and RTSV. Dahal *et al.* (1992) compared varietal reactions of three isolates from the Philippines, Malaysia and India. There were variations in symptom severity and transmission profile among the three isolates. Variety TKM6, which was previously identified as resistant to the Philippine type isolate of RTSV, reacted differently to RTSV isolates from India and the Philippines. An RTSV strain Vt6 that can infect TKM6 was found in the Philippines (Cabauatan *et al.* 1995). Four strains of RTBV, designated L, G1, G2, and Ic, were also isolated from the Philippines on the basis of their characteristic symptoms on rice cultivars FK135 and Taichung Native 1 (Cabauatan *et al.* 1995). On FK135, strain Ic caused severe stunting, reduced tillering, narrow leaves, and distinct interveinal chlorosis (striping and/or mottling) on the leaf blades, while G1 and G2 caused only mild stunting and foliage was a normal green. On Taichung Native 1, G1 and Ic caused mild stunting but no discoloration of the foliage; G2 caused severe stunting and yellow to orange discoloration similar to that caused by the type strain L. Genomic sequences of these strains were determined (Cabauatan *et al.* 1999).

Substantial works on genomic and serological variations among isolates of RTBV and RTSV have been done (Fan *et al.* 1996; Dahal *et al.* 1996; Druka and Hull, 1998; Rekha and Dasgupta, 2001). These studies revealed that RTBV isolates mainly fell into 2 groups, those from the Indian subcontinent and those from Southeast Asian countries. The genomes of isolates from the Indian subcontinent contained a deletion of 64 bp when compared with those from Southeast Asia (Fan *et al.* 1996). Although RTSV variations were not so conclusive as those of RTBV because of the high degree of micro-variation in RNA genomes, 3 genomic variants of RTSV were identified. The geographic distribution, of these RTSV variants, does not correlate with that found for strains of RTBV (Druka and Hull, 1998).

11.2.2 Diagnosis, transmission and epidemiology

The simplest way of field diagnosis of tungro is by symptoms. Generally rice viruses induce symptoms specific to each virus. Yellow to orange color degradation from the tip of leaves and plant stunting are characteristics of tungro. The double-sandwich enzyme-linked immunosorbent assay (DAS-ELISA) is routinely used for the detection of both rice tungro viruses (Bajet *et al.* 1985). Other serological methods such as rapid immunofilter paper assay (Cabauatan and Koganezawa, 1997) and tissue printing (Nath *et al.* 2000) were also applicable. A biological assay using *O. glaberrima* (International Rice Germplasm Center Acc. No. 100139) can be applied when laboratory equipment for serological test is not available. *O. glaberrima* plants doubly infected with RTBV and RTSV show severe necrosis and die within 4 weeks after inoculation. Plants infected with RTBV show similar but milder symptoms and survive longer. Plants infected with RTSV alone show pale green and elect leaves (Cabauatan *et al.* 1993). Polymerase chain reaction was applied to detect RTBV in tolerant rice varieties (Takahashi *et al.* 1993). Tungro viruses are not mechanically transmissible. They are transmitted by several leafhoppers including *Nephotettix virescens*, *Recilia dorsalis* and other *Nephotettix* species. Among them, *N. virescens* is the most efficient vector, and its distribution coincides with that of tungro. Transmission manner of tungro viruses by leafhoppers is unique among rice viruses. Leafhoppers transmit RTBV and RTSV in semi-persistent manner, and the minimum acquisition and inoculation feeding periods for *N. virescens* are 5 min and 7 min, respectively. *N. virescens* retains tungro infectivity for 2 to 6 days but loses infectivity after molting (Ling, 1972). Leafhoppers readily acquire RTSV from plants infected with RTSV alone, but do not acquire RTBV from plants infected with RTBV alone. Leafhoppers acquire RTBV only when exposed to RTSV-infected plants before feeding on RTBV-infected plants (Hibino *et al.* 1979; Hibino, 1983). RTBV absolutely depends on RTSV for transmission. The helper factor coded in RTSV genome remains to be clarified.

Tungro can become widespread within a short period because of its semi-persistent transmission manner, and its relatively short latent period in plants. Soon after transplanting, the leafhoppers migrate from surrounding fields. The infection of plants with RTSV occurs at a more rapid rate than RTBV in fields, and then RTBV infection increases subsequently (Tiongco *et al.* 1993b). Vector leafhoppers are considered sedentary and less migratory than the delphacids. Tungro incidence was low 100 m away from the disease source (Hasanuddin *et al.* 1999). Holt *et al.* (1996) demonstrated that if fields were more than 1 km from the disease source, the risk of tungro would be low. However, migratory flight of *N. virescens* up to 30 km has been observed (Riley *et al.* 1987). Flying for almost 7 hours was recorded for tethered *N. virescens* females (Cooter *et al.* 2000). The moderate-distance migration of vectors plays a role in tungro spread over wide areas.

The main factors affecting tungro incidence are the varieties, asynchronous cropping, abundance of vector and inoculum (Savary *et al.* 1993; Holt *et al.* 1996). In several studies, asynchronous cropping was associated with tungro epidemics (Bottenberg *et al.* 1990; Holt *et al.* 1996; Cabunagan *et al.* 2001). Often tungro is most severe in areas where irrigation facilities are well developed and water is available even in the dry season. This allows the farmers grow rice continuously and various stages of rice are growing side by side. In these staggered planting areas during the rainy season, peak vector population densities are reached by mid growing season (Widiarta, *et al.* 1990; Chancellor *et al.* 1996). At this peak stage inoculum sources become maximum, and leafhoppers move from older to younger plants that are more preferred by them and are also susceptible to infection. Moreover, continuous cropping enables the monophagous leafhoppers to complete their life cycle and produce several generations of offspring, spreading the disease over a wide production area.

During the rainy season of 2001 in the Isavela province of the Philippines, about 2200 ha of susceptible rice varieties that were transplanted late in the season became heavily infected with tungro. This year, the peak-planting season was longer than normal because early release of water prompted farmers to plant early, and there was carryover of the disease and the vectors from the previous dry season. (H. Koganezawa, unpublished observation).

11.2.3 Control

Synchronized cropping together with the creation of fallow period is considered to be a major component of control strategies against tungro.

Tungro incidence and vector populations are high in continuous and asynchronous cropping areas. The disease source is limited to infected rice plants including stubbles and ratoons. Major tungro vector *N. virescens* is basically monophagous. Viruliferous green leafhoppers lose infectivity within a week. If a fallow period can be implemented over a wide area, it is possible to reduce disease sources and vector populations. In several areas tungro was successfully controlled by introducing fallow period and synchronized cropping (Koganezawa, 1998). Although synchronous cropping is the most practical means of tungro control, its large-scale implementation has socioeconomic constraints. It requires a great deal of labor in planting seasons, extensive irrigation network, community-wide basis cooperation and leadership of local government or authorities. In addition, it is adversely possible that natural enemy populations are more drastically reduced than vectors during a fallow period. Therefore, vector populations and other insects especially planthoppers may increase rapidly in the crop season after a fallow period.

In the past, roguing of affected plants was recommended in farmer's brochures to reduce disease sources, but experimental data indicated that roguing was not effective under high disease pressure (Tiongco *et al.* 1998).

Chemical control by spraying insecticides is considered by many researchers not to be enough to control tungro when vector populations are high. Root zone application and soil incorporation of insecticides were effective in reducing tungro (Satapathy, 1998). In Japan, a transplanting machine achieves root zone application of residual pesticides. With the spread of this practice, rice virus diseases in Japan decreased progressively. It is expected that the pesticide application by transplanting machines would also be effective to control tungro in tropical area.

Breeding for resistance to tungro disease using resistance to the vectors has been attempted. However, resistance to leafhoppers breaks down after 4 –5 years, owing to change in the vector's virulence or adaptation (Dahal *et al.* 1990). For example, IR36, IR42 and IR64 showed resistance to tungro when released, but became susceptible after wide distribution of these varieties. Some varieties like IR62 and IR74 are still resistant in some areas. After the breakdown of vector resistance was recognized, researchers focused on breeding virus-resistant varieties. After long-term screening of more than 40,000 rice accessions for virus-resistance, several promising sources were selected as donors. Among them are Utri Merah, Balimau Putih and Utri Rajapan, ARC11554 and *O. rufipogon* (Hibino *et al.* 1990; Kobayashi *et al.* 1993; Koganezawa and Cabunagan, 1997). The first three, which originated in Indonesia, always showed tolerance. RTBV multiplication in these plants is low (Takahashi *et al.* 1993; Santa Cruz *et al.* 1999). In addition, Utri Merah and Utri Rajapan have resistance to RTSV infection. ARC 11554 had a low infection rate of the Philippine isolate of RTBV and RTSV due to unusual vector resistance and resistance to RTSV (Sebastian *et al.* 1996). A wild rice, *O. rufipogon*, showed resistance to infection to both RTBV and RTSV (Kobayashi *et al.* 1993). Efforts to incorporate resistance gene(s) in these varieties to improved varieties have being made at IRRI and several national research institutions (Cabunagan *et al.* 1999; Sebastian *et al.* 1999). In 1998, two varieties Tukud Petanu and Tukud Unda were released in Indonesia (Azzam and Chancellor, 2002). Tukud Petanu was derived from Utri Merah that is resistant to RTSV and tolerant to RTBV. Tukud Unda has tolerance to RTBV derived from Balimau Putih (Cabunagan *et al.* 1999). In the Philippines, eight 'Matatag' lines were released with resistance derived from Utri Merah and ARC11554 (Ravelo, 2000). These new advanced varieties or lines showed high level of resistance to tungro in farmers' fields. In the case of Matatag lines researchers consider them stopgap variety that can be used in tungro-endemic area, because of their low yield and susceptibility to other diseases and pests (E. M. Tiongco, personal communication). However, farmers tend to grow rice with better eating quality but without tungro resistance even in tungro endemic area.

Transgenic plants expressing the RTSV replicase or coat protein genes showed high level of resistance against RTSV infection (Huet *et al.* 1999; Sivamani *et al.* 1999). In contrast, transgenic plants transformed with RTBV genes did not show any protection against RTBV (Azzam *et al.* 1999).

11.2.4 Outlook and conclusions

Although introduction of modern high-yielding rice varieties contributed to increased rice production in developing countries, they are vulnerable to rice tungro viruses, and substantial economic losses caused by tungro are still seen in many areas of South and Southeast Asia. While cultrual control practices are effective in mitigating losses due to tungro, additional varieties that are acceptable to the farmer and with tungro resistance are needed.

11.3 *Rice yellow mottle virus* Genus *Sobemovirus* (RYMV)

Rice yellow mottle virus (RYMV) is present only in Africa (Abo *et al.* al. 1998). RYMV was first reported in Kenya in 1966 (Bakker, 1974) and in West Africa in 1976 (Fauquet and Thouvenel, 1977). Since the early 1990s, RYMV is present where rice is grown throughout the sub-Saharan Africa and in Madagascar. RYMV causes severe yield losses to rice production ranging from 25 to 100% and is considered to be the most important disease of rice in Africa (Abo *et al.* 1998). The natural host range of RYMV is restricted to a few members of the Graminae family, i.e. *Eragrostidae* and *Oryza* sp. The most commonly species are *O. sativa, O. longistaminata, O. barthii, Echinochloa colona, Ischaemum rugosum, Panicum repens.* A few additional *Poacea* have been infected experimentally (Bakker, 1974). Leaf mottling and yellowing are the typical symptoms of RYMV. However, a variety of symptoms are observed on some cultivars or under specific growing conditions including streaks and whitening on leaves and in severe cases necrosis of the plants. Infection results in stunting of the plant, reduced tillering, poor panicle exertion and sterility. Death of the plants of susceptible cultivars occurred after early infection (Bakker, 1974).

For purification RYMV particles are easily extracted from infected tissues and yields can reach 0.2 to 0.8 mg of virus per g of fresh leaves (Fauquet and Thouvenel, 1977).

The principal vectors of RYMV are chrysomelidae, although other biting insects including the grasshopper *Conocephalus* occasionally transmit the virus in a non-presistent manner (Bakker, 1974). Seed transmission does not occur (Konaté *et al.* 2001). Transmission through plant residues, irrigation water, contact between plants, and by infected agricultural tools is suspected. Mechanical transmission is easily achieved (Abo *et al.* 2000).

11.3.1 Viral characteristics

The structure of RYMV was determined to 2.8 Å resolution by X-ray crystallography (Qu *et al.* 2000). The capsid contains 180 copies of the coat protein subunit that consists of 239 amino acid residues each arranged with $T = 3$ icosahedral symmetry. Each subunit adopts a jellyroll sandwich fold with an internal domain R that is disordered in the structure and the S domain that forms the viral shell. The RYMV particle is stabilized by 180 calcium-binding sites and by a phenomenon called "3D domain swapping" which introduces long-range interactions in the icosahedral surface lattices and enhances the stability of virions.

The genome of the RYMV is a single stranded, linear, positive-sense RNA non-polyadenylated with a mol. wt of 1.47 10^6 (Yassi *et al.* 1994). Previously, RYMV genomic organization was classified as an SCPMV-like virus with 4 ORFs and a polyprotein encoded by a large continuous ORF2 (accession number L20893). However, sequencing data from Nigerian (accession number U23142) and other strains (Fargette *et al.* 2002c) showed that the RYMV genomic organization is similar to that of *Cocksfoot mottle virus* (CfMV). There are two overlapping ORFs (ORF2a and ORF2b), and the polyprotein is expressed through a –1 ribosomal frame shift mechanism. The genomic length is 4452 nt. The coding sequences are as follows: ORF1 encodes a protein (P1) with molecular masses of 18 to 19 kDa observed in translation *in vitro* (Bonneau *et al.* 1998). This protein is required for virus spread. When expressed in transgenic plants, P1 strongly enhances infection (Voinnet *et al.* 1999) and is an inhibitor of gene silencing. There are at least three functional domains in the polyproteins encoded by ORF2a and b. ORF2a encodes a protein with a serine protease-like domain and a VPg, and ORF2b encodes a protein with a RNA-dependent RNA polymerase (RdRp)-like domain. The ORF3 encodes a capsid protein of 26 kDa via sub-genomic RNA. A subgenomic RNA of 1010 nt was identified from RYMV infected plants but not from particles and was determined to encode for CP after hybridization with an ORF3 sequence probe.

A low molecular weight RNA (220 nt) was isolated from the virion of some RYMV isolates (Seghal *et al.* 1993). This associated small RNA shows a strong structural homology with satellite RNA of the *Canadian lucerne transient streak* sobemovirus (Collins *et al.* 1998; Seghal *et al.* 1993).

Several strains have been distinguished on the basis of their immunological properties and of their sequences (Konaté *et al.* 1997; N'Guessan *et al.* 2000, 2001; Traoré *et al.* 2001). They were related to the geographic origins of the isolates, strains from East-Africa differed from West Africa ones and in West Africa, strains from savannah regions differed from forest regions (Pinel *et al.* 2000). East-African strains were more variable than West/Central African strains and phylogenetic studies suggested an earlier diversification in East Africa and a latter radiation in West/Central Africa (Abubakar *et al.* 2003).

11.3.2 Diagnosis and epidemiology

Initial diagnois is based on symptoms of mottle and yellowing. However, symptoms are often variable and can be confounded with mineral deficiencies. Serological detection of RYMV is reliable (Mansour and Bailis, 1994; Konaté *et al.* 1997), as the specific antisera do not cross-react with other known virus. Monoclonal antibodies allowed the distinction of different serotypes (Konaté *et al.* 1997; N'Guessan *et al.* 2000). The coat protein gene is a good marker of diversity, and variations in it are used to distinguish strains (Pinel *et al.* 2000). There is a good correspondence between the serological and the molecular typing (Fargette *et al.* 2002b).

RYMV is an emergent disease. It is thought that the virus originated in wild Graminae and only recently infected cultivated rice. The rapid and intense spread of the virus was associated with changes of agricultural practices. High performing but susceptible *O. sativa indica* cultivars were introduced from Asia. The associated use of irrigation allows two crop cycles a year, resulting in maintenance of inoculum during the entire the year. A model of the spread of the virus was recently proposed (Konaté and

Fargette, 2001b). (1) Volunteers, of wild Graminae or plant residues acts as the primary inoculums; (2) Direct contamination of the new rice fields occur after mechanical transmission by man or infection by insects; (3) Rice seed-beds can also be infected by insect vectors; (4) Additional spread occurs during transplanting from infected seed-beds, (5) Further spread in the field by man, animals, wind, irrigation water, and vectors.

There are few epidemiology studies and many uncertainties about the importance of biotic and abiotic factors, in the transmission of RYMV. At our current level of knowledge, forecasting for RYMV is not possible.

11.3.3 Control

Phytosanitary measures are sometimes advised even though their effect in reducing virus spread and economic impact are unknown. They include protection of seedbeds by nets, disinfections of the tools used at replanting, destruction of volunteers and rice residues (Konaté and Fargette, 2002a). Chemical control of the vectors is not economically feasible, and the large number of vector species indicates that chemical control is unlikely to be successful. Biological control of the vectors has not been studied.

Several breeding programs to obtain resistant cultivars are in progress. The genotype and phenotype of two kinds of natural resistance were characterized. Partial resistance, where there is a delayed virus multiplication and symptom expression, was found in cultivars of *O. sativa japonica* (Ioannidou *et al.* 2000). Partial resistance is polygenic and a major quantitative trait locus has been identified on chromosome 12 acting in epitasis with other QTLs (Ghesquiere *et al.* 1997; Albar *et al.* 1998; Pressoir *et al.* 1998; Ahmadi *et al.* 2001). A high level of resistance was identified in a limited number of *O. glaberrima* and *O. sativa indica* species. In these lines, no symptoms are apparent, and the virus was not detectable (Ndjiondjop *et al.* 2001). This resistance is monogenic and recessive (Ndjiondjop *et al.* 1999).

Studies to develop transgenic plants with resistance to RYMV are in progress. Transgenic plants expressing a portion of the protein of the RYMV ORF2 were produced (Pinto *et al.* 1999). The level of the transgenic resistance in the transgenic plants is similar to that of partial natural resistance. Like natural resistance, it can be overcome by some isolates (Fargette *et al.* 2002a).

11.3.5 Outlook and conclusions

There has been a great increase in the knowledge of the genomic organisation, particle structure, strain diversity of RYMV, however, there is still a lack of information on the ecology of the disease. Knowledge of the transmission means and of the ecology of the disease will help to forecast the epidemics, that will allow appropriate phytosanitary measures and deployment resistances to be more efficiently. Both natural and transgenic sources of resistance to RYMV have been identified and characterised. These resistances are being introgressed into popular cultivars. However, there is now evidence that these resistances can be overcome by isolates of RYMV. Understanding resistance sources and pathogen variability are needed to pyramiding different sources of resistances in order to develop cultivars with durable resistance.

11.4 *Rice hoja blanca virus* (RHBV)

The first report of rice hoja blanca was in Colombia in 1935 . In the period of 1954 to 1957, there were additional epidemics of the disease in Cuba, Colombia, Costa Rica, Panama and Venezuela (Garces-Orejuela *et al.* 1958). The disease is prevalent in the tropical countries of South America and is present in Peru, Ecuador, Colombia, Venezuela, Guyana, Suriname and French Guiana (Morales and Niessen, 1985). There have been documented RHBV epidemics since 1935. Epidemics of RHBV are cyclic, and when outbreaks of RHBV occur, yield losses at the country level have been estimated to be between 25% and 50%. There were major epidemics in the early 1980's, and Colombia suffered large losses during in 1981-82 (Vargas, 1985). During the last decade, Peru, Costa Rica, Colombia, Dominican Republic and Venezuela experienced outbreaks of RHBV. Through the 90s, severe outbreaks of RHBV were reported in the mid-altitude rice grown in Northern Peru. Approximately 32,000 hectares of rice is grown in Huallaga region, and the loss was estimated to be 50% of the total crop.

Besides direct damage, RHBV infection predisposes rice to *Helminthosporium oryzae* (Lamey and Everett, 1967), and discoloration of the grain. Often farmers spray pesticides at the first signs of RHBV. This leads to pesticide abuse and destroys the natural enemies of the insect pests of rice.

The host range of RHBV is restricted to the Graminae. The most important host of hoja blanca virus is rice, *O. sativa*. It can also infect red rice and other cultivated species of rice, *O. glaberrima. Avena sativa, Digitaria horizontalis*, *Hordeum vulgare*, *Leptochloa filiformis*, and *Secale cereale* are hosts for the virus (Gálvez *et al.* 1961; Galvez, 1968b).

The symptoms in rice are chlorotic streaks that can coalesce and cause the leaves to turn white. When young plants become infected, they are stunted and in severe cases the leaves turn necrotic and the plants die.

Infections that occur before the emergence of the panicle can reduce seed set and quality.

The planthopper *Tagosodes orizicolus* is both a host and vector of RHBV (Galvez *et al.,* 1961). The virus is acquired through feeding on an infected plant, and there is a period of 17-22 days of viral propagation before the planthopper become viruliferous (Webber *et al.* 1971). There is a high rate of transovarian transmission to the progeny, and these nymphs can transmit the virus shortly after they emerge (Galvez et al, 1968). It was reported that RHBV causes a disease in the planthopper (Jennings and Pineda, 1971). Viruliferous vectors lay one-third as many eggs and hatch fewer nymphs than non-viruliferous insects. In populations of *T. orizicolus*, there is resistance to RHBV (Ziegler & Morales, 1990). The filamentous viral particles can easily be purified from either the roots or young leaves that show typical symptoms of hoja blanca disease. The extraction is with organic solvents, low speed centrifugation and polyethylene precipitation. This is followed by resuspension and two cycles of cesium chloride centrifugation (Morales and Niessen, 1983).

11.4.1 Virus characteristics

Rice hoja blanca virus is in the genus *tenuivirus*, which are still not classified in a family (Van Regenmortel *et al.* 2000). The *tenuiviruses* are related to the *Bunyaviridae* with the major difference being the particle structure. The virions consist of filamentous particles 3-8 nm wide that contain 12 % ribonucleic acid and 88% of a nucleocapsid protein of 34 kDa (Morales and Niessen, 1983; Espinosa *et al.* 1992). There is a hypothesis that these are defective forms of enveloped virions, and there exist particles similar to other *Bunyaviridae*. Despite numerous studies, no enveloped virions have been observed for any *tenuivirus* including in situ studies for RHBV (Espinosa *et al.* 1992). The genome consists of single stranded linear RNA and the total genome is estimated to be 17.6 kb (Ramirez *et al.* 1992; Calvert *et al.* 1994). The genome consists of four species. The RNA 1 is 9.8 kb and encodes for a RNA dependent polymerase. RNA 2 is 3.5 kb and encodes in an ambisense manner two proteins, which have homology with the glycosylated coat proteins of tomato spotted wilt virus (Miranda *et al.* 1996a). The RNA 3 consists of 2.3 kb and encodes for two proteins in an ambisense manner. The nucleoprotein gene (*N*) is on the virus complementary strand of RNA 3. The N protein is predicted to consist of 319 amino acids with a molecular weight of 35 kDa (Miranda *et al.* 1994). The RNA 4 consists of 1.9 kb and encodes for two proteins in an ambisense manner (Ramirez *et al.* 1993). The protein encoded by the virus strand of RNA 4 accumulates in the plant cells but cannot be detected in the insect host. It is 23 kDa and is referred to as the major non-structural protein. There are no known strains of the virus either on the basis of symptoms or using serological methods. Molecular characterization of RHBV (Miranda *et al.* 1997) showed that there are only minor differences between the RNA 3 and 4 of the RHBV strains in Colombia and Costa Rica. A virus that infects the Graminaceous weed *Echinochloa* was though to be a strain of RHBV. These viruses have different insect hosts (vectors), and the molecular characterization have been used to classify this virus as *echinochloa hoja blanca virus* (Miranda *et al.* 1996b).

11.4.2 Diagnosis, epidemiology and control

In the field, infected young plants can be diagnosed by the chlorotic streaks on the leaves. The disease was named because the infected leaves turn white, as they grow older. Even looking from a distance, one can observe white leaves that contrast to the normal green leaves. The purified virus is a good immunogen and ELISA based tests for the detection of RHBV have been developed. These are effective in detecting virus in both the plant and the insect host (vector).

RHBV epidemics are cyclic and the peaks of epidemics are usually separated by several years. During the peak of RHBV epidemics, yield losses have been estimated to be between 25% and 50% (Jennings, 1963; Vargas, 1985). It was estimated that during epidemics that between 5 and 25% of the planthoppers are able to transmit the virus (Galvez, 1968b, Reyes *et al.*1997). The low fecundity of the planthoppers infected with RHBV and genetic resistance are the probable factors responsible for the cyclic nature of the RHBV epidemics (Ziegler and Morales, 1990). Apparently, in the absence of selective pressure for resistance, the percentage of the genetic capability of the insect to transmit the virus tends to increase. Epidemics of RHBV occur when there are sufficient number of infected plants and vectors.

Cultural control methods are not sufficient to control the virus or the vector. Nevertheless, ratoons of rice should be destroyed as they encourage the build up of high populations of *T. orizicolus* and other pests and diseases.

Chemical control can be used to reduce populations of *T. orizicolus*. To be effective, these must be applied early in the growing cycle. Since there are many natural enemies of *T. orizicolus*, indiscriminate use of pesticides can led to an increase in the populations of *T. orizicolus* and to epidemics. If pesticides are to be used, selective pesticides that have the minimal effect on the beneficial insects are recommended.

Resistance breeding. While resistant materials have been identified (Lamey, 1969; Zeigler *et al.* 1994), screening for resistance to RHBV depends on maintaining highly viruliferous colonies of *T. orizicolus*. A field method to screen for resistance to RHBV was developed and thousands of lines are tested each year at CIAT (Zeigler *et al.* 1988). This method is effective for eliminating the most susceptible lines, but is not sufficient to select highly resistant varieties. Additional screening using different disease pressures and age of infestation have led to the development of varieties that are more resistant than the parents used in the crosses. Varieties and advance lines are classified as susceptible, intermediate resistant and highly resistant. No rice varieties in Latin America are immune to the virus, and resistance can be overcome with high disease pressure. Nevertheless even varieties that are classified as intermediate confer adequate field resistance.

Using the N-gene from RHBV, transgenic rice was developed that confers a level of resistance similar to the highly resistant varieties (Lentini *et al.* 2003). The mechanism of resistance appears to be gene silencing. The progeny of crosses of these materials with intermediate resistant varieties are superior in resistance than the parents and the transgenic plants are a potential source of additional resistance to RHBV.

11.4.3 Outlook and conclusions

In the mid to late 1990s, outbreaks of RHBV became more frequent and it appeared that widespread epidemics would occur. There were some rice growing regions that were affected by local epidemics, but at the regional level, the losses were much less than in the previous cycle of epidemics. This was partially attributed to less but more judicious use of pesticides including ones those that were more selective. The use of intermediate resistant varieties has been highly successful and eliminates the need for chemical control. In the last few years, varieties with even a better level resistance were developed.

11.5 Other tenuiviruses

Rice stripe virus (RSV) and *rice grassy stunt virus* (RGSV) are in the tenuivirus group and are related to RHBV. The biology is similar to RHBV, and they also cause periodic epidemics (Gingery, 1988). RSV is found in Japan, Korea, China, and Russia. The principal vector is *Laodelphax striatellus*. There are several alternative hosts including wheat, oat, barley and millet. Control can be achieved with the judicious use of pesticides, but resistant varieties should be planted in areas where this disease is a problem. Like RHBV, RGSV is a disease in tropical zones, and it is found in many areas of tropical Asia. There have been significant epidemics of RGSV in India, the Philippines, Thailand and Indonesia. Many modern varieties have a source of resistance to RGSV, which was derived from the wild species *O. nivara*.

11.6 Reoviridae viruses

Rice dwarf virus (RDV), rice gall dwarf virus (RGDV), rice ragged stunt virus (RRSV), and rice black streak virus (RBSDV) are members of the reoviridae and have similar physiological properties. The typical symptoms of these viruses include stunting and some form of gall or tumor. Also, the tillering is reduced and panicle develops poorly. The virions are 60-70 nm in diameter and consist of an inner and outer shell. The genomes contain 10-12 species of double stranded RNA. The viruses tend to be very labile and difficult to purify. The viruses are phloem limited and the concentrations of virus in the plants tend to be low. A range of planthoppers and leafhoppers transmits these viruses. There are many Graminaceous weeds that are host to these viruses. The transmission is in a propagative manner, but the epidemiology of these viruses is diverse. For example, RBSDV causes most damage on rice that is planted at the beginning of the season. High populations of the vectors can move from the overwintering hosts, and the following generations tend to have smaller populations. This contrasts with RDV in China, where the highest incidence of disease is in late-planted rice. Cultural practices that remove alternative hosts and rice residuals tend to reduce the incidence of these diseases. In the tropical areas insecticides tend not to be effective in controlling the vector populations, but in the temperate zone judicious use of pesticides can reduce disease incidence. Resistant varieties are available for several of the diseases including RBSDV, RRSV, and RDV. When the resistance is to the virus, it tends to be stable. While resistance to the vector confers some resistance, it tends to breakdown after several years.

11.6.1 Rice stripe necrosis virus (RSNV)

Rice stripe necrosis virus is caused by a virus transmitted by the soil-borne fungus *Polymyxa graminis*. It is a species of the *benyvirus* genus with rod-shaped particles of ca. 20 nm in diameter, with bimodal length 260 and 360 nm (Fauquet et al. 1983; Morales et al. 1999). The disease is characterised by seedling death, leaf stripes and necrosis, and plant malformation. Cystosori, characteristic of plasmodiophori fundal vectors of plant viruses, were consistently observed in the roots of diseased rice plants. PCR and sequence analyses of amplified fungal DNA products from infected roots revealed that the putative vector was *P. graminis* (Morales *et al.* 1999).

The disease was reported in 1977 in Côte d'Ivoire (Louvel and Bidaux, 1977) and latter in several other countries of West-Africa (Fauquet et al. 1988). The virus was isolated in 1983 (Fauquet and Thouvenel, 1983). At sites in Sierra-Leone and Côte d'Ivoire, where the disease has become a local problem, rice production was mechanised and relatively intensive. With the intensification of rice-based farming systems in West-Africa, the disease threatens to become increasingly prevalent (Johnson *et al.* 1998). In 1991, the disease appeared in Colombia (Morales *et al.* 1999). The disease spread rapidly and yield losses associated with the presence of RSNV averaged 20% in the severely infected fields. Even though the disease is present within a field, it is sporadic and difficult to predict the severity in any given season. Host resistance to RSNV exists within *O. glaberrima* and is being exploited through interspecific hybridization with *O. sativa* to produce resistant cultivars.

11.7 Conclusion

Viruses cause several important diseases of rice. The epidemics tend to be periodic but very severe. Understanding disease complexes like rice tungro has been difficult but has led to a better understanding of the disease. This is permitting a range of strategies to be developed in order to control the disease.

Cultural practices and chemical control can reduce the economic losses that these viruses cause. Abuse of pesticides can lead to increased disease incidence as well as to new pest problems. A striking example occurred in Peru when chemical control for *T. orizicolus*, the vector of RHBV, led to an attack of whiteflies on rice. An integrated pest management approach needs to be used to control viral diseases.

Resistant varieties are the most cost effective means to control the viral diseases of rice. Unfortunately, most of these viruses are not mechanically transmitted. The lack of consistent disease pressure and the inability to mechanically inoculate plants has made breeding resistant rice varieties a challenge. Many screening programs rely on the use of colonies of vectors to assure high disease pressure in order to develop resistant varieties. Several breeding programs are successfully developing varieties with resistance to either the vectors and/or the viruses, but this type of screening is labor intensive and costly. There is a need to develop molecular markers for the resistance genes to the major viral diseases. Marker aided selection is being used successfully to breed for RMYV and the development of molecular markers for RHBV is underway.

References

AbAbo, M., Sy, A. and Alegbejo, M.1998. Rice yellow mottle virus (RYMV) in Africa: evolution, distribution, economic significance and sustainable rice production andmanagement strategies. *J. Sustainable Agric.* 11: 85-111.

Abo, M., Alegbejo, M., Sy, A. and Misari, S. 2000. An overview of the mode of transmission, host plants and methods of detection of rice yellow mottle *virus. J. Sustainable Agric.* 17: 19-36.

Abubakar, Z., Ali, F., Pinel, A., Traoré, O., N'Guessan, P., Notteghem, JL., Kimmins, F., Konaté, G. and Fargette, D. 2003. Phylogeography of Rice yellow mottle virus in Africa. *J. Gen. Virol.*, in press.

Ahmadi, N., Albar, L., Pressoir, G., Pinel, A., Fargette, D. and Ghesquière, A. 2001. Genetic basis and mapping of the resistance to Rice yellow mottle virus. III. Analysis of the QTLs efficiency in introgressed progenies confirmed the hypothesis of complementary epistasis between two resistance QTLs. *Theor. Appl. Gen.* 103: 1084-1092.

Albar, L., Lorieux, M., Ahmadi, N., Rimbault I., Pinel A., Sy A., Fargette, D. and Ghesquière, A. 1998. Genetic basis and mapping of the resistance to rice yellow mottle virus. I. QTLs identification and relationship between resistance and plant morphology. *Theor. Appl. Gene.* 97: 1145-1154.

Azzam, O. and Chancellor, T.C.B. 2002. The biology, epidemiology, and management of rice tungro disease in Asia. *Plant Dis.* 86: 88-100.

Azzam, O., Klöti, A., Sta. Cruz, F., Füttere, J., Coloquio, E.L., Potrykus, I. and Hull, R. 1999b. Genetic engineering of rice for tungro resistance. In: *Rice Tungro Management.* T.C.B Chancellor, O. Azzam and K.L Heong, (eds.) Intern. Rice Res. Inst., Los Baños, Philippines. pp. 39-44.

Bajet, N.B., Daquioag, R.D. and Hibino, H. 1985. Enzyme-linked immunosorbent assay to diagnose rice tungro. *J. Plant Prot. Trop.* 2: 125-129.

Bakker, W. 1974. Characterisation and ecological aspects of rice yellow mottle virus in Kenya. *Agric. Res. Rept.* Wageningen 829: 1-152.

Bonneau, C., Brugidou, C., Chen, L., Beachy, R. and Fauquet, C. 1998. Expression of the Rice yellow mottle virus P1 protein in vitro and in vivo and its involvement in virus spread. *Virology* 244: 79-86.

Bottenberg, H., Litsinger, J.A., Loevinsohn, M.E. and Kenmore, P. 1991. Impact of cropping intensity and asynchrony on the epidemiology of rice tungro virus in Malaysia. *J. Plant Prot. Trop.* 7: 103-116.

Cabauatan, P.Q. and Hibino, H. 1988. Isolation, purification, and serology of rice tungro bacilliform and rice tungro spherical viruses. *Plant Dis.* 72: 526-528.

Cabauatan, P.Q. and Koganezawa, H. (1997). Alternative diagnostic methods for detection of rice tungro viruses. In: *Epidemiology and Management of Rice Tungro Disease.* T.C.B. Chancellor and J.M. Thresh, (eds.) Natural Resources Institute, U.K. pp. 1-4.

Cabauatan, P.Q., Cabunagan, R.C. and Koganezawa, H. 1995. Biological variants of rice tungro viruses in the Philippines. *Phytopathology* 85: 77-81.

Cabauatan, P.Q., Kobayashi, N., Ikeda, R. and Koganezawa, H. (1993). Oryza glaberrima: an indicator plant for rice tungro spherical virus. *Intl. J. Pest Manage.* 39:273-276.

Cabauatan, P.Q., Melcher, U., Ishikawa, K., Omura, T., Hibino, H., Koganezawa, H. and Azzam, O. 1999. Sequence changes in six variants of rice tungro bacilliform virus and their phylogenetic relationships. *J. Gen. Virol.* 80: 2229-2237.

Cabunagan, R.C., Angeles, E.R., Villareal, S., Azzam, O., Teng, P.S., Khush, G.S., Chancellor, T.C.B., Tiongco, E.R., Truong, X.H., Mancao, S., Astika, I.G.N., Muis, A.,

Chowdhury, A.K., Narasimhan, V., Ganapathy, T. and Subramanian, N. 1999. Multilocational evaluation of advanced breeding lines for resistance to rice tungro viruses. In: *Rice Tungro Disease Management.* T.C.B. Chancellor, O. Azzam and K.L. Heong, (eds.) International Rice Research Institute, Los Baños, Philippines. pp. 45-57.

Cabunagan, R.C., Castilla, N., Coloquio, E.L., Tiongco, E.R., Truong, X.H., Fernandez, J., Du, M.J., Zaragosa, B., Hozak, R.R., Savary, S. and Azzam, O. 2001. Synchrony of planting and proportions of susceptible varieties affect rice tungro disease epidemics in the Philippines. *Crop Protect.* 20: 499-510.

Calvert, L.A., Lozano, I. and Morales, F.J. 1994. Caracterización molecular del virus de la hoja blanca del arroz (RHBV). *Fitopat. Colombiana* 16: 139-147.

Chancellor, T.C.B., Cook, A.G. and Heong, K.L. 1996. The within-field dynamics of rice tungro disease in relation to the abundance of its major leafhopper vectors. *Crop Protect.* 15: 439-449.

Collins, R., Gellatly, D., Seghal, O. and Abouhaidar, G. 1998. Self-cleaving circular RNA associated with Rice yellow mottle virus is the smallets viroid-like RNA. *Virology* 241 : 269-275.

Cooter, R.J., Winder, D. and Chancellor, T.C.B. 2000. Tethered flight activity of Nephotettix virescens (Hemiptera: Cicadellidae) in the Philippines. *Bull. Entomol. Res.* 90: 49-55.

Dahal, G., Dasgupta, I., Lee, G. and Hull, R. 1992. Comparative transmission of, and varietal reaction to, three isolates of rice tungro virus disease. *Ann. appl. Biol.* 120: 287-300.

Dahal, G., Druka, A., Burns, T.M., Villegas, L.C., Fan, Z., Shrestha, R.B. and Hull, R. 1996. Some biological and genomic properties of rice tungro bacilliform badnavirus and rice tungro spherical waikavirus from Nepal. *Ann. appl. Biol.* 129: 267-287.

Dahal, G., Hibino, H., Cabunagan, R.C., Tiongco, E.R., Flores, Z.M. and Aguiero, V.M. 1990. Changes in cultivar reaction to tungro due to changes in "virulence" of the leafhopper vector. *Phytopathology* 80: 659-665.

Davis, R I., Parry, J.N., Geering, A.D.W., Thomas, J.E., and Rahamma, S. 2000. Confirmation of the presence of Rice tungro bacilliform virus in Papua (formerly Irian Jaya), Indonesia. *Austral. Plant Pathol.* 29: 223.

Druka, A. and Hull, R. 1998.Variation of rice tungro viruses: further evidence of two rice tungro bacilliform virus strains and possibly several rice tungro spherical virus variants. *J. Phytopathol.* 146: 175-178.

Espinoza, A.M., Hernandez, M., Pereira, R., Falk, B. and Medina, V. 1992. In situ immunogold labeling analysis of the rice hoja blanca virus nucleoprotein and major noncapsid protein. *Virology* 191: 619-627.

Espinoza, A.M., Pereira, R., Macaya-Lizano, A.V., Hernandez, M., Goulden, M. and Rivera, C. 1993. Comparative light and electron microscopic analyses of tenuivirus major noncapsid protein (NCP) inclusion bodies in infected plants, and of the NCP in vitro. *Virology* 195: 156-166

Fan Z.-F., Dahal, G., Dagupta, I., Hay, J. and Hull, R. 1996. Variation in the genome of rice tungro bacilliform virus: molecular characterization of six isolates. *J. Gen. Virol.* 77: 847-854.

Fargette, D., Pinel, A., Halimi, H., Brugidou, C., Fauquet, C. and Van Regenmortel, M. 2002b. Comparison of molecular and immunological typing of isolates of Rice yellow mottle virus. *Arch. Virology* 147: 583-596.

Fargette, D., Pinel, A., N'Guessan, P., Traoré, O., Abubakar, Z., Brugidou, C., Hebrard, E., and Konaté, G. 2002c. Rice yellow mottle virus and its RNA satellite: genomic organisation, diversity and evolution. *XIIth Inter. Cong. Virology* 27/7-1-8, Paris, Abstract addendum, p 10.

Fargette, D., Pinel, A., Traoré, O., Ghesquière, A. and Konaté, G. 2002a. Emergence of resistance-breaking isolates of Rice yellow mottle virus during serial inoculations. *Europ. J. Plant Pathol.* 108: 585-591.

Fauquet C, Thouvenel J-C, Fargette D, Fishpool L. 1988. Rice stripe necrosis virus : a soil borne rod-shaped virus. In : *Developments in Applied Biology 2, Viruses with Fungal Vectors.* Eds. J. I. Cooper & M. J.C. Asher. Assoc. Appl. Biol. Wellesbourne, pp 71-82.

Fauquet C, Thouvenel J-C. 1983. Association d'un nouveau virus en bâtonner avec la maladie de la nécrose à rayures du riz en Côte d'Ivoire. *Compt.-Rendus l'Acad. Sci.* Paris 296, 575-580.

Fauquet, C and Thouvenel, JC. 1977. Isolation of the rice yellow mottle virus in Ivory Coast. *Plant Dis Rept* 61: 443-446.

Galvez, G.E. 1968. Purification and characterization of rice tungro virus by analytical density-gradient centrifugation. *Virology* 35: 418-426.

Galvez, G.E. 1968. Transmission studies of the hoja blanca virus with highly active, virus-free colonies of Sogatodes oryzicola. *Phytopathology* 58: 818-821.

Galvez, G.E., Thurston, H.D. and Jennings, P.R. 1961. Host range and insect transmission of the hoja blanca disease of rice. *Plant Dis. Rept.* 45: 949-953.

Garces-Orejuela, C., Jennings, P.R., and Skiles, R.L. 1958. Hoja blanca of rice and the history of the disease in Colombia. *Plant Dis. Rept.* 42: 750-751.

Ghesquière, A., Albar, L., Lorieux, M., Ahmadi, N., Fargette, D., Huang, N., McCouch, S. and Notteghem, JL. 1997. A major quantitative trait locus for rice yellow mottle virus resistance maps to a cluster of blast resistance genes on chromosome 12. *Phytopathology* 12: 1243-1249.

Gingery, R. E. 1988. The rice strip virus group. Pages 297-329. in: *The Plant Viruses: The Filamentous Plant Viruses.* R. G. Milne. Plenum Press. New York.

Harper, G., Hull, R., Lockhart, B. and Olszewski, N. 2002. Viral sequences integrated into plant genomes. *Annu. Rev. Phytopathol.* 40: 119-136.

Hasanuddin, A. and Hibino, H. 1989. Grain yield reduction, growth retardation, and virus concentration in rice plants infected with tungro-associated viruses. *Trop. Agric. Res. Ser.* 22: 56-73.

Hasanuddin, A., Widiarta, I.N. and Yulianto 1999. Improving IPM technology for rice tungro disease in Indonesia. In: *Rice Tungro Disease Management.* T.C.B. Chancellor, O. Azzam and K.L. Heong, (eds.) International Rice Research Institute, Los Baños, Philippines. pp. 129-137.

Herdt, R. W. 1991. Research priorities for rice biotechnology. In: *Rice Biotechnology.* G.H. Toenniessen and G.S. Khush, (eds.) CAB International, Wallingford, UK. pp.19-54.

Hibino, H. 1983. Transmission of two rice tungro-associated viruses and rice waika virus from doubly or singly infected source plants by leafhopper vectors. *Plant Dis.* 67: 774-777.

Hibino, H., Daquioag, R.D., Mesina, E.M. and Aguiero, V.M. 1990. Resistances in rice to tungro-associated viruses. *Plant Dis.* 74: 923-926.

Hibino, H., Saleh, N. and Roechan, M. 1979. Transmission of two kinds of rice tungro associated viruses by insect vectors. *Phytopathology* 69: 1266-1268.

Hibino, H., Usugi, T., Omura, T., Tsuchizaki, T., Shohara, K. and Iwasaki, M. 1985. Rice grassy stunt virus: a planthopper-borne circular filament. *Phytopathology.* 75: 894-899.

Hino, T., Wathanakul, L., Nabheerong, N., Surin, P., Chaimongkol, U., Distahporn, S., Putta, M., Kerdchocchai, D. and Surin, A. 1974. Studies on rice yellow orange leaf virus disease in Thailand. *Tech. Bull., Tropical Agriculture Research Center*, Japan 7:1-67.

Holt, J., Chancellor, T.C.B., Reynolds, D.R. and Tiongco, E.R. 1996. Risk assessment for rice planthopper and tungro disease outbreaks. *Crop Protect.* 15: 359-368.

Huet, H., Mahendra, S., Wang, J., Sivamani, E., Ong, C.A., Chen, L., Kochko, A. de., Beachy, R.N. and Fauquet, C. 1999. Near immunity to rice tungro spherical virus achieved in rice by a replicase-mediated resistance strategy. *Phytopathology* 11: 1022-1027.

Hull, R. 1996. Molecular biology of rice tungro viruses. *Annu. Rev. Phytopathol.* 34: 275-297.

Ioannidou, D., Lett, J., Pinel, A., Assigbetse, K., Brugidou, C., Ghesquière, A., Nicole, M. and Fargette, D. 2000. Responses of Oryza sativa japonica sub-species to infection with Rice yellow mottle virus. *Physiol. Molec. Plant Pathol.* 57: 177-188.

Isogai, M., Cabauatan, P.Q., Masuta, C., Uyeda, I. and Azzam, O. 2000. Complete nucleotide sequence of the rice tungro spherical virus genome of the highly virulent strain Vt6. *Virus Genes* 20: 79-85.

Jennings, P.R. 1963. Estimating yield loss in rice caused by hoja blanca. *Phytopathology* 53:492.

Jennings, P.R. and Pineda A. 1971. The effect of the hoja blanca virus on its insect vector. *Phytopathology* 61: 142-143.

Johnson D, Dingkuhn M, Fomba S. 1998. Incidence of rice stripe necrosis virus in upland rice in West-Africa. *Intern. Rice Res. Notes* 3, 25-26.

Kobayashi, N. and Ikeda, R. 1992. Necrosis caused by rice tungro viruses in Oryza glaberrima and O. barthii. *Japan. J. Breed.* 42: 885-890.

Kobayashi, N., Ikeda, R. and Vaughan, D.A. 1993. Resistance to infection of rice tungro viruses in wild species of rice (Oryza spp.). *Japan. J. Breed.* 14: 247-255.

Koganezawa, H. 1998. Present status of controlling rice tungro virus. In: *Plant Virus Disease Control*. A. Hadidi, R.K. Khetarpal and H. Koganezawa, (eds.) APS Press, USA. pp. 459-469.

Koganezawa, H. and Cabunagan, R.C. 1997. Resistance to rice tungro virus disease. In: *Epidemiology and Management of Rice Tungro Disease.* T.C.B. Chancellor and J.M. Thresh, (eds.) Natural Resources Institute, U.K. pp. 54-59.

Konaté, G. and Fargette, D. 2001. Rice yellow mottle virus : an overview. *Plant Virology in Sub-Saharan Africa.* 4-8 juin, Ibadan, Nigéria, p 2.

Konaté, G., Sarra, S. and Traoré, O. 2001. Rice yellow mottle virus is seed-borne but not seed transmitted in rice seeds. *Eur. J. Plant Pathol.* 107: 361-364.

Konaté, G., Traoré, O. and Coulibaly, M. 1997. Characterisation of rice yellow mottle virus isolates in Sudano-Sahelian areas. *Arch. Virol.* 142: 1117-1124.

Lamey, H.A. 1969. Varietal resistance to hoja blanca. In: *The virus diseases of the rice plant.* p.293-311. Johns Hopkins Press, Baltimore, Md.

Lentini Z., Lozano I, Tabares E., Fory L., Domínguez J., Cuervo M., Calvert L. 2003. Expression and inheritance of hypersensitive resistance to rice hoja blanca virus mediated by the viral nucleocapsid protein gene in transgenic rice. *Theor. Appl. Genetics* 106: 1018-1026.

Lin Q.Y., Xie L.H., Wu Z.J., Zhou Z.J. and Xue L.Y. 1994. Comparative study on two kinds of rice virus diseases between China and Philippines 1. Pathogen of rice tungro virus disease. *Sci. Agric. Sinica* 27: 1-6 (in Chinese).

Ling, K.C. 1972. *Rice Virus Diseases.* Inter. Rice Rese. Inst., Los Baños, Philippines.

Louie R. 1995. Vascular puncture of maize kernels for the mechanical transmission of maize white line mosaic virus and other viruses of maize. *Phytopathology* 85:139-143.

Louvel D, Bidaux J-M. 1977. Observations de nouveaux symptômes sur des variétés précoces de riz en Côte d'ivoire. *Agronomie Tropicale* 32 : 257-261.

Mansour, A. and Baillis, K. 1994. Serological relationships among rice yellow mottle virus isolates. *Ann. appl. Biol.* 125:133-140.

Marmey P., Bothner B., Jacquot E., de Kochko A., Ong C.A., Yot P., Siuzdak G., Beachy R.N. and Fauquet C.M. 1999. Rice tungro bacilliform virus open reading frame 3 encodes a single 37-kDa coat protein. *Virology* 253: 319-326.

Mayo, M.A., Murant, A.F., Reavy, B., Vetten, H.J. and Reddick, B.B. 2000. Family Sequiviridae. In: *Virus Taxonomy, Seventh Rep. Intern. Comm.Taxonomy of Viruses.* M.H.V van Regenmortel, C.M. Fauquet, D.H.L. Bishop et al., (eds.) Academic Press, SanDiego, pp. 685-689.

Miranda, G.I., Yambao, M.L.M., Cabauatan, P.Q., Venkitesh, S.R. and Koganezawa, H. 1995. Degradation of a coat protein of rice tungro spherical virus during purification. *Ann. Phytopathol. Soc. Jpn.* 61: 41-43.

Miranda, J. de, Ramirez, B.C., Muñoz, M., Lozano, I., Wu, R., Haenni, A.L., Espinoza A.M. and Calvert, L.A. 1997. Comparison of Colombian and Costa Rican strains of rice hoja blanca tenuivirus. *Virus Genes* 15: 191-193.

Miranda, J.R. de, Hernandez, M., Hull, R. and Espinoza, A.M. 1994. Sequence analysis of rice hoja blanca virus RNA 3. *J. Gen. Virol.* 75: 2127-2132.

Miranda, J.R. de, Munoz, M., Wu,R., Hull,R. and Espinoza,A.M., 1996a. Sequence of rice hoja blanca tenuivirus RNA-2 *Virus Genes* 12: 231-237.

Miranda,J.R. de, Munoz,M., Wu,R. and Espinoza,A.M. 1996b. Sequence of Echinochloa hoja blanca tenuivirus RNA-4. *Virus Genes* 13: 61-64.

Morales F, Ward E, Castano, Arroyave J, Lozano I, Adams M. 1999. Emergence and partial characterization of rice stripe necrosis virus and its fungus vector in South America. *Eur. J.Plant Pathol.* 105, 643-650.

Morales, F.J. and Niessen, A.I. 1983. Association of spiral filamentous viruslike particles with rice hoja blanca. *Phytopathology* 73: 971-974.

Morales, F.J. and Niessen, A.I. 1985. Rice hoja blanca virus. *AAB Desc. Plant Viruses*, No. 299.

N'Guessan, P., Pinel, A., Caruana, M., Frutos, R., Sy, A., Ghesquière, A. and Fargette, D. 2000. Evidence of the presence of two serotypes of rice yellow mottle sobemovirus in Côte d'Ivoire. *Eur. J. Plant Pathol.* 106: 167-178.

Nath, P.D., Kenyon, L., Bartolome, V.I., McLaren, G. and Azzam, O. 2000. Simple serological assays for detecting rice tungro viruses. *Food Agric. Immunol.* 12: 139-151.

Ndjiondjop, MN., Albar, L., Fargette, D., Fauquet, C. and Ghesquière, A. 1999. The genetic basis of high resistance to rice yellow mottle virus (RYMV) in cultivars of two cultivated rice species. *Plant Dis.* 83: 931-935.

Ndjiondjop, MN., Brugidou, C., Zang, S., Fargette, D., Ghesquière, A., Fauquet, C. 2001. High resistance to Rice yellow mottle virus (RYMV) in two cultivated rice cultivars is correlated to the failure of cell-to-cell movement. *Physiol. Mol. Plant Pathol.* 59: 309-316.

Ngon A Yassi, M., Ritzenthaler, C., Brugidou, C., Fauquet, C. and Beachy, R (1994) Nucleotide sequence and genome characterisation of rice yellow mottle virus RNA. *J. Gen. Virol.* 75: 249 -257.

N'Guessan, P., Pinel, A., Sy, A., Ghesquière, A. and Fargette, D. 2001. Distribution, pathogenicity and interactions of two strains of Rice yellow mottle virus in forested and savanna zones of West-Africa. *Plant Dis.* 85: 59-64.

Ou, S.H. 1965. Rice diseases of obscure nature in tropical Asia with special reference to "mentek" disease in Indonesia. *Intl. Rice Com. Newsletter* 14: 4-10.

Ou, S.H. 1985. *Rice Diseases,* 2nd ed. Commonwealth Mycological Institute, Kew, UK.

Pinel, A., N'Guessan, P., Bousalem, M. and Fargette, D. 2000. Molecular variability of geographically distinct isolates of Rice yellow mottle virus in Africa. *Arch. Virol.* 145: 1621-1638.

Pinto, Y., Kok, R. and Baulcombe, D. 1999. Resistance to rice yellow mottle virus (RYMV) in cultivated African rice varieties containing RYMV transgenes. *Nature Biot.* 17: 702-707.

Pressoir, G., Albar, L., Ahmadi, N., Rimbault, I., Lorieux, M., Fargette, D. and Ghesquière A. 1998. Genetic basis and mapping of the resistance to rice yellow mottle virus. II. Evidence of a complementary epistasis between two QTLs. *Theor. App. Gen.* 97: 1155-1161.

Qu, C., Lars, L., Opalka, N., Brugidou, C., Yeager, M., Beachy, R., Fauquet, C., Johnson, J. and Lin, T. 2000. 3D domain swapping modulates the stability of members of an icosahedral virus group. *Structure* 8: 1095-1103.

Ramirez, B.C., Lozano, I., Constantino L.M., Haenni, A.L. and Calvert, L.A. 1993. Complete nucleotide sequence and coding strategy of rice hoja blanca virus RNA4. *J. Gen. Virol.* 74: 2463-2468.

Ramirez, B.C., Macaya, G., Calvert, L.A. and Haenni, A.L. 1992. Rice hoja blanca virus genome characterization and expression in vitro. *J. Gen. Virol.* 73: 1457-1464.

Ravelo, G.B. 2000. Matatag lines: farmer's new allies against tungro. *PhilRice Newsl.* 13: 14-15.

Rekha J. and Dasgupta I. 2001. Cloning and molecular analysis of DNA of rice tungro bacilliform viruses from different rice growing regions of India. *Indian Phytopathol.* 54: 469-475.

Reyes, L., Yencho, C., Velasco, A., Calvert L. 1997. Análisis de poblaciones de Tagosodes orizicolus (Muir) vector del virus de la hoja blanca en zonas arroceras de Colombia. *Revista Colombiana Entom.* 23 : 165-169.

Riley, J.R., Reynolds, D.R. and Farrow, R.A. 1987. The migration of Nilaparvata lugens (Stål) (Delphacidae) and other Hemiptera associated with rice during the dry season in the Philippines: a study using radar, visual observations, aerial netting and ground trapping. *Bull. Entomol. Res.* 77: 145-169.

Rivera, C.T. and Ou, S.H. 1965. Leafhopper transmission of "tungro" disease of rice. *Plant Dis. Reptr.* 49: 127-131.

Saito, Y., Roechan, M., Tantera, D.M. and Iwaki, M. 1975. Small bacilliform particles associated with Penyakit habang (tungro-like) disease of rice in Indonesia. *Phytopathology* 65: 793-796.

Santa Cruz, F.C., Koganezawa, H. and Hibino, H. 1993. Comparative cytology of rice tungro viruses in selected rice cultivars. *J. Phytopathol.* 138: 274-282.

Santa Cruz, F.C., Boulton, M.I., Hull, R. and Azzam, O. 1999. Agroinoculation allows the screening of rice for resistance to rice tungro bacilliform virus. *J. Phytopathol.* 147: 653-659.

Satapathy, M.K. 1998. Chemical control of insect and nematode vectors of plant viruses. In: *Plant Virus Disease Control.* A. Hadidi, R.K. Khetarpal and H. Koganezawa, (eds.) APS Press, USA. pp.188-195.

Satapathy, M.K. and Anjaneyulu, A. 2001. *Rice Tungro Disease Epidemic. An Analysis.* Discovery Publishing House, New Delhi.

Savary, S., Fabellar, N., Tiongco, E.R. and Teng, P.S. 1993. A characterization of rice tungro epidemics in the Philippines from historical survey data. *Plant Dis.* 77: 376-382.

Sebastian, L. S., Tiongco, E.R., Tabanao, D.A., Maramara, G.V., Abdula, S. and Tabelin, E.R. 1999. Breeding for rice tungro disease resistance at PhilRice. In: *Rice Tungro Disease*

Management. T.C.B Chancellor, O. Azzam and K.L Heong, (eds.) International Rice Research Institute, Los Baños, Philippines. pp. 23-29.

Sebastian, L.S., Ikeda, R., Huang, N., Imbe, T., Coffman, W.R., Yano, M., Sasaki, T. and McCouch, S.R. 1996. Genetic mapping of resistances to rice tungro spherical virus (RTSV) and green leafhopper (GLH) in ARC11554. In: *Rice Genetics III* (Proc. Third Inter. Rice Genetics Symp.). G.S. Khush, (ed.) International Rice Research Institute, Los Baños, Philippines. pp. 560-564.

Seghal, O., Abouhaidar, M., Gellatly, D., Ivanov, I. and Thottappilly, G. 1993. An associated small RNA in rice yellow mottle sobemovirus is homologous to the satellite RNA of lucerne transient streak. *Phytopathology* 83: 1309-1311.

Serrano, F.B. 1957. Rice "accep na pula" or stunt disease - a serious menace to the Philippine rice industry. *Philipp. J. Sci.* 86: 203-230.

Shinkai, A. 1977. Rice waika, a new virus disease, and problems related to its occurrence and control. *JARQ* (Japan Agric. Res. Q.) 11: 151-155.

Sivamani, E., Huet, H., Ping, S., Ong C.-A., Kochko, A. de., Fauquet, C. and Beachy, R.N. 1999. Rice plant (Oryza sativa L.) containing rice tungro spherical virus (RTSV) coat protein transgenes are resistant to virus infection. *Mol. Breed.* 5: 177-185.

Sogawa, K. 1976. Rice tungro virus and its vectors in tropical Asia. *Rev. Plant Protec. Res.* 9: 21-46.

Takahashi, Y., Tiongco, E.R., Cabauatan, P.Q., Koganezawa, H., Hibino, H. and Omura, T. 1993. Detection of rice tungro bacilliform virus by polymerase chain reaction for assessing mild infection of plants and viruliferous vector leafhoppers. *Phytopathology* 83:655-659.

Tiongco, E.R., Cabauatan, R.C., Flores, Z.M., Hibino, H. and Koganezawa, H. 1993b. Serological monitoring of rice tungro disease development in the field: its implication in disease management. *Plant Dis*. 77: 877-882.

Tiongco, E.R., Cabauatan, R.C., Koganezawa, H. and Teng, P.S. 1993a. Inoculum sources of rice tungro viruses. *Philippine Phytopathol.* 29:30-41.

Tiongco, E.R., Chancellor, T.C.B., Villareal, S., Magbanua, M. and Teng, P.S. 1998. Roguing as a tactical control for rice tungro virus disease. *J. Plant Prot. Trop.* 11: 45-52.

Traoré, O., Pinel, A., Fargette, D. and. Konaté, G. 2001. First report and characterization of Rice yellow mottle virus in Central Africa. *Plant Dis* 85: 920.

Van Regenmortel, M., Fauquet, C., Bishop D., Cartens, E., Estes, M., Lemon, S., Maniloff, J., Mayo, M., McGoeoch, D., Pringle, C. and Wickner, R. 2000. *Virus Taxonomy.* Seventh report Intern. Comm. Taxonomy Viruses. San Diego: Academic Press 1162 pp.

Vargas, J. P. (1985). La hoja blanca: descalabro de CICA-8. *Arroz, Bogota Colombia* 34:18-19.

Voinnet, O., Pinto, Y. and Baulcombe, D. 1999. Suppression of gene silencing: a general strategy used by diverse DNA and RNA viruses of plants. *Proc Natl Acad Sci* U S A. 96:14147-52.

Webber, A. J., Damsteeg, V. D., and Graban, C. L. 1971. Studies on the relationship of the hoja blanca virus to the planthopper vector, Sogatodes oryzicola. *Phytopahtology* 61:916.

Widiarta, I.N., Suzuki, Y., Sawada, H. and Nakasuji, F. 1990. Population dynamics of the green leafhopper, Nephotettix virescens Distant (Hemiptera: Cicadellidae) in synchronized and staggered transplanting areas of paddy fields in Indonesia. *Res. Popul. Ecol.* 32: 319-328.

Zeigler, R. and Morales, F. J. 1990. Genetic determination of replication of rice hoja blanca virus within its planthopper vector, Sogatodes oryzicola. *Phytopathology* 80:559-566.

Zeigler, R. S., Pantoja, A., Duque, M. C., and Weber, G. 1994. Characteristics of resistance in rice to rice hoja blanca virus (RHBV) and its vector. *Ann. appl. Biol.* 124:429-440.

Zeigler, R., Rubiano, M., and Pineda, A. 1988. A field screening method to evaluate rice breeding lines for resistance to the hoja blanca virus. *Ann. appl. Biol.* 112:151-158.

Table 11.1. Virus disease of rice

Virus	**Family**	**Genus**	**Geographic Distribution**
Black-streaked Dwarf	Reoviridae	Fijivirus	China & Japan
Dwarf	Reoviridae	Phytoreovirus	North Asia
Gall Dwarf	Reoviridae	Phytoreovirus	Southeast Asia
Barley Yellow Dwarf	Luteoviridae	Luteovirus	Italy
Grassy Stunt		Tenuivirus	Asia
Hoja Blanca		Tenuivirus	Latin America
Necrosis Mosaic		Bymovirus (tentative)	India & Japan
Ragged Stunt	Reoviridae	Oryzavirus	Asia
Stripe		Tenuivirus	North Asia
Stripe Necrosis		Benyvirus	Africa & Latin America
Transitory Yellowing	Rhabdoviridae		Asia
Tungro Bacilliform	Caulimoviridae	RTBV-like viruses	Asia
Tungro Spherical	Sequiviridae	Waikavirus	Asia
Yellow Mottle		Sobemovirus	Africa

12. Maize and Sorghum

D.T. Gordon* and G. Thottappilly**
**Department of Plant Pathology, The Ohio State University, Wooster, Ohio, USA, E-Mail: gordon.4@osu.edu and ** Mahyco Research Foundation, Road No. 12, Banjara Hills, Hyderabad-500 034, India, E-Mail: execdirector@satyam.net.in*

12.1 Introduction

Maize (*Zea mays*) is of tropical Mexican origin, used worldwide as human food, animal and poultry feed and finds diversified uses in a large number of industrial products. Maize is the third most important crop after rice (*Oryza sativa*) and wheat (*Triticum aestivum*). It is grown over an area of 139 million ha. The USA is the leading producer followed by China, Brazil, Mexico, France, Argentina, Italy, India, Indonesia, South Africa and Canada (FAOSTAT, 2001). Maize is also important in subsistence-level farming.

Sorghum (*Sorghum bicolor*), the fifth major cereal in the world, is grown in an area of 44 million ha with a production of 55 mt (FAOSTAT, 2001). The USA is the leading producer followed by Nigeria, India, Mexico, Argentina, Sudan, China, Australia, Ethiopia and Burkino Faso. In the developed world sorghum is used for feed, but in Africa and Asia it is mostly used for food, especially in the semiarid zones of the world.

Of more than 57 viruses reported worldwide from maize and 18 from sorghum (Damsteegt, 1981; McGee, 1988; Thottappilly *et al.* 1993; Tsai and Falk, 1993; Brunt *et al.* 1996; Web sites: http://image.fs.uidaho.edu/vide/famly064.htm and http://www.ictvdb.iacr.ac.uk/ICTVdB/index.htm), only the most important ones of maize and sorghum are described here. The viruses include members of the mastreviruses, potyviruses, fijiviruses, tenuiviruses, marafiviruses and nucleorhabdoviruses.

12.2 Mastreviruses

Probably the most destructive viral disease of maize is maize streak caused by the *Maize streak virus* (MSV), genus *Mastrevirus*, family *Geminiviridae* (Thottappilly, 1992; Rybicki and Pietersen, 1999; Bosque-Pérez, 2000). Mastreviruses mostly infect grasses and are transmitted by leafhoppers in a persistent manner (Rybicki *et al.* 2000). MSV is related to the less economically important streak viruses *Panicum streak virus* (PanSV), *Sugarcane streak virus* (SSV), *Sugarcane streak Réunion virus* (SSRV), *Sugarcane streak Egypt virus* (SSEV) and *Digitaria streak virus* (DSV) (Donson *et al.* 1987; Hughes *et al.* 1992; Bigarré *et al.* 1999). The maize streak disease was first named 'mealie variegation' by Fuller (1901) and renamed 'maize streak' by Storey (1925) in a demonstration that the disease agent was transmitted by the leafhopper *Balclutha*, subsequently named *Cicadulina mbila*. The Maize streak virus site http.//www.uct.ac.za/microbiology/mastrevirus.htm gives further information.

MSV occurs extensively throughout sub-Saharan Africa (Rose, 1978; Thottappilly *et al.* 1993; Rybicki and Petersen, 1999), from the Sudan to South Africa and from the east to west coasts, and also occurs on the adjacent islands of Madagascar, Mauritius, Réunion, Sao Tomé and Principe, and in Egypt (Ammar *et al.*1987) and Yemen (Alhubaishi *et al.* 1987). Its occurrence in India on pearl millet (*Pennisetum glaucum*) and wheat (Seth *et al.* 1972 a, b) is uncertain (Thottappilly *et al.* 1993). Thus, the occurrence of MSV appears limited to the African continent and the adjacent islands. MSV is ecologically versatile and is found throughout sub-Saharan Africa in both forest and savannah and from sea level to an elevation of 2000 m (Efron *et al.* 1989).

Serious epidemics are known to have occurred in many parts of Africa in the early 1980s, and streak is now recognized as an important constraint on yield, especially in late-planted and irrigated maize (Thresh and Fargette, 2001). MSV epidemics occur sporadically throughout much of sub-Saharan Africa and may cause major losses (Bosque-Pérez, 2000). Epidemics occurred in Nigeria during 1966, 1971, 1973 and 1976 and more recently in 1983 and 1984. Epidemics have also been reported from several other African countries (Efron *et al.* 1989; Bosque-Pérez, 2000). Yield losses of up to 100 % are known if infection of susceptible lines occurs at an early stage of growth (Fig.12.1). The sporadic nature of epidemics in any country or region is likely due to variation in disease incidence from season to season and to the growth stage at the time of infection. In Zimbabwe, expansion of irrigated winter-wheat production over the past 25 years has resulted in maize streak becoming a major disease of maize in the irrigated farming districts (Caulfield, 1994).

While precise estimates of the losses in different countries are not available, experimentally determined estimates are known. In Nigeria yield reductions of as much as a 71% were reported for a susceptible variety compared to 1.5 and 10% reductions for a resistant variety and hybrid, respectively (Bosque-Pérez *et al.*1998). In other experiments average yield reductions due to MSV infection were about 70% and for another test 71-93 % under controlled infection with viruliferous *Cicadulina* (Efron *et al.*1989). With respect to resistant varieties, incidence as well as severity of maize streak are frequently less and these varieties may yield well even when infected. Also, susceptible varieties may not be much affected if infected late.

12.2.1 Host range and symptoms

MSV has a broad host range within the *Poaceae*. Of 152 species tested 95 species are susceptible, including *Z. mays,* sorghum, pearl millet, finger millet (*Eleusine coracana* ssp. *africana*), rice, oat (*Avena sativa*), barley (*Hordeum vulgare*), wheat and sugarcane (*Saccharum officinarum*) (Gorter, 1953; Damsteegt, 1983). Among the susceptible weed grasses, *Axonopus compressus*, *Brachiaria lata*, and *Setaria barbata* are the ones most likely to perpetuate MSV between maize plantings in West Africa (Mesfin *et al.*1992).

The first symptoms of MSV on young maize seedlings appear 3-7 days after inoculation as small, spherical, chlorotic spots, 0.5 mm in diameter, on the lowest exposed portions of the youngest leaves (Bock, 1974). As leaves expand, conspicuous chlorotic streaks appear that range from a few mm to several cm in length and from 0.5 to1 mm in width (Fig. 12. 2 a, b). Streaks develop along the veins on most of the leaf blade and frequently fuse laterally. In highly susceptible varieties, chlorotic streaking may develop into uniform chlorosis covering the entire leaf blade. On resistant genotypes, the chlorotic stripes may appear at low to very low density (Fig.12. 2b), and on highly resistant genotypes they may be absent. Although maize plants are vulnerable to the virus throughout their growth, plants infected within 3 weeks of planting become stunted and may die prematurely or produce small and poorly filled ears, whereas plants infected at the 6-8 weeks usually show little or no visible effect on growth or yield.

12.2.2 Transmission

MSV is transmitted by leafhoppers in the genus *Cicadulina* (Fig. 12.3). Twenty-two species are recognized of which 18 occur in Africa (Webb, 1987). Nine of the latter are vectors of MSV and include *C. arachidis*, *C. bipunctata* (=*bipunctella*), *C. ghaurii*, *C. latens*, *C. mbila*, *C. niger*, *C. parazeae*, *C. similis* and *C. storeyi* (=*triangula*) (Storey, 1925, 1936; Okoth *et* *al.*1988; Bigirwa *et al.* 1995). MSV is transmitted in a persistent, circulative manner without virus replication in the vector (Boulton and Markham, 1986). The virus persists in the leafhopper throughout its life. In Ethiopia, *C. mbila* has been identified as the most important vector (Mesfin *et al.* 1991). Both virus transmitters and non-transmitters occur within populations of *C. mbila*, *C. storeyi* (*triangula*), *C. ghaurii*, and *C. arachidis* (Asanzi *et al.* 1995). The ability to transmit the virus is a genetically inherited trait that is dominant and sex linked with males being heterozygous and is related to the permeability of the insect's gut wall to MSV. *C. mbila* acquires MSV from infected maize plants within 15 sec and inoculates in as short as 5 min with a latent period (LP) of 6-12 hrs (Storey, 1925, 1928, 1933). C. *storeyi* has a minimum acquisition access period (AAP) of 30 sec, minimum LP of 14-18 hrs and an inoculation access period (IAP) of 2 hrs (Okoth *et al.*1988) . *C. arachidis* and C. *ghaurii* have minimum AAP's of 15 min and 1 hr, respectively, LP 50s of 14-24 and 12-14 hrs, respectively, and both species have a minimum IAP of 1 hr (Asanzi *et al.* 1995). Natural populations of *C. storeyi* (*triangula*) and *C. mbila* have more active transmitters than populations of *C. ghaurii* and *C. arachidis* and for all species the percentage of active transmitters among the females is 2-3 times higher than that of males (Efron *et al.*1989). *C. mbila*, *C. storeyi* (*triangula*) and *C. ghaurii* transmit MSV more efficiently than *C. arachidi*s (Asanzi *et al.* 1995). The latter has a maximum transmission of 15%, whereas the former have efficiencies of 40-50% (Efron *et al.* 1989; Asanzi *et al.*1995).

Natural transmission of MSV is exclusively mediated by *Cicadulina* vectors. MSV is not seed- or sap-transmissible. Experimentally, MSV or its cloned genomic DNA is transmissible to germinating maize seedlings by vascular puncture inoculation (VPI) (Louie, 1995; Redinbaugh, 2003). Experimental transmission also is possible using cloned virus and delivery by a procedure called "Agroinfection" or " Agroinoculation" (Grimsley *et al.* 1986, 1987; Boulton *et al.* 1989; Martin *et al.* 1999; Martin and Rybicki, 2000). Martin *et al.* (1999) used agroinfection to rapidly screen for MSV resistance in various maize genotypes. Normally, screening for resistance is done either in the field depending upon natural infection by the leafhoppers or in the screen house with viruliferous leafhoppers. Depending solely upon natural infection under field conditions may lead to erroneous results, as there can be escapes, which can be mistaken as resistant plants. Further, artificial rearing of leafhoppers is tedious and requires appropriate facilities. Agroinfection and VPI provide versatile and efficient alternatives to artificial inoculation by leafhoppers (Martin and Rybicki, 2000; C. Changa, D.T. Gordon and R. Louie, *unpublished*).

12.2.3 Diagnosis

Symptoms generally give a reliable indication of MSV infection in maize plants. Verification is achieved by transmission by *C. mbila* or other vector *Cicadulina* species with resulting characteristic symptoms. Serological and molecular assays also are available for virus identification. For virus detection, polyclonal and monoclonal antibodies have been used (Dekker *et al.* 1988; Pinner *et al.* 1988; Pinner and Markham, 1990). Peterschmitt *et al.* (1991b) characterized MSV isolates from 11 African countries and regarded them as the same serotype. Molecular hybridization assays involving polymerase chain reaction (PCR) provide more reliable results. A PCR-RFLP technique was also developed and used successfully to type MSV and related viruses in maize and wild grasses (Willment *et al.* 2001).

12.2.4 Purification of the virus

MSV may be purified from maize leaves with newly developed symptoms. One protocol (Bock *et al.* 1974) recommends homogenization of 100 g infected leaf tissue in 200 ml 0.01 M phosphate buffer containing 0.1 % thioglycollic acid, pH 3.9-4.1; clarification with butanol (7 ml/100 ml extract); concentration by ultracentrifugation, and resuspension of virus in 0.01 M phosphate, pH 7.7.

12.2.5 Virion characteristics

MSV has quasi-icosahedral, twinned or geminate particles of 18x30 nm. Nucleocapsids appear angular with 22 capsomers per nucleocapsid. The nucleic acid of MSV is single-stranded (ss) DNA existing predominantly as closed circular molecules of mol. wt. 7.1×10^5 Da (Harrison *et al.* 1977; Harrison, 1985). The coat protein subunit of MSV is reported to be a single species of mol. wt. 28,000 Da (Bock *et al.* 1977).

The MSV genome has been extensively analyzed involving the complete sequences of more than 20 isolates (Howell, 1984; Mullineaux *et al.* 1984; Lazarowitz, 1988; Martin *et al.* 2001). All mastreviruses have monopartite genomes between 2.6 and 2.8 kb and an almost identical arrangement of genes (Rybicki *et al.* 2000). Full genomic sequences of three MSV isolates from wild annual grass species and wheat also have been determined (Schnippenkoetter *et al.* 2001; Willment *et al.* 2002).

The genome of MSV contains four major open reading frames (ORFs) encoding proteins greater than 10 kDa. These ORFs diverge from a long intergenic region (LIR) and converge on the opposite side of the genome at a short intergenic region (SIR). Both the LIR and the SIR are non-coding regions. Transcription of the viral genome is bi-directional with transcripts independently controlled and initiated from the LIR. Open reading frames

V1 and V2 are transcribed from the virion sense strand, whereas C1 and C2 are expressed from the complementary strand. Four functional proteins are encoded from these ORFs (Liu *et al.* 2001; Boulton, 2002). These proteins include a replication-initiator protein [Rep protein (~41.3 kDa) translated from a C1/C2 ORF splice event], replication-associated protein [RepA protein (31.5-31.6 kDa) from an unspliced RNA of the C1 ORF], a movement protein [MP (10.9 kDa); V1 ORF] and a coat protein [CP (26.9-27.0 kDa); V2 ORF].

The other non-coding region, the SIR, is associated with a DNA primer of about 80 nts and 5' terminal ribonucleotides. The SIR functions as the complementary-sense origin of replication and the production of the ds DNA is primed by this short stretch of ribonucleotides complementary to the SIR, which also contains consensus termination and polyadenylation signals and may control gene expression.

12.2.6 Strains

Before molecular techniques came into use, MSV isolates from various African grasses and other gramineous crops were designated as host-adapted isolates or strains of MSV. These strains, some of which differed in their pathogenicity on maize, were classified as maize adapted and non-adapted strains. Based on genome sequence analyses, some of these strains are now considered distinct mastreviruses and include PanSV, SSV, SSRV and SSEV. These distinct species share between 63 to 66% genome sequence identity with typical maize-infecting MSV (termed MSV-A) strains (Martin *et al.* 2001). Currently recognized groupings of MSV strains include MSV-A comprised of strains and isolates with approximately >95% sequence identity, which induce more severe streak symptoms on maize than isolates of groups MSV-B and MSV-C. MSV-B includes strains from wheat and a number of wild grass species that share 89% genome sequence identity with MSV-A isolates. MSV-C is currently represented only by MSV-Set, an isolate from a *Setaria* sp. that shares 80% genome sequence identity with MSV-A and MSV-B isolates. Based on 20% nucleotide sequence divergence between MSV-Set and other completely sequenced MSV isolates, MSV-Set may be eventually classified as a new African streak virus species.

Although MSV has a wide host range among African *Poaceae* species, individual isolates show considerable host adaptation. Examples are the severe and mild isolates of MSV from maize, a form from sugarcane (Uba-virus), and three forms from wild grasses (*Sporobolus*, *Eleusine* and *Paspalum*) transmissible to maize but not to sugarcane. Four streak-disease viruses (those occurring naturally on maize, sugarcane, *Panicum* and *Eleusine*) have identical particle morphology and are serologically related, though distinguishable. It seems that some strains adapted to wild hosts, such as those occurring naturally in *Eleusine* and *Sporobolus*, are highly pathogenic on maize. A geminivirus from *Digitaria setigera* (reported earlier as *D. sanguinalis*) was described from Vanuatu (South Pacific) (Dollet *et al.* 1986) and reported as a strain of MSV (Pinner *et al.* 1988). However, the causal agent, DSV, while serologically related, is not identical to MSV.

Genomic sequence analyses have revealed substantial evidence for widespread recombination among geminiviruses (Martin *et al.* 2001). Recombination between divergent geminivirus genomes is believed to be a major mechanism for generation of novel virus genotypes (Padidam *et al.* 1999; Sanz *et al.* 2000). Evidence of recombination among MSV isolates has been published (Schnippenkoetter *et al.* 2001).

12.2.7 Epidemiology

In recent years in Africa, the shift to monoculture of maize involving increased agricultural inputs, introduction of new, higher-yielding but MSV-susceptible genotypes, and the increased area planted to maize have been associated with increased MSV incidence and *Cicadulina* populations (Bosque-Pérez, 2000). Further, year-round cultivation under irrigation has also increased disease incidence, with leafhoppers occurring year around. However, where cool temperatures or dry conditions interrupt maize cultivation, alternate hosts for the virus and its vector are required for perpetuation. While weed grass hosts are considered sources of MSV, virus isolates occurring in many grasses in Nigeria, are not readily transmissible to susceptible field maize. Perpetuation of MSV and its vectors during the dry season in West Africa occurs mainly along riverbanks and low-lying fields with residual moisture where both maize and grass weeds occur. MSV outbreaks appear only when favorable weather conditions allow vector build-up along with growth of grasses infected with MSV. Such weeds in Nigeria include the native annuals *S. barbata* and *B. lata*. The perennial *A. compressus* also serves as an MSV host. In Zimbabwe where winter cereal crops, including wheat, are grown, the latter serve as hosts for leafhoppers and virus that later move to early-planted maize (Rose, 1973).

For maize cultivated in forest locations, the initial virus source appears to be located outside the field, but further spread is mainly from plant-to-plant within the field. Incidence of MSV varies with late season plantings, showing higher virus incidence than early season ones.

12.2.8 Control

Various cultural practices and insecticides are able to prevent maize streak disease. The most important are: planting "barrier" crops between early and late-planted maize fields, crop rotation, and avoidance of maize plantings down wind from earlier-planted, MSV-susceptible cereal crops (Rose, 1978).

Control by intercropping is not always consistent or substantial, when maize is interplanted with sorghum or cowpea (*Vigna* sp.). Insecticides have been used to control vectors and hence the disease (Barrow, 1992). However, resistance is the most practical and cost-effective means for the control of MSV.

The cultivar 'Révolution' bred against a highly virulent MSV isolate in Réunion is resistant in most African countries, whereas highly resistant maize lines from continental Africa are susceptible in Réunion (Rodier *et al.* 1995). Maize varieties that combine resistance to MSV with other desirable characters have been developed at the International Institute of Tropical Agriculture (IITA) in Nigeria and, more recently, at Harare, Zimbabwe at the International Maize and Wheat Improvement Center (CIMMYT). Both high-yielding varieties and varieties traditionally grown in various countries in Africa have been converted into MSV-resistant lines at IITA in cooperation with CIMMYT and National Programs in Africa (Efron *et al.* 1989). Breeding lines of maize with resistance to MSV have also been developed by IRAT in Réunion (Rodier *et al.* 1995) and at the Mauritius Sugar Industry Research Institute (Govinden and Rummun, 1996). MSV-resistant hybrids have been released by the Pannar seed company of South Africa and the Seed Coop in Zimbabwe. By the 1990s, in most African countries streak-resistant maize germplasm had been released and was grown by farmers (Barrow, 1992, 1993).

Resistance to MSV in IITA maize germplasm is controlled by two or three major gene pairs, with the possible involvement of minor genes (Kim *et al.* 1989). A single major gene (designated *msv*1) controls tolerance to MSV in Tzi 4, a tolerant line from IITA (Kyetere *et al.* 1999). CML202 from CIMMYT also contains a major QTL for MSV resistance (Welz *et al.* 1998). Resistance to streak virus in D211 is provided by one region on chromosome 1, with a major effect, and four other regions on chromosomes 2, 3 (two regions) and 10, with moderate or minor effects. The major QTL on chromosome 1 of line CIRAD390 from Réunion seems to be common to the other sources of resistance, namely Tzi4 and CML202 (Pernet *et al.* 1999). However, the distribution of the other QTLs on chromosomes 2, 3, 5 (two QTLs), 6, 8, and 10 reveal differences in Réunion germplasm and across these other resistance sources. This diversity is of great importance when considering the durability of the resistance.

In continental Africa and especially in Nigeria, where the resistant lines were deployed over large areas, the streak-resistant maize has not become susceptible. More aggressive viral strains/isolates do not appear to have been 'selected' by the resistant varieties and, thus resistance 'breakdown' has not happened.

12.3 Potyviruses

Viruses in the genus *Potyvirus* infecting maize and/or sorghum include: *Guinea grass mosaic virus* (GGMV), *Johnson grass mosaic virus* (JGMV), *Maize dwarf mosaic virus* (MDMV), *Pennisetum mosaic virus* (PenMV), *Sorghum mosaic virus* (SrMV), *Sugarcane mosaic virus* (SCMV) and *Zea mosaic virus* (ZeMV). Of these, MDMV and SCMV are the most widespread in developing countries. *Wheat streak mosaic virus* (WSMV), a species in the family *Potyviridae*, genus *Tritomovirus*, also infects maize but is not known to cause a disease of maize in developing countries. Only MDMV and SCMV will be described in detail, while PenMV, SrMV, GGMV, JGMV and ZeMV will be described in brief.

MDMV and SCMV infect maize or sorghum worldwide. Among developing countries, JGMV, sometimes mistakenly identified as MDMV, occurs in maize mainly in South America, notably in Venezuela. SrMV is likewise of limited occurrence, mainly in India where it infects sorghum. Of the remaining maize infecting potyviruses, GGMV occurs in the Ivory Coast, PenMV in China and ZeMV in Israel.

12.3.1 Crop losses, host range, symptoms and transmission

In single infections these potyviruses generally cause only moderate yield reductions but experimental losses of 54% have been reported for MDMV in susceptible maize hybrids (Gordon *et al.* 1981). Infections of MDMV or SCMV-MDB with *Maize chlorotic mottle virus* (MCMV), genus *Machlomovirus*, cause corn lethal necrosis, a serious maize disease with major yield losses (Uyemoto and Claflin, 1981). In China, losses caused by MDMV-B (SCMV-MDB) have been estimated at 20-80%, making it the most important virus disease of maize in that country (Chen *et al.* 2002).

Losses due to MDMV infection in sorghum range from 100% for susceptible cultivars to less than 5% for tolerant ones (Toler, 1985). Losses caused by SCMV-MDB in various sorghum accessions ranged from 2 to 75%, with greatest losses in cultivars showing red leaf and necrosis (Alexander *et al.* 1984). For the most susceptible sorghums, losses are as high as 95%. In experimental infection of sorghum by *Sorghum red stripe virus* (SRSV), a synonym for MDMV, reduction of grain yield is as much as 53–67%.

The important hosts of MDMV are maize and sorghum and the overwintering host Johnson grass (*Sorghum halepense*) and for SCMV-MDB maize and sorghum; SCMV-MDB does not infect Johnson grass. JGMV like MDMV infects maize, sorghum and Johnson grass and in addition oat, which is not susceptible to MDMV (McDaniel and Gordon, 1985). SrMV mainly infects sorghum and sugarcane but not Johnson grass or oat. Hosts of GGMV are guinea grass (*Urochloa maxima* syn. *Panicum maximum*), *Setaria italica* and maize (Lamy *et al.* 1979) and for ZeMV - maize, sorghum and Johnson grass (Seifers *et al.* 2000). PenMV infects maize and sorghum but not Johnson grass or oat (Shi *et al.* 1986).

Leaf symptoms incited by MDMV on maize are mosaic or mottle including chlorotic flecks, streaks and rings. Symptoms of MDMV on sorghum include systemic mosaic, red leaf, red stripe and necrosis of leaves, plant stunting, delayed flowering, and reduction in head length, number of heads, seed size, number of seeds per head and grain yield (Tosic *et al.* 1990). SCMV-MDB incites systemic mosaic on maize and mosaic, streaks, stripes and/or necrosis on sorghum. Symptoms of JGMV infection are mosaic, ring spots and chlorosis on maize and mosaic, necrotic red leaf, necrotic red stripe, yellow mosaic or necrosis and sometimes stunting on sorghum. On maize SrMV causes systemic mosaic and on sorghum either mosaic or short, yellow-green streaks that develop into reddish-brown necrotic streaks and stripes followed by extensive reddish-brown necrosis of leaves (red-leaf symptoms).

These potyviruses are transmitted by sap inoculation and probably by VPI of germinating maize kernels, although only VPI transmission of MDMV is known (Louie, 1995). In addition to virion transmission, VPI transmits MDMV genomic RNA (Redinbaugh *et al.* 2001).

All maize- and sorghum-infecting potyviruses are non-persistently transmitted by aphids (Nault and Knoke, 1981). Several aphid species vector one or more of these potyviruses. MDMV and SCMV-MDB are also transmitted to uninfected young maize plants by uredospores of *Puccinia sorghi* from infected maize plants (Von Wechmar *et al.* 1992).

Seed transmission at low rates is reported - MDMV in dent maize at frequencies up to 0.5% (Mikel *et al.* 1984) and SCMV-MDB in maize at similar rates (Hill *et al.* 1974). However, MDMV is not transmitted in sorghum seed (Toler, 1985).

12.3.2 Purification, diagnosis and virion characteristics

The number of different published purification protocols (Rao *et al.* 2001) attests to the difficulty of obtaining purified virus in sufficient quantity for use in studies of their characteristics and raising antisera of sufficient antibody titer and minimal reactivity with host antigens.

Polyclonal antibodies have been raised to each of the above four potyviruses and to GGMV, PenMV and ZeMV (Bond and Pirone, 1971; Mali, 1999; Seifers *et al.* 2000), allowing identification of the viruses, mainly by ELISA. Microprecipitin assays, immunosorbent electron microscopy and electro-blot immunoassay have also been used. Antibodies to these viruses are moderately cross-reactive with heterologous potyviral antigens and cross-absorption of antisera with heterologous antigens may be needed for greater specificity. A monoclonal antibody specific to MDMV strain A has been raised and used in capture indirect ELISA for identifying MDMV (Jones *et al.* 1988).

A molecular procedure, as yet not widely employed, differentially detects MDMV, SCMV, JGMV and SrMV from infected maize (Marie-Jeanne *et al.* 2000). The procedure involves extraction of total RNA, reverse transcription (RT) and PCR using potyvirus-group specific primers to produce a 327-nucleotide fragment from the capsid gene of the four viruses. Enzymatic restriction of the fragment with the two endonucleases, AluI and DdeI, gives virus-specific patterns. This method differentiates well-characterized strains, field collected isolates, co-infections and new potyviruses infecting maize.

Sequence determinations followed by analysis of the nucleotide and/or deduced amino acid sequences of the NIb or coat protein genes or the nucleotide sequences of the 3'-untranslated region (UTR), allow identification of the viruses and in some cases even of strains (Chen *et al.* 2002; Cheng *et al.* 2002). Other molecular techniques for identifying the four potyviruses included HPLC peptide profiling of tryptic digests (McKern *et al.* 1990, 1991), molecular hybridization with probes corresponding to the 3' UTR (Frenkel *et al.* 1992) and immunocapture RT-PCR (Jiang and Zhou, 2002).

JGMV, MDMV, SCMV and SrMV have flexuous, filamentous particles of 700-755 x 12-16 nm (Rao *et al.* 2001). Their genomes are ss, positive-sense (+) RNA with Mr. of 2.7-3.41 kDa for the first three viruses (Berger *et al.* 1989). Complete genome sequences are known for JGMV (Gough and Shukla, 1993), MDMV (Kong and Steinbiss, 1998; T. Muelia and D. T. Gordon, *unpublished*), SCMV (Chen *et al.* 2002; Cheng *et al.* 2002; Fan *et al.* 2003) and SrMV (Chen *et al.* 2002). The sequences for SCMV and SrMV are for isolates occurring in China. The genomes are all monopartite.

The genomic RNA of JGMV is 9,766 nucleotides (nts) in length, excluding the poly (A) tail, and contains a large ORF that encodes a polyprotein of 3,052 amino acids (Gough and Shukla, 1993). The ORF is flanked by a 5' UTR of 135 and a 3' UTR of 475 nts. MDMV strain A (type strain) and MDMV-Bg (Bulgarian isolate) genomic RNAs, respectively, are 9418 and 9499 nts in length, excluding the 3' terminal poly(A) tail, and contain single ORFs of 9048 and 9126 nts that encode polyproteins of 3014 and 3042 amino acid residues (Kong and Steinbiss, 1998; T. Muelia and D. T. Gordon, *unpublished*). The ORFs are flanked by 5'- and 3'-UTRs of 139 and 235 nucleotides, respectively. The three SCMV genome sequences of isolates from maize are 9595 or 9596 nts in length with ORFs of 9192 nts encoding polyproteins of 3063 amino acids (Chen *et al.* 2002; Cheng *et al.* 2002; Fan *et al.* 2003). Their 5' UTRs and 3' UTRs are 149 nts and 255 nts, respectively.

The organization of the genomes of JGMV, MDMV, SCMV and SrMV is similar to that of other members of the genus *Potyvirus*. Their polyproteins contain potyvirus cleavage sites for ten encoded proteins which are 5' to 3' P1, HC-PRO, P3, $6K_1$, CI, $6K_2$, NIa-VPg, NIa-Pro, NIb and CP.

12.3.3 Strains

Characterized strains of JGMV include JGMV-JG and JGMV-Krish, from Australia (Suranto *et al.* 1998) and JGMV-KS1 (McKern *et al.*, 1990) and JGMV-O (McDaniel and Gordon, 1985) from the USA. JGMV-Krish is able to infect sorghum cultivars with Krish resistance (Suranto *et al.* 1998).

Characterized strains of MDMV are MDMV-A, MDMV-C, MDMV-D, MDMV-E, and MDMV-F from the USA (Louie and Knoke, 1975). Strains A, D, E, and F differ in biological, serological and nucleotide sequence properties (Lenardon *et al.* 1993). MDMV isolates from Bulgaria (Bg), Illinois (I) and Spain (S) differ similarly in nucleotide and derived amino acid sequence identities between 93 to 99% (S. Lenardon and D. T. Gordon, *unpublished*) but have not been designated as strains. Another possible strain is an isolate from sorghum in Venezuela that infects sorghum cultivars with MDMV resistance, but not those with Krish resistance (Toler *et al.* 1982). Another Venezuelan MDMV isolate infects Johnson grass from Venezuela but not that from the USA (Garrido and Trujillo, 1988). The isolate was designated MDMV-V. A putative MDMV strain, designated MDMV-D, occurring on maize and Johnson grass in Israel, caused markedly more severe symptoms on maize and sorghum than typical MDMV (Antignus, 1987). MDMV-D is serologically distantly related to MDMV but not to SRSV or SrMV (SCMV-H). It was subsequently characterized as ZeMV (R. Salomon, *personal communication*; Seifers *et al.* 2000)

The most frequently recognized strains of SCMV are: SCMV-A, SCMV-B, SCMV-D, and SCMV-E from the USA (Yang and Mirkov, 1997); SCMV-BC, SCMV-Brisbane, SCMV-Bundaberg, SCMV-Isis, SCMV-Sabi and SCMV-SC from Australia (Handley *et al.* 1998); and SCMV-MDB initially from the USA (MacKenzie *et al.* 1966), but now apparently worldwide. Sequenced SCMV strains and isolates show diversity with isolates from maize forming a distinct phylogenetic group (Alegria *et al.* 2003). SCMV-MDB shows the greatest divergence, prompting the suggestion that SCMV-MDB be considered a distinct potyvirus species (Handley *et al.* 1998). Likewise, maize isolates from Germany are distinct from other SCMV isolates in phylogenetic analysis and a new strain designation, SCMV-GER, has been proposed (Oertel *et al.* 1999). Although not given a strain designation, a SCMV isolate infecting maize, sorghum and Johnson grass in Iran is serologically related to SCMV-D but not to JGMV-O, MDMV-A, SCMV-MDB, or SrMV-H (Afsharifar and Izadpanah, 1994). This isolate differs from SCMV-D by naturally infecting Johnson grass. Confirmation of this finding is awaited.

Strains of SrMV are SrMV-SCH, SrMV-SCI, and SrMV-SCM from the USA (Yang and Mirkov, 1997) and SrMV-IBS from India (Mali, 1999). The latter, possibly the same as SrMV-IBH and/or SrMV-RBH, is a QL3 sorghum immunity breaking strain (Mali, 2001). QL3 sorghum carries the Krish sorghum resistance.

12.3.4 Epidemiology and control

The aphid vectors of MDMV, JGMV, SCMV and SrMV generally retain acquired virus for short periods of only minutes. Thus, viruliferous aphids are presumably able to spread the viruses for only short distances from the virus source plants to seedling maize or sorghum (Louie and Knoke, 1991). The closer the latter are to the source, the greater the probability that they will be infected. Further, the proximity of infected Johnson grass to maize appears to have a greater impact on MDMV incidence than the number of vector aphids (Vangessel and Coble, 1993).

Resistant maize or sorghum cultivars most frequently control JGMV, MDMV, SCMV and SrMV. Sources of resistance have been identified for these viruses (Mali and Thakur, 2001) and the genetics of resistance are described, especially for maize (Dussle *et al.* 2000; Quint *et al.* 2002; Simcox *et al.* 1995; Wu *et al.* 2002).

Among numerous maize inbreds identified as tolerant or resistant to MDMV in the USA, an especially resistant one is the inbred Pa405 (Louie *et al.* 1991). Pa405 is similarly resistant to MDMV-A, -D, -E and -F and SCMV-A, -MDB and -Jg (JGMV) (Kovacs *et al.* 1994). A major dominant gene for resistance, designated *Mdm*1, is located on the short arm of chromosome 6 near the centromere of Pa405 (Louie *et al.* 1991).

Other maize inbreds showing resistance to MDMV and SCMV under both field and greenhouse conditions are D21, D32 and FAP1360A (Kuntze *et al.* 1997). These early-maturing European dent maize inbreds are also resistant to JGMV and SrMV under greenhouse conditions (Kuntze *et al.* 1995). Two major genes, *Scmv*1 and *Scmv*2, conferring resistance to SCMV, map to the short arm of chromosome 6 and the long arm of chromosome 3, respectively, of maize with resistance derived from FAP1360A (Dussle *et al.* 2000). *Scmv*1 and *Scmv*2 are not present in Pa405 and *Scmv*2 is not confirmed for resistant inbreds D21 and D32. The *Scmv*1 region may contain two closely linked resistance genes (Dussle *et al.* 2003). In Chinese evaluations of resistance to MDMV, probably SCMV, maize varieties Guangyou 5, F09-2 and Bai 197 are highly resistant, DJ29 resistant and Baizong moderately resistant (Chen *et al.* 1996). Further, Tainung 2, a Chinese forage hybrid, is immune to MDMV-B (SCMV-MDB) infection (Shieh *et al.* 1994).

Resistance of the Chinese maize inbred Huangzaosi to MDMV-B (SCMV-MDB) is conditioned by a major resistance gene, *mdm*1(t), located on the long arm of chromosome 6 (Wu *et al.* 2002). *Mdm*1(t) is different from *mdm*1 since the latter is located near the centromere on the short arm of chromosome 6.

Pathogen-derived resistance (PDR) has been developed for MDMV-B (SCMV-MDB) involving the CP gene introduced into maize to provide cross-protection/resistance to MDMV-A as well as to MDMV-B (Murry *et al.* 1993). The extent of the deployment of this PDR is unknown.

For control of MDMV in grain sorghum tolerant, field resistant, resistant to infection and immune sorghum lines are available (Toler, 1985). Lines with greatest tolerance are RS 621, RS 625 and Tx 414. Field-resistant lines are IS-2549C (SC-228), IS2816C (SC-120), IS-12612C (SC-112), IS-12666C (SC-175), Rio, TAM2566 and Q-7539. The field-resistance of the latter line was identified in Australia and of the former lines in the USA. Line Tx2786 is resistant to infection by MDMV-A and immune to MDMV-B and SCMV-H when mechanically or naturally inoculated (Toler, 1985). Finally, Krish sorghum and lines derived from it are immune to most isolates of MDMV, SCMV, JGMV and SrMV (Giorda *et al.* 1986). Immune sorghum lines include SC0120-14E, 73-C9-31-32, TAM-B51 and TAM-B52 (Toler, 1985) and 15 parental lines, QL6-17 and QL20-22, are homozygous for Krish resistance (Henzell *et al.* 1982).

Further control of these potyviruses is by early planting and for MDMV, and possibly JGMV eradication of Johnson grass by post-emergence application of herbicides (Vangessel and Coble, 1993), or weeding, where sufficient labor is available. Eradication of Johnson grass within and adjacent to maize fields early in the season before aphids build up reduces the incidence of MDMV infection and subsequent disease occurrence within the field. However, herbicide eradication of Johnson grass may cause increased disease incidence because of increased movement of insect vectors from dying Johnson grass to the maize crop (Eberwine and Hagood, 1995). Crop rotation with a non-gramineous crop insusceptible to MDMV coupled with eradication of Johnson grass may be required to eliminate the latter from fields with a persistent Johnson grass problem and a history of the virus disease (Vangessel and Coble, 1993).

12.4 Fijiviruses

Maize-infecting fijiviruses include *Maize rough dwarf virus* (MRDV), *Rice black streaked dwarf virus* (RBSDV) and *Maize Rio Cuarto virus* (MRCV). Sometimes these names are used synonymously (Fang *et al.* 2001; Zhang *et al* 2001b). However, lack of cross-hybridization between MRCV and MRDV for at least 7 of the 10-genome segments and of sufficient sequence homology indicate that the two viruses are distinct (Distefano *et al.* 2002).

Maize-infecting fijiviruses have been reported from Europe, the Middle East (Conti, 1983), the Far East (Zhang *et al.* 2001b), and South America (Milne *et al.* 1983). A dwarfing disease of maize, subsequently named maize rough dwarf, was first reported from Italy in 1949, and later from Israel, France, Spain, Switzerland, Czechoslovakia and Yugoslavia (Conti, 1983). RBSDV was first reported in 1952 from Japan (Shikata, 1974) and in the 1970s from China (Zhang *et al.* 2001b), although it probably first appeared there in the 1960s (Bai *et al.* 2002). A similar disease of maize was reported from Korea. Since 1990 maize rough dwarf has occurred each year in China, causing very severe damage (Li *et al.* 1999). RBSDV infects sorghum in China causing dwarfing of plants (Bai *et al.*2002).

A disease, named "Enfermedad de Rio Cuarto" (MRCV), was recorded for the first time in 1970s from Argentina (Distefano *et al.* 2002). The causal agent was initially identified as MRDV (Milne *et al.* 1983), but lately as a distinct *Fijivirus* species (Distefano *et al.* 2002). In addition to Argentina (Distefano *et al.* 2002) MRCV occurs in Uruguay (Ornaghi *et al.* 1999) and southern Brazil.

12.4.1 Crop losses, host range, symptoms and transmission

Maize Rio Cuarto (MRC) is the most important viral disease affecting maize in Argentina with yield losses higher than 70% (Rodriguez Pardina *et al.* 1998). China also experiences widespread maize losses (Fang *et al.* 2001) and since 1990 the disease, usually referred to as maize rough dwarf, has occurred each year in northern, northwestern and northeastern China, causing very severe damage (Li *et al.*, 1999, as cited by Zhang *et al.*, 2001b).

The host range of fijiviruses is limited to species of the *Poaceae* (*Gramineae*), including maize, rice, wheat, oat and barley (Milne and Lovisolo, 1977). Natural hosts of MRDV, in addition to maize, are *Digitaria sanguinalis* and *Echinochloa crus-galli*. RBSDV naturally infects rice, maize and sorghum, whereas MRDV infects rice and sorghum with difficulty (Bai *et al.* 2002). Additional hosts of MRCV are sorghum, millet (*Panicum milliaceum*), foxtail millet (*Setaria italica*), rye (*Secale cereale*), and several species of grasses (Rodriguez Pardina *et al.* 1998; Truol *et al.* 2001). Early infection by fijiviruses, particularly MRDV, induces symptoms within *ca.* 3 weeks after emergence. Leaves are darker green, unusually erect and stiff. Minute swellings of veins appear on the lower leaf surfaces. Subsequently, dwarfing becomes more conspicuous with whitish enations, on leaf sheaths and ears. Leaves are unusually dark green, and may die prematurely. Plants infected later are dwarfed to varying degrees with enations appearing only on the under surfaces of top leaves and on ear and sheath leaves.

Maize plants infected with RBSDV are dwarfed with dark green leaves (Zhang *et al.* 2001b). Early infection shows white streaks and white waxy swellings on the veins on the lower surface of leaves. Margins of old infected leaves are split.

Symptoms of MRCV on maize appear within 4-5 weeks following inoculation (S. L. Lenardon, *personal communication*; Nome *et al.* 1981). Plants are dwarfed, leaves show fine chlorotic flecks on secondary and tertiary veins, developing into vein swelling and enations on the lower surfaces of leaves and on leaf sheaths and husks. Tassels are atrophied and sterile and the ears are reduced and malformed with few kernels. Sorghum plants infected when young are stunted with enations on leaf undersurfaces (S.L. Lenardon, *personal communication*).

Laodelphax striatellus is the natural vector of both MRDV and RBSDV. Several other Delphacid species experimentally transmit the two viruses (Shikata, 1974; Milne and Lovisolo, 1977). The natural vector of MRCV is *Delphacodes kuscheli* (Ornaghi *et al.* 1993). The three fijiviruses are persistent in their vectors and MRDV, RBSDV and possibly MRCV replicate in the vectors. RBSDV is transmitted transovarially (Shikata, 1974). Also, MRDV, RBSDV and MRCV are transmitted to maize by VPI of germinating kernels (Louie, 1995; R. Louie and D. T. Gordon, *unpublished*). None of the three viruses is transmitted by leaf rubbing or through seed.

12.4.2 Purification, diagnosis and virion characteristics

Purification protocols for MRDV, RBSDV and MRCV are found in Wetter *et al.* (1969), Shikata, (1974) and Rodriguez Pardina *et al.* (1998), respectively.

Antisera to each of these fijiviruses are available and serological identification is by immune electron microsopy and ELISA (Caciagli *et al.* 1985; Isogai *et al.* 1998; Lenardon *et al.* 1998). However, their close serological relationships, especially between MRDV and RBSDV, make serological identification of the viral species problematic.

Sequences of several of genome segments of each of these maize-infecting fijiviruses are available, allowing virus identification by molecular techniques. The latter for RBSDV detection include RT-PCR and dot-blot hybridization with PCR-generated probes (Isogai *et al.* 2001). For MRCV, SDS-PAGE (Rodriguez Pardina *et al.* 1998) and dot blot hybridization assay with MRCV probes (Arneodo *et al.* 2002a) are used, while for MRDV PAGE separation of genomic segments (Marzachi *et al.*1999) is used. Of these, the RT-PCR method is the most sensitive.

MRDV virions are spherical (63-70 nm diam.), with a double capsid and probably 92 morphological units in the outer capsid that includes 12 projecting (A) spikes, one at each 5-fold symmetry axis (Milne *et al.* 1973).

Complete virions of RBSDV are 75-85 nm in diameter (Shikata, 1974). The outer coat of MRDV and RBSDV may be removed by purification or PTA staining, yielding particles of 50-60 nm diam. MRCV virions are icosahedral, measuring 60-70 nm in diameter with a dark core (Nome *et al.* 1981; Arneodo *et al.* 2002b). All three maize infecting fijiviruses have a multipartite genome consisting of ten linear segments (S1–S10) of double-stranded (ds) RNA. The 10 segments of these fijiviruses range between approximately 4.5 to 1.8 kbp with a total genome size of about 29 kbp (Distefano *et al.* 2002; Zhang *et al.* 2001a, b).

For RBSDV the nucleotides sequences for the ten-segment genome that together contain 29,141 nucleotides are published (Zhang *et al.* 2001a, b). Sequences of several dsRNA segments of MRDV and MRCV are available from the GenBank and for MRDV include genome segments S6, S7, S8 and S10 (Marzachi *et al.* 1991,1995) and for MRCV segments S4 and S8 (Distefano *et al.*, 2002).

12.4.3 Strains, epidemiology and control

MRDV and RBSDV have an 85-87.5% nucleotide sequence identity (Azuhata *et al.*1993), suggesting that RBSDV is a geographical variant of MRDV (Marzachi *et al.*1995), although they are classified as distinct viral species. Likewise, sequence variation among fijivirus genomes for isolates from China and Japan suggest the existence of geographical variants of RBSDV. While MRCV was initially thought to be a geographic strain of MRDV, it is now considered to be a new species of the genus *Fijivirus* (Distefano *et al.* 2002).

The epidemiologies of the three maize-infecting fijiviruses have a common pattern but with variations for each virus. For MRDV, the vector, *L. striatellus*, overwinters possibly on *Cynodon dactylon* (Conti, 1983). The vectors of RBSDV and MRCV overwinter on wheat, oats, rye or barely, as well as some grass species (Laguna *et al.* 2000; Rodriguez Pardina *et al.* 1998; Zhang *et al.* 2001b). Infective vectors migrate early in the season to the surrounding young maize (March *et al.* 1995) and inoculate the seedling plants.

Application of agrochemicals, adjusting the time of planting (Li *et al.* 1999), and planting resistant or tolerant lines, limit virus spread. Likewise, systemic insecticides like carbofuran or imidacloprid experimentally applied as a seed coating provide good control of *D. kuscheli*, decreased MRCV severity and increase grain yield in field trials (March *et al.* 2002). Since the vectors do not breed on maize and feed only occasionally, herbicide application to maize fields prior to sowing to eliminate weed grasses prevents vectors from becoming resident within these fields sufficiently long thereby hindering virus spread early in the season (Conti, 1983). While most commercial hybrids are apparently susceptible to these viruses, tolerant or resistant maize lines are available (Lenardon *et al.*1998). Until highly virus resistant maize lines are developed and made available to farmers, control is probably best achieved by combining some or all of the above control procedures.

12.5 Tenuiviruses

Two maize infecting tenuiviruses or tenui-like viruses are known. These are *Maize stripe virus* (MSpV) (Gingery *et al.* 1981) and *Maize yellow stripe virus* (MYSV) (Ammar *et al.* 2003). MSpV and MYSV also infect sorghum causing sorghum stripe disease (SStD) (Peterschmitt *et al.* 1991a; Garud *et al.* 2000).

MSpV belongs to the genus *Tenuivirus*. MYSV is currently listed as a tentative species of the genus *Tenuivirus*. However, as indicated below, MYSV may be a tenui-like virus.

MSpV occurs only in subtropical and tropical countries of both the Eastern and Western hemispheres, occasionally causing serious losses of maize (Falk and Tsai, 1998). MSpV has been reported from Africa, Central, North and South America, Asia, and Australia (Thottappilly *et al.* 1993). Its planthopper vector, *Peregrinus maidis*, is limited to subtropical and tropical maize-growing regions and MSpV occurs where maize and *P. maidis* both occur. In India MSpV infects sorghum (Peterschmitt *et al.* 1991a). Earlier reports of maize stripe in India involved a virus with rhabdovirus morphology, possibly maize mosaic virus (Sharma and Payak, 1983). MYSV infecting maize and sorghum has been reported only from Egypt (Ammar *et al.* 1990).

MSpV causes losses in maize with experimental grain loss of up to 80% (Roca de Doyle and Autrey, 1992). However, relative to other maize-infecting viruses, it is not considered to be a major virus of maize. Natural infection of sorghum by MSpV causes reductions in ear head weight (70-93%) and in thousand-grain weight of 23-25% and a marked reduction in number of grains/ear head (Revuru and Garud, 1998). No major outbreaks of the disease in sorghum are reported.

MSpV naturally infects maize, sorghum and itchgrass (*Rottboellia exaltata*) (Gingery, 1985). Other natural hosts include *Sorghum verticilliflorum*, *S. halepense*, Sudan grass (*S. bicolor* ssp. *drummondii*), and Sudan grass X *S. bicolor* hybrids. Experimental hosts are barley, rye, triticale (*Triticum aestivum* X *Secale cereale*), oat and several teosintes (*Zea* spp.). MSpV from infected sorghum in India does not infect wheat or maize (Peterschmitt *et al.* 1991a). MYSV infects maize, sorghum, wheat, barley, oat and several grasses (Mahmoud *et al.* 1996; Thouvenel *et al.* 1996).

MSpV-infected maize plants are partially stunted and have yellow to light-green uninterrupted stripes or bands which start at the base of leaves and extend almost to the tip during early stages of the disease (Fig. 12.4). Characteristic systemic symptoms of MSpV start as chlorotic spots and streaks on the youngest leaves within 4-7 days after inoculation. Subsequently, the spots and streaks fuse to form chlorotic bands or stripes and leaves may become entirely chlorotic or whitish (hence the name maize hoja blanca for occurrences in Latin America) in late infections. Infected plants are stunted with a characteristic marked bending of the apex and may die when infected early. When infected young, plants produce little or no grain.

MYSV is associated with three types of symptoms on infected maize and sorghum that include fine stripe, coarse stripe and chlorotic stunt. These may be present on different leaves of the same plant (Ammar *et al.* 1990). When infected young, plants are stunted with reduced yields.

12.5.1 Transmission

The vector of MSpV is the corn planthopper, *P. maidis* (Gingery *et al.* 1979). The virus propagates and persists in the planthopper (Nault and Gordon, 1988) but may be intermittently transmitted (Narayana and Muniyappa, 1996a, b). Nymphs transmit MSpV more efficiently than adults and retain ability to inoculate for a much longer time (Tsai and Zitter, 1982). The mean latent period in the vector varies from 10 to 15 days with transmission by individuals beginning as soon as 4 days to as long as 22 days. The first instar nymphs are the most efficient transmitters (64%) compared to second to fourth instars (50%) and adults (33%). MSpV is transmitted through the egg (Gingery *et al.* 1981; Tsai and Zitter, 1982) to a small percentage of progeny.

MYSV is transmitted by the leafhopper *C. chinai* but not by *P. maïdis* (Mahmoud, 2001). Nymphs and adults of the former transmit MYSV in a persistent manner. Molecular and serological evidence indicate that MYSV multiplies in its vector (Khalifa, 2001; Mahmoud, 2001).

Neither virus is transmitted mechanically, although MSpV is transmitted by VPI of germinating maize kernels (R. Louie, *personal communication*). Mechanical transmission was reported for a Venezuelan isolate of MSpV (Lastra and Carballo, 1985).

12.5.2 Diagnosis and virion characteristics

Antisera to MSpV allow detection and identification by ELISA in both plants and insects (Ammar *et al.* 1987; Nault and Gordon, 1988). Further, antiserum to the noncapsid protein (NCP) or p4 protein found in abundance in MSpV infected leaves but not in vectors provides another means of identification when tested by indirect ELISA (Falk and Tsai, 1983b).

Antisera to the nucleoprotein of MYSV also are used for its detection in host plants and vector leafhoppers (Mahmoud *et al.*1996; Thouvenel *et al.* 1996) by ELISA. MYSV also may be detected with ^{32}P labelled DNA probes. *In lieu* of these tests, characteristic symptoms on maize or sorghum, persistent transmission by *C. chinaï* and tenuivirus-like filaments in purified preparations may serve for diagnosis.

MSpV virions are fine-stranded, flexuous nucleoprotein filaments that are 3-4 nm wide with undefined lengths and no envelope (Gingery *et al.* 1981). While filaments are usually linear, circular and helical forms also exist. Purified MYSV from naturally infected maize and sorghum leaves consists of fine filaments 4-8 nm in diameter, similar to those of MSpV (Ammar *et al.* 1990).

MSpV contain 5-6% nucleic acid and 94-95% protein by weight (Gingery *et al.* 1981). The nucleic acid is linear ss RNA. One structural virion protein subunit is present in the virion with a size of 32,700 Da. A virus-coded non-structural protein (=NCP) of 16,000 Da in size is present in infected plants.

Likewise, MYSV has an RNA genome and a single capsid protein (CP) with mol. wt. of 34-35.6 kDa (Thouvenel *et al.* 1996; Mahmoud, 2001). A NCP, mol. wt. 14-14.7 kDa, similar to that found for MSpV, is present in infected plants.

MSpV has a multipartite genome of five segments of ss RNA with estimated lengths of *c.* 8.3, 3.3, 2.4, 2.2 and 1.3 kb for RNAs 1–5, respectively, and with relative masses (Mr) of 3.01, 1.18, 0.81, 0.78, and 0.52 X 10^6 (Huiet *et al.* 1992; Estabrook *et al.* 1996). Apparently, negative and positive sense RNAs are separately encapsidated in different ribonucleoprotein particles (RNPs). The different sized genomic segments are also separately encapsidated, with the smallest RNA (RNA 5) encapsidated in the slowest sedimenting RNP and the larger RNAs in the faster sedimenting RNPs. Ds RNAs presumably arise *in vitro* by hybridization of complementary strands for each RNA. However, unequal amounts of the complementary molecules for all five MSpV RNAs are present and each of the five RNAs is represented primarily by one polarity with smaller amounts of the complementary RNA (Falk *et al.* 1989). MYSV contains five RNA segments that range in size from *c.* 9.5 to 1.6 kb (Mahmoud, 2001).

12.5.3 Strains, epidemiology and control

No strains of MSpV have been designated although differences among isolates suggest that strains exist. As to whether MSpV and MYSV are strains, serologically the two viruses are at most distantly related when tested with antisera to the CP and NCP of both viruses (Mahmoud, 2001). Further, the absence of sequence similarities between a cDNA of the MYSV genome and the sequences available in GenBank, the failure of the cDNA of the MYSV genome to hybridize with the genome of MSpV and of the RNA3 segment of MSpV to hybridize with the genome of MYSV, indicate that MYSV and MSpV are distinct viruses. The similarities between MSpV and MYSV in symptomatology, morphology of the virions, circulative propagative transmission by vectors, the CP and NCP proteins, the tenuivirus-like 5' and 3' terminal sequences of the RNA genomic segments, and the distant serological relationship suggest that MYSV is a related but distinct virus possibly belonging to a new genus related to the genus *Tenuivirus*.

The epidemiology of MSpV while not fully elucidated undoubtedly is dependent on its vector *P. maidis*. The latter is dimorphic, having both short- and long-winged adult forms. The former spreads MSpV over short distances by walking from plant to plant and the latter over distances of up to several kilometers by flight. *P. maidis* populations fluctuate seasonally and year-to-year.

In the epidemiology of MYSV, wheat, barley and possibly graminaceous weeds may serve as alternate hosts for infection of young maize in Egypt (Ammar *et al.*1989). Like with MSpV, incidence of MYSV is very high in some years. MYSV is especially high during September and October on late-sown maize in Giza and other parts of Egypt after the build-up of vector populations.

Little information is available on the control of maize stripe. It seems that eradication of alternate hosts of virus and vector during periods when populations of the latter are increasing would serve to control the disease. Several sources of resistance to MSpV are known. One such source, identified by IRAT in Réunion, is the variety Revolution that has been considered a useful resistance source in their breeding program (Thottappilly *et al.* 1993). Further, some IITA MSV-resistant varieties are also resistant to MSpV. Little is also known for the control of maize yellow stripe. But as for maize stripe, the epidemiology of MYSV suggests means of potential control. These include control of alternate grass hosts of the virus and vector, especially when build-up of vector populations occurs on these hosts in September and October in Egypt. Resistance to MYSV has not been reported.

12.6 Marafiviruses

The principal maize-infecting marafivirus is *Maize rayado fino virus* (MRFV). Synonyms include *Brazilian corn streak virus* (Kitajima *et al.* 1976) and *virus del rayado colombiano del maiz* (Martinez-Lopez, 1977).

Rayado fino was initially described in El Salvador as a third "strain" of corn stunt. Viral identity was established following its accidental discovery in field-collected maize leafhoppers (*Dalbulus maidis*) in Costa Rica (Gamez, 1969). MRFV occurs in the USA, Mexico, Guatemala, Honduras, El Salvador, Nicaragua, Costa Rica, Panama, Colombia, Peru, Bolivia, Brazil, Argentina and Uruguay. The geographical range of MRFV is limited by that of its vector, *D. maidis* (Gamez and Leon, 1988).

Incidences of rayado fino in various tropical areas and at diverse planting times range from 0-40% in most fields to nearly 100% in others (Gamez *et al.* 1979), depending on the cultivar. Total losses may occur due to wilting and quick death of infected plants. Rayado fino is not as prevalent or economically important as the corn stunt disease, which occurs where maize rayado fino is found. Corn stunt is now known to involve the corn stunt spiroplasma (CSS) and/or maize bushy stunt phytoplasma (MBSP) (Bedendo, 1999).

MRFV has a limited host range with only genera *Zea*, *Tripsacum*, and *Rottboellia* reported susceptible (Nault *et al.* 1980). Host species in the genus *Zea* include maize, *Z. mays* ssp. *mays*, and the wild annual and perennial teosintes [*Z. diploperennis* (diploperennial teosinte), *Z. luxurians* (teosinte), *Z. mexicana* (Mexican teosinte that is limited to parts of Mexico and northern Guatemala) and *Z. perennis* (perennial teosinte)]. Recently barley proved susceptible when inoculated by VPI of germinating kernels (Madriz Ordenana *et al.* 2000).

The vector of MRFV, *D. maidis*, also has as its host only maize, the annual teosintes, *Z. mays* ssp. *mexicana* and ssp. *parviglumis*, and the perennial teosintes, *Z. diploperennis* and *Z. perennis* (Gamez and Leon, 1983). *D. maidis* is considered a maize specific leafhopper since it is found principally on maize.

12.6.1 Symptoms and transmission

Disease symptoms on maize incited by MRFV isolates found in Costa Rica appear 8 to 14 days after inoculation as a few small chlorotic dots or short stripes at the base and along the veins of young leaves. As symptoms develop, the dots become more numerous in new leaves and some fuse to form the characteristic chlorotic or yellow stipple striping of the veins. Leaves of tolerant lines develop only a few long continuous stripes, which fade as the plants mature.

Disease symptoms associated with infection of maize by virus del rayado colombiano del maiz (a distinct isolate of MRFV) in Colombia are generally more severe. In some plants death of cells in the area of the lesion causes holes to develop in the leaf blade at these locations. These symptoms, including wilting and even plant death, are associated with infection of the most susceptible cultivars. Root systems are poorly developed and ears on diseased plants have very few or no grains.

MRFV is transmitted by the cicadellid leafhopper, *D. maidis*, and by VPI of developing maize kernels, but not by leaf rub nor through seeds of infected plants (Madriz Ordenana *et al.* 2000). Experimental vectors include *Baldulus tripsaci, D. elimatus, Graminella nigrifrons and Stirellus bicolor* (Nault *et al.* 1980). These vectors probably have limited importance due to preference for hosts other than maize for feeding, inefficient transmission, low population densities, and narrow geographical distributions (Gamez and Leon, 1988).

D. maidis leafhoppers persistently transmit MRFV after incubation periods of eight to 32 days. However, transmission is intermittent and leafhoppers eventually lose the ability to transmit. Only up to one third of the leafhoppers transmit the virus, while nearly 80% have detectable virus. Transmission of MRFV by *D. maidis* appears to be genetically dependent on a recessive gene. The virus replicates in *D. maidis* with no cytopathological effects and no apparent effect on leafhopper longevity and reproduction (Rivera and Gamez, 1986).

12.6.2 Diagnosis, purification and viral characteristics

Diagnosis of maize rayado fino requires identification of MRFV since symptoms alone are not sufficiently distinct and MRFV infection of maize may occur in combination with those by CSS and MBSP, whose symptoms may mask those of the former (Gamez, 1980 a, b). MRFV may be initially identified by transmission with *D. maidis* in a persistent relationship. In addition the virus cannot be transmitted by leaf rub inoculation.

MRFV can be identified by serological procedures, particularly ELISA (Rivera and Gamez, 1986), immune electron microscopy and immune rate-zonal centrifugation (Bradfute *et al.* 1985). The specificity of MRFV antisera plus ample titers allows serological identification of MRFV without initial identification by the above biological assays. A more recent assay for identification is with specific cRNA oligonucleotide primers designed to amplify the coat protein gene of MRFV (Kogel *et al.* 1996). Purification protocols for MRFV were presented by Falk and Tsai (1986). In these protocols the source of virus is freshly harvested or frozen virus-infected leaves. After several steps, MRFV is concentrated from clarified extracts by high-speed centrifugation. The virus is further purified by rate-zonal sucrose density gradient centrifugation followed by isopycnic centrifugation in CsCl.

MRFV virions have icosahedral symmetry and are of two sizes: 33 nm diameter for particles lacking nucleic acid and 31.5 nm diameter for those containing nucleic acid (Leon and Gamez, 1981). Virions have two major capsid proteins with molecular weights of 20-22 and 24-29 kDa found in molar ratios of approximately 3:1, respectively, and which contain common peptide sequences (Gamez and Leon, 1988).

The nucleic acid of MRFV is RNA with a molecular mass of 2.0 - 2.4 x 10^6 Da, which comprises 25-36 % of the virion (Gamez and Leon, 1988). The MRFV genome is a capped, ss, monopartite, +-sense RNA that lacks a terminal VPg and 3' poly(A) tail. The complete nucleotide sequence is 6305 nts in length and contains two putative open reading frames (ORFs).

12.6.3 Strains

Evidence for the existence of strains of MRFV has been weak (Hammond *et al.* 1997). While MRFV isolates with differing properties are reported from several countries (Rivera and Gamez, 1986), their differences do not suggest that they are strains. Possible exceptions are the Brazilian corn streak virus (Kitajima *et al.* 1976) and maize rayado colombiano virus (MRCV) (Martinez-Lopez, 1977). The latter viruses are serologically related to, but distinguishable from, MRFV, and so may be strains (Gamez, 1980 a, b).

12.6.4 Epidemiology and control

The epidemiology of MRFV is described in several publications based on studies done primarily in a tropical moist forest location (Gamez and Saavedra, 1986). Survival of MRFV depends on both maize and *D. maidis*. Multiplication in the latter provides survival between maize plantings.

MRFV and its vector are found from deserts and lowland rain forests through savannahs to deciduous and montane forests suggesting a wide ecological adaptability while having a very narrow host adaptation. Maize is considered the only natural host of both vector and virus throughout these zones, further indicating that it provides a satisfactory habitat for both wherever they survive year round.

Measures for control of MRFV are limited. Few maize lines show resistance and none immunity to infection by MRFV when inoculated by viruliferous *D. maidis*. Locally adapted Central American landraces appear in general to be more tolerant to the virus and show lower incidences within a population (Gamez and Leon, 1988). Susceptibility among maize cultivars to MRFV is variable (Bustamante *et al.* 1998). North American maize accessions show no immunity or resistance but only different degrees of susceptibility (Toler *et al.* 1985). Two genotypes from the CIMMYT germplasm collection (accession nos. 2980-93 and 3974) show mild and delayed symptoms and low MRFV concentrations as determined by ELISA (Bustamante *et al.* 1998). These resistant genotypes will need to be utilized in national programs to develop maize hybrids with increased resistance to MRFV.

Among the close relatives of maize, the perennial diploid teosinte *Z. diploperennis* show few or no symptoms while the perennial tetrapolid teosinte, *Z. perennis*, is immune (Nault *et al.* 1980). Only *Z. diploperennis* will naturally hybridize with maize and thus might serve as a source of MRFV resistance. *Z. perennis* does not readily cross with diploid maize and, therefore, its resistance may be difficult to incorporate into hybrids.

As to other means of control, in Colombia insect control using systemic insecticides did not reduce the incidence of MRFV but crop rotation and adjustment of planting dates did , with reduction from 80% to less than 20% incidence (Martinez-Lopez, 1977).

12.7 Nucleorhabdoviruses

Among the rhadoviruses infecting maize, *Maize mosaic virus* (MMV) is important in developing countries (Jackson *et al.* 1981). Other rhabdoviruses of maize are *Barley yellow striate mosaic virus* (BYSMV) (Lockhart and Elyamani, 1983), *Iranian maize mosaic rhabdovirus* (IMMRV) (Izadpanah, 1989) [= *Maize Iranian mosaic virus* (MIMV)] *and Maize fine streak virus* (MFSV) (Redinbaugh *et al.* 2002). Rhabdoviruses of sorghum are MMV (Naidu *et al.* 1989) and *Sorghum stunt mosaic virus* (Mayhew and Flock, 1981).

MMV occurs worldwide in moist or irrigated lowlands in tropical and subtropical areas (Brewbaker, 1981) such as North, Central and South America and the adjacent Caribbean country islands; Africa; islands of the Western Indian Ocean; India and Pacific Ocean islands. In some countries, notably India, diseases named maize mosaic may not be associated with rhabdovirus infection but with maize-infecting potyviruses. MMV infection of sorghum was reported from Africa (Traore and Konate, 1997) and India (Naidu *et al.* 1989).

Early infection of maize by MMV causes serious crop loss of up to 100% (Brewbaker, 1981). MMV is also severe on sorghum (Naidu *et al.* 1989). MMV is economically important primarily when maize is grown continuously. In addition to maize, the virus also occurs naturally in reservoir hosts *Rottboellia exaltata*, *Setaria vulpiseta* and *Sorghum verticilliflorum*, and has been experimentally transmitted to *Z. mays mexicana*, *Axonopus compressus*, barley and sorghum (Autrey, 1983).

The principal symptoms of MMV are dwarfing of internodes and husks, chlorotic striations adjacent to leaf veins and extending to about 10 cm on the leaf, and chlorotic mosaic on sheath and stalk (Brewbaker, 1981). The extent of internode dwarfing depends on plant age at the time of infection. The type of chlorotic striping on leaves depends on the MMV strain (Autrey, 1983). Strain F (MMV-F) or raya fina (MMV-RF) incites fine chlorotic striations, strain C (MMV-C) coarse or broad striations and strain B (MMV-B) discontinuous or broken striping. Interveinal areas on leaf blades rarely become chlorotic (Brewbaker, 1981). These different striation patterns may overlap to make strain recognition difficult.

Initial symptoms on sorghum are chlorotic streaks between the veins, which later may become necrotic (Naidu *et al.* 1989). In severe infections, the leaves turn chlorotic and plants become severely stunted. Panicles on such plants contain very few seeds.

12.7.1 Transmission

MMV is mechanically transmissible by VPI of germinating maize seed (Louie, 1995) but not by leaf-rub inoculation. Natural transmission is by the corn planthopper, *P. maidis*, the only vector of MMV (Autrey, 1983). Nymphs and adults, both males and females transmit the virus with an efficiency of about 5% to 23%. MMV strains vary in their ability to be transmitted (Autrey, 1983), and transmission seems to be influenced by the presence of a latent *Peregrinus* virus. Nymphs appear to be more efficient vectors than adults. The minimum AAP is 15 min but effective acquisition takes 24 hrs (Autrey, 1983). The incubation period within the vector ranges between 11 days and 7 weeks (Herold, 1972). The virus is transmitted within 15 min of feeding, but transmission is more efficient after 24 hr (Autrey, 1983). MMV multiplies in *P. maidis* (Falk and Tsai, 1985), but is intermittently transmitted (Ammar and Nault, 1985). No seed transmission of the virus has been reported.

12.7.2 Diagnosis, purification and virion characteristics

Symptoms of maize mosaic are often confused with those due to other maize infecting viruses (Brewbaker, 1981), *viz.*, MRFV and possibly MSV. Confirmation of maize mosaic is by transmission of the virus with *P. maidis*, detection of bacilliform or bullet-shaped particles in infected maize leaf cells, and reactivity of extracts from infected tissue with MMV antiserum using ELISA (Naidu *et al.* 1989) or immunosorbent electron microscopy. Light microscopy may be used to identify MMV-distinctive intracellular inclusions in stained epidermal strips (Overman *et al.* 1992).

Several purification protocols have been published for MMV isolates (Naidu *et al.* 1989). Purification of rhabdoviruses poses special problems due to their relatively low concentration and instability.

Both bullet shaped and bacilliform rhabdovirus particles are observed for MMV from various regions and measured by different investigators (Ammar and Nault, 1985; McDaniel *et al.* 1985). Their lengths range from 204 to *c.* 300 nm and diameters from 48 to 90 nm. Particles consist of an envelope with thread-like or knob-like protrusions, a helical structure composed of beaded units, a hollow cylinder and an inner core.

MMV has a ss RNA of mol. wt. c. 4.2 X 10^6 (Falk and Tsai, 1983a). The genome of MMV has yet to be sequenced and its organization described. By analogy with the genome of *Sonchus yellow net virus* (SYNV), another nucleorhabdovirus, the genome is probably negative sense and encodes six proteins designated beginning at the 3'-terminus N, P, sc4, M, G and L (Jackson *et al.* 1999). Putative L, N, Ns (P), Ml (sc4), M2 (M) and G proteins are demonstrated for MMV isolates MMV-CR, MMV-FL and MMV-HI.

12.7.3 Strains, epidemiology and control

Several strains of MMV are known. In Venezuela, two symptomatological strains were recognized. In Mauritius three strains designated MMV-Fine (MMV-F), MMV-Coarse (MMV-C), and MMV-Broken (MMV-B) have been partially characterized (Autrey, 1983). Isolates from the USA designated MMV-FL (Florida), MMV-HI (Hawaii) and MMV-CR (Costa Rica) differ in their location of occurrence but not in viral characteristics and are not considered strains.

The occurrence and spread of MMV is mostly limited to the warm humid tropics and subtropics (Brewbaker, 1981). Environmental conditions in these areas favor survival and development of the vector, *P. maidis*. When large numbers of *P. maidis* develop under these conditions, spread of the virus may occur.

Maize mosaic is controlled primarily by planting MMV-resistant cultivars. Several sources of resistance are known and have been introduced by direct backcrossing into more than 100 inbreds and cultivars and 130 genetic stocks (Brewbaker, 1981). Resistance was first recognized in Cuban flint evaluated in Hawaii. Maize from the Western Hemisphere shows four distinct and apparently unrelated sources of resistance. The resistance locus, designated *mv1*, is located on chromosome 3 near the centromere (Ming *et al.* 1997)

Pure lines from local cultivars in Mauritius and Rodrigues and hybrids from them are also highly resistant and do not become infected with MMV even after a second or third exposure to viruliferous insects, whereas all foreign hybrids are susceptible (Autrey, 1984). The genetic relationship between this resistance and that identified in Western Hemisphere has not been reported.

12.8 Conclusions

Many of the viruses of maize and sorghum crops have been reasonably characterized but very little has so far been done to systematically survey for their actual distribution and incidence and likewise assess their actual economic importance. With the increasing virus research in the tropics, one can expect to see additional literature from the developing countries. It should be realized that with continuing crop improvement, movement of germplasm, "commercial" seed production, further intensified cropping and the change of cropping systems, may lead previously unimportant problems and/or new emerging viruses to become or cause new important diseases. This, together with prevailing insufficient knowledge of existing viruses and of their distribution and incidence, emphasizes the need of continuing efforts to study the viruses in maize and sorghum.

The infection of the same plant by two or more viruses in the field as well as similar mottling and mosaic symptoms caused by different viruses have led to confusion in disease identification. Reliable identification requires a combination of various diagnostic methods like host range, vector transmission, electron microscopy, serology and molecular methods.

Use of insecticides to prevent virus transmission by vectors has only been moderately successful because the available insecticides will not kill the insects quickly enough before they can transmit the virus. Successful vector control may be enhanced by knowing the vector and the mode of virus transmission. A certain degree of control can also be gained by good cultural practices such as removal of diseased plants where practical, crop rotation, weeding to remove alternate virus and vector hosts, possible use of barrier crops to prevent movement of vectors, etc.

Perhaps the best single control method now available for viruses is the use of resistant varieties. The first and often most promising step towards crop improvement and virus control is large scale screening and selection for resistant and tolerant lines to local virus isolates under local conditions. Transgenic resistance is being applied in a few cases.

Acknowledgments

The authors wish to thank Drs. V.T. John, Margaret Boulton and Stephan Winter for all their help in finalizing this chapter.

References

Afsharifar, A.R. and Izadpanah, K. 1994. A type of sugarcane mosaic virus infecting maize, sorghum and johnsongrass in Iran. *Iran Agric. Res.* 13: 33-47.

Alegria, O.M., Royer, M., Bousalem, M., Chatenet, M., Peterschmitt, M., Girard, J.-C. and Rott, P. 2003. Genetic diversity in the coat protein coding region of eighty-six sugarcane mosaic virus isolates from eight countries, particularly from Cameroon and Congo. *Arch. Virol.* 148: 357-372.

Alexander, J.D., Toler, R.W., and Miller, F.R. 1984. The effect of maize dwarf mosaic virus strain B on the growth and yield of susceptible sorghum. *Phytopathology* 74: 625.

Alhubaishi, A.A., Walkey, D.G.A., Webb, M.J.W., Bolland, C.J., and Cook, A.A. 1987. A survey of horticultural plant virus diseases in the Yemen Arab Republic. *FAO Plant Prot. Bull.* 35: 135-143.

Ammar, E. D. Mahmoud A. and Peterschmitt M. 2003. Description of maize yellow stripe virus. In: *Viruses and virus diseases of Poaceae*. H. Lapierre and P.A. Signoret (eds.). INRA Editions, Centre de Versailles, route de St Cyr, 78026 Versailles cedex. In press.

Ammar, E. D., Elnagar, S., Aboul-Ata, A. E., and Sweify, G. H. 1989. Vector and host plant relationships of the leafhopper-borne maize yellow stripe virus. *J. Phytopathol.* 126: 246-252.

Ammar, E. D., Gingery, R. E., Gordon, D. T., and Aboul-Ata, A. E. 1990. Tubular helical structures and fine filaments associated with the leafhopper-borne maize yellow stripe virus. *Phytopathology* 80: 303-309.

Ammar, E.D. and Nault, L.R. 1985. Assembly and accumulation sites of maize mosaic virus in its planthopper vector. *Intervirology* 24: 33-41.

Ammar, E.D., Abul-Ata, A.E., El-Sheikh, M.A., and Sewify, G.H. 1987. Incidence of virus and viruslike disease syndromes on maize and sugarcane in Middle and Lower Egypt. *Egypt. J. Phytopathol.* 19: 97-107.

Antignus, Y. 1987. Comparative study of two maize dwarf mosaic virus strains infecting corn and johnsongrass in Israel. *Plant Dis.* 71: 687-691.

Arneodo, J.D., Guzman, F.A., Conci, L.R., Laguna, I.G., and Truol, G.A. 2002a. Transmission features of Mal de Rio Cuarto virus in wheat by its planthopper vector Delphacodes kuscheli. *Ann. appl. Biol.* 141: 195-200.

Arneodo, J.D., Lorenzo, E., Laguna, I.G., Abdala, G., and Truol, G.A. 2002b. Cytopathological characterization of Mal de Rio Cuarto virus in corn, wheat and barley. *Fitopatologia Brasileira* 27: 298-302.

Asanzi, M.C., Bosque-Perez, N.A., Nault, L.R., Gordon, D.T., and Thottappilly, G. 1995. Biology of *Cicadulina species* (Homoptera: Cicadellidae) and transmission of maize streak virus. *African Entomol.* 3: 173-179.

Autrey, L.J.C. 1983. Maize mosaic virus and other maize virus diseases in the islands of the Western Indian Ocean. Pages 167-181 In: *Proceedings International Maize Virus Disease Colloquium and Workshop.* D.T. Gordon, J.K. Knoke, L.R. Nault, and R.M. Ritter, eds. 2-6 Aug., 1982, Ohio State Univ., Ohio Agric. Res. Dev. Cent., Wooster. 261 pp.

Autrey, L.J.C. 1984. On assessing yield losses due to maize mosaic virus in the field. *Maize Virus Dis. Newsl.* 1: 39-45.

Azuhata, F., Uyeda,I.,Kimura,I. And Shikata, E. 1993. Close similarity between genome structures of rice black streaked dwarf and maize rough dwarf viruses. *J. Gen. Virol.* 74: 1227-1232.

Bai, F. W., Yan, J., Qu, Z. C., Zhang, H. W., Xu, J., Ye, M. M., and Shen, D. L. 2002. Phylogenetic analysis reveals that a dwarfing disease on different cereal crops in China is due to rice black streaked dwarf virus (RBSDV). *Virus Genes* 25: 201-206.

Barrow, M.R., 1992. Development of maize hybrids resistant to maize streak virus. *Crop Protect.* 11: 267–271.

Barrow, M.R., 1993. Increasing maize yields in Africa through the use of maize streak virus resistant hybrids. *Afr. Crop. Sci. J.* 1: 139-144.

Bedendo I.P. 1999. Corn stunt and maize bushy stunt: symptomatology, etiology and techniques for detection and identification of the causal agents. *Summa Phytopathologica* 25: 190-196.

Berger, P.H., Luciano, C.S., Thornbury, D.W., Benner, H.I., Hill, J.H., and Zeyen, R.J. 1989. Properties and *in vitro* translation of maize dwarf mosaic virus RNA. *J. Gen. Virol.* 70: 1845-1851.

Bigarré, L., Salah, M., Granier, M., Frutos, R., Thouvenel, J.-C., and Peterschmitt, M. 1999. Nuleotide sequence evidence for three distinct sugarcane streak mastreviruses. *Arch. Virol.* 144: 2331-2344.

Bigirwa, G., Gibson, R.W., Page, W.W., Hakiza, J.J., Kyetere, D.T., Kalule, T.M., Baguma, S.D., 1995. A new maize disorder in Uganda caused by *Cicadulina niger*. In: Jewell, D.C., Waddington, S.R., Ransom, J.K., Pixley, K.V. (Eds.), *Maize Research for Stress Environments.* Proceedings of the Fourth Eastern and Southern Africa Regional Maize Conference, Harare, Zimbabwe. CIMMYT, Mex-ico, D.F., pp. 202–204.

Bock, K.R. 1974. Maize streak virus. No. 133 In: *Descriptions of Plant Viruses.* Commonw. Mycol. Inst., Assoc. Appl. Biol., Kew, Surrey, England. 4 pp.

Bock. K.R., Guthrie, E.J. and Woods, R.D. 1974. Purification of maize streak virus and its relationship to viruses associated with streak diseases of sugan cane and *Panicum maximum. Ann. appl. Biol.* 77: 289-296.

Bock, K.R., Guthrie, E.J., Merideth, G., Barker, H., 1977. RNA and protein components of maize streak and cassava latent viruses. *Ann. appl. Biol.* 85: 305–308.

Bond, W.P. and Pirone, T.P. 1971. Purification and properties of sugarcane mosaic virus strains. *Phytopathol. Z.* 71: 56-65.

Bosque-Pérez, N.A. 2000. Eight decades of maize streak virus research. *Virus Res.* 71: 107-121.

Bosque-Pérez, N.A., Olojede, S.O., and Buddenhagen, I.W. 1998. Effect of maize streak virus disease on the growth and yield of maize as influenced by varietal resistance levels and plant stage at time of challenge. *Euphytica* 101: 307-317.

Boulton, M.I. 2002. Functions and interactions of mastrevirus gene products. *Physiol. and Molec. Plant Path.* 60: 243-255.

Boulton, M.I. and Markham, P.G. 1986. The use of squah-blotting to detect plant pathogens in insect vectors. Pp. 55-69, In: *Developments in applied biology 1. Developmen and applications in virus testing*, R.A.C. Jones and L. Torrance (eds). Assoc. appl. Biologists, WellesbournUK.

Boulton, M.I., Buchholz, W.G., Marks, M.S., Markham, P.G.and Davies, J.W. 1989. Specificity of *Agrobacterium*-mediated delivery of maize streak virus DNA to members of the Graminae. *Plant Mol. Biol.* 12: 31-40.

Bradfute, O.E., Tsai, J.H., and Falk, B.W. 1985. Maize rayado fino and maize dwarf mosaic viruses in Ecuador. *Plant Dis.* 69: 1078-1080.

Brewbaker, J.L. 1981. Resistance to maize mosaic virus. Pages 145-151 In: *Virus and Viruslike Diseases of Maize in the United States.* D.T. Gordon, J.K. Knoke, and G.E. Scott, eds. *Southern Coop. Ser. Bull.* 247. 210 pp.

Brunt, A.A., Crabtree, K., Dallwitz, M.J., Gibbs, A.J. and Watson, L. 1996. *Viruses of plants. Descriptions and lists from the vide database.* CAB International, Oxow, 1484pp.

Bustamante, P.I., Hammond, R., and Ramirez, P. 1998. Evaluation of maize germ plasm for resistance to maize rayado fino virus. *Plant Dis.* 82: 50-56.

Caciagli, P., Roggero, P., and Luisoni, E. 1985. Detection of maize rough dwarf virus by enzyme-linked immunosorbent assay in plant hosts and in the planthopper vector. *Ann. appl. Biol.* 107: 463-471.

Caulfield, M.J. 1994. A field selection method to screen for resistance to maize streak virus. In: *Proceedings of the Tenth South African Maize Breeding Symposium, Potchefstroom, South Africa, 17-19 March 1992.* J.G du Plessis, J.B.J. van Rensburg, N.W. McLaren, , and B.C. Flett (eds), Pages 46-49.

Chen, J., Chen, J. and Adams, M.J. 2002. Characterisation of potyviruses from sugarcane and maize in China. *Arch. Virol.* 147: 1237-1246.

Chen, Y.T., Guo, M.K., Zhu, X.Y., Gao, W.D., Da, F.C., and Li, L. 1996. Evaluation of resistance to germplasm dwarf mosaic virus in maize. *Plant Protection* 22: 13-15.

Cheng, Y., Chen, J., and Chen, J. 2002. The complete sequence of a sugarcane mosaic virus isolate causing maize dwarf mosaic disease in China. *Sci. China, Series C: Life Sci.* 45: 322-330.

Conti, M. 1983. Maize viruses and virus diseases in Italy and other Mediterranean countries. In: *Proceedings International Maize Virus Disease Colloquium and Workshop*. D. T. Gordon, J. K. Knoke, L. R. Nault, and R. M. Ritter, eds. 2-6 Aug., 1982. Ohio State Uni.., Ohio Agric. Res. Dev. Cent., Wooster. Pp 103-112.

Damsteegt, V.D. 1981. Exotic virus and viruslike diseases of maize. In: *Virus and virus-like diseases of maize in the United States*, D.T. Gordon, J.K. Knoke and G.E. Scott (eds), *Southern Co-operative Series Bull. 247*, 110-123, Ohio Agricultural Research and Development Center, Wooster, Ohio.

Damsteegt, V. D. 1983. Maize streak virus: I. Host range and vulnerability of maize germ plasm. *Plant Dis.* 67: 734-737.

Dekker, E.L., Pinner, M.S., Markham, P.G., van Regenmortel, M.H.V., 1988. Characterization of maize streak virus isolates from different plant species by polyclonal and monoclonal antibodies. *J. Gen. Virol.* 69: 983-990.

Distefano, A. J., Conci, L. R., Munoz Hidalgo, M., Guzman, F. A. Hopp, H E. and Del Vas. M. 2002. Sequence analysis of genome segments S4 and S8 of Mal de Rio Cuarto virus (MRCV): evidence that the virus should be a separate Fijivirus species. *Arch. Virol.* 147: 1699-1709.

Dollet, M., Accotto, G.P., Lisa, V., Menissier, J., Boccardo, G., 1986. A geminivirus, serologically related to maize streak virus, from *Digitaria sanguinalis* from Vanuatu. *J.Gen. Virol.* 67: 933–937.

Donson, J., Accotto, G.-P., Boulton, M.I., Mullineaux, P.M. and Davies, J.W. 1987. The nucleotide sequence of a geminivirus from *Digitaria sanguinalis. Virology* 161: 160-169.

Dussle, C.M., Melchinger, A.E., Kuntze, L., Stork, A., and Lubberstedt, T. 2000. Molecular mapping and gene action of Scml and Scm2, two major QTL contributing to SCMV resistance in maize. *Plant Breeding* 119:299-303.

Dussle, C.M., Quint, M., Melchinger, A.E., Xu, M.L., and Luebberstedt, T. 2003. Saturation of two chromosome regions conferring resistance to SCMV with SSR and AFLP markers by targeted BSA. *Theor. Appl. Genetics* 106: 485-493.

Eberwine, J.W. Jr. and Hagood, E.S. Jr. 1995. Effect of johnsongrass (*Sorghum halepense*) control on the severity of virus diseases of corn (*Zea mays*). *Weed Technology* 9: 73-79.

Efron, Y., Kim, S.K., Fajemisin, J.M., Mareck, J.H., Tang, C.Y., Dabrowski, Z.T., Rossel, H.W., and Thottappilly, G. 1989. Breeding for resistance to maize streak virus: a multidisciplinary team approach. *Plant Breed.* 103: 1–36.

Estabrook, E.M., Suyenaga, K., Tsai, J.H., and Falk, B.W. 1996. Maize stripe virus Tenuivirus RNA2 transcripts in plant and insect hosts and analysis of pvc2, a protein similar to the Phlebovirus virion membrane glycoproteins. *Virus Genes* 12: 239–247.

Falk, B. W. and Tsai, J.H. 1983a. Physicochemical characterization of maize mosaic virus. *Phytopathology* 73:1536-1539.

Falk, B.W., and Tsai, J.H.. 1983b. Assay for maize stripe virus-infected plants by antiserum produced to a purified noncapsid protein. *Phytopathology* 73: 1259–1262.

Falk, B. W. and Tsai, J. H. 1985. Serological detection and evidence for multiplication of maize mosaic virus in the planthopper, Peregrinus maidis. *Phytopathology* 75:852-855.

Falk, B.W. and Tsai, J.H. 1986. The two capsid proteins of maize rayado fino virus contain common peptide sequences. *Intervirology* 25: 111-116.

Falk, B.W. and Tsai, J.H. 1998. Biology and molecular biology of viruses in the genus Tenuivirus. *Annu. Rev. Phytopathol.* 36: 139–163.

Falk, B.W., Klaassen, V.A., and Tsai, J.H.. 1989. Complementary DNA cloning and hybridization analysis of maize stripe virus RNAs. *Virology* 173: 338–342.

Fan, Z.F., Chen, H.Y., Liang, X.M., and Li, H F. 2003. Complete sequence of the genomic RNA of the prevalent strain of a potyvirus infecting maize in China. *Arch. Virol* 148: 773–782.

Fang, S., Yu, J., Feng, J., Han, C., Li, D., and Liu, Y. 2001. Identification of rice black-streaked dwarf fijivirus in maize with rough dwarf disease in China. *Arch. Virol.* 146: 167-170.

FAOSTAT, 2001. www.fao.org

Frenkel, M.J., Jilka, J.M., Shukla, D.D., and Ward, C.W. 1992. Differentiation of potyviruses and their strains by hybridization with the 3' non-coding region of the viral genome. *J. Virol. Methods* 36: 51-62.

Fuller, C. 1901. Mealie variegation. Pages 17-19 In: *First Rep. Gov. Entomol., Natal 1899-1900.*

Gamez, R. 1969. A new leafhopper-borne virus of corn in Central America. *Plant Dis. Rep.* 53:929-932.

Gamez, R. 1980a. Maize rayado fino virus. No. 220 In: *Descriptions of Plant Viruses.* Commonw. Mycol. Inst., Assoc. Appl. Biol. , Kew, Surrey, England. 4 pp.

Gamez, R. 1980b. Rayado fino virus disease of maize in the American tropics. *Trop. Pest Manage.* 26:26-33.

Gamez, R. and Leon, P. 1983. Maize rayado fino virus: Evolution with plant host and insect vector. Pages 149-168 In: *Current Topics in Vector Research*, Vol. 1. K.F. Harris, ed. Praeger Publ., New York, N.Y.

Gamez, R. and Leon, P. 1988. Maize rayado fino and related viruses. Pages 213-233 In: Plant Viruses, Vol. 3. *Polyhedral Virions with Monopartite RNA Genomes.* R. Koenig, ed. Plenum Press, NY. 294 pp.

Gamez, R. and Saavedra, F. 1986. Maize rayado fino: A model of a leafhopper-borne virus disease in the Neotropics. Pages 315-326 In*: Plant Virus Epidemics: Monitoring, Modeling and Predicting Outbreaks.* G.D. McLean, R.G. Garrett and W.G. Ruesink, eds. Academic Press Australia, North Ryde, N.S.W. 550 pp.

Gamez, R., Kitajima, E.W., and Lin, M.T. 1979. The geographical distribution of maize rayado fino virus. *Plant Dis. Rep.* 63:830-833.

Garrido, M.J. and Trujillo, G.E. 1988. Identification of a new strain of maize dwarf mosaic virus (MDMV) in Venezuela. *Fitopatologia Venezolana* 1: 73-81.

Garud, T.B., Syed, I., Shinde, B.M., and Ismail, S. 2000. Sorghum stripe disease in Maharashtra: incidence and sources of resistance. *Internatl. Sorghum Millets-Newsletter* 41: 50-51.

Gingery, R. E. 1985. Maize stripe virus. No. 300 In: Descriptions of Plant Viruses. Assoc. Appl. Biol., Natl. Veg. Res. Stn., Wellesbourne, Warwick, U.K. 4 pp.

Gingery, R.E., Nault, L.R., and Bradfute, O.E. 1981. Maize stripe virus: characteristics of a new virus class. *Virology* 182: 99–108.

Gingery, R.E., Nault, L.R., Tsai, J.H., and Lastra, R.J. 1979. Occurrence of maize stripe virus in the United States and Venezuela. *Plant Dis. Rep.* 63: 341-343.

Giorda, L.M., Toler, R.W. and Miller, F.R. 1986. Identification of sugarcane mosaic virus strain H isolate in commercial grain sorghum. *Plant Dis.* 70: 624-628.

Gordon, D.T., Bradfute, O.E., Gingery, R.E., Knoke, J.K., Louie, R.,. Nault, L.R and Scott, G.E. 1981. Introduction: History, geographical distribution, pathogen characteristics, and economic importance. Pages 1-12 In. *Virus and viruslike diseases of maize in the United States.* D.T. Gordon, J.K. Knoke, and G.E. Scott, eds. *South. Coop. Ser. Bull.* 247. 210 pages

Gorter, G.J. M.A., 1953. Studies on the spread and control of the streak disease of maize. *Union So. Afr. Dept. Agric. Sci. Bull.* 341: 1–20.

Gough, K.H. and Shukla, D. D. 1993. Nucleotide sequence of Johnsongrass mosaic potyvirus genomic RNA. *Intervirology* 36: 181–192.

Govinden, N. and Rummun, K. 1996. Registration of MSIRI 3B streakresistant maize germplasm. *Crop Sci.* 36: 824.

Grimsley, N., Hohn, B., Hohn, T.and Walden, R.M. 1986. Agroinfection, an alternative route for plant virus infection by using Ti plasmid. *Proc. Natl. Acad. Sci. USA*, 83: 3282-3286.

Grimsley, N., Hohn, T., Davies, J.W. and Hohn, B. 1987. *Agrobacterium*-mediated delivery of infectious maize streak virus into maize plants. *Nature* 325: 177-179.

Hammond, R.W., Kogel, R., and Ramirez, P. 1997. Variability of geographically distinct isolates of maize rayado fino virus in Latin America. *J. Gen. Virol.* 78: 3153-3159.

Handley, J.A., Smith, G.R., Dale, J.L., and Harding, R.M. 1998. Sequence diversity in the coat protein coding region of twelve sugarcane mosaic potyvirus isolates from Australia, USA and South Africa. *Arch. Virol.* 143: 1145-1153.

Harrison, B.D., 1985. Advances in geminivirus research. *Annu. Rev. Phytopathol.* 23, 55–82.

Harrison, B.D., Baker, H., Bock, K.R., Guthrie, E.J., Me-rideth, G., Atkinson, M., 1977. Plant viruses with circular single-stranded DNA. *Nature* 270: 760–762.

Henzell, R.G., Persley, D.M., Greber, R.S., Fletcher, D.S., and Van Slobbe, L. 1982. Development of grain sorghum lines with resistance to sugarcane mosaic and other sorghum diseases. *Plant Dis.* 66: 900-901.

Herold, F. 1972. Maize mosaic virus. No. 94 In: *Descriptions of Plant Viruses.* Commonw. Mycol. Inst., Assoc. Appl. Biol., Kew, Surrey, England. 4 pp.

Hill, J.H., Martinson, C.A., and Russell, W.A. 1974. Seed transmission of maize dwarf mosaic and wheat streak mosaic viruses in maize and response of inbred lines. *Crop. Sci.* 14: 232-235.

Howell, S.H., 1984. Physical structure and genetic organisation of the genome of maize streak virus (Kenyan isolate). *Nucleic Acids Res.* 12:7359-7362.

Hughes, F.L., Rybicki, E.P. and Von Wechmar, M.B. 1992. Genome typing Southern African subgroup-1 geminiviruses. *J. Gen. Virol.* 73: 1031-1040.

Huiet, L., Tsai, J.H., and Falk, B.W. 1992. Complete sequence of maize stripe virusRNA4 and mapping of its subgenomic RNAs. *J. Gen. Virol.* 73: 1603–1607.

Isogai, M., Uyeda, I.., and Lee, B.C. 1998. Detection and assignment of proteins encoded by rice black streaked dwarf fijivirus S7, S8, S9 and S10. *J. Gen. Virol.* 79: 1487-1494.

Isogai, M., Uyeda, I., Choi J.K., and Choi J.K. 2001. Molecular diagnosis of Rice black-streaked dwarf virus in Japan and Korea. *Pl. Pathol. J.* 17: 164-168.

Izadpanah, K. 1989. Purification and serology of the Iranian maize mosaic rhabdovirus. *J. Phytopathol.* 126: 43-50.

Jackson, A.O., Milbrath, G.M., and Jedlinski, H. 1981. Rhabdovirus diseases of the Gramineae. Pages 51-76, In: *Virus and Viruslike Diseases of Maize in the United States.* D.T. Gordon, J.K. Knoke, and G.E. Scott,(eds). *Southern Coop. Ser. Bull.* 247. 210 pp.

Jackson, A.O., Goodin, M., Moreno, I., Johnson, J., and Lawrence, D.M. 1999. Plant Rhabdoviruses. In: *Encyclopedia of Virology*, Second Edition, Allan Granoff and Robert G. Webster (eds). Academic Press. doi:10.1006/rwvi.1999.0251.

Jiang, J.X. and Zhou, X.P. 2002. Maize dwarf mosaic disease in different regions of China is caused by sugarcane mosaic virus. Brief Report. *Arch. Virol.* 147: 2437-2443.

Jones, F.E., Hill, J.H., and Durand, D.P. 1988. Detection and differentiation of maize dwarf mosaic virus, strains A and B, by use of difference class immunoglobulins in a double-antibody sandwich enzyme-linked immunosorbent assay. *Phytopathology* 78: 1118-1124.

Khalifa, E.A., 2001. Studies on the relationship between maize yellow stripe virus and its leafhopper vector. Ph.D. Thesis, Faculty Agric., Cairo Univ. 82 pp.

Kim, S.K., Efron, Y., Fajemisin, J.M., Buddenhagen, I.W., 1989. Mode of gene action for resistance in maize to maize streak virus. *Crop Sci.* 29: 890–894.

Kitajima, E.W., Yano, T., and Costa, A.S. 1976. Purification and intracellular localization of isometric viruslike particles associated with Brazilian corn streak virus infection. *Cienc. Cult.* (Sao Paulo) 28: 427-430.

Kogel, R., Hammond, R.W., and Ramirez, P. 1996. Incidence and geographic distribution of maize rayado fino virus (MRFV) in Latin America. *Plant Dis.* 80: 679-683.

Kong, P. and Steinbiss, H.-H. 1998. Complete nucleotide sequence and analysis of the putative polyprotein of maize dwarf mosaic virus genomic RNA (Bulgarian isolate). *Arch. Virol.* 143: 1791-1799.

Kovacs, G., Gaborjanyi, R., and Toldi, E. 1994. Inheritance of resistance to maize dwarf mosaic virus and sugarcane mosaic virus in maize. *Cereal Res. Communications* 22: 361-368.

Kuntze, L., Fuchs, E., Gruntzig, M., Schulz, B., Henning, U., Hohmann, F., and Melchinger, A.E. 1995. Evaluation of maize inbred lines for resistance to sugarcane mosaic virus (SCMV) and maize dwarf mosaic virus (MDMV). *Agronomie* 15: 463-467.

Kuntze, L., Fuchs, E., Gruntzig, M., Schulz, B., Klein, D., and Melchinger, A.E. 1997 . Resistance of early-maturing European maize germplasm to sugarcane mosaic virus (SCMV) and maize dwarf mosaic virus (MDMV). *Plant Breeding* 116: 499-501.

Kyetere, D.T., Ming, R., McMullen, M.D., Pratt, R.C., Brew-baker, J., Musket, T., 1999. Genetic analysis of tolerance to maize streak virus in maize. *Genome* 42: 20–26.

Laguna, I.G., Gimenez Pecci, M.P., Herrera, P.S., Borgogno, C., Ornaghi J., and Pardina P.R. 2000. Rol de los cereales en la epidemiologia del virus del Mal de Rio Cuarto en Argentina. *Fitopatologia* 35: 41-49.

Lamy, D., Thouvenel, J.C., and Fauquet, C. 1979. A strain of guinea grass mosaic virus naturally occurring on maize in the Ivory Coast. *Ann. appl. Biol.* 93: 37-40.

Lastra, R., and Carballo, O. 1985. Mechanical transmission, purification and properties of an isolate of maize stripe virus from Venezuela. *Phytopathol. Z.* 114: 168–179.

Lazarowitz, S.G., 1988. Infectivity and complete nucleotide sequence of the genome of a south African isolate of MSV. *Nucleic Acids Res.* 16: 229-249.

Lenardon, S.L., Gordon, D.T., and Gingery, R.E. 1993. Serological differentiation of maize dwarf mosaic potyvirus strains A, D, E, and F by electro-blot immunoassay. *Phytopathology* 83: 86-91.

Lenardon, S.L., March, G.J., Nome, S.F., and Ornaghi, J.A. 1998. Recent outbreak of "Mall de Rio Cuarto" virus on corn in Argentina. *Plant Dis.* 82:448 (Abstr.)

Leon, P. and Gamez, R. 1981. Some physicochemical properties of maize rayado fino virus. *J. Gen. Virol.* 56: 67-75 .

Li, J., Li, G., and Yuan, .F 1999. Control of the maize rough dwarf virus disease. *Beijing Agric. Sci.* 17: 23–25.

Liu, H., Boulton, M.I., Oparka, K.J. and Davies, J.W. 2001. Interaction of the movement and coat proteins of Maize streak virus: implications for the transport of viral DNA. *J. Gen. Virol.* 82: 35-44.

Lockhart, B.E.L. and Elyamani, M. 1983. Virus and viruslike diseases of maize in Morocco. Pages 127-129 In: *Proceedings International Maize Virus Disease Colloquium and Workshop.* D.T. Gordon, J.K. Knoke, L.R. Nault, and R.M. Ritter, eds. 2-6 Aug.,1982, Ohio State Univ., Ohio Agric. Res. Dev. Cent., Wooster. 261 pp.

Louie, R. 1995. Vascular puncture of maize kernels for the mechanical transmission of maize white line mosaic virus and other viruses of maize. *Phytopathology* 85: 139-143.

Louie, R. and Knoke, J.K. 1975. Strains of maize dwarf mosaic virus. *Plant Dis. Rep.* 59: 518-522.

Louie, R. and Knoke, J.K. 1991. Detection of maize dwarf mosaic onset in northern Ohio. *Phytopathology* 81: 760-765.

Louie, R., Findley, Knoke, J.K. and McMullen, M.D. 1991. Genetic basis of resistance in maize to five maize dwarf mosaic virus strains. *Crop Sci.* 31: 14-18.

MacKenzie, D.R., Wernham, C.C., and Ford, R.E. 1966. Differences in maize dwarf mosaic virus isolates of the northeastern United States. *Plant Dis. Rep.* 50: 814-818.

Madriz Ordenana, K., Rojas Montero, R., Lundsgaard, T., Ramirez, P., Thordal Christensen, H., and Collinge, D.B. 2000. Mechanical transmission of maize rayado fino marafivirus (MRFV) to maize and barley by means of the vascular puncture technique. *Plant Pathol.* 49: 302-307.

Mahmoud, A. 2001. Biological and molecular characteristics of Maize yellow stripe virus and its relationship with the leafhopper vector. PhD. Thesis, Faculty Agric. Cairo Univ.156 pp.

Mahmoud, A., Thouvenel, J.C., Abol Ela, S.E., Sewify, G.H., and Ammar, E.D. 1996. Detection of maize yellow stripe tenui-like virus by ELISA and dot-blot tests in host plants and leafhopper vector in Egypt. *Phytopathol. Medit.* 35: 19-23.

Mali, V.R. 1999. An immunity (QL3) breaking strain of sorghum mosaic potyvirus in India. *Plant Dis.* 83: 877.

Mali, V.R. 2001. Current status of sugarcane mosaic subgroup of potyviruses causing sorghum red stripe in India. *Sugarcane Pathol.*2: 99-154.

Mali, V.R. and Thakur, R.P. 2001. Reactions of promising sorghum genotypes to a distinct strain of sorghum mosaic potyvirus (SrMV-IBS) prevalent in peninsular India. *Acta Phytopathol. Entomol. Hungarica* 36: 317-327.

March, G.J., Balzarini, M., Ornaghi, J.A., Beviacqua, J.E., and Marinelli, A. 1995. Predictive model for "Mal de Rio Cuarto" disease intensity. *Plant Dis.* 79: 1051-1053.

March, G.J., Ornaghi, J.A., Beviacqua, J.E., Giuggia, J., Rago, A., and Lenardon, S.L. 2002. Systemic insecticides for control of *Delphacodes kuscheli* and the Mal de Rio Cuarto virus on maize. *International J. Pest Manage.* 48: 127-132.

Marie-Jeanne, V., Ioos, R., Peyre, J., Alliot, B., and Signoret, P. 2000. Differentiation of poaceae potyviruses by reverse transcription-polymerase chain reaction and restriction analysis. *J. Phytopathol.* 148: 141-151.

Martin, D.P. and Rybicki, E.P. 2000. Improved efficiency of *Zea mays* agroinoculation with *Maize streak virus*. *Plant Dis.* 84: 1096-1098.

Martin, D.P., Willment, J.A., Rybicki, E.P. 1999. Evaluation of maize streak virus pathogenicity in differentially resistant *Zea mays* genotypes. *Phytopathology* 89: 695–700.

Martin, D.P., Willment, J.A., Billharz,R., Velders, R., Odhiambo, B., Njuguna, J., James, D and Rybicki, E.P. 2001. Sequence diversity and virulence in *Zea mays* of *Maize streak virus* isolates. *Virology* 288: 247-255.

Martinez-Lopez, G. 1977. New maize virus diseases in Colombia. Pages 20-29 In: *Proceedings International Maize Virus Disease Colloquium and Workshop*. L.E. Williams, D.T. Gordon, and L.R. Nault, eds. 16-19 Aug.,1976, Ohio Agric. Res. Dev. Cent., Wooster. 145 pp.

Marzachi, C., Boccardo, G., and Caciagli, P. 1999. Heterogeneity of maize rough dwarf fijivirus genomic segments isolated from 1997 field infection. *Zeitschrift Pflanzenkrankheiten Pflanzenschutz* 106: 431-436.

Marzachi, C., Boccardo, G., and Nuss, D.L. 1991. Cloning of the maize rough dwarf virus genome: Molecular confirmation of the plant-reovirus classification scheme and identification of two large non-overlapping coding domain within a single genomic segment. *Virology* 180: 518-526.

Marzachi, C., Boccardo, G., Milne, R., Isogai, M., and Uyeda, I. 1995. Genome structure and variability of Fijiviruses. *Seminars in Virology* 6: 103-108.

Mayhew, D.E. and Flock, R.A. 1981. Sorghum stunt mosaic. *Plant Dis.* 65: 84-86.

McDaniel, L.L. and Gordon, D.T. 1985. Identification of a new strain of maize dwarf mosaic virus. *Plant Dis.* 69: 602-607.

McDaniel, L.L., Ammar, E.D., and Gordon, D.T. 1985. Assembly, morphology and accumulation of a Hawaiian isolate of maize mosaic virus in maize. *Phytopathology* 75: 1167-1172.

McGee, D.C. 1988. Maize diseases. A reference source for seed technologists, American Phytopathological Society, APS Press, St. Paul, Minnesota, USA, 150pp.

McKern, N.M., Whittaker, L.A., Strike, P.M., Ford, R.E., Jensen, S.G., and Shukla, D.D. 1990. Coat protein properties indicate that maize dwarf mosaic virus-KS1 is a strain of Johnson grass mosaic virus. *Phytopathology* 80: 907-912.

McKern, N.M., Shukla, D.D., Toler, R.W., Jensen, S.G., Tosic, M., Ford, R.E., Leon, O., and Ward, C.W. 1991. Confirmation that the sugarcane mosaic virus subgroup consists of four distinct potyviruses by using peptide profiles of coat proteins. *Phytopathology* 81: 1025-1029.

Mesfin, T., Bosque-Pe´rez, N.A., Buddenhagen, I.W., Thottappilly, G., Olojede, S.O., Den Hollander, J., Markham, P.G., 1991. *Cicadulina* species and maize streak virus in Ethiopia. *Trop. Pest Management* 37: 240–244.

Mesfin, T., Bosque-Perez, N.A., Buddenhagen, I.W., Thottappilly, G., and Olojede, S.O. 1992. Studies on maize streak virus isolates from grass and cereal hosts in Nigeria. *Plant Dis.* 76: 789-795.

Mikel, M.A., D'Arcy, C.J., and Ford, R.E. 1984. Seed transmission of maize dwarf mosaic virus in sweet corn. *Phytopathol. Z.* 110: 185-191.

Milne, R.G. and Lovisolo, O. 1977. Maize rough dwarf and related viruses. *Adv. Virus Res.* 21: 267-341.

Milne, R.G., Conti, M. and Lisa, V. 1973. Partial purification, structure and infectivity of complete maize rough dwarf virus particles. *Virology* 53: 130-141.

Milne, R. G., Boccardo, G., Dal Bo. E., and Nome, F. 1983. Association of maize rough dwarf virus with Mal de Rio Cuarto in Argentina. *Phytopathology* 73: 1290-1292.

Ming, R., Brewbaker, J. L., Pratt, R. C., Musket, T. A., and McMullen, M. D. 1997. Molecular mapping of a major gene conferring resistance to maize mosaic virus. *Theor. Appl. Genet.* 95: 271-275.

Mullineaux, P.M., Donson,J., Morris-Krinich, B.A., Boulton, M.I., and Davies, J.W. 1984. The nucleotide sequence of maize streak virus DNA. *EMBO journal* 3: 3063-3068.

Murry, L.E., Elliott, L.G., Capitant, S.A., West, J.A., Hanson, K.K., Scarafia, L., Johnston, S., DeLuca-Flaherty, C., Nichols, S. *et al.* 1993. Transgenic corn plants expressing MDMV strain B coat protein are resistant to mixed infections of maize dwarf mosaic virus and maize chlorotic mottle virus. *Bio/Technology* 11: 1559-1564.

Naidu, R.A., Harikrishnan, R., Manohar, S.K., Reddy, D.V.R., Ratna, A.S., King, S.B., and Bandyopadhyay, R. 1989. The occurrence of maize mosaic virus on sorghum in India. *Ann. appl. Biol.* 114: 301-310.

Narayana, Y. D., and Muniyappa, V 1996a. Virus-vector relationships of a planthopper (Peregrinus maidis) borne sorghum stripe tenuivirus. *Int. J. Pest Manage.* 42: 165-170.

Narayana, Y.D. and Muniyappa, V. 1996b. Evaluation of techniques for the efficient transmission of sorghum stripe virus by vector (Peregrinus maidis) and screening for disease resistance. *Trop. Agric.* 73: 119-123.

Nault, L.R. and Knoke, J.K. 1981. Maize Vectors. Pages 77-84 In: *Virus and Viruslike Diseases of Maize in the United States.* D.T. Gordon, J.K. Knoke and G.E. Scott, eds. *Southern Coop. Ser. Bull.* 247. 210 pp.

Nault, L.R., and Gordon, D.T. 1988. Multiplication of maize stripe virus in Peregrinus maidis. *Phytopathology* 78: 991–995.

Nault, L.R., Gingery, R.E., and Gordon, D.T. 1980. Leafhopper transmission and host range of maize rayado fino virus. *Phytopathology* 70: 709-712.

Nome, S. L., Lenardon, S. L., Raju, B. C., Laguna, I. G., Lowe, S. K. and Docampo, D. 1981. Association of reoviruslike particles with "enfermedad of Río Cuarto" of maize in Argentina. *Phytopathol. Z.* 101: 7-15.

Oertel, U., Fuchs, E., and Hohmann, F. 1999. Differentiation of isolates of sugarcane mosaic potyvirus (SCMV) on the basis of molecular, serological and biological investigations. *Z. Pflanzenkrankheiten Pflanzenschutz* 106: 304-313.

Okoth, V.A.O., Dabrowski, Z.T., Thottappilly, G., Van Em-den, H.F., 1988. Comparative analysis of some parameters affecting maize streak virus (MSV) transmission of various *Cicadulina* spp. populations. *Inst. Sci. Applic.* 8: 445–461.

Ornaghi, J. A., Boito, G., Sanchez, G., March, G. J., and Beviacqua, J. E. 1993. Studies on the populations of *Delphacodes kuscheli* Fennah, in different years and agricultural areas. *J. Gen Breed.* 47: 277–282.

Ornaghi, J.A,, Beviacqua, J.E., Aguirrezabala, D.R., March, G.J., and Lenardon, S.L.. 1999. Detection of Mal del Rio Cuarto virus in Uruguay. *Fitopatologia Brasileira* 24: 471.

Overman, M.A., Ko, N.J., and Tsai, J.H. 1992. Identification of viruses and mycoplasmas in maize by use of light microscopy. *Plant Dis.* 76: 318-322.

Padidam, M., Sawyer, S. 7 Fauquet, C.M. 1999. Possible emergence of new geminiviruses by frequent recombination. *Virology* 265: 218-225.

Pernet, A., Hoisington, D., Dintinger, J., Jewell, D., Jiang, C., Khairallah, M., Letourmy, P., Marchand, J.-L., Glaszmann, J.-C., and De Leon, D. Gonzalez. 1999. Genetic mapping of maize streak virus resistance from the Mascarene source. II. Resistance in line CIRAD390 and stability across germplasm. *Theor. Appl. Genet.* 99: 540-553.

Peterschmitt, M., Ratna, A.S., Sacks, W.R., Reddy, D.V.R., and Mughogho, L.K. 1991a. Occurrence of an isolate of maize stripe virus on sorghum in India. *Ann. appl. Biol.* 118: 57-70.

Peterschmitt, M., Reynaud, B., Sommermeyer, G., Baudin, P.,1991b. Characterization of maize streak virus isolates using monoclonal and polyclonal antibodies and by transmission to a few hosts. *Plant Dis.* 75: 27–32.

Pinner, M.S., Markham, P.G., 1990. Serotyping and strain identification of maize streak virus isolates. *J. Gen. Virol.* 71: 1635–1640.

Pinner, M.S., Markham, P.G., Markham, R.H., Dekker, E.L., 1988. Chacterisation of maize streak virus: description of strains; symptoms. *Plant Pathol.* 37: 74–87.

Quint, M., Mihaljevic, R., Dussle, C.M., Xu, M.L., Melchinger, A.E., and Luebberstedt, T. 2002. Development of RGA-CAPS markers and genetic mapping of candidate genes for sugarcane mosaic virus resistance in maize. *Theor. Appl. Genetics* 105: 355-363.

Rao, G.P., Jain, R.K., Tosic, M., and Ford, R.E. 2001. Advancements in purification of sugarcane mosaic potyvirus. *Sugarcane Pathol.* 2: 247-264.

Redinbaugh, M.G. 2003. Transmission of *Maize streak virus* by vascular puncture inoculation with unit-length genomic DNA. J. Virol. Methods 109: 95-98.

Redinbaugh, M.G., Louie, R., Ngwira, P., Edema, R., Gordon, D.T., and Bisaro, D.M. 2001. Transmission of viral RNA and DNA to maize kernels by vascular puncture inoculation. *J. Virol. Methods* 98: 135-143.

Redinbaugh, M.G., Seifers, D.L., Meulia, T., Abt, J J., Anderson, R.J., Styer, W.E., Ackerman, J., Salomon, R., Houghton, W., Creamer, R., Gordon, D.T., and Hogenhout, S.A. 2002. Maize fine streak virus, a new leafhopper-transmitted rhabdovirus. *Phytopathology* 92: 1167-1174.

Revuru, S.S. and Garud, T.B. 1998. Effect of chlorotic stripe stunt disease on plant growth and grain yield of different sorghum cultivars. *J. Maharashtra Agric. Univ. Publ.* 23: 253-255.

Rivera, C. and Gamez, R. 1986. Multiplication of maize rayado virus in the leafhopper vector *Dalbulus maidis*. *Intervirology* 25: 76-82.

Roca de Doyle, M.M., and Autrey, L.J.C. 1992. Assessment of yield losses as a result of co-infection by maize streak virus and maize stripe virus in Mauritius. *Ann. appl. Biol.* 120: 443–450.

Rodier, A., Assie, J., Marchand, J.-L., Herve, Y., 1995. Breeding maize lines for complete and partial resistance to maize streak virus (MSV). *Euphytica* 81: 57–70.

Rodriguez Pardina, P. E., Giménez Pecci, M. P., Laguna, I. G., Dagoberto, E., and Truol, G. 1998. Wheat: A new natural host for the Mal de Río Cuarto virus in the endemic disease area, Río Cuarto, Córdoba Province, Argentina. *Plant Dis.* 82: 149-152.

Rose, D.J.W. 1973. Field studies in Rhodesia on *Cicadulina* spp. (Homoptera: Cicadellidae), vectors of maize streak disease. *Bull. Entomol. Res.* 62: 477-495.

Rose, D.J.W., 1978. Epidemiology of maize streak disease. Annu. Rev. Entomol. 23: 259–282.

Rybicki, E.P. and Pietersen, G. 1999. Plant virus disease problems in the developing world. *Adv. Virus Res.* 53: 127-175.

Rybicki, E.P., Briddon, R.W., Brown, J.E., Fauquet, C.M., Maxwell, D.P., Harrison, B.D., Markham, P.G. and Stanley, J. 2000. *Geminviridae.* In: Virus tTaxonomy, M.H.V. Van Regenmortel, C.M. Fauquet, D.H.L. Bishop, E. Carstens, M.K. Estes, S. Lemon, J. Maniloff, M.A. Mayo, D. McGeoch, C.R. Pringle and R.B. Wickner (ed), *Seventh Report of the International Committee on Taxonomy of Viruses.* Academic Press, New York, pp. 285-297.

Sanz, A.I., Fraile, A., Gracía-Arenal, F., Zhou, X., Robinson, D.J., Khalid, S., Butt, T., and Harrison, B.D. 2000. Multiple infection, recombination and genome relationships among begomovirus isolates found in cotton and other plants in Pakistan. *J. Gen. Viol.* 81: 1839-1849.

Schnippenkoetter, W.H., Martin, D.P., Willment, J.A. and Rybicki, E.P. 2001. Forced recombination between distinct strains of Maize streak virus. *J. Gen. Viol.* 82: 3081-3090.

Seifers, D.L., Salomon, R., Marie-Jeanne, V., Alliot, B., Signoret, P., Haber, S., Loboda, A., Ens, W., She, Y.-M., and Standing, K.G. 2000. Characterization of a novel potyvirus isolated from maize in Israel. *Phytopathology* 90: 505-513.

Seth, M.L., Raychaudhuri, S.P., and Singh, D.V. 1972a. Bajra (pearl millet) streak: A leafhopper-borne cereal virus in India. *Plant Dis. Rep.* 56: 424-428.

Seth, M.L., Raychaudhuri, S.P., and Singh, D.V. 1972b. Occurrence of maize streak virus on wheat in India. *Curr. Sci.* (Bangalore) 41: 684.

Sharma, R.C. and Payak, M.M. 1983. An overview of virus and viruslike diseases of maize in India. Pages 186-190 in: *Proceedings International Maize Virus Disease Colloquium and Workshop.* D.T. Gordon, J.K. Knoke, L.R. Nault, and R.M. Ritter (eds). 2-6 Aug.,1982, Ohio State Univ., Ohio Agric. Res. Dev. Cent., Wooster. 261 pp.

Shi, Y., Zhang, Q., Wang, F., Xi, Z., and Xu, S. 1986. Identification of strains of maize dwarf mosaic virus. (in Chinese, English summary). *Acta Phytopathol. Sin.* 16: 99-104.

Shieh, G.J., Ho, C.L., Lee, T., Liu, K.S. and Lu, H.S. 1994. The development of Tainung No. 2 maize - a hybrid cultivar for forage uses. *J. Agric. Res. China* 43: 354-372.

Shikata, E. 1974. Rice black streaked dwarf virus. CMI/AAB Descriptions of Plant Viruses No. 135. Descriptions of Plant Viruses. *Commonw. Mycol. Inst., Assoc. Appl. Biol.* Kew, Surrey, England. 4 pp.

Simcox, K.D., McMullen, M..D., and Louie, R. 1995. Co-segregation of the maize dwarf mosaic virus resistance gene, *Mdm1*, with nucleolus organizer region in maize. *Theor. Appl. Genetics* 90: 341-346.

Storey, H.H., 1925. The transmission of streak disease of maize by the leafhopper *Balclutha mbila* Naude´. *Ann. appl. Biol.* 12: 422–439.

Storey, H.H., 1928. Transmission studies of maize streak disease. *Ann. appl. Biol.* 59: 429–436.

Storey, H.H., 1933. Investigations of the mechanism of the transmission of plant viruses by insect vectors. I. *Proc. R. Soc. Lond.* (Ser. B) 113: 463–485.

Storey, H.H., 1936. Virus diseases of East African plants IV. A survey of the viruses attacking the Gramineae. *East Afr. Agric. J.* 1: 333–337.

Suranto, Gough, K.H., Shukla, D.D., and Pallaghy, C.K. 1998. Coat protein sequence of Krish-infecting strain of Johnson grass mosaic potyvirus. *Arch. Virol.* 143: 1015-1020.

Thottappilly, G. 1992. Plant virus diseases of importance to African agriculture. *J. Phytopathol.* 134: 264-288.

Thottappilly, G., Bosque Perez, N.A., and Rossel, H.W. 1993. Viruses and virus diseases of maize in tropical Africa. *Plant-Pathol.* 42: 494-509.

Thouvenel, .J.C., Abol Ela, S., Sewify, G.H., El Hariry, M.., Hussein, A., and Ammar, E.D. 1996. Characterization and serology of the leafhopper-borne maize yellow stripe virus in Egypt. Bull. *Faculty Agric. Univ. Cairo* 47: 179-189.

Thresh, J.M. and Fargette, D. 2001. Virus diseases of tropical crops. *Encyclopedia of Life Sciences,* pages 1-9, Nature Publishing Group (www.els.net)

Toler R.W. 1985. Maize dwarf mosaic, the most important virus disease of sorghum. *Plant Dis.* 69: 1011-1015.

Toler, R.W., Rosenow, D.T., Riccelli, M., and Mena, H.A. 1982. Variability of Venezuelan isolate of maize dwarf mosaic virus in sorghum. *Plant Dis.* 66: 849-850 .

Toler, R.W., Skinner, G., Bockholt, A.J., and Harris, K.F. 1985. Reactions of maize (*Zea mays*) accessions to maize rayado fino virus. *Plant Dis.* 69: 56-57.

Tosic, M., Ford, R.E., Shukla, D.D., and Jilka, J. 1990. Differentiation of sugarcane, maize dwarf, johnsongrass, and sorghum mosaic viruses on reactions of oats and some sorghum cultivars. *Plant Dis.* 74: 549-552.

Traore, O. and Konate, G. 1997. Maize viruses and their reservoir hosts in Western and Central Africa. *Cahiers Agric.* 6: 257-262.

Truol, G.A., Usugi, T., Hirao, J., Arneodo, J.D., Gimenez Pecci, M.P., and Laguna, I.G. 2001. Experimental transmission of Mal del Rio Cuarto virus by *Delphacodes kuscheli. Fitopatologia Brasileira* 26: 39-44.

Tsai, J.H. and Falk, B.W. 1993. Viruses and mycoplasma agents affecting maize in the Tropics. *Plant Pathol. Bull.* 2: 203-217.

Tsai, J.H., and Zitter, T.A. 1982. Characteristics of maize stripe virus transmission by the corn delphacid. *J. Econ. Entomol.* 75: 397–400.

Uyemoto, J.K. and Claflin, L.E. 1981. Maize chlorotic mottle virus and corn lethal necrosis disease. Pages 163-165 in: *Virus and Viruslike Diseases of Maize in the United States.* D.T. Gordon, J.K. Knoke, and G.E. Scott, eds. Southern Coop. Ser. Bull. 247. 210 pp.

Vangessel, M.J. and Coble, H.D. 1993. Postemergence control of johnsongrass and its effect on maize dwarf mosaic virus incidence and vectors in corn. *Plant Dis.* 77: 613-618.

Von Wechmar, M.B., Chauhan, R., and Knox, E. 1992. Fungal transmission of a potyvirus: uredospores of *Puccinia sorghi* transmit maize dwarf mosaic virus. *Arch. Virol. Supplementum* 5: 239-250.

Webb, M.D., 1987. Species recognition of *Cicadulina* leafhoppers Hemiptera: Cicadellidae) vectors of pathogens of ramineae. *Bull. Entomol. Res.* 77: 683–712.

Welz, H.G., Schechert, A., Pernet, A., Pixley, K.V., Geiger,H.H., 1998. A gene for resistance to maize streak virus in the Arican CIMMYT maize inbred line CML202.*Mol. Breed* 4, 147–154.

Wetter, C., Luisoni, E., Conti, M., and Lovisolo, O. 1969. Purification and serology of maize rough dwarf virus from plant and vector. *Phytopathol. Z.* 66: 197-212.

Willment, J.A., Martin, D.P. and Rybicki, E.P. 2001. Analysis of the diversity of African streak mastreviruses using PCR-generated RFLP's and partial sequence data. *J. Virol. Methods* 93: 75-87.

Willment, J.A., Martin, D.P., Van der Walt, E. and Rybicki, E.P. 2002. Biological and genomic sequence characterization of *Maize streak virus* isolates from wheat. *Phytopathology* 92: 81-86.

Wu, J.Y., Tang, J.H., Xia, Z.L., and Chen, W.C. 2002. Molecular tagging of a new resistance gene to maize dwarf mosaic virus using microsatellite markers. *Acta Phytopathol. Sinica* 44: 177-180.

Yang, Z.N. and Mirkov, T.E. 1997. Sequence and relationships of sugarcane mosaic and sorghum mosaic virus strains and development of RT-PCR-based RFLPs for strain discrimination. *Phytopathology* 87: 932-939.

Zhang, H.M., Chen, J.P., and Adams, M.J. 2001a. Molecular characterisation of segments 1 to 6 of Rice black-streaked dwarf virus from China provides the complete genome. *Arch. Virol.* 146: 2331–2339.

Zhang, H.M., Chen, J.P.,Lei, J.L., and Adams, M.J. 2001b. Sequence analysis shows that a dwarfing disease on rice, wheat and maize in China is caused by *Rice black-streaked dwarf virus*. *European J. Plant Pathol.* 107: 563-567.

MAIZE : Figure Legends

Fig. 12.1: A maize field in Nigeria infected with MSV. Please note the severely affected plants in the front and resistant plants in the background.

Fig. 12.2 a and b: Symptoms due to MSV. Symptoms are severe on susceptible lines, while very mild symptoms are observed on tolerant lines.

Fig. 12.3 : MSV vector (*Cicadulina* sp.) on maize plants

Fig. 12.4: Symptoms due to Maize stripe virus: Characterized by chlorotic bands of increasing width.

Source: Courtesy IITA (Nigeria) and H.W. Rossel

12.1
12.2a
12.2b
12.3
12.4

13. Other Cereals

Monique Henry and Michael J Adams
CIMMYT, Mexico and *Rothamsted Research, UK*

13.1 Introduction

Cereals are the major source of food for the world population. In the developing countries, the main cereals grown are rice, wheat, maize, sorghum and barley. Rice, maize, sorghum and millet have been treated in chapter 12. Other cereals include barley, oat, rye, triticale and wheat (bread and durum) (Table 13.1).

Wheat is the most widely grown cereal grain, occupying approx. 17% of the total cultivated land in the world. It is planted on about 100 million hectares in the developing world, excluding the countries of Central Asia and the Caucasus. It is the staple food for nearly 35% of the world population and demand for wheat will grow faster than for any other major crop (Braun *et al.* 1998). A big proportion of the wheat grown in the developing world is bread wheat, where the most important producers are China, India, and Turkey.

Table 13.1: World areas under cereals and average yields.

	Developing countries		World	
Crop	Area (Millions ha)	Yield (Tonnes/ha)	Area (Millions ha)	Yield (Tonnes/ha)
Barley	15.4	1.49	54.5	2.62
Oat	1.6	1.58	13.1	2.06
Rye	0.8	1.37	9.8	2.36
Triticale	0.5	1.88	2.9	3.63
Wheat	98.2	2.70	213.0	2.77

Source: FAO, 2001

Bread wheat dominates the wheat area of South Asia, eastern and southern Africa, the Southern Cone of South America, Central Asia, and Mexico/Guatemala. Durum wheat production is concentrated in the Middle East, Central India, and the Mediterranean region of West Asia and North Africa (WANA).

Cultivation of barley traces back to the earliest remains of agricultural activity in the Old World about 10,000 years ago. Nowadays, barley is mainly used as animal feed (85%) or in the malting industry and it only plays a minor role in food consumption except in a few countries such as the West of China, Tibet, Ethiopia and the Andean highlands (Fischbeck, 2002).

Oat, triticale and rye are of much less economic importance in the developing world. Oat and triticale are mainly used for animal feed in some parts of the Southern Cone, China, North and East Africa and parts of Central Asia.

13.2 The Main virus diseases

Small grain cereals (not including rice) are hosts for a large number of viruses. However, of these, only a few have been reported to cause extensive damage and an economic effect on small grain cereal production in the developing world. Wiese (1991) and Mathre (1997) reported up to 30 and 25 viruses that naturally infect wheat and barley, respectively.

The most important viruses affecting wheat and barley in the developing world are the aphid-transmitted viruses in the family *Luteoviridae* that cause Barley yellow dwarf disease and the fungally-transmitted viruses in the genera *Bymovirus* and *Furovirus*. Other viruses that have been occasionally reported include: *Wheat streak mosaic virus* (genus *Tritimovirus*), *Mal de Rio Cuarto virus* (genus *Fiivirus*), Nariño dwarf virus (family *Reoviridae*), *Barley yellow striate mosaic virus* (genus *Cytorhabdovirus*), *Barley stripe mosaic virus* (genus *Hordeivirus*), *Peanut clump virus* and *Indian peanut clump virus* (genus *Pecluvirus*).

13.2.1 *Luteoviridae* of cereals

Several members of the *Luteoviridae* family infect small grain cereals, (oat, wheat, barley, rye). A complex of viruses, previously named *Barley yellow dwarf virus* (BYDV) was first reported on barley by Oswald and Houston (1951) and subsequently found on other cereals such as oat and wheat (Oswald and Houston, 1953). Five variants of the virus were characterized in New York (Rochow, 1969; Rochow and Muller, 1971) based on their transmissibility by aphids. In China, other variants were found to be different from the NY variants based on their transmission properties and named GPV, GAV, DAV, GPDAV (Zhang *et al.* 1983, Zhou *et al.* 1984), although only GPV seems to differ significantly from other known variants. The viruses causing Barley yellow dwarf disease (BYD) were divided into two groups (1: PAV, MAV and SGV; 2: RPV and RMV), based on serology (Waterhouse *et al.* 1988), cytopathological alterations in the host plants (Gill and Chong, 1979) and nucleic acid sequences (Martin and D'Arcy, 1990). Recently, the viruses were reclassified into two genera: *Luteovirus* (BYDV-PAV, BYDV-MAV) and *Polerovirus* (*Cereal yellow dwarf virus*, CYDV-RPV) (D'Arcy *et al.* 1999). Subsequently, it was suggested that the Mexican RPV isolate should be considered as a different virus from CYDV-RPV and named CYDV-RPS (Mayo, 2002). BYDV-GPV, -RMV and -SGV are recognised as different viruses but were not yet assigned to a genus (D'Arcy *et al.* 1999).

BYDV and CYDV are the most common and widespread viruses of small grain cereals in the world. They have been reported from most countries (for review, see Lister and Ranieri, 1995) and are of economic importance in China, the Eastern African Highlands, the Mediterranean region, such as Algeria, Tunisia and Morocco (El Yamani *et al.* 1990), the Southern Cone of South America and the Andean Highlands.

Surveys of CIMMYT nurseries and other locations between 1988 and 89 indicated the preponderance of BYDV-MAV in the Andean zone of Latin America (Bolivia, Colombia, Ecuador and Peru). In Central America (Guatemala and Mexico), all types except SGV were found; BYDV-PAV and BYDV-MAV were predominant. In the Southern Cone (Brazil, Chile, Paraguay, Uruguay), BYDV-PAV was predominant, followed by BYDV-MAV and SGV (Webby *et al.*, 1993). In Africa (Ethiopia, Rwanda, Tanzania, Zaire and Zimbabwe), BYDV-PAV was the preponderant species followed by RMV. In the Mediterranean region (Jordan, Syria, Tunisia and Turkey) BYDV-PAV and RMV were the most important. In Asia (India, Iran and Pakistan), all types except SGV were identified, PAV again being predominant (Webby *et al.* 1993).

BYDV and CYDV affect principally barley, oat and to a lesser extent wheat production and can be a limiting constraint to small grain cereal production in many countries. Disease severity and yield losses in a crop depend on various factors, including the viruses and cultivars, the time of infection and environmental conditions (Lister and Ranieri, 1995; Pike, 1990). In wheat, losses of 47% and 26% have been reported after artificial inoculation with PAV in Kenya (Wangai, 1990) and Mexico (Burnett and Mezzalama, 1992). Losses around 11-12% due to natural infection have been reported in Morocco (El Yamani and Hill, 1990) and Chile (Ramirez et al. 1992). Average yield losses of 16 and 25% were reported by Pike (1990) on barley and oat respectively.

BYDV and CYDV are restricted to the Poaceae (Graminae). Cultivated hosts include all the major cereal crops: barley, maize, oat, rice, rye, triticale and wheat as well as many annual and perennial cultivated and wild grasses.

Symptoms caused by BYDV and CYDV differ with the host species and cultivar, the age and the physiological condition of the host plant at the time of infection, the strain and the environmental conditions and can be easily confused with nutritional and abiotic disorders. Symptoms include leaf discoloration from tip to base and from margin to centre, in shades of yellow for barley and triticale, yellow or sometimes red in wheat and deep purpling in oat. Plants are usually stunted, with a decrease in tiller numbers and biomass and a weak root system. Suppressed heading, sterility and failure to fill grains occur in the most severe cases. In the field, symptoms appear usually as yellow or red patches of stunted plants. Symptoms vary considerably between the different BYDVs or CYDV. In Mexico, PAV-Mex tends to induce a lot of stunting and little chlorosis while MAV induces intense yellowing and little dwarfing. RPV-Mex induces severe leaf notching (corkscrew symptoms) on some wheat cultivars, in most cases with spike sterility.

BYDV and CYDV are phloem-limited and transmitted in a persistent manner by several species of cereal aphids. At least 25 species have been reported to be vectors of BYDV and CYDV (Halbert and Voetglin, 1995). As reported in Power and Gray (1995), transmission patterns of BYDV-CYDV isolates in the New-York experimental system follow a trend, but cannot be generalized to all BYDV-CYDV isolates. BYDV-PAV is transmitted non-specifically by several species including *Rhopalosiphum padi* , *Sitobion avenae* and *Schizaphis graminum* . BYDV-MAV is efficiently transmitted by *S. avenae.* BYDV-SGV is transmitted mainly by *S. graminum*, RMV mainly by *Rhopalosiphum maidis* but also by *S. graminum* and *R. padi.* CYDV-RPV is transmitted by *R. padi* and *S. graminum* (Power and Gray, 1995). The Chinese isolate GAV (serologically related to MAV) is transmitted by *S. graminum* and *S. avenae* (Zhou *et al.* 1984) and GPV by *S. graminum* and *R. padi* (Wang *et al.* 1998). In Mexico, BYDV-PAV is efficiently transmitted by *Metopolophium dirhodum* and *Rhopalosiphum rufiabdominalis* (Henry, unpublished data).

The viruses can be detected and discriminated by ELISA or PCR. ELISA using polyclonal antibodies allows the differentiation of 5 of the viruses (PAV, MAV, RPV, RMV, SGV) and is used routinely to detect the presence of the virus in fresh, frozen or dried-leaf samples (Lister *et al.* 1985). PCR with *Luteoviridae* specific primers combined with reverse transcriptase and restriction digest has been used to identify BYDV and CYDV in field samples (Robertson *el al.* 1991). The viruses can be detected in single aphids or pools of five aphids by ELISA with enzymatic amplification (Torrance, 1987) and PCR (Kulemeka *et al.* 2002)

The virus can be purified by freezing and grinding tissue (root or shoot) to a powder in liquid nitrogen and extraction in 0.5 M Phosphate buffer, pH 6.0, maceration with industrial grade enzyme (celluclast, extractase P20X), clarification with cloroform:butanol, PEG precipitation, clarification through low speed centrifigation, ultracentrifigation through a 30% sucrose pad and passage through a sucrose gradient (Hewings, 1995; D'Arcy *et al.* 1989; Webby and Lister, 1992).

BYDV and CYDV have single isometric particles 25-30 nm in diameter (Waterhouse *et al.* 1988), with a single-stranded positive RNA genome of approximately 5,600 nucleotides. The genome consists of 6 open reading frames (numbered 1 to 6 for BYDVs and 0 to 5 for CYDV-RPV). The genome of CYDV-RPV has a protein (VPg) linked to its 5'end and terminates in GU, while BYDV-PAV lacks a 5'cap and terminates with the sequence CCC (Miller *et al.* 2002).

The epidemiological cycle includes two phases. Primary infection occurs when aphids colonise the crop by flying from neighbouring crops or from a longer distance, usually at an early growth stage. Secondary infection is due to movement of aphids inside the crop, disseminating the virus from one plant to the other. As a consequence, typical symptoms appear in patches of dwarfed and yellowed plants. Many factors intervene in the epidemiological cycle, including number and infectivity of aphid vectors, strains of virus, weather conditions that are favourable or not to aphid flight and their survival during winter or summer and the availability of alternate hosts, including grasses, maize and other small grain cereal crops.

As there is no direct control of the viruses, management of the disease relies on control of the vectors through insecticide application (organosphosphates or pyrethroids). It is critical to control infection at the early growth stages, when plants are more vulnerable to infection and are damaged most (Plumb and Johnstone, 1995). Effects of Imidachloprid on the disease have been well documented. Yield benefits of 36-43% on late sown crops seed-treated with Imidachloprid were reported in Kenya (Wangai *et al.* 2000). Delaying planting dates, thus breaking the cycle of aphid survival and virus infection, can be used to moderately reduce virus incidence. The most economically and sustainable control method for BYDV and CYDV would be the use of resistant/tolerant germsplasm. In oat, tolerance has been reported to be heritable and due to 2-4 genes (Burnett *et al.* 1995). In barley, the major semi-dominant gene *Yd2* (Rasmusson and Schaller, 1959), located on chromosome 3 (Schaller *et al.* 1964) was identified in Ethiopian landraces and has been extensively used in the breeding programs around the world. It provides tolerance and resistance to BYDV-PAV and BYDV-MAV but it less effective against CYDV-RPV (Herrera, 1989). The gene *Bdv1* (Singh *et al.* 1993) has been reported to bring moderate tolerance to BYDV-MAV and BYDV-PAV in wheat and has been used extensively in the CIMMYT breeding program. True resistance (immunity) was reported for wheat relatives such as *Thinopyrum intermedium* and has been successfully introgressed in wheat (Banks *et al.* 1995; Larkin *et al.* 1995; Francki *et al.* 2001) and is used in several breeding programs in the world. The first cultivar containing such resistance was released recently (Larkin et al. 2002). The gene *Bdv2* is associated with reduction in virus titres and infection rate (Ayala *at al.* 2001). Coat protein-mediated resistance to BYDV has been reported in transgenic barley, oat and wheat (McGrath *et al.* 1997; Koev *et al.* 1998; Cheng *et al.* 2002) but the effectiveness of this resistance is still unclear. Some results indicate delay in symptom development or moderate tolerance. More recently, immunity was reported from barley plants transformed with a transgene designed to produce hairpin (hp)RNA containing BYDV-PAV sequences (Wang *et al.* 2000).

13.2.2 Bymoviruses of wheat

Wheat yellow mosaic virus (WYMV), genus *Bymovirus*, family *Potyviridae*, was first reported from Japan in 1927 and is now known to be widespread on winter wheat in China (Chen *et al.* 2000b, Han *et al.* 2000). *Wheat spindle streak mosaic virus* (WSSMV) is a serologically related virus that occurs in North America and Europe and which has recently been reported from Zambia (Kapooria *et al.* 2000) (although the methods used may not have discriminated between the two viruses). Since the 1970s, WYMV has been particularly severe in the winter wheat growing areas of China, including Shandong, Jiangsu, Zhejiang, Anhui, Shaanxi, Henan, Hubei and Sichuan provinces. Co-infection with a furovirus related to *Soil-borne wheat mosaic virus* occurs in Shandong province. Yield losses have been estimated commonly at 10-30% and sometimes up to 70%. It has been estimated that a total of c.3,300,000 ha of wheat are affected, resulting in yield losses of 5 million tonnes (Jianping Chen, personal communication). Losses in Africa have not been estimated. The virus is restricted to cultivated wheat. Symptoms appear as yellow patches in the field during the winter and individual plants have a yellow, mosaic, streaking especially on the youngest (unfurling) leaves. Infected plants may remain stunted although symptoms often disappear as temperatures rise and the stem extends to flowering. The root-infecting zoosporic parasite *Polymyxa graminis*, often considered to be a fungus but probably more closely related to protists, transmits the virus. The vector is an obligate parasite of plant roots and acquires/transmits the virus while growing in the root epidermal or cortical cells. It survives between crops as long-lived resting spores and the virus survives inside the spore. The mosaic symptoms on the young leaves are very suggestive of a bymovirus and/or furovirus infection and the particular virus can be confirmed by serological methods (e.g. ELISA; Usugi *et al.* 1984) or by RT-PCR (Li *et al.* 1997). The virus can be purified by extraction in citrate buffer, differential centrifugation and caesium chloride density gradient separation (Usugi and Saito, 1976). Its particles are filamentous, approx 13nm in diameter and of two modal lengths, around 550 and 275nm. The WYMV genome consists of two segments of linear single-stranded RNA of about 7.6 (RNA1) and 3.6 (RNA2) kb each of which contains a single open reading frame that encodes a polyprotein in the virion (+) sense. Each polyprotein is subsequently processed into 8 (RNA1) or 2 (RNA2) functional proteins by virus-encoded proteases (Namba *et al.* 1998). WSSMV is similar but with a larger RNA2 (about 4.0 kb) (Anke Sohn personal communication). The viral coat protein is encoded at the C-terminus of the RNA1 polyprotein. Epidemiology is typical of a soil-borne disease, with irregular patches in the field that are enlarged slowly by cultivation or, more rapidly, by irrigation or flooding. The only practical control measure is to grow wheat cultivars that are resistant to the virus and efficient resistance genes have been successfully used to combat the disease in China although their response sometimes varies between sites suggesting some strain variation (Chen *et al.* 2000a). It is likely that genetic resistance will remain the principal control strategy. Genes introgressed from foreign germplasm and/or transgenic resistance based on the sequence of the virus are likely to be most promising. There is little experimental evidence that some wheat cultivars are relatively resistant to the vector but this has not been actively exploited in breeding programmes.

13.2.3 Bymoviruses of barley

Barley yellow mosaic virus (BaYMV), genus *Bymovirus*, family *Potyviridae*, was first recognised in Japan in 1940. It is a disease of winter barley that occurs in Asia (China, Korea, Japan) as well as in Europe (Ukraine to France and the UK). Losses in heavily infected susceptible crops are typically 30-40% but can be greater. The virus is restricted to cultivated barley and a few closely related wild *Hordeum* species. Symptoms, transmission and epidemiology, diagnosis, purification and virus properties are very similar to those for the closely related *Wheat yellow mosaic virus* and are therefore not repeated here. BaYMV is the type member of the genus. Strains of the virus able to overcome the resistance of several individual genes have been identified in Japan (Kashiwazaki *et al.* 1989) and Western Europe and there is some evidence that Chinese isolates also differ in specific virulence (Chen *et al.* 1998, Chen *et al.* 1992, Chen *et al.* 1996). The strains do not differ serologically but only in cultivar response. The only practical control measure is to grow barley cultivars that are resistant to the virus and efficient resistance genes conferring immunity were successfully used to control the disease in eastern China. Genetic resistance will probably remain the principal control strategy. Genes introgressed from foreign germplasm and/or transgenic resistance based on the sequence of the virus may also be promising. There is little convincing evidence that effective resistance to the vector is available in barley lines.

Barley mild mosaic virus (BaMMV) is a distantly related member of the same genus, which has many properties in common with BaYMV (Kashiwazaki *et al.* 1998). The two viruses often occur together in Europe, where BaMMV was first recognised. BaMMV has been reported from Korea (Lee *et al.* 1996) and China (Zheng *et al.* 1999) but does not appear to be widespread. The disease can be controlled by using resistant cultivars, even though different pathotypes have been recognised in Japan (Nomura *et al.* 1996).

13.2.4 Furoviruses of wheat

Soil-borne wheat mosaic virus (SBWMV) is the type member of the genus Furovirus. It was first reported from the USA where it causes major losses, and was the first recognised virus disease of wheat (Brakke and Langenberg, 1988). SBWMV or related viruses are now known to occur on winter wheat in many other parts of the world including Western Europe, Poland, Turkey (Kose and Ertunc, 1999), Brazil (Barbosa et al. 2001), Argentina, Iran (Sahandpour and Izadpanah, 1999), China, Japan, Egypt and Zambia (Kapooria et al. 2000). Co-infection with the bymoviruses Wheat yellow mosaic virus or Wheat spindle streak mosaic virus is common in some regions and the symptoms and effects are not always easy to separate. Yield losses of 50% or more may occur on susceptible varieties (Barbosa et al. 2001; Dalbosco et al. 2002). In China, the virus has been named Chinese wheat mosaic virus because it is genetically rather distant from SBWMV but it appears to be restricted to the coastal regions of Shandong province and always occurs in association with WYMV (Diao et al. 1999; Yang et al. 2001; Ye et al. 1999). In South America, the disease is particularly severe in the northern part of the state of Rio Grande do Sul, Brazil. The host range of the virus includes wheat, barley, rye and some grass species but natural infection of barley appears to be rare. Symptoms appear as yellow patches in the field during the winter and individual plants have a yellow, mosaic, streaking. Infected plants may remain stunted although symptoms often disappear as temperatures rise and the stem extends to flowering. The root-infecting zoosporic parasite *Polymyxa graminis*, often considered to be a fungus but probably more closely related to protists, transmits the virus. The vector is an obligate parasite of plant roots and acquires/transmits the virus while growing in the root epidermal or cortical cells. It survives between crops as long-lived resting spores and the virus survives inside the spore. The mosaic symptoms on the young leaves are very suggestive of a furovirus and/or bymovirus infection and the particular virus can be confirmed by serological methods (e.g. ELISA) or by RT-PCR. The virus can be purified by extraction in borate buffer, differential centrifugation and caesium chloride density gradient separation. Its particles are rod-shaped about 20nm in diameter with a central canal and are of two size classes. The longer virions (about 280nm) encapsidate the larger of the two genomic RNAs (RNA1); shorter virions encapsidate RNA2 and are about 140nm long, but shorter forms accumulate and predominate later in the growing season because of spontaneous deletions in the RNA. The full-length genome consists of two segments of linear single-stranded RNA of about 7.1 (RNA1) and 3.6 (RNA2) kb which encode proteins in the virion (+) sense (Shirako and Wilson, 1993). RNA1 encodes a large protein involved in replication and a 37 kDa cell-to-cell movement protein. RNA2 encodes the 19 kDa coat protein near its 5'-end and this can be extended by partial suppression of the termination codon to produce a much larger (84 kDa) protein that has a role in transmission. A 19 kDa protein encoded towards the 3'-end of RNA2 is rich in cysteine residues but its function is unknown. Virus isolates from different parts of the world may have nucleotide sequences that are only about 70% identical and it has not been agreed whether these represent strains of SBWMV or a group of related furovirus species. Epidemiology is typical of a soil-borne disease, with irregular patches in the field that are enlarged slowly by cultivation or, more rapidly, by irrigation or flooding. The only practical control measure is to grow wheat cultivars that are resistant to the virus and efficient resistance genes are being used to control the disease in many parts of the world including Brazil and Turkey. It is likely that genetic resistance will remain the principal control strategy. Genes introgressed from foreign germplasm and/or transgenic resistance based on the sequence of the virus are likely to be most promising. There is experimental evidence that some wheat cultivars are relatively resistant to the vector but this has not been actively exploited in breeding programmes.

13.2.5 Pecluviruses on cereals

Peanut clump virus and *Indian peanut clump virus* are the two members of the genus *Pecluvirus* and are genetically related but serologically distinct. They are mostly significant for their effects on groundnuts (see chapter 15). However, the viruses can occur on cultivated crops of wheat, barley and sorghum (Delfosse *et al.*1999). This may reduce yields particularly in some sorghum cultivars in Africa and is important in their epidemiology (Philippe Delfosse, personal communication).

13.2.6 *Wheat streak mosaic virus* Genus ***Tritimovirus*** (WSMV) has been reported from Czech Republic, Hungary, Iran (Izadpanah and Behjatnia, 1989), Jordan, Mexico (Sànchez Sànchez *et al.* 2001), Romania, Russia and Yugoslavia (French and Stenger, 2002). WSMV causes a severe mosaic in winter wheat and less frequently in spring wheat. Infected plants are stunted, with a rosette appearance (French and Stenger, 2002). Symptoms range from mild mosaic to severe chlorosis resulting in death of tillers; reduced seed set and shrivelled kernels. WSMV infect most varieties of wheat, oats, barley, rye, triticale, certain varieties of maize, sorghum and millets. It is persistently transmitted by the wheat curl mite (*Eryophyidae*), formally identified as *Aceria tulipae* and recently re-named as *A. tosichella* (Amrine and Stasny, 1994). It is detected in leaves samples by DAS-ELISA, western blotting and filter paper immunobinding assay (Sherwood, 1987) and in its mite vector using fluorescein-labelled antiserum (Mahmood *et al.* 1997). It is purified according to French and Stenger (2002). Virions are flexuous rods c.15 nm in diameter and c. 700 nm in long. Virus particles contain a single species of ssRNA of 9,339-9,384 nucleotides that encodes a single ORF of about 350kDa, which is processed into proteins analogous to those of other poytviruses (French and Stenger, 2002). Isolates of WSMV differ in serological properties (Montana *et al.* 1996) and nucleotide sequences (Choi *et al.* ; Rabenstein *et al.* 2002). Control of WSMV involves reducing the risk of vectors colonizing the crop through delayed planting and removal of reservoirs (volunteer wheat or weeds) as chemical control of the vector is not feasible. No effective source of resistance has been found in wheat, but resistance to WSMV or the vector has been introgressed from wild relatives, such as *Th. intermedium* (Chen *et al.* 1999; Baley *et al.* 2001).

13.3 Other viruses

13.3.1 *Brome mosaic virus* Genus ***Bromovirus*** (BMV) was first reported to cause economic damage in wheat in South Africa and to mimic BYDV infection (von Wechmar and Rybicki, 1985). It has been reported to naturally infect wheat in the ex- Soviet Union, Yugoslavia, Hungary (Pocsai *et al.* 1991) and Brazil (Caetano *et al.* 1990). Some wheat cultivars are symptomless carriers of the disease. Symptoms on leaves are yellow or white spots and streaks that turn into a yellow mosaic pattern. Infected plants can be slightly stunted and produce shrivelled grains. It is transmitted mechanically to cereals and grasses and has been experimentally transmitted with low efficiency by beetles (Lane, 1979). BMV has three types of icosahedral particle with identical diameters (26 nm), but with different RNA components (Lane, 1979).

13.3.2 ***Barley stripe mosaic virus*** Genus ***Hordeivirus*** (BSMV) occurs in North America, Europe, Japan, Australia, the Soviet Union, China and Korea (Wiese, 1987) but is not an economically important disease of wheat. It is transmitted through the seed and pollen and is a quarantine disease in several countries. BSMV causes yellow to white mottling, spotting and streaking on leaves, severe mosaic, dwarfing, excessive tillering and necrosis. Plants grown from BSMV infected seeds may show symptoms as early as the 2-3-leaf stage.

13.3.3 Reoviruses

Cereals are hosts for several Reoviruses: *Nariño dwarf virus* (Enanismo, possible Oryzavirus) has been associated with strong symptoms in wheat and barley in Colombia, (Uyeda and Milne, 1995) and Mal de Rio Cuarto has been reported to cause extensive damage to wheat and maize in Argentina (Rodriguez Pardina *et al.* 1998). Symptoms are deformed leaves, spikes and spikelets, shortened internodes, leaves with serrated borders and sterile spikelets. "Mal de Rio Cuarto" was classified as a strain of *Maize rough dwarf virus*, genus *Fijivirus* (Brunt *et al.* 1996) but recent molecular data (Rodriguez Pardina *et al.* 1998) suggest that it is a separate member of the genus. Reoviruses are exclusively transmitted by leafhoppers (Phytoreoviruses) and planthoppers (Fijiviruses). They propagate in the vector and the plant. They have icosahedral double-shelled particles 65-75 nm in diameter, containing 10-12 segments of double-stranded RNA and at least six different structural polypeptides (Boccardo and Milne, 1984).

13.3.4 Rhabdoviruses

There are several rhabdoviruses that infect wheat but none of them is economically important. Recent reports mention the occurrence of *Barley yellow striate mosaic virus* in Turkey (Makkouk *et al.* 1996) and Uzbekistan (Makkouk *et al.* 2001). It is transmitted in a propagative manner by the planthopper *Laodelphax striatellus*. Particles are bacilliform, measuring 55 x 330 nm or 55 x 270 nm depending on the staining method (Milne and Conti, 1986). The disease causes chlorotic stripes or mosaic in wheat.

13.4 Conclusion

With the constantly rising human population and the increasing demand for food, cereal production will need to increase significantly in the coming years. One among many limiting factors to high yields is the impact of virus diseases.

Despite our current knowledge and understanding of viruses affecting small grain cereals, much remains to be done to find effective and sustainable ways of control, both for the fungally-transmitted mosaic viruses and for the members of the Luteoviridae causing barley yellow dwarf disease. As more knowledge is acquired of the viruses causing barley yellow dwarf disease, more questions will arise on the variability of the isolates and the interactions between aphid species and virus.

New strains of viruses and new viruses are being reported from various parts of the world, and the present distribution of viruses might change with changes in environmental conditions.. Viruses that did not occur in some regions might well become of importance, while others might become less significant. Predicting particular changes is impossible but continued monitoring is essential.

References

Amrine, J. W. and Stasny, T. A. 1994. Catalog of Eriophyoidea (Acarina: Prostigmata) of the World. Indira Publishing Co., West Bloomfield, MI.

Ayala, L., Van Ginkel, M., Khairallah, M., Keller, B., and Henry, M. 2001. Expression of *Thinopyrum intermedium*-Derived Barley yellow dwarf virus Resistance in Elite Bread Wheat Backgrounds. *Phytopathology*. 91: 55-62.

Baley, G.J., Talbert, L.E., Martin, J.M., Young, M.J., Habernicht, D.K., Berg, J.E., Lanning, S.P. and Bruckner, P.L. 2001. Agronomic and end-use qualities of Wheat streak mosaic virus resistant spring wheat. *Crop. Sci.* 41: 1779-1784.

Banks, P.M., Larkin, P.J., Bariana, H.S., Lagudah, E.S., Appels, R., Waterhouse, P.M., Brettell, R.I.S., Chen, X., Xu, H.J., Xin, Z.Y., Qian, Y.T., Zhou, X.M., Cheng, Z.M., and Zhou, G.H. 1995. The use of cell culture for subchromosomal introgressions of barley yellow dwarf virus resistance from *Thinopyrum intermedium* to wheat. *Genome* 38: 395-405.

Barbosa, M. M., Goulart, L. R., Prestes, A. M. and Juliatti, F. C. 2001. Genetic control of resistance to soilborne wheat mosaic virus in Brazilian cultivars of *Triticum aestivum* L. Thell. *Euphytica* 122: 417-422.

Boccardo, G. and Milne, R.G. 1984. The plant Reovirus group. AAB Descriptions of Plant Viruses no. 294.

Brakke, M. K. and Langenberg, W. G. 1988. Experiences with soil-borne wheat mosaic virus in North America and elsewhere. In: Developments in Applied Biology II. Viruses with Fungal Vectors. (eds) J. I. Cooper and M. J. C. Asher, pp. 183- 203. Association of Applied Biologists, Wellesbourne, Warwicks.

Braun, H. J., Payne, T.S., Morgounov, A.I., van Ginkel, M. and Rajaram, S. 1998. Germplasm development and breeding. In: *Proc. 9th Intern. Wheat Symp.*, Saskatoon, Saskatchewan, Canada, 2-7 August 1998, pp. 33-40.

Brunt, A. A., Crabtree, K., Dallwitz, M. J., Gibbs, A. J. and Watson, L. Eds. 1996. *Viruses of plants - Descriptions and lists from the VIDE database.* CAB International, Wallingford, UK, 1484 pp.

Burnett, P. A. and Mezzalama, M. 1992. The CIMMYT barley yellow dwarf program. In: *Barley yellow dwarf in West Asia and North Africa,* A. Comeau and K. Makkouk (eds). ICARDA, Syria, pp. 33-46.

Burnett, P. A., Comeau, A. and Qualset, C. A. 1995. Host plant tolerance or resistance for control of barley yellow dwarf. In: *Barley Yellow Dwarf, 40 Years of Progress.* C.J. D'Arcy and P. Burnett, (eds) APS Press, St. Paul, Minnesota, pp. 321-343.

Caetano, V. R., Marinho, V. L. A., Lin, M. T., Formiga, L. C. M. and Kitajima, E. W. 1990. Ocorrencia do virus do mosaico do capim bromo (brome mosaic virus) em trigo, no estado do Rio Grande do Sul. *Fitopatol. Bras.*, 15: 363-365.

Chen, J. P., Adams, M. J., Zhu, F. T., Shi, C. and Chen, H. 1992. Responses of some Asian and European barley cultivars to UK and Chinese isolates of soil-borne barley mosaic viruses. *Ann. appl. Biol.* 121: 631-639.

Chen, J. P., Adams, M. J., Zhu, F. T., Wang, Z. Q., Chen, J., Huang, S. Z. and Zhang, Z. C. 1996. Response of foreign barley cultivars to barley yellow mosaic virus at different sites in China. *Plant Pathol.* 45:1117-1125.

Chen, J., Adams, M. J., Zhu, F. and Chen, J. 1998. Preliminary identification of strains of barley yellow mosaic virus by cultivar responses in China. *Sci. Agric. Sin.* 31: 40-45.

Chen, J., Chen, J. P., Yang, J. P., Cheng, Y., Diao, A., Adams, M. J. and Du, J. 2000a. Differences in cultivar response and complete sequence analysis of two isolates of wheat yellow mosaic bymovirus in China. *Plant Pathol.* 49: 370-374.

Chen, J., Du, J. and Adams, M. J. 2000b. Sequence diversity in the coat protein coding region of wheat yellow mosaic bymovirus isolates from China. *J. Phytopathol.* 148: 515-521.

Chen, Q., Conner, R.L., Laroche, A., Fedak, G and Thomas, J.B. 1999. Genomic origins of *Thinopyrum* chromosomes specifying resistance to wheat streak mosaic virus and its vector, *Aceria tosichella. Genome* 42: 289-295.

Cheng, Z., Wu, M., He, X., Xu, L., Zhou, G., and Waterhouse, P. M. 2002. Nucleotide Sequence Analysis of the BYDV-GPV Isolate Genome, and Transgenic Wheat Obtained via Pollen Tube Pathway. In: *Barley Yellow Dwarf Disease: Recent Advances and Future Strategies.* M. Henry, and A. McNab, (eds) CIMMYT, Mexico D.F., pp. 29-31.

Choi, I.-R., Hall, J.S., Henry, M., Zhang, L., Hein, G.L., French, R. and Stenger, D.C. 2001. Contributions of genetic drift and negative selection on the evolution of three strains of wheat streak mosaic tritimovirus. *Arch. Virol.* 146: 619-628.

Dalbosco, M., Schons, J., Prestes, A. M. and Cecchetti, D. 2002. Effects of soil-borne wheat mosaic virus on yield in wheat and triticale cultivars. *Fitopatol. Bras.* 27: 53-57.

D'Arcy, C. J., Martin, R. R., and Spiegel, S. 1989. A comparative study of luteovirus purification methods. *Can. J. Pl. Path.* 11: 251-255.

D'Arcy, C. J., Domier, L. L., and Mayo, M. A. 1999. Luteoviridae. In: *Virus Taxonomy. Classification* and *Nomenclature of Viruses. Seventh Report of the International Committee on Taxonomy of Viruses.* M.H.V van Regenmortel, C.M. Fauquet, D.H.L.

Bishop, E.B. Carstens, M.K Estes, S.M Lemon, J. Maniloff, M.A. Mayo, D.J. McGeoch, C.R. Pringle and R.B. Wickner, (eds) Academic Press, pp. 775-784.

Delfosse, P., Reddy, A. S., Legrève, A., Devi, P. S., Devi, K. T., Maraite, H. and Reddy, D. V. R. 1999. Indian peanut clump virus (IPCV) infection on wheat and barley: symptoms, yield loss and transmission through seed. *Plant Pathol.* 48: 273-282.

Diao, A., Chen, J., Ye, R., Zheng, T., Yu, S., Antoniw, J. F. and Adams, M. J. 1999. Complete sequence and genome properties of Chinese wheat mosaic virus, a new furovirus from China. *J. Gen. Virol.* 80: 1141-1145.

El Yamani, M. and Hill, J. H. 1990. Identification and importance of barley yellow dwarf virus in Morocco. *Plant Dis.* 74: 291-294.

FAO, 2001. Food and Agriculture Organization of the United nations, FAOSTAT, Agriculture database, http://apps.fao.org/default.htm

Fischbeck, G. 2002. Contribution of barley to agriculture: a brief overview. In: *Barley Science, Recent Advances from Molecular Biology to Agronomy of Yield and Quality.* G.A Slafer, J.L Molino-Cano, R. Savin, J.L. Araus, and I. Romagosa (eds). Food Product Press, New York, USA, pp. 1-14.

Francki, M. G., Ohm, H. W., and Anderson, J. M. 2001. Novel germplasm providing resistance to barley yellow dwarf virus in wheat. *Aust. J. Agric. Res.* 52: 1375-1382.

French, R. and Stenger, D. C. 2002. Wheat streak mosaic virus. AAB Descriptions of Plant Viruses no. 393. http://www.dpvweb.net/dpv/showdpv.php?dpvno=393.

Gill, C. C. and Chong, J. 1979. Cytopathological evidence for the division of barley yellow dwarf virus isolates into two subgroups. *Virology* 95: 59-69.

Halbert, S. and Voegtlin, D. 1995. Biology and taxonomy of vectors of Barley yellow dwarf viruses. In: *Barley Yellow Dwarf, 40 Years of Progress.* C.J. D'Arcy and P. Burnett, (eds) APS Press, St. Paul, Minnesota, pp. 217-258.

Han, C., Li, D., Xing, Y., Zhu, K., Tian, Z., Cai, Z., Yu, J. and Liu, Y. 2000. Wheat yellow mosaic virus widely occurring in wheat (Triticum aestivum) in China. *Plant Dis.* 84: 627-630.

Herrera, M. G. F. 1989. Interactions between host plants and British isolates of barley yellow dwarf virus, Thesis - University of London, 232 pp.

Hewings, A. D. 1995. Purification and virion characterization of barley yellow dwarf viruses. In: *Barley Yellow Dwarf, 40 Years of Progress.* C.J. D'Arcy and P. Burnett, (eds) APS Press, St. Paul, Minnesota, pp. 165-179.

Izadpanah, K. and Behjatnia, A.A. 1989. Natural hosts of wheat streak mosaic virus and its mite vector in Fars. In: *Proc. Ninth Plant Protection Congr. Iran*, 163.

Kapooria, R. G., Ndunguru, J. and Clover, G. R. G. 2000. First reports of Soilborne wheat mosaic virus and Wheat spindle streak mosaic virus in Africa. *Plant Dis.* 84: 921.

Kashiwazaki, S., Huth, W. and Lesemann, D. 1998. Barley mild mosaic virus. AAB Description of Plant Viruses no. 356.

Kashiwazaki, S., Ogawa, K., Usugi, T., Omura, T. and Tsuchizaki, T. 1989. Characterization of several strains of barley yellow mosaic virus. *Ann. Phytopath. Soc. Japan* 55: 16-25.

Koev, G., Mohan, B. R., Dinesh-Kumar, S. P., Torbert, K. A., Somers, D. A., and Miller, W. A. 1998. Extreme Reduction of Disease in Oats Transformed with the 5' Half of the Barley Yellow Dwarf Virus-PAV Genome. *Phytopathology.* 88: 1013-1019.

Kose, A. and Ertunc, F. 1999. Virus diseases of wheat and barley in Eskisehir province. *J. Turk. Phytopathol.* 28: 55-62.

Kulumeka B., Bliss, W., Harness, C., Sutula, C., Bandla, M. and Gray, S.M. 2002. Diagnosis of Barley yellow dwarf virus and Cereal yellow dwarf virus isolates in infected plants. In:

Barley Yellow Dwarf Disease: Recent Advances and Future Strategies. M. Henry, and A. McNab, (eds) CIMMYT, Mexico D.F., p. 100.

Lane, L.C. 1979. Bromovirus group. AAB Descriptions of Plant Viruses no. 215.

Larkin, P. J., Banks, P. M., and Chen, X. 1995. Registration of Six Genetic Stocks of Wheat with Rust and BYDV Resistance: Z1, Z2, Z3, Z4, Z5, and Z6 Disomic Addition Lines with *Thinopyrum intermedium* Chromosomes. *Crop Sci.* 35: 604.

Larkin, P.J., Kleven, S., and Banks, P.M. 2002. Utilizing Bdv2, the *Thinopyrum intermedium* Source of BYDV Resistance, to Develop Wheat Cultivars. In: *Barley Yellow Dwarf Disease: Recent Advances and Future Strategies.* M. Henry, and A. McNab, (eds) CIMMYT, Mexico D.F., pp. 60-63.

Lee, K., Kashiwazaki, S., Hibi, T. and So, I. 1996. Properties and capsid protein gene sequence of a Korean isolate of barley mild mosaic virus. *Ann. Phytopath. Soc. Japan* 62: 397-401.

Li, D., Han, C., Xing, Y., Tian, Z., Yu, J., Cai, Z. and Liu, Y. 1997. Identification of the wheat yellow mosaic virus occurring in China by RT-PCR. *Acta Phytopathol. Sinica* 27: 303-307.

Lister, R. M. and Ranieri, R. 1995. Distribution and economic importance of barley yellow dwarf. In *Barley yellow dwarf, 40 years of progress.* C.J. D'Arcy and P. Burnett, (eds) APS Press, St. Paul, pp. 29-53.

Lister, R. M., Clement, D., Skaria, M., McFatridge, J. A. 1985. Stability of ELISA activity of barley yellow dwarf virus in leaf samples and extracts. *Plant Dis.* 69: 854-857.

Mahmood, T., Hein, G.L., and French, R.C. 1997. Development of serological procedures for rapid and reliable detection of wheat streak mosaic virus in a single wheat curl mite. *Plant Dis.* 81: 250-253.

Makkouk, K. M., Bertschinger, L., Conti, M., Bolat, N. and Dusunceli, F. 1996. Barley yellow striate mosaic rhabdovirus naturally infects cereal crops in the Anatolian Plateau of Turkey. *J. Phytopathol.*, 144: 413-415.

Makkouk, K.M., Kumar, S.G., Kadirova, Z. and Zueva, A. 2001. First record of Barley yellow striate mosaic virus infecting wheat in Uzbekistan. *Plant Dis.* 85: 1122.

Martin, R. R. and D'Arcy, C. J. 1990. Relationships among luteoviruses based on nucleic acid hybridization and serological studies. *Intervirology* 31: 23-30.

Mathre, D. E. 1997. Compendium of barley diseases (4th edition). APS Press, The American Phytopathology Society, 90 pp.

Mayo, M. A.2002. Virology Division news: Virus taxonomy-Houston. 2002. *Arch.Virol.* 147: 1072-1076

McGrath, P. F., Vincent, J. R., Lei, C. H., Pawlowski, W. P., Torbert, K. A., Gu, W., Kaeppler, H. F., Wan, Y., Lemaux, P. G., Rines, H. R., Somers, D. A., Larkins, B. A., and Lister, R. M. 1997. Coat protein-mediated resistance to isolates of barley yellow dwarf in oats and barley. *Euro. J. Plant Pathol.* 103: 695-710.

Miller, W. A., Beckett, R., and Liu, S. 2002. Structure, Function, and Variation of the Barley Yellow Dwarf Virus and Cereal Yellow Dwarf Virus Genomes. In: *Barley Yellow Dwarf Disease: Recent Advances and Future Strategies.* M. Henry, and A. McNab, (eds) CIMMYT, Mexico D.F., pp. 1-8.

Milne, R.G. and Conti, M. 1986. Barley yellow striate mosaic virus. AAB Descriptions of Plant Viruses no. 312.

Montana, J.R., Hunger, R.M., and Sherwood, J.L. 1996. Serological characterization of Wheat streak mosaic virus isolates. *Plant Dis.* 80: 1239-1244.

Namba, S., Kashiwazaki, S., Lu, X., Tamura, M. and Tsuchizaki, T. 1998. Complete nucleotide sequence of wheat yellow mosaic bymovirus genomic RNAs. *Arch. Virol.* 143: 631-643.

Nomura, K., Kashiwazaki, S., Hibino, H., Inoue, T., Nakata, E., Tsuzaki, Y. and Okuyama, S. 1996. Biological and serological properties of two strains of barley mild mosaic virus. *J. Phytopathol.* 144: 103-107.

Oswald, J. W. and Houston, B. R. 1953. Host Range and Epiphytology of the Cereal Yellow Dwarf Disease. *Phytopathol.*. 43:309-313.

Oswald, J.W. and Houston, B. R. 1951. A New Virus Disease of Cereals, Transmissible by Aphids. *Plant. Dis. Rep.* 35:471-475.

Pike, K. S. 1990. A review of barley yellow dwarf virus grain yield loses. In: *World Perspectives on Barley Yellow Dwarf.* P. Burnett, (ed) CIMMYT, Mexico, pp. 356-361.

Plumb, R. T. and Johnstone, G. R. 1995. Cultural, chemical and biological methods for the control of barley yellow dwarf. In: *Barley Yellow Dwarf, 40 Years of Progress.* C.J. D'Arcy and P. Burnett, (eds) APS Press, St. Paul, Minnesota, pp. 307-319.

Pocsai, E., Kobza, S., Murányi, I. and Szunics, L. 1991. Brome mosaic virus infection in different cereal breeding materials. *Acta Phytopathol.Entomol. Hungarica*, 26: 207-212.

Power, A. G. and Gray, S.M. 1995. Aphid transmission of barley yellow dwarf viruses: interactions between viruses, vectors, and host plants. In: *Barley Yellow Dwarf, 40 Years of Progress.* C.J. D'Arcy and P. Burnett, (eds) APS Press, St. Paul, Minnesota, pp. 259-289.

Rabenstein, F., Seifers, D.L., Schubert, J., French R. and Stenger, D.C. 2002. Phylogenetic relationships, strain diversity and biogeography of tritimoviruses. *J. Gen. Virol.* 83: 895-906.

Ramirez, I., Zerene, M. and Cortazar, R. 1992. The barley yellow dwarf virus program in Chile. In: *Barley yellow dwarf in West Asia and North Africa,* A. Comeau and K. Makkouk (eds). ICARDA, Syria, pp. 47-54

Rasmusson, D. C. and Schaller, C. W. 1959. The Inheritance of Resistance in Barley to the Yellow-Dwarf Virus. *Agronomy Journal* 51: 661-664.

Robertson, N. L., French, R., and Gray, S. M. 1991. Use of group-specific primers and the polymerase chain reaction for the detection and identification of luteoviruses. *J. gen. Virol.* 72: 1473-1478.

Rochow, W. F. 1969. Biological properties of four isolates of barley yellow dwarf virus. *Phytopath.* 59: 1580-1589.

Rochow, W.F. and Muller, I. 1971. A fifth variant of Barley yellow dwarf virus in New York. *Plant Dis. Rep.* 55: 874-877.

Rodriguez Pardina, P. E., Giménez Pecci, M. P., Laguna, I. G., Dagoberto, E. and Truol, G. 1998. Wheat: A new natural host for the Mal de Río Cuarto virus in the endemic disease area, Río Cuarto, Córdoba Province, Argentina. *Plant Dis.*, 82: 149-152.

Sahandpour, A. and Izadpanah, K. 1999. Report of Polymyxa graminis and a virus resembling wheat soilborne mosaic virus from Fars. *Iranian J. Plant Pathol.* 35: 61.

Sánchez-Sánchez H., Henry M., Cárdenas-Soriano E., Alvizo-Villasana H.F. 2001. Identification of Wheat streak mosaic virus and its vector Aceria Tosichella in Mexico. *Plant Dis.*, 85: 13-17.

Schaller, C. W., Qualset, C. O. and Rutger, J.N. 1964. Inheritance and Linkage of the Yd2 Gene Conditioning Resistance to the Barley Yellow Dwarf Virus Disease in Barley. *Crop Sci.* 4: 544-548.

Sherwood , J.L. 1987. Comparison of filter paper immunobinding assay, western blotting and an enzyme linked immunosorbent assay for the detection of wheat streak mosaic virus. *J. Phytopathol.* 118: 68-75.

Shirako, Y. and Wilson, T. M. A. 1993. Complete nucleotide sequence and organization of the bipartite genome of soil-borne wheat mosaic virus. *Virology* 195: 16-32.

Singh, R.P., Burnett, P.A., Albarrán, M., and Rajaram, S. 1993. *Bdv1* : A gene for tolerance to barley yellow dwarf virus in bread wheats. *Crop Sci.* 33: 231-234.

Torrance, L. 1987. Use of enzymatic amplifucation in an ELISA to increase sensitivity of detection of Barley yellow dwarf virus in oats and in individual vector aphids. *J. Virol. Methods* 15: 131-138.

Usugi, T. and Saito, Y. 1976. Purification and serological properties of barley yellow mosaic virus and wheat yellow mosaic virus. *Ann. Phytopath. Soc. Japan* 42: 12-20.

Usugi, T., Kuwabara, T. and Tsuchizaki, T. 1984. Serological detection of barley yellow mosaic virus, wheat yellow mosaic virus and soil-borne wheat mosaic virus by ELISA enzyme linked immunosorbent assay. *Ann. Phytopath. Soc. Japan* 50: 63-68.

Uyeda, I. and Milne, R. G. 1995. Introduction: Genomic organization, diversity and evolution of plant reoviruses. *Virology*, 6: 85-88

von Wechmar, M. B. and Rybicki, E. P. 1985. Brome mosaic virus infection mimics barley yellow dwarf virus disease symptoms in small grains. *J. Phytopathol.*, 114: 332-337.

Wang, M. B., Cheng, Z. M., Keese, P. K., Graham, M.W., Larkin, P. J., and Waterhouse, P.M. 1998. Comparison of the coat protein, movement protein and RNA polymerase gene sequences of Australian, Chinese and American isolates of barley yellow dwarf virus transmitted by *Rhopalosiphum padi. Arch. Virol.* 143: 1005-1013.

Wang, M. B., Abbott, D. C. and Waterhouse, P. M. 2000. A single copy of a virus-derived transgene encoding hairpin RNA gives immunity to barley yellow dwarf virus. *Mol. Plant Path.* 1: 347-356.

Wangai, A.W. 1990. Effects of barley yellow dwarf virus on cereals in Kenya. In: *World Perspectives on Barley Yellow Dwarf.* P. Burnett, (ed) CIMMYT, Mexico, pp. 391-393.

Wangai, A.W., Plumb, R.T., and Van Emden, H.F. 2000. Effects of Sowing Date and Insecticides on Cereal Aphid Populations and Barley Yellow Dwarf Virus on Barley in Kenya. *J. Phytopathol.* 148: 33-37.

Waterhouse, P. M., Gildow, F. E. and Johnstone, G. R. 1988. Luteovirus Group. AAB Descriptions of Plant Viruses, no. 339.

Webby, G. N. and Lister, R. M. 1992. Purification of the NY-RMV and NY-SGV Isolates of Barley Yellow Dwarf Virus and the Production and Properties of Their Antibodies. *Plant Dis.* 76: 1125-1132.

Webby, G. N., Lister, R. M. and Burnett, P. A. 1993. The occurrence of barley yellow dwarf viruses in CIMMYT bread wheat nurseries and associated cereal crops during 1988-1990. *Ann. appl. Biol.* 123: 63-74.

Wiese, M. V. 1991. Compendium of wheat diseases (2nd edition). APS Press, The American Phytopathology Society, 112 pp.

Yang, J., Chen, J., Jiang, H., Zhao, Q. and Adams, M. J. 2001. Sequence of a second isolate of Chinese wheat mosaic furovirus. *J. Phytopathol.* 149: 135-140.

Ye, R., Zheng, T., Chen, J., Diao, A., Adams, M. J., Yu, S. and Antoniw, J. F. 1999. Characterization and partial sequence of a new furovirus of wheat in China. *Plant Pathol.* 48: 379-387.

Zhang, Q. F., Guan, W. N., Ren, Z. Y., Zhu, X. S. and Tsai, J. H. 1983. Transmission of barley yellow dwarf virus strains from northwestern China by four aphid species. *Plant Dis.* 67: 895-899.

Zheng, T., Cheng, Y., Chen, J. P., Antoniw, J. F. and Adams, M. J. 1999. The occurrence of barley mild mosaic virus (BaMMV) in China and the nucleotide sequence of its coat protein gene. *J. Phytopathol.* 147: 229-234.

Zhou, G-H, Cheng, Z-M., Zhang, S-X. and Qian, Y-T. 1984. Serological identification of luteoviruses of small grains in China. *Plant Dis.* 68: 710-713.

14. Cowpea

R. O. Hampton[1] and G. Thottappilly[2]
[1]Research Plant Pathologist, U. S. Department of Agriculture, Agriculture Research Service, and Professor (Courtesy), Department of Botany and Plant Pathology, Oregon State University, Corvallis, OR 97331-2902, USA [2]Executive Director, Mahyco Research Foundation, Road No. 12, Banjara Hills, Hyderabad- 500 034, India

14.1 Introduction

Cowpea (*Vigna unguiculata*) is an important crop, contributing to the health and livelihood of millions of people in tropical and sub-tropical countries, particularly Africa (Nigeria, Niger, Burkina Faso, Ghana, Kenya, Uganda, Tanzania, Senegal, Togo) and Asia (India, Pakistan, Bangladesh, Sri Lanka, Burma, Thailand, Nepal, China, Malaysia) (Rachie, 1985). Young leaves, immature pods, and dry-stored seeds from the plant serve as food, particularly for rural families. The crop also provides cash income and animal forage/fodder, and residue from the crop is valued for soil enrichment. Cowpea crops have also become important in southern states of the U.S. and in South America, particularly Brazil, as advanced, multi-adapted cultivars (Quin, 1997).

Historically, many factors have limited cowpea production in regions where the crop is best adapted; including a complex of variously transmitted viral diseases. Prior summaries of this information, up-dated by this chapter, include (i) viral taxonomic identities and relationships, (ii) the means of their origin and spread in cowpea plantings, (iii) conventional methods for their control, and (iv) references to viral molecular biology, nucleotide sequences of viral genes and their function (Regenmortel, *et al.* 2000) and to the nature and potential of producing anti-viral transgenes for direct introduction into plant tissues through molecular-engineering methods (Lomonossoff, 1995; Monti *et al.* 1997; Karasawa *et al.*1999; Bertens *et al.* 2000; Gopinath *et al.* 2000; Assis *et al.* 2002). Unfortunately, ill-informed, fear-based activist groups in developed countries have widely propagandized against genetically modified crop plants, disregarding food shortages and famine among peoples subject to subsistence agriculture. But fortunately, there have been concurrent off-setting factors: (1) persevering and prevailing counter-activist "molecular engineers" whose work will benefit millions of needy people, e.g., by developing beta-carotene enhanced rice (Beyer and Potrykus, 1999) and pest-resistant sweet potato (Wambubu, 1999) (press releases), and (2) development of elite cultivars by pyramiding genes from new pest-resistant sweet potato (Wambubu, 1999) (press releases), and (2) development of elite cultivars by pyramiding genes from new virus-resistant cowpea cultivars (Singh *et al.* 1997) and by utilizing international sources of virus-resistant *V. unguiculata* germplasm (see Table 14.2) and other *Vigna* spp.

Simultaneous with the development of information on viral genetics (i.e., viral gene structure and viral gene functions) has begun increased molecular characterization of cowpea (Galasso *et al.* 1997; Kononowitz *et al.* 1997). The continuing development of such understanding can facilitate prioritized, genetically engineered cowpea resistance against the most damaging viruses. Continuing synthesis of this information can facilitate genetically engineered cowpea resistance to the most damaging viruses; thereby improving cowpea productiveness and the prosperity of families dependant on this crop.

14.2 Cowpea viruses

This chapter complements and/or updates previous reviews by Allen (1983), Mali and Thottappilly (1986), Thottappilly and Rossel (1985, 1992), Thottappilly *et al.* 1995, Shoyinka *et al.* (1997), Hampton *et al.* (1997), and Allen *et al* (1998). Viruses that are seedborne in cowpea are, at the same time, (a) insidious and severely damaging and (b) the most simply and preclusively controlled (by production and use of certified, virus-free seedstocks). These are: *Blackeye cowpea mosaicvirus* Genus *Potyvirus* (BlCMV), *Cowpea aphid-borne mosaic virus* Genus *Potyvirus* (CABMV), *Cowpea mosaic virus* (CPMV) and *Cowpea severe mosaic virus* (CPSMV) comoviruses, *Cowpea mottle virus* Genus *Carmovirus* (CPMoV), *Cucumber mosaic virus* Genus *Cucumovirus* (CMV), and *Southern bean mosaic virus* Genus *Sobemovirus* (SBMV) (Table 14.1). Because of their seed-transmissibility, these viruses have now been disseminated to, and established in, most cowpea-producing areas of the world *via* commercial seedlots, variety trials, dispensed germplasm, etc. Some virus combinations (e.g., BlCMV + CMV; BlCMV + CPSMV; and CMV + CPSMV + SBMV) can cause drastically worsened disease symptoms and crop losses (Kuhn 1990; Anderson *et al.* 1994). Other detected cowpea mixed-infections include CMV + CPSMV and SBMV + CPSMV (Hampton *et al.* 1992, and *unpublished data*).

Non-seedborne viruses include *Cowpea golden mosaic virus* Genus *Geminivirus* (Winter *et al.* 2002), which causes one of the most destructive cowpea diseases in the world, and *Cowpea chlorotic mottle virus* Genus *Bromovirus*, which causes disease losses either alone or in combination with other viruses (reviewed by Kuhn 1990).

Recently developed, ultra-sensitive methods for detecting/identifying viruses in plants or seeds (Gillaspie *et al.* 1999; Guia *et al.* 2001; Lima *et al.* 2001), enhanced monitoring and control of viral diseases.

Blackeye cowpea mosaic virus Genus ***Potyvirus*** (BlCMV) occurs wherever cowpeas are extensively grown. It is transmitted non-persistently by several aphid species, including *Aphis craccivora* (Purcifull and Gonsalves, 1985). Particularly in combination with other viruses (Pio Ribeiro *et al.* 1980; Collins *et al.* 1985; Kuhn 1990), BlCMV can cause severe losses to cowpea crops. Distinct BlCMV strains exist, but strain variants may be less decisive in disease epidemics than is notable for CABMV (Bashir, 1992; Bashir and Hampton, 1992).

The work of Taiwo *et al.* (1982a) partitioned potyviruses seedborne in cowpea into two distinct kinds. With differing results and interpretations, Dijkstra *et al.* (1987) distinguished two potyviruses, but recommended that both be called BlCMV. The relationship between BlCMV and CABMV was discussed at a potyvirus taxonomy workshop (Barnett, 1992), with clear indications that BlCMV and CABMV are distinct potyviruses and that separate viral terminology should be retained.

Bashir (1992) biologically and serologically characterized some 140 cowpea potyvirus isolates seedborne in cowpea seedlots from various countries (Bashir 1992; Bashir and Hampton 1992, 1993), in comparison with type isolates BlCMV-Georgia (BlCMV-GA) and CABMV-Morocco CABMV-Mor). This work clearly partitioned the two viruses, determined that CABMV-Kenya (Bock 1973; Dijkstra *et al* 1987) was instead BlCMV, and verified much of the work of Taiwo *et al.* (1982a). Key isolates characterized by Bashir (1992) were also instrumental in definitive monoclonal antibody distinctions of BlCMV and CABMV by Huguenot *et al.* (1993, 1994). The Florida isolate of BlCMV was considered by McKern *et al.* (1992) to be a strain of bean common mosaic virus. The sequence of the BlCMV genome had not yet been published, as of November 2002.

Cowpea aphid-borne mosaic virus Genus ***Potyvirus*** (CABMV). First described by Lovisolo and Conti (1966), CABMV is endemic in Africa. It is now widely disseminated in the world through infected cowpea seedlots, and causes severe crop damage either alone (Ndiaye *et al.* 1993) or in combination with other viruses. Like BlCMV, it is transmitted nonpersistently by several aphid species, including *Aphis craccivora*. The virus comprises numerous distinct strains (Fischer and Lockhart 1976; Bashir 1992; Ndiaye *et al.* 1993).

CABMV and BlCMV produce indistinguishable symptoms on cowpea genotypes susceptible to them, typically consisting of veinal chlorosis, interveinal chlorosis, or dark-green vein banding (Bock and Conti 1974; Purcifull and Gonsalves 1985). The Morocco isolate (Fischer and Lockhart 1976), CABMV-Mor, has been widely used as a quasi type isolate, but it is extremely virulent and poorly representative of 80 separate seedborne CABMV isolates that were evaluated at Corvallis, Oregon, USA (Bashir 1992; Ndiaye *et al.* 1993).

The potyvirus designated PTY+ by Ndiaye *et al.* (1993) was later determined to be a distinct, virulent strain of CABMV (Hampton, *unpublished data*). This CABMV pathotype clearly differed from CABMV-Mor, and sources of cowpea genetic resistance were identified (Ndiaye *et al.* 1993). A sizeable but unknown number of pathogenic variants exist in nature, some of which respond to separate cowpea genes/alleles for resistance. Because none of the 14 elite cowpea selections were resistant to all 10 selected CABMV isolates, Van Boxtel *et al.* (2000) proposed complementing natural cowpea resistance with genetically engineered resistance to BlCMV and CABMV. The complete nucleotide sequence of the CABMV genome was published by Mlotshwa *et al.* (2002).

Cowpea chlorotic mottle virus Genus *Bromovirus* (CCMV). CCMV was not accepted as a distinct virus until the definitive work of Kuhn (1964a) and Bancroft *et al.* (1968). Kuhn (1964b) also developed differential hosts for distinguishing CCMV, SBMV, CMV, and BYMV (actually BlCMV). In susceptible cowpea cultivars, CCMV can cause severe crop damage, alone or in mixed infections with SBMV (Kuhn and Dawson, 1973). Once assumed to be confined to North and South America, CCMV was later isolated from *Desmodium heterocarpon* and *Clitoria ternatae* in Nigeria (Thottappilly *et al.* 1993). The occurrence of CCMV in natural hosts outside of the Americas suggests that it may persist in native legumes of other cowpea- producing regions of the world. The genomic RNA of CCMV was sequenced and compared to that of other bromoviruses by Allison *et al.* (1988, 1989). Nobumitsu *et al* (2001) analyzed codon-change patterns in

CCMV adaptive mutants, and determined that single amino acid changes to Lys or Arg in the movement protein were associated with the virus' adaptation to a non-host. Mutational analysis of CCMV-T (type strain) RNA-3 by Assis *et al.* (2002) suggested that typical chlorotic symptoms were determined by amino acid residue 151 of the coat protein, not the associated nucleic acid sequence.

Cowpea golden mosaic virus Genus ***Geminivirus*** (CGMV). Wherever this virus occurs in cowpea crops, principally Africa and Asia, but also in Brazil (Faria *et al.* 2000), it is capable of causing a very destructive disease. Thottappilly (1992) and Thottappilly and Rossel (1992) reported the occurrence of CGM-like diseases in at least seven African countries. These diseases were considered related to those reported as "cowpea yellow fleck" in India, by Sharma and Varma (1976), and as "cowpea bright yellow mosaic" in Pakistan, by Ahmed (1978), and as "mungbean yellow mosaic" in Pakistan by Bashir and Bashir (1988). Cowpea samples from Nigeria with CGM-like symptoms produced weak reactions with monoclonal antibodies reactive to whitefly-transmitted geminiviruses (Thottappilly and Rossel 1992), suggesting that the CGM symptoms were caused by geminivirus(es). The demonstration of cowpea resistance to CGM was initiated by Anno-Nyako 1980, and the dynamics of whitefly (*Bemisia tabaci*)-transmission of CGM were also investigated (Anno-Nyako *et al.* 1983). CGM-resistant cowpea breeding lines have since been developed (Singh *et al.* 1997)

The CGM disease, as recognized in Pakistan (M. Bashir, *personal communication*), has caused increasingly severe damage to cowpea plantings in that country, since 1988. Although no sources of genetic resistance to CGM were identifiable in Pakistan *V. unguiculata* germplasm resources, Anno-Nyako (1980) reported that 23 of 47 IITA (Nigeria) cowpea accessions and 8 of 11 other *Vigna* spp. accessions were resistant to the CGM disease, based on non-detectability of the virus in symptomless whitefly-inoculated plants. The extent to which CGMV comprises divergent pathotypes or even variable genetic components could limit development of durable resistance in cowpea genotypes. Prerequisite to elucidation of cowpea genes that confer durable CGMV resistance are (i) controlled inoculations of (ii) standardized indicator cowpea genotypes, using (iii) defined geminivirus isolates that (iv) consistently produce typical golden mosaic symptoms/responses in (v) proven-susceptible genotypes and none in proven-resistant (Singh *et al.* 1997). And, as geminivirus-resistant cultivars come into wider use, the threat of serious losses by CGMV may be reduced locally by removing CGMV-susceptible weedy plants (natural inoculum reservoirs), such as *V. unguiculata* var. dekindtiana, from the vicinity of cowpea fields.

The complete sequencing of CGMV DNAs from India (Malathi, *et al.* 2002), from Nigeria (Winter *et al.* 2000) and Taiwan (Tsai *et al.* 2000) geminivirus isolates is currently being concluded, pending final biological trials with isolates comprising cloned viral DNA. Interestingly, sequence comparisons between the DNA A of cowpea golden mosaic geminivirus isolates from India and Nigeria are only 62% homologous, indicating unexpectedly great viral diversity in the same crop, between geographic locations (Winter *et al.* 2000). Similar biological diversity existed among geminiviruses affecting tomato crops in North, Central and South America (Polston and Anderson, 1997), and among those affecting cowpea and other crops in Brazil (Faria *et al.* 2000).

Cowpea mosaic virus Genus ***Comovirus*** (CPMV). CPMV, originally described as cowpea yellow mosaic virus (Chant 1959), reportedly occurred in the Americas before 1964, since an isolate from Surinam (Agrawal, 1964) was identified as CPMV. It has since been reported from several African countries (Thottappilly and Rossel, 1985), and non-standard names for CPMV continue to be published (Sangwan *et al.* 2000). Though its identity and existence in older cowpea landraces/varieties in both West Africa (Chant 1959; Patel and Kuwite 1982) and India (Hampton *et al.* 1992) are generally accepted, CPMV was not detected recently in either Senegal (Ndiaye *et al.* 1993) or Pakistan (Bashir and Hampton 1993). Some CPMV isolates appear to be marginally seed transmissible (Gilmer *et al.* 1974 suspected 1-5%); seed-transmission of CPMV could not be confirmed in other cowpea genotypes (Thottappilly and Rossel, 1982).

Owing to its common occurrence, pathogenicity, and epidemic potential, CPMV is one of the most important viruses in Africa. Most locally grown varieties (large, white, rough-seeded) continue to be susceptible and highly sensitive to infection. The virus also occurs in pigeon pea (Bock, 1971), soybean, and bambara groundnut [*Vigna subterranean*](Thottappilly and Rossel, 1992). As with other viral diseases of cowpea, the most practical control of CPMV continues to be use of resistant cultivars (Robertson 1965; Williams 1975, 1977; Singh *et al.* 1997).

The RNA genome of CPMV, type member of the comovirus group, was sequenced and defined in a series of classic investigations by van Kammen and colleagues (e.g. Sijen *et al.* 1995; Lekkerkerker *et al.* 1996; Bertens *et al.* 2000; Gopinath *et al.* 2000). Continuing research is likely to identify means by which transgenic CPMV-resistant cowpea genotypes can be generated.

Cowpea severe mosaic virus Genus ***Comovirus*** (CPSMV). CPSMV was characterized by Shepherd (1964) as "Arkansas cowpea mosaic virus". Its host range was very extensive, in contrast to that of standard isolates of cowpea mosaic virus (CPMV). Isolates of this virus, however, were transmitted by Chrysomelid beetles and were separated from CPMV by Agrawal (1964) and named CPSMV by de Jager (1979). The isolate of de Jager did indeed induce very severe symptoms on some cowpea cultivars; however, symptoms induced by CPSMV on other cowpea cultivars were no more severe than those induced by CPMV. Crop losses caused by CPSMV were sometimes severe (50-80%, Debrot and De Rojas 1967; Valverde *et al.* 1982); however, losses depend largely on specific interactions between CPSMV strains and cowpea genotypes. CPSMV is seed transmissible and also efficiently transmitted by several beetle species, including *Cerotoma ruficornis* and *C. trifurcata* (Walters and Barnett, 1964; Debrot and De Rojas, 1967), which can retain infective virus for more than 7 days.

The virus has probably assumed worldwide distribution via movement of infected seedlots, appears to be more common than CPMV in the cowpea cultivars of southern Europe and the Americas, and is less common in old-world cowpea-growing regions (Bashir and Hampton, 1993; Ndiaye *et al.* 1993). CPSMV was reported to occur in five legume species, in Brazil (Lin 1982), and cowpea resistance to Brazil isolates was reported by Umaharan *et al.* (1997), by Paz, da *et al.* (1999), and by Lima *et al.* (2001). Resistance to a Trinidad isolate of the virus was also reported by Umaharan *et al.* (1999). Lima *et al.* (2001) and Guia, *et al.* (2001) reported simplified serological methods for diagnosing CPSMV and for identifying CPSMV-resistant cowpea genotypes.

CPSMV comprises at least nine serotypes (J.H. Hill, isolate donations to The American Type Culture Collection; Di *et al.* 1993) and an unknown number of pathogenic variants.

Cowpea mottle virus Genus ***Carmovirus*** (CPMoV). Originally isolated in Nigeria (Shoyinka *et al.* 1978; Bozarth and Shoyinka, 1979), CPMoV readily cross-reacts with antiserum to bean mild mosaic carmovirus (Gillaspie *et al.* 1994) and is probably abiotically transmitted in soil for >2 months after infected plants are removed (R. Hampton, *unpublished data*). Thouvenel *et al.* (1990) characterized an Ivory Coast isolate of CPMoV and reported thatit caused a 65% reduction in yield there. The virus was also reported from the Republic of Benin (Thottappilly and Rossel 1988), Togo (Gumedzoe *et al.* 1990), and Pakistan (Bashir and Hampton 1993). The virus was also detected by DAC-ELISA in cowpea seedlots from Botswana and Senegal (R. Hampton, *unpublished data*). An RT-PCR-based assay that greatly exceeded the sensitivity and virus-specificity of DAC-ELISA for detecting CPMoV in cowpea seedlings was since reported by Gillaspie *et al.* (1999)

Only tolerance to CPMoV has ever been reported in cowpea, e.g., Allen *et al.* (1982). However, Ogundawin (2002) identified resistance to CPMoV in two IITA accessions of *Vigna unguiculata* subsp. *dekindtiana* that was conferred by a single dominant gene. Because this subspecies is cross-incompatible with cowpea, Ogundawin proposed that the gene be cloned and transferred to cowpea by molecular engineering. This work also opens opportunity to define the exact mechanism by which this gene precludes infection by CPMoV, facilitating yet further research.

The capsid protein gene of CPMoV was sequenced by Kim and Bozarth (1992), and the entire CPMoV genome was sequenced by You (1995) and You *et al.* (1995). The genomes of four other carmoviruses were sequenced, as reviewed by Hacker *et al.* (1992) and Skotnicki *et al.* (1993). Both the identification of a CPMoV-resistant Vigna gene and of CPMoV RNA coding regions invite genetic transformation efforts for producing CPMoV-resistant cowpea lines.

Cucumber mosaic virus Genus ***Cucumovirus*** (CMV). CMV is one of the most broadly adapted of all plant viruses (Francki *et al* 1979), and is one of the most commonly seedborne viruses in cowpea seedlots. Despite its common and widespread occurrence, through both seed- and aphid transmission, CMV is considered a mild cowpea pathogen except in a minority of infection-sensitive genotypes. However, CMV, when combined with BlCMV (and some CABMV isolates), typically causes severe symptom and yield reductions, a disease referred to as cowpea stunt (Pio-Ribeiro *et al.* 1978, 1980). A selection of cowpea germplasm accession PI 441918 was found to be "stunt resistant" by virtue of its restriction of CMV replication (Gillaspie 2001). Interestingly, cowpea resistance to CMV-Y, a strain unable to infect most cowpea genotypes, was conditioned by a single dominant gene (Nasu *et al.* 1996), and Karasawa *et al.* (1999) determined that the capacity of CMV-Y to infect cowpea cv. Kurodane-Sanjaku was attributable to a single amino acid of CMY-Y RNA polymerase. Both of these host-pathogen attributes could help facilitate cowpea resistance to CMV, through molecular transformation.

The extent to which cowpea isolates of CMV differ from other legume-infecting forms is not well defined. Legume-infecting isolates CMV-Pg and CMV-Le (Hampton and Francki, 1992) are distinguishable from CMV-CP biologically but have antigenic determinants in common with CMV-CP. Antisera/IgG to either CMV-Pg or CMV-Le react with, but also differentiate CMV-CP (R. Hampton, *unpublished data*).

The tripartite RNA genome of CMV was sequenced by Symons and colleagues (Gould and Symons, 1982; Rezaian *et al.* 1984, 1985) and cloned, transcribed and tested for infectivity by Hayes and Buck (1990). Several pathological traits have been ascribed to genomic RNA-1, 2, and 3 (Rao and Francki, 1982; Edwards *et al.* 1983; Lakshman *et al.* 1985). CMV-mediated transgenic resistance appears plausible, if needed, as a CMV control measure.

Southern bean mosaic virus Genus ***Sobemovirus*** (SBMV). The cowpea strain of SBMV (SBMV-C) was discovered as a seedborne isolate in a seedlot of cv. Wilt Resistant Early Ramshorn cowpea (Shepherd and Fulton, 1962). It often occurs in mixtures with other beetle-transmissible viruses, including CCMV (Kuhn 1990) and CPSMV (R. Hampton, *unpublished data*). Like other seedborne viruses, SBMV is becoming distributed to most cowpea-producing regions of the world. Reports of SBMV from India and many locations in Africa since 1974 were reviewed by Thottappilly and Rossel (1992).

SBMV-induced symptoms are exceptionally variable among cowpea genotypes (Kuhn, 1990), ranging from symptomless infection to severe mottle/mosaic with leaf deformity. Kuhn (1990) reviewed several forms of SBMV resistance in cowpea, including infection localization and inhibition of virus synthesis. A separate resistance mechanism in cv. Bountiful bean, associated with the formation of abnormal SBMV-C virions, apparently prevented systemic spread of the virus to noninoculated trifoliolate leaves (Fuentes and Hamilton, 1993). However, resistance to intercellular SBMV-C movement in inoculated primary bean leaves was overcome by co-infection with sunnhemp mosaic tobamovirus (Fuentes and Hamilton 1991). A resistance-breaking mutant of SBMV (SBMV-S) was characterized by Lee and Anderson (1998).

The micro-structure of the SBMV virion was elucidated by Rossman and colleagues (e.g., Silva and Rossman 1987) and has perhaps received more attention for fine-structure than any other plant virus. Antigenic determinants of the SBMV capsid were defined with monoclonal antibodies by Tremaine *et al.* (1985), and the SBMV RNA genome was sequenced and elaborated by Wu *et al.* (1987).

Other viruses. Viruses isolated from cowpea but of undetermined or minor significance include *Alfalfa mosaic alfamovirus* (Jaspers and Bos, 1980), *Cowpea mild mottle carlavirus* (Brunt and Kenton, 1973; Naidu *et al.* 1998), *Peanut mottle potyvirus* (Demski *et al.* 1983), *Peanut stunt* cucumo*virus* (Abdelbagi and Ahmed, 1990), *Sunnhemp mosaic tobamovirus* (Chant and Gbaja, 1987; Fuentes and Hamilton, 1991), and *Tobacco ringspot* nepo*virus* (Zeeuw and Ballard, 1959; Mali and Ganacharya, 1984). *Beet curly* top *geminivirus (*Regenmortel, *et al.* 2000) has been observed and identified in cowpea, in California (Hampton and A. Hall, 1990, *unpublished*). *Tomato spotted wilt tospovirus* is infectious to cowpea, has caused increasing damage to susceptible crops in semitropical regions (Brunt *et al.* 1996), and could become a threat to cowpea, when tomato crops are grown near by.

14.3 Genes conferring resistance to cowpea viruses

Cowpea genetics was reviewed by Fery (1985) and Fery and Singh, (1997). Basic concepts of virus-resistance breeding were thoroughly reviewed from three perspectives (Kyle and Provvident, 1993; Provvidenti, 1993; Scully and Federer, 1993). At least 25 genes or genetic resources providing resistance or tolerance have been reported since 1955 (Table 14.2). Virus resistance should be preferred wherever feasible, since virus-tolerant plants are potential reservoirs from which viruses can be spread to susceptible cowpea cultivars (Allen *et al.* 1982), e.g., tolerance to *Cowpea chlorotic mottle virus* (Kuhn *et al.* 1981) potentiated virus multiplication and further dissemination of the virus by insect-vectors. Other examples of this risk are: CMV (Khalf-Allaha *et al.* 1973; Gillaspie, 2002) and CPMoV (Allen *et al.* 1982)(see Table 14.2).

Of the gene-source cowpea genotypes reported to be resistant to cowpea-infecting viruses (Table 14.2), resistance was conferred in 10 by dominant genes (e.g., BlCMV: By and "D", PEPH-BVR and WA-BVR) and in nine by recessive genes, (e.g., *bcm*, TVu 2480 and *blc*, Worthmore). Genes in 12 other resistant genotypes were not specified or named.

Resistance to BlCMV is distinct from that of CABMV (Bashir, 1992) and independently inherited (Taiwo *et al.* 1982b). The two independent genes conferring resistance were also demonstrated for a Tanzanian isolate of CABMV (BlCMV): one recessive and one "partially dominant" (Patel *et al.* 1982). Partial dominance in this case was probably attributable to lower-than-normal virulence of the virus isolate, after successive local-lesion passages through *Chenopodium amaranticolor*. This virus isolate was later reported to be BlCMV, rather than CABMV (Bashir, 1992; P.N. Patel, personal communication, 1992). Both dominant and recessive genes were also reported to govern resistance to CPMV (Patel, 1982a; Ponz *et al.* 1988) and to SBMV (Brantley and Kuhn, 1970; Melton *et al.* 1987; Hobbs *et al.* 1987). Recently published BlCMV-resistant cowpea genotypes and cultivars include those of Gumedzoe *et al.* 1998, Putaraju *et al.* 2000, and Ferry, 2002.

Resistance sources are also indicated for CPMV, CPSMV, CPMoV (i.e., in *V. vexillata*), CMV, SBMV, *Peanut mottle virus* (PMV), TRSV, and CCMV (Table 14.2). *Cowpea mild mottle virus* (Table 14.1, footnote), while not considered serious in cowpea crops, can be damaging to other crops, e.g., eggplant crops in Jordan (Mansour *et al.* 1998).

In retrospect, three separate research efforts since 1980 succeeded in identifying IITA cowpea lines with genes conferring resistance to two or more viruses: (a) TVu lines 612, 1948, and 2480 were resistant to BlCMV (*sic.*CABMV), to CPMV (Patel, 1982; 1982 *et al.*[a]; 1982 *et al.* [b]) and also to CSMV (Fulton and Allen, 1982), and (b) genes of TVu 1509 conferred resistance to both CGMV (Anno-Nyako, 1980) and CPMV (Patel, 1982)

Although resistance to cowpea golden mosaic virus has now been incorporated into advanced breeding lines (Singh *et al.* 1997), identities of resistance-conferring genes still require (1) establishment of accepted CGMV type isolates, (2) choices of resistant and susceptible cowpea parental genotypes, and (3) classical cowpea-genetic studies.

14.4 Cultural practices for reducing viral diseases

Pest management methods (e.g., Jackai and Adalla,, 1997), and even those of non-cowpea crops (e.g., Fondong and Thresh, 2002) are obviously relevant, wherever they pertain directly or in principle to controlling insect vectors of viruses or other cowpea pests. However few practicable cultural measures, other than establishing and planting virus-free seed stocks, can significantly reduce the incidence or spread of or damage by viruses to cowpea crops in tropical and subtropical regions (Sharma, 1984; Orozco *et al.* 1994). Rather, the principal disease-determinant factors are (1) proximity of virus-inoculum reservoirs in crop or non-crop plant species, (2) prevalence of insect pests capable of transmitting viruses from infected to healthy plants, and (3) levels of virus-susceptibility of cowpea cultivars planted within cowpea-producing areas or regions (ref. Table 14.2).

14.5 Cowpea germplasm resources

Drs Quat Ng and B.B. Singh, with others at the International Institute for Tropical Agriculture (IITA), Ibadan, Nigeria, acquire, preserve, utilize, and distribute germplasm of cowpea and wild (native) species of Vigna (see books Singh and Rachie, 1985, and Singh, Mohan Raj, Dashiell and Jackai, 1997). IITA safe-stores the world's largest collection of *V. unguiculata* germplasm (~15,000 accessions, ~85% of which are primative cultivars [landraces]; remainder, commercial cultivars and breeder lines). Dr. Singh and cooperating scientists developed tropic-adapted, high-yielding cultivars with resistance to major diseases, insect pests, nematodes, and parasitic weeds (Singh *et al.* 1997). Seeds of such cultivars have been released for evaluation and utilization to more than 60 countries.

Likewise, Dr. Henry Shands, Director of the U.S. Department of Agriculture, Agricultural Research Service (USDA-ARS) National Center for Genetic Resources Preservation (NCGRP), Fort Collins, Colorado, safe-stores more than 8,000 cowpea accessions from some 64 countries where *Vigna* spp. are native or historically cultivated, principally in Africa, Asia, and the Americas. Taxonomic terms and relationships endorsed for USDA-

ARS *Vigna* spp. germplasm accessions were published in "Legume (Fabaceae) Nomenclature in the USDA Germplasm System", Technical Bulletin No. 1757 (1990). Such cowpea accessions from numerous countries were screened under the supervision of Dr. Graves Gillaspie.. Once introduced, this cowpea germplasm is then stored at both the Griffin and Fort Collins facilities, and made available internationally.

Dr. Richard Fery, USDA-ARS Research Geneticist, Charleston, South Carolina, reported that the combined efforts of USDA-ARS and state-university breeders produced 42 new cowpea cultivars during 1960 to 1985. Cooperative cowpea breeding efforts during that time included scientists in Arkansas, Alabama, California, Louisiana, Mississippi, Minnesota, South Carolina, and Texas (Fery, 1985). These research efforts were supported and supplemented by leading U.S. cowpea-producing and/or cowpea processing-companies.

Finally, the International Board of Plant Genetic Resources (IBPGR) is committed to the exploration, documentation, and preservation of land races of valued plant species. The contributions of IBPGR have supplemented those of indigenous scientists in developed and developing countries, sometimes facilitating *in situ* preservation of species related to internationally grown crop plants. Collections and addresses, reported in the 1989 IBPGR Directory of Food Legume Germplasm Collections, including *Vigna* spp., have been supplemented by numerous other Vigna collections reported in their Plant Genetic Resources Newsletter, e.g., issues 73/74 (1988), 77, 78/79, 80, 83/84, 101, Mededelingen LandbHoogesch, Wageningen 64:1-53and 107 (1996).

References

Abdelbagi, A.O. and Ahmed, A.H. 1990. Effect of the Sudanese strain of peanut stunt virus on the growth, nodulation, and yield of cowpea. *Trop.Agric.* 67:66-68.

Agrawal, H.W. 1964. Identification of cowpea mosaic virus isolates. ?????

Ahmed, M. 1978. Whitefly (*Bemisia tabaci)* transmission of the yellow mosaic disease of cowpea (*Vigna unguiculata). Plant Dis. Reptr.* 62:224-226.

Allen, D.J. 1983. *The pathology of tropical food legumes. Disease resistance in crop improvement.* John Wiley and Sons, Chichester, UK. 413 pp.

Allen, D.J., Thottappilly, G., and Rossel, H.W. 1982. Cowpea mottle virus: field resistance and seed transmission in virus-tolerant cowpea , Vigna unguiculata. *Ann. appl. Biol.* 100:331-336.

Allen, D.J., Thottappilly, G., Emechebe, A.M., and Singh, B.B. 1998. Diseases of Cowpea, Pages 267-324 *in The Pathology of Food and Pasture Legumes.* Editors D.J. Allen and J.M. Lenne, pages 267-324, CAB International, Oxon, UI, 750 pp.

Allison, R.F., Janda, M, and Ahlquist, P. 1988. Infectious in vitro transcripts from cowpea clhlorotic mottle virus cDNA clones and exchange of individual RNA components with brome mosaic virus. *J. Virol.* 62:3581-3588.

Allison, R.F., Janda, M., and Ahlquist. P. 1989. Sequence of cowpea chlorotic mottle virus RNAs 2 and 3 and evidence of a recombination event during bromovirus evolution. *Virology* 172:321-330.

Anno-Nyako, E.O. 1980. Studies on the vector relationships of whitefly-transmitted golden mosaic diseases of cowpea and lima bean. *M.Sc. thesis, Univer. Science* Technol., Kumasi, Ghana. 122pp.

Anno-Nyako, E.O., Vetten, H.J., Allen, D.J., and Thottappilly, G. 1983. The relation between cowpea golden mosaic and its vector, Bemisia tabaci. Genn. *Ann. appl. Biol.* 102:319-323.

Assis, F.F.M., de, Paguio, O.R., Sherwood, J.L., and Deom, C.M. 2002. Symptom induction by cowpea chlorotic mottle virus on Vigna unguiculata is determined by amino acid residue 151 in the coat protein. *J. Gen. Virol.* 83:897-883.

Bancroft, J.B., Hiebert, E., Rees, M.W., and Markham, R. 1968. Properties of cowpea chlortic mottle virus, its protein and nucleic acid. *Virology* 34:224-239.

Barnet, O.W. (ed.) 1992. Potyvirus taxonomy. *Arch. Virol.* Supplement 5. Springer-Verlag, Wien, New York. 450 pp.

Bashir, M. 1992. Serological and biological characterization of seed-borne isolates of blackeye cowpea mosaic and cowpea aphid-borne mosaic potyviruses in Vigna unguiculata (L.) Walp. *PhD Thesis, Oregon State Univ.* 222 pp.

Bashir, M. and Bashir. A.M. 1988. Diseases of major pulse crops in Pakistan: A review. *Trop. Pest Manag.* 34:309-314.

Bashir, M. and Hampton. R.O. 1993. Natural occurrence of five seedborne cowpea viruses in Pakistan. *Plant Dis.* 77:948-951.

Bashir, M., Ahmad, Z., and Ghafoor, A. 2002. Cowpea aphid-borne mosaic potyvirus: a review. *Int. J.. Pest Mgmt.* 48:155-168.

Bertens, P., Wellink, J., Goldbach, R., and van Kammen, A. 2000. Mutational analysis of the cowpea mosaic virus movement protein. *Virology* 267:199-208.

Bijaisoradat, M., Kuhn, C.W., and Benner, C.P. 1988. Disease reactions, resistance, and viral antigen content in six legume species infected with eight isolates of peanut mottle virus. *Plant Dis.* 72:1042-1046.

Bock, K.R. 1971. Notes on East African plant virus diseases. I. Cowpea mosaic virus. *East African Agric. Forest. J* 37:60-62.

Bock, K.R. 1973. East African strains of cowpea aphid-borne mosaic virus. *Ann. appl. Biol.* 74:75-83.

Bock, K.R., and Conti. M. 1974. Cowpea aphid-borne mosaic virus. Descript. plant virus. No. 134. *Commonw. Mycol. Inst. & Assoc. Appl. Biol.*, Kew, Surrey, UK. 4 pp.

Bozarth, R.G. and Shoyinka S.A. 1979. Cowpea mottle virus. Descriptions of plant viruses, No. 212. *Commonw. Mycol. Inst. & Assoc. Appl. Biol.* Kew, Surrey, UK. 3 pp.

Brantley, B.B. and Kuhn, C.W. 1970. Inheritance of resistance to southern bean mosaic virus in southern pea (Vigna unguiculata*). J. Amer. Soc. Hort. Sci.* 95:155-158.

Brunt, A.A., and Kenton, R.H. 1973. Cowpea mild mottle virus. Descriptions of plant viruses No. 140. *Commonw. Mycol. Inst. & Assoc. Appl. Biol.* Kew, Surrey, UK. 4 pp.

Brunt, A.A., Crabtree, K., Dallwitz, M.J., Gibbs, A.J., and Watson, L. 1996. Viruses of plants. *Descriptions and lists from the VIDE database.* CAB International, Wallingford. UK. 1484 pp.

Chant, S.R. 1959. Viruses of cowpea, Vigna unguiculata (L.) Walp. in Nigeria. *Ann. appl..* 47:565-573.

Chant, S.R., and Gbaja, I.S. 1987. Further studies of co-infection of cowpea by sunnhemp mosaic virus and cowpea mosaic virus. *Trop. Agric.* 64:115-118.

Chen, X. and Bruening G. 1992a. Cloned DNA copies of cowpea severe mosaic virus genomic RNAs: infectious transcripts and complete nucleotide sequences of RNA 1. *Virology* 191:607-618.

Chen, X. and Bruening G. 1992b. Nucleotide sequence and genetic map of cowpea severe mosaic virus RNA 2 and comparisons with RNA 2 of other comoviruses. *Virology* 187:682-692.

Collins, M.H., Witcher, W., Barnett, O.W., and Ogle, W.L. 1985. Reactions of 16 cowpea cultivars to six viruses. *Plant Dis.* 69:18-20.

Debrot, C.E., and Rojas, C.E.B., de. 1967. El virus del mosaico del frijol. Vigna sinensis Endl. en Venezuela. *Agronomia Tropical* 17:3-16.

Demski, J.W., Alexander, A.T., Stefani, M.A., and Kuhn, C.W. 1983. Natural infection, disease reactions, and epidemiological implications of peanut mottle virus in cowpea. *Plant Dis.* 67:267-269.

Di, R., Hill, J.H., and Deusen, R.A., van. 1993. Antigenic signature analysis reflects differences among plant virus isolates. *J.. Virol. Meth.* ?????:281-292.

Dijkstra, J., Bos. L., Bouwmeester, H.J., Hadiastono, T., and Lohuis, H. 1987. Identification of blackeye cowpea mosaic virus from germplasm of yard-long bean and from soybeans, and the relationship between blackeye cowpea mosaic virus and cowpea aphid-borne mosaic virus. *Neth. J. Pl. Path.* 93:115-133.

Eastwell, K.C., Kiefer, M.C., and Bruening, G. 1983. Immunity of cowpeas to cowpea mosaic virus. Pages 201-211 in Plant molec. biol., edited by R.B. Goldberg. UCLA Symposia of Molec. & Cell Biol., New Series Vol XII. Alan R. Liss, New York, NY, USA.

Edwards, M.C., Gonsalves, D., and Provvidenti, R. 1983. Genetic analysis of cucumber mosaic virus in relation to host resistance: location of determinants for pathogenicity to certain legumes and Lactuca salingna. *Phytopathology* 73: 269-273.

Faria, J.C., Bezerra, I.C., Zerbini, F.M., Ribeiro, S.G., and Mirtes, M.F. 2002. Current status of diseases caused by geminiviruses in Brazil. *Fitopatol. Brasil.* 25: 125-137.

Fery, R. 1985. The genetics of cowpea: a review of the world literature. Pages 25-62 in *Cowpea research, production and utilization,* (eds) S.R. Singh and K.O. Rachie. John Wiley and Sons, Chichester, UK.

Fery, R. 2002. Notice of release of 'KnuckleHull-VNR', a blackeye cowpea mosaic virus and root-knot nematode resistant, crowder-type southernpea. USDA, ARS, Washington, D.C.

Fischer, H.U. and Lockhart, B.E. 1976. A strain of cowpea aphid-borne mosaic virus isolated from cowpeas in Morocco. *Phytopath. Z.* 85:43-48.

Fondong, V.N., Thresh, J.M. and Zok, S. 2002. Spatial and temporal spread of cassava mosaic virus disease in cassava grown alone and when intercropped with maize and/or cowpea. *J. Phytopath.* 150 :365-374.

Francki, R.I.F., Mossop, D.W., and Hatta, T. 1979. Cucumber mosaic virus. *Descriptions of plant viruses,* No. 213. Commonw. Mycol. Inst. & Assoc. Appl. Biol., Kew, UK. 6 pp.

Fuentes, A.L., and Hamilton, R.I. 1991. Sunnhemp mosaic virus facilitates cell-to-cell spread of southern bean mosaic virus in nonpermissive host. *Phytopathology* 81:1302-1305.

Fuentes, A.L., and Hamilton, R.I. 1993. Failure of long-distance movement of southern bean mosaic virus in a resistant host is correlated with lack of normal virion formation. *J. Gen. Virol.* 74: 1903-1910.

Fulton, J.P., and Allen, D.J. 1982. Identification of resistance to cowpea severe mosaic virus, in Vigna unguiculata. *Trop. Agric.* 59:66-68.

Galasso, I., Heslop-Harrison, J.S., Perrino, P., and Pignone, D. 1997. Location and organization of major repetitive DNA sequence families in Vigna unguiculata [L.] Walp. Pages 372-375 in *Adv. Cowpea Res.* (eds.) B.B. Singh, D.R. Mohan Raj, K.E. Dashiell, and L.E.N. Jackai. Copublication Intern. Inst. Trop. Agric. (IITA) and Japan Intern. Res. Center Agric. Sciences (JIRCAS). IITA, Ibadan, Nigeria.

Gillaspie, A.G., Jr. 2001. Resistance to cucumber mosaic virus in cowpea and implications for control of cowpea stunt disease. *Plant Dis.* 85:1004-1005.

Gillaspie, A.G., Jr., Hopkins, M.S., Pinnow, D.L., and Hampton, R.O. 1994. Seedborne viruses in pre-introduction cowpea seedlots and establishment of virus-free accessions. *Plant Dis.* 79:388-391.

Gillaspie, A.G., Jr., Mitchell, S.E., Stuart, G.W., and Bozarth, R.F. 1999. RT-PCR method for detecting cowpea mottle carmovirus in Vigna germplasm. Plant Dis. 83:639-643.

Gilmer, R.M., Whitney, W.K., and Williams, R.J. 1974. Epidemiology and control of cowpea mosaic in western Nigeria. Page 269 in Proc. First IITA Grain Legume Workshop, IITA, Ibadan, Nigeria.

Gopinath, K., J. Wellink, C. Porta, K.M. Taylor, G.P. Lomonossoff, and A. van Kammen. 2000. Engineering cowpea mosaic virus RNA-2 into a vector to express heterologous proteins in plants. *Virology* 267:159-173.

Gould, A.R. and Symons, R.H. 1982. Cucumber mosaic virus RNA 3. *Eur. J. Biochem.* 126: 217-226.

Guia da, S.L.M., Florindo, G.M.I., Farias, A.M.E., Marinho, S.A.C., Oliveira, O.M.L., and Fernandes, M.D. 2001. Specific antibody production in mice orally fed with cowpea severe mosaic virus. *FASEB J.* 15(5). A1229.

Gumedzoe, M.Y. 1993. Major virus diseases of medicinal and aromatic plants for human welfare. *Acta Horticul.* 331: 307-310.

Gumedzoe, M.Y., Sunu, D.Y., Thottapilly, G., and Asselin, A. 1990. Importance of cowpea mottle virus and cowpea yellow mosaic virus in Togo. *Phytoprotec.* 71: 85-91.

Gumedzoe, M.Y., Rossel, H.W., Thorrappilly, G., Asselin, A., and Huguenot C. 1998. Reaction of cowpea (Vigna unguiculata L. Walp.) to six isolates of blackeye cowpea mosaic virus, cowpea aphid-borne mosaic virus (CABMV), two potyviruses infecting cowpea in Nigeria. *Intern. J. Pest Mgmt.* 44: 11-16.

Hacker, D.L., Petty, I.T.D., Wei, N., and Morris, T.J. 1992. Turnip crinkle virus genes required for RNA replication and virus movement. *Virology* 186: 1-8

Hampton, R.O. and Francki, R.I.B. 1992. RNA-1 dependent seed transmissibility of cucumber mosaic virus in Phaseolus vulgaris. *Phytopathology* 82: 127-130.

Hampton, R.O., Albrechtsen, S.E., and Mathur, S.B. 1992. Seed health (viruses) of Vigna unguiculata selections from developing countries. *Seed Sci. & Technol.* 20:23-38.

Hampton, R.O., Thottappilly, G., and Rossell, H.W. 1997. Viral diseases of cowpea and their control by resistance-conferring genes. *Adv. Cowpea Res.* Pages 159-175 (see publication details, Galasso et al. 1997)

Hayes, R.J., and Buck, K.W. 1990. Infectious cucumber mosaic virus RNA transcribed in vitro from clones obtained from cDNA amplified from cDNA amplified using the polymerase chain reaction. *J. Gen. Virol.* 71: 2503-2508.

Hobbs, H.S., Kuhn, C.W., Papa, K.E., and Brantley, B.B. 1987. Inheritance of non-necrotic resistance to southern bean mosaic virus in cowpea. *Phytopathology* 77: 1624-1629.

Huguenot, C., Furneaux, M.T., and Hamilton, R.I. 1994. Capsid protein properties of cowpea aphid-borne mosaic virus and blackeye cowpea mosaic virus confirm the existence of two major subgroups of aphid-transmitted, legume-infecting potyviruses. *J. Gen. Virol.* 75: 3555-3560.

Huguenot, C., Furneaux, M.T., Thottappilly, G., Rossel, H.W., and Hamilton, R.I. 1993. Evidence that cowpea aphid-borne mosaic and blackeye cowpea mosaic viruses are two different potyviruses. *J. Gen. Virol.* 74: 335-340.

Jackai, L.E.N. and Adalla, C.B. 1997. Pest management practices in cowpea: a review in *Adv. Cowpea Res.* (see publication details, Galasso, et al. 1997).

Jager, C.P., de. 1979. Cowpea severe mosaic virus. *Descriptions of plant viruses.* No. 209. Commonw. Mycol. Inst. & Assoc. Appl. Biol., Kew, Surrey, UK. 5 pp.

Jaspars, E.M.J., and Bos, L. 1980. Alfalfa mosaic virus. *Descriptions of plant viruses*, No. 229. Mycol. Inst. & Assoc. Appl. Biol., Kew, Surrey, UK. 7 pp.

Karasawa, A., Itaru, O., Kayoko, A., Ytaka, C., Shuu, H., Yoshiko, N-N., Akiko, I. and Yoshio, E. 1999. One amino acid change in cucumber mosaic virus RNA polymerase determines virulent/avirulent phenotypes on cowpea. *Phytopathology* 89: 1186-1192.

Khalf-Allah, A.M., Faris, F.S., and Nassar, S.H. 1973. Inheritance and nature of resistance to cucumber mosaic virus in cowpea, Vigna sinensis. *Egyp. J. Genet. & Cytol.* 2: 274-282.

Kim, J.W., and R.F. Bozarth. 1992. Mapping and sequence analysis of the capsid protein gene of cowpea mottle virus. *Interviology* 33: 135-147.

Kononowicz, A.K., Cheah, K.T., Narasimhan, M.S., Murdock, L.L., Shade, R.E. Chrispeels, M.J., Filippone, E., Monti, L.M., Bressan, R.A., and Hasegawa, P.M. 1997. Developing a transformation system for cowpea (Vigna unguiculata [L.] Walp.) Pages 361-371 in Adv. *Cowpea Res*. (see publication details, Galasso et al. 1997).

Kuhn, C.W. 1964a. Purification, serology, and properties of a new cowpea virus. *Phytopathology* 54: 853-857.

Kuhn, C.W. 1964b. Separation of cowpea virus mixtures. *Phytopathology* 54: 739-740.

Kuhn, C.W. 1990. Cowpea virus diseases in the United States: A status report. Pages 7-23 *in Cowpea Research*: a U.S. Perspective, edited by J.C. Miller, J.P.Miller, and R.L. Fery. Texas Agricultural Experiment Station, College Station, TX, USA.

Kuhn, C.W., and Brantley, B.B. 1963. Cowpea resistance to the cowpea strain of southern bean mosaic virus. *Plant Dis. Reptr.* 47: 1094-1096.

Kuhn, C.W., and Dawson, W.O. 1973. Multiplication and pathogenesis of cowpea chlorotic mottle virus and southern bean mosaic virus in single and double infections in cowpea. *Phytopathology* 63: 1380-1385.

Kuhn, C.W., Benner, C.P., and Hobbs, H.A. 1986. Resistance responses in cowpea to southern bean mosaic virus based on virus accumulation and symptomatology. *Phytopathology* 76: 795-799.

Kuhn, C.W., Brantley, B.B., Demski, J.W., and Pio-Ribeiro, G. 1984. 'Pinkeye Purple Hull-BVR', 'White Acre-BVR', and Corona' cowpeas. *HortSci.* 19: 592.

Kuhn, C.W., Wyatt, S.D., and Brantley, B.B. 1981. Genetic control of symptoms, movement, and virus accumulation in cowpea plants infected with cowpea chlorotic mottle virus. *Phytopathology* 71: 1310-1315.

Kyle, M.M., and Provvidenti, R. 1993. Genetics of broad spectrum viral resistance in bean and pea. Pages 153-166 in *Resistance to viral diseases of vegetables: genetics and breeding*, edited by M.M. Kyle. Timber Press, Portland, OR, USA.

Laksman, D.K., Gonsalves, D., and Fulton, R.W. 1985. Role of Vigna species in the appearance of pathogenic variants of cucumber mosaic virus. *Phytopathology* 75:751-757.

Lee, L. and Anderson, E.J. 1998. Nucleotide sequence of a resistance breaking mutant of southern bean mosaic virus. *Arch. Virol.* 143:2189-2201.

Lekkerkerker, A., Wellink, J., Yuan, P., van Lent, J., Goldbach, R., and van Kammen, A. 1996. Distinct functional domains in the cowpea mosaic virus movement protein. *J. Virol.* 70: 5658-5661.

Lima, J.A.A., Lima, R.C.A., and Goncalves, M.F.B. 2001. Production of polyclonal antisera specific to plant viruses by rabbit oral immunization. *Fitopat. Brasil.* 26:774-777.

Lomonossoff, G.P. 1995. Pathogen-derived resistance to plant viruses. Annu. Rev. Phytopathol. 33:323-343.

Lomonossoff, G.P., and Shanks, M. 1983. The nucleotide sequence of cowpea mosaic virus B RNA. EMBO Jour. 2:2253-2258.

Lovisolo, O., and Conti, M. 1966. Identification of an aphid-transmitted (sic.) cowpea mosaic virus. Neth. Jour. Pl. Path. 72:265-269.

Malathi, V.V., Naghma, A., and Surendranath, B. 2002. Nucleotide sequence update, bases 1 to 2746 of cowpea golden mosaic geminivirus DNA, Direct Submission.

Mali, V.R., and Ganacharya, N.M. 1984. Comparative studies on two isolates of tobacco ring spot virus from cowpea. Indian Phytopath. 37:630-632.

Mali, V.R., and Thottappilly, G. 1986. Virus diseases of cowpea in the tropics. Pages 361-403 in Reviews of tropical plant diseases, Vol. 3, edited by S.P. Raychaudhri and J.P. Verma. Today & Tomorrow Publishers, New Delhi, India.

Mali, V.R., Patil, F.S., and Gaushal, D.H. 1981. Immunity and resistance to bean yellow mosaic, cowpea aphid borne mosaic and tobacco ringspot viruses in cowpea, Vigna sinensis. Indian Phytopath. 34:521-522.

Mansour, A., Al-Musa, A., Vetten, H.J., and Lesemann, D-E. 1998. Properties of a cowpea mild mottle virus (CPMMV) isolate from eggplant in Jordan and evidence for biological serological differences between CPMMV isolates from leguminous and solanaceous host. Jour. Phytopath. (Berlin)146:539-547.

McKern, N.M., Mink, G.I., Barnett, O.W., Mishra, L.A. , Whittaker, L.A., Silbernagel, M.J., Ward, C.W., and Shukla, D.D. 1992. Isolates of bean common mosaic virus comprising two distinct potyviruses. *Phytopathology* 82:923-929.

Melton, A., Ogle, W.L., Barnett, O.W., and Caldwell, J.D. 1987. Inheritance of resistance to viruses in cowpea. *Phytopathology* 77:642.

Mlotshwa, S., Verver, J., Sithole-Niang, I., Van Kampen, T., Van Kammen, A., and Wellink, J. 2002. The genomic sequence of cowpea aphid-borne mosaic virus and its similarities to other potyviruses. *Arch. Virol.* 147:1043-1052.

Monti, L.M, Murdock, L.L., and Thottappilly, G. 1997. Opportunities for biotechnology in cowpea. Pp. 341-351 in *Adv. Cowpea Res.* (see publication details, Galasso et al. 1997)

Naidu, R.A., Gowda, S., Satyanarayana, T., Boyko, V., Reddy, A.S., Dawson, W.O., and Reddy, D.V.R. 1998, Evidence that whitefly-transmitted cowpea mild mottle virus belongs to the genus Carlavirus. Arch. *Virol.* 143: 769-780.

Nasu, Y., Karasawa, A., Hase, S., and Ehara, Y. 1996. Cry, the resistance locus of cowpea to cucumber mosaic virus strain Y. *Phytopathology* 86: 946-951.

Ndiaye, M., Bashir, M., Keller, K.E., and Hampton, R.O. 1993. Cowpea viruses in Senegal, West Africa: identification, distribution, seed-transmission, and sources of genetic resistance. *Plant Dis.* 77:999-1003.

Negri, V., Tosti, N., Falcinelli, M., and Veronesi, F. 2002. Characterisation of thirteen cowpea landraces from Umbria (Italy). Strategy for their conservation and promotion. *Genetic Resources & Crop Evolution* 47: 141-146.

Nobumitsu, S., Yasunari, F., Kazuyiki, M., and Iwao, Furusawa. 2001. Site-specific single amino acid changes to Lys or Arg in the central region of the movement protein of a hybrid Bromovirus are required for adaptation to a nonhost. *Virology* 279: 47-57.

Odu, B.O. 1999. Isolation, characterization and identification of a potyvirus from Doscorealata L. (water yam) in Nigeria. *Ann. appl. Biol.* 134: 65-71.

Ogundiwin, E.A. 2002. Genetic dissection of cowpea mottle virus disease resistance in Vigna species. PhD thesis, Dept of Agron., Biotech. Research Unit, Univ. of Ibadan, Nigeria.

Orozco, S.M., Lopez, A.O., Perez, O., Delgadillo, S.F. 1994. Effect of transparent mulch, floating row covers and oil sprays on insect populations, virus diseases and yield of cantaloupe. (Mexico) *Biological Agric. & Hortic.* 10: 229-234.

Ouattara, S., and Chambliss, O.L. 1991. Inheritance of resistance to blackeye cowpea mosaic virus in 'White Acre-BVR' cowpea. *HortSci.* 26: 194-196.

Patel, P.N. 1982. Reactions of cowpeas to two strains of cowpea mosaic virus from Tanzania. *Indian Phytopath.* 35: 461-466.

Patel, P.N., Mligo, J.K., Leyna, H.K., Kuwite, C., and Mmbaga, E.T. 1982a. Sources of resistance, inheritance, and breeding of cowpeas for resistance to a strain of cowpea aphid-borne mosaic virus from Tanzania. *Indian J. Genet.* 42: 221-229.

Patel, P.N., Mligo, J.K. Leyna, H.K., Kuwite, C., and Mmbaga, E.T. 1982b. Multiple disease resistance cowpea breeding program in Tanzania. *Indian J. Genet.* 42: 230-239.

Paz C.D., da., Lima, A.A., Pio-Ribeiro, G., Assis, F.F.M., Andrade, G.P., and Goncalves, M.F.B. 1999. Purification of an isolate of cowpea severe mosaic virus, obtained in Pernambuco, production of antiserum, and determination of sources of resistance in cowpea. *Summa Phytopath.* 25: 285-288.

Pio-Ribeiro, G., Kuhn, C.W., and Brantley, B.B. 1980. Cowpea stunt: inheritance pattern of the necrotic synergistic reaction. *Phytopathology* 70: 250-252.

Pio-Ribeiro, G., Wyatt, S.D., and Kuhn. C.W. 1978. Cowpea stunt: a disease caused by a synergistic interaction of two viruses. *Phytopathology* 68: 1260-1265.

Ponz, F., Russell, M.S., Rowhani, A., and Bruening, G. 1988. A cowpea line has distinct genes for resistance to tobacco ringspot virus and cowpea mosaic virus. *Phytopath.ology* 78: 1124-1128.

Provvidenti, R. 1993. Genetics of resistance to viral diseases of bean. Pp. 112-152 in *Resistance to viral diseases of vegetables: genetics and breeding,* edited by M.M. Kyle. Timber Press, Portland, OR, USA.

Purcifull, D., and Gonsalves, D. 1985 Blackeye cowpea mosaic virus. *Descriptions of plant viruses,* No. 305. Commonw. Mycol. Inst. & Assoc. Appl. Biol., Kew, Surrey, UK. 5 pp.

Puttaraju, H.R., Prakash, H.S., and Shetty, H.S. 2000. Field incidence, seed-transmission and susceptibility of cowpea varieties with reference to blackeye cowpea mosaic potyvirus. *Seed Research* (New Delhi) 28: 196-202.

Quin, F.M. 1997. Introduction Pp ix- xv in *Adv. Cowpea Res.* (see publication details in Galasso *et al.)*

Rachie, K.O. 1985. Introduction. Pages xxi – xxviii in *Cowpea Research, Production, and Utilization.* (Eds S.R. Singh and K.O. Rachie. John Wiley & Sons Ltd. 460 pp.

Rao, A.L.N. and Francki, R.I.B. 1982. Distribution of determinants for symptom production and host range of the three RNA components of cucumber mosaic virus. *J. Gen. Virol.* 61:197-205.

Reeder, B.D., Norton, J.D., and Chambliss, O.L. 1972. Inheritance of bean yellow mosaic virus resistance in southern pea, Vigna sinensis. *J. Amer. Soc. Hort. Sci.* 98:62-63.

Regenmortel, M.H.V, van, and 10 Co-Editors. 2000. *Virus Taxonomy* (Seventh Report of the, International Committee on Taxonomy of Viruses).

Rezaian, M.A.I., Williams, R.H.V. Gordon, J.H.K, Gould, A.R., and Symons, R.H. 1984. Nucleotide sequence of cucumber mosaic virus RNA 2 reveals a translation product significantly homologous to corresponding proteins of other viruses. *Eur. J. Biochem.* 143:277- 284.

Rezaian, M.A.I., Williams, R.H.V., A.R., and Symons, R.H. 1985. Nucleotide sequence of cucumber mosaic virus RNA 1: presence of a sequence complementary to part of the viral satellite RNA and homologies with other viral RNAs. *Eur. J. Biochem.* 150: 331-339.

Robertson, D.G. 1965. The local lesion reaction for recognizing cowpea varieties immune from and resistant to cowpea yellow mosaic virus. *Phytopathology* 55: 923-925.

Rogers, K.M., Norton, J.D., and Chambliss. 1973. Inheritance of resistance to cowpea chlorotic mottle virus in southern pea, Vigna sinensis. *J. Amer. Soc. Hort. Sci.* 98: 62-63.

Sangwan, R.S., Lodhi, G.P., and Jaglan, R.S. 2000. Stability of resistance to yellow mosaic virus over environments in cowpea. *Indian J. Genet. & Pl. Breeding* 60: 553-555.

Santos, F.M.L., Lima, J.A.A., Santos, A.A. 1999. Single and mixed virus infections in cowpea in Ceara, Brazil. Fito*pat. Brasil.* 24: 518-522.

Scully, B.T., and Federer, W.T. 1993. Application of genetic theory in breeding for multiple viral resistance. Pages 167-195 in *Resistance to viral diseases of vegetables: genetics and breeding,* edited by M.M. Kyle. Timber Press, Portland,OR, USA.

Sharma, S.R. 1984. Effect of cultural practices on virus infection in cowpea. (German) Journal of Agronomy and *Crop Science* 153: 23-31.

Sharma, S.R., and Varma, A. 1976. Cowpea yellow fleck, a whitefly-transmitted disease of cowpea. *Indian Phytopath.* 29: 421-423.

Shepherd, R.J. 1964. Properties of a mosaic virus of cowpea and its relationship to the bean pod mottle virus. *Phytopathology* 54: 466-473.

Shepherd, R.J., and Fulton J.P. 1962. Identity of a seedborne virus of cowpea. *Phytopathology* 52: 489-493.

Shoyinka, S.A., Bozarth, R.F., Rees, J., and Rossel, H.W. 1978. Cowpea mottle virus: a seedborne virus with distinctive properties infecting cowpeas in Nigeria. *Phytopathology* 68: 693-699.

Shoyinka, S.A., Thottappilly, G., Adebayo, G.G., and Anno-Nyako, F.O. 1997. Survey of cowpea virus incidence and distribution in Nigeria. *Intern. J. Pest Manag.* 43: 127-132.

Sijen, T., Wellink, J., Hendriks, J., Verver, J. and van Kammen, A. 1995. Replication of cowpea mosaic virus RNA1 or RNA2 is specifically blocked in transgenic Nicotiana benthamiana plants expressing the full-length replicase or movement protein genes. *Molec. Plant-Microbe Interac.* 8: 340-347.

Silva, A.M., and Rossman, M.R. 1987. Refined structure of southern bean mosaic virus at 2.9 angstrom resolution. *J. Molec. Biol.* 197:69-87.

Singh, B.B., Chambliss, O.L., and Sharma, B. 1997. Cowpea genetics: a review of the recent literature. Pages 13-29 in *Adv. Cowpea Res.* (see publication details, Galasso *et al.* 1997)

Skotnicki, M.S., Mackenzie, A.M., Torronen, M., and Gibbs, A.J. 1993. The genomic sequence of cardamine chlorotic fleck carmovirus. *J. Gen. Virol.* 74. 1933-1937.

Strniste, P.B. 1987. The inheritance and assessment of a second qualitative gene for blackeye cowpea mosaic virus resistance in southern pea, Vigna unguiculata (L.)Walp. *M.S. Thesis*, Auburn Univ., Auburn, AL (USA).

Taiwo, M.A., Gonsalves, D., Provvidenti, R., and Thurston, H.D. 1982a. Partial characterization and grouping of isolates of blackeye cowpea mosaic and cowpea aphidborne mosaic virus. *Phytopathology* 72: 590-596.

Taiwo, M.A., Provvidenti, R. and Gonsalves, D. 1982b. Inheritance of resistance to blackeye cowpea mosaic virus in Vigna unguiculata. *J. Hered.* 72: 433-434.

Thottappilly, G. 1992. Plant virus diseases of importance to African agriculture. *J. Phytopath.* 134: 265-268.

Thottappilly, G., and Rossel, H.W. 1982. Seed transmission of cowpea (yellow) mosaic virus unlikely in cowpea. *Tropical Grain Bull.* 34: 27-28.

Thottappilly, G., and Rossel, H.W. 1985. Worldwide occurrence and distribution of virus diseases. Pages 155-171 in *Cowpea research, production and utilization*, eds. S.R. Singh and K.O. Rachie, John Wiley and Sons, Chicester, UK.

Thottappilly, G., and Rossel, H.W. 1988. Occurrence of cowpea mottle virus and other viruses (cowpea yellow mosaic virus, southern bean mosaic virus) in cowpea. *FAO Plant Protection Bull.* 36: 1984-1985.

Thottappilly, G., and Rossel, H.W. 1992. Virus diseases of cowpea in tropical Africa. *Tropical Pest Managm.* 38: 337-348.

Thottappilly, G., Hamilton, R.I., Huguenot, C., Rossel, H.W., Furneaux, M.T., Gumedzoe, M.Y., Shoyinka, S.A., Naik, D.M., Konate, G. Atcham-Agneroh, T., Haciawa, H.C., Anno-Nyako, F.O., Saifodine, N., Wangai, A. Lamptey, P. Gubba, A., Mbwaga,

A.M., Neya, J, Offei, S.K. 1993. Identification of cowpea viruses and their strains in tropical Africa. *FAO Plant Protection Bull.* 41: 65-71.
Thottappilly, G., ibid. 1995. Identification of cowpea viruses and their strains in tropical Africa. *IITA Research* 10: 12-15.
Thottappilly, G., Sehgal, O.P., and Rossel, H.W. 1993. Characteristics of a cowpea chlorotic mottle virus isolate from Nigeria. *Plant Dis.* 77:60-63.
Thouvenel, J.C., Tia, E., and Fishpool, L.D.C.. 1990. Characterization of cowpea mottle virus on cowpea (Vigna unguiculata) in the Ivory Coast and identification of a new vector. *Tropical Agriculture* 67: 280-282.
Tremaine, J.H., Ronald, W.P., and Mackenzie, D.J. 1985. Southern bean mosaic virus monoclonal antibodies: reactivity with virus strains and with the virus antigen in different conformations. *Phytopathology* 75: 1208-1212.
Tsai, W.S., Green, S.K., and Deshpande, R.R. 2000. Direct Submission of bases 1 to 2728, nucleotide sequence of cowpea golden mosaic geminivirus DNA.
Umaharan, P., Ariyanayagam, R.P., and Haque, S.Q. 1997. Resistance to cowpea severe mosaic virus, determined by three dosage dependent genes in [Vigna unguiculata (L.) Walp. *Euphytica* 95: 49-55.
Umaharan, P., Haque, S.Q., and Ariyanayagam R.P. 1997. Identification of resistance to cowpea severe mosaic virus (Trinidad isolate) in cowpea [Vigna unguiculata (L.) Walp.]. *Trop. Agric.* (Trinidad) 74: 324-328.
Valverde, R.A., Moreno, R., and Gamez, R. 1982. Yield reduction in cowpea Vigna unguiculata [L.] Walp.) infected with cowpea severe mosaic virus in Costa Rica. *Turriabla* 32: 89-90.
Van Boxtel, J., Lesley, T.C., and Maule, J.A. 2000. Phylogenetic analysis of two potyvirus pathogens of commercial cowpea lines: Implications for obtaining pathogen-derived resistance. *Virus Genes* 20:71-77.
Walker, C.A., and Chambliss, O.L.. 1981. Inheritance of resistance to blackeye cowpea mosaic virus in Vigna unguiculata (L.) Walp. *J. Amer. Soc. Hort. Sci.* 106:410-412.
Walters, H.J., and Barnett, O.W. 1964. Bean leaf beetle transmission of Arkansas cowpea mosaic virus. *Phytopathology*. 54:911.
Williams, R.J. 1975. The control of cowpea diseases in the IITA Grain Legume Improvement Program. Pages 139-146 in *Tropical diseases of legumes*, eds. J. Bird and K. Maramorosch. Academic Press, New York, NY, USA.
Williams, R.J. 1977. The identification of multiple disease resistance in cowpea. *Tropic.* Agric. (Trinidad) 54:53-60.
Winter, S., Butgereitt, A., and Thottappilly, G. 2002. Cowpea golden mosaic virus and related geminiviruses associated with Vigna spp. in Nigeria. Poster presentation*, Intern Virology Congress,* Sydney, AU, 1999.
Wu, S., Rinehart, C.A. and Kaesberg, P. 1987. Sequence and organization of southern bean mosaic virus genomic RNA. *Virology* 161:73-80.
You, X.J. Rinehart, C.A., and Kesberg, P. 1987. Sequence and organization of southern bean mosaic virus genomic RNA. *Virology* 161:73-80.
You, X.L., Kim, J.W., Stuart, G.W., and Bozarth, R.F. 1995. The nucleotide sequence of cowpea mottle virus and its sequence homology to carmoviruses. *J.. Gen. Virol.* 76:2841-2845.
Zeeuw, D.J. de, and Ballard, J.C. 1959. Inheritance in cowpea of resistance to tobacco ringspot virus. *Phytopathology* 49:332-334.
Zeeuw, D.J. de, and Crum, R.A. 1963. Inheritance of resistance to tobacco ringspot and cucumber viruses in black cowpea crosses. *Phytopathology* 53:337-340.
Zheng, H., Chen, Jiong., and Chen, Jianping. 2002. Bean common mosaic virus isolates causing different symptoms in asparagus bean in China differ greatly in the 5'-parts of their genomes. *Arch. Virol.* 147:1257-1262.

Table 14.1. S ome properties of viruses causing principle diseases of cowpea.

	Transmission			Coat Protein		Genome		
					Mol Wt		Nucleotides	
Virus	Mech.	Vector	Seed (%)	Particle	x 10^4	Parts	x 10^3	Key References
Blackeye cowpea mosaic virus (BlCMV), *Potyvirus*	Yes	Aphid	3-55	Filament	3.4	One	9.5	Purcifull & Gonsalves, 1985; Taiwo *et al*. 1982a
Cowpea aphid-borne mosaic virus (CABMV) *potyvirus*	Yes	Aphid	0-40	Filament	3.4	One	9.5	Bock & Conti, 1974; Taiwo *et al*. 1982a
Cowpea chlorotic mottle virus (CCMV), *Bromovirus*	Yes	Beetle	0	Icosahedron	2.0	Three	8.2	Kuhn, 1964a; Allison *et al*. 1989
Cowpea golden mosaic virus (CGMV), *Geminivirus*	No	Whitefly	0	Duplex	31	ssDNA	5.0	Thottappilly & Rossel, 1992; Winter *et al*. 2002; Malathi *et al*. 2002
Cowpea mosaic virus (CPMV), *Comovirus*	Yes	Beetle	0-5	Icosahedron	2.3, 3.7	Two	9.4	Agrawal, 1964; van Kammen & de Jager, 1978; Lomonossoff & Shanks, 1983
Cowpea severe mosaic virus (CPSMV), *Comovirus*	Yes	Beetle	3-10	Icosahedron	2.3, 3. 7	Two	9.4	Jager, 1979; Chen & Bruening, 1992a,b; Paz, da *et al*. 1999
Cowpea mottle virus (CPMoV), Carmovirus	Yes	Beetle	0-10	Icosahedron	4.4	One	4.1	Bozarth & Shoyinka, 1979; Thouvenel *et al*. 1990
Cucumber mosaic virus (CMV), *Cucumovirus*	Yes	Aphid	4-26	Icosahedron	2.4	Three	8.3	Francki et al. 1979; Rezaian *et al*. 1984
Southern bean mosaic virus (SBMV), *Sobemovirus*	Yes	Beetle	3-4	Icosahedron	3.1	One	4.2	Shepherd& Fulton, 1962; Wu *et al*. 1987

See "other viruses", p. 9 (reported in cowpea; less significant).

Table 14. 2. Cowpea genes or genetic sources providing virus resistance/tolerance.

Virus[1]	Gene(s)	Source[2]	Reference
(BYMV)[3] BlCMV	By	PI 297562	Reeder *et al.* 1972
(CABMV)[4]	1-r[5]	TVu 612, TVu 1948	
BlCMV	1-Dp[6]	TVu408-P, TVu 410 (many others also)	Patel *et al.* 1982a
BlCMV	bcm	TVu 2480	Taiwo et al. 1982b
	blc	'Worthmore'	Walker & Chambliss 1981
	-	PEPH-BVR, WA-BVR, Corona	Kuhn *et al.* 1984
	1-r[5]	'Mississippi Silver'	Melton *et al.* 1987
	1-D[6]	PEPH-BVR	Strniste 1987
	1-D[6]	WA-BVR	Ouattara & Chambliss 1991
	-	Knuckle Hull-VNR[9]	Fery 2002
CABMV	-	(sources unknown)[10]	Lapido & Allen 1979
	-	TVu 401, TVu 1582	Bashir 1992
CCMV	cc	PI 255811	Rogers *et al.* 1973
	1-r[5]	PI 255811 & 186465 (tolerance)	Khun *et al.* 1997
CGMC	-	Vita 1 & 3; TVu 143, 274, 283, 393, 410, 470; other TVu's	Anno-Nyako, 1980
		Ife, TVx, & many IT# lines	Singh *et al.* 1997
CPMV	-	'Arlington', 'Blackeye', others	Robertson 1965
	mvs	TVu 227, TVu 345	
		TVu 612, and TVu 2331	Patel 1982; Patel *et al.* 1982b
	-	'Arlington'	Eastwell *et al.* 1983
	1-D[6]	'Arlington'	Ponz *et al.* 1988
		TVu 1948, TVu 2480, Macaido	Fulton & Allen, 1982
		Piaui, Mulato, Gurgeia, others	Paz, da, *et al.* 1999
CPMoV		TVu 3901 (tolerance)	Allen *at al.* 1982
	1-D[6]	Tvnu 390, and 1561	Ogundawin *et al.* 2002
CMV	1-D[6]	Selection from 'Black'	Zeeuw, de & Crum 1963
	1-D[6]	'Fetriat' (tolerance)	Khalf-Allah *et al.* 1973
	Cry	PI 189375 (model)	Nasu *et al.* 1995
SBMV	-	'Iron', 'Clay', others	Kuhn & Brantley, 1963
	Sbm[3]	'Clay'	Brantley & Kuhn ,1970
	-	PI 147562, PI 186465	Kuhn *et al.* 1986
	sbc-1, sbc-2	'Mississippi Silver'	Melton *et al.* 1987
	sbm-2	(PI 186465)	Hobbs *et al.* 1987
PMV[7]	-	'Corona', 'Early Pinkeye'	
		'Iron', 'Worthmore'	Bijaisoradat *et al.* 1988
TRSV[8]	Tr[3]	'California Blackeye #5'	Zeeuw, de & Ballard, 1959
	-	(sources unknown)[10]	Mali *et al.* 1981
	1-D[7]	'Arlington'	Ponz *et al* 1988a

1. See Table 14. 1 for virus names.
2. PI = USDA-ARS Plant Introduction lines; TVu = IITA accessions of V. unguiculata; PEPH = Pink Eye Purple Hull; BVR = BlCMV-resistant.
3. Reported as BYMV; actually BlCMV (O.L. Chambliss, personal communication). Term, By, assigned by Fery; previously reviewed (Fery 1985).
4. Reported as CABMV; actually BlCMV (P.N. Patel, personal communication).
5. Resistance apparently conferred by a single recessive gene; no term assigned.
6. Resistance reportedly conferred by a single dominant gene; no term assigned. 1-Dp = partial dominance reported.
7. PMV = peanut mottle virus; typically causes minor disease in cowpea (Table 14.1 footnote).
8. TRSV = tobacco ringspot virus; ibid. (Table 14.1 footnote).
9. VNR = virus and nematode resistant
10. Published resistance source not available to authors.

15. Soybean

John H. Hill
Dept. of Plant Pathology, Iowa State University, Ames, Iowa.50011, USA. E-mail: johnhill@iastate.edu

15 .1 Introduction

Soybean, *Glycine max*, is the world's foremost provider of protein and oil. Oilseeds constitute the world's fourth most important component of the world's food supply behind wheat and rice, coarse grains, and potatoes and cassava. Soybeans are the leading oilseed crop and constitute 50% – 55% of the world's oilseed production. For the most part, soybeans are grown in developed countries with the United States producing 43% of the global production of 6751 million bushels in 2001 – 2002. Brazil followed with 24% of the world's production followed by Argentina, China, India, Paraguay and others with 16%, 8%, 3%, 2% and 4%, respectively (United Soybean Board, 2002).

The first written record of soybean occurred as early as 5,000 years ago in China when Emperor Cheng-Nung included soybeans in Chinese books that described plants of China. Before that time, the origin and history are unknown. Subsequently, the plant was repeatedly mentioned as a most-important cultivated legume crop throughout Asia and particularly in China. The first planted soybeans in the United States were grown on a farm in Thunderbolt Georgia by Henry Younge in 1765. During the 1800's, they were brought to the United States, sometimes as inexpensive ballast that was later dumped to make room for cargo. They were grown variously for soy sauce or as "coffee berries" to brew coffee during the Civil War, when real coffee was scarce.

The concept of soybeans grown as a forage crop changed when, in 1904 at the Tuskegee Institute in Alabama, George Washington Carver began studies of the soybean. His discoveries showed soybeans provided valuable protein and oil. By 1929 soybean production in the United States had grown to nine million acres. However, soybean farming in the United States did not really come to fruition until soybean fields in China were devastated by World War II and an internal revolution in China in the 1940's. Henry Ford developed a soybean-based plastic in the 1940's and the impetus for the use of soybean meal for livestock feed began in the 1950's.

Today, soybean is grown primarily for protein content and secondarily for oil. However, that represents only a portion of their attraction since processing of soybeans leads to a wide variety of uses that range from from tofu, soy milk and soy sauce, meat substitutes, margarine, cooking and salad oils, cereals, plywood, particle board, printing inks, soap, paint, industrial lubricants, candy products, cosmetics, and biodiesel. Although this represents only a small portion of how soybeans are utilized, it illustrates the diversity of uses for the crop.

Soybeans constitute a significant part of the world's agricultural trade. For example, in the United States, soybeans constitute the second largest crop in cash sales and the number one crop in terms of the value of exports. World exports in 2001/02 totaled 2056 million bushels. Of that, 52%, 27%, 12%, 4%, and 5% were exported by the United States, Brazil, Argentina, Paraguay, and others, respectively (United Soybean Board, 2002). For the United States, China constitutes the largest export customer followed by Mexico, Japan, The Netherlands, Taiwan, Spain, Indonesia, Korea, Canada, and Thailand. Soybean yields in Iowa, generally the number one soybean producing state in the United States, have grown steadily from 14 bushels per acre in the decade of the 1920s to 40-50 bushels/acre in the decade of the 1990s. Yields produced by some progressive farmers can occasionally reach 60 or more bushels per acre. Nevertheless, there is concern that a relatively narrow germplasm base in North American soybean cultivars may limit future yield increases.

Global yield loss caused by plant disease during 1998 was estimated to be 28,522.3 thousand metric tons in the top ten soybean-producing countries (Wrather *et al.* 2001). Although loss caused by soybean viruses is difficult to discern because of interactions between soybean cultivar, time of inoculation and virus strain, approximately 5% was estimated to be caused by virus disease. However, due to the recent soybean virus epidemic in the primary soybean-growing region of the United States, the estimate is probably now extremely conservative (Giesler *et al.* 2002). Different viruses are known to infect soybean (Table 15.1; Hartman *et al.* 1999; Thottapilly and Rossel, 1987) and their relative importance depends upon soybean cultivar, virus strain, propensity for seed transmission, presence of potential vectors, characteristics of the agroecosystem, and geographic location. However, on a global basis major soybean virus diseases are believed to include soybean mosaic, bean pod mottle, bud blight, and yellow mosaic. In this chapter, we discuss soybean mosaic, bean pod mottle and bud blight.

Colored illustrations of symptoms caused by these viruses on soybeans are readily available (North Central Soybean Program, 2003).

15. 2 The Main Viruses

Soybean mosaic virus Genus *Potyvirus* (SMV) has historically been the most common soybean virus worldwide and can be present wherever soybeans are grown. The wide global distribution of the virus is probably a result of seed transmission, which in some soybean varieties can approach 50%. However, seed transmission rates in modern commercial cultivars now approximate 0% - 5%. In addition, the virus is experimentally transmitted mechanically and in nature by at least 32 different aphid species belonging to 15 different genera. Some of the most important species include *Acayrhosiphon pisum, Aphis fabae, A. glycines, Myzus persicae*, and *Rhopalosipum maidis* (Edwardson and Christie, 1991). As with most potyviruses, aphid transmission is dependent upon interaction of a DAG sequence at the C-terminus end of the coat protein and the virus encoded HC-Pro gene product (Jayaram *et al.* 1998).

The most economically important host is soybeans. Plant hosts reportedly infected by various SMV isolates are limited to six plant families (hosts marked with an asterisk have also been reported as nonsusceptible to SMV, suggesting differences in strain susceptibility). Amaranthaceae (*Gomphrena globosa**); Chenopodiaceae (*Chenopodium album, C. amaranticolor*, C. quinoa*); Fabaceae (*Astragalus monspessulanus, Canavalia ensiformis, Cassia laevigata, Crotalaria spectabilis, Cyamopsis tetragonologa, Dolichos biflorus, D. falcatus, Galactia* sp., *Glycine max, G. soja, G. usuriensis, Hippocrepis multisiliquosa, Indigofera hirsuta, Kummerowia stipulacea, K. striata, Lablab purpureus, Lotus tetragonolobus, Lourea vespertilionis, Lupinus albus, L. angustifolius,* L. luteus,* Macroptilum lathyroides, Macrotyloma uniflorum, Mucuna pruriens* var. *utilis, Neonotonia wightii, Phaseolus acutifolius, P. lunatus, P. nigricans, P. speciosus, P. vulgaris,* Pisum sativum,* Scorpiurus sulcata,*

Senna occidentalis, Sesbania exaltata, Stizolobium deeringianum, Trigonella carulea, T. foenum-gracum, Vicia faba, V. narboensis, Vigna mungo,* V. ungiculata* subsp. *cylindrica**); Passifloraceae (*Passiflora edulis* var. *flavicarpa,* P. ligularis,* P. quadrangularis,* P. edulis**); Schropulariaceae (*Antirrhinum majus*); Solonaceae (*Nicandra physalodes, Nicotiana benthamiana, N. tabacum,* Petunia* x *hybrida**). Local lesion hosts include *Chenopodium album, Cyamopsis tetgagonoloba, D. biflorus, L. purpureus* and *P. vulgaris* cv. Top Crop. Latent infection has been reported in *H. multisiliquosa, L. tetragonolobus. P specious,* and some cultivars of *P. vulgaris* and *S. sulcata* (Hill, 1999). The most useful local lesion assay utilizes *Phaseolus vulgaris* cv. Top Crop in a detached leaf assay conducted under low light at 30^0 C (Milbrath and Soong, 1976).

Numerous SMV strains have been characterized on the basis of differential reaction on a series of differential soybean cultivars. In the United States, these include strains G1 – G7 and G7a, (the existence of G7a may be in doubt; Hajimorad and Hill, unpublished) (Buzzel and Tu, 1984; Cho and Goodman, 1979) and in Japan strains A – E (Takahashi *et al.* 1980). Studies to investigate the correspondence between the Japanese and United States strains suggest that Japanese strain B is most similar to the U.S. strain G2 and that strain G3 may correspond to the Japanese strain A (Kanematsu *et al.* 1998). Other countries, including China, Korea, and Brazil also report differentiation of SMV isolates using soybean cultivars (Almeida, 1981; Anjos *et al. 1985;* Chen *et al.* 1986; Cho *et al.* 1977; Lee *et al.* 1992; Pu *et al.* 1982; Xu *et al.* 1986)

Virus symptoms, often confused with herbicide damage or with symptoms caused by other viruses that cause similar or identical symptoms, vary with host genotype, virus strain, plant age at infection, and environment. Most infected cultivars are slightly stunted and have fewer pods. Pods that do develop are sometimes dwarfed and flattened and lack hairs and seeds. Trifoliolate leaves have a mosaic of light and dark green areas that may later become raised or blistered, particularly along the main veins. Chlorosis may develop between the dark green areas. Leaf margins may be wavy or curled downward. Leave veins do not exhibit parallel veination or grow together, a symptom also typical of herbicide injury. Primary leaves of some cultivars may develop necrotic local lesions, which emerge into veinal necrosis followed by yellowing and leaf abscission. The G7 strain overcomes resistance induced by the *Rsv1* gene and can cause severe stunting, systemic necrosis, leaf yellowing, petiole and stem necrosis, terminal necrosis, and defoliation, usually leading to death. Different strains often produce different symptoms on the same soybean cultivar. Rugosity is most severe in plants grown at about 18^0 C. Symptoms are less severe at 24^0 – 25^0 C and are largely masked at temperatures above 30^0 C (Hill, 1999).

Seeds from infected plants may be mottled brown or black, depending upon hilum color. They may also be smaller than seeds from noninfected plants, and germination levels may be reduced. Mottling does not indicate that the virus is present in seeds, since not all mottled seeds contain virus, not all seeds from virus-infected plants are mottled, and plants not infected with SMV may also produce low percentages of mottled seed (Hill *et al.* 1980). The correlation between the percentage of infected mottled seeds and the severity of mottling varies from year to year and from cultivar to cultivar (Bryant *et al.* 1982).

A synergistic reaction occurs when plants are infected with both SMV and *bean pod mottle virus* (BPMV) (Calvert and Ghabrial, 1983). Reduction in yield may be as high as 66% - 88% in susceptible cultivars with mixed infections compared with reductions of 8% - 25% in cultivars infected with SMV. Symptoms of double infection may include severe dwarfing, foliar distortion, necrosis, and mottling of leaves. Doubly infected plants have fewer root nodules than singly infected plants. The ultimate effect on yield, seed mottling, and seed transmission of SMV is dependent upon infection by either virus, cultivar, and virus strain (Hill, 1999).

SMV particles are flexuous rods averaging approximately 15 – 18 nm x 750 nm. The virions consist of single-stranded positive sense RNA (ca. 10 kb), which constitutes 5.3% of the particle, and have helical symmetry with a pitch of 34^0.The radial location of the RNA is 3.7 nm. Repeating monomeric coat protein subunits have a molecular weight of approximately 30 kDa (Hill and Benner, 1980a, b). The complete nucleotide sequence of the viral RNA has been determined for strains G2 and G7. A genetic map has been constructed, and nine virus-encoded proteins have been predicted through proteolytic cleavage of a single polyprotein precursor by three virus-encoded proteases (Jayaram *et al.* 1992). Infections cDNA clones have been prepared from the G2 and G7 strains. Nucleotide sequence data is available through GeneBank for selected regions of other SMV strains.

SMV cross-reacts serologically with watermelon mosaic virus 2, bean common mosaic virus, bean yellow mosaic virus, clover yellow vein virus, and lettuce mosaic virus, but highly specific antisera have also been prepared. Serological differences among SMV isolates have been difficult to show with polyclonal antisera because of high sequence identity of coat protein among strains; nevertheless the use of monoclonal antibodies has allowed serological differentiation of isolates characteristic of SMV strains (Hill *et al.* 1989; Hill *et al.* 1994). Coat protein trypsin peptide profiles of 14 strains are nearly identical (Jain *et al.* 1992). Antisera are available for detection of virus in both soybean leaves and seed.

Reported thermal inactivation points are 55^0 – 60^0 C. The dilution end point varies from 10^{-3} to 10^{-5}, and reported longeivity in vitro is 2 - 3 days or 4 – 5 days. The virus remains infectious in desiccated leaves for 7 days at 25^0 – 33^0 C but may be easily stored for long periods in leaves frozen at -80^0 C. The virus loses infectivity below pH 4.0 and above pH 9.0. It degrades below pH 3.0 (Hill, 1999).

Virus can be purified from soybeans but the ease of purification and yield of purified virus is dependent upon virus strain due to propensity of the virions to aggregate. The virus can be purified by homogenizing soybean leaves 3 – 6 weeks after inoculation, according to Hill and Benner (1980a).

Plants grown from infected seed often constitute the primary inoculum source with secondary spread by aphids, characterized by a logistic model, occurring at a moderately fast rate (Nutter *et al.* 1998). The virus moves both up and down within the plant and can be detected in all parts of a systemically infected plant. At least three naturally occurring resistance genes to SMV have been identified. The first dominant resistance gene identified in the soybean line PI 96983 was designated *Rsv1*. Single resistance genes in other cultivars, which confer differential reactions to strains G1 to G7, are alleles at the *Rsv1* locus and have been designated *Rsv1y, Rsv1m, Rsv1t, Rsv1k, Rsv1s*, and *Rsv1n*. Subsequently, *Rsv3* and *Rsv4* have been assigned to resistance loci independent of the *Rsv1* locus (Ma *et al.* 2002). Both the *Rsv1* and *Rsv4* loci have been mapped (Yu *et al.* 1994; Hayes *et al.* 2000). Very recent evidence suggests that, in contrast to previous genetic data indicating resistance to SMV is conferred by the single dominant gene *Rsv1* derived from PI 96983, it is likely that several tightly linked genes besides *Rsv1* may condition resistance to SMV. Further, both genes *Rsv1* and *Rpv1*, for resistance against the soybean-infecting *peanut mottle potyvirus,* are tightly linked as a distance of 1.1 cM (Gore *et al.* 2002).

Recent studies have focused on the mechanism of *Rsv1*-mediated resistance. SMV strain G7 overcomes the resistance conferred by *Rsv1* and results in a systemic necrosis of virus infected plants; however, the resistance gene confers resistance against other SMV strains, including SMV strain G2. This suggests matched specificity between a disease resistance gene *(R)* and pathogen avirulence *(avr)* gene. The resistance against strain G2 has been characterized as extreme resistance (ER) in which no virus can be detected in the inoculated leaf. By comparison with SMV G7, the location of the apparent *avr* gene resides at the 3'- and 5' – ends of the viral HcPro and P3 genes, respectively (Eggenberger *et al.* 1997; Hajimorad *et al.* 2001). The SMV G2 -*Rsv1* interaction also has the potential to induce a hypersensitive (HR)-like defense response that is characterized by elevated transcription of a soybean pathogenesis-related (PR-1) protein gene (Hajimorad *et al.* 2001). The features of this unusual SMV-*Rsv1* interaction suggest the resistance resembles that of *Rx* in potato (Adams *et al.* 1986; Bendahmane *et al.* 1999).

In work that constitutes the first report of stable transgenic resistance against any pathogen of soybean, coat protein-mediated resistance has been developed by using a non-aphid transmissible isolate of SMV to reduce the risk of spreading new pathogenic strains by recombination (Cervera *et al.* 1993; Revers *et al.* 1996). Tests showed the resistance to be effective against numerous SMV strains (Wang *et al.* 2001). In a tactic intended to maximize the durability of resistance, the resistance could be successfully deployed in the field as rate-reducing resistance (Steinlage *et al.* 2002).

In order to preclude losses caused by SMV, several approaches are now available. Both naturally occurring resistance and transgenic resistance are potentially viable approaches. In addition, the use of SMV-free seed to reduce primary inoculum significantly reduces disease incidence, since in most soybean growing areas alternate plant hosts are not prevalent. Careful attention to seed quality coupled with resistance to virus disease will reduce global problems associated with this virus disease.

Bean pod mottle virus Genus *Comovirus* (BPMV) has recently assumed increasing importance, particularly in the North Central soybean-growing region of the United States (Giesler *et al.* 2002). This is of major importance because this region of the United States provides approximately 80% of the production in the largest soybean producing country of the world. However, presence of the virus is not limited to the United States as it has recently become problematic in Canada and is reported from Brazil and Ecuador (Zettler *et al.* 1991; Anjos *et al.* 1999; Michelutti *et al.* 2002).

The recent outbreak of the disease in the North Central United States is associated with, and believed to be caused by, a significant increase in the population of bean leaf beetles (*Cerotoma trifurcata*), which are believed to be the principal vectors of the virus. This is presumably linked to warm winters during the past few years. Several other vectors, including leaf-feeding beetles (Coleoptera) in the families Chrysomelidae, Coccinellidae, and Meliodae can also transmit BPMV. These include *Colaspis brunnea, C. lata, Diabrotica balteata, D. undercmpunctata howardi, Epicauta vittata, and Epilachna varvivestis* (Gergerich, 1999). More recently, *Diabrotica virgifera virgifera* and *Odontota horni* have been identified as potential vectors (Werner *et al.* 2002).

The most economically important host of the virus is soybeans. Although numerous host plants have been experimentally identified, only *Desmodium canadense, D. paniculatum, Glycine max,* and *Phaseolus vulgaris* have been demonstrated to be naturally infected. Other hosts, demonstrated through mechanical inoculation, include *Glycine* spp., *Lespedeza cundata, L. striata, L. stipulacea, Phaseolus lunatus, Pisum sativum, Stizolobium deeringianum, Trifolium incarnatum, Vigna sinensis, and V. unguiculata* (Krell, 2002). All are members of the Fabaceae. Probably the most useful local lesion host is *Phaseolus vulgaris* cv. Pinto.

At least two genetically distinct BPMV subgroups, I and II, can clearly be distinguished by nucleic hybridization. Furthermore, naturally occurring reassortants between the two subgroups have also been isolated and appear to be associated with increased symptom severity (Gu *et al.* 2002).

Virus symptoms on soybeans are similar to those described for SMV and, on the basis of phenotypic symptoms alone, cannot clearly be differentiated from SMV. Plant symptoms range from a mild chlorotic mottling of foliage to a severe mosaic, with the most obvious symptoms appearing on younger leaves. Severe strains may cause puckering and distorted leaves. Depending on the soybean variety, the virus may cause terminal necrosis and death. BPMV delays maturity of soybean, and has been associated with a "green stem" syndrome (Schwenk and Nickell, 1980). However, virus presence is not absolutely correlated with green stem and other factors also appear to be involved. The pod-mottling symptom that is prominent in snap beans is not prominent in many soybean cultivars due to pubescence, but does appear in some. Infection by BPMV also predisposes soybeans to *Phomopsis* spp. seed infection, a major cause of poor seed quality in soybean (Schmitthenner and Kmetz, 1980; Stuckey *et al.* 1982).

Seeds from infected plants may be mottled as described for SMV. It is not possible, based upon visual examination, to discern the seed coat mottling caused by BPMV from that caused by SMV. As with SMV, mottling does not indicate that the virus is present in seeds since not all mottled seeds contain BPMV and not all seeds from virus-infected plants are mottled. Further, mottled seeds do not always contain BPMV. As with SMV, the correlation between the percentage of infected mottled seeds and the severity of mottling varies from year to year and from cultivar to cultivar (Krell, 2002).

The synergistic reaction that can occur between SMV and BPMV is described in the section on SMV. In soybean plants that are infected only with BPMV, yield losses can range from 3% to 52%. Over a broad geographic range, yield reductions between 10% and 40% have been reported (Giesler *et al.* 2002). Impact of BPMV on yield depends upon the time of virus infection relative to plant development with early infection giving the highest yield reduction (Gergerich and Scott, 1996). Reports from Louisiana suggest the BPMV incidence level needs to be between 20% and 40% of the plant population to cause economic loss (Horn *et al.* 1973).

BPMV consists of isometric particles 28 nm in diameter. The virus has a bipartite positive-strand RNA genome consisting of RNA-1 and RNA-2 which are separately encapsidated in isometric particles. Virions can be separated by density gradient centrifugation into three components designated top (T), middle (M), and bottom (B). The T component contains empty particles, and the M and B components contain single molecules of RNA-1 (ca. 6.0 kb) or RNA-2 (ca. 3.6 kb). The three components have identical protein composition, consisting of 60 copies each of a large (L) and small (S) coat protein of 41 kDa and 22 kDa, respectively. The S-coat protein occurs in two major size classes, the intact protein and a C-terminus truncated version. Therefore, BPMV virions have two electrophoretic forms, a slow- and a fast-migrating form, each containing both M and B nucleoprotein components. Intact S-coat protein converts to the C-terminus truncated form with ageing of the virions. The genomic RNAs are polyadenylated, and each has a small basic protein, VPg, covalently linked to the 5' terminus. (Geisler *et al.* 2002). The complete nucleotide sequences of the two genomic RNAS of BPMV strain KY-G7 has been reported and a genetic map has been constructed that suggest RNA-1 and RNA-2 encode 5 and 3 proteins, respectively (Di *et al.*1999; MacFarlane *et al.* 1991).

BPMV is strongly immunogenic and cross reacts with radish mosaic, squash mosaic and red clover mottle viruses, all members of the Comoviridae. Antisera are available for detection of virus in both soybean leaves and seed. The reported thermal inactivation point is $70^0 - 75^0$ C. The dilution end point varies from 10^{-4} to 10^{-5}, and the reported longeivity in vitro is 62 – 93 days (Bruening, 1978).

Virus can be readily purified from soybeans by homogenizing leaves 3 – 4 weeks after inoculation in 0.2 M potassium phosphate, pH 7.0. After passing the homogenate through cheesecloth, chloroform-butanol is added to 10%. Following stirring for 60 minutes at 4^0 C, two or three differential cycles of centrifugation (5,000 x g for 10 minutes; 44,000 x g for 3 hours) and resuspension in 0.1 M potassium phosphate, pH 7.0, results in purified virus (modified from Semancik and Bancroft, 1965).

Potentially, three primary inoculum sources exist. These include perennial host plants of which few have been shown to be naturally occurring, potential for overwintering in insect vectors, or seedlings emerging from infected seeds. Because the disease causes the most damage when infection occurs at early stages of soybean growth, elimination of or reduction in primary inoculum would facilitate disease management. Although the virus can be detected in overwintered beetles at high levels and it is mechanically transmissible, the level of transmission by beetles is low. Similarly, transmission through seed is less than 0.1% and its occurrence is probably cultivar specific. (Giesler *et al.* 2002; Krell, 2002). Nevertheless, the high beetle populations that currently exist in the North Central United States may allow the low rates of transmission through seed and through overwintered beetles to be important. Although the relative importance of the primary inoculum sources is currently unclear, it is possible that all three sources of BPMV inoculum play more-or-less important roles in BPMV epidemiology depending upon location and environment. Seasonal variations that influence putative virus reservoirs are expected to influence BPMV incidence.

There are currently no commercial BPMV-resistant soybean cultivars. However, resistance to BPMV has been identified in the genus *Glycine* that may permit introduction of BPMV resistance with interspecific crosses (Scott *et al.* 1974). Also, genetically engineered resistance to one strain of BPMV has been described but field tests have not yet been reported (Reddy *et al.* 2001). Several studies also suggest that tolerance may exist in commercial soybean cultivars (Giesler *et al.* 2002). In the absence of resistance, vector management provides the best option for disease control at the present time. Recent studies have suggested that protection of the soybean plant from emergence until bloom time with long-lasting insecticide applications can increase yield, reduce virus incidence, and reduce seed coat mottling. Further, planting dates that are timed to reduce bean leaf beetle populations and reduce colonization may enhance disease control (Krell, 2002).

Development of management tactics that will achieve effective disease control will necessitate an interdisciplinary approach through the efforts of virologists, pathologists, entomologists, and agronomists. Vector management through insecticide application can only be a short-term approach to disease control. Longer-term solutions must be based on host plant resistance.

Tobacco ringspot virus Genus *Nepovirus* (TRSV) was first described in the United States in 1941 and has been reported in Canada, Cuba, Brazil, India, Australia, the former Soviet Union, and the Peoples Republic of China (Almeida, 1980; Demski *et al.* 1999; Fernandez-Suarez, 1984; Gupta, 1978). It causes a disease known as bud blight, which should not be confused with Brazilian bud blight, caused by the *Ilarvirus tobacco streak virus*. Global distribution of the virus probably occurs through seed transmission.

Although seed transmission can occasionally range as high as 100%, the potential of harvesting seed fields with such high infection rates is unlikely because infected plants remain green until killed by frost. This makes mechanical harvesting almost impossible (Athow and Bancroft, 1959; Athow and Laviolette, 1961). Because plants must become infected before bloom for seed transmission to be significant (Athow and Bancroft, 1959; Demski and Harris, 1974) and because plants infected before bloom produce few or no seeds, seed transmission is not believed to provide a major source of primary inoculum in the field.

The dagger nematode, *Xiphinema americanum,* is an efficient vector of TRSV to some host species such as *Cucumis sativus*. However, transmission efficiency to soybean is low (McGuire and Douthit, 1978). Observational reports, based upon rapid downwind directional spread, suggest the potential of efficient natural transmission by arthropod vectors (Hill *et al.* 1973). Several arthropod vectors have been reported including aphids *Myzus persicae* and *Aphis gossypii;* the grasshoppers *Melanoplus differentialis, M. mexicanus,* and *M. femr-rubrum;* the tobacco flea beetle, *Epitrix hirtipennis;* and the thrips, *Thrips tabaci* and probably *Frankliniella* sp. (summarized in Granillo and Smith, 1974). However, the reported arthropod vectors are not efficient vectors of TRSV. Additional research is required to discern how apparent virus can spread in soybeans more rapidly than expected through activity of inefficient arthropod or nematode vectors.

The virus, characterized by numerous strains, has been reported to infect a large number of plant hosts in at least 30 different plant families shown here. These include the Amaranthaceae, Aquifoliaceae, Asteraceae, Aizozceae, Balsaminaceae, Begoniaceae, Caricaceae, Caryophyllidae, Chenopodiaceae, Commelinaceae, Compositae, Cornaceae, Crucifereae, Cucurbitaceae, Ericaceae, Fabaceae, Geraniaceae, Hydrangeaceae, Iridaceae, Labiatae, Lilaceae, Malvaceae, Nyctaginaceae, Oleaceae, Pedaliaceae, Polygonaceae, Rosaceae, Scropholariaceae, Solonaceae and Umbellifareae .Additional potential hosts are found in the compilation of De Zeeuw and Hooker (1965), which includes 54 plant families. One of the most useful local lesion hosts is *Vigna sinensis* cv. California Blackeye.

Plants infected at early vegetative growth stages (Fehr *et al.* 1971) are severely stunted and will likely remain green after a killing frost. The most characteristic symptom is the 'shepherd's crook' caused by the curving of the terminal bud to form a crook. Later, other buds on the plant become brown, necrotic, and brittle. Adventitious leaf and floral buds may proliferate excessively. Petioles of the youngest trifoliolate leaves are often shortened, thickened, and may be curved to distort the shoot tips. Leaflets are dwarfed and may cup or roll. Leaf blades become more or less rugose and bronzed. Pods are generally underdeveloped and often aborted. Those that do set often develop dark blotches, generally do not produce viable seeds, and drop early (Demski *et al.* 1999).

Of the many diseases caused by TRSV, budblight of soybeans is the most severe and can cause the most significant yield loss. In general, most significant yield loss occurs when plants become infected before flowering. Yield of early inoculated soybeans has been reported to be reduced by as much as 79% (Demski *et al.* 1971) and yield loss as a result of natural infection has been reported as greater than 50% (Crittenden *et al.* 1966).

TRSV virions are isometric averaging 25-29 nm in diameter (virions may swell at pH 8.0). The genome consists of positive sense single-stranded RNA with a total genome size of 11.2 kb. The genome is bipartite and consists of two polyadenylated RNAs, both of which are required for infection and encapsidated separately in particles surrounded by identical coat protein subunits of molecular weight 58,000. Sedimentation profiles show three components designated top (T), middle (M), and bottom (B). The T components consists of empty particles, while the M and B components contain RNA-1 and RNA-2 of 2.6 x 10^6 and 1.2 x 10^6 kb, respectively. A small VPg is covalently linked at the 5' terminus (Brunt *et al.*1996). Comparison of cloned coat protein genes of three isolates from the United States, Iran, and the United Kingdom showed nucleotide sequence identity ranging from 92% to 98% (Zadeh and Foster, 2001). Several isolates of TRSV contain a satellite RNA of about 359 nucleotides in a protein shell that is identical to the helper virus. The satellite cannot replicate on its own and interferes with the replication of TRSV. The satellites either decrease or have little effect on symptom severity caused by TRSV (Demski *et al.* 1999).

TRSV is moderately immunogenic and cross reacts with potato black ringspot virus. Isolates can be grouped on the basis of serological strain specificity (Gooding, 1970). Antisera can be effective for disease diagnosis. The reported thermal inactivation point is 55^0 – 65^0 C. The dilution end point is 10^{-4} and the longeivity in vitro is 22 days (Brunt *et al.* 1996). The virus may be easily stored in infected leaf tissue at -80^0 C.

Virus can be purified in relatively high quantity by homogenizing french bean leaves or cucumber cotyledons infected for 5 days in 1 ml of 0.1 M phosphate buffer, pH 7.0, and 1 ml of chloroform per gram of tissue. After centrifugation at 10,000 x g for 30 minutes, polyethylene glycol (MW 8000) and NaCl are added to 6% and 0. 3 M, respectively. Following incubatiion of the extract for 30 minutes at 4^0 C, the precipitate is collected by centrifugation and resuspended overnight at 4^0 C in phosphate buffer containing 0.01 M EDTA. After centrifugation at 160,000 x g for 50 minutes, the virus is resuspended in phosphate buffer and clarified by centrifugation at 10,000 x g for 10 minutes (Rezaian and Francki, 1973).

The primary inoculum source of TRSV in soybeans is unclear. Because transmission in seed used for commercial production is low and the unequivocal demonstration of efficient transmission by an arthropod vector remains elusive, the epidemiology of the virus remains vague. Demonstration of a disease gradient on the downwind side of mixed legume (alfalfa and red clover) fields suggests spread more rapid than that associated with nematode transmission (Hill *et al.* 1973). Resistance to TRSV has been identified in *Glycine soja* but is not known in the cultivated soybean, *Glycine max* (Orellana, 1981). Although not clear that it contributes in a major way to significant soybean disease problems, the use of virus-free soybean seed seems appropriate for disease control. Also, producers should avoid fields that are known to be infested with the dagger nematode (or treat them with appropriate nematacides) as well as avoid planting soybeans adjacent to mixed legume fields.

15. 3 Outlook and Conclusions

Numerous virus diseases of soybeans have been documented. However, difficulty in correct identification of the causal agent of virus-like diseases of soybeans is common because symptoms caused by widely different viruses are often not unique. The large variety of soybean cultivars, and symptoms that vary with cultivar and time of inoculation, makes this problem more vexing. Little advancement has been made in identification of resistance sources for many of the virus-based diseases that infect soybean. Also, little research progress has been made that identifies sources of primary inoculum or provides understanding of the basis for disease epidemics that occur. This lack of knowledge is not surprising upon realization that, until the last 20 - 30 years, disease problems in soybeans were virtually nonexistent. However, due to changes in farming practices, changes in climate, and a narrow genetic base of commercially grown soybeans, soybean disease problems have become important. Research input has been unable to keep up with mounting problems. Further efforts, on a global scale, will be necessary to provide knowledge that will allow development of science-based tactics for virus disease control in soybeans. This will require interdisciplinary effort.

Acknowledgments

This chapter was supported, in part, by the Iowa Soybean Promotion Board and the North Central Soybean Research Program. This chapter of the Iowa Agriculture and Home Economics Experiment Station, Ames, IA, U.S.A., Project 2428, was supported by the Hatch Act and State of Iowa funds.

References

Adams, S. E., Jones, R. A. C., and Coutts, R. H. A. 1986. Expression of potato virus X resistance gene *Rx* in potato leaf protoplasts. *J. Gen. Virol.* 67:2341-2345.

Almeida, A. M. R. 1981. Identification of strains of soybean common mosaic virus in Parana State. *Fitopatol. Brasil.* 6:131-136.

Almeida, A. M. R. 1980. Survey of soybean common mosaic and bud blight viruses in different regions of Parana State. *Fitopatol. Brasil.* 5:125-128.

Anjos, J. R. N., Brioso, P. S. T., and Charchar, M. J. A. 1999. Partial characterization of bean pod mottle virus in soybeans in Brazil. *Fitopatol. Brasil.* 24:85 – 87.

Anjos, J. R. M., Lin, M. T., and Kitajima, E. W. 1985. Characterization of an isolate of soybean mosaic virus. *Fitopatol. Brasil* 10:143-157.

Athow, K. L., and Bancroft, J. B. 1959. Development and transmission of tobacco ringspot virus in soybean. *Phytopathology* 49:697-701.

Athow, K. L., and Laviolette, F. A. 1961. The relation of seed-transmitted tobacco ringspot virus to soybean yield. *Phytopathology* 51:341-342.

Bendahmane, A., Kanyuka, K., and Baulcombe, D. C. 1999. The *Rx* gene from potato controls separate virus resistance and cell death responses. *Plant Cell* 11:781-791.

Brunt, A.A., Crabtree, K., Dallwitz, M.J., Gibbs, A.J., Watson, L., and Zurcher, E.J. (eds.) (1996 onwards). `Plant Viruses Online: Descriptions and Lists from the VIDE Database. Version: 20th August 1996.' URL http://biology.anu.edu.au/Groups/MES/vide/

Bryant, G. R., Hill, J. H., Bailey, T. B., Tachibana, H., Durand, D. P., and Benner, H. I. 1982. Detection of soybean mosaic virus in seed by solid-phase radioimmunoassay. *Plant Dis.* 66:693-695.

Bruening, G. 1978. Bean pod mottle virus. In: *CMI/AAB Descr. Pl. Viruses* no. 199, 4 pp.

Buzzel, R. I., and Tu, J. C. 1984. Inheritance of soybean resistance to soybean mosaic virus. *J. Hered.* 75:82.

Calvert, L. A., and Ghabrial, S. A. 1983. Enhancement by soybean mosaic virus of bean pod mottle virus titer in doubly infected soybean. *Phytopathology* 73:992-997.

Cervera, M. T., Reichmann, J. L., Martin, M. T., and Garcia, J. A. 1993. 3'-terminal sequence of the plum pox PS and 06 isolates: evidence for RNA recombination within the potyvirus group. *J. Gen. Virol.* 74:329-334.

Chen, Y., Xue, B., Hu, Y., and Fang, Z. 1986. Identification of two new strains of soybean mosaic virus. *Acta. Phytophyl. Sinica* 13:222-226.

Cho, E.-K., Chung, B. J., and Lee, S. H. 1977. Studies on identification and classification of soybean virus diseases in Korea. II. Etiology of a necrotic disease of *Glycine max*. *Plant Dis. Rep.* 61:313-317.

Cho, E.-K., and Goodman, R. M. 1979. Strains of soybean mosaic virus: classification based on virulence in resistant soybean cultivaars. *Phytopathology* 69:467-470.

Crittenden, H. W., Hastings, K. M., and Moore, D. M. 1966. Soybean losses caused by tobacco ringspot virus. *Plant Dis. Reptr.* 50:910-913.

Demski, J.W., and Harris, H. B. 1974. Seed transmission of viruses in soybean. *Crop Sci.* 14:888-890.

Demski, J. W., Haris, H. B., and Jellum, M. D. 1971. Effects of time of inoculation with tobacco ringspot virus on the chemical composition and agronomic characteristics of soybean. *Phytopathology* 61:308-311.

Demski, J. W., Kuhn, C. W., and Hartman, G. L. 1999. Tobacco ringspot. In: *Compendium of soybean diseases, 4th edition.* (eds) G. L. Hartman, J. B. Sinclair, and J. C. Rupe, pp. 66 – 68. APS Press, St. Paul, Minnesota.

De Zeeuw, D. J., and Hooker, W. J. 1965. Additional suscepts of the tobacco ringspot virus. *Quarterly Bulletin of the Michigan Agr. Expt. Sta.* 48:76-80.

Di. R., Hu, C.-C., and Ghabrial, S. A. 1999. Complete nucleotide sequence of *Bean pod mottle virus* RNA1: sequence comparisons and evolutionary relationships to other comoviruses. *Virus Genes* 18:129-137.

Edwardson, J. R., and Christie, R. G. 1991. Soybean mosaic virus. *The potyvirus group, vol. III. University of Florida Monograph Series No. 16-III.* pp. 821-835.

Eggenberger, A. L., and Hill, J. H. 1997. Analysis of resistance-breaking determinants in soybean mosaic virus. *Phytopathology* 87:S27.

Fehr, W. R., Caviness, C. E., Burmood, D. T., and Pennington, J. S. 1971. Stage of development descriptions for soybeans, Glycine max (L.) Merrill. Crop Sci. 11:929-931.

Fernandez-Suarez, R. 1984. Bibliographic review of some virus diseases of soya bean (Glycine max). Reporte de Investigacion del Instituto de Investigaciones Fundamentales. *Agricultural Tropical* No. 16. 31pp.

Geisler, L. J., Ghabrial, S. A., Hunt, T. E., and Hill, J. H. 2002. Bean pod mottle virus. A threat to U.S. soybean production. *Plant Dis.* 86:1280-1289.

Gergerich, R. C. 1999. Comoviruses: Bean pod mottle comovirus. In: *Compendium of soybean diseases, 4th edition.* (eds) G. L. Hartman, J. B. Sinclair, and J. C. Rupe, pp. 61 – 62. APS Press, St. Paul, Minnesota.

Gergerich, R. C., and Scott, H. A. 1996. Comoviruses: Transmission, epidemiology, and control. In: *The Plant Viruses 5: Polyhedral Virions and Bipartite RNA Genomes.* pp. 77-98. Plenum Press, New York.

Gooding, Jr., G. V. 1970. Natural serological strains of tobacco ringspot virus. *Phytopathology* 60:708-713.

Gore, M. A., Hayes, A. J., Jeong, S. C., Yue, Y. G., Buss, G. R., and Saghai Maroof. M. A. 2002. Mapping tightly linked genes controlling potyvirus infection at the *Rsv1* and *Rpv1* region in soybean. *Genome* 45: 592-599.

Granillo, C. R., and Smith, S. H. 1974. Tobacco and tomato ringspot viruses and their relationships with Tetranychus urticae. *Phytopathology* 64: 494-499.

Gu, H., Clark, A. J., de Sa, P. B., Pfeiffer, T. W., Tolin, S., and Ghabrial, S. A. 2002. Genetic diversity among isolates of *Bean pod mottle virus*. *Phytopathology* 92: 446-452.

Gupta, V. K. 1978. Further studies on bud blight disease of soybean. 1978. *Acta Botanica Indica* 6 Suppl: 169-170.

Hajimorad, M. R., and Hill, J. H. 2001. *Rsv1*-mediated resistance against *soybean mosaic virus*-N is hypersensitive response-independent at inoculation site, but has the potential to initiate a hypersensitive response-like mechanism. *Mol. Plant Micro. Interact.* 14:587-598.

Hartman, G. L., Sinclair, J. B., and Rupe, J. C. (eds). 1999. Diseases caused by viruses. In: *Compendium of soybean diseases, 4th edition,* pp. 57-73. APS Press, St. Paul, Minnesota.

Hayes, A. J., Ma, G., Buss, G. R., and Saghai Maroof, M. A. 2000. Molecular marker mapping of *RSV*4, a gene conferring resistance to all known strains of soybean mosaic virus. *Crop Sci.* 40:1434-1437.

Hill, J. H. 1999. Soybean mosaic. In: *Compendium of soybean diseases, 4th edition.* (eds) G. L. Hartman, J. B. Sinclair, and J. C. Rupe, pp. 70 – 71. APS Press, St. Paul, Minnesota.

Hill, J. H., and Benner, H. I. 1980a. Properties of soybean mosaic virus and its isolated protein. *Phytopath. Z.* 97:272-281.

Hill, J. H., and Benner, H. I. 1980b. Properties of soybean mosaic virus ribonucleic acid. *Phytopathology* 70:236-239.

Hill, J. H., Benner, H. I., Permar, T. A., Bailey, T. B., Andrews, R. E., Durand, D. P., and Van Deusen, R. A. 1989. Differentiation of soybean mosaic virus isolates by one-dimensional trypsin peptide maps immunoblotted with monoclonal antibodies. *Phytopathology* 79:1261-1265.

Hill, J. H., Benner, H. I., and Van Deusen, R. A. 1994. Rapid differentiation of soybean mosaic virus isolates by antigenic signature analysis. *J. Phytopath.* 142:152-162.

Hill, J. H., Epstein, A. H., McLaughlin, M. R., and Nyvall, R. F. 1973. Aerial detection of tobacco ringspot virus-infected soybean plants. *Plant Dis. Reptr.* 57:471-472.

Hill, J. H., Lucas, B. S., Benner, H. I., Tachibana, H., Hammond, R. B., and Pedigo, L. P. 1980 Factors associated with the epidemiology of soybean mosaic virus in Iowa. *Phytopathology* 70: 536-540.

Horn, N. S., Newsom, L. D., and Jensen, R. L. 1973. Economic injury thresholds of bean pod mottle and tobacco ringspot virus infection of soybeans. *Plant Dis. Reptr.* 57: 811-813.

Jain, R. K., McKern, N. M., Tolin, S.A., Hill, J. H., Barnett, O. W., Tosic, M., Ford, R. E., Beachy, R. N., Yu, M. H., Ward, C. W., and Shukla, D. D. 1992. Similarity of coat protein peptide profiles of fourteen potyvirus isolates from soybean confirms that they are strains of the one virus. *Phytopathology* 82: 294-299.

Jayaram, Ch., Hill, J. H., and Miller, W. A. 1992. Complete nucleotide sequences of two soybean mosaic virus strains differentiated by response of soybean containing the *Rsv* resistance gene. *J. Gen. Virol.* 73: 2067-2077.

Jayaram, Ch., Van Deusen, R. A., Eggenberger, A. L., Schwabacher, A. W., and Hill, J. H. 1998. Characterization of a monoclonal antibody recognizing a DAG-containing epitope conserved in aphid transmissible potyviruses: evidence that the DAG motif is in a defined conformation. *Virus Res.* 58: 1-11.

Kanematsu, S., Eggenberger, A. L., and Hill, J. H. 1998. Comparison of soybean mosaic virus strains G2 and G3 with Japanese strains A and B. *Ann. Phytopathol. Soc. Japan* 64:607.

Krell, R. K. 2002. Bean pod mottle virus ecology and management in Iowa. Ph.D. thesis. Iowa State University. 153 pp.

Lee, Y. C., Kim, J. J., and Cho, E.-K. 1992. Classification of seed-borne SMV strains and resistance to SMV in leading soybean cultivars. *Korean J. Breed.* 23:53-58.

Ma., G., Chen, P., Buss, G. R., and Tolin, S. A. 2002. Complementary action of two independent dominanat genes in Columbia soybean for resistance to soybean mosaic virus. *J. Hered.* 93: 179-184.

MacFarlane, S. A., Shanks, M., Davies, J. W., Zlotnick, A., and Lomonossoff, G. P. 1991. Analysis of the nucleotide sequence of bean pod mottle virus middle component RNA. *Virology* 183: 405-409.

McGuire, J. M., and L. B. Douthit. 1978. Host effect on acquisition and transmission of tobacco ringspot virus by Xiphinema americanum. *Phytopathology* 68:457-459.

Michelutti, R., Tu., J. C., Hunt, W. A., Gagnier, D., Anderson, T. R., Welacky, T. W., and Tenuta, A. U. 2002. First report of bean pod mottle virus in Canada. *Plant Dis.* 86:330.

Milbrath, G. M., and Soong, M.-M. 1976. A local lesion assay for soybean mosaic virus using *Phaseolus vulgaris* L. cv. Top Crop. *Phytopath. Z.* 87:255-259.

North Central Soybean Research Program. 2003. Plant Health Initiative. URL http://.planthealth.info/

Nutter, F. W., Jr., Schultz, P. M., and Hill, J. H. 1998. Quantification of within-field spread of soybean mosaic virus in soybean using strain-specific monoclonal antibodies. *Phytopathology* 88: 895-901.

Orellana, R. G. 1981. Resistance to bud blight introductions from the germ plasm of wild soybean. *Plant Dis.* 65:594-595.

Pu, Z., Cao, Q., Fang, D., Xi, B., and Fang, C. 1982. Identification of strains of soybean mosaic virus. *Acta Phytophyl. Sinica* 9:15-20.

Reddy, M. S. S., Ghabrial, S. A., Redmond, C. T., Dinkins, R. D., and Collins, G. B. 2001. Resistance to *Bean pod mottle virus* in transgenic soybean lines expressing the capsid polyprotein. *Phytopathology* 91:831-838.

Revers, F., Le Gall, O., Candresse, T., Le Romancer, M., and Dunez, J. 1996. Frequent occurrence of recombinant potyvirus isolates. *J. Gen. Virol.* 77:1953-1965.

Rezaian, M. A., and Francki, R. I. B. 1973. Replication of tobacco ringspot virus. I. Detection of a low molecular weight double-stranded RNA from infected plants. *Virology* 56:238-249.

Schmitthenner, A. F., and Kmetz, K. 1980. Role of *Phomopsis* sp in the soybean seed rot problem. In: *Proc. World Soybean Res. Conf. 2nd*. pp. 355–366. Westview Press, Boulder CO.

Schwenk, F. W., and Nickell, C. D. 1980. Soybean green stem caused by bean pod mottle virus. *Plant Dis.* 64:863-865.

Scott, H. A., Van Scyoc, J. V., and Van Scyoc, C. E. 1974. Reaction of *Glycine* spp. to bean pod mottle virus. *Plant Dis. Rept.* 58:191-192.

Semancik, J. S., and Bancroft, J. B. 1965. Stability differences between the nucleoprotein components of bean pod mottle virus. *Virology* 27:476-483.

Steinlage. T. A., Hill, J. H., and Nutter, Jr., F. W. 2002. Temporal and spatial spread of soybean mosaic virus (SMV) in soybeans transformed with the coat protein gene of SMV. *Phytopathology* 92:478-486.

Stuckey, R. E., Ghabrial, S. A., and Reicosky, D. A. 1982. Increased incidence of *Phomopsis* spp. in seeds from soybean infected with bean pod mottle virus. *Plant Disease* 66:826-829.

Takahashi, K., Tanaka, T., Wataru, I., and Tsuda, T. 1980. Studies on virus diseases and causal viruses of soybean in Japan. Bull. Tohoku Natl. Agric. Exp. Stn. 62:1-130.

Thottappilly, G., and Rossel, H. W. 1987. Viruses affecting soybean. In: *Soybeans in the tropics.* (eds) S. R. Singh, K. Rachie, and K. Dashiell, pp. 53-68. John Wiley and Sons, New York.

Tolin, S. A. 1999. Diseases caused by viruses. In: *Compendium of soybean diseases, 4th edition.* (eds) G. L. Hartman, J. B. Sinclair, and J. C. Rupe, pp. 57 – 59. APS Press, St. Paul, Minnesota.

United Soybean Board Annual Soybean Statistics Guide. 2002. URL http://www.unitedsoybean.org/

Wang, X., Eggenberger, A. L., Nutter, Jr., F. W., and Hill, J. H. 2001. Pathogen-derived transgenic resistance to soybean mosaic virus in soybean. *Mol. Breeding* 8:119-127.

Werner, B. J., Krell, R. K., and Pedigo, L. P. 2002. New host plant and vector relationships for *Bean pod mottle virus.* Iowa State University. North Central Branch of the

Entomological Society of America. On-line, publication D108. URL http://esa.ent.iastate.edu/progsearch?m=35&s=1&q=Werner

Wrather, J. A., Anderson, T. R., Arsyad, D. M., Tan, Y., Ploper, L. D., Porta-Puglia, A., Ram, H. H., and Yorinori, J. T. 2001. Soybean disease loss estimates for the top ten soybean-producing countries in 1998. *Can. J. Plant Pathol.* 23:115-121.

Xu, Z., Polston, J. E., and Goodman, R. M. 1986. Identification of soybean mosaic, southern bean mosaic and tobacco ringspot viruses from soybean in the People's Republic of China. *Ann. appl. Biol.* 108:51-57.

Yu, Y. G., Saghai Maroof, M. A., Buss, G. R., Maughan, P. J., and Tolin, S. A. 1994. RFLP and microsatellite mapping of a gene for soybean mosaic virus resistance. *Phytopathology* 84:60-64.

Zadeh, A. H., and Foster, G. D. 2001. Cloning and sequencing of the coat protein gene of tobacco ringspot virus isolates from UK and Iran. *Acta Virologica* 45:319-326.

Zettler, F. W., Stanaly, P. A., Elliot, M. S., Peralta, A., Carranza, C., and Morales, F. J. 1991. Bean pod mottle virus (BPMV) in Ecuador and its transmission by *Cerotoma facialis maculata*. *Phytopathology* 81:695.

Table 15.1. **Virus diseases of soybean.**

Genus type Family Genus	Viruses in Field	By Inoculation
Single-stranded RNA, positive sense		
Comoviridae		
Comovirus	Bean pod mottle	Broad bean stain
	Broad bean true mosaic	Cowpea mottle
	Cowpea mosaic	Glycine mosaic
	Cowpea severe mosaic	Pea mild mosaic
	Quail pea mosaic	Red clover mottle
Fabavirus	Broad bean wilt	
Nepovirus	Soybean severe stunt	Arabis mosaic
	Tobacco ringspot	Cacao necrosis
	Tomato ringspot	Cherry leaf roll
		Grapevine fanleaf
		Mulberry ringspot
		Raspberry ringspot
		Tomato blackring
Potyviridae		
Potyvirus	Azuki bean mosaic	Beet mosaic
	Bean common mosaic	Cassia yellow spot
	Bean yellow mosaic	Chickpea distortion mosaic
	Blackeye cowpea mosaic	Clover yellow vein
	Passion fruit woodiness	Cowpea aphid-borne mosaic
	Peanut chlorotic ring mottle	Kennedya Y
	Peanut mottle	Peanut green mosaic
	Peanut stripe	Watermelon mosaic
	Soybean mosaic	Wisteria vein mosaic
Unamed		
Sobemovirus	Southern bean mosaic	Subterranean clover mottle
Unamed		

Luteovirus	Bean leaf roll Beet western yellows Indonesian soybean dwarf Soybean dwarf (subterranean clover red leaf)	Chickpea stunt Pea leafroll
Unamed		
Enamovirus	Pea enation mosaic	
Unamed		
Umbravirus		Groundnut rosette
Tombusviridae		
Carmovirus	Black gram mottle	Bean mild mosaic Cowpea mottle Glycine mottle Pea stem necrosis Pelargonium line pattern Tephrosia symptomless
Tombusvirus		Cymbidium ringspot
Unamed		
Necrovirus		Tobacco necrosis
Unamed		
Dianthovirus		Red clover necrotic mottle
Unamed		
Tobamovirus	Tobacco mosaic Sunn-hemp mosaic	
Unamed		
Tobravirus	Tobacco rattle (soybean fleck)	Pea early browning
Unamed		
Furovirus	Soybean yellow vein	Peanut clump
Bromoviridae		
Alfamovirus	Alfalfa mosaic	
Bromovirus	Cowpea chlorotic mottle (bean yellow stipple)	Broad bean mottle Cassia yellow bloth Spring beauty latent
Cucumovirus	Cucumber mosaic (soybean stunt) Peanut stunt	Tomato aspermy
Ilarvirus	Tobacco streak	Asparagus virus 2 Sunflower ringspot
Closterovirus		Cassia severe mosaic
Unamed		
Tymovirus		Abelia latent Clitoria yellow vein Kennedya yellow mosaic Ononis yellow mosaic Plantago mottle
Unamed		
Carlavirus	Cowpea mild mottle	Cassia mild mottle Cole latent Pea streak

		Poplar mosaic
		Red clover vein mosaic
Unamed		
Potexvirus		Asparagus 3 virus
		Clover yellow mosaic
		Commelina X
		Crotalaria yellow mosaic
		Foxtail mosaic
		Pea wilt
		Narcissus mosaic
		White clover mosaic
Single-stranded RNA, ambisense		
Bunyaviridae		
Tospovirus	Tomato spotted wilt	Peanut yellow spot
Single-stranded DNA		
Geminiviridae		
Begomovirus	Abutilon mosaic	Bean calico mosaic
	African soybean dwarf	Cotton leaf crumple
	Bean golden mosaic	Tobacco leaf curl
	Euphorbia mosaic	
	Horsegram yellow mosaic	
	Mung bean yellow mosaic	
	Rhynchosia mosaic	
	Soybean crinkle leaf	
	Soybean yellow mosaic	
Double-stranded DNA, reverse transcribing		
Caulimoviridae		
Caulimovirus	Soybean chlorotic mottle	Peanut chlorotic streak
Not known		
Unamed		
Unamed	Soybean mild mosaic	

Data from Tolin, 1999.

16. Peanuts

D.V.R. Reddy[1] and K. Thirumala-Devi [2]
[1]*Donald Danforth Plant Science Center, 975. North Warson Road, St. Louis, MO, USA; and* [2]*Department of Plant pathology, Russell labs, 1630 Linden Drive, University of Wisconsin, Madison, WI, USA*

16.1 Introduction

Peanut or groundnut (*Arachis hypogea*) is a member of Papilionoideae, family Leguminosae. It is an important oil, food and forge crop cultivated mainly in tropical and subtropical regions of nearly 100 countries in all the six continents, with latitudes 40° N and 40° S. It is believed to have originated in southern Bolivia and northwestern Argentina, grown for many centuries in south and Central America and the Caribbean. It was introduced into West Africa from Brazil in the sixteenth century and subsequently to Asia including India and China. Peanut is highly nutritious food (25 to 34% protein) and the fifth most important oil seed crop (44% to 56% oil). Additionally peanut haulms (vegetative growth after harvest) and peanut cake (by product of oil extraction) are used as livestock feed. India is currently the largest producer of peanut accounting for nearly 33% of total world's production. India and China together contribute to nearly 60% of world's production. In Africa and Asia peanuts are consumed as fresh boiled or roasted peanuts and additionally used as an ingredient in many cereal-based foods. In Asia peanut is mainly cultivated for its edible oil.

Varieties can be broadly classified into two main groups, which differ in their branching pattern, erect brunch types and spreading or runner types. Peanuts are grown as a sole crop, and in multiple-cropping systems, including that of a relay crop after harvesting rice.

The current area under peanut production is estimated to be 22.65 million hectares and world's production in 2002 was about 33.24 million metric tons in shell. Nearly 50% of world's production is from crops grown in the developing countries (China is not included under this category) (Golbitz, 2003). We describe in this review various aspects of six economically important peanut virus diseases.

16.2 Peanut Bud Necrosis (PBND) and Spotted Wilt Diseases (PSWD)

Synonyms: Bud blight, groundnut mosaic, groundnut ring mosaic, bunchy top, chlorosis, ring mottle, ring spot and spotted wilt.

Tomato spotted wilt virus (TSWV) on peanut was first reported by Costa (1941) in Brazil. The occurrence of bud necrosis was first reported by Reddy *et al.* (1968) from India. Latter bud necrosis was reported under several different names from India (Reddy, 1988)

Peanut bud necrosis disease (PBND) was reported from South and Southeast Asia including China, India, Indonesia, Nepal, Pakistan, Philippines, Sri Lanka and Thailand. Peanut spotted wilt disease (PSWD) has been reported from South and North America (Argentina, Brazil, USA), and several African Countries (South Africa, Nigeria, Kenya, Malawi and Uganda)

PBND and PSWD have become increasingly important in the production of peanut in Southeast Asia and in the southeastern United States (Reddy, 1991; Black, 1991; Bertrand, 1996). Crop losses up to 89 million dollars from India were reported.

The viruses causing PBND and PSWD have a wide host range infecting more than 800 plant species in 82 different families which include many agricultural, horticultural and green house crops and weeds (Cho *et al.* 1987; Latham and Jones, 1997; Peters 1998; Reddy and Wightman 1988).

Symptoms of PBND caused by *Peanut bud necrosis tospovirus* (PBNV) and PSWD caused by *Tomato spotted wilt tospovirus* (TSWV) in peanut are similar. Early symptoms appear on young leaflets as chlorotic spots or mild mottling that may develop into chlorotic and necrotic rings and streaks until the whole plant is affected. Petioles bearing the fully expanded leaflets with initial symptoms usually become flaccid and droop more frequently with PBNV infected than with TSWV. This is followed by terminal bud necrosis, especially when temperatures are above 30°C during the day. Early infection causes stunting and sometimes proliferation of axillary shoots. Leaflets produced on axillary shoots are reduced in size and may show puckering, chlorosis, mosaic and distortion of lamina (Fig. 16.3 and Fig. 16.4). Entire plant shows bushy appearance and often confused with infection by *Peanut stunt cucumovirus* and *groundnut chlorotic rosette virus*. Any seed produced on early-infected plants are small, shriveled and testae have red, brown or purple mottling. Late-infected plants may produce seed of normal size; however the testae of such seed are often mottled (Reddy, 1991)

16.2.1 Transmission

PBNV and TSWV are mechanically transmissible. Only chilled extracts containing antioxidants (thioglycerol or mercaptoethanol) permit sap transmission. Factors that affect mechanical transmission of TSWV to peanut were investigated by Mandal *et al.* (2000).

Both the viruses are transmitted by thrips in a persistent manner. Vectors of TSWV are *Frankliniella fusca, F. intonsa, F. occidentalis, F. schultzei, Thrips setosus* and *T. tabaci* (Mound 1996; Ullman *et al.* 1997). PBNV is known to be transmissible only by *Thrips palmi*. Thrips can acquire the virus only during larval stages and larvae as well as adults transmit the virus (Reddy 1991; Peters *et al.* 1996). Both TSWV and PBNV have been shown to multiply in their vectors (Ullman *et al.* 1993; Wijkamp *et al.* 1993; DVR Reddy and KLN Reddy, unpublished; Pappu *et al.* 1998). TSWV and PBNV are not transmitted through peanut seed (Reddy *et al.* 1991; Pappu *et al.* 1999)

16.2.2 Diagnosis

Biological. Both the viruses are mechanically transmissible. Terminal bud necrosis is also caused by *Tobacco streak virus* and therefore should not be used to diagnose infections by PBNV and TSWV. Local lesions produced on cowpea are similar for TSWV and PBNV. The lesions are concentric chlorotic and necrotic type and distinguishable from lesions produced by other peanut viruses. *Petunia* produces necrotic lesions within 2 to 3 days. Thrips transmission is a good diagnostic tool especially for research workers who have no access to antisera, as vectors of TSWV and PBNV are not the same. PBNV is unique in that it is only transmitted by *T. palmi*, while TSWV is transmitted by a number of thrips species as listed above.

Serological methods. Good quality polyclonal and monoclonal antisera are currently available for PBNV and TSWV. Both antisera are produced for whole virions as well as for viral proteins cloned *in vitro*, which include structural and non-structural proteins (Bandla *et al.* 1994). These antisera have been used in different ELISA systems (Reddy *et al.* 1992; Sherwood *et al.* 1995).

Nucleic acid based. RT-PCR has been applied for rapid detection of TSWV (Jain, 1998) and PBNV infections (DVR Reddy and KLN Reddy unpublished). Immuno-capture allows rapid and reliable amplification of TSWV sequences from peanut and other crops from a range of different tissues. Primers derived from the nucleocapsid protein gene as well as from the large RNA gene, which codes for polymerase, are able to amplify the target sequences in a specific and reproducible manner. This technique has also the potential to distinguish TSWV infection from other tospoviruses infection in peanut including that of PBNV.

16.2.3 Purification

Detailed purification procedure adopted for PBNV were given by Satyanarayana *et. al.* (1996) and should be applicable for TSWV

16.2.4 Viral Characteristics

PBNV and TSWV are members of the Bunyavirideae, genus *Tospovirus*. Virions are spherical or pleiomorphic, 80-120nm in diameter. The particle contains a core with three nucleocapsids, each containing genomic RNA tightly associated with nucleoprotein (N). The outer layer contains surface glycoprotein projections (G1 and G2 of molecular mass 58 kd and 78 kd, respectively) of 5-10 nm, which are embedded in a lipid envelope. TSWV and PBNV also contain a nucleocapsid (N) protein (29 kd for TSWV and 31 kd for PBNV) and a large (L) transcriptase protein of over 200 kd. Both viruses encapsidate 3 species of RNA small (S) medium (M) and large (L) and the complete nucleotide sequences have been determined. The nucleotide sequences of PBNV are 3057 for S RNA, 4801 for M RNA and 8911 for L RNA (Gowda *et al.* 1998; Satyanarayana *et al.* 1996). The consensus terminal sequences of L, M, and S segments are UCU CGU UA.... at the 3' and AGA GCA AU.... at the 5' end. The L RNA of TSWV has 8897 nucleotides, M RNA has 4821 and S RNA has 2916 nucleotides. The terminal nucleotides of each genome RNA segments are base paired forming non-covalently closed circular RNAs (ribonucleic capsids). The terminal nucleotide sequences of the genome segments are conserved. Viral mRNAs are polyadenylated at the 3'end and are truncated relative to the genome RNAs. mRNA's have 5'methylated caps and 10-18 non templated nucleotides which are derived from host cell mRNAs. RNA 1 encodes the viral RNA polymerase L protein and the S RNA codes for the N protein. A single continuous ORF in the m RNA encodes the glycoproteins, G1 and G2. Non-structural proteins of 34 kd and 29 kd are encoded by M and S RNAs, respectively (van Regenmortel *et al.* 2000; Moyer 1999).

16.2.5 Epidemiology

TSWV and PBNV are transmitted under natural conditions by thrips in a persistent manner. The primary sources of inoculum comprise a range of hosts, which include weeds as well as crop plants, which can sustain virus infection, and support thrips vector multiplication. The incidence and spread are dependent on many factors, which include those that contribute to multiplication and spread of the vector thrips. They include transmission efficiency of the individual thrips species, temperature, and wind velocity and crop canopy. For PSWD risk index for peanuts has been determined by evaluating various factors that contribute to the disease occurrence. Risk factors that affect the severity of TSWV and PBNV include peanut variety, planting date, plant population, intercropping with cereal crops and application of insecticides (Reddy *et al.* 1983; Culbreath *et al.* 1999). The most important vectors for TSWV under field conditions are *F. fusca* and *F. occidentalis*. For PBNV *Ageratum conyzoides* has been shown to support the virus and vector multiplication (KLN Reddy and DVR Reddy unpublished). Many weeds which include spiny sowthistle, common chickweed and dandelion support large vector thrips population (Groves *et al.* 2002). Secondary spread from infected peanut plants with in a peanut field is considered to be negligible. Incidence continues to depend on infection by viruliferous thrips that acquire the virus from alternative hosts. Therefore it is a monocyclic disease.

16.2.6 Control

Management of TSWV as well as PBNV depends on the control of the thrips vectors. Insecticide use is not recommended as this can lead to no significant effect or in some instances can even increase the disease and may encourage other pest problems. If information is available on the time of arrival of principal thrips vectors and if the growing season permits adjustment of sowing dates so that plants as well as the crop canopy develop before infection occurs, the disease can be reduced (Reddy, 1998). Cultural practices can reduce incidence of PSWD and PBND. These include sowing of crops to coincide with the lowest population of vectors, and use of seed of high quality treated with seed protectants to avoid seedling fungal diseases, with adequate soil moisture supply. These measures will lead to maintenance of good plant stand (2-3 million plants per hectare is recommended) and vigorous crop growth will ensure close canopy, leading to reduction in incidence of PBNV and TSWV. Though many weeds can provide the primary source of inoculum their eradication to reduce the primary spread is not considered to be practical especially in tropical countries. Removal of infected plants in peanut fields, especially those infected during early stages of crop growth, can create gaps in the field and can lead to increased incidence. Additionally intercropping with quick growing cereals as pearl millet or sorghum or maize can reduce the incidence by acting as a barrier for thrips colonization.

Peanut sources with good field-resistance have been identified for TSWV and PBNV. They were tested under laboratory as well as field conditions

(Fig. 16.5). For PBNV they are ICGV 86029, 86031, 86388, 91239, 91245, 91246 and 91249 for PBNV. Several cultivars with resistance to PBNV such as ICGS 44 and ICGS 11 have been released in India. Cultivars with field resistance to TSWV have been released and currently being grown in the southeastern USA. These include Georgia Green, Virugard, C99R and Southern Runner (Culbreath *et al.* 2000; Gorbet *et al.* 1999). Advanced breeding lines with high levels of resistance to TSWV were identified (Culbreath *et al.* 2000; Mandal *et al.* 2002). Plant biotechnological approaches are being used to produce TSWV resistant plants (Pappu 1997). Peanut plants transformed with NP gene of TSWV have been tested under glasshouse and field conditions (Magbanua *et al.* 2000). Based on the nucleotide sequences of several TSWV isolates from Georgia (Pappu *et al.* 1998; Yang *et al.* 2001) a highly conserved region of the NP gene has been used to generate transgenic peanut plants. However to our knowledge peanut transgenic plants with resistance to TSWV have not yet been released (H. R. Pappu, personal communication).

16.3 Groundnut Rosette Virus Disease

Synonyms: groundnut chlorotic rosette, groundnut green rosette and groundnut mosaic rosette

Groundnut rosette disease (GRD) is by far the most destructive disease of peanuts throughout Africa, south of the Sahara. It was first reported from Tanzania. Rosette has also been reported from other countries in Asia, but these are now regarded as erroneous (Reddy, 1991). Since peanut is not a native African plant, it is presumed that GRD may be indigenous to other native African species.

It is currently known to be present in the entire African Continent, south of Sahara and in Madagascar, and in some years occurs in epidemic proportions. One of the most severe epidemics was recorded in 1975 when the disease destroyed more than 0.7 million hectares of peanut in Nigeria (Yayock, 1977), a value of $ 250 million. In 1995 more than 43,000 hectares were affected in eastern Zambia resulting in losses up to $ 5 million. Epidemics in Malawi during 1994-1995 lead to reduction in area under peanut cultivation by 23%, from 89,000 to 69,700 ha (Taliansky *et al.* 2000).

Peanut is the only known natural host of the rosette disease complex. Hosts, which are infected under laboratory conditions, are listed by Taliansky *et al.* (2000)

Three major forms of rosette disease have been reported: Chlorotic rosette (Storey and Bottomley, 1928), green rosette (Hayes, 1932) and mosaic rosette (Storey and Ryland, 1955; Reddy, 1991).

Chlorotic rosette was recorded throughout Africa south of the Sahara. Considerable variation has been observed within chlorotic rosette (Fig.

16.1). Faint chlorotic mottling appears on young leaflets and subsequent leaflets show bright yellow color interspersed with green islands. As a result of shortening of internodes early-infected plants are severely stunted. Few pods are formed and these may not contain marketable seeds. Another form of chlorotic rosette, referred to as green mottle, is characterized by mild chlorotic mottling with isolated flecks. Older leaflets are dark green, remain small and show outward rolling of their margins. If the older plants are infected the symptoms may be restricted to apical branches.

Green rosette is known to occur mainly in West Africa, Uganda and Angola. Younger leaflets show mild mottling and isolated flecks. Older leaflets are reduced in size, show downward rolling and are not distorted. Plants are severely stunted and are darker green than healthy plants. Plants infected early with green rosette may show striking resemblance to a completely different virus disease, peanut clump disease.

Mosaic rosette was recorded only in East and Central Africa. Young leaflets show conspicuous mosaic symptoms. Later symptoms resemble those of chlorotic rosette. Stunting is less pronounced than for chlorotic rosette.

16.3.1 Transmission

The GRD complex is transmitted by *Aphis craccivora* in a persistent manner (circulative, non propagative). Aphids retain the ability to transmit for at least 15 days, probably for life. Rosette disease is caused by a complex of three agents, in which the groundnut rosette umbravirus (GRV) depends on groundnut rosette assistor luteovirus (GRAV) for transmission by *Aphis craccivora*. The main cause of rosette disease symptoms in plants is due to a third component, a satellite RNA, which depends on GRV for its replication but on GRAV for its encapsidation and transmission by aphids. The aphid may transmit GRAV only, GRV and satellite RNA, but not GRV alone. In peanut the ability of GRAV to assist aphid transmission of GRV depends on the packaging of GRV RNA in the coat protein of GRAV. The dependence of GRV on GRAV, therefore, is for acquisition as well as inoculation by *A. craccivora,* but not necessarily for transmission to a peanut plant (Taliansky *et al.* 2000).

16.3.2 Diagnosis

GRD can be identified in the field by its characteristic symptoms. Green rosette disease may be more difficult to identify based on symptoms because of its resemblance to peanut clump disease. GRV is mechanically transmissible. GRV with or without satellite RNA, gives symptoms on *Nicotiana benthamiana* ranging from almost none to necrotic spots, rings or target spots on inoculated leaves. On *N. clevelandii* GRV induces local chlorotic or necrotic sports or rings followed by systemic chlorosis or necrosis. *Chenopodium amaranticolor* is the best local lesion host.

The GRD complex is readily transmitted by *Aphis craccivora*, distinguishing it from peanut clump. GRV and sat RNA can be identified by RT-PCR and dot blot hybridization assays.

16.3.3 Purification

A purification procedure has been developed for GRAV (Rajeshwari *et al.* 1987). ss RNA can be purified by following the procedure described by Murant *et al.* (1988).

16.3.4 Viral Characteristics

Groundnut rosette virus (GRV): GRV is a member of genus *Umbravirus,* which comprises viruses that depend on a helper virus from the Luteoviridae (GRAV) for transmission by aphids. In plants infected with GRV alone no nucleoprotein particles typical of those observed for plant viruses can be detected by electron microscopy. The genome of GRV has 4019 nucleotides and contains four large ORFs. GRV-infected plants contain, in addition to the full length genomic RNA, two subgenomic RNA species of similar size, which include the sequences of ORFs 3 and 4 and the 3'-terminal untranslated region (Taliansky *et al.* 1996).

GRV satellite RNA: Satellite RNAs from five different isolates have been sequenced. Their length ranged from 895 to 902 nucleotides and was at least 87% identical to one another in sequence. The satellite RNA contained up to 5 short ORFs, although only one ORF was present in 10 clones. It is unlikely that any of these ORFs are translated *in vivo*.

The nucleotide sequences of the overlapping ORFs 3 and 4 of GRV were highly conserved from isolates within a geographic region but less conserved between isolates from the two distinct geographic regions. On the basis of comparison of ORFs 3 and 4, Malawian GRV isolates are considered to be distinct from Nigerian GRV isolates. The satellite RNA sequence from isolates from different geographic regions showed clustering within the country of origin as well as within isolates from two distinct regions of Malawi (Deom *et al.* 2000)

16.3.5 Epidemiology

Aphis craccivora is the principal vector of the groundnut rosette disease complex. This aphid is ubiquitous in Africa. Females of this aphid reproduce parthenogenetically throughout the year. The rate of reproduction is dependent on such climatic factors as temperature and the nutritional status of the plant host. The aphid can transmit the virus up to two weeks and probably for life (Misari *et al.* 1988; Naidu *et al.* 1999). Till now it is not known where the primary source of inoculum comes from. Possible sources are infected peanut plants surviving between cropping seasons and peanut crops raised in off-season under irrigated conditions. An alternative host other than peanut is suspected to be present. Therefore GRD is regarded as a polycyclic disease. None of the agents involved in GRD complex are seed-transmitted. The secondary spread of rosette within the crops is well documented. Early infected plants form foci for aphids to spread the disease. The disease spread can be influenced especially by the crop density, age of crop, aphid population and the efficiency with which they can transmit the disease complex (Naidu *et al.* 1999).

16.3.6 Control

Cultural practices: Infected plants, especially those of green rosette, survive longer than healthy plants. These are not harvested and thus can form a primary source of inoculum. Additionally self-sown peanut plants that survive through the dry season also provide a source of inoculum. Alternative hosts of the virus and the vector exist in Africa. These sources have not been identified (Gibbons, 1977; Reddy, 1998). It is likely that viruliferous aphids disseminated by the moving rainfall fronts are responsible for early infection. Therefore ground-keepers and self sown peanut plants after harvesting the main crop must be destroyed. Early sowing and uniform spacing can reduce rosette incidence (Booker, 1963; A'Brook, 1964 and Farrell, 1976). Several insecticides have been shown to be efficient in reducing aphid population and in containing the spread of the rosette disease. However the cultural practices as well as the application of insecticides are often not adopted by marginal farmers. The best approach to control rosette disease lies in the development of cultivars resistant to one or more of the causal agents and/or their aphid vector (Reddy, 1998).

Rosette resistant cultivars: The majority of the rosette resistant cultivars developed to date were derived from genotypes located in the border region between Cote d' Ivoire and Burkina Faso. The plants were highly resistant and resistance is controlled by two independent recessive genes and is effective against both chlorotic and green forms of rosette (Nigam and Bock 1990; Olorunju *et al.* 1992). Unfortunately, the cultivars developed were late maturing (69-101, RMP 12, RMP 40 and RG1 and Spanish types KH 149-A, KH 241-C, KH 241-D, CN 94C, QH-243C) and therefore not suitable for planting in most parts of sub-Saharan Africa, where seasons are short and rainfall erratic. It is essential to develop rosette resistant varieties with early maturity and high yield suitable for small holders in Africa. The following early maturing lines ICG 7638, ICG 7727, 7632, 7636, 7625, 7637 were found to be resistant (Subrahmanyam *et al.*1998; Olorunju *et al.* 2001).

All these genotypes were found to be resistant only to GRV and its satellite RNA. Many wild *Arachis* species resistant to all the 3 components of the disease were recently identified (Subrahmanyam *et al.* 2001). One peanut genotype, EC 36892, was also found to be resistant to the aphid vector. It would be desirable to combine virus resistance with vector resistance.

16.4 *Peanut clump virus* Genus *Pecluvirus*

Synonyms: Peanut clump pecluvirus; Indian peanut clump pecluvirus

Peanut clump virus was first reported in 1927 in peanut in India. Subsequently in 1931, a similar disease was reported from West Africa.

The virus has been found in West Africa and in the Indian subcontinent. The clump virus that occurs in the Indian subcontinent is referred to as the *Indian peanut clump virus* (IPCV) and the virus that occurs in Africa is called *Peanut clump virus* (PCV). PCV occurs in Burkina Faso, Niger, Mali, Cote d'Ivoire, Gambia and Senegal. In India it is known to occur in the states of Andhra Pradesh, Gujarat, Punjab, Rajasthan and Tamil Nadu and from Pakistan in the states of Punjab and Sindh. The annual losses due to clump on a global scale have been estimated to exceed 38 million US dollars.

IPCV and PCV have wide host ranges. They can infect monocotyledonous and dicotyledonous plants. These include peanut, wheat, barley, maize, sorghum, pearl millet, finger millet and pigeonpea.

Symptoms induced by IPCV isolates in several hosts may differ, as, in *Canavalia ensiformis, Nicotiana benthamiana* and *N. clevelandii* and the hybrid N. *clevelandii* X *N. glutinosa.* PCV isolates from Senegal, Burkina Faso and Niger differed markedly in symptoms produced in *Chenopodium amaranticolor*.

Symptoms produced by IPCV and PCV in peanuts are similar. Symptoms of clump disease can be confused to those by green rosette. Plants affected by clump disease stand out in the field because of their severe stunting and dark green appearance. These plants often occur in patches and the disease recurs in the same area of the field in successive peanut crops. Initial symptoms appear on young leaflets as mottling, mosaic and chlorotic rings. However, when these leaflets mature, they turn dark green with or without faint mottling. Plants infected when they are young become severely stunted. Late infected plants may not show conspicuous stunting but appear dark green with faint mottling on young leaflets. In late-infected plants clumping may be only restricted to few branches. Infected plants can produce flowers and if pods are formed, they are not well developed. Yield loss up to 60% has been recorded even from late-infected plants.

16.4.1 Transmission

All IPCV and PCV isolates can be transmitted by mechanical sap inoculation, though not always successfully. PCV and IPCV are transmitted through seed. IPCV is transmitted through seed of pearl millet, finger millet, foxtail millet, wheat and maize. IPCV has been shown to be transmitted by *Polymyxa graminis*. However, transmission of PCV by *Polymyxa* so far has not been reported.

16.4.2 Diagnosis

PCV can be identified in the field by its symptoms, especially in countries where green rosette is not present The occurrence of severely stunted dark green plants in well defined patches in the field especially in light soils is characteristic for peanut clump disease. *Chenopodium amaranticolor* is a good diagnostic host for PCV. IPCV and PCV can be identified by serological tests. Both the viruses exist as serologically distinct isolates. IPCV isolates can be grouped into three distinct serotypes namely IPCV-H (Hyderabad), IPCV-D (Durgapura), and IPCV-L (Ludhiana). All IPCV isolates are distinct from PCV isolates, and *vice versa*. By using monoclonal antibodies, a number of PCV isolates have been placed in five distinct species. None of seven monoclonal antibodies produced against PCV reacted in triple antibody sandwich ELISA with IPCV-D (Manohar *et al.* 1995; Huguenot *et al.* 1989).

16.4.3 Purification

IPCV can be purified by using the procedure described by Reddy *et al.* (1985).

16.4.4 Viral Characteristics

Peanut clump virus is rod shaped with lengths of about 245 nm and 190 nm and diameter of about 21 nm.

RNA1: The 5' ORF encodes a large protein of 131 kd. The ORF is followed by another that is in frame with the termination codon of the 131 kd polypeptide and which is expressed by read-through of the UGA termination codon to generate a fusion polypeptide of 190 kd, that contains the 130 kd polypeptide at its 5'end. A third ORF that encodes a 15 kd polypeptide is downstream of the 190 kd ORF and separated from it by a noncoding region of about 60 nucleotides.

RNA2: The RNA 2 contains 5 ORFs. The ORF near the 5' end codes for the coat protein gene. The second ORF overlaps the first ORF by 2 nucleotides and encodes a protein of 39 kd. This protein is suspected to be involved in the transmission of virus by its fungal vector. Further downstream, separated by a 135 nucleotide intergenic region, is a triple gene block sequence that codes for proteins of 51 kd, 14 kd, and 17 kd. These are implicated in the movement of the virus from cell to cell. The 3' noncoding regions for RNA 1 are 2998 nucleotides and 275 nucleotides for RNA 2. The last 96 nucleotides are identical in both RNAs. The noncoding regions differ in size among isolates from the serotypes. Sequences of RNA2 are more diverse, between 390 and 500 nucleotides. RNA1 and RNA2 have similar 5' noncoding region of six to seven nucleotides and these sequences are shared between *Pecluvirus* species. The 3' noncoding regions are 300 nucleotides in length and the 100 terminal nucleotides are almost identical among all *Pecluvirus* RNAs sequenced so far. It should be possible to develop a nucleic acid probe, which can detect all the currently known *Pecluviruses*. The sequence of RNA 1 of IPCV H is similar to that of PCV and corresponding encoded polypeptides are 88%, 95% and 75% identical (Herzog *et al.* 1994).

16.4.5 Epidemiology

PCV is soil-borne, transmitted by *Polymyxa graminis*. The nature of the vector and its survival as highly resistant resting spores, or sporosori, is responsible for the patchy appearance of the disease and its recurrence year after year. Although the fungus can transmit the virus to dicotyledonous plants it does not extensively colonize their roots and resting spores are seldom detected in such plants. Also, roots of naturally virus-infected peanut plants failed to induce the disease when incorporated into sterile sand, whereas infected sorghum and pearl millet roots could. For these reasons, dicotyledonous plants are considered as "fortuitous" hosts that are unlikely to contribute to build-up of virus inoculum. In contrast monocotyledonous hosts such as maize, pearl millet and sorghum are regarded as "preferred" hosts for *P. graminis* because of its high incidence and multiplication in them (Delfosse *et al.* 1996; Legreve 1999). Additionally, the virus is also transmitted through the seed of cereal crops. Peanut clump disease often occurs on crops raised during the rainy season. Ambient temperatures during this season range from 25 to 40°C. The optimum temperature for infection by *P. graminis* is between 27 and 30°C. Below 23° C infection does not occur and fungal development is delayed. This explains the reason for the absence of peanut clump disease on peanut crops raised during post-rainy season in India, when ambient temperatures range between 15 to 30 C. Rainfall appears to play a significant role in natural infection by *P. graminis* and IPCV. *P. graminis* is known to be favored by alternate watering and drainage. Soils where peanut clump disease occurs are generally high-textured permitting excellent drainage. In such soils adequate levels of free water for *Polymyxa* zoospore movement becomes available with the occurrence of high rainfall. Continuous cropping with fortuitous hosts such as peanut, cowpea and pigeonpea is likely to reduce the inoculum in the soil. Intercropping with susceptible cereal crops can contribute to increase in incidence and spread of clump. Apparently, pecluviruses are not typical peanut viruses but actually graminaceous viruses that possibly coevolved in tropical and subtropical areas with wild grasses and cereal crops such as millets and sorghum (Delfosse *et al.* 2002).

16.4.6 Control

Peanut clump is a persistent soil-borne virus disease, difficult to control. Satisfactory control can only be achieved by adopting cultural practices over a period of time. The results from epidemiological studies (Delfosse *et al.* 1999, 2002) led in India to formulation of the following cultural practices:

1. Early sowing of the peanut crops before the onset of monsoon rains
2. Use of bait crops, as pearl millet. These are planted soon after the onset of monsoon preferably under irrigation and up rooted in three weeks after germination. This permits the infection by *P. graminis* but not the development into sporosori. As a result, inoculum load was reduced. Peanut crops grown subsequently showed lower incidence of clump disease than those where trap cropping was not adopted.
3. Sowing peanut during the post-rainy season, continuous rotation with dicotyledonous crops and soil solarization during hot summer months. Soil solarization is achieved by covering well-irrigated soils with a transparent polyethylene sheet for at least three weeks.

Host plant resistance: No resistance to IPCV was found in 9000 *Arachis* germplasm tested. Recently, resistance was identified in wild *Arachis* species (P. Delfosse personal communication). However, it has yet to be incorporated into cultivated peanut. Transgenic approaches utilizing polymerase genes or other nucleotide sequences may be an answer in the future. The report on the incorporation of IPCV-H coat protein gene into peanut by non-conventional methods (Sharma and Anjayya, 2000) has yet to be confirmed. The transgenic peanut plants generated were tested under laboratory conditions and were found to be susceptible for IPCV-H (DVR Reddy Personal communication).

16.5 *Tobacco streak virus* Genus *Ilarvirus* (TSV)

TSV is known to occur in peanuts in India, Pakistan, South Africa and Brazil. The virus is economically important in India and caused over US$ 65 million worth of crop losses in 2000, when it occurred in an epidemic on 225,000 ha of peanuts in Andhra Pradesh, (Reddy *et al.* 2002)

Tobacco streak virus has a wide host range, and is economically important in soybeans, tobacco, sunflower, marigold, cowpea, mung beans, green gram and black gram.

Initial symptoms on peanut appear a week after inoculation. Young leaflets show large necrotic lesions, which subsequently coalesce and cover the entire lamina (Fig. 16.2). This leads to the total necrosis of the leaflets followed by stem necrosis. In severe cases the entire plant becomes necrotic (Fig. 16.3). Necrotic spots also occur on the shells of the pods. In early infections the plants are severely stunted and the yield is substantially reduced. In late infected plants only terminal leaflets or a single branch may show necrosis.

16.5.1 Transmission

The virus is readily transmitted by mechanical sap inoculations and by a range of thrips species. These include *Scirtothrips dorsalis*, *Frankliniella schultzeii*, *F. fusca, Thrips palmi, Megalurothrips usitatus* (Prasada Rao *et al.* 2003). The virus is transmitted through infected pollen. The virus is seed transmitted in soybean, but seed transmission in peanut has not been reported.

16.5.2 Diagnosis

Field symptoms by TSV resemble those caused by PBNV in that both show necrosis of terminal leaflets. TSV often induces total stem necrosis and necrotic lesions on pods., PBNV causes axillary shoot proliferation and distorted leaflets. This has not been observed in TSV infections.

TSV can be readily identified using serological techniques. High quality polyclonal antisera have been produced. The virus can be detected by nucleic acid hybridization technique in leaf and pollen extracts.

16.5.3 Purification

A purification procedure has been reported for TSV (Reddy *et al.* 2002).

16.5.4 Viral Characteristics

Virus particles are quasi-isometric or occasionally bacilliform, 25-35 nm in diameter. The molecular mass of capsid protein is 28 kd. All purified virus preparations also contain a minor polypeptide of 57 kd which presumably is a dimer of 28 kd. The genome of TSV consists of three linear positive sense ssRNAs with 5'-terminal cap structures. The 3' termini are not polyadenylated, generally are highly conserved within a species or isolate, and form strong secondary structures. They are either tRNA-like and can be aminoacylated. The total genome length is approximately 8 kb (van Regenmortel *et al.* 2000).

16.5.5 Epidemiology

TSV is efficiently transmitted by several thrips species. Transmission can only occur through infected pollen. It has been shown in India that , *Parthenium,* a ubiquitous weed, harbors the virus and the thrips that visit their flowers carry the virus inoculums. Secondary spread does not occur in peanut because the pollen from peanut flowers is not an efficient source of inoculum. TSV incidence on peanut could be correlated with the proportion of infected parthenium near the fields (Prasada Rao *et al.* 2003)

16.5.6 Control

Removal of *Parthenium* plants from the vicinity of peanut fields is expected to reduce the disease incidence and in field trials in India, removal of *Parthenium* decreased TSV incidence. However, *Parthenium* removal is not a practical measure. Border cropping with 10-12 lines of a quick growing cereal such as pearl millet or sorghum or maize can reduce the TSV incidence. Additionally intercropping with these crops also reduces the TSV incidence. Cultivation of peanut near sunflower and marigold should be discouraged. Currently no peanut genotype resistant to TSV has been reported. In initial laboratory tests ICG 94379 was found to be resistant (DVR Reddy, AS Reddy and SN Nigam unpublished data). However, these findings need further confirmation. Resistance to TSV can be incorporated through non-conventional approaches. Both CP genes as well as movement protein genes can be used very efficiently (Cooper *et.al.* 1995). However, transgenic plants are currently not available for cultivation.

16.6 *Peanut mottle virus Genus Potyvirus* (PMV)

Synonyms: Peanut mottle, peanut mosaic

Peanut mottle was first reported from United States in 1961, and is present in all major peanut-growing countries. PMV is considered to be economically important on a global scale (Demski and Reddy 1997). In Georgia losses due to PMV are estimated at 5-6% and in India in susceptible cultivars crop losses may reach 40%.

PMV produces a range of symptoms on peanut. The younger leaflets may show mild mottle or mosaic of irregular dark-green islands. In older leaflets mosaic symptoms are not obvious but can be seen in transmitted light. In some genotypes, inter-veinal depression and inward curling of the margins of leaflets can occur. Infected plants, including those infected early, are not severely stunted. The number and size of pods are reduced.

16.6.1 Transmission

The virus is mechanically transmissible. It is also transmitted in a non-persistent manner by *Aphis craccivora, A. gossypii, Myzus persicae, Hyperomyzus lactucae, Rhopalosiphum padii* and *R. maides.*

16.6.2 Diagnosis

PMV can be detected by its characteristic reddish-brown lesions produced on *Phaseolus vulgaris* (Topcrop). ELISA is commonly used to detect PMV in leaves as well as in seed. An immunocapture-RT-PCR technique that detects the virus in seed lots and can distinguishing it from PStV has been reported (Gillaspie *et al.* 2000). This method has been shown to be more sensitive than ELISA for virus detection in seed.

16.6.3 Purification According to Rajeshwari *et. al.* (1983).

16.6.4 Viral Characteristics Particles are flexuous rods about 750 nm in length and 12 nm in diameter. The coat protein has an apparent molecular mass of 32 kd. The nucleic acid is approximately 9500 nucleotides in size (Teycheney and Dietzgen, 1994).

Strains: The mild mottle isolate reported is the most widely distributed one. Additionally 5 more isolates of PMV on the basis of particle morphology, serological relations and host range were reported. The 3' terminal of 1247 nucleotides of an Australian isolate of PMV has been sequenced. Comparison of nucleotide sequences of PMV and PStV revealed a similarity of 64.4% for the coat protein gene and 34.6% for the 3' non-coding region (Teycheney and Dietzgen, 1994).

16.6.5 Epidemiology

PMV is transmitted under field conditions by a number of aphids. The natural host-range of virus includes several legume crops (cowpea, phaseolus bean, soybean, lupine) and weeds that occur in peanut cropping systems. The virus is seed transmissible at a low rate, less than 1.0%. PMV is seed-transmitted also in cowpea, mung bean and common bean. Therefore these hosts are likely sources of inoculum. There is evidence that seed-transmisson in peanuts can be a source of secondary spread. The primary source of inoculum has been shown to come from seed (Demski *et al.* 1975)

16.6.6 Control

No resistance to PMV was found in cultivated peanuts. However, the genotypes ICG 5043 and NCAC 2240 are tolerant with no significant effects on yield after infection. One high yielding peanut genotype ICG 89336 with tolerance to PMV is currently available. Since the seed can provide the primary source of inoculum, genotypes, which do not transmit the virus to seed, are useful in containing the spread of virus. Peanut genotypes ICG 2716 (EC 76446-292), ICG 7013 (NCAC 17133) and ICG 1697 (NCAC 17090) do not transmit the virus through seed. They have been used in conventional breeding to transmit this trait to high yielding peanut cultivars. The seed of advanced breeding lines from these crosses have been tested for frequency of PMV transmission. ICGS 65 and ICGS 76 were identified as non-seed transmitting high yielding peanut genotypes. *Arachis chacoense* and *A. pusilla* and wild *Arachis* species have been found to be resistant to PMV. This resistance has yet to be transferred to cultivated peanuts.

16.7 *Peanut stripe virus* Genus Potyvirus (PStV)

A virus resembling peanut stripe was first reported by Xu *et al.* (1983) from China. Subsequently Demski *et al.* (1984) reported a potyvirus isolated from peanut seed imported from China and named it "peanut stripe potyvirus". Peanut stripe virus (PStV) is endemic in East- and Southeast Asia. It is widely distributed in all peanut growing areas in China. Currently PStV is known to occur in the majority of peanut growing countries including India.

PStV can severely affect yields of peanut in Southeast Asia. Yield losses can reach up to 70% in early-infected plants. An annual yield loss in north China was estimated at 200,000 tons of pods (Mc Donald *et al.* 1998).

Susceptible natural hosts are *Centrosema pubescens*, *C. macrocarpum, Calopogonium caeruleum, Crotalaria striata, Desmodium siliquosum,* and *Pueraria phaseoloides* (Wongkaew, 1986). The virus causes stripes and green banding symptoms along lateral veins (Demski *et al.* 1993). PStV isolates from different regions of the world can produce distinct symptoms. The stripe isolate produces discontinuous stripes along the lateral veins on young quadrifoliates. Older leaflets show striping, mosaic in the form of green islands, and an oak leaf pattern. For most other PStV isolates initial symptoms appear as chlorotic flecks followed by mild mottle, blotch or chlorotic ring mottle symptoms. An isolate that produces necrotic symptoms was reported.

16.7.1 Transmission

The virus is readily transmitted by mechanical sap inoculation and by many aphids. *Aphis craccivora, A. gossypii* and *Myzus persicae* are efficient vectors. PStV transmission through peanut seed can reach 37%. However, in naturally infected plants seed transmission rates are lower (0-7%), generally less than 4%.

16.7.2 Diagnosis

PStV produces chlorotic or necrotic lesions on *Chenopodium amaranticolor*. This virus does not infect *Phaseolus vulgaris* (cv Topcrop) and peas unlike PMV. High quality sera detect the virus in leaves and seed. Antisera to the virus react also strongly with blackeye cowpea mosaic potyvirus, soybean mosaic potyvirus, clover yellow vein mosaic potyvirus and adzuki bean mosaic potyvirus (Demski *et al.* 1993).

A RT-PCR assay was developed, detecting pgs of virus (Dietzgen *et al.* 2001). An immunocapture RT-PCR for the detection of PStV in seed lots was shown to be more sensitive than ELISA utilizing polyclonal antibodies (Gillaspie *et al.* 2000).

16.7.3 Purification: Accordong to Demski *et al.* 1993.

16.7.4 Viral Characteristics

Virus particles are flexuous rods 12nm wide and 752 nm long. Several PStV isolates have been described from different countries. These were based on a wide range of symptoms produced on peanut which include stripe, mild mottle, blotch and chlorotic rings. Currently PStV is considered to be a peanut infecting strain of bean common mosaic virus (BCMV). A full-length cDNA clone of a blotch isolate of PStV was sequenced and shown to contain 9852 nucleotides (Flasinski *et al.* 1996). PStV isolates from Thailand considered to be strains of BCMV that can be distinguished from bean infecting strain of BCMV and blackeye cowpea mosaic virus through differences in the nucleotide sequence and host range (Higgins *et al.* 1998). The 3' region of a necrotic strain (1356 nucleotides) of PStV was sequenced and found to have 92.95% homology to a blotch strain at the nucleotide sequence level. The polymorphism that exist in the coat protein gene could be utilized in designing oligonucleotides which are specific for the necrotic strain (Pappu *et al.* 1998). Coat protein genes and a portion of the 3' non-coding region of several PStV isolates have been sequenced.

These strains are related to each other despite the variation in the symptoms caused by them (Higgins *et al.* 1998; Pappu *et al.*1998).

16.7.5 Epidemiology

PStV is naturally transmitted by aphids in a non-persistent manner, probably the only way of spread from its primary source. PStV can be transmitted through seed up to 37%, but generally less than 0.7%. The primary sources of inoculum are seed (Demski *et al.* 1993)

16.7.6 Control

PStV occurs in peanut crops throughout Southeast Asia and China and is seed transmitted.. Resistance to PStV could not be found in cultivated peanut., Attempts to locate genotypes which do not transmit PStV through seed have not been successful. Since the primary source of inoculum is seed, planting should be done with seed lots obtained from disease free areas. Several wild *Arachis* sp. have been shown to be resistant, but this resistance has not been transferred to cultivated peanuts (Reddy, 1998). Transgenic peanuts, which are resistant to PStV, have been developed (R. Dietzgen personal communication). To our knowledge they are not yet been commercialized.

16.8 Future Thrust

In this review we dealt with the major aspects of six economically important viruses of peanut. Peanut is a crop most important for the welfare of resource poor farmers in developing countries. Therefore disease control measures must be practical, affordable and suitable for situations in developing countries. In the case of PBNV sources of resistance are available in peanut germplasm and they have been fully exploited for incorporating resistance into widely adapted peanut cultivars. Resistance recently identified in wild Arachis species is yet to be transferred to cultivated peanut. Many peanut genotypes have been identified with resistance to TSWV and several advanced breeding lines with field resistance to TSWV are available. Therefore in the case of PBNV and TSWV incorporation of resistance by non-conventional methods does not require high priority. Application of acibenzolar-s-methyl (ASM) to tobacco seedlings prior to transplanting resulted in significant reduction in TSWV under field conditions in Georgia (Pappu *et al.* 2000). ASM is known to trigger systemic acquired resistance and it is possible that it may have similar suppressive effect on TSWV and PBNV in peanut. Rosette is known to occur for nearly 100 years. Excellent progress has been made in identifying the causal agents and in developing sensitive diagnostic tools for the three agents that cause the disease. The epidemiology of rosette disease is an enigma and is yet to be solved. Research in Africa in 50s, 60s and 70s was focused on understanding the epidemiology and in formulating cultural methods for disease control. During these three decades lack of diagnostic tools hampered the efforts to identify alternative sources of inoculum. Despite the availability of sensitive diagnostic tools, efforts during the last decade have failed in identifying the alternative source of inoculum. If the alternative sources of inoculums are found, it may be possible to contain the severity of epidemics, which continue to occur in Africa.

Peanut clump is a restricted disease and the virus cannot establish in new areas through inoculum from peanut seed. Efforts by ICRISAT in collaboration with Universite Catholoque de Louvain (UCL), Belgium, resulted in understanding the epidemiology, have lead to the development of effective cultural methods of disease control. An important constraint has been the lack of sources of resistance to this virus except in wild Arachis species. For generating host plant resistance to clump disease, non-conventional approaches are expected to yield encouraging results. Viral polymerase or those genes, which can offer broad-based resistance to this virus, are most suitable for incorporating transgenic resistance. In our opinion coat protein mediated resistance may be of limited value because of the diversity that occurs among various clump virus isolates. Nonetheless it is important to mention that epidemics due to clump disease does not occur due to the soil borne nature. Additionally spread of the disease even under ideal conditions was found to be negligible in observations spanning over nine years on ICRISAT farm.

TSV offers the greatest challenge for research workers in Asia. The virus can be devastating and can destroy large areas of peanut crop in a short time. Control methods, which can be adopted by small-scale farmers, are yet to be developed. A major stumbling block appears to be that the virus found shelter in a pernicious weed, parthenium, which is very difficult to eliminate. Although resistance to the virus has been identified in one peanut genotype, it has to be confirmed under field tests at multi-locations. Therefore high priority must be given to develop peanut cultivars with resistance to TSV. We believe the best option is to concentrate on developing transgenic peanut lines with TSV resistance, by using coat protein or movement protein genes. This approach is expected to yield very encouraging results.

PMV and PStV occur in all major peanut-producing countries. They are seed transmitted and can therefore move to new locations. They are often present in germplasm collections and highly sensitive methods are available for their detection in seed. The limited economic impact observed for these two viruses do not warrant concentrated research efforts towards the development of strategies for disease management.

Acknowledgement: We thank Dr. H.R. Pappu, Associate Professor, Washington State University, Pullman, for his valuable comments.

References

A'Brook, J. 1964. The effect of planting date and spacing on the incidence of groundnut rosette disease and of the vector, *Aphis craccivora* Koch, at Mokwa, Northern Nigeria. *Ann Appl Biol.* 54: 199-208.

Bandla, M.D., Westcot, D.M., Chenault, K.D., Ullman, D.E., German, T.L., Sherwood, J.L. 1994. Use of monoclonal antibody to the nonstructural protein encoded by small RNA of tomato spotted wilt tospovirus to identify viruliferous thrips. *Phytopathology*. 84: 1427-1431.

Bertrand, P. F. 1996. "*Georgia plant disease loss estimates*." Plant Pathology Pub. No. 96-007, Univ. Georgia Coop. Ext. Serv., Coll. Agric. Envir. Sciences, Athens, GA.

Black, M.C. 1991. Effects of spotted wilt in selected peanut cultivars. *Proc. Am. Penaut. Res. Educ. Soc.* 23:52

Booker, R.H. 1963. The effect of sowing date and spacing on rosette of groundnut in Northern Nigeria, with observations on the vector, *Aphis craccivora. Ann. Appl. Biol.* 52: 125-131.

Brown, S., Todd, J., Culbreath, A., Baldwin, J., Beasley, J. and Pappu, H. 2000. Tomato spotted wilt of peanut: Identifying and avoiding high-risk situations. *Univ. Georgia Coop. Ext. Serv. Bull.* 1165. pp 11.

Cho, J.J., Mau, RFL, Mitchell, WC Gonsalves, D. and Yudin L.S. 1987. Host list of plants susceptible to tomato spotted wilt virus (TSWV*). Univ. Hawaii Coll. Trop. Agric. Human Resources. Res Ext Ser. 078*

Cooper, B., Lapidot, M., Heick, J.A.,Dods, J.A., and Beachy, R.N. 1995. A defective movement protein of TMV in transgenic plants confer resistance to multiple viruses whereas the functional analog increases susceptibility. *Virology* 206: 307-313

Costa, A.S. 1941. Uma molestia de virus da amendoim (Archis hypogea L.). *A monchanector. O. Biologica* 7: 249-251, 1941

Culbreath, A.K., Todd, J.W., Brown, S.L., Baldwin, J.A. and Pappu, H.R. 1999. A genetic and cultural "package" for management of tomato spotted wilt virus in peanut. *Biological and Cultural Tests.* 14:1-8.

Culbreath, A.K., Todd, J.W., Demski, J.W. and Chamberlin, J.R. 1992. Disease progress of spotted wilt in peanut cultivars florunner and southernrunner *Phytopathology* 82: 766-771

Culbreath, A.K., Todd, J.W., Gorbet, D.W., Brown, S.L., Baldwin, J., Pappu, H.R., and Shokes, F.M. 2000. Reaction of peanut cultivars to spotted wilt. *Peanut Sci.* 27:35-39.

Delfosse, P., Devi, P.S., Reddy, A.S., Risopoulos, J., Doucet, D., Legreve, A., Maraite, H. and Reddy, D.V.R. 1996. Epidemiology of Indian peanut clump virus transmitted by polymyxa sp.

Delfosse, P., Reddy, A.S., Legreve, A., Devi, P.S., Thirumala Devi, K., Maraite, H. and Reddy, D.V.R. 1999. Indian peanut clump virus infection on wheat and barley: Symptoms, yield loss and transmission through seed. *Plant Pathol.* 48: 273-282

Delfosse, P., Reddy, A.S., Thirumala Devi, K., Legreve, A., Risopoulos, J., Doucet, D., Maraite, H. and Reddy, D.V.R. 2002. Dynamics of *Polymyxa graminis* and Indian peanut clump virus (IPCV) infection on various monocotyledonous crops and groundnut during the rainy season. *Plant Pathol.* 51: 546-560

Demski, J.W. and Reddy, D.V.R. 1997. Diseases caused by viruses. In *Compendium on Peanut Diseases*, Eds: N.K.B. La, D.M. Porter, R. Rodriguez-Kabana, D.H. Smith and P. Subrahmanyam. Amer. Phytopath. Soc., pp 53-59

Demski, J.W., Deom, M., Olorunju, P., Wongkaew, S. 1993. Peanut viruses: etiology, epidemiology, and nature of resistance. In *Annu.Rep.Peanut.Coll.Res.SupportProgram-CRSP*. Griffin, Ga. Univ. Georgia, Georgia Exp. Sta. p. 119-138.

Demski, J.W., Reddy, D.V.R., Sowell, G., Bays, D. 1984 Peanut stripe virus-a new seed-borne potyvirus from China infecting groundnut (Arachis hypogea) *Ann Appl Biol* 10 : 495-501

Demski, J.W., Smith, D.H. and Kuhn, C.W. 1975. Incidence and distribution of peanut mottle virus in peanut in the United States. *Peanut Science* 2: 91-93

Deom, C.M., Naidu, R.A., Chiyembekeza, A.J., Ntare, B.R., Subrahmanyam, P. 2000. Sequence diversity within the three agents of groundnut rosette disease. *Phytopathology* 90 : 214-219.

Dietzgen, R.G., Callaghan, B.,Higgins, C.M., Birch, R.G.,Chen and K., Xu,-Z. 2001. Differentiation of peanut seedborne potyviruses and cucumoviruses by RT-PCR. *Plant Dis*. 9: 989-992.

Dyer, RA. 1949. Botanical surveys and control of plant disease. *Farming in S. Africa* 24:121

Farrell, J. A. K. 1976. Effect of groundnut crop density on the population dynamics of *Aphis craccivora* Koch (Hemiptera, Aphididae) in Malawi. *Bull. Entomol. Res.* 66 : 317-329.

Flasinski, S., Gunasinghe, U.B., Gonzales, R.A., Cassidy, B.G. 1996. The cDNA sequence and infectious transcripts of peanut stripe virus. *Gene.* 71: 299-300.

Gibbons, R.W. 1977. Groundnut rosette virus. *In: Diseases of Tropical Corps*. J. Kranz, J. Schmutterer, and W. Koch (eds.) Paul Parey, Berlin. pp 19-21.

Gillaspie, A.G. Jr., Pittman, R.N., Pinnow, D.L. and Cassidy, B.G. 2000. Sensitive method for testing seed lots for peanut stripe and peanut mottle viruses by immunocapture reverse transcription-polymerase chain reaction. *Plant Dis.* 84: 559-561

Golbitz, P. 2003. World Major Oilseeds: Area, Yield & Production by Main Producers. *In Soya & Oilseed Bluebook* , Bar Harbor , M.E.: Soyatech, Inc. pp 23.

Gorbet, D.W., Shokes, F.M., Culbreath, A.K., Todd, J.W., and Whitty, E.B. 1999. C-99R - A new multiple disease resistant peanut cultivar. *Marianna North Florida Res. Educ. Center Res. Rep.* 99-2.

Gowda, S., Satyanarayana, T., Naidu, R.A., Mushegian, A., Dawson, W.O., Reddy, D.V. R. 1998. Characterization of the large (L) RNA of peanut bud necrosis tospovirus. *Arch. Virol.* 143:2381-90.

Groves, R. L., Walgenbach, J.F., Moger, J.W. and Kennedy, G.G. 2002. The role of weed hosts and tobacco thrips, *Franklinella fusca*, in the epidemiology of tomato spotted wilt virus. *Plant Dis.* 86: 573-582.

Halliwell, R.S. and Philley, G. 1974. Spotted wilt of peanut in Texas. *Plant Dis. Rept.* 58: 23-25

Hayes, T. R. 1932. Groundnut rosette disease in the Gambia. *Trop. Agric. Trinidad.* 9:211-217.

Herzog, E., Guilley, H., Manohar, S.K., Dollet, M., Richards, K., Fritsch, C., Jonard, G. 1994. Complete nucleotide sequence of peanut clump virus RNA 1 and relationships with other fungus-transmitted rod-shaped viruses. *J. Gen. Virol.* 75:3147-55.

Higgins, C.M., Cassidy, B.G., Teycheney, P.Y., Wongkaew, S., Dietzgen, R.G. 1998. Sequences of the coat protein gene of five peanut stripe virus (PStV) strains from Thailand and their evolutionary relationship with other bean common mosaic virus sequences. *Arch.Virol.* 143: 1655-1667.

Huguenot, C., Givord, L., Sommermeyer, G., and Van Regenmortel, M.H.V. 1989. Differentiation of peanut clump virus serotypes by monoclonal antibodies. *Res. Virol.* 140 : 87-102.

Jain, R.K., Pappu, H.R., Pappu, S.S., Reddy, M.K. and Vani, A. 1998. Watermelon bud necrosis tospovirus is a distinct virus species belonging to serogroup IV. *Arch. Virol..* 143: 1637-44.

Latham, L.J. and Jones, R.A.C. 1997. Occurrence of tomato spotted wilt tospoviuses in native flora, weeds and horticultural crops. *Aust. J. Agric. Res*. 48: 359-369

Legreve, A., Vanpee, B., Delfosse, P. and Maraite, H. 1999. High temperature during storage favors infection potential of resting spores of *Polymyxa graminis* of Indian origins. *Ann. Appl. Biol.* 134: 163-169.

Magbanua, Z.V., H.D. Wilde, J.K. Roberts, K. Chowdhury, J. Abad, J.W. Moyer, H.Y. Wetzstein, and W.A. Parrott. 2000. Field resistance to tomato spotted wilt virus in transgenic peanut (*Arachis hypogaea* L.) expressing an antisense nucleocapsid gene sequence. *Mol. Breed.* 6:227-236

Mandal, B., H. R. Pappu, A. K. Culbreath, C. C. Holbrook, D. W. Gorbet and J. W. Todd. 2002. Differential response of selected peanut genotypes to mechanical inoculation by Tomato spotted wilt virus. *Plant Dis.* 86: 939-944.

Mandal, B., Pappu, H. R., and Culbreath, A. K. 2001. Factors affecting mechanical transmission of *Tomato spotted wilt virus* to peanut (*Arachis hypogaea*). *Plant Dis.* 85:1259-1263.

Manohar, S.K., Dollet, M., Dubern. J. and Gargani, D. 1995. Studies on variability of peanut clump virus: symptomalogy and serology. *J. Phytopath* 143: 233-238.

McDonald, D., Reddy, D.V.R., Sharma, S.B., Mehan, V.K., Subrahmanyam, P. 1998. Diseases of groundnut. *The Pathology of Food and Pasture Legumes.* CAB International in association with the International Crops Research Institute for the Semi-Arid Tropics (ICRISAT), 63-124.

Misari, S. M., Abraham, J. M., Demski, J. W., Ansa, O. A. and Kuhn, C.W. Casper R and Breyel E. 1988. Aphid transmission of the viruses causing chlorotic rosette and green rosette diseases of peanut in Nigeria. *Plant Dis.* 72: 250-253.

Morwood, R. B. 1954. Peanut diseases. *Queensland J. Agric* 79: 267-270

Mound, L. A. 1996. The thysanoptera vector species of tospoviruses. *Acta Hortic*, 431: 298-309

Moyer, J. W. 1999. *Tospoviruses* (*Bunyaviridae*). Pages 1803-1807 In*: Encyclopedia of Virology.* A. Granoff and R. G. Webster, eds, Academic Press, San Diego, CA.

Murant, A.F., Rajeshwari, R., Robinson, D. J., and Raschke, J. H., 1988. A satellite RNA of groundnut rosette virus that is largely responsible for symptoms of groundnut rosette disease. *J Gen Virol.* 69: 1479-1486.

Naidu, R.A., Kimmins, F.M, Holt Robinson, D.J., Deom, C.M. and Subrahmanyam P. 1999. Spatio-temporal separation of groundnut rosette virus disease agents *Phytopathology* 89: 934-941.

Nigam S.N. and Bock K. R. 1990. Inheritance of resistance to groundnut rosette virus in groundnut (*Arachis hypogea* L.). *Ann appl. Biol.* 117: 553-560

Olorunju, P. E., Ntare, B. R., Pande, S. and Reddy, S. V. 2001. Additional sources of resistance to groundnut rosette disease in groundnut germplasm and breeding lines. *Ann. Appl.. Biol..* 139: 259-268

Olorunju, P.E., Kuhn, C.W., Demski, J.W., Misari, S.M. and Ansa, O.A. 1992. Inheritance of resistance in peanut to mixed infections of groundnut rosette virus (GRV) and groundnut rosette assistor virus and a single infection of GRV. *Plant Dis.* 76:95-100.

Pappu, H.R. 1997. Management of Emerging Virus Threats of Crops: Impact of Biotechnology in Controlling Tospoviruses. *Biotechnology and Development Monitor.* 32: 14-17. URL: http://www.biotech-monitor.nl/3207.htm

Pappu, H.R., Csinos, A.S., McPherson, R.M., Jones, D.C. and Stephenson, M.G. 2000. Effect of acibenzolar-S-methy and imidaclopridon suppression of tomato spotted wilt tospovirus in flue-cured tobacco. *Crop Protect.* 19 : 349-354.

Pappu, H.R., Todd, J.W., Culbreath, A.K., Bandla, M.D. and Sherwood, J. L. 1998. First report on the multiplication of tomato spotted wilt tospovirus in tobacco thrips, *Frankliniella fusca*. *Plant Dis.* 82: 1282.

Pappu, H.R., Pappu, S.S., Jain, R.K., Bertrand, P.F., Culbreath, A.K. McPherson, R and Csinos, A.S. 1998. Sequence characteristics of natural populations of tomato spotted wilt tospovirus infecting flue-cured tobacco in Georgia. *Virus Genes.* 17167-175.

Pappu, S.S., H.R. Pappu, A.K. Culbreath, and J.W. Todd. 1999. Localization of tomato spotted wilt tospovirus in peanut pod. *Peanut Science* 26: 98-100.

Peters D., Wijkamp, F Van de Wetering F., and Goldbach R. 1996 Vector relations in the transmission and epidemiology of tospoviruses *Acta Hortic.* 431: 29-43

Peters, D. 1998. An updated list of plant species susceptible to tospoviruses. Pages 107-110. *In: Recent Progress in Tospovirus and Thrips Research.* D. Peters and R. Goldbach eds. Wageningen, Netherlands.

Prasad Rao, R. D. V.J., Reddy, A.S., Reddy, SV., Thirumala-Devi, K., Chander Rao, S., Manoj Kumar, V., Subrahmanyam, K., Yellamanda Reddy, T., Nigam, S. N. and Reddy D V R (2003). Epidemiology of tobacco streak virus in India: Host range and transmission by thrips. *Ann. Appl. Biol.* (In press)

Rajeshwari, R., Iizuka, N., Nolt, B.L. and Reddy, D.V.R. 1983. Purification, serology and physico-chemical properties of a peanut mottle virus isolate from India. *Plant Pathol.* 32: 197-205

Rajeshwari, R., Murant, A.F., and Massalski, P.R. (1987). Use of monoclonal antibody to potato leafroll virus for detecting groundnut rosette assistor virus by ELISA. *Ann. appl. Biol.* 111: 353-358.

Reddy, A.S., Prasada Rao, R.D.V.J., Thirumala-Devi, K., Reddy, S.V., Subrahmanyam, K., Satyanarayana, T., Mayo, M.A., Roberts, I. and Reddy, D.V.R. (2001) First record of *tobacco streak ilarvirus* occurrence on peanut in India. *Plant Dis.* 86: 173-178

Reddy, A.S., Reddy, L.J., Mallikarjuna, N., Abdurahman, M.D., Reddy, Y.V., Bramel, P.J., and Reddy D.V.R. 2000. Identification of resistance to Peanut bud necrosis virus (PBN) in wild *Arachis* germplasm. *Ann. Appl. Biol.* 137:135-139.

Reddy, D.V.R. 1988. Groundnut virus diseases occurring in India. In *Monograph of Groundnut.* ed. By P.S. Reddy. Indian Coun. Agric. Res. New Delhi, India. pp 508-525

Reddy, D.V.R. 1991. Groundnut viruses and virus diseases: Distribution, identification and control. *Rev. Plant Pathol.* 70: 665-678.

Reddy, D.V.R. 1998. Control measures for the economically important peanut viruses. *In Plant Virus Disease Control.* Eds A. Hadidi, R.K. Khetarpal and A. Koganezawa. American Phytopathological Society Press, USA. pp 541-546

Reddy, D.V.R. and Wightman, J.A. 1988. Tomato spotted wilt virus, thrips transmission and control. Harris KR ed. *Adv. Disease Vector Res.* 5: 203-220

Reddy, D.V.R., Amin, P.W., Mc Donald, D. and Ghanekar, A.M. 1983. Epidemiology and control of groundnut bud necrosis and other diseases of legume crops in India caused by tomato spotted wilt virus. *In:Plant Virus Epidemiology.* Eds. R.T. Plumb and J.M. Thresh. Blackwell Scientific Publications, UK pp 93-102

Reddy, D.V.R., Demski, J.W., Wong Kaew, S. and Middleton, K. 1998. Occurrence of isolates of peanut stripe virus. In*: Proceed. First Coordinator Meeting Peanut Stripe Virus*. 1987, Malang, Indonesia

Reddy, D.V.R., Ratna, A.S., Sudarshana, M.R., Poul, F. and Kumar, I.K. 1992. Serological relationships and purification of bud necrosis virus, a tospovirus occurring in peanut (Arachis hypogaea L) in India. *Ann. Appl. Biol.* 120, 279–286.

Reddy, D.V.R., Robinson, D.J., Roberts, I.M. and Harrison, B.D. 1985. Genome properties and relationships of Indian peanut clump virus. *J. Gen Virol*, 66: 2011-2016.

Reddy, Sugunakar M., Reddy, D.V.R. and Appa Rao, A. 1968. A new record of virus disease on peanut. *Plant Dis. Reprt.* 52: 494-495

Satyanarayana, T., Mitchell, S.E., Reddy, D.V.R., Brown, S., Kresovin, S., Jarret, R., Naidu, R.A. and Demski, J.W. 1996. Peanut bud necrosis tospovirus RNA: complete nucleotide sequence, genome organization and homology to other tospoviruses. *Arch.Virol.*141, 85-98.

Satyanarayana T, Mitchell, S.E., Reddy, D.V.R., Kresovich, S., Jarret, R., Naidu, R.A., Gowda, S., Demski, J.W. 1996. The complete nucleotide sequence and genome organization of the M RNA segment of peanut bud necrosis tospovirus and comparison with other tospoviruses. *J Gen Virol.* 77: 2347-52.

Sharma, K.K. and Anjayya, V.V. 2000. An efficient method for the production of transgenic plants of peanut (Arachis hypogaea L.) through Agrobacterium tumefaciens-mediated genetic transformation. *Plant Sci.* 159: 7-19.

Sherwood, J.L., Bandla, M.D., Chenault, K.D., Ullman, D.E., Westcot, D.M. and German, T.L. 1995. Utility of antibodies to explore and control tomato spotted wilt virus. *In* Recent *Studies on Peanut Bud Necrosis Disease.* Ed by A.A.M. Buiel, J.E. Parlevliet and J.M. Lenne. ICRISAT publication, India. pp 25-33

Sherwood, J.L., German, T.L. Moyer, J.W., Ullman, D.E. and Whitfield, A.E. 2000. Tomato spotted wilt. pp. 1030-1031 *In: Encyclopedia of Plant Pathology*, O.C. Maloy and T. D. Murray, eds. John Wiley & Sons, New York.

Storey, H. H and Ryland, A.K. 1955. Transmission of groundnut rosette virus. *Ann. Appl. Biol.* 43: 423-432.

Storey, H.H. and Bottomley, A.M. 1928. The rosette disease of peanuts (*Arachis hypogaea* L.). *Ann. Appl. Biol.* 15: 26-45.

Subrahmanyam, P., Hildebrand, G.L., Naidu, R.A., Reddy, L.J. and Singh, A. K. 1998. Sources of resistance to groundut rosette disease in global groundnut germplasm. *Ann. Appl. Biol.* 132: 473-485.

Subrahmanyam, P., Naidu, R.A., Reddy, L.J., Kumar, P.L. and Ferguson, M. 2001. Resistance to groundnut rosette disease in wild Arachis species. *Ann. appl. Biol.* 139: 45-50.

Taliansky, M.E. Robinson, D.J. and Murant, A.F. 2000. Groundnut rosette disease virus complex: Biology and molecular biology. *Adv. Virus Res.* 55: 357-400.

Taliansky, M.E., Robinson D.J. and Murant, A.F. 1996. Complete nucleotide sequence and organization of the RNA genome of groundnut rosette umbravirus *J Gen. Virol.* 77: 2335-2345.

Teycheney, P.Y. and Dietzgen, R.G. 1994. Cloning and sequence analysis of the coat protein genes of an Australian strain of peanut mottle and an Indonesian 'blotch' strain of peanut stripe potyviruses. *Virus Res.* 31:235-44.

Ullman, D. E., Sherwood, J. L. and German, T. L. 1997. Thrips as vectors of plant pathogens. *In Thrips as Crop Pests* (Lewis, T.L., ed.). London: CAB International, pp. 539-565.

Ullman, D.E., German, T.L., Sherwod, J.L., Westcot, D.M. and Cantone, F.A. 1993. Tospovirus replication in insect vector cell: Immunochemical evidence that the non-structural protein encoded by the s RNA of tomato spotted wilt virus is present in thrips vector cells. *Phytopathology.* 82: 1333-1342.

van Regenmortel, M.H.V. Fauquet, D.H.L., Carstens, E.B., Estes, M.K., Lemon, S.M., Maniloff, J., Mayo M.A., McGeoch, DJ., Pringle, CR., Wickner, RB. 2000. Virus taxonomy. Classification and nomeclature of viruses. *Seventh Rep. Intern.Comm. Taxonomy of Viruses.* Academic press. pp 923-926.

Wijkamp, I., Van lent, J., Kormelink, R., Goldbach, R. and Peters, D. 1993. Multiplication of tomato spotted wilt virus in its insect vector *Franklinella occidenrtalis*. *J. Gen. Virol.* 74: 341-349.

Wongkaew, S. 1986. Peanut stripe and other viruses in Thailand. In: *Proc. Peanut CRSP Workshop,* Khon Kaen, Thailand pp 86-90

Woodward, J. W. 2001. 2000 Georgia plant disease loss estimates. *University of Georgia, Cooperative Extension Service Publication.* Path 01-001. May 2001, 19 pp.

Xu, Z., Yu, Z., and Jialing, L. 1983. A virus causing peanut mild mottle in Hubei province, China. *Plant Disease.* 67: 1029-1032

Yang, H.Y., Pappu, H.R. and Ozias-Akins, P. 2001. Transformation of peanut with truncated nucleocapsid gene of Tomato spotted wilt virus in cultivated peanut (*Arachis hypogaea* L.) using particle bombardment. *Proc. 2001 Congress In Vitro Biology*. June, 2001. St. Louis, MO.

Yayock, J.Y. 1977. An epidemic of rosette disease and its effect on growth characteristics and yield of groundnuts in Nigeria. *Oleagineux.* 32: 113-115.

Zimmerman, G. 1907. Uber eine Krankheit der Erdnusse (*Arachis hypogea*) *Der Pflanze*r. 3:1

Lergends to figures:

Fig. 16.1 – Symptoms of chlorotic rosette on peanut plants. Note chlorotic and deformed leaflets especially at the apical positions.

Fig. 16.2 - Typical field symptoms caused by *Tobacco streak virus* on peanut. Note stunting of plants and necrotic blotches on leaflets.

Fig. 16.3 – Field symptoms induced by *Tobacco streak virus* (TSV) (right) and Peanut bud necrosis disease (PBND) (left). TSV caused complete necrosis and death. PBND caused severe stunting and axillary shoot proliferation.

Fig. 16.4 – Typical symptoms induced by Peanut bud necrosis disease. Note severe stunting and axillary shoot proliferation.

Fig. 16.5 – Evaluation of field-resistant cultivars to Peanut bud necrosis disease in North India. Susceptible cultivar is on the left side and a field-resistant one is on the right side.

17. Common Bean

Francisco J. Morales
Virology Reseach Unit, International Center for Tropical Agriculture, Palmira, Colombia

17.1 Introduction

The common bean (*Phaseolus vulgaris*) is one of the most important food legumes in the world (over 26 million hectares worldwide), particularly in Latin America, its center of origin (Singh *et al.*, 1991), and East Africa. In Latin America, Brazil is the main producer of common beans, with approximately 2.8 million tons, followed by Mexico with 1.2 million tons, and Central America with 335,700 tons. The estimated production of common beans in East Africa, is 1.5 million tons. In Latin America, *per capita* annual consumption of common beans ranges from 10-18 kg, whereas in East Africa, common bean annual consumption may be as high as 50 kg *per capita*. As a food staple, common bean can contribute up to 35% of the protein and 340 calories/100 grams to the daily diet of resource poor urban and rural families. Common beans are also grown and consumed in various Asian countries, such as China and the Philippines, and in West Asia and

North Africa, primarily in Turkey, Egypt, Iran, Lebanon, Pakistan, Sudan, and Syria.

The common bean is very susceptible to a large number of plant viruses, which are partially responsible for the low yields observed in developing countries. These viruses are often transmitted via the seed or by insect vectors, mainly aphids, beetles and whiteflies.

The viral pathogens and vectors of viruses that infect common bean in developing countries are adequately characterized, and the majority have been sufficiently investigated in order to manage them. However, small-scale farmers still grow common bean landraces that are highly susceptible to the majority of bean viruses, and which act as permanent reservoirs of these viruses.

The control of viruses affecting common beans has been accomplished mainly by genetic improvement, including genetic engineering. Insecticides are still widely used to control insect vectors of common bean viruses in developing countries.

17.2 Bean Common Mosaic Disease

Bean common mosaic was first described in the United States, ca. 1917. This disease was first referred to as 'bean mosaic', but it was later named 'bean common mosaic' to distinguish it from bean yellow mosaic. Bean common mosaic reduces yield by as much as 80%. Susceptible plant species infected under natural conditions include: common bean, *Phaseolus vulgaris*, *Rhynchosia minima*, and some wild species of *Phaseolus* (Morales and Bos, 1988). Most common bean landraces in the American center of origin are susceptible to the causal virus.

On the foliage of infected common bean plants well-defined light and dark green areas (mosaic) with or without malformation (e.g. blisters), leaf roll, or yellow dots may be observed (Fig.17.1). The virus usually causes dwarfing. Seed-borne infections often produce healthy-looking common bean plants with mild mosaic symptoms only on the primary and first true leaves (Morales and Bos, 1988).

Bean common mosaic is caused by two species of the genus *Potyvirus*: *Bean common mosaic virus* (BCMV) and *Bean common mosaic necrosis virus* (BCMNV). These viruses have flexuous filaments, ca. 15 X 750 nm, and contain a single molecule of ssRNA, ca. 10 kb in size. The capsid protein is formed from a single protein subunit of M_r 31,000, but some potyvirus coat protein sub-units yield different molecular weights as observed in SDS-PAGE (Regenmortel *et al.* 2000).

Ten major strains (pathotypes) have been described (Drijfhout, 1978). NLI/USI, NL7, US5 (Florida), US2 (NYI5), and NL4/US6 (Mexican) are non-necrosis-inducing BCMV strains. NL2, NL6, US3 (Idaho), and US4

(Western) are temperature-dependent necrosis-inducing BCMV strains. NL3, NL5, and NL8 are temperature-independent necrosis-inducing BCMNV strains. Recently, Azuki bean mosaic, Black*eye* cowpea mosaic and Peanut stripe viruses have been classified as strains of BCMV based on close similarities in their genomic characteristics (Berger *et al.* 1997). However, these new BCMV strains do not induce characteristic common mosaic symptoms in *Phaseolus vulgaris*.

Bean host differentials are divided into two main groups: those with and those without gene *I*. Eleven resistance groups have been found. Lines IVT 7214 and IVT 7233 are resistant to all of the original 10 strains of the virus. A gene-for-gene relationship has been found between strain-specific resistance genes *bc-l*, *bc-* 1^2, *bc-2*, *bc-*2^2, and members of strains with the same numerical codes (Drijfhout, 1978). The original virus strains can be diagnosed using these common bean differentials; by serology; nucleic acid hybridization and polymerase chain reaction.

Virus purification depends on the virus strain (some BCMV strains are easier to isolate than others). The use of organic solvents, differential and density gradient centrifugation using sucrose or, preferably, cesium chloride, yields adequate amounts of most BCMV strains. Buffers are usually supplemented with anti-oxidants (Morales and Bos, 1988).

Epidemiology

BCMV is found wherever common beans are cultivated, due to the relatively high rates (average 35%) of seed transmission. BCMNV is also seed-borne in common bean, but its geographical distribution is more restricted and its incidence is considerably lower in most common bean production regions of the world. BCMV predominates in the Western World, whereas BCMNV is more common in eastern Africa. The most important means of secondary spread during the growing season is by aphids. These viruses can also be transmitted in pollen and are readily transmitted by mechanical means (Morales and Bos, 1988).

Control

The use of virus-free seed is an effective measure. The most widely used control method is the use of resistant cultivars. Gene combinations of *bc-u* (strain-nonspecific epistatic gene) plus any of the *bc-1*, *bc-*1^2, *bc-2*, or *bc-*2^2 genes confer strain-specific, recessive resistance. But since *bc-1* and *bc-*1^2 or *bc-2* and *bc-*2^2 are allelic pairs, it is not possible to have resistance to all strains of the virus in one plant genotype. The combination of *bc-u* and *bc-3* gives recessive resistance to all known strains of BCMV and BCMNV.

Similarly, the dominant gene *I* inhibits all known strains of the virus. However, this gene can be challenged by necrosis-inducing strains of BCMV and BCMNV, unless protected by the *bc-2*^{2} or *bc 3* and *bc-u* genes, in which case only restricted necrotic local lesions are produced. Another difficulty with the *I* gene is that it has a darkening effect on red and yellow seed, among other seed colors. The combination of the recessive *bc-u, bc-2*^{2}*, bc-3* genes, and the dominant *I* gene, provides durable multiple resistance to all known strains of BCMV and BCMV (Drijfhout, 1978).

17.3 Black Root Disease

'Black root' is a lethal systemic necrosis that affects common bean cultivars possessing monogenic dominant resistance (necrosis *I* gene) to bean common mosaic (Ali, 1950).

The 'black-root' reaction starts as a local lesion that extends through the vascular system of the leaf in a star-like fashion. The systemic necrosis spreads to the vascular system, causing wilt and subsequent plant death. When infection occurs late in plant development, pods may show necrotic symptoms (Morales and Bos, 1988). The type and severity of symptoms depend on host genotype, virus strain, and environmental conditions. Systemic necrosis characteristic of the 'black root' syndrome does not occur in common bean genotypes devoid of the dominant necrosis (*II*) gene (Drijfhout, 1978).

The 'black root' syndrome is caused by temperature-dependent, necrosis-inducing *Bean common mosaic virus* (BCMV) strains NL2, NL6, US3 and US4, and by the temperature-independent necrosis-inducing *Bean common mosaic necrosis virus* (BCMNV) strains NL3, NL5, and NL8 (Drijfhout, 1978; Regenmortel *et al*, 2000). Some legume potyviruses, such as *Soybean mosaic virus* and *Peanut mottle virus* can also induce black root symptoms in *I* gene common bean genotypes (Provvidenti, 1993). These viruses have flexuous filaments, ca. 15 X 750 nm, and contain a single molecule of ssRNA, ca. 10 kb in size. The capsid protein is formed from a single protein subunit of M_r 31,000 (Regenmortel *et al.* 2000). BCMNV strains are easier to purify than most BCMV strains. These viruses can be isolated using organic solvents, such as chloroform and carbon tetrachloride, differential centrifugation, and cesium chloride gradients. The detection and diagnosis of these viruses can be accomplished as described for bean common mosaic.

Disease Cycle and Epidemiology

The necrosis-inducing strains of BCMV and BCMNV are transmitted via the seed of susceptible common bean genotypes devoid of the dominant *I* or recessive *bc 3* genes. The simultaneous planting of dominant *I* gene and recessive I^+ gene common bean cultivars has caused major epidemics of 'black root' in the presence of BCMNV and aphid species capable of transmitting the virus in nature. Dominant *I* gene cultivars cannot transmit BCMV or BCMNV via the seed. Some wild legumes found in East Africa, have been shown to act as reservoirs of necrosis-inducing strains of BCMNV in nature (Spence and Walkey, 1993).

Control

The most effective control measure against 'black root' is the cultivation of BCMV/BCMNV-resistant common bean varieties. In the absence of virus-resistant common bean cultivars, the use of virus-free common bean seed minimizes the risk of outbreaks. It is equally important to avoid co-cultivation of dominant *I* gene and recessive I^+ gene common bean cultivars in adjacent fields. Genetic resistance to 'black root' can be incorporated in dominant *I* gene genotypes by hybridization with sources of recessive genes (e.g. $bc\text{-}2^2$ and *bc-3*), which protect the *I* gene against necrosis-inducing strains of BCMV and BCMNV (Drijfhout, 1978). Resistant common bean genotypes show only pin-point local lesions (*I bc-22*) or immunity (*I bc-3*). The *bc 3* recessive gene can protect common bean genotypes against 'black root' by itself, but it is usually combined with the necrosis dominant gene to prevent the emergence of a virus strain capable of challenging this unique recessive gene.

17.4 *Bean southern mosaic virus* Genus *Sobemovirus* (SBMV)

This disease was first observed in 1941 (Zaumeyer and Harter, 1943), affecting common bean in southern United States (Louisiana), hence the misnomer 'southern bean mosaic'. This disease is distributed worldwide but its characteristic mild symptoms are often overlooked. BSMV may act synergistically with other legume viruses to induce very noticeable plant malformation or necrotic symptoms (Brunt *et al.* 1996). The disease may also be particularly severe in related *Phaseolus* species, such as *P. acutifolius*, or in *P. vulgaris* x *P. acutifolius* hybrids.

Bean southern mosaic symptoms on common bean are usually mild, consisting of a light olive-green color and down curling of the infected foliage. Witches' broom and necroses are often associated with mixed infections (Zaumeyer and Harter, 1943), or infection of genotypes derived from inter-specific crosses.

Virions are ca. 30 nm in diameter, and contain a ssRNA molecule 4.1 kb in size, and one capsid protein subunit of M_r 29,000 (Regenmortel *et al.* 2000). The virus is very stable and tolerates organic solvents. Differential and density gradient centrifugation in sucrose gradients are recommended for SBMV, due to the high concentration of this virus in systemically-infected plants. SBMV exhibits pathogenic variability. The type (B) strain infects most common bean varieties systemically and induces local lesions in others. The B strain can infect a limited number of other legumes but does not infect cowpea. The Mexican and Ghanaian strains of SBMV infect both common bean and cowpea (Grogan and Kimble, 1964).

SBMV is a good immunogen and various serological tests, such as the Ouchterlony gel diffusion and ELISA tests, have been used to detect and diagnose this virus. The high concentration of this virus in infected common bean plants facilitates its observation by electron microscopy.

Disease Cycle and Epidemiology

SBMV is a ubiquitous contaminant because of its *in vitro* stability and high concentration in infected plant tissue. The virus is seed-borne in common beans and pods usually contain a considerable amount of virus both in the seeds and pod. However, most of the virus particles found in pods are inactivated during the post-harvest, seed drying process.

Chrysomelid beetles are efficient vectors of SBMV, and in some cases the virus is retained by beetles for as long as 2 weeks. Beetle vectors include *Cerotoma, Diabrotica* and *Epilachna* species.

Control

No major effort has been made to introduce genetic resistance to SBMV in common bean. It is possible that highly susceptible common bean genotypes have been consistently eliminated under field conditions. A single gene governs resistance to SBMV in common bean, and the local lesion reaction is dominant over the systemic infection. Immunity against the type strain of SBMV in bean is not known, but local-lesion bean cultivars are considered resistant (Zaumeyer and Harter, 1943).

The dissemination of SBMV could be reduced with insecticides or other methods that control beetle vectors. However, the virus is highly contagious and can be easily spread in common bean fields by workers and contaminated tools.

17.5 ***Bean severe mosaic virus*** Genus ***Comovirus*** (BSMV)

Bean severe mosaic was originally referred to as 'bean curly dwarf mosaic', when it was first described in El Salvador, in 1977 (Meiners *et al.*

1977). BSMV is present in Honduras, Guatemala, Brazil and Venezuela (Morales, 1992). Yields may be greatly reduced and total losses can occur, particularly in common bean cultivars that react with systemic necrosis.

Most common bean varieties show mosaic, leaf malformation and curling (Fig. 17.2) Dwarfing and shoot proliferation may be observed in mixed infections with other viruses, such as *Bean common mosaic virus* or *Bean southern mosaic virus*. Some common bean genotypes react with top necrosis. Restricted or systemic necrosis occurs in common bean cultivars that possess monogenic dominant resistance to *Bean common mosaic virus* . The development of necrosis, however, is influenced by environmental factors. Mechanically inoculated plants may exhibit local lesions in the inoculated leaves (Morales and Singh, 1997).

Bean severe mosaic (bean curly dwarf) was originally described (Meiners *et al.* 1977) as a disease caused by a strain of *Quail pea mosaic virus* (QPMV). Further research (Morales, 1992) has shown that this disease can be induced by different comoviruses related to QPMV and/or *Cowpea severe mosaic virus* (CPSMV). These viruses belong to the *Comovirus* genus, family *Comoviridae*, and, consequently, they have three types of isometric particles, 25 nm in diameter, one empty and the other two containing ssRNA molecules, 6.0 (RNA1) and 3.7 (RNA2) kb in size (Regenmortel *et al.* 2000). These viruses can be isolated using organic solvents and differential centrifugation. Cesium chloride or sulphate density gradients can be used for the final steps of purification. Three capsid protein subunits of M_r 39,000, 22,000, and 4,000 are associated with comoviruses, which make them good immunogens. ELISA can easily detect comoviruses. Comoviruses are also readily observed by negative staining and electron microscopy.

Disease Cycle and Epidemiology

Comoviruses are sap-transmitted, but do not seem to be seed-borne in common bean. Legume comoviruses are transmitted by different species of chrysomelid beetles, such as *Diabrotica undecimpunctata* and *D. balteata, Epilachna varivestis, Cerotoma ruficornis,* and species of *Gynandrobrotica* and *Paranapiacaba.* These chrysomelid beetles retain the virus for a few days following its acquisition from susceptible plant species, mainly weeds growing in the vicinity of bean plantings (Hobbs, 1981).

Control

Although none of the bean genotypes tested so far has shown immunity to any of the comoviruses associated with bean severe mosaic, there are some common bean genotypes that are not appreciably affected by the causal viruses (Morales and Singh, 1997).

Chemical control of beetle vectors usually reduces the incidence of bean severe mosaic comoviruses. Corn, a crop frequently grown in association with common bean in Latin America, is a host of some chrysomelid vectors and, consequently, bean severe mosaic incidence is higher in maize-bean cropping systems.

17.6 *Bean golden mosaic virus* Genus ***Begomovirus*** (BGMV)

Bean golden mosaic was first described in the state of São Paulo, Brazil, as a disease of minor importance (Costa, 1965). This disease soon reached epidemic proportions throughout the main bean growing regions of Brazil (Costa, 1975); and eventually reached northwestern Argentina and the southeastern Bolivian Plains. Bean golden mosaic is considered the main biotic constraint of common bean production in South America, particularly during the warmer months of the year. Yield losses depend on the time and frequency of virus inoculation, and the common bean genotype affected. If infection occurs within the first two weeks after planting, yield losses may be total in susceptible cultivars (Costa, 1978).

BGMV causes intense systemic yellowing (Fig. 17.3) in most common bean landraces grown in South America. Systemically infected plants usually abort their flowers or produce distorted pods with few small and damaged seeds. Flower abortion is aggravated under high temperature conditions. All known common bean genotypes are susceptible to the causal virus, but genotypes differ in their level of susceptibility.

The causal virus of bean golden mosaic has a relatively narrow host range. The virus infects *Calopogonium mucunoides*, *Macroptilium lathyroides*, *M. longepedunculatum*, *Phaseolus lunatus*, and *P. vulgaris* (Galvez and Morales, 1989).

BGMV is the type species of the *Begomovirus* genus, family *Geminiviridae*. BGMV has geminate, near-icosahedral (18 X 30 nm) particles containing two circular ss-DNA molecules of 2,617 and 2,580 nucleotides in size (Gilbertson *et al.* 1993). BGMV is not mechanically transmissible, but its DNA can be mechanically inoculated using biolistic methods. The purification of begomoviruses is difficult due to their relatively low concentration in infected plants. Moreover, begomoviruses are restricted mainly to the phloem, from which virions have to be extracted with the use of enzymes capable of dissolving phloem fibers and their lignified secondary walls (e.g. driselase). Differential centrifugation can be used for clarification and precipitation, followed by centrifugation in a log-linear sucrose density gradient. The virus band is then pelleted by ultracentrifugation, resuspended and layered again in a cesium sulphate density gradient to obtain a cleaner suspension (Morales *et al.* 1990). Begomoviruses can be detected by polyclonal antisera prepared to most virus species in this genus. For specific detection, monoclonal antibodies have been prepared against BGMV. Nucleic acid hybridization tests and specific primers could be used to diagnose BGMV, but the characteristic symptoms induced by this virus often suffices for identification purposes.

Disease Cycle and Epidemiology

BGMV is transmitted by different biotypes of the whitefly *Bemisia tabaci.* Biotype B of *B. tabaci* is sometimes referred to as *B. argentifolii.* Adult *B. tabaci* may acquire the virus from systemically infected common bean plants in few minutes, but longer virus acquisition-feeding times are usually required to achieve high transmission rates. Viruliferous whiteflies can transmit the virus for periods ranging from a few days to life. BGMV does not seem to propagate in *B. tabaci* or be transovarially transmitted (Costa, 1978).

BGMV epidemics depend on the presence of a suitable reproductive host for the whitefly vector and dry and warm environmental conditions that favor whitefly reproduction and vector movement. Pesticide abuse, particularly in the crops where the whitefly vector reproduces, is another factor associated with BGMV epidemics (Morales and Anderson, 2000).

Control

Planting beans during the rainy, cooler seasons of the year can reduce BGMV incidence. Only systemic insecticides should be used and applied at sowing time, and the crop has to be protected as much as possible during the first month. Genetic resistance has been found in black-seeded common bean varieties of Mesoamerican origin, as well as in Andean and Mexican genotypes (Morales and Niessen, 1988). Genetically modified genotypes possessing some level of partial resistance to BGMV have been developed.

17.7 ***Bean golden yellow mosaic virus*** Genus ***Begomovirus*** (BGYMV)

BGYMV and BGMV were considered diseases caused by the same virus for over two decades (Bird *et al.* 1972; Goodman, 1981). With the advance of molecular techniques, it was shown that these diseases are caused by related but different virus species (Gilbertson *et al.* 1993). BGYMV was first observed in the late 1960s (Bird *et al.* 1972), and it spreads mainly in Central America, southern Mexico and the Caribbean region (Gamez, 1970). Yield losses depend on the time of infection and the common bean genotype affected. Susceptible common bean genotypes infected at the seedling stage usually sustain 100% yield losses.

Symptoms associated with BGYMV vary depending on the common bean genotype, time of infection and environmental conditions. Susceptible bean genotypes infected soon after emergence, generally display an intense systemic yellowing. Plant malformation, including dwarfing, is more apparent in common bean genotypes of Mesoamerican origin. Flower abortion is common in infected Mesoamerican common bean genotypes, particularly under high temperature conditions (Morales and Niessen, 1988).

BGYMV is a species of the genus *Begomovirus*, family *Geminiviridae*. Virions are geminate, 18-20 nm in diameter and 30 nm in length for the dimer (Goodman, 1981). The BGYMV isolates characterized so far have two molecules of ssDNA ca 2,640 and 2600 nucleotides in size, and a coat protein subunit of M_r 27,000 (Faria *et al.* 1994). The purification procedure described above for BGMV is suitable for BGYMV. The advantage with the latter virus is that it can be transmitted mechanically, thus, facilitating the production of virus-infected tissue for purification. The virus can be diagnosed using polyclonal or monoclonal antibodies. PCR amplification of different regions of the viral genome, followed by partial sequencing, are commonly used to diagnose BGYMV strains or isolates. BGYMV symptoms are very characteristic as well.

Disease Cycle and Epidemiology

BGYMV is transmitted by the whitefly *Bemisia tabaci.* Adult *B. tabaci* may acquire the virus from systemically infected bean plants in few minutes, but longer virus acquisition-feeding times are required before whiteflies become efficient vectors. Viruliferous whiteflies can transmit the virus for periods ranging from a few days to life. BGYMV has not been shown to either propagate or be transovarially transmitted in *B. tabaci* (Gamez, 1970).

BGYMV epidemics depend on the presence of suitable reproductive hosts for the whitefly vector, and dry/warm climatic conditions that favor whitefly reproduction and vector activity. High value horticultural and industrial crops are constantly treated with a mixture of insecticides that greatly reduce whitefly predators and generate pesticide-resistant whitefly populations (Morales and Anderson, 2000).

Control

Genetic resistance has been the most effective control method used to date, although all known common bean cultivars can be systemically infected. The first sources of resistance identified were black-seeded

Mesoamerican bean genotypes, which were at best tolerant and had the ability to escape infection under field conditions. The discovery of other BGYMV resistant genes in *P. vulgaris* races, such as the *bgm-1* gene in genotypes of the Mexican 'Durango' race, or the *bgm-2* gene of Andean origin (Morales and Singh, 1991). Systemic insecticide applications have been shown to be effective when applied at planting time, but *B. tabaci* has developed resistance to most of the insecticides used in the past. Some of the new chemistries developed in the past decade, have shown to be effective in controlling *B. tabaci*. Cultivation of common bean during the rainy seasons, when whitefly populations are generally low, has been another effective control practice (Morales and Anderson, 2000). Genetically modified bean genotypes have been produced following different strategies to control BGYMV, but there is little adoption of these materials so far.

17.8 *Bean dwarf mosaic virus* Genus *Begomovirus* (BDMV)

BDMV is a ubiquitous disease of common bean in Latin America, where it is recognized under various names, such as chlorotic mottle, mottled dwarf, mosaico anão (Portuguese) and *achaparramiento* (Spanish). This disease is usually found at a low incidence (<5%) in most countries where it occurs, but major epidemics have affected extensive bean-producing areas in northwestern Argentina, causing total yield losses in susceptible bean genotypes (Morales *et al.*1990). The main reservoirs of the causal viruses in nature are various species of malvaceous weeds, particularly *Sida* species, commonly found in the American lowland tropics (Costa, 1965).

Bean plants affected by BDMVat the cotyledonary leaf stage generally exhibit marked stunting (Fig.17.4), shortening of internodes, foliar malformation, flower abortion, and pod distortion. Plants affected at a later stage,,show irregular mottling or mosaic symptoms. Chlorotic areas of variegated leaves grow more slowly than the adjacent tissues, which result in leaf malformation. Chlorotic lesions may develop into yellow patches that are irregularly distributed in the foliage. Flower abortion is commonly observed in systemically affected plants, but if pods are formed, they become severely distorted and do not produce normal seeds (Costa, 1965; Morales *et al.* 1990).

BDMV is a species of the *Begomovirus* genus, family *Geminiviridae* Morales *et al.* 1990). Virions consist of two incomplete icosahedral particles, each one approximately 20 nm in diameter. These geminate virions contain each a single molecule of ssDNA, 2.6 kb in size (Hidayat, 1990). BDMV is closely related to *Abutilon mosaic virus*. BDMV is purified as described above for BGMV (Morales *et al.* 1990). Selective amplification and partial sequencing of the viral genome are necessary to diagnose this virus, given the existence of different strains of BDMV in Latin America.

Disease Cycle and Epidemiology

BDMV is mechanically transmitted with some difficulty. In nature, the virus is transmitted by the whitefly *Bemisia tabaci* in a persistent manner, but no evidence has been found of virus multiplication in the vector or transovarial transmission. BDMV does not seem to be seed-transmitted in either common bean or *Sida* spp. Epidemics of BDMV are associated with the occurrence of large populations of the whitefly vector at the beginning of the bean planting season, favored by the presence of suitable whitefly reproductive hosts, pesticide abuse and warm/dry conditions in the affected regions (Morales *et al.* 1990).

Control

Avoidance of host plants favored by *B. tabaci* is highly recommended, especially cotton, soybean, tobacco, tomato, potato, and eggplant. The vector can be controlled with systemic insecticides, but chemical control is expensive. Resistant varieties are available and should be used whenever possible. Some of the resistant cultivars include Porrillo Sintetico, DOR 41, Red Mexican UI 35, and Pinto UI 114. Most of the sources of resistance to bean golden mosaic and bean golden yellow mosaic are also effective against BDMV (Morales and Niessen, 1988).

17.9 *Bean calico mosaic virus* Genus *Begomovirus* (BcaMV)

BCaMV affects common bean in northwestern Mexico since the early 1970s. At that time, the disease was thought to be caused by either BYMV or BGYMV. It was not until the early 1990s, when the causal agent was shown to be a different whitefly-transmitted *Geminivirus*, that the name of this disease was changed to bean calico mosaic (Brown *et al.* 1999).

BCaMV symptoms resemble those induced by BGMV and BGYMV, but the causal virus tends to induce a progressive chlorosis and yellowing, that eventually causes bleaching and necrosis of the affected foliage.

BCaMV is considered a distinct species of the *Begomovirus* genus, family *Geminiviridae* (Brown *et al.* 1999). However, BCaMV is most likely a pathogenic variant of *Squash leaf curl virus* (SLCV), adapted to common bean in northwestern Mexico. SLCV is occasionally isolated in this region from common bean plants showing the characteristic chlorosis and yellowing associated with bean calico mosaic (F.J. Morales, *unpublished data*). BCaMV has a bipartite genome (DNA-A and DNA-B) each ca. 2.6

kb. Virions consist of two incomplete icosahedra, ca. 18X30 nm (Brown *et al.* 1999). Nucleic acid hybridization, and use of specific primers followed by partial sequencing, can be used to diagnose BCaMV and related SLCV strains

Disease Cycle and Epidemiology

BCaMV is transmitted in a persistent manner by the whitefly *B. tabaci*, particularly during the dry and warm months of the year. The intensive and continuous mixed cropping systems found in the desert southwest of the United States and northwestern Mexico, favor the development of large populations of *B. tabaci* and, consequently, the dissemination of different *Begomoviruses* that attack horticultural crops throughout that region. The intensive use of pesticides has also led to the emergence of insecticide-resistant populations of the whitefly vector (Morales and Anderson, 2000).

Control

The control of BCaMV has been primarily achieved through the use of tolerant common bean cultivars, which yield well despite their systemic infection by BCaMV. The best sources of resistance to BCaMV in *P. vulgaris* have come from the Andean center of origin. The isolation of common bean plantings from reproductive hosts of the whitefly vector, and cucurbits harboring SLCV strains, is recommended to prevent crop losses by this begomovirus.

17.10 *Bean yellow mosaic virus* Genus *Potyvirus* (BYMV)

BYMV was first identified in 1931 near Madison, Wisconsin (Pierce, 1934). The virus was observed to infect common bean, pea, and a number of other leguminous species, mainly in temperate regions of the World. BYMV can cause major yield losses in susceptible common bean cultivars, depending on factors such as: virus strain, common bean cultivar, plant age, and environmental conditions.

Typical mosaic symptoms are observed as contrasting yellow and green areas in systemically infected foliage (Fig. 17.5). Some strains incite only a mild and diffuse chlorotic mottle and limited plant stunting, whereas others cause a striking yellow mosaic, rugosity, malformation, and severe stunting. Some cultivars may react to some virus strains with necrosis, wilting, and premature death. Plant maturity is delayed and the quality and quantity of seed production is often affected (Bos, 1970).

BYMV is a species of the genus *Potyvirus*. Virus particles are long, flexuous rods 750 X 15 nm and contain a single strand of RNA, ca. 10 kb in size. BYMV is serologically related to a number of other potyviruses, such as *Bean common mosaic virus*, *Clover yellow vein virus*, *Soybean mosaic virus*, and *Watermelon mosaic virus 2*. BYMV can infect most species of the *Leguminosae* family and non-legume species, such as *(Alpinia, Amaranthus, Babiana, Chenopodium, Freesia, Gladiolus, Sparaxis,* and *Tritonia* (Bos, 1970). BYMV can be purified in the presence of organic solvents, but sometimes EDTA must be added to the extraction buffers to avoid aggregation of particles. Differential and density gradient centrifugation in cesium chloride can be used to purify most BYMV strains. ELISA with specific antisera, permits diagnosis of BYMV strains. Partial sequencing of the viral genome, using potyvirus-specific primers, and inoculation of diagnostic host plants, such as pea, are complementary tests to identify BYMV strains.

BYMV is spread from cultivated and wild host reservoirs *(Lathyrus* spp., *Melilotus* spp., and *Trifolium* spp.) by several aphid species in a non-persistent manner. The virus is transmitted mechanically, but no seed transmission has been demonstrated for BYMV in common bean. However, BYMV can be seed-borne in other legumes, such as *Lupinus albus*, *L. luteus*, *Trifolium pratense*, and *Vicia faba* (Bos, 1970). Dissemination of BYMV may also occur through the use of infected, vegetatively-propagated species, such as *Gladiolus* spp.

Control

Eradication of BYMV hosts is a difficult task, and application of insecticides for the control of non-persistent, aphid-borne viruses is not effective. The use of resistant cultivars is the most viable control method, but some BYMV strains have a broad host range within the species *P. vulgaris*. Interspecific crosses between *Phaseolus coccineus* and *P. vulgaris* have revealed the existence of a single dominant gene, By-2, which confers resistance to most BYMV strains. Another single dominant gene *(By)* has been detected, but it confers resistance only to the pea strain of the virus. A second source of strain-specific resistance was found in the cultivar Great Northern 31. Resistance in this cultivar is associated with three complementary recessive genes (Dickson and Natti, 1968).

17.11 Bean Yellow Stipple Disease

Bean yellow stipple was first reported in Illinois in 1948 (Zaumeyer and Thomas, 1950). This disease has caused severe yield losses in the Caribbean and Central America regions, particularly in Cuba and Costa Rica, respectively (Morales and Gámez, 1989). In Costa Rica, the causal agent has also been associated with the 'male sterility' ('amachamiento') disease of beans (Morales *et al.* 1999).

The disease affects only leguminous species, such as: *Macroptilium lathyroides*, *Phaseolus aconitifolius*, *P. acutifolius*, *P. calcaratus* and *P. lunatus* and several cultivars of *Cajanus indicu*, *Cyamopsis tetragonoloba*, *Glycine javanica*, G. *max*, *Pisum sativum*, *Vigna hirta*, *V. mungo*, *V. sesquipedalis*, *V. umbellata* and *V. unguiculata* (Morales and Gámez, 1989).

Infected bean plants exhibit a yellow mottle that gradually develops into distinct small spots, unevenly distributed in the trifoliolate leaves, forming the yellow stippling characteristic of the disease. Severity of symptoms varies according to environmental conditions, time of infection and bean genotype. Reduction of leaf size, rugosities and deformations has been observed in certain common bean genotypes grown under highland or cool glasshouse conditions. The most damaging effect of the virus is its capacity to reduce pod formation and seed production (Morales *et al.* 1999).

Bean yellow stipple is caused by *Cowpea chlorotic mottle virus* (CCMV), a species of the genus *Bromovirus* (Fulton *et al.* 1975). Virions arc icosahedral, 26-30 nm in diameter, and contain 23.7% ssRNA in three parts. A subgenomic mRNA (RNA-4) molecule corresponds to the coat protein (Regenmortel *et al.* 2000). CCMV may be serologically related and it is often confused with some cucumoviruses that infect common bean. The virus can be isolated in the presence of organic solvents, such as chloroform and n-butanol. Differential and density gradient centrifugation in cesium chloride or caesium sulphate gradients yield highly purified virus particles. The virus is best diagnosed by ELISA or serologically-specific electron microscopy using polyclonal antiserum.

Disease Cycle and Epidemiology

CCMV is easily transmitted by sap inoculations but not through seed of infected plants. Natural spread of the virus occurs through chrysomelid beetles; the two main and most abundant vectors in Central America being *Cerotoma ruficornis* and *Diabrotica balteata.* Insects may transmit the virus after acquisition access periods of less than 24 hr. The virus is retained for 3-6 days by C. *ruficornis* and for 1-3 days by *D. balteata* (Gilmez, 1976).

Control

Control of CCMV in bean plantings is achieved by reducing the populations of its chrysomelid vectors. The frequent association of common bean and maize in Central America favors the development of high populations of certain chrysomelid vectors, which transmit the virus in common bean. All common bean cultivars tested to date, have shown susceptibility to CCMV.

17.12 *Cucumber Mosaic Virus* Genus *Cucumovirus* (CMV)

CMV is one of the most ubiquitous viral diseases of plants, affecting some 775 plant species in 85 families, including *Phaseolus vulgaris*. Several strains of the causal virus can induce different symptoms in common bean, ranging from mild mosaic to severe plant malformation. Yield losses vary from 5% to 75% depending on the cultivar, age of infection, virus strain and environmental conditions (Bird *et al.*1974).

In early stages of growth, plants show systemic symptoms consisting of a mosaic or mottle, often accompanied by rugosity of leaves (Fig.17.6). Leaf curling, blisters, and dark green veinbanding along the main veins, may also be part of the syndrome. Some common bean cultivars respond with an extreme leaf deformity (shoe stringing) that can be confused with herbicide toxicity. Pods are usually malformed and do not produce much seed (Bird *et al.* 1974).

CMV a species of the *Cucumovirus* genus consist of 29-nm icosahedral particles, with a tripartite genome that is encapsidated in three distinctive particles. Subgenomic mRNA (for coat protein) or satellite RNA may be found in certain CMV strains. The capsid protein is made up of many copies of one protein subunit of M_r 24,200. Numerous strains of the virus are recognized (Brunt *et al.,* 1996). Often, common bean infections attributed to CMV are caused by a different cucumovirus species, *Peanut stunt virus* (PSV). CMV/PSV hybrids have been detected in Chile under natural conditions, causing severe plant malformation in common bean cultivars (White *et al.* 1995). Most cucumoviruses can be isolated using chloroform and differential centrifugation. To further purify the virus, sucrose density gradient centrifugation is recommended. Polyclonal antisera are widely used for diagnosis, but the significant pathogenic variability of CMV, complicates the selection of the correct antiserum. Monoclonal antibodies have been produced for some strains, but they cannot be used to detect most CMV strains. PCR is another diagnosis technique that can be readily applied for the diagnosis of CMV. Finally, the inoculation of diagnostic hosts, such as cucurbits and tobacco, can complement other diagnostic tests.

Disease Cycle and Epidemiology

CMV strains are easily transmitted mechanically. Various bean-infecting strains of CMV are seed-borne in bean and can thus be disseminated over long distances in seed shipments. CMV strains infecting leguminous plants are transmissible in a non-persistent manner by different aphid species from a variety of plant reservoirs (Bos and Maat, 1974).

Control

In certain countries (e.g., Chile, Turkey, Iran) CMV causes significant yield losses and for best control measures, the use of virus-free seed, is recommended. The destruction of wild reservoir hosts or isolation of common bean fields from other cultivated hosts (e.g. cucurbits) may also contribute to lower disease incidence (Horvath, 1983). Useful levels of resistance or tolerance to CMV transmission through seed do exist in common bean and are available for breeding programs. Several species of *Phaseolus* are resistant to cucumber mosaic, including *P. acutifolius*, *P. adenanthus*, *P. leptostachyus.*, *P. polyanthus*, *P. trilobus*, and some accessions of *P. coccineus L.* (Provvidenti, 1976). The mechanism of resistance is not known.

17.13 Present and Future of Common Bean Viruses in Developing Countries

Currently, the main threat to common bean production in developing countries comes from whitefly-transmitted begomoviruses. These viruses cause significant yield losses in the lowlands and mid-altitude valleys of tropical and subtropical America, despite considerable success in the development of common bean cultivars possessing high levels of genetic resistance. Unfortunately, traditional common bean breeding programs in national agricultural research institutions and international agricultural development centers in Latin America, have been largely downsized due to financial stress. This trend has resulted in a significant reduction in the number of common bean breeding materials available to bean improvement programs in developing countries.

Beetle-transmitted viruses are endemic in developing countries where common beans are predominantly grown in association with maize, as in many regions of Central and South America. Fortunately, the control of beetle vectors is more feasible than the control of whiteflies, although it implies the use of chemicals. Diseases such as bean severe mosaic and the 'male sterility', caused by strains of *Cowpea severe mosaic virus* and *Cowpea chlorotic mottle virus*, respectively, have re-emerged in Latin America in response to changes in the genetic make-up of some of the new common bean cultivars.

The main aphid-transmitted viruses of common bean, *Bean common mosaic virus* and *Bean common mosaic necrosis viruses*, are currently under control in Latin America and Africa. However, the higher incidence of necrosis-inducing strains of the latter virus (BCMNV) in East Africa, poses a constant threat to improved common bean cultivars not resistant to these necrotic strains. Bean yellow mosaic, cucumber mosaic and peanut stunt are a potential threat to common bean production in developing nations located in the temperate region.

References

Ali, M. A. 1950. Genetics of resistance to the common bean mosaic virus (bean virus 1) in the bean *(Phaseolus vulgaris* L.). *Phytopathology* 40:69-79.

Berger, P.H., Wyatt, S.D., Shiel, P.J., Silbernagel, M.J., Druffel, K, and Mink, G.I. 1997. Phylogenetic analysis of the Potyviridae with emphasis on legume-infecting potyviruses. *Arch. Virol.* 142: 1979-1999.

Bird, J., Perez, J.E., Alconero, R., Vakili, N.G., and Melendez, P.E. 1972. A whitefly-transmitted golden-yellow mosaic of *Phaseolus lunatus* in Puerto Rico. *J. Agric. Univ.Puerto Rico* 56: 54-74.

Bird, J., Sanchez, J., Rodriguez, L., Cortes-Monllor, A., and Kaiser, W. J. 1974. A mosaic of beans (Phaseolus vulgaris L.) caused by a strain of common cucumber mosaic virus. *J. Agric. Univ. Puerto Rico* 53:151-161.

Bos, L. 1970. Bean yellow mosaic virus. *Descriptions Plant Viruses*, No.40. Comm. Mycological Inst. and Association of Applied Biologists, Kew, Surrey, England.

Bos, L., and Maat, D. Z. 1974. A strain of cucumber mosaic virus seed transmitted in bean. *Neth. J. Plant Pathol.* 80:113-123.

Brown, J.K., Ostrow, K.M., Idris, A.M., and Stenger, D.C. 1999. Biotic, molecular, and phylogenetic characterization of *Bean calico mosaic virus*, a distinct begomovirus species with affiliation in the *Squash leaf curl virus* cluster. *Phytopathology* 89: 273-280.

Brunt, A.A., Crabtree, K., Dallwitz, M.J., Gibbs, A.J., and Watson, L. 1996. Viruses of Plants: *Descr.and Lists from the VIDE Database.* CAB Intern., Wallingford, UK. 1484.

Costa, A. S. 1965. Three whitefly-transmitted virus diseases of beans in Sao Paulo, Brazil. *FAO Plant Prot. Bull.* 13:121-130.

Costa, A. S. 1975. Increase in the populational density of *Bemisia tabaci,* a threat of widespread virus infection of legume crops in Brazil. Pages 27-49 in: *Tropical Diseases of Legumes*. J. Bird and K. Maramorosch, eds. *Academic Press, New York. 171 pp.*

Costa, A. S. 1978. Whitefly-transmitted plant diseases. *Annu. Rev. Phytopathol.* 14:429-449.

Debrot, E. A., and Centeno, F. 1986. Occurrence of quail pea mosaic virus (QPMV) infecting beans and soybeans in Venezuela. (Abstr.) *Phytopatho)ogy* 76:1089-1090.

Dickson, M. H., and Natti, J. J. 1968. Inheritance of resistance of *Phaseolus vulgaris* to bean yellow mosaic virus. *Phytopathology* 58:1450.

Drijfhout, E. 1978. Genetic interaction between *Phaseolus vulgaris* and bean common virus with implications for strain identification and breeding for resistance. *Agric. Res. Rep.* 872. Centre Agric. Publ. Doc. Wageningen.

Faria, J.C., Gilbertson, R.L., Hanson, S.F., Morales, F.J., Ahlquist, P., Loniello, A.O., and Maxwell, D.P. 1994. Bean golden mosaic geminivirus type II isolates from the Dominican Republic and Guatemala: nucleotide sequences, infectious pseudorecombinants, and phylogenetic relationships. *Phytopathology* 84: 321-329.

Fulton, J. P., Gilmez, R., and Scott, H. A. 1975. Cowpea chlorotic mottle and bean yellow stipple viruses. *Phytopathology* 65:741-742.

Galvez, G. E., and Morales, F. J. 1989. Whitefly-transmitted viruses. Pages 379-406 in: *Bean Production Problems in the Tropics*, 2nd ed. H. F. Schwartz and M. A. Pastor-Corrales, eds. Centro Internacional de Agricultura Tropical, Cali, Colombia. 726 pp.

Gámez, R. 1970. Los virus del frijol en Centroamérica. I. transmisión por moscas blancas (*Bemisia tabaci* Genn.) y plantas hospedantes del virus del mosaico dorado. *Turrialba* 21: 22-27.

Gilbertson, R.L., Faria, J.C., Ahlquist, P., and Maxwell, D.P. 1993. Genetic diversity in geminiviruses causing bean golden mosaic disease: the nucleotide sequence of the infectious cloned DNA components of a Brazilian isolate of bean golden mosaic geminivirus. *Phytopathology* 83: 709-715.

Gilmez, R. 1976. Los virus del frijol en Centroamerica, IV: Algunas propiedades y transmision por insectos crisomelidos del virus del moteado amarillo del frijol. *Turrialba* 26:160-166.

Goodman, R. M. 1981. Geminiviruses. Pages 879-910 in: *Handbook of Plant Virus Infections: Comparative Diagnosis*. E. Kurstak, ed. Elsevier, Amsterdam. 943 pp.

Grogan, R. G., and Kimble, K. A. 1964. The relationship of severe bean mosaic virus from Mexico to southern bean mosaic virus and its related strain in cowpea. *Phytopathology* 54:75-78.

Hidayat, S.H. 1990. The construction and application of DNA probes for the detection of bean dwarf mosaic geminivirus. *Thesis, University of Wisconsin*, Madison, WI, 53 p.

Hobbs, H. A. 1981. Transmission of bean curly dwarf mosaic virus and bean mild mosaic virus by beetles in Costa Rica. *Plant Dis.* 65:491-492.

Meiners, J. P., Waterworth, H. E., Lawson, R. H., and Smith, F. F. 1977. Curly dwarf mosaic disease of beans from El Salvador. *Phytopathology* 67:163-168.

Morales, F. J., and Niessen, A. I. 1988. Comparative responses of selected *Phaseolus vulgaris* germ plasm inoculated artificially and naturally with bean golden mosaic virus. *Plant Dis.* 72:1020-1023.

Morales, F., Niessen, A., Ramirez, B., and Castano, M. 1990. Isolation and partial characterization of a geminivirus causing bean dwarf mosaic. *Phytopathology* 80:96-101.

Morales, F.J. 1992. Increased disease severity induced by some comoviruses in bean genotypes possessing monogenic dominant resistance to bean common mosaic potyviruses. *Plant Dis.* 76: 570-573.

Morales, F.J., Araya, C.M., Hernandez, J.C., Arroyave, J., Cuervo, M., Velasco, A.C., and Castaño, M. 1999. Etiology of "amachamiento" ("sterility") in common bean in Costa Rica. *Manejo Integrado de Plagas*. 52: 42-48.

Morales, F.J., and Anderson, P.K. 2000. The emergence and dissemination of whitefly-transmitted geminiviruses in Latin America. *Arch. Virol.* 146: 415-441.

Morales, F.J., and Bos, L. 1988. Bean common mosaic virus. *Description of Plant Viruses,* No.337. Assoc. Appl. Biol., Kew, Surrey, England.

Morales, F.J., and Singh, S.P. 1991. Genetics of resistance to bean golden mosaic virus in *Phaseolus vulgaris* L. *Euphytica* 52: 113-117.

Morales, F.J., and Singh, S.P. 1997. Inheritance of the mosaic and necroses reactions induced by comoviruses in *Phaseolus vulgaris* L. *Euphytica* 93: 223-226.

Pierce, W. H. 1934. Viroses of the bean. *Phytopathology* 24:87-115.

Provvidenti, R. 1976. Reaction of *Phaseolus* and *Macroptilium* species to a strain of cucumber mosaic virus. *Plant Dis. Rep.* 60:289-293.

Provvidenti, R. 1993. Genetics of resistance to viral diseases of bean. Pp. 112-152, In: *Resistance to Viral Diseases of Vegetables: Genetics and Breeding* (M.M. Kyle, Ed.). Timber Press, Oregon.

Regenmortel, M.H.V., Fauquet, C.M., Bishop, D.H.L., Carstens, E.B., Estes, M.K., Lemon, S.M., Maniloff, J., Mayo, M.A., McGeoch, D.J., Pringle, C.R., and Wickner, R.B. 2000. *Virus Taxonomy: Classification and Nomenclature of Viruses:* Seventh Rep. Intern. Comm. Taxonomy of Viruses. Academic Press, New York, New York, 1162 p.

Singh, S.P., Gepts, P., and Debouck, D.G. 1991. Races of common bean (*Phaseolus vulgaris*, Fabaceae). *Economic Botany* 45: 379-396.

Spence, N., and Walkey, D.G.A. 1993. Bean common mosaic virus and related viruses in Africa. NRI Bulletin 63, Chatham, U.K. 167

White, P.S., Morales, F., and Roossinck, M.J. 1995. Interspecific reassortment of genomic segments in the evolution of cucumoviruses. *Virology* 207: 334-337.

Zaumeyer, W. J., and Harter, L. L. 1943. Two new virus diseases of beans. *J. Agric. Res.* 67:305-328.

Zaumeyer, W. J., and Thomas, H. R. 1950. Yellow stipple, a virus disease of bean. *Phytopathology* 40:847-859.

Legends to figures:
Fig. 17.1 –Symptoms induced Bean common mosaic disease.
Fig. 17.2 – Beans infected with Bean severe mosaic virus.
Fig. 17.3 – Beans infected with Bean golden mosaic virus.
Fig. 17.4 – Beans infected with Bean dwarf mosaic virus.
Fig. 17. 5 – Beans infected with Bean yellow mosaic virus.
Fig. 17.6 – Beans infected with Cucumber mosaic virus.

18. Other legumes

Faba bean, chickpea, lentil, pigeonpea, mungbean, blackgram, lima bean, horsegram, bambara groundnut and winged bean

K.M. Makkouk[1], S.G. Kumari[1], J. d'A. Hughes[2], V. Muniyappa[3] and N.K. Kulkarni[3]

[1]Virology Laboratory, Germplasm Program, International Center for Agricultural Research in the Dry Areas (ICARDA), P.O. Box 5466, Aleppo, Syria; [2]Virologist, International Institute of Tropical Agriculture, (IITA), Ibadan, Nigeria (Address for correspondence: c/o L W Lambourn & Co Ltd., 26 Dingwall Road, Croydon, CR9 3EE, UK.); [3]Department of Plant Pathology, University of Agricultural Sciences, GKVK, Bangalore-560 065, India

18.1 Introduction

The ten legumes discussed in this chapter (faba bean, chickpea, lentil, pigeonpea, mungbean, blackgram, lima bean, horsegram, bambara groundnut and winged bean) are economically important food crops in the arid and semi arid tropics of Asia and Africa. All are susceptible to numerous viruses, and our understanding of the ecology of these viruses and the impact they have on crop yields is continuously improving. This chapter will focus on the main features of the most economically important of those viruses that attack these legumes. For more details on these viruses, the reader is referred to previously published reviews (Bos *et al.* 1988; Chenulu and Varma, 1988; Cockbain, 1983).

Grain yield losses due to virus diseases can vary greatly, from almost complete crop failure to little or no loss. In some cases the effect is indirect and not clearly visible—for example reduced faba bean root nodulation following infection with *Bean yellow mosaic virus* (Frowd and Bernier, 1977). Quality of produce may also be affected, as with *Broad bean stain virus*, which renders faba bean seeds unusable for canning.

18.2 Faba bean (*Vicia faba*)

Faba bean is the fourth most important pulse crop in the world. Of the area cropped with faba bean globally, more than 50% is located in China. Faba bean has multiple uses in China. Consumed as dry seeds, a green vegetable, or a processed food, its products are a cheap source of high-quality protein in the human diet, while its dry seeds, green haulm and dry straw are used as animal feed. Faba bean also serves as a break crop in continuous cereal rotations, improving the productivity of the soil. Its green haulms are used for green manuring. The major viruses reported to infect faba bean worldwide are *Bean leafroll virus, Bean yellow mosaic virus, Beet western yellows virus, Broad bean mottle virus* and *Faba bean necrotic yellows virus* (Tables 18.1 and 18.22). In the following paragraphs summaries of the main features of these viruses are presented.

18.2.1 *Bean leafroll virus* Family *Luteoviridae* (BLRV)

Bean leafroll virus is the most important virus known to infect faba bean. It was first isolated from faba bean by Quantz and Volk (1954) in Germany. The main symptoms produced by the virus are interveinal chlorosis, yellowing, stunting, leaf rolling, reddening and thickening of the leaves, suppression of flowering and pod setting. In faba bean, old leaves become leathery. Because the symptoms produced by BLRV can vary slightly among the species infected, the common disease names vary based the crop species involved. Hence, the names Pea leafroll, Pea yellow top, Chickpea stunt and Lentil yellows reflect the main symptoms produced in these species in response to infection by the same virus. BLRV and the other luteoviruses, can have a marked effect on yield. Yield losses of 50-90% have been reported in Europe (Schmutterer and Thottappilly, 1972). In Iran, faba bean plants inoculated with BLRV either during or after flowering yielded, respectively, 89% and 59% less than uninoculated plants; plants inoculated before flowering failed to set any seeds (Kaiser, 1973b).

Bean leafroll virus can be purified following the procedure described by D'Arcy *et al.* (1989). A purified preparation of the virus has an A_{260}:A_{280} of 1.83. The virus particles are isometric, *ca* 27 nm in diameter with an ssRNA genome of 6 kb. BLRV can be diagnosed in infected tissue by using ELISA or Tissue-blot immunoassay (TBIA) (using either polyclonal or monoclonal antibodies).

The spread of BLRV within crops can sometimes be checked by aphicides (Schmutterer and Thottappilly, 1972), seed treatment with Imidacloprid before planting (Makkouk and Kumari, 2001) and manipulation of sowing date (Johnstone and Rapley, 1979). Genotypes resistant to BLRV have been reported in faba bean (Makkouk *et al.* 2002b) and in lentil (Makkouk *et al.* 2001).

18.2.2 *Bean yellow mosaic virus* Genus *Potyvirus* (BYMV)

BYMV was reported in faba bean by Böning (1927) in Germany. The number of fields infected with BYMV can vary greatly among locations. A high incidence, up to 100%, has been noted in some regions of Egypt, Iraq and Sudan. Two of these countries (Egypt and Sudan) are known for their relatively warm winters, which favour increased aphid populations and movement. The effect of BYMV on yield depends, to a great extent, on the time of infection and on the strain of the virus. In Syria, faba bean plants inoculated with BYMV, 11, 15 and 20 weeks after sowing, resulted in yield losses of 81%, 56% and 39%, respectively (Makkouk *et al.* 1988). In Canada, faba bean inoculated with a mild isolate 5, 7 and 9 weeks after sowing yielded, respectively, 59%, 48% and 17% less than uninoculated plants, while plants inoculated with a severe isolate yielded 96%, 70% and 17% less (Frowd and Bernier, 1977).

The symptoms that result from BYMV infection are variable and are greatly affected by the strain of the virus and the genotype of the plant involved. Symptoms produced by the virus include mosaic, mottling, green vein-banding and chlorosis. Severe strains of the virus can produce stem and tip necrosis and early death in some faba bean genotypes.

TBIA and ELISA are widely used for the large-scale testing of many samples and for the purposes of quarantine (in order to certify the absence of the virus). PCR has also been advocated for seed testing (Bariana *et al.* 1994).

BYMV can be purified following the procedure described by Bos (1970). A purified preparation of the virus has an A_{260}:A_{280} of 1.21. The virus particles sediment as one component *ca* 151*S* (varying from 140-166*S*), which is composed of 5.5-6% nucleic acid and 94-94.5% of one type of protein. The nucleic acid is unipartite, single-stranded, positive-sense RNA *ca* 10 kb in size.

Measures to control BYMV include manipulation of sowing date (Abu Salih *et al.* 1973), spraying with mineral oils (Proeseler *et al.* 1976), soil mulching with reflective polyethylene sheets (Tachibana, 1981) and ensuring that crops are not grown in the vicinity of known overwintering disease sources. In regions where seed-borne infection is a problem, it is clearly advisable to ensure that seed crops are kept as virus-free as possible. In addition, repeated inoculation and continuing re-selection in faba bean has led to the development of resistant genotypes in Canada (Gadh and Bernier, 1984) and in Syria (Makkouk and Kumari, 1995).

18.2.3 *Beet western yellows virus* (BWYV)

(See details given under 'chickpea,' below)

18.2.4 *Broad bean mottle virus* Genus *Bromovirus* (BBMV)

Broad bean mottle virus was first described in faba bean by Bawden *et al.* (1951). The symptoms produced in faba bean are, mainly, mottling, marbling or diffuse mosaic, which is often associated with leaf malformation and sometimes with plant stunting. Some faba bean genotypes may show necrosis in response to BBMV infection. A faba bean grain yield loss of 35-55%, depending on time of BBMV infection has been reported (Makkouk *et al.* 1988). The virus has been found to reach high levels of incidence in faba bean fields in Morocco, Sudan and Tunisia (Fortass and Bos, 1991; Makkouk *et al.* 1988; Najar *et al.* 2000).

For a long time, BBMV attracted academic interest because of its structural characteristics; its high concentration of infected cells and is an easy virus to purify. The virus was successfully purified as described by Lane (1974). A purified preparation of the virus has an A_{260}:A_{280} ratio of 5.4. The virus particles have an ssRNA genome which is *ca* 8.25 kb in size.

The use of healthy seed, and beetle control reduce the spread of BBMV. No faba bean cultivars resistant to BBMV infection are available.

18.2.5 Faba bean necrotic yellows virus Genus *Nanavirus* (FBNYV)

Faba bean necrotic yellows virus was first isolated from faba bean near Lattakia, Syria (Katul *et al.* 1993). The virus can be very damaging in those years in which an epidemic occurs, as happened in Middle Egypt during 1992, 1997 and 1998 and in Tunisia in 2001. In Middle Egypt, losses due to FBNYV infection reached 80-90% during 1992 (Makkouk *et al.* 1994). In West Asia and North Africa, FBNYV is considered to be the virus to most damaging faba bean.

One-week-old faba bean plants show retarded growth as early as five days after inoculation. At two weeks after infection, the plants are usually severely stunted. The leaves become thick and brittle and show interveinal chlorotic blotches, which begin at the leaf margins. Those young leaves remain very small and are cupped upwards, whereas older leaves are rolled downward. New shoots, leaves and flowers develop poorly. About 3-4 weeks after infection, interveinal chlorosis usually turns necrotic and infected plants die within 5-7 weeks after infection.

The virus can be diagnosed in infected tissue using ELISA or TBIA (using either polyclonal or monoclonal antibodies) and by dot-blot hybridization. These tests are essential to the differentiation of FBNYV from viruses in the family *Luteoviridae,* which induce similar symptoms in legume species such as BLRV, BWYV and *Soybean dwarf virus* (SbDV), or viruses of the family *Geminiviridae,* such as Chickpea chlorotic dwarf virus (CpCDV).

The virus has been successfully purified Katul *et al.* (1993). A purified preparation of the virus has an A_{260}:A_{280} ratio of 1.4. The genome consists of circular ssDNA, with at least seven components. Each is approximately 1 kb in size and appears to be encapsidated in a separate particle.

The most effective measure to control FBNYV in faba bean crop is an integrated virus management approach consisting of: (i) planting at an appropriate time, to avoid peak numbers of viruliferous aphids coming from susceptible summer legume crops or wild hosts, (ii) well-timed application (either once or twice) of aphicide, (iii) seed treatment with Imidacloprid before planting (Makkouk and Kumari, 2001), (iv) roguing of infected plants early in the season, and (v) use of an appropriate seeding rate. Such measures are essential to reach an economic yield in regions where FBNYV epidemics are likely to occur. In addition, genotypes resistant to this virus have been reported in lentil (Makkouk *et al.* 2001).

18.3 Chickpea (*Cicer arietinum*)

Chickpea is the second most important pulse crop in the world. It is a major food legume in West Asia and North Africa. It is an important source of protein in human diets and plays a significant role in farming systems. Four major chickpea production regions can be distinguished: the Indo-Pakistan subcontinent, East Africa, the Mediterranean Region and the Americas. The former two regions mostly grow the type known as 'desi'; the latter two grow, almost exclusively, the type known as 'kabuli'. Much of the research effort focused on the desi type is concentrated in the Indo-Pakistan subcontinent, while that on the kabuli type is mainly concentrated in the Mediterranean region. The major viruses reported to infect chickpea worldwide are *Alfalfa mosaic virus, Bean leafroll virus, Beet western yellows virus, Chickpea chlorotic dwarf virus*, *Cucumber mosaic virus* and *Faba bean necrotic yellows virus* (Tables 18.1 and 18.2). In the following paragraphs summaries of the main features of these viruses are given.

18.3.1 *Alfalfa mosaic virus* Genus *Alfamovirus* (AlMV)

Alfalfa mosaic virus was first reported on faba bean from Sudan (Nour and Nour, 1962). Symptoms induced by AlMV generally take the form of systemic mosaic or mottling. Under some conditions, symptoms can disappear or become extremely mild. In some species (such as pea and faba bean) infection with this virus leads to stem necrosis. Symptoms of AlMV infection in chickpeas include chlorosis, proliferation of axillary shoots and plant dwarfing. In lentil, symptoms of infection with this virus include chlorosis, mottling, deformation, twisting of leaflets and dwarfing. Mechanical inoculation of *P. vulgaris* with many virus strains causes necrotic local lesions, and this can be used as an assay for the virus.

Lucerne (*Medicago sativa*) is probably the main source of the virus in many countries, but other perennial legumes, including red and white clovers (*Trifolium pratense* and *T. repens*), are also known sources (Hull, 1969). In Western Australia, the most important legume hosts for virus carry-over outside the growing season through seed transmission (i.e. oversummering) are annual medics (*Medicago* spp.) (Jones and Pathipanawat, 1989).

Alfalfa mosaic virus can be purified as described by Van Vloten-Doting and Jaspars (1972). A purified preparation of the virus has an A_{260}:A_{280} ratio of 1.7-1.8. It contains bacilliform particles, which form a sediment composed of four major components, having a constant diameter of 18 nm and varying in form from 30 nm to 57 nm in length, depending on the nucleic acid species encapsidated. In most strains, each particle contains one species of single-stranded (+) RNA, the size of which varies from 1.0 to 3.4 kb. Virions are readily separated into their components by sucrose density gradient centrifugation.

Virus incidence in most fields is extremely low, and the virus is not considered of economic importance in cool-season legumes. However, losses in terms of grain yield and quality can, in some cases, be substantial in both lentil and chickpea (Bosque-Perez and Buddenhagen, 1990; Kaiser, 1973a).

Since AlMV is both seed and aphid transmitted, a range of control measures are applied: (i) use of virus-free seed, (ii) control of virus spread from overwintering hosts by spatial separation, and (iii) aphid vector control. Insecticides have limited use, as the virus is non-persistent. No source of resistance to the virus in legume crops has been reported, and there are no reports of attempts being made to control the virus using aphicides.

18.3.2 ***Bean leafroll virus*** (BLRV)

(See details given under 'faba bean,' above)

18.3.3 ***Beet western yellows virus*** Genus ***Polerovirus*** (BWYV)

Beet western yellows virus was first isolated by Duffus (1961) in California. Synonyms for BWYV are Malva yellows virus and Turnip mild yellows virus. The main symptoms produced in chickpea upon infection with BWYV are yellowing and stunting with brown phloem discoloration. Chickpea grain yield losses can vary widely based on time of infection and variety, and have been found to range from 8% to 90% when infection occurs 4-5 months after sowing (Bosque-Perez and Buddenhagen, 1990).

Beet western yellows virus can be purified as described by D'Arcy *et al.* (1989). A purified preparation of the virus has an A_{260}:A_{280} ratio of 1.8. Virus particles have a diameter of *ca* 26 nm with an ssRNA genome of 5.6 kb; the coat protein Mr of the virus is around 24 kd. BWYV polyclonal antibodies cross react with many viruses of the family *Luteoviridae*. Specific serological detection of BWYV can be made using monoclonal antibodies.

Controlling BWYV in chickpea involves the use of cultural practices, which reduce virus incidence in the crop. Examples of such practices are: planting date, sowing seed density, rouging of infected plants early in the season, and site selection (select chickpea field away from a source of virus infection). No chickpea varieties resistant to BWYV are available.

18.3.4 Chickpea chlorotic dwarf virus Genus *Mastrevirus* ? (CpCDV)

Chickpea chlorotic dwarf virus was first reported on chickpea in India (Horn *et al.* 1993). The main symptoms in chickpea are reddening or yellowing of leaflets with severe stunting. Similar symptoms are observed in faba bean, but without reddening of leaflets.

Crop loss is proportional to virus incidence in the field, as CpCDV-infected plants produce little or no grain. Losses of 5-10% are estimated to occur in some regions of India and Pakistan (Horn *et al.* 1996), whereas losses of up to 50% have been reported in some areas of Sudan (Hamed and Makkouk, 2002). CpCDV is, potentially, a serious virus with regard to chickpea production. It has been found to reach a high level of incidence in some chickpea fields in Iran (Makkouk *et al.* 2002a).

Purified virus was obtained by Horn *et al.* (1993). A purified preparation of the virus has an A_{260}: A_{280} ratio of 1.4. The virus particles have a single stranded, circular DNA genome of around 2.9 kb. The coat protein has a Mr of 32 kd. CpCDV is serologically related to *Tobacco yellow dwarf virus,* but not to the *African cassava mosaic, Squash leaf curl, Chloris striate, Maize streak* or *Wheat dwarf viruses*.

Work in Sudan has suggested that a delay of 3-4 weeks in the date of planting coupled with short irrigation intervals significantly reduces CpCDV incidence. No chickpea cultivars resistant to the virus are available at present, but some local cultivars (e.g. 'Shendi') are less infected than others (Hamed and Makkouk, 2002).

18.3.5 *Cucumber mosaic virus* Genus *Cucumovirus* (CMV)

CMV was first recorded infecting faba bean in Iran (Kaiser *et al.* 1968). The symptoms induced by CMV infection range from almost none to clear, systemic mottling or mosaic and, sometimes, stunting and leaf malformation.

It causes economic losses in chickpea and lentil crops in Western Australia (Jones and Coutts, 1996).

CMV can be purified as described by Lot *et al.* (1972). A purified preparation of the virus has an A_{260}:A_{280} ratio of 1.7. Virus particles have an ssRNA genome. CMV particles contain three genomic RNA components (RNA1, RNA2 and RNA3), each encapsidated separately. The subgenomic RNA4, which is not needed for infectivity, is encapsidated with RNA3.

18.3.6 Faba bean necrotic yellows virus (FBNYV)

(See details given under 'faba bean,' above)

18.4 Lentil (*Lens culinaris*)

Lentil is a staple pulse in developing countries, where it complements the cereal-rich diet of the people (particularly those in the low-income bracket). It is consumed either as a whole seed or after decortication. The milling wastes and its straw provide nutritious feed for animals. The major viruses reported to infect lentil worldwide are *Bean leafroll virus, Bean yellow mosaic virus, Broad bean stain virus, Pea enation mosaic virus* and *Pea seed-borne mosaic virus* (Tables 18.1 and 18.2).

18.4.1 *Bean leaf roll virus* (BLRV)

(See details given under 'faba bean,' above)

18.4.2 *Bean yellow mosaic virus* (BYMV)

(See details given under 'faba bean,' above)

18.4.3 *Broad bean stain virus* Genus *Comovirus* (BBSV)

Broad bean stain virus was first reported in faba bean in the UK (Lloyd *et al.* 1965) displaying systemic mottling and, occasionally, leaf shape distortion. In lentils, a very mild mottling occurs which, often, cannot be easily recognized. Grain yield losses varied from 14% to 61% when 19 lentil genotypes were inoculated at the flowering stage (Makkouk and Kumari, 1990) and from 18% to 20% when faba bean plants were inoculated with the virus at other different stages (Makkouk *et al.* 1988). Seed transmission rates of BBSV vary widely based on host genotype and virus isolates, and were found to range from 0.2% to 32.4% when 19 lentil genotypes were evaluated (Makkouk and Kumari, 1990).

Broad bean stain virus can be purified as described by Makkouk *et al.* (1987). A purified preparation of the virus has an A_{260}:A_{280} ratio of 1.64. Virus particles have two ssRNA components of 6.75 and 4.5 kb. BBSV has three components, which have sedimentation coefficients of 127 *S* (B), 100 *S* (M) and 60 *S* (T).

The use of healthy seed in conjunction with beetle control seems to be the logical approach with regard to reducing BBSV spread. Makkouk and Kumari (1990) found that the lentil cv. Red Chief had the lowest seed transmission rate (0.2%) and yield loss (14%) among the 19 lentil genotypes evaluated. Seed transmission was reduced to zero when the seeds were exposed to 70 °C for 28 days; however this treatment caused a 57% reduction in seed germination (Kumari and Makkouk, 1996). Such an approach could be useful to eliminate seed infection in germplasm accessions deposited in gene banks, but is not appropriate for commercial production.

18.4.4 *Pea enation mosaic virus-1* Genus ***Enamovirus*** (PEMV)

What is commonly referred to as *Pea enation mosaic virus* is a mixture of two viruses, *Pea enation mosaic virus-1* (PEMV-1) and *Pea enation mosaic virus-2* (PEMV-2, genus *Umbravirus*). PEMV-1 is readily transmitted mechanically, a property dependent on its multiplication in cells co-infected with PEMV-2. PEMV-1 and PEMV-2 are transmitted by several aphid species.

The symptoms caused by PEMV-1 and PEMV-2 in lentil is growth reduction in conjunction with a rolling of leaves, accompanied by mottling and tip wilting or necrosis. In faba bean it causes yellow or hyaline streaks or spots on the leaves, and sometimes enations on the lower leaf surface. These symptoms became severe in lentil fields in Syria (Makkouk *et al.* 1999) and in chickpea and lentil fields in Eastern Washington (Klein *et al.* 1991).

The virus can be purified according to Mahmood and Peters (1973). A purified preparation of the virus has an A_{260}:A_{280} ratio of 1.63. Virus particles have a diameter of 28 nm, mostly with a hexagonal outline. However, a few are pentagonal, while others are more ovoid. Such particles are found more frequently in fractions rich in top component than they are in those rich in bottom component. The ssRNA genome of PEMV-1 is 5.7 kb, while that of PEMV-2 is 4.35 kb.

The virus can be detected by polyclonal antibodies, using ELISA or PCR (J. Skaf, personal communication). Serological tests cannot differentiate PEMV-1 from PEMV-2; however, PCR can differentiate between the two by using specific primers.

18.4.5 Pea seed-borne mosaic virus Genus ***Potyvirus*** (PSbMV)

Pea seed-borne mosaic virus was first reported in faba bean in Japan (Inouye, 1967). Symptoms in faba bean consist mostly of mild chlorosis and mottling with downward curling and distortion of leaf margins. Brownish necrotic lesions sometimes develop on the seed surface. In peas, the seed coat of seeds from infected plants often cracks at maturity.

Seed transmission rates of PSbMV can vary widely, based on host and virus isolates, ranging between 0 and 1.5% when 20 lentil genotypes were evaluated (Kumari and Makkouk, 1995); and can be as high as 100% in peas, especially when the testa is split (Brunt *et al.* 1996a).

Three strains of PSbMV (P1, P2 and P4) have been identified, and polyclonal antisera produced against PSbMV P1 and P2 can be used in DAS-ELISA for the detection of all three strains (Kohnen *et al.* 1995). Several serological methods have been used to test for PSbMV in seeds or tissue. These include ELISA, TBIA, dot-blot, immunospecific electron microscopy (ISEM) and dot immuno binding (DIB). In addition, RT-PCR has been used for pathotype-specific detection of P1 and P4 (Kohnen *et al.* 1995).

Pea seed-borne mosaic virus can be purified according to Alconero *et al.* (1986). A purified preparation of the virus has an A_{260}:A_{280} ratio of 1.14-1.18. PSbMV particles sediment as one component *ca* 154S and contain a single positive-sense strand RNA of *ca* 10 kb.

Good PSbMV control can be achieved if virus-free seed is planted. The use of genotypes that are resistant to the virus or resistant to seed transmission is a control measure that should also be considered. Kumari and Makkouk (1995) found that the lentil cvs Red Chief and Palouse had the lowest seed transmission rate and yield loss of the 20 lentil genotypes they evaluated. Controlling aphid vectors on plants will reduce the spread of the virus in the field. Insecticides, in general are less effective in controlling the spread of viruses vectored by aphids in a non-persistent manner. However, because seed-borne viruses create a 'within-field' inoculum source, insecticides could reduce the spread of the virus (Matthews, 1991). Reflective mulches have been used to reduce the incidence of PSbMV by as much as 78% in faba bean in Japan (Tachibana, 1981).

18.5 Pigeonpea (*Cajanus cajan*)

Pigeonpea is an important drought-resistant food legume in the Indian sub-continent. It is grown on an area of 3.8 million hectares, and has an annual production of more than 3 million tonnes. India contributes nearly 90% of the pigeonpea produced globally. It is mainly cultivated for its seed, which provides vital protein in the diets of an estimated 1.1 billion people around the world. Fifteen viruses have been reported to infect pigeonpea, but Pigeonpea sterility mosaic virus is the most devastating.

18.5.1 Pigeonpea sterility mosaic disease (PpSMD)

Pigeonpea sterility mosaic disease is becoming a potential threat to pigeonpea cultivation in the Indian subcontinent (Ghanekar *et al.* 1992). This disease was first reported from Pusa, Bihar state, India (Mitra, 1931). A susceptible genotype, infected at an early stage of crop growth, became completely sterile, leading to a 100% yield loss. However, as the plant matures, susceptibility to the virus decreases; such plants show only partial sterility (Muniyappa and Nangia, 1982). PpSMD has been reported to cause an annual loss, which exceeds 205,000 tons (valued at more than US$300 million) in India alone (Kannaiyan *et al.* 1984).

Symptoms are conspicuous mosaic, a bushy and pale green plant appearance, a drastic reduction in leaf size, leaf distortion, an increased number of secondary and tertiary branches from leaf axils and complete or partial cessation of reproductive structures (which leads to complete or partial sterility). Late infections may not show clear symptoms but, when ratooned, the new growth shows clear mosaic symptoms and sterility. The symptoms of pigeonpea sterility mosaic disease vary with the genotype of the plant and are of three types (a) severe mosaic on leaves with complete sterility, (b) mild mosaic with partial sterility and (c) chlorotic ring spots without any sterility (Reddy *et al.* 1998).

The causal agent of PpSMD remained unidentified for more than seven decades. Recently, however, a breakthrough was made using a new purification method. A novel virus was identified and provisionally named *Pigeonpea sterility mosaic virus* (PpSMV) (Kumar *et al.* 2000, 2001). The virus particles contain a 32 kDa protein and up to 6 RNA species of 0.8-3.5 kb (Kumar *et al.*, 2001). ELISA and RT-PCR-based diagnostic tools for the precise and sensitive detection of the virus in infected plants have been developed (Kumar *et al.*, 2001). The main characteristics of this virus are summarized in Table 18.2.

High summer rainfall, a low maximum temperature and a higher relative humidity (RH) in summer favour a PpSMD outbreak in the following crop season. The mite vector population is found throughout the year, but reaches its peak in April/May and is at a minimum in the winter season (Singh and Rathi, 1997). Temperatures in the range of 25-35 °C, low rainfall and RH are found to be highly favourable for mite multiplication (Singh, 1992).

Various insecticides and acaricides were used to manage PpSMD through vector control. PpSMD incidence may be reduced by removing perennial and volunteer plants of pigeonpea in the vicinity of pigeonpea fields in advance of the sowing season (Raychaudhary and Nariani, 1977). No significant reduction in disease incidence was observed either as a result of altered dates of sowing (Lakshmikantha *et al.* 1997) or of intercropping with non-host crops such as sorghum (Bhatnagar *et al*., 1984) or pearl millet (Zote *et al*., 1988). Pigeonpea cultivars resistant to PpSMV are available (Singh *et al*., 1995; Srinivas and Reddy, 1995).

18.6 Mungbean (*Vigna radiata*)

Greengram (generally known as mungbean) is an important short duration legume crop which provides easily digestible, high-quality protein in the human diet. Mungbean is grown in India, Myanmar, Pakistan, Thailand, Sri Lanka, Indonesia and China. Global production of mungbean is 2.5 to 3 million tonnes, grown on an area of 5 million hectares. The major viruses reported to infect mungbean worldwide are Mungbean leaf curl virus disease and *Mungbean yellow mosaic* (Tables 18.1 and 18.2).

18.6.1 Mungbean leaf curl virus disease (MLCV)

Leaf curl is an important viral disease affecting mungbean. Infection with MLCV may lead to yield losses of up to 40% (Amin *et al.* 1985). Characteristic symptoms are downward curling of leaves, veinal necrosis, chlorotic areas on the leaf lamina, and overall stunting of the plants and death of affected plants due to apical necrosis (Ghanekar *et al.* 1979). The virus produces characteristic ring spots on cowpea and necrotic lesions on Petunia (Amin *et al.* 1985).

The virus can easily be detected serologically by ELISA and reacts with antiserum produced against *Peanut bud necrosis virus* (PBNV) in ELISA and hemagglutination tests (Amin *et al.* 1985). MLCV is transmitted by thrips, *Frankliniella schultzei* in a persistent manner and also by sap (Amin *et al.* 1985). The virus is not transmitted by either seed or soil. The information available (Amin *et al.* 1985) propose a serological relationship between MLCV and TSWV, suggesting that MLCV is likely to be a Tospovirus

Though insecticides have been found to suppress whitefly populations, thereby reducing the spread of Mungbean yellow mosaic virus (MYMV), farmers do not use pesticides on mungbean, or on other pulses, because of the low returns obtained from these crops. Though sources of resistance are available in wild accessions, few resistant cultivars are available for cultivation. Therefore, efforts should be made to develop high yielding cultivars resistant to MYMV and MLCV.

18.6.2 ***Mungbean yellow mosaic virus*** Genus ***Begomovirus*** (MYMV)

Mungbean yellow mosaic virus is a severe disease of mungbean. The disease was first reported on mungbean in Delhi, India, by Nariani (1960). It can cause 100% yield loss if infection occurs three weeks after planting. However, seed loss will be small if infection occurs eight weeks after planting (Ahmad and Ahmad, 1991).

The disease can be characterized by the presence of bright yellow patches on the leaves interspersed with green areas. The first symptoms appear on young leaves in the form of mild scattered yellow specks or spots. Emerging trifoliate leaves show irregular yellow and green patches, with slight puckering and malformation in some cases. Diseased plants usually mature late and bear very few flowers, whilst their pods are of a reduced size.

The virus particles have a single stranded, circular DNA genome (Honda *et al.* 1986), which consists of two split DNA species of about 2.7 kb each. These types of DNA were identified as DNA-A and DNA-B by nucleic acid hybridization tests (Lazarowitz, 1992; Varma *et al.* 1991). Blackgram has been systemically infected with cloned DNAs of the blackgram isolate of MYMV through agroinoculation (Mandal *et al.* 1998).

The MYMV isolates from India, Pakistan and Thailand differed in their host range. The host range of the Thai isolate is restricted to *Leguminosae* (Honda *et al,* 1983), while the Indian isolates can infects species from other families, as *Brachiaria ramosa* (*Graminae*) and *Eclipta alba* (*Compositae*) (Nene *et al.* 1971). Moreover, the Indian isolates differed from the Thai isolate in the restriction pattern of their DNA-B. At least five variants of MYMV have been reported in India (Biswas and Varma, 2000; Mandal and Varma, 1996).

The incidence and spread of yellow mosaic disease are generally high in summer crops, as compared with rainy or winter season crops. Weeds serve as a reservoir of the virus, and are a source of primary inoculum. The virus starts spreading with the onset of the monsoon in India (Grewal, 1988), reaching a high incidence in areas where the temperature ranges between 31 and 35 °C and the RH is about 70% (Sharma *et al.* 1993). These conditions favour the development and multiplication of the vector *Bemisia tabaci.*

Management of yellow mosaic virus disease is mainly through host resistance and vector control. Spraying the crop 15 and 30 days after germination with dimethoate (0.03%) was found to reduce both whitefly populations and disease incidence. Monocrotophos was also effective in controlling whiteflies and MYMV spread (Sethuraman *et al.* 2001). Use of cotton as a trap crop (sown one month before the mungbean, between the mungbean rows), and spraying once with dimethoate (0.03%) 15 days after germination of mungbean, controlled the spread of both *B. tabaci* and

MYMV (Borah *et al.* 1996). Mungbean cultivars and breeding lines, which are resistant to MYMV, have been located (Raje and Rao, 2002; Sandhu *et al.*, 1996).

18.7 Blackgram (*Vigna mungo*)

Blackgram (known in India as urdbean) is widely grown in the Indian subcontinent. Viral diseases are a major constraint to blackgram production (Chenulu and Varma, 1988). Blackgram is being naturally infected by Blackgram mottle virus (BMoV), *Cowpea severe mosaic virus* (CPSMV), *Horegram yellow mosaic vuirs* (HgYMV), Mungbean leaf curl virus (MLCV), Mungbean mosaic virus (MbMV), *Mungbean yellow mosaic virus* (MYMV), *Tomato spotted wilt virus* (TSWV) and Urdbean leaf crinkle virus (ULCV) (Table 18.1). Their characteristics are summarized in Table 18.2.

Leaf crinkle virus is a major problem in all the urdbean growing areas in India (Brar and Rataul, 1989; Williams *et al.* 1968) and Pakistan (Bashir *et al.* 1991), causing losses in the range of 62-100%, depending upon the stage at which the plants become infected (Brar and Rataul, 1989). Hot water treatment of seeds at 55 °C for 30 min (Sharma and Dubey, 1984) or dry heat treatment at 70 °C for 10 min (Beniwal *et al.* 1983) resulted in elimination of ULCV without affecting seed germination. Spraying of mustard, rapeseed, sesame or groundnut oils at 1% and emulsions of rape seed and sesame oils at 2%, prevented the transmission of ULCV (Bharadwaj and Dubey, 1988). Urdbean cultivars resistant to ULCV are available (Sharma and Dubey, 1984).

18.8 Lima bean (*Phaseolus lunatus*)

P. lunatus (syn. *P. limensis,* are also known as Sieva bean, butter bean, sugar bean, and towe bean) is a native to Central America.. Lima beans are susceptible to over 30 different viruses when transmitted experimentally. The following viruses infect lima bean: *Bean calico mosaic virus* (BcaMV) in North America and Mexico, *Bean common mosaic virus* (BCMV) is widespread in lima bean (Morales and Bos, 1988), *Bean golden mosaic virus* (BGMV) in Latin America and in the Caribbean (Goodman and Bird 1978). *Bean yellow mosaic virus* (BYMV) is distributed worldwide, *Cucumber mosaic virus* (CMV) in USA (Harter, 1938), *Limabean golden mosaic virus* (LGMV), *Cowpea aphid-borne mosaic virus* (CABMV), Blackeye cowpea mosaic virus (BlCMV), *Cowpea mild mottle virus* (CPMMV) and CMV in Nigeria (Brunt *et al.* 1996b). Their main characteristics are summarized in Table 18.2. Lima bean is the only natural host for LGMV and there are no other experimental hosts reported. In Nigeria, LGMV was the major constraint. There was a mandated breeding program at the International Institute for Tropical Agriculture in Ibadan, Nigeria (IITA), but because of lima LGMV, IITA could not progress with the crop. As soon as the breeders planted their lines in the field at Ibadan, they became totally infected with LGMV. It was therefore decided to remove this crop from IITA's mandate.

18.9 Horsegram (*Macrotyloma uniflorum*)

Horsegram (often referred to as poor man's pulse) is considered to be the most drought tolerant and least risky of the pulse crops. It can do well in poor soils and in areas where other crops fail. Horsegram is mainly grown in India. It is relatively rich in the amino acid lysine, in comparison with pigeonpea and chickpea. It is also grown as a fodder and cover crop. Infection with *Horsegram yellow mosaic virus* Genus *Begomovirus* (HgYMV) is considered to be a threat to the cultivation of horsegram in India.

HgYMV was first reported by Williams *et al.* (1968) in India. Incidence of the disease ranges from 50 to 100% in both summer and early rainy season crops (Muniyappa, 1983).

The disease is characterized by typical, yellow mosaic symptoms on leaves, coupled with reduction of leaf size and plant height in severely infected plants. The leguminous weed *Indigofera hirsuta* serves as a natural reservoir of HgYMV (Muniyappa and Reddy, 1979). The main characteristics of HgYMV are summarized in Table 18.2.

HgYMV infection is severe in early summer reaching an incidence level which can be as high as 100%. The whitefly vector populations are most affected by rainfall, humidity and temperature. Severe HgYMV epidemic can be avoided by sowing horsegram during July-October when rainfall and humidity are high and temperatures are low (Muniyappa, 1983; Muniyappa *et al.* 1978b).

The best way to control HgYMV is through host resistance, and a number of horsegram genotypes tolerant/resistant to HgYMV have been identified (Muniyappa *et al.* 1978a). In future, HgYMV-resistant cultivars may control the disease if the immunity exhibited by *Macrotylum axillare* can be transferred to cutivars that mature early. This will permit the successful cultivation of horsegram in seasons when whitefly populations are high.

18.10 Bambara groundnut (*Vigna subterranea*)

Bambara groundnut (also known as ground-bean, Madagascar groundnut, Congo goober and kacang Bogor) is grown throughout the world, but is particularly important in Africa (e.g. Namibia, Botswana, Swaziland and Ghana). It is grown in the drier areas, due to its ability to withstand drought. It can be eaten roasted and salted, and is consumed as *akara* and bambara milk in Nigeria. Viruses can be local problems; many viral diseases have been reported, but rarely has the causal agent been identified.

Seven viruses have been reported previously from bambara groundnut in Nigeria, namely *Cowpea aphid-borne mosaic virus* (CABMV), Blackeye cowpea mosaic virus (BlCMV), *Cowpea mottle virus* (CPMoV), *Cowpea mosaic virus* (CPMV), *Cowpea mild mottle virus* (CPMMV), *Cucumber mosaic virus* (CMV) and *Southern bean mosaic virus* (SBMV) (Thottappilly and Rossel, 1997). The following viruses have been reported from bambara groundnut elsewhere: *Peanut mottle virus* (PeMoV) from Kenya (Bock *et al.* 1978) and West Africa (Li *et al.* 1991), CPMoV from Burkina Faso (Drabo *et al.* 2002) and Ghana (Doku, 2002), *Voandezia necrotic mosaic virus* (VNMV), *Voandezia distortion mosaic virus* (VDMV) and SBMV from Côte d'Ivoire (Fauquent and Thouvenel, 1987) (Table 18.1). The main characteristics of these viruses are summarized in Table 18.2.

18.11 Winged bean (*Psophocarpus tetragonolobus*)

P. tetragonolobus (syn: *Botor tetragonolobus*, *Dolichos tetragonolobus*; also known as winged bean, Goa bean, princess bean, and pois carré) is widely cultivated in the tropics, especially in Asia (including India, Myanmar, Malaysia, the Philippines and Indonesia), West Africa, the West Indies and parts of the USA. Winged bean is used to make 'milk' and flour or the beans are cooked and eaten; both the shoots and the tubers can be eaten as vegetables. In Sri Lanka, the yield obtained from winged bean is about 10 t/ha (Crop Protection Programme, Ministry of Agriculture and Livestock, Sri Lanka). The following viruses have been reported from winged bean: *Cucumber mosaic virus* (CMV) reported to cause ringspots of winged bean in West Africa and Fiji (Brunt and Phillips, 1981; Fauquet *et al.* 1979; Rossel and Thottappilly, 1985). *Cowpea severe mosaic virus* (CPSMV) spreads in North, South and Central America, and similar symptoms may be caused by this virus in East Africa (Allen, 1983; Kitajima *et al.*, 1979). In addition, *Cowpea aphid-borne mosaic virus* (CABMV), Blackeye cowpea mosaic virus (BlCMV), *Cowpea mild mottle virus* (CPMMV) and CMV were detected routinely in winged bean in Nigeria (Table 18.1). The main characteristics of these viruses are summarized in Table 18.2.

References

Abu Salih, H. S., Ishag, H. M. and Siddig, S. A. (1973). Effects of sowing date on incidence of Sudanese broad bean mosaic virus in, and yield of, *Vicia faba*. *Ann. appl. Biol.* 74: 371-378.

Ahmad, Q. and Ahmad, Q. (1991). Growth attributes and grain yield of mungbean plants affected by mungbean yellow mosaic virus in field. *Indian Phytopathology* 43: 559-560.

Alconero, R., Provvidenti, R. and Gonsalves, D. (1986). Three pea seedborne mosaic virus pathotypes from pea and lentil germ plasm. *Plant Dis.* 70: 783-786.

Allen, D. J. (1983). *The Pathology of Tropical Food Legumes: Disease Resistance in Crop Improvement.* John Wiley & Sons, Chichester, UK, 413 pp.

Amin, P. W., Ghanekar, A. M., Rajeshwari, R. and Reddy, D. V. R. (1985). Tomato spotted wilt virus as the causal pathogen of leaf curl of mungbean and urdbean in Andhra Pradesh. *Indian J. Plant Protection* 13: 9-13.

Ashby, J. W. 1984. Bean leaf roll virus. *CMI/AAB Descriptions of Plant Viruses* No. 286.

Bashir, M., Mughal, S. M. and Malik, B. A. (1991). Assesment of yield lossess due to leaf crinkle virus in urdbean. *Pakistan J. Botany* 23: 140-142.

Bariana, H. S., Shannon, A. L., Chu, P. W. G. and Waterhouse, P. M. (1994). Detection of five seedborne legume viruses in one sensitive multiplex polymerase chain reaction test. *Phytopathology* 84:1201-1205.

Bawden, F. C., Chaudhuri, R. P. and Kassanis, B. (1951). Some properties of broad bean mottle virus. *Ann. appl. Biol.* 38: 774-784.

Beniwal. S. P. S., Bharathan, N. and Chaubey, S. N. (1983). Two cucurbit hosts of ULCV. *Indian Phytopathology* 36: 577-579.

Bhardwaj, S. V. and Dubey, G. S. (1988). Effect of oil on transmission of urdbean (*Vigna mungo*) leaf rinkle virus by *Aphis craccivora*. *Indian J. Virology* 4: 82-86.

Bhatnagar, V. S., Nene, Y. L. and Jadhav, D. R. (1984). Sterility mosaic disease of pigeonpea in a sorghum based cropping system. *Intern. Pigeonpea News Letter* 3: 37-38.

Bird, J., Sanchez, J., Rodriguez, R. L. and Julia, F.J. (1975). Rugaceous (whitefly transmitted) viruses in Puerto Rico. In: *Tropical Diseases of Legumes*, p. 3. (ed. J. Bird and K. Maramorosch). Academic Press, New York.

Biswas, K. K. and Varma, A. (2000). Identification of variants of mungbean yellow mosaic geminivirus by host reaction and nucleic acid spot hybridization. *Indian Phytopathology* 53: 134-141.

Bock, K. R. and Conti, M. (1974). Cowpea aphid-borne mosaic virus. *CMI/AAB Descriptions of Plant Viruses* No. 134.

Bock, K. R. and Kuhn, C. W. (1975). Peanut mottle virus. *CMI/AAB Descriptions of Plant Viruses* No. 141.

Bock, K. R., Guthrie, E. J. and Meredith, G. (1978). Viruses occurring in East Africa that are related to peanut mottle virus. *Ann. appl. Biol.* 89: 423-428.

Böning, K. (1927). Die Mosaikkrankheit der Ackerbohne (*Vicia faba*). *Forsch. Geb. Pflanzenkrankh. Immunität Pflanzenr.* 4: 43-111.

Borah, R. K., Nath, P. D. and Deka, N. (1996). Effect of insecticides and crop trap on the incidence of white fly Bemisia tabaci and yellow mosaic virus in greengram, *Vigna radiata*. *Indian J. Virology* 12: 75-77.

Bos, L. (1970). Bean yellow mosaic virus. *CMI/AAB Descriptions of Plant Viruses* No. 40.

Bos, L., Hampton, R. O. and Makkouk, K. M. (1988). Viruses and virus diseases of pea, lentil, faba bean and chickpea. In: *World Crops: Cool Season Food Legumes*, pp. 591-615 (ed. R. J. Summerfield). Kluwer Academic Publishers, Dordrecht, The Netherlands.

Bosque-Perez, N. A. and Buddenhagen, I. W. (1990). Studies on epidemiology of virus disease of chickpea in California. *Plant Dis.* 74: 372-378.

Bozarth, R. F. and Shoyinka, S. A. (1979). Cowpea mottle virus. *CMI/AAB Descriptions of Plant Viruses* No. 212.

Brar, J. S. and Rataul, H. S. (1989). Incidence and extent of losses due to leaf crinkle virus in urdbean (*Vigna mungo*). *PAU J. Research* 26: 65-70.

Brunt, A. A. and Phillips, S. (1981). *Rept. Glasshouse Crops Res. Inst. 1979*, pp. 157-158.

Brunt, A. A. and Kenten, R. H. (1974). Cowpea mild mottle virus. *CMI/AAB Descriptions of Plant Viruses* No. 140.

Brunt, A. A., Crabtree, K., Dallwitz, M. J., Gibbs, A. J. and Watson L. (1996a). Viruses of Plants; Descriptions and Lists from the VIDE Database. CAB Interna. Wallingford, UK.

Brunt, A. A., Crabtree, K., Dallwitz, M. J., Gibbs, A. J., Watson, L. and Zurcher, E. J. (eds.) (1996b onwards). `Plant Viruses Online: Descriptions and Lists from the VIDE Database. Version: 20th August 1996.' URL: http://biology.anu.edu.au/Groups/MES/vide/

Chenulu, V. V. and Varma, A. (1988). Virus and virus like diseases of pulse crops commonly grown in India. In: *Pulse Crops,* pp. 338-370 (eds. B. Baldev, S. Ramanujan and H. K. Jain). Oxford and IBH Publishing Company, Pvt. Ltd. New Delhi, India.

Cockbain, A. J. (1983). Viruses and virus-like disease of *Vicia faba* L. In: *The Faba Bean (*Vicia faba *L.),* pp. 421-461 (ed. P.D. Hebblethwaite). Butterworths, London, UK.

D'Arcy, C. J., Martin, R. R. and Spiegel, S. (1989). A comparative study of luteovirus purification methods. *Can. J. Plant Pathol.* 11: 251-255.

De Jager, C. P. (1979). Cowpea severe mosaic virus. *CMI/AAB Descriptions of Plant Viruses* No. 209.

Dickson, B. T. (1921). Plant diseases of 1920-21. In: *13th Ann. Rept. Quebec Soc. Protection Plants*, pp. 66-67.

Doku, E. V. (2002). *Proc. 2nd Internal. bambara groundnut network workshop, Accra,* September 23-25, 1998, pp. 30-32.

Drabo, I., Sérémé, P. and Dabire, C. (2002). *Ibid.* pp. 19-26.

Dubey, G. S., Sharma, I. and Prakash, N. (1983). Some properties of urdbean leaf crinkle virus. *Indian Phytopathology* 36: 762-764.

Duffus, J. E. (1961). Economic significance of beet western yellows (radish yellows) on sugar beet. *Phytopathology* 51: 605-607.

Duffus, J. E. (1972). Beet western yellows virus. *CMI/AAB Descriptions of Plant Viruses* No. 89.

Fauquet, C. and Thouvenel, J. C. (1987). Plant viral diseases in the Ivory Coast. ORSTOM, Paris, Documentation Techniques, 46: 243 p.

Fauquet, C., Lamy, D. and Thouvenel, J. C. (1979). *FAO Plant Protection Bulletin* 27: 81-87.

Fauquet, C., Monsarrat, A. and Thouvenel, J.C. (1984). Voandzeia necrotic mosaic virus. *CMI/AAB Descriptions of Plant Viruses* No. 279.

Fortass, M. and Bos, L. (1991). Survey of faba bean (*Vicia faba* L.) for viruses in Morocco. *Neth. J. Plant Path.* 97: 369-380.

Francki, R. I. B., Mossop, D. W. and Hatta, T. (1979). Cucumber mosaic virus. *CMI/AAB Descriptions of Plant Viruses* No. 213.

Frowd, J. A. and Bernier, C. C. (1977). Virus diseases of faba beans in Manitoba and their effects on plant growth and yield. *Can. J. Plant Science* 57: 845-852.

Gadh, I. P. S. and Bernier, C. C. (1984). Resistance in faba bean (*Vicia faba*) to bean yellow mosaic virus. *Plant Dis.* 68: 109-111.

Ghanekar, A. M., Reddy, D. V. R. and Amin, P. W. (1979). Leaf curl disease of mungbean and urdbean caused by tomato spotted wilt virus. *Indian Phytopathology* 32: 163.

Ghanekar, A. M., Sheila, V. K., Beniwal, S. P. S., Reddy, M. V. and Nene, Y. L. (1992). Sterility mosaic of pigeonpea. In: *Plant Diseases of International Importance: Diseases of Cereals and Pulses vol I,* pp. 415-428 (eds. U. S. Singh, A. N. Mukhopadhyay, J. Kumar, and H. S. Chaubey). Prentice Hall, New Jersey, USA.

Gibbs, A. J. (1972). Broad bean mottle virus. *CMI/AAB Descriptions of Plant Viruses* No. 101.

Gibbs, A. J. and Paul, H. L. (1970). Echtes ackerbohnemosaik-virus. *CMI/AAB Descriptions of Plant Viruses* No. 20.

Gibbs, A. J. and Smith, H. G. (1979). Broad bean stain virus. *CMI/AAB Descriptions of Plant Viruses* No. 29.

Goodman, R. M. and Bird, J. (1978). Bean golden mosaic virus. *CMI/AAB Descriptions of Plant Viruses* No. 192.

Grewal, J. S. (1988). Diseases of pulse crops an over view. *Indian Phytopathology* 41: 1-14.

Hamed, A. A. and Makkouk, K. M. (2002). Occurrence and management of chickpea chlorotic dwarf virus in chickpea fields in northern Sudan. *Phytopath. Medi.* 41: 193-198.

Hampton, R. O. and Mink, G. I. (1975). Pea seed-borne mosaic virus. *CMI/AAB Descriptions of Plant Viruses* No. 146.

Harrison, B. D. (1973). Pea early-browning virus. *CMI/AAB Descriptions of Plant Viruses* No. 120.

Harter, L. L. (1938). Mosaic of lima beans (*Phaseolus lunatus macrocarpus*). *J. Agric. Res.* 56: 895-906

Honda, Y. and Ikegami, M. (1986). Mung bean yellow mosaic virus. *CMI/AAB Descriptions of Plant Viruses* No. 323.

Honda, Y., Iwaki, M., Saito, Y., Thongmeearkom, P., Kittisak, K. and Deema, N. (1983). Mechanical transmission, purification and some properties of whitefly borne mungbean yellow mosaic virus in Thailand. *Plant Dis.* 67: 801-804.

Honda, Y., Iwaki, M., Thongmeearkom, P., Kiratiya, A. K., Kiratiya, A. S., Srithongchai, W., Prommin, M., Kittipakorn, K., Sarindu, N., Deema, N., Syamananda, R., Hibi, T. and Saito. Y. (1986). Mungbean yellow mosaic virus isolated from mungbean in Thailand. *Tech. Bull. Trop. Agri. Res. Center* 21: 189-202.

Horn, N. M., Reddy, S. V., Roberts, I. M. and Reddy, D. V. R. (1993). Chickpea chlorotic dwarf virus, a new leafhopper-transmitted geminivirus of chickpea in India. *Ann. appl. Biol.* 122: 467-479.

Horn, N. M, Reddy, S. V., van den Heuvel, J. F. J. M. and Reddy, D. V. R. (1996). Survey of Chickpea (*Cicer arietinum* L.) for Chickpea Stunt Disease and Associated Viruses in India and Pakistan. *Plant Dis.* 80: 286-290.

Hull, R. (1969). Alfalfa mosaic virus. *Adv. Virus Rese.* 15: 265-433.

Ie, T. S. (1970). Tomato spotted wilt virus. *CMI/AAB Descriptions of Plant Viruses* No. 39.

Inouye, T. (1967). A seed-borne mosaic virus of pea. *Ann. Phytopath. Soci. Japan* 33: 38-42.

Jaspars, E. M. J and Bos, L. (1980). Alfalfa mosaic virus. *CMI/AAB Descriptions of Plant Viruses* No. 229.

Johnston, G. R. and Rapley, P. E. L. (1979). The effect of time of sowing on the incidence of subterranean clover red leaf virus infection in road bean (*Vicia faba*). *Ann. appl. Biology* 99: 135-41.

Jones, R. A. C. and Coutts, B. A. (1996). Alfalfa mosaic and cucumber mosaic virus infection in chcikepa and lentil: incidence and seed transmission. *Ann. appl. Biol.* 129: 491-506.

Jones, R. A. C. and Pathipanawat, W. (1989). Seed-borne alfalfa mosaic virus infecting annual medics (*Medicago* spp.) in western Australia. *Ann. appl. Biol.* 115: 263-277.

Kaiser, W. J. (1973a). Etiology and biology of viruses affecting lentil (*Lens esculenta* Moench) in Iran. *Phytopath. Mediter.*, 12: 7-14.

Kaiser, W. J. (1973b). Biology of bean yellow mosaic and pea leaf roll viruses affecting *Vicia* faba in Iran. *Phytopath. Z.* 78: 253–263.

Kaiser, W. J., Danesh, D., Okhovat, M. and Mossahebi, H. (1968). Diseases of pulse crop (edible legumes) in Iran. *Plant Dis. Rept.* 52: 687-691.

Kannaiyan, J., Nene, Y. L., Reddy, M. V., Ryan, J. G. and Raju, T. N. (1984). Prevalence of Pigeonpea diseases and associated crop losses in Asia, Africa, and America. *Tropical Pest Management* 30: 62-71.

Katul, L., Vetten, H. J., Maiss, E., Makkouk, K. M., Lesemann, D. E. and Casper, R. (1993). Charecteristics and serology of virus-like particles associated with faba bean necrotic yellows. *Ann. appl. Biol.* 123: 629-647.

Kitajima, E. W., Noda, H., Lin, M. T. and Costa, C. (1979). Um mosaico em feijρo de asa)*Psophocarpus tetragonolobus*) causado por um isolado do sub-grupo severo do virus do mosaico de Vigna. *Fitopatologia Brasiliera* 4: 519-524.

Klein, R. E., Larsen, R. C. and Kaiser, W. J. (1991). Virus epidemic of grain legume in Eastern Washington. *Plant Dis.* 75: 1186.

Kohnen, P. D., Johansen, I. E. and Hampton, R. O. (1995). Characterization and molecular detection of the P4 pathotype of pea seedborne mosaic potyvirus. *Phytopathology* 85: 789-793.

Kumar, P. L., Jones A. T., Sreenivasulu, P. and Reddy D. V. R. (2000). Breakthrough in identifying the causal virus of Pigeonpea sterility mosaic disease. *J. Mycology Plant Pathol.* 30: 249.

Kumar, P. L., Jones, A. T., Duccan, G., Roberts, I. M. and Reddy, D. V. R. (2001). Characterization of a novel mite transmitted virus associated with pigeonpea sterility mosaic disease. *Phytopathology* 91: 51.

Kumari, S. G. and Makkouk, K. M. (1995). Variability among twenty lentil genotypes in seed transmission rated and yield loss induced by pea seed-borne mosaic potyvirus infection. *Phytopath. Mediter.* 34: 129-132.

Kumari, S. G. and Makkouk, K. M. (1996). Inactivation of broad bean stain comovirus in lentil seeds by dry heat treatment. *Phytopath. Mediter.* 35: 124-126.

Lakshmikantha, B .P., Prabhuswamy, H. P. and Jagadish, K. S. (1997). Effect of staggered sowing on the incidence of sterilty mosaic and its mite vector, *Aceria cajani* Channabasavanna. *Mysore J. Agric. Scie.* 31: 29-32.

Lane, L. C. (1974). The Bromovirus group. *Adv. Virus Res.* 19: 151-220.

Lazarowitz, S. G. (1992). Geminiviruses: Genome structure and function. *Critical Review of Plant Science* 11: 327-349.

Li, R. H., Zettler, F. W., Elliott, M. S., Peterson, M. A., Still, P. E., Baker, C. A. and Mink, G. I. (1991). A strain of peanut mottle virus seed-borne in bambara groundnut. *Plant Dis.* 75: 130-133.

Lloyd, A. T. E., Smith, H. G. and Jones, L. H. (1965). Evesham stain-a virus disease of broad beans (*Vicia faba* L.). *Hort. Res.* 5: 13-18.

Lot, H., Marrou, J., Quiot, J. B. and Esvan, C. (1972). Contribution l' etude du virus mosaique du concombre (CMV). II. Methode de purification rapide du virus. *Annal. Phytopathologie* 4: 25-38.

Mahmood, K. and Peters, D. (1973). Purification of pea enation mosaic virus and the infectivity of its components. *Neth. J. Plant Pathol.* 79: 138-147.

Makkouk, K. M. and Kumari, S. G. (1990). Variability among 19 lentil genotypes in seed transmission rates and yield loss induced by broad bean stain virus infection. *LENS Newsletter* 17: 31-33.

Makkouk, K. M. and Kumari, S. G. (1995). Screening and selection of faba bean (*Vicia faba* L.) germplasm for resistance to bean yellow mosaic potyvirus. *J. Plant Disease Protection* 102: 461-466.

Makkouk, K. M. and Kumari, S. G. (2001). Reduction of spread of three persistently aphid-transmitted viruses affecting legume crops by seed-treatment with Imidacloprid (Gaucho®). *Crop Protection* 20: 433-437.

Makkouk, K. M., Bos, L., Azzam, O. I., Katul, L. and Rizkallah, A. (1987). Broad bean stain virus: identification detectability in faba bean leaves and seeds, occurrence in West Asia and North Africa and possible wild hosts. *Neth. J. Plant Pathol.* 93: 97-106.

Makkouk, K. M., Bos, L., Azzam, O. I., Kumari, S. and Rizkallah, L. (1988). Survey of viruses affecting faba bean in six Arab countries. *Arab J. Plant Protection* 6: 53-61.

Makkouk, K. M., Rizkallah, L., Madkour, M., El-Sherbeiny, M., Kumari, S. G., Amriti, A. W. and Solh, M. B. (1994). Survey of faba bean (*Vicia faba* L.) for viruses in Egypt. *Phytopath. Mediter.* 33: 207-211.

Makkouk, K. M., Kumari, S. G. and Bayaa, B. (1999). First report of pea enation mosaic virus affecting lentil (*Lens culinaris* Medik.) in Syria. *Plant Dis.* 83: 303.

Makkouk, K. M., Kumari, S., Sarker, A. and Erskine, W. (2001). Registration of six lentil germplasm lines with combined resistance to viruses. *Crop Science* 41: 931-932.

Makkouk, K. M., Fazlali, Y., Kumari, S. G. and Farzadfar, S. (2002a). First record of *Beet western yellows, Chickpea chlorotic dwarf, Faba bean necrotic yellows* and *Soybean dwarf viruses* affecting chickpea and lentil crops in Iran. *Plant Pathology* 51(3): 387.

Makkouk, K. M., Kumari, S. G. and van Leur, J. A. G. (2002b). Screening and selection of faba bean (*Vicia faba* L.) germplasm resistant to *Bean leafroll virus. Austr. J. Agric. Res.* 53: 1077-1082.

Mandal, B. and Varma, A. (1996). Differentiation of natural variants of mungbean yellow mosaic geminivirus by host reactions and DNA-DNA hybridization. *Internat. J. Tropical Plant Dis.* 14: 189-202.

Mandal, B., Varma, A. and Malathi, V. G. (1998). Some biological and genomic properties of pigeonpea isolate of mungbean yellow mosaic geminivirus. *Indian Phytopathology* 51: 121-129.

Matthews, R. E. F. (1991). *Plant Virology*. Academic Press, Inc. (London) LTD. Berkeley Square House, London W1X 6BA, UK. 813 pp.

Mitra, M. (1931). Report of Imperial Mycologist. Sci. Rept. Imp. Inst. Agric. Res. Pusa 1929-30, 45-71.

Morales, F. J. and Bos, L. (1988). Bean common mosaic virus. *CMI/AAB Descriptions of Plant Viruses* No. 337.

Muniyappa,V. (1983). Epidemiology of Yellow mosaic disease of horsegram (Macrotyloma uniflorum) in southern India. In: *Plant Virus Epidemiology*, pp. 331-335 (eds. R.T Plumb and J. M. Thresh). Blackwell Scientific Publications, Oxford, UK.

Muniyappa V., Chandrashekariah, S. C. and Shivashankar, G. (1978a). Horsegram cultures tolerant to yellow mosaic. *Ind. J. Genetics and Plant Breeding* 38: 148.

Muniyappa V., Reddy, H. R. and Mustak Ali, T. M. (1978b). Studies on the yellow mosaic disease of horsegram (*Dolichos biflorus*). IV. Epidemiology of the disease. *Mysore J. Agric. Sciences* 12: 277-279.

Muniyappa, V. and Nangia, N. (1982). Pigeonpea cultivars and selection for resistance to sterilty mosaic in relation to prevalence of eriophyid mite (*Aceria cajani* Channabasavanna). *Trop. Grain. Legume Bull.* 25: 28-30.

Muniyappa, V. and Reddy H. R. (1979). Indigofera hirsutum A Natural Reservoir of horsegram yellow Mosaic Virus. *Madras Agric J.* 66: 350.

Najar, A., Makkouk, K. M., Boudhir, H., Kumari, S. G., Zarouk, R., Bessai, R. and Ben Othman, F. (2000). Viral diseases of cultivated legume and cereal crops in Tunisia. *Phytopath. Mediter.* 39: 423-432.

Nariani, T. K. (1960). Yellow mosaic of mung. *Indian Phytopathology* 13: 24-29.

Nene, Y. L., Naresh, J. S. and Nair, N. G. (1971). Additional hosts of mungbean yellow mosaic virus. *Indian Phytopathology* 24: 415-417.

Nour, M. A. and Nour, J. J. (1962). A mosaic disease of *Dolichos lablab* and diseases of other crops caused by alfalfa mosaic virus in the Sudan. *Phytopathology* 52: 427-432.

Peters, D. (1982). Pea enation mosaic virus. *CMI/AAB Descriptions of Plant Viruses* No. 257.

Proeseler, G., Fritzsche, R. and Schimanski, B. (1976). Laboratory trials with insecticides, mineral oil and combinations of them for the reduction of aphid transmission of non-persistent viruses. *Arch. Phytopathologie, Pflanzenschutz* 12: 19-26.

Purcifull, D. and Gonsalves, D. (1985). Blackeye cowpea mosaic virus. *CMI/AAB Descriptions of Plant Viruses* No. 305.

Quantz, L. and Volk, J. (1954). Die Blattrollkrankheit der Ackerbohne und Erbse, eine neue Viruskrankheit bei Leguminosen. *NachrBl. dt. PflSchtzdienst* 6: 177-182.

Raje, R. S. and Rao, S. K. (2002). Screening of mungbean (*Vigna radiata*) germplasm for yellow mosaic virus, Cercospora leaf spot and Powdery mildew. *Legume Res.* 25: 99-104.

Raychaudhary, S. P. and Nariani, T. K. (1977). Sterility mosaic of pigeonoea. In: *Virus and Mycoplasma Diseases of Plants in India,* pp. 19-20. Oxford & IBH Publishing Co., New Delhi, India.

Reddy, M. V., Raju, T. N. and Lenne, J. M. (1998). Diseases of pigeonpea. In: *The Pathology of Food and Pasture Legumes*, pp 517-558 (eds D. J. Allen and J. M. Lenne). CAB International, Wallingford, UK.

Rossel, H. W. and Thottappilly, G. (1985). *Virus Diseases of Important Food Crops in Tropical Africa.* Intern. Inst. Tropical Agric. (IITA) Publications Series, Ibadan, Nigeria, 67 pp.

Sandhu, J. S., Dhiman, J. S., Sharma, S. K and Brar, J. S. (1996). Screening of mungbean genotypes for resistance against mungbean diseases. *Crop Improvement* 23: 293-294.

Schmutterer, H. and Thottappilly, G. (1972). Zur wirtschaftlichen Bedeutung und Ausbreitung des Erbsenblattrollvirus im Ackerbohnenbestand sowie zur chemischen Bekampfung der Vektoren. *Z. PflanzenKrankh. PflanzenSchutz* 79: 478–484.

Scott, H. A. and Hoy, J. W. (1981). Blackgram mottle virus. *CMI/AAB Descriptions of Plant Viruses* No. 237.

Sethuraman, K., Manivannan, N. and Natarajan, S. (2001). Management of yellow mosaic disease of urdbean using neem products. *Legume Res.* 24: 197-199.

Sharma, I. and Dubey, G. S. (1984). Control of leaf crinkle virus through heat treatment, chemotherapy and resistance. *Indian Phytopathology* 37: 26-30.

Sharma, S., Yadav, O. P., Thareja, R. K., Kaushik, J. C. and Sharma, S. (1993). Nature and extent of losses due to mungbean yellow mosaic virus and its epidemiology. *HAU J. Res.* 23: 51-53.

Singh, A. B., Singh, D. P., Singh, N. P. and Kumar, R. (1995). Evaluation of pigeonpea germplasm for resistance to sterility mosaic. *Intern. Chickpea Pigeonpea News Letter* 2: 69-70.

Singh, A. K. (1992). Sterility mosaic of pigeonpea: etiology, epidemiology and management. Ph. D. Thesis, GBPU & AT, Pantanagar, U. P., India, 151 pp.

Singh, A. K. and Rathi, Y. P. S. (1997). Epidemiology of pigeonpea sterility mosaic virus. *Indian J. Virology* 13: 143-145.

Srinivas, T and Reddy, M. V. (1995). Pigeonpea lines resistant to old and new strains of sterility mosaic disease. *Intern. Chickpea Pigeonpea News Letter* 2: 67-69.

Tachibana, Y. (1981). Control of aphid-borne viruses in faba bean by mulching with silver polyethylene film. *FABIS Newsletter* 3: 56.

Thottappilly, G. and Rossel, H. W. (1997). Identification and characterization of viruses infecting bambara groundnut (*Vigna subterranean*) in Nigeria. *Intern. J. Pest Manag.* 43: 177-185.

Tomlinson, J. A. (1970). Lettuce mosaic virus. *CMI/AAB Descriptions of Plant Viruses* No. 9.

Tremaine, J.H. (1983). Southern bean mosaic virus. *CMI/AAB Descriptions of Plant Viruses* No. 274.

Van Kammen, A. and de Jager, C.P. (1978). Cowpea mosaic virus. *CMI/AAB Descriptions of Plant Viruses* No. 197. .

Van Vloten-Doting, L. and Jaspars E. M. J. (1972). The uncoating of alfalfa mosaic virus by its own RNA. *Virology* 48: 699-708.

Varma, A., Dhar, A. K. and Malathi, V. G. (1991). Cloning and restriction analysis of mungbean yellow mosaic virus. In*: Intern/ Conf. Virology in the Tropics*. Lucknow, India, 114 pp.

Williams, F. J., Grewal, J. S. and Amin, K. S. (1968). Serious and new disease of pulse crops in India in 1966. *Plant Dis. Reptr.* 52: 300-304.

Zote, K. K., Dandanaik, B. P., Khalikar, P. V. and Raut, K. G. (1988). Effect of intercropping with sorghum and pearl millet on incidence of pigeonpea sterility mosaic. *Intern. Pigeonpea News Letter* 14: 19-21.

Table 18.1. Economically important viruses from naturally-infected faba bean, chickpea, lentil, pigeonpea, mungbean, blackgram, lima bean, horsegram, bambara groundnut and winged bean.

Crop/ Virus	Crop/ Virus
Faba bean	**Mungbean (cont.)**
* *Bean leafroll virus*	Mungbean mosaic virus
* *Bean yellow mosaic virus*	Mungbean mottle virus
* *Beet western yellows virus*	* *Mungbean yellow mosaic virus*
* *Broad bean mottle virus*	Urdbean leaf crinkle virus
Broad bean true mosaic virus	**Blackgram**
* *Faba bean necrotic yellows virus*	* Blackgram mottle virus
Pea early-browning virus	*Cowpea severe mosaic virus*
Pea enation mosaic virus	*Horsegram yellow mosaic virus*
Tomato spotted wilt virus	Mungbean leaf curl virus
Chickpea	Mungbean mosaic virus
* *Alfalfa mosaic virus*	*Mungbean yellow mosaic virus*
* *Bean leafroll virus*	*Tomato spotted wilt virus*
* *Beet western yellows virus*	* Urdbean leaf crinkle disease
Chickpea bushy dwarf virus	**Lima bean**
* Chickpea chlorotic dwarf virus	* *Bean calico mosaic virus*
Chickpea filiform virus	* *Bean common mosaic virus*
Chickpea stunt disease associated virus	* *Bean golden mosaic virus*
* *Cucumber mosaic virus*	* *Bean yellow mosaic virus*
* *Faba bean necrotic yellows virus*	Blackeye cowpea mosaic
Lettuce mosaic virus	*Cowpea aphid-borne mosaic virus*
Tomato spotted wilt virus	*Cowpea mild mottle virus*
Lentil	* *Cucumber mosaic virus*
Alfalfa mosaic virus	*Horsegram yellow mosaic virus*
* *Bean leafroll virus*	* *Limabean golden mosaic virus*
* *Bean yellow mosaic virus*	*Peanut mottle virus*
* *Broad bean stain virus*	**Horsegram**
Cucumber mosaic virus	* *Horsegram yellow mosaic virus*
* *Pea enation mosaic virus*	**Bambara groundnut**
* *Pea seed-borne mosaic virus*	* *Bean common mosaic virus*
Pigeonpea	Blackeye cowpea mosaic virus
Cowpea severe mosaic virus	* *Cowpea aphid-borne mosaic virus*
Horsegram yellow mosaic virus	* *Cowpea mild mottle virus*
Mungbean yellow mosaic virus	* *Cowpea mottle virus*
Pigeonpea proliferation virus	*Cowpea mosaic virus*
* Pigeonpea sterility mosaic virus	*Cucumber mosaic virus*
Rhynchosia mosaic virus	* *Peanut mottle virus*
Mungbean	*Southern bean mosaic virus*
Alfalfa mosaic virus	* *Voandezia distortion mosaic virus*
Bean common mosaic virus	* *Voandezia necrotic mosaic virus*
Blackgram mottle virus	**Winged bean**
Cowpea severe mosaic virus	Blackeye cowpea mosaic virus
Cucumber mosaic virus	*Cowpea aphid-borne mosaic virus*
Horsegram yellow mosaic virus	*Cowpea mild mottle virus*
* Mungbean leaf curl virus	* *Cowpea severe mosaic virus*
	* *Cucumber mosaic virus*

*Viruses which are more important and causes epidemics in some regions.

Table 18. 2. The main features of viruses reported from naturally infected faba bean, chickpea, lentil, pigeonpea, mungbean, blackgram, lima bean, horsegram, bambara groundnut and winged bean.

Virus	Natural Legume crops infected[1]	Manner of transmission[2]	Virus particle Shape[3] and Size (nm) Diameter/ length	Geographical distribution[4]	Host range (non legumes)	Reference[5]
Alfalfa mosaic virus (AlMV, genus *Alfamovirus*, family *Bromoviridae*)	Cp*[6], FB*, L*, M, OL	Sa, Se[7], Aphids NP[8]	Ba, 30-57	Af, Am, As, Au, Eu	167 genera in 71 families	Jaspars and Bos, 1980 (229)
Bean calico mosaic virus (BcaMV, genus *Begomovirus*, family *Geminiviridae*)	LB	Sa, Whiteflies	Ge, 20 x 30	Am	2 species in 2 families	Brunt *et al.* 1996b
Bean common mosaic virus (BCMV, genus *Potyvirus*, family *Potyviridae*)	Bg, LB, M, OL	Sa, Se, Aphids NP	Fi, 750	Af, Am, As, Au, Eu	6 species in 4 families	Morales and Bos, 1988 (337)
Bean golden mosaic virus (BGMV, genus *Begomovirus*, family *Geminiviridae*)	LB, OL	Sa, Whiteflies	Ge, 18-20 x 30	Af, Am, As	2 species in 2 families	Goodman and Bird, 1978 (192)
Bean leafroll virus (BLRV, family *Luteoviridae*)	Cp, FB, L, OL	Aphids P[8]	Is, 27	Af, Am, As, Au, Eu	None	Ashby, 1984 (286)
Bean yellow mosaic virus (BYMV, genus *Potyvirus*, family *Potyviridae*)	Cp, FB*, L*, LB*, OL	Sa, Se, Aphids NP	Fi, 750	Af, Am, As, Au, Eu	24 species in 12 families	Bos, 1970 (40)
Beet western yellows virus (BWYV, genus *Polerovirus*, family *Luteoviridae*)	Cp, FB, L, OL	Aphids P	Is, 26	Af, Am, As, Au, Eu	Over 100 species in 21 families	Duffus, 1972 (89)

Blackeye cowpea mosaic virus (BlCMV, genus *Potyvirus*, family *Potyviridae*)	Bg, LB, Wb	Sa, Se, Aphids NP	Fi, 750	Af, Am, As	36 species in 7 families	Purcifull and Gonsalves, 1985 (305)
Blackgram mottle virus (BMoV, family *Tombusviridae*)	B*, M*, P, OL	Sa, Se, Beetles	Is, 28	As, Au	None	Scott and Hoy, 1981 (237)
Broad bean mottle virus (BBMV, genus *Bromovirus*, family *Bromoviridae*)	Cp*, FB*, L, OL	Sa, Se, Beetles	Is, 26	Af , As, Eu	None	Gibbs, 1972 (101)
Broad bean stain virus (BBSV, genus *Comovirus*, family *Comoviridae)*	FB*, L*, OL	Sa, Se, Beetles	Is, 25	Af, As, Au, Eu	None	Gibbs and Smith, 1979 (29)
Broad bean true mosaic virus (BBTMV, genus *Comovirus*, family *Comoviridae*)	FB*	Sa, Se, Beetles	Is, 25	Af, As, Eu	None	Gibbs and Paul, 1970 (20)
Chickpea bushy dwarf virus (CpBDV, genus *Potyvirus*, family *Potyviridae*)	Cp	Sa, Aphids NP	Fi, 700-720	As	5 species in 3 families	Brunt *et al.*, 1996b
Chickpea chlorotic dwarf virus (CpCDV, family *Geminiviridae*)	Cp, FB, L, OL	Leafhoppers	Ge, 15 x 25	Af, As, Au	Few species in 2 families	Horn *et al.* 1993
Chickpea filiform virus (CpFV, genus *Potyvirus*, family *Potyviridae*)	Cp	Sa, Aphids NP	Fi, 770	Am, As	4 species in 3 families	Brunt *et al.* 1996b
Chickpea stunt disease associated virus (CpSDaV, family *Luteoviridae*)	Cp	Aphids P	Is, ≈ 25	As	None	Brunt *et al.* 1996b

Cowpea aphid-borne mosaic virus (CABMV, genus *Potyvirus*, family *Potyviridae*)	Bg, OL	Sa, Se, Aphids NP	Fi, 750	Af, As, Eu	None	Bock and Conti, 1974 (134)
Cowpea mild mottle virus (CPMMV, genus *Carlavirus*)	Bg, LB, Wb, OL	Sa, Se, Whiteflies	Fi, 650	Af, Am, As	11 species in 5 families	Brunt and Kenten, 1974 (140)
Cowpea mottle virus (CPMoV, genus *Carmovirus*, family *Tombusviridae*)	Bg*, OL	Sa, Se, Beetles	Is, 30	Af	3 species in 2 families	Bozarth and Shoyinka, 1979 (212)
Cowpea mosaic virus (CPMV, genus *Comovirus*, family *Comoviridae)*	Bg	Sa, Se, Beetles	Is, 24	Af, Am	Few non-legumes species	Van Kammen and de Jager, 1978 (197)
Cowpea severe mosaic virus (CPSMV, genus *Comovirus*, family *Comoviridae*)	B, M, P, Wb, OL	Sa, Se, Beetles	Is, 25	Am	11 species in 4 families	De Jager, 1979 (209)
Cucumber mosaic virus (CMV, genus *Cucumovirus*, family *Bromoviridae*)	Cp*, FB, L*, LB, M, Wb, OL	Sa, Se, Aphids NP	Is, 28	Af, Am, As, Au, Eu	800 species in 85 families	Francki *et al.*, 1979 (213)
Faba bean necrotic yellows virus (FBNYV, genus *Nanovirus*, family *Nanoviridae*)	Cp, FB, L, OL	Aphids P	Is, 18	Af, As, Eu	None	Katul *et al.* 1993
Horsegram yellow mosaic virus (HgYMV, genus *Begomovirus*, family *Geminiviridae*)	Bg, B, H, LB, M, P, OL	Whitefly	Ge, 15-18x30	As	None	Brunt *et al.* 1996b
Lettuce mosaic virus (LMV, genus *Potyvirus*, family *Potyviridae*)	Cp	Sa, Se, Aphids NP	Fi, 750	Af, Am, As, Au, Eu	26 species in 9 families	Tomlinson, 1970 (9)

Limabean golden mosaic virus (LGMV, genus *Begomovirus*, family *Geminiviridae*)	LB	Whiteflies	Ge, 18 x 30	Af	None	Brunt *et al.* 1996b
Mungbean leaf curl virus (MLCV)	B, FB, M, P	Sa, Thrips	No information	As	None	Amin *et al.* 1985
Mungbean mosaic virus (MbMV, family *Potyviridae*)	B, M, P	Sa, Aphids NP	Fi, 750	As	None	Kaiser *et al.* 1968
Mungbean mottle virus (MMoV, family *Potyviridae*)	B, M	Sa, Aphids NP	Fi, 750	As	3 species in 2 families	Brunt *et al.* 1996b
Mungbean yellow mosaic virus (MYMV, genus *Begomovirus*, family *Geminiviridae*)	B, H, M, P	Sa, Whitefly	Ge, 18 x 30	As	4 species in 2 families	Honda and Ikegami, 1986 (323)
Pea early-browning virus (PEBV, genus *Tobravirus*)	FB	Sa, Se, Nematodes	Short rod 105-215	Af, Eu	Few species in 2 families	Harrison, 1973 (120)
Pea enation mosaic virus-1 (PEMV-1, genus *Enamovirus*, family *Luteoviridae*) and *Pea enation mosaic virus-2* (PEMV-2, genus *Umbravirus*).	Cp, FB, L, OL	Sa, Se, Aphids P	Is, 28	Am, As, Eu	None	Peters, 1982 (257)
Pea seed-borne mosaic virus (PSbMV, genus *Potyvirus*, family *Potyviridae*)	Cp*, FB*, L*, OL	Sa, Se, Aphids NP	Fi, 770	Af, Am, As, Au, Eu	47 species in 12 families	Hampton and Mink, 1975 (146)
Peanut mottle virus (PeMoV, genus *Potyvirus*, family *Potyviridae*)	Bg*, LB, OL	Sa, Se, Aphids NP	Fi, 740-750	Af, Am, As, Au, Eu	7 species in 5 families	Bock and Kuhn, 1975 (141)

Pigeon pea proliferation virus (PPPV, family *Rhabdoviridae*)	P	Leafhoppers	Rhabdo	As	None	Brunt *et al.* 1996b
Pigeonpea sterility mosaic disease (PpSMD)	P, OL	Sa, Mite	F	As	Few species in 2 families	Kumar *et al.* 2000, 2001
Rhynchosia mosaic virus (RhMV, genus *Begomovirus*, family *Geminiviridae*).	P	Sap, Whiteflies	Ge, 18 x 30	Am	5 species in 3 families	Bird *et al.* 1975
Southern bean mosaic virus (SBMV, genus *Sobemovirus*)	Bg, OL	Sa, Se, Beetles	Is, 30	Af, Am, As, Eu	2 non-legumes species	Tremaine, 1983 (274)
Tomato spotted wilt virus (TSWV, genus *Tospovirus*, family *Bunyaviridae*)	Cp, FB	Sa, Se, Thrips	Is, 70-90	Af, Am, As, Au, Eu	940 species in 90 families	Ie, 1970 (39)
Urdbean leaf crinkle virus (ULCV)	B*, M, P, OL	Sa, Se, Aphids, Beetles, Whitefly	Is, 32	As	Only two cucurbitous species	Williams *et al.* 1968; Dubey *et al.* 1983
Voandezia distortion mosaic virus (VDMV, genus *Potyvirus*, family *Potyviridae*)	Bg	Sa, Apids NP	Fi, 744	Af	None	Fauquet and Thouvenel, 1987
Voandezia necrotic mosaic virus (VNMV, genus *Tymovirus*)	Bg	Sa	Is, 28	Af	1 species (*Chenopodium*)	Fauquet *et al.* 1984 (279)

[1] Bg= bambara groundnut, B= blackgram, Cp= chickpea, FB= faba bean, H= horsegram, L= lentil, LB= lima bean, M= mungbean, P= pigeonpea, Wb= winged bean, OL=Other legume

[2] Sa= Sap, Se= seeds

[3] Ba = Bacilliform, Fi= Filamentous, Ge= Gemini, Is= Isometric

[4] Af= Africa, Am= America, As= Asia, Au= Australia, Eu= Europe

[5] Reference of Descriptions of Plant Viruses by Commonwealth Mycological Institute/Association of Applied Biologists (CMI/AAB Description numbers is in brackets).

[6] * Indicates seed transmission in that species

[7] Indicates seed transmission of the virus in at least one of its host species.

[8] NP= non-persistent transmission; P= persistent transmission; SP= semi-persistent transmission

Fig. 18.1
(A) *Horsegram yellow mosaic virus* induced bright yellow mosaic symptoms on the leaves of Horsegram, (B) *Bean leaf roll virus* induced interveinal chlorosis develops to full chlorosis on faba bean plants, (C) Severe crinkling of leaves of Blackgram due to the infection by the causal agent of urd bean leaf crinkle disease, (D) *Bean yellow mosaic virus* induced mosaic symptoms on faba bean leaves, (E) *Pigeonpea sterlity mosaic virus* induced mosaic symptoms, paling and reduction of leaves of pigeonpea, (F) *Mungbean yellow mosaic virus* produced bright yellow mosaic symptoms on Greengram, , (G) Chickpea plant in the field infected with *Beet western yellows virus* (left) as compared to healthy looking plant (right), (H) Lentil plant in the field infected with *Faba bean necrotic yellows virus* (left) as compared to healthy looking plant (right), (I) *Faba bean necrotic yellows virus* induced leaf cupping and chlorosis of growing point of faba bean plants.

19. Banana and Plantain

J. E. Thomas[1], A. D. W. Geering[1], G. Dahal[2], B. E. L. Lockhart[2] and G. Thottappilly[3]

[1]Agency for Food and Fibre Sciences, Department of Primary Industries, Queensland, 80 Meiers Road, Indooroopilly, Qld 4068, Australia; [2]University of Minnesota, Department of Plant Pathology, St Paul, MN 55108, USA; [3]Mahyo Research Foundation, Banjara Hills, Road No. 12, Hyderabad 500 034, India

19.1 Introduction

Banana and plantain (*Musa* spp.) are grown in more than 125 countries (Anon., 2001) and are a major source of carbohydrate for more than 400 million people (Swennen *et al,* 1995). The progenitors of almost all domesticated *Musa* are two wild species, *Musa acuminata* and *M. balbisiana*, whose genomes are represented using the letter codes A and B, respectively. The most commonly cultivated cultivars have the triploid genotypes AAA, AAB or ABB.

Only five viruses, namely *Banana bunchy top virus*, *Cucumber mosaic virus*, *Banana streak virus*, *Banana bract mosaic virus* and Banana mild mosaic virus, are known to cause disease in banana. Abacá mosaic virus is reported to be a significant constraint to abacá (*Musa textilis*) production in the Philippines (Diekmann and Putter, 1996) and has been experimentally transmitted to banana, but there are as yet no field records of this virus in this host. Recently, Agindotan (2001) detected bacilliform virus particles

(60-70 nm wide and 267-300 nm long) in *Musa* samples from Nigeria, suspected to be a new virus belonging to the genus *Rhabdovirus*.

As each of the viruses known to infect banana differ in molecular and biological properties, the techniques used for their effective management and control also differ. In addition, successful control of these viral diseases depends on the availability of reliable virus detection methods and appreciation of the risk factor that each of these diseases poses . Each of the viruses is discussed separately in the following sections.

19.2 ***Banana bunchy top virus*** Genus ***Babuvirus*** (BBTV)

Banana bunchy top virus (BBTV) is an aphid-transmitted virus occurring in some parts of Africa, Australia, Asia, and the Pacific Islands, but is absent from the Americas (Diekmann and Putter, 1996). BBTV is the most devastating of viruses known to infect banana.

Severely infected plants are easily recognizable from a distance by their stunted and 'choked' appearance (bunchy top). The other diagnostic symptom of infection is the Morse code pattern of green dots and dashes running along the leaf veins and hooking into the midrib. These green dots and dashes are also evident along the midrib, on the leaf sheaths, petioles, and on the bracts of the male flower. Other foliar symptoms include a narrower, stiffer leaf lamina, which is slightly upturned and yellow at the margin and maybe ripped. Foliar symptoms, especially the green dots and dashes, are most prominent in dessert bananas belonging to the Sucrier and Cavendish subgroups (*Musa* AA and AAA groups), and less distinctive in *cv.* Gros Michel (*Musa* AAA group), and in plantains and cooking bananas (*Musa* AAB and ABB groups). Infection severely affects yield, and plants infected at an early stage generally do not produce a bunch.

BBTV, the type member of genus *Babuvirus* (Mayo, 2002), has isometric particles, 18 nm in diameter, and a circular, ssDNA genome, divided amongst at least six different components (Burns *et al.* 1995; Karan *et al.* 1997; Thomas and Dietzgen, 1991). DNA-1 encodes a replication-associated protein (Rep) protein and a second smaller protein of unknown function (Beetham *et al.* 1997; Hafner *et al.* 1997). Coat protein, movement protein and a nuclear shuttle protein are encoded by DNA-3, -4 and -6, respectively (Wanitchakorn *et al.* 1997; Wanitchakorn *et al.* 2000). DNA-5 encodes a retinoblastoma-binding protein, involved in converting the cell cycle to the S-phase, a state favoring virus replication (Wanitchakorn *et al.* 2000). The function of the protein encoded by DNA-2 is currently unknown.

BBTV isolates from around the world fall into two groups, the Asian and the South Pacific groups, which are on average 9.6%, 11.9% and 14.5% different at the nucleotide level for DNA-1, DNA-3, and DNA-6, respectively (Karan *et al.* 1994, 1997; Wanitchakorn *et al.* 2000). The biological significance of these groups is not known. Three different satellite components (DNA-S1, -S2 and –Y), all encoding a Rep protein, have been found associated with virus isolates from the Asian group (Horser *et al.* 2001). These satellite components are able to initiate self-replication, but in contrast to DNA-1, are unable to direct replication of the other BBTV genomic components (Horser *et al.* 2001). The effects of these satellite components on symptom production are unknown.

19.2.1 Detection and Identification

Serological and nucleic acid-based assays have been developed for BBTV. Monoclonal and polyclonal antibodies to BBTV are commercially available, and can be used in a variety of formats such as ELISA and western immunoblot (Dietzgen and Thomas, 1991; Geering and Thomas, 1996). A range of PCR assays have been developed, which are either specific for a single component (Hafner *et al.* 1995; Hu *et al.* 1996; Karan *et al.* 1997; Thomson and Dietzgen, 1995), or capable of amplifying all components (Burns *et al.* 1994). BBTV can also be detected simultaneously with *Cucumber mosaic virus* and *Banana bract mosaic virus* in a multiplex PCR assay (Sharman *et al.* 2000).

19.2.2 Disease Cycle and Epidemiology

The epidemiology of BBTV is relatively simple, as there is only one known vector, *Pentalonia nigronervosa* (banana aphid), and in most cases, one host plant, the banana. BBTV is transmitted in the circulative, persistent manner, after a minimum acquisition-access period of 4 hours and a minimum inoculation-access period of 15 minutes and is retained for the life of the aphid (Hafner *et al.* 1995; Hu *et al.* 1996). The narrow host range of *P. nigronervosa*, which is restricted to families of *Araceae, Cannaceae, Commelinaceae, Heliconiaceae, Musaceae, Strelitzeaceae* and *Zingiberaceae* (Blackman and Eastop, 1984; Wardlaw, 1961), limits potential alternative hosts of the virus. Studies in Australia and Hawaii suggest that *Alocasia brisbanensis*, *Canna indica*, *Colocasia esculenta*, *Hedychium coronarium*, *Heliconia psittacorum* and a range of *Alpinia* spp. are non-hosts of the virus (Geering and Thomas, 1997; Hu *et al.* 1996). Local spread is often via the aphid vector. Allen (1978, 1987) showed that over 99% of secondary spread was within 86 m of the source of infection. Average distance of spread was only 15.5-17.2 m.

BBTV is transmitted through all forms of vegetative planting material, including suckers, corms, bits, and tissue culture plantlets, and it is in this form that spread between regions and countries usually occurs.

All banana cultivars, and their wild progenitors, *Musa acuminata* and *Musa balbisiana*, are susceptible to BBTV. Some dessert bananas such as cv. Gros Michel (*Musa* AAA group), and many plantains and cooking bananas (*Musa* AAB, ABB groups) appear to have partial resistance to BBTV in the form of lower susceptibility to infection, less severe symptoms and reduced virus titres (Jose 1981; Muharam 1984, ADW Geering and JE Thomas, unpublished)). Other than domesticated bananas, the only other known hosts of BBTV are from within family *Musaceae*, and include *Musa coccinea*, *M. jackeyi*, *M. ornata*, *M. textilis*, and *M. velutina*, and their close relative, *Ensete ventricosum* (Thomas and Iskra-Caruana, 2000).

19.2.3 Control

The Australian banana industry was nearly destroyed by BBTV in the early twentieth century, but was later able to minimize the disease problem by implementing strict domestic quarantine regulations and a program of inspection and rouging infected plants. Movement of planting material out of BBTV-affected regions was prohibited, and planting material could only be obtained from on-farm sources, or from plantations that were free of disease. Laws were enacted to enforce these control measures (Magee, 1927, Dale, 1987). Based on computer simulations of epidemics, Allen (1987) recommended inspections at intervals equating to the period taken for four leaves to emerge. This equated to the time from infection until the plant was itself a source of infection. Eradication of diseased plants at these intervals prevented most secondary infection. BBTV is now rare in Australia, and has been successfully eradicated from some districts in New South Wales and Queensland.

Currently there are no other control measures for BBTV other than rouging and sanitation programs similar to those described above. BBTV can sometimes be eliminated from germplasm using meristem tip culture (Thomas *et al.* 1995). Use of systemic insecticides to control the vector is considered ineffective, and no resistant (immune) cultivars are available.

19.3 *Cucumber mosaic virus* Genus *Cucumovirus* (CMV)

Banana mosaic, caused by *Cucumber mosaic virus* (CMV), occurs in most countries where bananas are grown. The disease is also known as infectious chlorosis, heart rot, virus sheath rot, and cucumber mosaic (Magee, 1930; Stover, 1972; Wardlaw, 1961). This aphid-transmitted virus can be of economic importance especially when heart rots symptoms occur (Bouhida and Lockhart, 1990).

Symptoms are often sporadic and usually consist of mosaics and line patterns and are usually more severe at temperatures below 24°C. Leaf distortion may also occur. Some strains can produce necrosis of the cigar leaf and heart rot (necrosis of the internal pseudostem tissue). Previously, symptoms were sometimes confused with those caused by *Banana streak virus* (Stover, 1972; Wardlaw, 1961).

Banana mosaic disease is caused by CMV, type member of genus *Cucumovirus*. It has isometric virions, about 28 nm in diameter and a tripartite ssRNA genome and an encapsidated subgenomic RNA (Palukaitis *et al.* 1992). Some isolates also contain encapsidated satellite RNA and in banana this is known to ameliorate symptoms (Gafney *et al.* 1996).

Most isolates fall into one of two subgroups that are consistently grouped by serology, nucleic acid hybridisation and sequence analysis (Devergne *et al.* 1973; Owen *et al.* 1990; Piazzolla *et al.* 1979). These subgroups are designated I (or DTL, WT) and II (or ToRS, S). Subgroup I isolates predominate in the tropics and are more commonly found in banana (Hu *et al.* 1995; Singh *et al.* 1995; Gafny *et al.* 1996).

19.3.1 Detection and Identification

Antisera and ELISA kits for either subgroup or all strains of CMV are commercially available. A number of PCR assays have also been described (Hu *et al.* 1995; Sharman *et al.* 2000; Singh *et al.* 1995).

19.3.2 Disease Cycle and Epidemiology

CMV has a very wide host range, with over 800 host species recorded. It is transmitted in the non-persistent manner by at least 75 different aphid species (Palukaitis *et al.* 1992) and is also seed transmitted in many hosts, including banana (Gold, 1972). Infection of banana is thought to occur usually via non-colonizing aphids moving from adjacent infected weed or crop hosts. The only aphid species colonizing banana in significant numbers is *Pentalonia nigronervosa* (Blackman and Eastop, 1984) and recent observations suggest that it may act as a vector to spread CMV from *Commelina diffusa*, a commonly infected weed in the tropics (Lockhart and Jones, 2000).

Banana mosaic has been recorded from a wide range of cultivars and genotypes and it is not known whether any resistance to the virus is present in *Musa*.

19.3.3 Control

Alternative weed and crop hosts of the virus within and around banana plantings should be avoided or eradicated if possible. In Taiwan, improved control was achieved by growing bananas next to non-hosts (e.g. rice) instead of susceptible vegetable crops (Tsai *et al.* 1986).

As CMV can be transmitted in all forms of vegetative planting material, it is important that virus-free material is used for propagation. Meristem tip culture (Berg and Bustamante, 1974) and cryopreservation (Helliot *et al.* 2002) have been used to eliminate CMV from infected germplasm.

19.4 *Banana streak virus* Genus *Badnavirus* (BSV)

Banana streak disease, also known as '*la mosaïque à tirets*', was first reported from Cote d'Ivoire in 1974 (Lassoudière, 1974). A similar leaf streak disease was observed in banana plantations in Morocco and was shown to be caused by *Banana streak virus* (BSV) (Lockhart, 1974). Subsequently the disease has been reported from 48 countries in Europe, Africa, the Middle East, Asia, Oceania and Tropical America (Diekmann and Putter, 1996; Lockhart and Jones, 2000).

Although banana streak disease is often reported as an economically important disease in many countries, experimental data on its effect on growth and yield are still scanty. The economic importance of banana streak stems from two factors: the direct effect of the disease on the crop yield and the restriction imposed on the deployment of improved hybrids due to the occurrence of the causal virus in-different genotypes. Disease incidence varies greatly; in some localities and countries the disease appears to be serious and in other places it is confined to a few plants of certain cultivars and not regarded as important (Lockhart and Jones, 2000). In a study conducted in Ivory Coast, BSV was reported to cause 7-90% yield loss depending on severity of symptoms (Lassoudière, 1994), while in Nigeria losses were less than 15% bunch weight loss (Dahal *et al.* 2000) and in Australia losses of 6-11 % were reported (Daniells *et al.* 2001).

The most common symptoms of BSV infection are narrow discontinuous and sometimes continuous, chlorotic or yellow streaks that run from the leaf midrib to the margin and occasional eye-shaped patterns or blotches. Symptoms can vary from an inconspicuous chlorotic flecking to lethal necrosis. Symptom expression varies, depending on the virus isolates, banana cultivars and the environment (Lockhart, 1994; Gauhl and Pasberg-Gauhl, 1995; Dahal *et al.* 1998b). Some naturally occurring isolates of BSV (e.g. the Rwanda isolate) cause severe symptoms such as die-back and internal necrosis of the pseudostem (Lockhart, 1994). Periodicity in symptom expression is commonly observed in BSV-infected plants, and has been associated with environmental and temperature changes (Lockhart, 1995, Dahal *et al.* 1998b), or possibly plant growth rate and development stage (Daniells *et al.* 2001). Other symptoms associated with BSV include leaf bases falling away from the pseudostem; abnormal arrangement of the leaves; streaks on the pseudostem; narrow, thicker leaves and constriction of bunch emergence (choking); stunting due to shortening of the internodes; internal necrosis of pseudostem, aberrant bunch emergence, reduced bunch size and distortion of fingers (Lassoudière, 1979, Gauhl and Pasberg-Gauhl, 1994; Daniells *et al.* 2001).

The foliar symptoms of BSV infection often resemble those caused by CMV as both viruses produce chlorotic streaks during the initial phase of the symptom development. However, common strains of CMV produce mosaic patterns that do not darken or become chlorotic with age (or temperature) and the heart-rot strain of CMV produces bright yellow chlorosis and characteristic internal necrosis and collapse of the pseudostem (Lockhart and Jones, 2000).

Banana streak disease of banana and plantain (*Musa* spp.) is caused by BSV (Lockhart, 1986). The virions are non-enveloped and bacilliform (130-150 nm x 30 nm) and contain circular double stranded DNA genome of 7.4 kb size (Lockhart, 1990; Harper and Hull, 1998). The genome contains three open reading frames (ORFs) in the arrangement typical of a badnavirus (Harper and Hull, 1998).

BSV isolates exhibit genomic, serological and biological heterogeneity (Lockhart and Olszewski, 1993; Geering *et al.* 2000). So far, serological and genomically distinct isolates of BSV from Morocco, Rwanda, Trinidad, Honduras, Philippines and Australia have been identified (Lockhart and Olszewski, 1993; Geering *et al.* 2000).

Recently, it has been shown that BSV sequences are integrated into genomic DNA of *Musa* and *Ensete* (LaFleur *et al.* 1996). While all *Musa* genotypes appear to contain integrated sequences, to date only one has been implicated in giving rise to *de novo* episomal infection in a wide range of *Musa* genotypes (Harper *et al.* 1999; Ndowora *et al.* 1999). These integrants are associated with the B genome, and include a significant number of tetraploid hybrids, which have been bred for improved yield and disease resistance (Geering *et al.* 2001). The development of episomal BSV infection from integrated viral sequences is linked to *in vitro* propagation and possibly other stress factors (Ndowora *et al.* 1999, Dallot *et al.* 2001), thus preventing on the deployment of *in vitro* propagated improved banana and plantain genotypes.

19.4.1 Detection and Identification

Current methods of detection of BSV include (i) visual inspection of symptoms, (ii) immunosorbent electron microscopy (ISEM), (iii) ELISA, and (iv) genome-based methods (nucleic acid hybridization, PCR amplification). Diagnosis of BSV based on visual symptoms is not reliable as banana streak can be readily mistaken for banana mosaic (Lockhart and Jones, 2000) and easily missed because of periodicity in symptom expression (Dahal *et al.* 1998 b,c).

The major obstacle to the reliable detection of BSV is the highly heterogeneous nature of BSV isolates (Lockhart and Olszewski, 1993; Geering *et al.* 2000). Immunoassay protocols do not reliably detect all BSV isolates. Despite the significant improvements in BSV detection by ELISA using polyclonal antibodies produced in two animal species (Ndowora, 1998), some virus isolates are still either poorly detected or not detected at all. ISEM detects these and all other BSV isolates from partially purified tissue extracts (Bouhida *et al.* 1993) but this method is expensive and laborious and requires specialized equipment and skills.

BSV detection by genome-based methods like PCR is highly sensitive but its reliability is seriously compromised by genomic heterogeneity among BSV isolates and presence of integrated BSV sequences in the *Musa* genome (LaFleur *et al.* 1996; Ndwora *et al.* 1999). However, some of these problems can be avoided by using an immunocapture-PCR protocol with either specific or degenerate primers (Thottappilly *et al.* 1998, Harper *et al.* 1999; Geering *et al.* 2000).

19.4.2 Disease Cycle and Epidemiology

Under experimental conditions, BSV is transmitted by citrus mealybug (*Planococcus citri*) and *Planococcus* spp. in the semi-persistent manner (Lockhart, 1995). *Dysmicoccus* spp., which are reported to colonize banana in Africa (Matile-Ferrero and Williams, 1995) and in South America, may also act as vectors (Lockhart and Jones, 2000). In Nigeria, *Planococcus musae*, a new species of mealybug, has been identified in the field where BSV incidence is high (Dahal *et al.* 1998a). In Ecuador, *Pseudococcus comstocki* is the most common mealybug found on banana, but its transmission of BSV has not been demonstrated (Lockhart and Jones, 2000). Ants seem to managing mealybugs in banana fields in Ecuador, thus accelerating the field spread of the disease. Pink sugarcane mealybug (*Saccharicoccus sacchari*) transmits *Sugarcane bacilliform virus* (SCBV) from sugarcane to banana (Lockhart and Autrey, 1991) and SCBV is present in all noble cane varieties of sugarcane (Lockhart, 1994a). In most tropical countries, sugarcane is often grown in close proximity to banana and thus it is highly likely that it may serve as inoculum source for banana (Jones, 1994, Lockhart, 1994a). However, field observations from many countries suggest that natural dissemination of BSV by mealybug vectors is limited (Lockhart, 1995, Daniells *et al.* 2001).

Attempts to transmit BSV by mechanical inoculation were not successful. Thus it is unlikely that BSV is transmitted mechanically by cutting tools or during cultural operations. It is not known to be soil-borne, but is transmitted through all forms of vegetative propagation, including tissue culture. Transmission through seed seems highly probable (Daniells *et al.* 1995). BSV has very limited natural and experimental host range. Under natural conditions, BSV occurs in cultivated landraces of banana (Dahal *et. al.* 1999, Lockhart and Jones, 2000). Under experimental conditions, BSV has also been transmitted to *M. balbisiana* (Lockhart, 1994).

BSV is reported to occur in a wide range of banana genotypes, including AA, AAA, AB, AAB, ABB groups (Dahal *et. al.* 1999, Lockhart and Jones, 2000). BSV infection arising from the integrated viral sequences is common amongst triploid (AAB) and tetraploid (AAAB) hybrids produced in breeding programs during the past ten years (Lockhart, 2000).

19.4.3 Control

Control of BSV is particularly difficult, primarily due to problems associated with detection, and the presence of integrated BSV sequences in the B genome of *Musa* hybrids. Dissemination of BSV occurs primarily in infected planting material, and in the absence of alternate hosts and mobile insect vectors, control of this disease is based on the use of virus-free planting materials. Effective control methods have not been established for areas where vector transmission is a significant means of spread.

Episomal BSV infections have been eliminated from Cavendish bananas using cryopreservation (Helliot *et al.* 2002), although this technique is unlikely to be successful for *Musa* hybrids containing integrated sequences.

19.5 *Banana bract mosaic virus* Genus *Potyvirus* (BBrMV)

Banana bract mosaic virus (BBrMV) is an aphid-transmitted potyvirus, apparently restricted to *Musa* spp. and known to occur in India, Sri Lanka and the Philippines. Records from other countries are not associated with typical symptoms and need to be clarified (Thomas *et al.* 2000). Little is documented of the economic impact of bract mosaic disease, but one study in the Philippines cites yield losses of up to 40% (Roperos and Magnaye, 1991).

The most characteristic symptom of infection by BBrMV is a mosaic pattern (usually dark red) on the flower bracts. When the older leaf sheaths are pulled away from the pseudostem, mosaic patterns are visible, the colour varying from green to red and brown, depending on the pigmentation of the cultivar. Leaf symptoms are erratic, occurring most often on newly emerged leaves, and consisting of spindle-shaped chlorotic lesions. Mosaic patterns also sometimes occur on the bunch stalk and fingers (Thomas *et al.* 2000).

BBrMV virions are 11 × 750 nm (Muñez, 1992) encapsidated in a coat protein of size *ca* 39 kDa. Virions have a buoyant density of 1.29-1.31 g cm^{-3} in caesium chloride and an $A_{260/280}$ of 1.17 (Thomas *et al.* 1997).

The 3' terminal region of the genome of several isolates has been sequenced (Bateson and Dale, 1995; Rodoni *et al.* 1997; Thomas *et al.* 1997) demonstrating that BBrMV is a novel, unique potyvirus. All isolates tested are serologically related and have a nucleotide identity in the coat protein region of greater than 87% (Thomas *et al.* 1997, 2000).

19.5.1 Detection and Identification

Serological and nucleic acid-based assays are available for BBrMV. A polyclonal antiserum has been produced and used in ELISA (Thomas *et al.* 1997) and immunosorbent electron microscopy, and a number of PCR assays have been reported, in both single virus (Bateson and Dale, 1995; Thomas *et al.* 1997) and multiplex formats (Sharman *et al.* 2000).

19.5.2 Disease Cycle and Epidemiology

Attempts to transmit BBrMV by sap inoculation have generally been unsuccessful, although one instance of transmission to banana has been observed (L V Magnaye, personal communication). The virus is transmitted through all forms of vegetative propagules, including micropropagated plantlets, suckers and corms.

Three aphid species are recorded as vectors. Transmission by *Pentalonia nigronervosa* was of the non-persistent type (Muñez, 1992), with a transmission efficiency of 10% (Caruana and Galzi, 1998). Other vector species include *Aphis gossypii* and *Rhopalosiphum maidis* (Magnaye and Espino, 1990).

All known natural and experimental hosts are in the genus *Musa*, and no sources of resistance are known. In addition to banana and plantain, abacá (*Musa textilis*) is also a host (Sharman *et al.* 2000).

19.5.3 Control

Virus-free planting material provides the most effective means of control. In the Philippines, some commercial plantations use roguing and sanitation programs for control (Magnaye, 1994).

19.6 Banana mild mosaic virus (BanMMV)

Banana mild mosaic virus (BanMMV) is a newly recognised virus with filamentous particles. It occurs widely throughout the world, often producing few, if any, symptoms. Little is known of its mode of spread or economic importance.

The symptoms caused by BanMMV seem to be variable. In most cases and in most cultivars, infections are symptomless. In cv. Ducasse (AAB group, syn. Pisang Awak), young leaves on suckers and tissue-cultured plantlets show a mild chlorotic mosaic, but subsequent leaves are symptomless. Broad chlorotic or fine silvery streaks have been observed on some cultivars (Thomas *et al.* 2000).

BanMMV commonly occurs as mixed infections with BSV or BBrMV and in such cases only symptoms of the latter viruses are present (Thomas *et al.* 2000). Additional leaf necrosis has been observed when BanMMV occurs as a mixed infection with CMV (Caruana and Galzi, 1998).

The virions of BanMMV are flexuous filaments about 580 × 14 nm (Thomas *et al.* 2000). The virus has been fully sequenced, and has a ssRNA genome of 7.4 kb, containing five ORFs, including a triple gene block, and encoding a coat protein of 27 kDa (Gambley and Thomas, 2001). The virus is phylogenetically related to potex-, carla- and foveaviruses, but is also distinct from all these genera.

19.6.1 Detection and Identification

A monoclonal antibody, and a number of polyclonal antisera have been developed (Thomas *et al.* 2000), and PCR assays are also available (M Sharman and J E Thomas, unpublished results).

19.6.2 Disease Cycle and Epidemiology

Vegetative propagation is the only confirmed means of transmission of BanMMV (Thomas *et al.* 2000). All attempts at sap transmission and experimental vector transmission (M Sharman and J E Thomas, unpublished) have been unsuccessful.

In Guadeloupe, virus-free micropropagated Cavendish bananas were frequently field infected with CMV and sometimes also BanMMV, implying that natural transmission of BanMMV was sometimes occurring simultaneously with the aphid-transmitted CMV (Caruana and Galzi, 1998).

The natural host range of BanMMV appears restricted to *Musa*, and all attempts at sap transmission to banana and herbaceous indicators have been unsuccessful (Thomas *et al.*2000). No sources of resistance are known.

19.6.3 Control

The use of virus-free planting material is the only option for control of BanMMV.

19.7 Outlook

Banana viruses continue to be both a major production constraint and a quarantine threat around the world. There are many gaps in our knowledge and understanding of these viruses, requiring input from both field and laboratory studies. Control still relies heavily on cultural practices. There are no reports of immunity to banana viruses in any cultivars, though the possibility of cultivars expressing tolerance or field resistance has received scant attention and would be worthy of investigation. Genetic engineering offers promise as an alternative source of resistance, though more likely in the medium to long term. As yet, there are no published reports of transgenic resistance to banana viruses in glasshouse or field experiments. Examples of transgenic resistance to viruses containing an RNA genome are well known, but obtaining resistance to viruses containing a DNA genome, such as BBTV and BSV, may require novel strategies.

BBTV continues to cause the single most devastating disease of banana. However, its impact is most serious on Cavendish cultivars, and other genotypes can show far milder symptoms. There is no evidence that different strains of the virus are involved. Preliminary glasshouse studies have shown that cultivars and genotypes can vary markedly in their susceptibility to aphid-mediated infection, their virus titre and the rapidity and degree of symptom expression (ADW Geering and JE Thomas, unpublished). The cultivar Gros Michel (AAA), for example, appeared to express partial resistance in these studies. Judicious selection of cultivars could play an important role in field control of the disease. There is a need to assess banana and plantain cultivars for field resistance to BBTV.

There is still a need for countries free of BBTV to ensure that imported germplasm has been indexed using currently available, validated ELISA and PCR assays (Sharman *et al.* 2000; Thomas and Iskra-Caruana 2000). The recent outbreaks in Pakistan (Khalid *et al.* 1993), Hawaii (Hu *et al.* 1993) and New Caledonia (Kagy *et al.* 2001) attest to the continuing quarantine threat posed by BBTV.

BSV has rapidly risen to prominence, especially because of its frequent infection of improved banana hybrids from the world's breeding programs. The integration of BSV into the B component of the *Musa* genome (Geering *et al.* 2001; Harper et al. 2002), its subsequent activation through tissue culture (Dallot *et al.* 2001), and its genomic and serological diversity (Geering *et al.* 2000; Lockhart and Olszewski 1993) caused many problems with indexing. There is still a need for economical, sensitive, robust and widely-available assays, capable of detecting all variants of episomal BSV and distinguishing them from integrated sequences. Some advances have been made with methods of rapid clonal propagation, avoiding tissue culture and resultant activation of integrated BSV (Auboiron 1997), but multiplication rates are still low compared with tissue culture.

BanMMV has only recently been characterised (Gambley and Thomas 2001) and little is known of its epidemiology. There is some circumstantial evidence of field spread in mixed infections with CMV (Caruana and Galzi 1998). However, there are no confirmed reports of transmission of this virus by any means other than vegetative propagation, despite its presence in a wide range of cultivars. The economic impact of this virus is unknown, and possible synergistic effects in co-infections with BSV and BBrMV have not been studied. Although infections with BanMMV are frequently symptomless, it should not be assumed that this equates to minimal yield losses.

It seems surprising that bract mosaic disease, caused by a non-persistently aphid-transmitted virus, occurs only in India, Sri Lanka and the Philippines. Systematic surveys have not been reported from elsewhere in south-east Asia, and although BBrMV has been reported from Thailand, Vietnam and Western Samoa (Rodoni *et al.* 1999), the virus was not associated with typical disease symptoms.

References

Agindotan, B. O. 2001. Production of monoclonal and polyclonal antibodies for the diagnosis of banana streak virus, genus Rhabdovirus. Ph. D. Thesis, Univ. Ibadan, Nigeria. 262 pp.

Allen, R. N. 1978. Spread of bunchy top disease in established banana plantations. *Aust. J. Agric. Res.* 29:1223-1233.

Allen, R. N. 1987. Further studies on epidemiological factors influencing control of banana bunchy top disease, and evaluation of control measures by computer simulation. *Aust. J. Agric. Res* 38:373-382.

Anon. (2001). www.fao.org

Auboiron, E. 1997. Fiche techniques. La multiplication sur souche decortique. Propagation rapide de materiel de plantation de bananiers at de plantains. Centre recherché regional sur bananiers et plantain. Douala, Cameroun.

Bateson, M. F. and Dale, J. L. 1995. Banana bract mosaic virus: characterization using potyvirus specific degenerate PCR primers. *Arch. Virol.* 140:515-527.

Beetham, P. R., Hafner, G.J., Harding, R. M. and Dale, J. L. 1997. Two mRNAs are transcribed from banana bunchy top virus DNA-1. *J. Gen. Virol.* 78:229-236.

Berg, Lloyd A., and Bustamante, Mario. 1974. Heat treatment and meristem culture for the production of virus-free bananas. *Phytopathology* 64:320-322.

Blackman, R. L., and Eastop, V. F.. 1984. *Aphids on the world's crops: an identification and information guide*. Chichester, UK: John Wiley & Sons.

Bouhida, M., and Lockhart, B. E. 1990. Increase in importance of cucumber mosaic virus infection in greenhouse-grown bananas in Morocco. *Phytopathology 80:981.*

Burns, T. M., Harding, R. M. and Dale, J. L. 1994. Evidence that banana bunchy top virus has a multiple component genome. *Arch. Virol.* 137:371-380.

Burns, T. M., Harding, R. M. and Dale, J. L. 1995. The genome organization of banana bunchy top virus: analysis of six ssDNA components. *J. Gen. Virol.* 76:1471-1482.

Caruana, M. L., and Galzi, S. 1998. Identification of uncharacterized filamentous viral particles on banana plants. *Act. Hortic.* 490:323-335.

Dahal, G., Hughes, J. d'A. and Lockhart, B. E. L. 1998a. Status of banana streak badnavirus in Africa: problems and future research needs. *Int. Pest. Mgt. Rev.* 3:85-97.

Dahal, G., Hughes, J. d'A, Thottappilly, G., and Lockhart, B. E. L. 1998b. Effect of temperature on symptom expression and reliability of banana streak *Badnavirus* detection in naturally infected plantain and banana (*Musa* spp.) *Plant Dis.* 82:16-21.

Dahal, G., Pasberg-Gauhl, C, Gauhl, F., Thottappilly, G. and Hughes, J. d'A. 1998c. Studies on a Nigerian isolate of banana streak badnavirus: II. Effect of intraplant variation on virus accumulation and reliability of diagnosis by ELISA. *Ann. appl. Biol.* 132:263-275.

Dahal, G., Gauhl, F., Pasberg-Gauhl, C., Hughes, J. d'A., Thottappilly, G. and Lockhart, B. E. L. 1999. Evaluation of micropropagated plantain and banana (*Musa* spp) for BSV incidence under field and screen house conditions. *Ann. appl. Biol.* 134:181-191.

Dahal, G., Ortiz, R., Tenkuano, A., Hughes, J. d'A., Vuylsteke, D. and Lockhart, B. E. L. 2000. Relationship between natural occurrence of banana streak badnavirus and symptom expression, relative concentration of virus antigens and yield characteristics of some micropropagated *Musa* spp. *Plant Pathol.* 49:68-79.

Dale, J. L. 1987. Banana bunchy top: an economically important tropical plant virus disease. *Adv. Virus Res.* 33:301-325.

Dallot, S., Acuña, P., Rivera, C., Ramírez, P., Côte, F., Lockhart, B. E. L. and Caruana M. L. 2001. Evidence that the proliferation stage of micropropagation procedure is determinant in the expression of *Banana streak virus* integrated into the genome of the FHIA 21 hybrid (*Musa* AAAB). *Arch. Virol.* 146:2179-2190.

Daniells, J., Thomas, J. E., and Smith, M. 1995. Seed transmission of banana streak *Badnavirus* confirmed. *InfoMusa* 4:7.

Daniells, J. W., Geering, A., D., W., Bryde, N., J., and Thomas, J. E. 2001. The effect of *Banana streak virus* on the growth and yield of dessert bananas in tropical Australia. *Ann. appl. Biol.* 139:51-60.

Devergne, J. C., and . Cardin, L.. 1973. Contribution a l'étude du virus de la mosaïque du concombre (CMV). IV. - Essai de classification de plusieurs isolats sur la base de leur structure antigénique. *Annal. Phytopath.* 73:409-430.

Diekmann, M., and Putter, C. A. J. 1996. *FAO/IPGRI Technical Guidelines for the Safe Movement of Germplasm. No. 15. Musa.* 2nd ed. Rome: Food and Agriculture Organization of the United Nations/International Plant Genetic Resources Institute.

Dietzgen, R. G., and Thomas, J. E. 1991. Properties of virus-like particles associated with banana bunchy top disease in Hawaii, Indonesia and Tonga. *Austr. Plant Pathol.* 20:161-165.

Gafney, R., A. Wexler, M. Mawassi, Y. Israeli, and M. Bar-Joseph. 1996. Natural infection of banana by a satellite containing strain of cucumber mosaic virus: nucleotide sequence of the coat protein gene and the satellite RNA. *Phytoparasitica* 24:49-56.

Gambley, C. F., and Thomas, J. E. 2001. Molecular characterisation of Banana mild mosaic virus, a new filamentous virus in *Musa* spp. *Arch. Virol.* 146:1369-1379.

Gauhl, F. and Pasberg-Gauhl, C. 1995. Temporal dynamics of banana streak *Badnavirus* (BSV) symptoms in *Musa* clones in Southeast Nigeria *Phytomedizin* 27

Geering, A. D. W., Olszewski, N. E., Dahal, G., Thomas, J. E., Lockhart, B. E. L. 2001. Analysis of the distribution and structure of integrated *Banana streak virus* DNA in a range of *Musa* cultivars. *Mol. Plant Path.* 2:207-213.

Geering, A. D. W., and Thomas, J. E. 1996. A comparison of four serological tests for the detection of banana bunchy top virus in banana. *Aust. J. Agric. Res.* 47:403-412.

Geering, A. D. W., and Thomas, J. E. 1997. Search for alternative hosts of banana bunchy top virus in Australia. *Aust. Plant Pathol.* 26:250-254.

Geering, A. D. W., McMichael, L. A., Dietzgen, R. G. and Thomas, J. E. 2000. Genetic diversity among *Banana streak virus* isolates from Australia. *Phytopathology* 90:921-927.

Geering, A.D.W., Olszewski, N.E. , Dahal, G., Thomas, J.E. and Lockhart, B.E.L. 2001. Analysis of the distribution and structure of integrated *Banana streak virus* DNA in a range of *Musa* cultivars. *Mol. Plant Pathol.* 2: 207-213.

Gold, A. H. 1972. Seed transmission of banana viruses. *Phytopathology* 62:760.

Hafner, G. J., Harding, R. M. and Dale, J. L. (1995). Movement and transmission of banana bunchy top virus DNA component one in bananas. *J. Gen. Virol.* 76:2279-2285.

Hafner, G. J., Stafford, M. R., Wolter, L. C, Harding, R. M., and Dale, J. L. 1997. Nicking and joining activity of banana bunchy top virus replication protein *in vitro*. *J. Gen. Virol.* 78:1795-1799.

Harper, G. and Hull, R. 1998. Cloning and sequence analysis of banana streak virus DNA. *Vir. Genes* 17:271-278.

Harper, G., Dahal, Thottappilly, G., and Hull, R. 1999. Detection of episomal banana streak *Badnavirus* by IC-PCR. *J. Virol. Methods* 79:1-8.

Harper, G., Osuji J.O., Heslop-Harrison, J. S. and Hull, R. and Pat Integration of banana streak badnavirus into the Musa genome:molecular and cytogenetic evidence.*Virology* 255:207-213.

Harper, G., Hull, R., Lockhart, B. and Olszewski, N. 2002. Viral sequences integrated into plant genomes. *Annu.Rev. Phytopath.* 40: 119-136.

Helliot, B., Panis, B., Poumay, Y., Swennen, R.,P. Lepoivre, and E. Frison. 2002. Cryopresevation for the elimination of cucumber mosaic and banana streak viruses from banana (*Musa* spp.). *Plant Cell Rep.* 20:1117-1122.

Horser, C L, Harding, R. M. and Dale, J. L. 2001. Banana bunchy top nanovirus DNA-1 encodes the 'master' replication initiation protein. *J. Gen. Virol.* 82:459-464.

Horser, C. L., Karan, M., Harding, R., M. and Dale, J. L. 2001. Additional Rep-encoding DNAs associated with banana bunchy top virus. *Arch. Virol.* 146:71-86.

Hu, J.S., Xu, M.Q., Wu, Z.C. and Wang, M. 1993. Detection of banana bunchy top virus in Hawaii. *Plant Dis.*77: 952.

Hu, J. S., Wang, M., Sether, D., Xie, W., and Leonhardt, K. W. 1996. Use of polymerase chain reaction (PCR) to study transmission of banana bunchy top virus by the banana aphid (*Pentalonia nigronervosa*). *Ann. appl. Biol.* 128:55-64.

Hu, J. S., Li, H. P., Barry, K., Wang, M., and Jordan, R. 1995. Comparison of dot blot, ELISA, and RT-PCR assays for detection of two cucumber mosaic virus isolates infecting banana in Hawaii. *Plant Dis.* 79:902-906.

Jones, D. R. 1994. Risks involved in the transfer of banana and plantain germplasm. In: Jones, D. R. (ed.) The Improvement and Testing of Musa: a Global Partnership, Proceedings of the First Global Conference of the International Musa Testing Program held at FHIA, Honduras, 27-30 April 1994. INIBAP, Montpellier, France.

Jose, P. C. 1981. Reaction of different varieties of banana against bunchy top disease. *Agri. Res. J. Kerala* 19.108-110.

Kagy, V., Thomas, J.E., Shrman, M. and Mademba-Sy, F. 2001. First record of banana bunchy top in New Caledonia. *Australasian Plant Pathol.* 30: 71.

Karan, M., Harding, R. M. and Dale, J. L. 1994. Evidence for two groups of banana bunchy top virus isolates. *J. Gen. Virol.* 75:3541-3546.

Karan, M., Harding, R. M. and Dale, J. L. 1997. Association of banana bunchy top virus DNA Components 2 to 6 with bunchy top disease. *Molecular Plant Pathology On-line [http://www.bspp.org.uk/mppol/]* 1997/0624karan.

Khalid, S.M., Somoro, M.H. and Stover, R.H. 1993. First record of banana bunchy top virus in Pakistan. *Plant Dis.* 77:101.

La Fleur, D.A., Lockhart, B. E. L. and Olszewski, N. E. 1996. Portion of the banana streak *Badnavirus* genome are integrated in the genome of its host *Musa* sp. *Phytopathology* 86:S 100.

Lassoudière, A. 1974. La mosaïque dite 'à tirets' du bananier 'Poyo' en Côte d'Ivoire. *Fruits* 29:349-357.

Lassoudière, A. 1979. Mise en évidence des répercussions économique de la mosaïque en tirets du bananier en Côte d'Ivoire. Possibilités de lutte par éradication. *Fruits* 43:3-34.

Lockhart, B. E. L. 1986. Purification and serology of a bacilliform virus associated with a streak disease of banana. *Phytopathology* 76:995-999.

Lockhart, B. E. L. 1990. Evidence for a double-stranded circular DNA genome in a second group of plant viruses. *Phytopathology* 80:127-131.

Lockhart, B. E. L. 1994. Banana streak Virus. In *Compendium of Tropical Diseases* (Ploctz R. *et al.,* eds). St Paul MN, USA: APS Press 88pp.

Lockhart, B. E. L. 1995. *Banana Streak Badnavirus Infection in Musa: epidemiology, diagnosis and control.* Food and Fertilizer Technology Center Technical bulletin 143, Food and fertilizer Technology Center, Taipei, Taiwan, 11 pp.

Lockhart, B.E.L. and Autrey, L .J. C. 1988. Occurrence in sugarcane of bacilliform virus related serologically to banana streak virus. *Plant Dis.* 72:230-233.

Lockhart, B. E. L. and Autrey, L. J. C. 1991. Mealybug transmission of sugarcane bacilliform and sugarcane closterolLike viruses. Third Sugarcane Path. *Workshop Intern. Soc. Sugarcane Technologists,* 22-26 July, 1991, Mauritius. Abstracts. 17.

Lockhart, B. E. L. and Olszewski, N .E. 1993. Serological and genomic heterogeneity of banana streak *Badnavirus* : Implications for detection in *Musa* germplasm. In *Breeding Banana and Plantain for Resistance to Diseases and Pests* (Ganry, J. Ed). Montpellier, France: INIBAP. Pages 105-113.

Lockhart, B. E. L., and Jones, D. R. 2000. Banana mosaic. In *Disease of banana, abacá and enset*, edited by D. R. Jones. Wallingford, UK: CABI Publishing.Magee, C. J. P. 1927. Bulletin No. 30. Investigation on the Bunchy Top Disease of the Banana. Council for Scientific and Industrial Research, Melbourne, 64pp.

Magee, C. J. P. 1930. A new virus disease of bananas. *Agric. Gaz. N. S. W.* 41:929.

Magnaye, L. V. 1994. Virus diseases of banana and current studies to eliminate the virus by tissue culture. In Towards making pest and disease management relevant to big and small banana growers. Proceedings of the 1st PPS-SMD National Symposium on Pests and Diseases in the Philippines, edited by N. G. Tangonan. Davao: Phytopath. Soc., Southern Mindanao Division.

Magnaye, L. V., and Espino, R. R. C. 1990. Note: banana bract mosaic, a new disease of banana I. symptomatology. *Phil. Agric.* 73:55-59.

Matile-Ferrero, D. and William, D. J. 1995. Recent outbreaks of mealybugs on plantain (Musa spp.) in Nigeria including a new record for Africa and description of a new species of Planococcus Ferris (Homoptera, Psuedococcidae). *Bull. Soc. Entomol. France* 100:445-449.

Mayo, M. A. 2002. A summary of taxonomic changes recently approved by the ICTV. *Arch. Virol.* 147:1655-1656.

Muharam, A. 1984. Test for resistance of some banana cultivars to banana bunchy-top disease. *Bull. Penel. Hort.* 11:16-19.

Muñez, Alejandro Rebucas. 1992. Symptomatology, transmission and purification of banana bract mosaic virus (BBMV) in 'giant cavendish' banana. Master of Science (Plant Pathology), Faculty of Graduate School, University of Philippines, Los Baños.

Muylle, H. 1998. Development of techniques for the elimination of virus diseases from *Musa*: Progress report. In: *Banana Streak Virus: A Unique Virus-Musa Interaction?* Frison, E.A and S.L.Sharrock (editors). Proc. workshop PRO*MUSA* Virology working group held in Montpellier, France, 19-21 January 1998.

Ndowora, T., Dahal, G., La Fleur, D., Harper, G., Hull, R., Olszewski, N. E. and Lockhart, B. E. L.. 1999. Evidence that badnavirus infection in Musa can originate from integrated pararetroviral sequences. *Virology* 255:214-220.

Ndowora, T. R. C., 1998. Banana Streak Virus: Development of an Immunoenzymatic Assay for Detection and Characterization of Sequences that are Integrated in the Genome of the Host. PhD. Thesis, University of Minnesota, USA. 90 pp.

Owen, J., Shintaku, M., Aeschleman, P.,Ben Tabar, S. and Palukaitis, P. 1990. Nucleotide sequence and evolutionary relationships of cucumber mosaic virus (CMV) strains. CMV RNA 3. *J. Gen. Virol.* 71:2243-2249.

Palukaitis, Peter, Roossinck, Marilyn J., Dietzgen, Ralf G., and Francki, Richard I. B. 1992. Cucumber mosaic virus.*Adv.Virus Res.* 41:281-348.

Piazzolla, P., Diaz-Ruiz, J., R., and Kaper, J., M. 1979. Nucleic acid homologies of eighteen cucumber mosaic virus isolates determined by competition hybridization. *J. Gen. Virol.* 45:361-369.

Rodoni, B. C., Ahlawat, Y. S., Varma, A., Dale, J. L. and Harding, R. M. 1997. Identification and characterisation of banana bract mosaic virus in India. *Plant Dis.* 81:669-672.

Rodoni, B. C., Dale, J.L. and Harding, , R.M. 1999. Characterization and expression of the coat protein-region of banana bract mosaic potyvirus, development of diagostic assays and derection of the virus in banana plants from five countries in southeast Asia. *Arch. Virol.* 144; 1725-1737.

Roperos, N. I., and Magnaye, L. V. 1991. Status of Banana Diseases in the Philippines. In *Banana Diseases in Asia and the Pacific*, edited by R. V. Valmayor, B. E. Umali and C. P. Bejosano. Montpellier, France: Intern. Network Improvement of Banana and Plantain.

Sharman, M., Thomas, J. E. and Dietzgen, R. G. 2000. Development of a multiplex immunocapture PCR with colourimetric detection for viruses of banana. *J. Virol. Methods* 89:75-88.

Sharman, M., Gambley, C. F., Oloteo, E. O., Abgona, R. V. J. and Thomas, J. E.. 2000. First record of natural infection of abaca (*Musa textilis*) with banana bract mosaic potyvirus in the Philippines. *Australas. Plant Patho.* 29:69.

Singh, Z., Jones, R. A. C., and Jones, M. G. K. 1995. Identification of cucumber mosaic virus subgroup I isolates from banana plants affected by infectious chlorosis disease using RT-PCR. *Plant Dis.* 79:713-716.

Stover, R. H. 1972. Virus diseases. In *Banana, Plantain and Abaca Diseases*. Kew: Commonwealth Mycological Institute.

Thomas, J. E., and Dietzgen, R. G. 1991. Purification, characterization and serological detection of virus-like particles associated with banana bunchy top disease in Australia. *J. Gen. Virol.* 72:217-224.

Thomas, J E, and Iskra-Caruana, M-L. 2000. Bunchy top. In *Diseases of banana, abacá and enset*, edited by D. R. Jones. Wallingford: CABI Publishing.

Thomas, J.E. and Iskr-Caruna, M.L. 2000. Bunchy top. In: *Diseases of banana, abuca and enset. Ed.* D.R Jones. Wallingford: CABI Publishing.

Thomas, J E, Iskra-Caruana, M.-L., Magnaye, L. V. and Jones, D. R. 2000. Bract mosaic. In *Diseases of banana, abacá and enset*, (ed) D. R. Jones. Wallingford: CABI Publishing.

Thomas, J. E., Lockhart, B. E. L. and Iskra-Caruana, M. L. 2000. Banana mild mosaic. In *Diseases of banana, abacá and enset*, e(d) D. R. Jones. Wallingford: CABI Publishing.

Thomas, J. E., Geering, A. D. W., Gambley, C. F., Kessling, A. F. and White, M.. 1997. Purification, properties and diagnosis of banana bract mosaic potyvirus and its distinction from abaca mosaic potyvirus. *Phytopathology* 87:698-705.

Thomson, D., and Dietzgen, R. G. 1995. Detection of DNA and RNA plant viruses by PCR and RT-PCR using a rapid virus release protocol without tissue homogenization. *J. Virol. Methods* 54:85-95.

Thottappilly, G., Dahal, G. and Lockhart, B. E. L. 1998. Studies on a Nigerian isolate of banana streak *Badnavirus*. I. Purification and enzyme-linked immunoassay. *Ann. appl. Biol.*:132:253-261.

Tsai, Y. P., Hwang, M. T., Chen, S. P. and Liu, S. S. 1986. An ecological study of banana mosaic. *Plant Prot. Bull. Taiwan* 28:383-387.

Wanitchakorn, R., Harding, R. M. and Dale, J. L. 1997. Banana bunchy top virus DNA-3 encodes the viral coat protein. *Arch. Virol.* 142:1673-1680.

Wanitchakorn, R., Harding, R. M. and Dale, J.L. 2000. Sequence variability in the coat protein gene of two groups of banana bunchy top isolates. *Arch. Virol.* 145:593-602.

Wanitchakorn, R., Hafner, G. J., Harding, R. M. and Dale, J. L. 2000. Functional analysis of proteins encoded by banana bunchy top virus DNA-4 to -6. *J. Gen. Virol.* 81:299-306.

Wardlaw, C. W. 1961. The Virus Diseases: Bunchy Top. In *Banana Diseases, including Plantains and Abaca.* London: Longmans.

Wardlaw, C. W. 1961. Mosaic, Infectious Chlorosis and other Virus Diseases. In *Banana Diseases, including Plantains and Abaca.* London: Longmans.

Legends for figures:

Fig. 19.1 - *Banana bunchy to virus*-infected banana, New Caledonia

Fig. 19.2 - *Banana bunchy top virus*-infected banana, New Caledonia

Fig. 19.3 - *Banana streak virus* – Chlorotic streaks produced on a leaf of cv. Williams banana

Fig. 19.4 - *Banana streak virus* - Necrotic streaks produced on a leaf of cv. Williams banana

Fig. 19.5 - *Banana bract mosaic virus* - Mosaic pattern on banana bracts.

Fig. 19.6 - *Banana bract mosaic virus* - Mosaic pattern on the pseudostem of banana, after removal of the outer leaf sheaths.

19.1
19.2
19.3
19.4
19.5
19.6

20. Papaya

Engineering Resistance against Papaya ringspot virus by Native, Chimeric and Synthetic Transgenes

Gustavo Fermin and Dennis Gonsalves
Pacific Basin Agricultural Research Center, 99 Aupuni St., Suite 204. Hilo, HI 96720, USA

20.1 Introduction

The case of transgenic papayas resistant to *Papaya ringspot virus* (PRSV) is a paradigm in the control of a plant pathogen: from the conventional strategies to the more sophisticated use of recombinant DNA technologies, the engineered papayas have served as an example in the use of novel approaches to protect crop plants against plant viruses, and as a model system to transfer the technology to several different countries. PRSV-resistant transgenic papayas were the first commercialized transgenic fruit on a worldwide basis. It is also the only case in which the genetic engineering of the coat protein (CP) gene has led to the development of plants with varying degrees of resistance through the use of different entire CP genes in different programs and countries around the world.

Instead of having an exhaustive literature review on PRSV and previous control efforts, this chapter will focus on describing the recent success in using PRSV resistant transgenic papaya to practically control PRSV in Hawaii, and in the analysis of the factors that affect resistance in transgenic papaya. The Hawaii papaya case has been covered in several recent reviews (Gonsalves, 1998; Gonsalves and Fermin, 2003). Strong emphasis has been put into the control of PRSV since it is the most economically important virus that attacks papaya. We will summarily describe, however, other viruses reported in the literature that can infect and/or cause disease in the plant. The later sections will focus on the approaches available to engineering papayas for single and multiple resistances against PRSV.

20.2 The papaya and the viruses

Papaya *(Carica papaya)* is an important fruit crop grown widely in tropical and subtropical lowland regions (Manshardt, 1992). Papaya is a herbaceous plant with a single stem bearing a crown of large palmately shaped leaves. Flowers are produced at the axils of the leaf petiole. The plant is polygamous, with male, female, and hermaphrodite plants. In the wild, dioecious plants predominate, whereas cultivated plants are both dioecious and hermaphrodites. In breeding cultivars, the latter can be inbred, which results in stable characteristics from generation to generation. Papaya is largely consumed as a fresh dessert fruit, and the green fruit is used in cooking and consumed as a green salad in many of the Southeast Asian countries. Commercial production of papain, which is commonly used as a meat tenderizer, is made by recovering the enzyme from latex of green fruits. The tree is widely planted in home gardens because it is relatively easy to grow from seed; the first mature fruits can be harvested nine months after sowing seeds, and fruit is produced continuously year-round. Commercially, when trees are grown at a density of 1500 to 2500 per hectare, annual production can range from 125,000 to 300,000 lbs per hectare. Fruits are harvested for 1 to 2 years, after which the trees are usually too tall for efficient harvesting. FAO estimated that about 6 million metric tons of fruit were harvested in 2002, almost doubling the 1980 harvest. Brazil (25.2%), Nigeria (12.6%), India (11.8%), Mexico (11.6%) and Indonesia (8.6%) are the largest producers of papaya.

Viruses belonging to very few plant virus groups can infect and cause diseases of varying economical importance in papaya. *Papaya ringspot virus* (PRSV, *Potyvirus*, *Potyviridae*), is the most important, and has been the subject of numerous studies in disease management control by classical approaches such as cross protection and the more sophisticated ones of genetic engineering. The complete nucleotide sequence of the PRSV genome has been reported for the following geographical isolates: Mexico (GeneBank AY231130), Hawaii (GeneBank NC_001785), Thailand strains

P (GeneBank AY162218) and W (GeneBank AY010722), and Taiwan strains P (GeneBank X97251) and W (GeneBank AY027810). The P strain of PRSV can infect papaya and cucurbits, while the W strain can infect only the latter. Other viruses infecting papaya are:

Papaya mosaic virus (PapMV, *Potexvirus*, *Closteroviridae*) was originally reported in Florida and Venezuela (Purcifull and Hiebert, 1971), and in Peru, Bolivia and the US (Brunt *et al.* 1996); it has been recently found in Jamaica (P. Tennant, personal communication), Mexico (Noa-Carrazana and Silva-Rosales, 2001), India (Bhaskar, 1983; Taya and Singh, 1995) and Sri Lanka (Rajapakse, 1981). PapMV can cause leaf mosaic and stunting in infected plants but the disease is not of economical importance. The full genome of the virus (GeneBank NC_001748) has been reported elsewhere (Sit *et al.* 1989).

*Papaya leaf distortion mosaic virus (*PLDMV, *Potyvirus*, *Potyviridae*) has a monopartite, positive polarity ssRNA genome (GeneBank BD171712). The virus was identified for the first time in Okinawa in 1954 (Kawano and Yanaha, 1992), and so far it has been found only in Japan and Taiwan (Kiritani and Su, 1999). In Japan PLDMV is the main cause of papaya's viral diseases. PLMDV shows the highest level of similarity to the potyvirus *Pea seed-borne mosaic virus* (ca.54%). PRSV and PLDMV differ in serology and symptomatology: plants infected with the latter show rosettes on the leaves, slender stems on the crown top and swollen ringspots on the fruits that are not observed in PRSV-infected papayas. Both viruses are transmitted by aphids (mainly *Aphis gosypii* and *Myzus persicae*) in a non-persistent manner. Roguing and eradication of alternate hosts are the common methods of control for PLMDV.

Papaya meleira virus (PMeV) has been recently described as a new virus (Maciel-Zambolim *et al.* 2003), and was purified from the latex of plants showing 'Meleira' or 'sticky disease' (Correa *et al.* 1988; Kitajima *et al.* 1993; Nakagawa *et al.* 1987). So far the virus has been reported only in Brazil. Purified PMeV virions are isometric particles; the virus is not sap-transmissible and the genome seems to consist of dsRNA molecules. Virions can be found only on the laticifers of infected papaya plants. The disease is characterized by a latex exudation in petioles, new leaves and fruits; when the exudate oxidizes it cause necrosis of the affected tissues. The latex of diseased plants loses viscosity and is easily dispersed. First symptoms appear on young leaves of 6 months old plants; eventually they become distorted and develop a burnt appearance. The presence of dark spots on the pulp of the fruit makes them unmarketable. The only effective way of control of PeMV is roguing and sanitation (Lima *et al.* 2001).

Papaya lethal yellowing virus (PLYV) seems to be present only in Brazil where it was initially reported (Loreto *et al.* 1983). PLYV shares similarities to the *Sobemovirus* group of plant viruses: 51% similarity with the polymerase and the VPg genes, and an average similarity of 44% with the CP genes of sobemoviruses. Infected plants show isometric particles in cell vacuoles. Lethal yellowing by PLYV starts with yellowing of the upper leaves that eventually fall off. As the disease advances, the leaves curl and become chlorotic, followed by wilting and senescence. Green spots appear on the surface of fruits that then turn yellow as they ripe (Lima *et al.* 2001). The virus can be transmitted mechanically, by grafting or wounding and can also be found in the soil and water (Camarco-Rosa *et al.* 1998). Control measures are limited to quarantine, roguing and sanitation.

Different geminiviruses (all of them *Begomovirus*) have been reported as able to infect and/or cause disease symptoms in papaya. For instance, *Squash yellow mottle virus* (SqYMoV, *Begomovirus*, *Geminiviridae*) is found in cucurbits in Costa Rica, but its DNA-B could also be detected in papaya (Karkashian *et al.* 2002). The virus is also known as *Squash yellow mild mottle virus* (GeneBank NC_003865 and NC_003860). SqYMoV shows less than 60% of nucleotide sequence similarity to *Croton yellow mosaic virus* (CYMV) and *Papaya leaf curl virus* (PalCV). CYMV and PalCV are 86% similar at the nucleotide sequence level. PaLCV (GeneBank NC_004147) has been reported in India (Saxena *et al.* 1998), Pakistan (Nadeem *et al.* 1997) and Taiwan (Chang *et al.* 2003). PaLCV-infected papayas show curled leaves, twisted petioles, vein enation and stunting; the affected fruits are small and tend to abscise prematurely. PaLCV is transmitted by whiteflies and is related to *Cassava mosaic virus*. On the other hand, CYMV (GeneBank NC_004300), has been reported only in India and it causes leaf curling in papaya like PaLCV. Different sources of evidence indicate that other geminiviruses are able to infect papaya: Valand and Muniyapra (1992) found that in India *C. papaya* can be an alternate host for *Tobacco leaf curl virus*. Based on the molecular analysis of different plant species Garzon-Tiznado *et al.* (2002) found that *C. papaya* grown in Central Mexico is an alternate host of the geminiviruses *Pepper huasteco virus* and *Pepper texas virus* .

Tomato spotted wilt virus (TSWV, *Tospovirus*, *Bunyaviridae*) can infect papaya in Hawaii (Gonsalves and Trujillo, 1986) and Brazil (Rezende and Costa, 1987), but only in one instance the virus was associated with a field disease. The disease occurs sporadically and causes severe chlorosis and necrosis of the apical leaves; plants defoliate prematurely and young infected plants show pronounced stem apex bending. Fruits can also show dark green rings and deformation; internally, the fruits can develop hard, brown necrotic spots and white lumps..

Papaya apical necrosis has been reported in Venezuela and seems to be caused by a *Rhabdovirus* (Lastra and Quintero, 1981; Marys *et al.* 2000).

The disease causes yellowing of the mature leaves and wilting of the younger ones, accompanied by necrosis of the plant apex. The virus is not mechanically transmissible but can be transmitted by the leafhopper *Empoasca papayae*. The virus was found to be associated to membrane-bound pockets inside the nuclei of parenchyma cells near the vascular bundles. Rhabdovirus type-2 particles detected in the perinuclear cavity of leaf cells were also reported in Brasil (Kitajima *et al.* 1991). The presence of the rhabdovirus in plant tissue was associated with mosaic and leaf distortion.

Marys *et al.* (1995) reported the presence of a thread-like virus in Venezuela and referred to it as *Papaya mild yellowing virus* (PMYV). Virus particles aggregate exclusively within xylem cells and mixed infections with PRSV were common (Marys *et al.* 2000).

C. papaya can be infected experimentally by *Tobacco rattle virus*, , *Cucumber mosaic virus,* and *Tobacco streak virus*,, (Rezende and Costa, 1987).

Papaya ringspot virus (PRSV)

PRSV is by far the most widespread and damaging virus that infects papaya (Purcifull *et al.* 1984). The name of the disease, papaya ringspot, is taken from the ringed spots that develop on fruits of infected trees (Jensen, 1949). Trees infected with PRSV develop mosaic and chlorosis of leaf lamina, water-soaked oily streaks on the petiole and upper part of the trunk, and distortion of young leaves that resembles mite damage. Infected plants lose vigor and become stunted. When infected at the seedling stage or within two months after planting, trees do not normally produce mature fruits. Production of fruit by trees infected at later stages is severely reduced and of poor quality, due to the presence of ringspots and generally lower sugar concentrations.

PRSV is transmitted by numerous species of aphids in a non-persistent manner to a limited host range of cucurbits and papaya. PRSV also produces local lesions on *Chenopodium quinoa* and *C. amaranticolor.* A number of investigations failed to demonstrate that PRSV is transmitted through seeds of papaya or cucurbits, except for one study that showed that 2 of 1355 papaya seedlings from PRSV-infected fruit (Bayot *et al.* 1990). However, seed transmission is not a significant way for spreading PRSV (Gonsalves personal observations).

PRSV is grouped into two types: Type P (PRSV-p) infects cucurbits and papaya, whereas type W (PRSV-w) infects cucurbits but not papaya (Purcifull *et al.* 1984). The latter was previously referred to as WMV-1. Although both types are serologically closely related, observations suggest that papaya is the major primary and secondary source for the spread of

PRSV-p.

Much progress has been made in the molecular characterization of PRSV. The genomic RNA consists of 10,326 nucleotides and has the typical array of genes found in potyviruses (Shukla, 1994). The genome is monocistronic and is expressed via a large polypeptide that is subsequently cleaved to yield all functional proteins. There are two possible cleavage sites, 20 amino acids apart, for the N terminus of the CP (Quemada *et al.* 1990; Yeh and Gonsalves, 1985). These two sites may be functional; the upstream site for producing a functional NIb protein (the viral replicase), and the other, to produce the CP present in aphid-transmissible virions. It is impossible to differentiate PRSV-p and PRSV-w types by their CP sequences. Within the p-types, however, the CP sequences can diverge by as much as 12%. The implication of these sequence differences is discussed later.

The 1991 discovery of PRSV infection of papaya in Southeast Queensland, Australia (Thomas and Dodman, 1993), has provided insights on the possible origins of PRSV-p. Bateson *et al.* (1994) proposed that PRSV-p from Australia could have originated through a mutation of the PRSV-w strain, which has been in Southeast Queensland since 1978. Furthermore, a phylogenetical analysis of PRSV, including its closest relative *Moroccan watermelon mosaic virus* (Lecoq *et al.* 2001), put the origin of PRSV in Asia but also confirmed that PRSV-p has emerged from PRSV-w in numerous occasions (Bateson *et al.* 2002).

20.3 *Control of PRSV in Hawaii and development of transgenic papaya*

The papaya industry in Hawaii started in the 1940s on the island of Oahu (Ferreira *et al* 1992); the virus was discovered in 1945, and the term *Papaya ringspot virus* coined by Jensen (1949). By the 1950s, production on Oahu was affected and the industry subsequently moved to Hawaii island into the area of Puna, which had no commercial production but was free of PRSV. Although Puna was still free of PRSV-p in the late 1970s, efforts to develop control measures for PRSV were started. Cross protection was the first approach to be tried with a mild nitrous acid mutant, designated HA 5-1 (Yeh and Gonsalves, 1984). Protection, however, was not sufficient to provide long lasting economic benefits because of the effort involved in producing mild strains and inoculating plants, and the fact that the mild strain produced significant symptoms on the popular cultivar 'Sunrise' (Ferreira *et al.* 1992).

The concept of parasite-derived resistance (PDR), conceived in the middle 1980s (Sanford and Johnston, 1985), offered a new approach for controlling PRSV. Parasite-derived resistance is a phenomenon whereby transgenic plants containing genes or sequences of a parasite (in this case, the CP gene of a virus) are protected against detrimental effects of the same or related pathogens. Our laboratory began utilizing the PDR concept in 1986 by cloning the CP gene of the PRSV mild strain HA 5-1 (Quemada *et al.* 1990). The gene was engineered (Ling *et al.* 1991) in such a way that could lead to the production of a chimeric protein containing 16 amino acids of *Cucumber mosaic virus* at the N terminus of the full-length CP gene of PRSV HA 5-1. A technique to produce highly embryogenic tissue starting from immature zygotic embryos was developed (Fitch *et al.* 1990), and in 1988-1989, embryogenic tissue was bombarded with tungsten particles coated with DNA of the PRSV HA 5-1 CP gene using the gene gun in Sanford's laboratory. Transgenic plants were obtained (Fitch *et al.* 1993; Fitch *et al.* 1992). Our transformation efforts were successful with the 'Sunset' variety, but not so with the 'Kapoho' variety, which was the dominant cultivar grown in Puna.

R0-micropropagated plants of the first line, designated as 55-1, showed excellent resistance to PRSV HA (Fitch *et al.* 1992). Line 55-1 was female and thus progenies could not be obtained directly from the R0 plants, as would be the case for a hermaphrodite. Research started to determine whether line 55-1 would be resistant to PRSV under field conditions and have suitable horticultural characteristics. First, field trials were conducted using R0 plants instead of waiting a year to obtain R1 plants; and second, R1 plants were obtained by crossing line 55-1 with nontransgenic 'Sunset' under greenhouse conditions, both at Cornell University and the University of Hawaii.

R1 plants of line 55-1 were tested against three PRSV isolates from Hawaii and 13 isolates from different parts of the world (Tennant *et al.* 1994). Genetic analysis of R1 plants suggested that they contained a single insert, since 50% of the R1 plants were resistant. The experiments illustrated several important points: *(a)* resistance of transgenic plants was not correlated to level of protein expression, *(b)* R1 plants were highly resistant to Hawaiian isolates, but (c) largely susceptible to non-Hawaiian isolates (i.e., resistance was homology-dependent, see later).

In 1991, the Animal Plant Health Inspection Service (APHIS) issued a permit for a field trial on Oahu island. The importance of a field evaluation at an early stage cannot be overemphasized because it allowed to appraise the resistance and horticultural characteristics of line 55-1, to bulk up seeds by crossing line 55-1 with nontransgenic plant cultivars, and to demonstrate the long-term resistance of a transgenic fruit crop to infection. A field trial (1992) was designed to determine the resistance of R0 plants to mechanical and aphid inoculations of PRSV. The transgenic papaya showed excellent resistance throughout the two-year trial (Lius *et al.* 1997). Nearly all (95%)

of the nontransgenic plants showed PRSV symptoms by 77 days after the start of the field trial, whereas none of the line 55-1 plants showed symptoms. Plants grew normally and fruit appearance and total soluble solids were within the expected range. Thus, by the mid-1993, the trial had provided convincing evidence that line 55-1 would be useful for controlling PRSV in Hawaii.

Field trial in Puna

The inevitable entry of PRSV into the Puna district on Hawaii island was discovered during the first week of May 1992 in a papaya field 1-3 miles from the major papaya growing areas (Fig. 20-1). The Hawaii Department of Agriculture (HDOA) intervened, and the initial spread of the virus was measurably slowed down. But ultimately, by October 1994, PRSV had spread so much that control efforts by HDOA were abandoned (Fig. 20-1). As expected, this led to an even more rapid spread of PRSV. By late 1994, APHIS allowed a field trial with the stipulation that *(a)* the field must be sufficiently isolated from commercial orchards to minimize the chance of transgenic pollen escape, *(b)* all abandoned trees in the area must be monitored for the introgression of the transgene, and (c) all fruits had to be buried on site.

As noted above, line 55-1 is a transgenic 'Sunset', which is a commercial red-fleshed cultivar. The dominant cultivar growing in Puna, however, is the yellow-fleshed 'Kapoho'. Unfortunately, no virus-resistant 'Kapoho' was developed as a substitute for this cultivar. Using the 1992 field trial as resource base, Manshardt (1998) developed two cultivars 'SunUp' (homozygous, red fleshed) and 'Rainbow' (hemyzigous, yellow fleshed) that were derived from line 55-1. The resistance and quality of these cultivars would be tested in the Kapoho field trial: none of the transgenic plants became infected, except for three plants that showed infection already at the beginning of the trial (Ferreira *et al.* 2002). In stark contrast, 50% of the nontransgenic control plants were infected within four and a half months; and by seven months all were infected (Fig. 20.2). The performance of 'Rainbow' was especially critical because it was targeted as an alternative to 'Kapoho'. Taste, production, color, size, and packing and shipping qualities of 'Rainbow' were analyzed. The consensus was that 'Rainbow', although slightly larger, was a more than adequate substitute for 'Kapoho'.

Deregulation and commercialization of transgenic papaya

To develop a transgenic papaya with virus resistance and excellent fruit qualities represents one hurdle, but to get it deregulated and commercialized is another. These were within the purview of private companies. The responsibility of getting the necessary licenses to commercialize the transgenic papaya was placed into the hands of the Papaya Administrative Committee (PAC), which consisted of Hawaii papaya growers, and was done through APHIS, the Food and Drug Administration (FDA), and the Environmental Protection Agency (EPA).

APHIS was largely concerned with the potential risk of transgenic papaya on the environment: heteroencapsidation of the incoming virus with CP produced by the transgenic papaya, and recombination of the transgene with incoming viruses. A third concern— the escape of the transgenic genes to wild relatives, which might make papaya or their relatives weedier because of resistance to PRSV— was of no consequence since there are no papaya relatives in Hawaii, nor is papaya considered a weed. In November 1996, transgenic line 55-1 and another recently tested line 63-1 (Tennant, 1996) and their derivatives were deregulated by APHIS (Strating, 1996).

According to the EPA, the CP transgenes are considered a pesticide, because they confer resistance to plant viruses. A pesticide is subjected to tolerance levels in the plant. We petitioned for an exemption from tolerance levels of the CP produced by the transgenic plant since the pesticide (the CP gene) was already present in many fruits consumed by the public, and much of the papaya eaten in the tropics is from PRSV-infected plants. In fact, we had earlier used cross protection (the deliberate infection of papaya with a mild strain of PRSV) to control PRSV. Fruit from these trees was sold to consumers. Furthermore, there is no evidence to date that the CP of PRSV or other plant viruses is allergenic or detrimental to human health in any way. Finally, measured amounts of CP in transgenic plants were much lower than those of infected plants. An exemption from tolerance to lines 55-1 and 63-1 was granted in August 1997.

The FDA is concerned with food safety of transgenic products. Several aspects of the transgenic papaya were considered: the concentration range of vitamins, including vitamin C; the presence of *gus* and *nptll* genes; and whether transgenic papaya had abnormally high concentrations of benzyl isothiocyanate. This latter compound has been reported in green papaya. FDA approval was granted in September 1997.

On May 1, 1998 seeds of the transgenic papaya were distributed free to growers who qualified by watching an educational video and signing an agreement that restricted growing of the transgenic papaya only in Hawaii. Fruits could be sold in any region, which would legally accept the transgenic papaya.

Adoption and impact of transgenic papaya in Hawaii

Farmer's adoption of the transgenic papaya was very high (Gonsalves, 2001). Adoption was defined as to whether the farmer had planted the seeds, and not on whether the farmer had signed up and obtained seeds. By September 1999, 90% of the farmers had obtained transgenic seeds and 76% of them had planted (adopted) the seeds. The transgenic 'Rainbow' papaya is now widely planted in Puna, with many formerly abandoned fields being replanted with 'Rainbow'. 'Rainbow' has remained resistant to PRSV and has produced high yields of marketable fruit. Papaya production in Puna has increased from a low of 26 million pounds in 1999 to 40 million in 2001. The percentage of transgenic papaya during this period has varied from 40-70%, and thus it is obvious that nontransgenic papaya is still being grown in Puna (Ferreira *et al.* 2002). There are several reasons for this, the most important being the Japan papaya market, since transgenic papaya cannot be sold in Japan. With the availability of the resistant 'Rainbow' variety, and with the increased demand for nontransgenic papayas for the Japan market, growers looked at options for strategically deploying the newly deregulated transgenic 'Rainbow' in order to assure economic yields of the nontransgenic 'Kapoho' variety. In the HDOA plan, a 1000-acre parcel of land in Kahuwai, which was isolated and predominantly upwind from the main planting areas in Puna, was targeted for nontransgenic fruit production destined for sale in Japan. About 600 acres were devoted to the nontransgenic 'Kapoho' variety and about 300 acres of the transgenic 'Rainbow' was planted to create a buffer of resistant plants that interrupted the movement of PRSV by viruliferous aphids into Kahuwai and their subsequent spread (Ferreira *et al.* 2002). Nontransgenic papaya was being grown with low incidences of PRSV. Results from the Kahuwai project suggest that a strategy of planting an "island" of a nontransgenic variety in an "ocean" of transgenic 'Rainbow' might be useful in allowing commercial production of the nontransgenic variety (Ferreira *et al.* 2002).

Although several other labs have engineered resistance against PRSV using the CP (Bau *et al.* 2003; Davis and Ying, 2002; Lines *et al.* 2002) or the replicase (Chen *et al.* 2001; Ming *et al.* 2003) genes of PRSV, we will emphasize here the work done at Gonsalves group because a) the transgenic papaya 'Rainbow' is the only one being deregulated and commercialized (Canada recently approved the importation of transgenic papaya into their country); b) more than one strategy and more isolates have been used to engineer single and multiple resistances with native, chimeric and synthetic genes; c) the technology has been transferred to other labs in countries pursuing similar goals.

20.4 Mechanisms of transgenic viral resistance

The concept of PDR has been successfully used for controlling plant viruses. Different genes coding for viral proteins have been tried, but the CP gene is, by far, the most widely used to engineering transgenic resistance

(Baulcombe, 1996; Beachy, 1990; Goldbach *et al.* 2003). CP-mediated resistance was hypothesized to lead to resistance by different mechanisms: a CP expressed *in planta* could potentially interfere with the uncoating of incoming virus particles (Wilson and Watkins, 1986), impede replication (Nelson *et al.* 1987; Osbourn *et al.* 1989), or movement of the virus (Wisnieswski *et al.* 1990). It has recently become clear, however, that only in a minority of cases is the gene product essential to confer resistance.

Occasionally, portions of the plant genome become silenced (i.e., not expressed) either at the transcriptional (TGS) or post-transcriptional (PTGS) levels (Matzke and Matzke, 1998). Gene silencing appears to be a universal phenomenon since it has been described in prokaryotes and in diverse eukaryotes. When a genome-integrated virus transgene is targeted for PTGS, degradation affects not only the corresponding transgene messenger but also the genome of an attacking virus with high sequence similarity to the transgene (Goldbach *et al.* 2003). In PTGS, thus, a transgene is efficiently transcribed in the nucleus of the transformed cell but it is targeted for degradation in the cytoplasm where the mRNA levels diminish or are undetectable. Upon infection with a similar virus the viral genome is also targeted for degradation and resistance ensues. Effective resistance is generally limited to virus isolates with greater than 90% similarity to the transgene (Baulcombe, 1996). Various factors can affect the efficiency of PTGS other than sequence similarity; they will be discussed later. It is worth mentioning, however, that certain viruses are able to suppress PTGS (Carrington and Whitham, 1998; Kasschau and Carrington, 1998), rendering transgenic plants susceptible to the homologous virus or even non-related ones.

Sequence similarity drives PTGS and hence homology-dependent resistance (HDR)

Sequence similarity between the transgene and the attacking virus is a fundamental requirement of PTGS-mediated virus resistance. For this reason it is also known as homology-dependent resistance, or HDR (Mueller *et al.* 1995; van den Boogaart *et al.* 1998; Wasseneger and Pelissier, 1998). In PTGS, the key player in the mechanism of RNA degradation is a dsRNA molecule. dsRNAs are produced in a transgenic plant showing PTGS because the transgene is engineered to self-anneal, because inverted repeats were created upon transformation, or simply due to the production of aberrant RNA molecules (Kooter *et al.* 1999; Waterhouse *et al.* 2001). Another potential source of dsRNA is the unintended production of antisense RNAs from neighbouring promoters (Kooter *et al.* 1999), or complementary RNAs (cRNA) synthesized by an RNA-dependent RNA polymerase RdRp (Lindbo *et al.* 1993). Regardless of their source, the dsRNA molecules are further processed to short interfering RNAs (siRNA) species 20-27 nt long (Hamilton and Baulcombe, 1999) by a putative Dicer homolog (Hannon, 2002), RNAseIII-like enzyme called carpel factory (CAF) or SIN1 in *Arabidopsis* (Golden *et al.* 2002; Jacobsen *et al.* 1999). The endogenously generated siRNAs give a degradation complex known as small interfering ribonucleoprotein particles (siRNPs) its target specificity. The siRNPs degrade the putative target mRNA by cutting it at the mid point of the region recognized by the siRNA (Elbashir *et al.* 2001).

Translatable coat protein resistance

The transgenic cultivars, 'SunUp' and 'Rainbow', contain the CP gene of PRSV HA 5-1, a mild mutant from Hawaii. Since the CP of PRSV emerges as a sub-proteolytic product of a polyprotein encoded in the virus genome, it does not contain a 5' untranslated region or a canonical start codon. These elements were added by cloning the PRSV HA 5-1 CP gene (Ling *et al.* 1991) into a *Cucumber mosaic virus* (CMV) CP-gene expresión vector (Quemada *et al.* 1991). The cloned PRSV CP gene is under the transcriptional control of the 35S double-enhancer promoter of CaMV, and flanked by the 5' untranslatable translational enhancer of CMV, and the 35S terminator. The start codon (ATG), and the nucleotides for the first 16 amino acids are from the CMV CP gene. The sequence covering the CP gene is the one corresponding from the Q/S cleavage site proposed by Quemada *et al.* (1990) for the CP protein up to the 3'UTR without the virus polyA tail. This expression cassette was sub cloned into the plant transformation vector pGA482GG, described elsewhere (An *et al.* 1988; Quemada *et al.* 1991). Resistance to PRSV in transgenic 'Rainbow' and 'SunUp' has been the target of numerous studies that can be summarized as follows:

1.- The CP can be detected in transgenic tissues (Tennant *et al.* 2001; Tennant *et al.* 1994). Other transgenic proteins detectable in transformed tissues include NPT-II and GUS.

2.- 'Rainbow', which is a hemyzigous CP/- hybrid, contains a single copy transgenic insertion (Fermin, 2002).

3.- Nuclear run-on analysis detected a very high level of expression of the CP gene in the nucleus (Tennant *et al.* 2001). Cytoplasmic levels of the CP mRNA, however, were quite much lower. The messenger, thus, undergoes PTGS and resistance is considered to be RNA-mediated.

4.- Resistance is better achieved when the challenging virus shares a high degree of similarity to the transgene (Tennant *et al.* 2001), i.e. resistance is homology-dependent.

5.- Although hemyzigous 'Rainbow' shows a very good level of resistance to the homologous strain of the virus, resistance is even higher and wider in homozygous 'SunUp'. It seems, thus, that the level and range of resistance depends on gene dosage (Tennant *et al.* 2001).

6.- Resistance is also dependent on the developmental stage when the transgenic plants were challenged with the non-homologous strain of the virus (Tennant *et al.* 2001). 'SunUp' for example, was resistant to the heterologous strain PRSV TH only when plants were challenged at older stages.

Other transgenic papaya plants have been created with different variants of the CP gene (Fermin *et al.* 2003; Souza, 1999; Tennant *et al.* 2002). Important differences exist between the transgene in 'Rainbow' and the engineered transgene in other transgenic papayas. Transgenic papayas from Brazil, Jamaica, Thailand and Venezuela, for example, are not fusion genes. The transgene in these cases includes only the PRSV CP gene sequence corresponding to the CP four aa downstream of the Q/S cleavage site. 'Rainbow' transgene contains the 5' UTR of CMV, while the others the translational enhancer of the white leaf-CMV. In all cases the transgenes are driven by the CaMV 35S promoter, but the terminator in 'Rainbow' derives also from the CaMV 35S DNA, while in the other transgenes is the *nos* terminator. Remarkably, the transgene in 'Rainbow' was isolated from a mild, mutant strain of PRSV HA while in all other cases the transgenes were isolated from aggressive local strains of the virus. Resistance in other transgenic papayas transformed with the translatable CP gene, however, shared with transgenic 'Rainbow' many features. It is RNA-mediated, homology-dependent and independent of the number of CP insertions. Resistance range, however, is wider than in 'Rainbow'. Similar results have been recently reported in Taiwan (Bau *et al.* 2003).

Non-translatable Coat Protein resistance

The lack of a viral protein in a transgenic crop has distinctive practical advantages over translatable CP-transgenic crops. For example, untranslatability of the CP gene precludes the possibility of transgenic protein production, and hence, virus transcapsidation and allergenicity of the engineered product (Bindslev-Jensen, 1998; Hammond *et al.* 1999). An untranslatable version of the PRSV HA 5-1 CP gene was engineered and used to transform somatic embryos of papaya (Cai *et al.* 1999; Gonsalves *et al.* 1998). A stop codon was introduced 6 nt downstream of the engineered start codon and if translated, the construct delivered to papaya embryos would lead to the synthesis of a tripeptide (Met-Ala-Val), but it could not be detected *in vitro*, or in transformed papaya.

To obtain transgenic papayas harboring untranslatable CP constructs somatic 'Sunrise' papaya embryos were bombarded with a non-linearized pGA482GG harboring the PRSV HA 5-1 untranslatable CP gene (Cai *et al.*

1999; Gonsalves *et al.*1998). Eighty-three R0 transgenic lines were sequentially challenged twice with the nearly homologous, severe PRSV HA (99.8% similarity to the transgene). Subsequent generations were challenged not only with the HA isolate but also with isolates from Bahamas, Brazil, Jamaica, Mexico and Thailand, and an additional heterologous Hawaiian isolate (C. Gonsalves, G. Fermin and D. Gonsalves, unpublished results). Untranslatable CP transgene-mediated resistance to PRSV was due to PTGS (Fermin, 2002):

1.- No transgene-derived protein is detected *in vitro*, nor *in planta*, which precludes the action of the CP product as the effector molecule of resistance.

2.- In a selected line (P2-175#13) showing a high level and wide range of resistance against diverse geographical isolates of PRSV, a high rate of CP transgene nuclear transcription was observed, but concomitant to a lower detection level of its cytoplasmic mRNA.

3.- Finally, as observed in 'SunUp' and 'Rainbow' transgenic papaya (Chiang *et al.* 2001; Tennant *et al.* 2001; Tennant *et al.* 1994) the level and range of resistance was largely homology-dependent.

Surprisingly, the NPTII protein was not detected in some lines (e. g., P2-175.13). Some of the engineered transgenic papayas, thus, produce neither CP nor NPTII as translational products of their respective transgenes. The absence of NPTII and CP improves the value of the engineered papayas since the risks associated with the potential allergenicity of transgenic proteins in the final good are avoided (Ruibal-Mendieta *et al.* 1997). The lack of a CP produced *in planta*, on the other hand, precludes the possibility of heterocapsidation (transcapsidation) – a source of environmental concern in transgenic plants transformed with viral sequences (Hammond *et al.* 1999; Robinson, 1996)

Transgenic papayas harboring an untranslatable CP transgene have been obtained elsewhere showing essentially the same attributes of ours (Lines *et al.* 2002).

Multiple resistance with coat protein segments

Evidence that viral transgenes could confer resistance only against closely related viruses stimulated new approaches to create transgenic plants that could be simultaneously resistant to different viruses. The first approach involved the use for plant transformation of multiple CP or nucleoprotein (*N*) genes regulated by independent promoters and terminators (Fuchs and Gonsalves, 1995; Prins *et al.* 1995; Tricoli *et al.* 1995). Two main disadvantages precluded the extensive use of this approach: concerns about the introduction of increasingly higher amounts of foreign DNA into crop plants, and the limited practicality of transforming plants with multiple CP genes. A second major step towards multiple virus resistance was taken when the minimal amount of transgenic DNA required to confer viral resistance by HDR was analyzed (Pang *et al.* 1997). Virtually any segment of DNA derived from the *N* gene of *Tomato spotted wilt virus* (TSWV) conferred resistance to TSWV in transgenic tobacco. These fragments could be as short as 100 bp if they were fused to a longer fragment of non-related DNA, such as the 720 bp green fluorescent protein (GFP) gene. Fragments 240 bp long of CP genes from different geographical isolates of PRSV have been held together in different combinations and with different orientations (Chiang, unpublished results). Resistance is predicted to be specific for all those isolates that share a high degree of similarity to any of the fragments the chimeric gene is composed of. The inconvenience of this approach is that not all known isolates of PRSV were targeted for degradation. Yet, this is the first rational approach to widen transgenic resistance without increasing the amount of foreign DNA delivered to the target plant. These plants are still under evaluation but the first virus challenges indicate that transgenic resistance has been widened for viruses whose transgenes are involved in the system here described.

Multiple resistance with synthetic genes

The next step to widen the range of resistance involved the creation of a single, short DNA segment (ca. 250 bp) able to confer resistance to different viruses. The approach did not involve cloning native CP genes, fusion of chimeras or multiple transformations with selected sequences. Instead, a single fragment was created by design and used for plant transformation avoiding isolate-specific resistance and diminishing the amount of foreign DNA inserted into the plant genome (Fermin and Gonsalves, 2003). In our first attempt we used transgenic *N. benthamiana* as a model system, and the *N* gene of TSWV as the protecting transgene against TSWV and other tospoviruses. We chose the third quarter of the *N* gene of TSWV because we already knew it was able to confer resistance (Pang *et al.* 1997). Based on this nucleotide sequence we designed and synthesized a synthetic new one that is highly similar (90%) to the corresponding fragments of three different tospoviruses-- TSWV, *Groundnut ringspot virus* (GRSV) and *Tomato chlorotic spot virus* (TCSV), which are distantly related (ca. 78%). The rationale of our approach was that by introducing nucleotide changes at specific locations we might create a transgene equally distant to all the sequences used for its design. This synthetic sequence allowed us to obtain transgenic *N. benthamiana* plants resistant to two different tospoviruses (GRSV and TSWV), and potentially to a third one (TCSV). The plants transformed with the synthetic transgene showed the same level of resistance to TSWV and GRSV (24%). It is worth mentioning that a transgene with 10% changes evenly scattered along the transgene only conferred resistance to TSWV in 9% of the transgenic plants, and none to GRSV (Fermin, 2002; Fermin and Gonsalves, 2003).

We have designed a PRSV-derived single sequence aimed at targeting for PTGS different isolates of the virus (Fermin, 2002). The engineered sequence is equally distant to all PRSV isolates sequences used. We targeted the variable (5') and conserved regions (3') of the PRSV CP. The two synthetic genes were cloned, independently, in tandems of 3 repetitions each, and transcriptionally fused to a silencer gene (Chiang and Wang, unpublished results). In this case, the silencer gene was the half *N* gene of *TSWV* (Jan, 1998; *Jan et al.* 2000a). The two different constructs were cloned in a plant expression cassette and somatic papaya embryos transformed by biolistic. Infection experiments will be performed by the summer of 2003. Results derived from these experiments might lead to the design and creation of a synthetic gene in which we would be able to adjust not only sequence similarity, but also the secondary and tertiary structures of the protecting transgene. The long-term goal of these experiments is to create a short, synthetic sequence able to confer universal and durable resistance to PRSV.

20.5 Concluding remarks

Transgenic papayas have offered a practical solution for the control of an economical important plant disease. The technology is being transferred to other countries where the application of the concept of PDR is proving to be as successful as the case of the transgenic papayas that helped to save the papaya industry in Hawaii. Beyond the practical implications we have also shown that the application of diverse approaches that keep pace with the development of new concepts in basic biology have served multiple purposes: a) it helped improve a technological good by widening the range of resistance against a particular pathogen, b) it contributed to the elucidation of some of the key aspects of HDR like transgene length, location and amount of changes etc., c) it provided a better understanding of the mechanisms of transgene-mediated resistance breakdown, d) it proved that it is possible to rely not only on native genes but that it is possible to design genes that suit particular needs.

References

An, G., Ebert, P., Mitra, A. and Ha, S. 1988. Binary vectors (In: *Plant Molecular Biology Manual).* S. Gelvin and Schilperoort, RA. Kluwer Academic Publishers, Boston

Bateson, M., Henderson, J., Chaleeprom, W., Gibbs, A. and Dale, J. 1994. *Papaya ringspot virus*: isolate variability and the origin of PRSV type W (Aust). *Arch. Virol.* 75:3547-3553.

Bateson, M., Lines, R., Revill, P., Chaleeprom, W., Ha, C., Gibbs, A. and Dale, J. 2002. On the evolution and molecular epidemiology of the potyvirus *Papaya ringspot virus*. *J. Gen. Virol.* 83:2575-2585.

Bau, H., Cheng, Y., Yu, T., Yang, J. and Yeh, S. 2003. Broad-spectrum resistance to different geographic strains of *Papaya ringspot virus* in coat protein gene transgenic papaya. *Phytopathology* 93:112-120.

Baulcombe, D. 1996. Mechanisms of pathogen-derived resistance to viruses in transgenic plants. *Plant Cell* 8:1833-1844.

Bayot, R., Villegas, V., Magdalita, P., Jovellana, M., Espino, T. and Exconde, S. 1990. Seed transmissibility of *Papaya ringspot virus*. Phillip. *J. Crop Sci.* 15:107-111.

Beachy, R. 1990. Coat protein-mediated resistance against virus infection. *Ann. Rev. Phytopathol*. 28:451-474.

Bhaskar, R. B. 1983. Confirmation of the etiology of *Papaya mosaic virus*. *Indian J. Agric. Sci.* 53:479-481.

Bindslev-Jensen, C. 1998. Allergy risks of genetically engineered foods. *Allergy* 53(Suppl. 45):58-61.

Brunt, A., Crabtree, K., Dallwitz, M., Gibbs, A. and Watson, L. 1996. *Viruses of Plants*. CAB International, UK.

Cai, W., Gonsalves, C., Tennant, P., Fermin, G., Souza Jr., M., Sarindu, N., Jan, F.-J., Zhu, H.-Y. and Gonsalves, D. 1999. A protocol for efficient transformation and regeneration of *Carica papaya* L. *In Vitro Cell. Dev.-Pl.* 35:61-69.

Camarco-Rosa, F., Lima, J. and Pio-Ribeiro, G. 1998. Transmission and presence in the soil of *Papaya lethal yellowing virus*. *Fitopatol. Bras.* 23:453-458.

Carrington, J. and Whitham, S. 1998. Viral invasion and host defense: strategies and counter-strategies. *Curr. Opin. Plant Biol.* 1:336-341.

Chang, L.-S., Lee, Y.-S., Su, H.-J. and Hung, T.-H. 2003. First report of *Papaya leaf curl virus* infecting papaya plants in Taiwan. *Plant Dis.* 87:204.

Chen, G., Ye, C., Huang, J., Yu, M. and Li, B. 2001. Cloning of the *Papaya ringspot virus* (PRSV) replicase gene and generation of PRSV-resistant papayas through the introduction of the PRSV replicase gene. *Plant Cell Rep.* 20:272-277.

Chiang, C.-H., Wang, J.-J., Jan, F.-J., Yeh, S.-D. and Gonsalves, D. 2001. Comparative reactions of recombinant *Papaya ringspot viruses* with chimeric coat protein (CP) genes and wild-type viruses on CP-transgenic papaya. *J. Gen. Virol.* 82:2827-2836.

Davis, M. and Ying, Z. 2002. Development of transgenic ringspot virus-resistant papaya for Florida. *Phytopathology* 92:S18.

Elbashir, S., Lendeckel, W. and Tuschl, T. 2001. RNA interference is mediated by 21- and 22-nucleotide RNAs. *Gene. Dev.* 15:188-200.

Fermin, G. 2002. Use, application, and technology transfer of native and synthetic genes to engineering single and multiple transgenic viral resistance. *Department of Plant Pathology, Cornell University*. 293.

Fermin, G. and Gonsalves, D. 2003. Multiple virus resistance to three non-related tospoviruses by a single and short non-native transgene. *Manuscript in preparation*

Fermin, G., Inglessis, V., Garboza, C., Rangel, S., Dagert, M. and Gonsalves, D. 2003. Technology transfer for engineered resistance against PRSV in Venezuelan transgenic papayas. *Plant Dis.* (accepted for publication).

Ferreira, S., Pitz, K., Manshardt, R., Zee, F., Fitch, M. and Gonsalves, D. 2002. Virus coat protein transgenic papaya provides practical control of *Papaya ringspot virus* in Hawaii. *Plant Dis.* 86:101-105.

Fitch, M., Manshardt, R., Gonsalves, D. and Slightom, J. 1993. Transgenic papaya plants from *Agrobacterium*-mediated transformation of somatic embryos. *Plant Cell Rep.* 12:245-249.

Fitch, M., Manshardt, R., Gonsalves, D., Slightom, J. and Sanford, J. 1990. Stable transformation of papaya via microprojectile bombardment. *Plant Cell Rep.* 9:189-194.

Fitch, M., Manshardt, R., Gonsalves, D., Slightom, J. and Sanford, J. 1992. Virus resistant papaya derived from tissues bombarded with the coat protein gene of *Papaya ringspot virus*. *Bio/Technol.* 10:1466-1472.

Fuchs, M. and Gonsalves, D. 1995. Resistance of transgenic hybrid squash ZW-20 expressing the coat protein genes of *Zucchini yellow mosaic virus* and *Watermelon mosaic virus 2* to mixed infections by both potyviruses. *Bio/Technol.* 13:1466-1473.

Garzón-Tiznado, J., Acosta-García, G., Torres-Pacheco, I., González-Chavira, M., Rivera-Bustamante, R., Maya-Hernández, V. and Guevara-González, R. 2002. Presencia de los Geminivirus, *Huasteco del Chile* (PHV), *Texano del Chile* variante Tamaulipas (TPV-T), y *Chino del Tomate* (VCdT), en los estados de Guanajuato, Jalisco y San Luis Potosi, México. *Rev. Mex. Fitopatol.* 20:45-52.

Goldbach, R., Bucher, E. and Prins, M. 2003. Resistance mechanisms to plant viruses. *Virus Res.* 92:207-212.

Golden, T., Schauer, S., Lang, J., Pien, S., Mushegian, A., Grossniklaus, U., Meinke, D. and Ray, A. 2002. Short integuments1/suspensor1/carpel factory, a Dicer homolog, is a maternal effect gene required from embryo development in *Arabidopsis*. *Plant Physiol.* 130:808-822.

Gonsalves, C. 2001. Transgenic virus-resistant papaya: farmer adoption and impact in the Puna area of Hawaii. Master in Liberal Arts. *Graduate School-Empire State College*, State University of New York. 170.

Gonsalves, C., Cai, W., Tennant, P. and Gonsalves, D. 1998. Effective development of *Papaya ringspot virus* resistant papaya with untranslatable coat protein gene using a modified microprojectile transformation method. *Acta Hortic.* 461:311-314.

Gonsalves, D. 1998. Control of *Papaya ringspot virus* in papaya: a case study. *Ann. Rev. Phytopathol.* 36:415-437.

Gonsalves, D. and Fermin, G. 2003. Transgenic papaya to control *Papaya ringspot virus* in Hawaii and technology transfer to other countries (In: *Handbook of Biotechnology).* John Wiley & Sons (accepted for publication).

Gonsalves, D. and Trujillo, E. 1986. *Tomato spotted wilt virus* in papaya and detection of the virus by ELISA. *Plant Dis.* 70:501-506.

Hamilton, A. and Baulcombe, D. 1999. A species of small antisense RNA in post-transcriptional gene silencing in plants. *Science* 286:950-952.

Hammond, J., Lecoq, H. and Raccah, B. 1999. Epidemiological risks from virus mixed infections and transgenic plants expressing viral genes. *Adv. Virus Res.* 54:189-314.

Hannon, G. 2002. RNA interference. *Nature* 418:244-251.

Jacobsen, S., Running, M. and Meyerowitz, E. 1999. Disruption of an RNA helicase/RNAseIII gene in *Arabidopsis* causes unregulated cell division in floral meristems. *Development* 126:5231-5243.

Jan, F.-J. 1998. Roles of non-target DNA and viral gene length in influencing multi-virus resistance through homology-dependent gene silencing. *Faculty of the Graduate School,* Cornell University. 286.

Jan, F.-J., Fagoaga, C., Pang, S.-Z. and Gonsalves, D. 2000a. A minimum length of *N* gene sequence in transgenic plants is required for RNA-mediated tospovirus resistance. *J. Gen. Virol.* 81:235-242.

Jan, F.-J., Fagoaga, C., Pang, S.-Z. and Gonsalves, D. 2000b. A single chimeric transgene derived from two distinct viruses confers multi-virus resistance in transgenic plants through homology-dependent gene silencing. *J. Gen. Virol.* 81:2103-2109.

Jensen, D. 1949. Papaya virus diseases with special reference to papaya ringspot. *Phytopathology* 39:191-211.

Kasschau, K. and Carrington, J. 1998. A counterdefensive strategy of plant viruses: suppression of post-transcriptional gene silencing. *Cell* 95:461-470.

Kawano, S. and Yanaha, T. 1992. The occurrence of *Papaya leaf distortion mosaic virus* in Okinawa. *FFTC Tech. Bull.* 132:12-23.

Kiritani, K. and Su, H. J. 1999. Papaya ringspot, banana bunchy top, and citrus greening in the Asia and Pacific region: Occurrence and control strategy. *Jarq - Jpn. Agric. Res. Q.* 33:23-30.

Kitajima, E., Ritzinger, C. and Rodrigues, M. 1991. Rhabdovirus-like particles in some compositae squash and papaya. *Fitopatol. Bras.* 16:141-144.

Kitajima, E., Rodrigues, C., Silveira, J., Alves, F., Ventura, J., Aragão, F. and Oliveira, C. 1993. Association of isometric virus-like particles, restricted to laticifers, with "meleira" ("sticky disease") of papaya (*Carica papaya*). *Fitopatol. Bras.* 8:118-122.

Kooter, J., Matzke, M. and Meyer, P. 1999. Listening to the silent genes: transgene silencing, gene regulation and pathogen control. *Trends Plant Sci.* 4:340-347.

Lastra, R. and Quintero, E. 1981. Papaya apical necrosis, a new disease associated with a rhabdovirus. *Plant Dis.* 65:439-440.

Lecoq, H., Dafalla, G., Desbiez, C., Wipf-Scheibel, C., Delécolle, B., Lanina, T., Ullah, Z. and Grumet, R. 2001. Biological and molecular characterization of *Moroccan watermelon mosaic virus* and a potyvirus isolate from Eastern Sudan. *Plant Dis.* 85:547-552.

Lima, R., Lima, J., Souza Jr., M., Pio-Ribeiro, G. and Andrade, G. 2001. Etiologia e estratégias de controle de viroses do mamoeiro no Brasil. *Fitopatol. Bras.* 26:689-702.

Lindbo, J., Silva-Rosales, L., Proebsting, W. and Dougherty, W. 1993. Induction of a highly specific antiviral state in transgenic plants-- implications for regulation of gene-expression and virus resistance. *Plant Cell* 5:1749-1759.

Lines, R., Persley, D., Dale, J., Drew, R. and Bateson, M. 2002. Genetically engineered immunity to *Papaya ringspot virus* in Australian papaya cultivars. *Mol. Breeding* 10:119-129.

Ling, K., Namba, S., Gonsalves, C., Slightom, J. and Gonsalves, D. 1991. Protection against detrimental effects of potyvirus infection in transgenic tobacco plants expressing the *Papaya ringspot virus* coat protein gene. *Bio/Technol.* 9:752-758.

Lius, S., Manshardt, R., Fitch, M., Slightom, J., Sanford, J. and Gonsalves, D. 1997. Pathogen-derived resistance provides papaya with effective protection against *Papaya ringspot virus*. *Mol. Breeding* 3:161-168.

Loreto, T., Vital, A. and Rezende, J. 1983. Ocorrência de um amarelo letal do mamoeiro solo no estado de Pernambuco. *O biologico* 49:275-279.

Maciel-Zambolim, E., Kunieda-Alonso, S., Matsuoka, K., de Carvalho, M. and Zerbini, F. 2003. Purification and some properties of *Papaya meleira virus*, a novel virus infecting papayas in Brazil. *Plant Pathol.* 52:389-394.

Manshardt, R. 1992. Papaya. Chapter 21 Chapter 21 in: *Biotechnology of Perennial Fruit Crops*. F. Hammerschlag and R. Litz. CAB /International Press, Cambridge

Manshardt, R. 1998. 'UH Rainbow' papaya. *University of Hawaii*, College of Tropical Agriculture and Human Resources Germplasm, Honolulu. G-1:2pp

Marys, E., Carballo, O. and Izaguirre-Mayoral, M. L. 1995. Properties of a previously undescribed supercoiled filamentous virus infecting papaya in Venezuela. *Arch. Virol.* 140:891-898.

Marys, E. E., Carballo, O. and Izaguirre-Mayoral, M. L. 2000. Occurrence and relative incidence of viruses infecting papaya in Venezuela. *Ann. appl. Biol.* 136:121-124.

Matzke, M. and Matzke, A. 1998. Epigenetic silencing of plant transgenes as a consequence of diverse cellular defence responses. *Cell. Mol. Life Sci.* 54:94-103.

Ming, Y., Dong, W., Hong, C., Yu, L. and Min, Z. 2003. Analyses of virus resistance and transgenes for transgenic papaya. *Yichuan* 25:181-184.

Mueller, E., Gilbert, J., Davenport, G., Brigneti, G. and Baulcombe, D. 1995. Homology-dependent resistance: transgenic virus resistance in plants related to homology-dependent gene silencing. *Plant J.* 7:1001-1013.

Nadeem, A., Mehmood, T. and Tahir, M. 1997. First report of papaya leaf curl disease in Pakistan. *Plant Dis.* 81:1333.

Nelson, R., Powell-Abel, P. and Beachy, R. 1987. Lesion and virus accumulation in inoculated transgenic tobacco plants expressing the coat protein gene of *Tobacco mosaic virus*. *Virology* 158:126-132.

Noa-Carrazana, J. and Silva-Rosales, L. 2001. First report of a Mexican isolate of *Papaya mosaic virus* in papaya (*Carica papaya*) and pumpkin (*Cucurbita pepo*). *Plant Dis.* 85:558.

Osbourn, J., Watts, J. and Beachy, R. 1989. Evidence that nucleocapsid disassembly and a later step in virus replication are inhibited in transgenic tobacco expressing TMV coat protein. *Virology* 172:370-373.

Pang, S.-Z., Jan, F.-J. and Gonsalves, D. 1997. Nontarget DNA sequences reduce the transgene length necessary for RNA-mediated tosposvirus resistance in transgenic plants. *Proc. Natl. Acad. Sciences*, USA 94:8261-8266.

Prins, M., de Haan, P., Luyten, R., van Veller, M., van Grinsven, M. and Goldbach, R. 1995. Broad resistance to tospoviruses in transgenic tobacco plants expressing three tospoviral nucleoprotein gene sequences. *Mol. Plant Microbe In.* 8:85-91.

Purcifull, D., Edwardson, J., Hiebert, E. and Gonsalves, D. 1984. *Papaya ringspot virus*. CMI/AAB descriptions of plant viruses 292 (No 84 Revis., July 1984):8.

Purcifull, D. and Hiebert, E. 1971. *Papaya mosaic virus*. CMI/AAB descriptions of plant viruses 56p.

Quemada, H., Gonsalves, D. and Slightom, J. 1991. Expression of coat protein gene from *Cucumber mosaic virus* strain-C coat in tobacco: protection against infection by CMV strains transmitted mechanically or by aphids. *Phytopathology* 81:794-802.

Quemada, H., Hostis, B., Gonsalves, D., Reardon, I., Heinrikson, R., Hiebert, E., Sieu, L. and Slightom, J. 1990. The nucleotide sequence of the 3' terminal regions of papaya ringspot virus strains W and P. *J. Gen. Virol.* 71:203-210.

Rajapakse, R. H. S. 1981. Host Susceptibility of the Papaya Mosaic-Virus in Sri-Lanka. Beitr. *Trop. Landwirt.* 19:429-432.

Rezende, J. and Costa, A. 1987. Four viruses infecting papaya plants experimentally. *Fitopatol. Bras.* 12:63-65.

Robinson, D. 1996. Environmental risks assessment of releases of transgenic plants containing virus-derived inserts. *Transgenic Res.* 5:359-362.

Ruibal-Mendieta, N., Nagy, A.-M. and Lints, F. 1997. The potential allergenicity of novel foods. *J. Sci. Food Agr.* 75:405-411.

Sanford, J. and Johnston, S. 1985. The concept of pathogen derived resistance: deriving resistance genes from the parasite's own genome. *J. Theor. Biol.* 113:395-405.

Saxena, S., Hallan, V., Singh, B. and Sane, P. 1998. Nucleotide sequence and intergeminiviral homologies of the DNA-A of papaya leaf curl geminivirus from India. *Biochem. Mol. Biol. Int.* 45:

Shukla, D. D., Ward, C.W. and Brunt, A.A 1994. The *Potyviridae*. CAB International, Wallingford.

Sit, T. L., Abouhaidar, M. G. and Holy, S. 1989. Nucleotide-Sequence of *Papaya mosaic virus* RNA. *J. Gen. Virol.* 70:2325-2331.

Souza Jr., M. 1999. Analysis of the resistance in genetically engineered papaya against *Papaya ringspot potyvirus*, partial characterization of the PRSV Brazil Bahia isolate, and development of transgenic papaya for Brazil. Doctor*al Thesis*. Department of Plant Pathology, Cornell University. 277.

Strating, A. 1996. Availability of determination of nonregulated status for papaya lines genetically engineered for virus resistance. *Fed. Regist.* 61:48663.

Taya, R. and Singh, J. 1995. Survey of *Papaya mosaic virus* disease in Haryana. *Agr. Sci. Dig.* 15:191-192.

Tennant, P. 1996. Evaluation of the resistance of coat protein transgenic papaya against *Papaya ringspot virus* isolates and development of transgenic papaya for Jamaica. *Faculty of Graduate Studies, Cornell University*. 317.

Tennant, P., Ahmad, M. and Gonsalves, D. 2002. Transformation of *Carica papaya* L. with virus coat protein genes for studies on resistance to *Papaya ringspot virus* from Jamaica. *Trop. Agr.* 79:105-113.

Tennant, P., Fermin, G., Fitch, M., Manshardt, R., Slightom, J. and Gonsalves, D. 2001. *Papaya ringspot virus* resistance of transgenic 'Rainbow' and 'SunUp' is affected by gene dosage, plant development, and coat protein homology. *Eur. J. Plant Pathol.* 107:645-653.

Tennant, P., Gonsalves, C., Ling, K., Fitch, M., Manshardt, R., Slightom, J. and Gonsalves, D. 1994. Differential protection against *Papaya ringspot virus* isolates in coat protein transgenic papaya and classically cross-protected papaya. *Phytopathology* 84:1359-1366

Thomas, J. and Dodman, R. 1993. The first record of *Papaya ringspot virus*-type P in Australia. *Aust. Plant Pathol.* 22:2-7.

Tricoli, D., Carney, K., Russell, P., McMaster, J., Groff, D., Haden, K., Himmel, P., Hubbard, J., Boeshore, M. and Quemada, H. 1995. Field evaluation of transgenic squash containing single or multiple virus coat protein gene constructs for resistance to *Cucumber mosaic virus*, *Watermelon mosaic virus* 2, and *Zucchini yellow virus*. *Bio/Technol.* 13:1458-1465.

Valand, G. and Muniyapra, V. 1992. Epidemiology of *Tobacco leaf curl virus* in India. *Ann. appl. Biol.* 120:257-267.

van den Boogaart, T., Lomonossoff, G. and Davies, J. 1998. Can we explain RNA-mediated virus resistance by homology-dependent gene silencing? *Mol. Plant Microbe In.* 11:717-723.

Wasseneger, M. and Pelissier, T. 1998. A model for RNA-mediated gene silencing in higher plants. *Plant Mol. Biol.* 37:349-362.

Waterhouse, P., Wang, M.-B. and Finnegan, E. 2001. Role of short RNAs in gene silencing. *Trends Plant Sci.* 6:297-301.

Wilson, T. and Watkins, P. 1986. Influence of exogenous viral coat protein on the co-translational disassembly of *Tobacco mosaic virus*. *Virology* 149:132-135.

Wisnieswski, L., Powell, P., Nelson, R. and Beachy, R. 1990. Local and systemic spread of *Tobacco mosaic virus* in transgenic tobacco. *Plant Cell* 2:559-567.

Yeh, S. and Gonsalves, D. 1985. Translation of *Papaya ringspot virus* RNA *in vitro*: detection of a possible polyprotein that is processed for capsid protein, cylindrical-inclusion protein, and amorphous-inclusion protein. *Virology* 143:260-271.

Yeh, S.-D. and Gonsalves, C. 1984. Evaluation of induced mutants of *Papaya ringspot virus* for control by cross protection. *Phytopathology* 74:1086-1091.

Figure 20.1 Top left, a commercial papaya orchard in the Puna district of Hawaii before PRSV was detected in the island. Top right, devastation caused by PRSV (1994) in the same area where commercial orchards flourished short time ago. Bottom left, after deregulation was granted seeds of transgenic papayas were sown and 13 months later non transgenic papayas (left row) were severely affected by the disease while 'Rainbow' (right row) was already harboring fruits. Bottom right, a commercial orchard 3 years later, with transgenic 'Rainbow' still bearing fruits and showing no symptoms of infection by PRSV.

21. Cocoa

M.J. Jeger[1], L.A. Ollenu[2], L.S. Hagen[3] and M. Jacquemond[3]
[1]*Department of Agricultural Sciences, Imperial College, Wye, Ashford, Kent TN25 5AH, U.K;* [2]*Cocoa Research Institute, P.O. Box 8, Tafo, Ghana;* [3]*INRA-Station de Pathologie, Végétale, Domaine Saint Maurice, BP94, F-84143 Montfavet Cedex, France.*

21.1 Introduction

Cocoa (*Theobroma cacao*) is a crop of ancient cultivation in central America and was introduced to the Caribbean, West Africa, South-East Asia and the Pacific overhe 16-19th centuries (Purseglove, 1977). Although grown over a wide latitudinal range (from the equator to as far as 18°S in Brazil) cocoa requires rather exacting climatic conditions to thrive (Williams *et al.* 1987), including suitable ranges for temperature and rainfall, and availability of shade.

21.2 The Main Viruses

Viruses affecting cocoa have been reported in Ghana, Nigeria, Cote d'Ivoire, Togo, Sierra Leone, Trinidad and Sri Lanka (Posnette, 1944; Brunt and Kenten, 1971; Blencowe, 1971; Carter, 1956). Symptoms suspected to be caused by virus infection have also been reported from Saba (Malaysia) (Thorold, 1975) Indonesia (Senagum, 1961; Turner, 1972), Venezuela (Posnette and Palma, 1944) as well as Colombia and Dominican Republic (Ciferri, 1948).

In West Africa three distinct diseases have been recognised in naturally infected cocoa: cocoa swollen shoot in Ghana, Nigeria, Togo, Sierra Leone and Côte d'Ivoire; cocoa necrosis in Ghana and Nigeria; and cocoa yellow mosaic in Sierra Leone. Cocoa swollen shoot is, by far, the most serious and economically important and has been for many years a major problem for the cocoa industries of Ghana and Nigeria (Padwick, 1956) and more recently Togo (Partiot *et al.* 1978). International movement of cocoa germplasm carries the inherent risk of moving pests, including viruses. Diekman (1999) suggested that germplasm be moved as seed from healthy pods, with budwood the next safest method. Cocoa budwood is tested for the presence of virus by grafting to sensitive West African Amelonado seedlings (Frison and Feliu, 1989).

21.3 ***Cocoa necrosis virus*** Genus ***Nepovirus*** (CNV)

Cocoa necrosis virus found in Ghana and Nigeria was originally reported in 1948 (Anon., 1948), but the causal agent was isolated much later (Owusu, 1971). The virus causes leaf necrosis, defoliation and dieback, and young seedlings may be killed. Plants that survive the acute infection later show leaves with irregular translucent lesions alongside the major veins and leaf distortion resembling that of zinc deficiency (Owusu, *et al. 1985*). CNV is sap transmissible to a wide range of hosts and is also transmitted through the seed of some hosts, but is not transmitted by mealybugs or other insects that commonly infest cocoa. It has been purified and partially characterised and shown to be a serotype of the nepovirus *Tomato black ring virus* (Adomako, *et al.* 1974; Kenten, 1972; 1977). However, the vector has not been found.

21.4 ***Cocoa yellow mosaic virus*** (CYMV)

Cocoa yellow mosaic virus was first found in Siera Leone inducing virus-like symptoms different from those induced by CSSV (Blencowe *et al.* 1963). Typically, infection of cocoa results in large chlorotic blotches, chlorotic mosaic or complete chlorosis of the leaves. Occasionally vein necrosis occurred in the chlorotic areas. The virus is readily sap transmissible to many woody and herbaceous dicotyledonous plants. Hosts of CYMV include plant species from the families of *Apocynaceae, Begoniaceae, Bombacaceae, Chenopodiaceae, Cucurbitaceae, Papillionaceae, Solonaceae and Sterculiacea.* The virus is not transmitted by mealybugs yet the vector is not known (Blencowe *et al.* 1963). The viral particles are spherical in shape about 25μ or 29μ depending on the medium.

21.5 *Cocoa swollen shoot virus* Genus *Badnavirus* (CSSV)

The disease occurs in West Africa (Ghana, Cote d'Ivoire, Nigeria, Sierra Leone, Togo) and was first reported in Ghana in 1936 (Steven, 1936). Its viral etiology was established in 1939 (Posnette, 1940). There is strong circumstantial evidence that swollen shoot virus originated from wild forest trees of the West African tropical rain forest areas. (Posnette, 1981). Forty plant species have proved to be susceptible to one or more isolates of CSSV. All of these fall within the *Tiliaceae, Sterculiaceae, Bombacaceae* and *Malvaceae* families. Of these forty species, only five have been found to be naturally infected. They are *Cola chlamydantha* K. Schum (Posnette *et al.* 1950; Todd, 1951; Tinsley, 1955), *Ceiba pentandra* Gaertn, *Cola gigantea var glabrescens* Brenman and Keay, *Adansonia digitata* Lindl. (Posnette, 1950b; Attafuah, 1965) and *Sterculia tragacantha* Lindl. (Legg and Agbodjan, 1969). Most of the alternative host plants were more tolerant to CSSV than cocoa. Some were symptomless carriers while others produce mild chlorotic symptoms (Dale, 1962). A morphologically similar virus has been isolated from the herbaceous plants *Asystesia sp* and *Commelina sp.* in Ghana and Togo (Hughes and Ollennu, 1994).

Virus strains originally distinguished as *Cocoa mottle leaf virus* (CMLV) (Posnette, 1947; Thresh and Tinsley, 1959) were found in the Volta Region in Ghana and Alaparun in Nigeria. They have similar particle morphology and are serologically closely related to CSSV strains (Sagemann *et al.* 1985). The virus is now included within the CSSV group (Owusu, 1988).

CSSV is one of the most economically important diseases of the world (Thresh, 1958). CSSV has undoubtedly contributed directly and also indirectly to the drastic decline in cocoa production that has occurred in Ghana. Since 1936, nearly 200 million cocoa trees have been cut out from about 130,000 hectares of land (Ampofo, 1997). From 1945 to 1950 yield of wet beans fell by 86.3% almost proportional to the killing of trees. In 1953 the 50 million infected trees represented a capital depreciation of £25 million since the discovery of swollen shoot (Padwick, 1956).

The main symptoms induced by CSSV isolates are leaf symptoms and swellings on the stems and roots (Fig. 21.1). Leaf symptoms may include red vein banding of flush leaves (Posnette, 1941) where the red coloration is attributed to an accumulation of anthocyanins in the mesophyll cells (Knight and Tinsley, 1958). Cocoa plants inoculated with a single DNA clone of the severe isolate Agou1 (Togo) showed all the diversity of symptoms characteristic of the disease syndrome (Hagen *et al.* 1994), thereby demonstrating that a single molecule can provoke the complexity of symptoms observed in the fields.

Posnette (1941) considered it likely that more than one virus or strain of the virus was responsible for the different swollen shoot disease symptoms observed. A naturally occurring virus complex was described by Posnette and Todd (1955). They investigated a CSSV infected-farm on which the trees had been coppiced but some of the infected stumps had regenerated. Some of the regenerated trees had developed symptoms characteristic of the severe isolates of CSSV while some exhibited only mild 'fern leaf' symptoms, which was described by Posnette (1947) as characteristic of attenuated CSSV 1A.

The same authors were further able to inoculate simultaneously mild and virulent isolates of CSSV to young trees. Some of the trees exhibited mild symptoms, some with severe symptoms and others with intermediate symptoms. These observations would indicate virus complexes existing in the doubly inoculated trees, with different strains dominating.

21.5.1 Virus particles and their genome

Cacao swollen shoot virus (CSSV) with bacilliform virions of 113 nm modal length and 28 nm modal width, containing one molecule of double-stranded DNA belongs to the *Badnavirus* genus in the family *Caulimoviridae*. Because these viruses encode an RNA-dependent DNA polymerase (RT) they are called pararetroviruses and also belong to the extensive Retroid agents (McClure, 2000).

The CSSV genome has a double-stranded DNA of about 7 kbp with two discontinuities, one on each strand (Lot *et al.* 1991). The nucleotide sequence of an infectious full-length genomic clone of the Togolese strain Agou1 of CSSV was determined (Hagen *et al.* 1993). The functions of the putative encoded proteins have been deduced from the presence of well-known conserved motifs and sequence comparisons with other badnaviruses.

CSSV genome is 7,161 bp long. As for all pararetroviruses, numbering begins at the first nucleotide of the $tRNA^{met}_{init}$ sequence which is the primer binding site for the minus strand (see below). Sequence analysis revealed an intergenic region and 5 open reading frames (ORFs) capable of encoding peptides greater than 10 kD on the plus strand (Fig.21.2). The intergenic region and ORF 1, ORF 2 and ORF 3 are found in all badnaviruses sequenced to date at similar places and with similar lengthsThe genomic organisation of viruses weighs heavily in their phylogenetic classification. Most similar to CSSV is *Citrus yellow mosaic virus* (CYMV). Huang and Hartung (2001), using the complete sequences, showed that CSSV and

CYMV were grouped together in all bootstrap replicates and were in the same clade as *Commelina yellow mottle virus* (CoYMV, the type member of the genus) and *Dioscorea bacilliform virus* (DBV).

The sequences of the C-termini of the badnavirus ORF 1 proteins are conserved. The CSSV protein is most closely related to that of CYMV by its size and their amino acid sequences can be aligned over their entire length. The role of badnavirus ORF1 proteins is still undefined.The ORF 2 protein of CSSV, which has significant sequence homologies to the CoYMV ORF 2 protein, was shown to possess non-specific sequence nucleic acid binding capabilities *in vitro* through a lysine and proline-rich region in its C-terminus, similar to that found in histone proteins (Jacquot et al. 1996). The ORF 2 protein of *Rice tungro bacilliform virus* (RTBV) binds to the capsid protein (Herzog *et al.* 2000). This suggests a role for the ORF 2 protein in encapsidation process and/or capside structure. The large ORF 3 encodes a predicted polyprotein of 211 kDa which would be cleaved to yield several functional proteins: putative cell-to-cell movement protein, capsid protein, aspartic protease, reverse transcriptase, and RNAse H. The order of the last four proteins is homologous to the *gag/pol* genes of the *gypsy* retrotransposon elements *gag/pr/RT-RNH/in* (McClure, 2000). However, this does not account for the entire coding region so other proteins may or may not be produced. This ORF is divided into three regions. The N terminal domain has amino acid sequence homology with viral plant movement proteins (Bouhid *et al.* 1993; Hagen *et al.* 1993). Moreover, the hydrophilic / hydrophobic profiles of the CSSV protein and the *Cauliflower mosaic virus* (CaMV, *Caulimovirus*) gene I movement protein share similarity. That this protein is involved in virus movement was shown for CoYMV (Tzafrir *et al.* 1997).

The region of CSSV ORF 3 protein from amino acids 507 to 1,040 has similarities with the same region of other badnaviruses. Particularly, there is a basic core rich in lysine residues preceding the zinc-finger motif or cysteine-histidine box ($CX_2CX_4HX_4C$), conserved in retrovirus *gag* genes coding for the capsid. Amino acid sequencing of the RTBV capsid confirmed that this region did encode the coat protein (Marmey *et al.* 1999). After the zinc finger motif in CSSV P211 are 2 histidine motifs (HX_4H) and three CX_2C motifs ($CX_2CX_{11}CX_2CX_4CX_2C$) conserved and unique to badnaviruses and of unknown function. The last part of ORF 3 corresponds to the replicase region and includes protease (PR), reverse transcriptase (RT), and RNAse H (RNH) domains. The viral PR is thought to process the ORF 3 polyprotein into its individual and functional components. By sequence conservation it is identified as an aspartic acid protease and this was confirmed experimentally for RTBV (Laco *et al.* 1995). CSSU has two ORFs, ORF X and ORF Y, which have a coding capacity of over 100 amino acids and a shorter ORF in the middle of ORF 3 (unnamed) with a coding capacity of 48 amino acids. Only CYMV has three ORFs in similar genomic positions but their biological significance, whether they are fortuitous or represent functional genes, is unknown.

21.5.2 Replication and translation

The following general model for replication is well supported for the caulimoviruses and support is accumulating for the badnaviruses as well. Discontinuities of the genomic DNA are repaired in the cell nucleus, giving rise to small 'chromosomes' which are the templates for transcription. A more than full-length, terminally redundant genomic mRNA transcript is synthesized in the nucleus by the host DNA-dependent RNA polymerase II, then exported towards the cytoplasm where it is used for both reverse transcription to yield the viral DNA progeny and for translation of the virus proteins (or at least some of them).

Transcription of the genomic RNA is driven from a promoter located in the intergenic region. The core promoter consists of a TATA box sequence and a putative CA cap site corresponding to the start of transcription. Concerning CSSV, a putative TATA box was identified at nt 6962 (TATAAA) as well as conserved motifs of some cis-acting elements. Efficiency and specificity of the promoter remain to be established.

Pararetroviruses use a consensus AATAAA polyadenylation signal for RNA 3'-end formation. Production of a more than full-length RNA transcript supposes the bypass of this polyadenylation signal at the first passage. A large viral mRNA of greater than 7 kb is present in infected cells in addition to other smaller uncharacterized RNAs (Hagen, 1994). Mapping of the genomic transcript's 3' and 5' termini indicated a 98 nt long redundancy (Hagen and Jacquemond, to be published elsewhere).

According to the model proposed for the caulimoviruses, synthesis of the minus strand by the RT is primed by the host cytosolic $tRNA^{met}{}_{init}$. The viral RNAse H degrades the RNA template generating a 500-600 nts long DNA, still bound to the primer, referred as strong-stop DNA. The strong stop DNA anneals to the 3' repeat of the genomic RNA, allowing the switch necessary for the production of a full-length minus DNA. Synthesis of the second (plus) strand occurs in a distinct region, which will determine the discontinuity on this strand.

The transcript of CSSV has a long leader of 609 nucleotides with seven small ORF and a predicted strong secondary stem-loop structure (Pooggin *et al.* 1999), both of which are unfavorable for efficient translation via Kozak's classical model for monocistronic gene transcripts. CaMV and RTBV, and the badnaviruses probably, use ribosome shunt to bypass the long leader (Hohn *et al.* 2001), then leaky scanning (Fütterer *et al.* 1997) and/or reinitiation to translate their proteins. In the leaky scanning model, translation of the first ORF would have to use an inefficient alternative start codon in order to allow some ribosomes to continue scanning. There are no start codons within CSSV ORF 1 and ORF 2 which would favor leaky scanning. Moreover, juxtaposition of the three ORFs would rather favor reinitiation of translating ribosomes. If the putative ORF Y was indeed translated, leaky scanning would not be plausible as there are many AUG codons in the preceding ORF 3. Alternatively, this ORF might be translated from a spliced genomic RNA as described for RTBV (Fütterer *et al.* 1994).

Integration of viral DNA into the host genome has been reported for different pararetroviruses. It is all the more preoccupying since integrated copies could give rise to an episomal infection as shown for the badnavirus *Banana streak virus* (BSV) (Ndowora *et al.* 1999). However, no integrase core motif has been found in badnaviruses and no hybridisation with a CSSV probe was detected in total DNA from healthy *Theobroma cacao* leaves (Lot *et al.* 1991). Thus, nothing presently suggests that CSSV is integrated in the cultivated cocoa genome.

21.5.3 Variability and detection

Several isolates of CSSV varying in virulence and symptomatology have been identified. Four variants of the virus in Ghana as 1A, 1B, 1C and 1D were identified by Posnette (1947) and Box (1945) used Roman numerals I – IV to describe them. This method of nomenclature was not satisfactory since some isolates in other countries could not be classified according to symptomatology. Particularly, the Nigerian CSSV isolates did not show any similarity to any of the above isolates (Anon., 1950) and isolates occurring in Sierra Leone though mild were slightly different from mild isolates of CSSV occurring in Ghana, Nigeria and Côte d'Ivoire (Thresh and Tinsley, 1959; Attafuah, *et al.*1963). Tinsley (1953a, b) and Thresh (1961) resorted to locality nomenclature (New Juaben isolate, Mampong isolate, Bisa isolate etc.) and probably because of the biological diversity of CSSV this nomenclature has persisted since.

Several serological detection methods have been used to detect and classify CSSV isolates: agar gel double diffusion test (Kenten and Legg 1971); ISEM (Adomako *et al.* 1983;Sagemann *et al.* 1985); ELISA (Sagemann *et al.* 1983; 1985) and a virobacterial agglutination test (VBA) (Hughes and Ollennu 1993; Hughes *et al.* 1995). Sagemann *et al.*(1985) distinguished five groups (A – E) of Ghanaian isolates of CSSV on the basis of leaf symptoms and serological properties using antiserum raised against CSSV 1A. However, some isolates were only poorly detected or not detected at all. Latter, VBA was reported to be more sensitive than ELISA and an easily portable method. Using a panel of 13 antisera raised against several mild and severe CSSV isolates, Hughes and colleagues were able to class 44 CSSV accessions into 8 groups. A combination of two antisera was able to detect all strains. Severe and mild strains were differentiated, which is essential for the use of cross-protection to control the CSSV disease.

Additionally, production of monoclonal antibodies in mice was developed by Hoffmann *et al.* (1999) for CSSV 1A, Nsaba and Kpeve isolates. Only 1A yielded a large number of 1A-specific monoclonal antibodies.

PCR has proven to be more sensitive than ELISA for other badnaviruses. Recently, PCR-based methods have been tested for CSSV detection. PCR performed on standard DNA extractions was either not more sensitive than ELISA (Jacquot *et al.* 1999) or totally unsuccessful (Hoffmann *et al.* 1997). Thus adaptation of standard protocols was needed. This was achieved by either further purification of total DNA (Muller *et al.* 2001) or by immunocapture (IC)-PCR (Hoffmann *et al.* 1997). In the former case, and using degenerate primers from the transcriptase region, CSSV was detected in asymptomatic tissues of experimentally infected plants and in trees naturally infected in and around the Agou region of Togo. However, detection was possible in only one out of three trees developing symptoms sampled in the Kpele location. In the latter, the IC-PCR test was 100 times more sensitive than ELISA. Several A1 group isolates from several areas of Ghana were detected reliably using non-degenerate primers from the capsid/ORF X region. However, the Kpeve and Nsaba isolates did not amplify in these conditions even when homologous antisera were used. Finally, sequencing of some amplified products from CSSV-1A isolates revealed a considerable level of variation.

It is evident from all these studies that significant variability between CSSV isolates exists. Complete sequencing of other strains than Agou1 will be needed to discover the extent of the diversity and to pinpoint genomic regions, which are conserved. These data will be essential in order to develop robust PCR assay(s) for indexing.

21.6 Experimental inoculation methods

Experimental inoculation of cacao plants is usually done either by viruliferous mealybugs transmission or by grafting. Mechanical inoculation with purified virus is also described. Each of the methods has its own drawbacks. Methods for infecting plants with cloned virus DNA offer controlled inoculation of cacao plants not only for research but also for applied purposes (cross-protection, plant-breeding resistance tests, etc.).

Both of these methods give high levels of infection, and use a known, controlled, cloned DNA inoculum which can be prepared in large quantities.

Microprojectile bombardment of cocoa beans with a full-length clone of CSSV was described by Hagen *et al.* (1994). Even if it requires a microparticle gun, the main advantage of this method is that only one copy of the genome is required, avoiding complicated clone construction.

Agroinfection uses the *Agrobacterium* to transfer DNA cloned within the borders of a T-DNA-containing plasmid (Grimsley *et a*l. 1986). This method requires a greater-than-genome-length clone including the viral promoter at the beginning of the construction and the polyadenylation signal downstream to allow transcription of the complete genomic RNA. It was successfully developed by Jacquot *et al.* (1999) for infecting young cacao seedlings. However, rigorous containment conditions are required for working with genetically modified *Agrobacterium* and the infected plants.

21.7 Disease Epidemiology

21.7.1 Vector ecology and behaviour

A range of mealybug species is known to vector CSSV in West Africa (Dufour, 1991) with *Planococcoides njalensis* the most important. *P. njalensis* exhibits much intraspecific morphological variation but a recent study (Padi and den Hollander, 1996) failed to find other *Planococcoides* species in fourteen populations sampled.

Cornwall (1958) showed that the mobile mealybug population in the canopy is almost entirely of first instar nymphs. Movement is enhanced with temperatures greater than 24°C, density of mobile insects is at a maximum close to the canopy top and nymphs can walk at least 28 ft under favorable conditions. Trees have to be spaced 12ft apart to have no contact between branches of adjoining trees. At 4ft spacing 40% of branches were in contact and about 8% of mobile mealybugs reached adjacent contact trees.

All vector species to a greater or lesser extent can be dislodged and dispersed by air currents (Cornwall, 1960). Mealybugs that drop from branches are carried by air currents when falling and become laterally dispersed. Gradients of deposition were recorded with a maximal recorded distance of aerial dispersal of 340ft. Over a five-month experimental period 22% (14/64) of young cocoa trees were infested by air-dispersed vectors. Mealybug vectors are found mostly in the canopy branches with peak colony numbers occurring with periods of leaf flushing, again stressing the importance of mealybug movement in the peripheral zone of the canopy and where contact is made with adjoining trees (Campbell, 1975). Techniques and sampling procedures for estimating mealybug and other Homoptera have been described (Campbell, 1983). The effect of shade on populations of the main mealybug vectors was investigated by Campbell (1984). An estimated 100, 99 and 83% of unshaded, lightly and heavily shaded trees were infested with mealybugs suggesting shade thinning over cocoa would lead to little change in the rate of virus spread.

In spite of the complexity of the cocoa ecosystem, cycles of many insect populations have been observed (Bigger, 1976), including the vector species *Phenacoccus hargreavesi* that shows cycles with a period of about 9 weeks. Time series of the mealybug vectors and their associated ant populations were studied in detail by Bigger (1993a, b).

21.7.2 Disease dynamics

Thresh (1958) gave the first comprehensive description of CSSV epidemiology, reviewing the formation of new outbreaks, the pattern and rate of spread of disease in relation to vector populations, tree factors, seasonal effects and virus-host-vector interactions. A series of coppicing experiments was started in Nigeria in the mid 50's to ascertain whether the large-scale destruction of apparently healthy trees surrounding outbreaks was justified (Thresh, 1959). The coppicing experiments indicated that cutting out of trees ('roguing') should be done according to size of outbreak. It is now known that many indigenous tree and shrub species are susceptible to the virus and many are commonly infested with the mealybug vectors. It seems likely that some tree species, notably *Adansonia digitata* and *Cola chlamydantha* were infected before cocoa but it seems impossible to provide an account of the origins of CSSV in West Africa.

Patterns and rates of spread of CSSV have been comprehensively described by Thresh *et al.* (1988b). The sequence of steps involved in spread are: (1) jump spread by windborne mealybugs to initiate outbreaks in new areas; (2) radial spread by mealybug movement on branches from tree to tree; (3) further enlargement of primary foci by radial spread, and to satellite foci by jump spread; (4) consolidation, by both mechanisms of spread, into an area of extensive infection. Gradients of infection from a source (infected plant or boundary) at different sites and sizes of outbreaks were compared. Gradients were steep with infection falling off within a few meters of the source. Nevertheless over relatively short periods of time disease incidence can increase from less than 10% to 80% over an 8-year period. Thus if CSSV had spread from a large number of 'indigenous' foci of diseased cocoa followed by relatively unrestricted spread then it is not surprising that CSSV has spread throughout Ghana and other areas of West

Africa. Thus following Vanderplank (1948), Thresh *et al.* (1988) termed CSSV an archetypal crowd disease.

Eradication measures have been mounted against CSSV in Ghana since 1946 (Thresh, 1988). The drastic measures adopted were difficult for small farmers to accept and comply with. In the areas of mass infection, infection was so prevalent that it would have been prohibitively expensive and unrealistic to bring all plantings under regular inspection and treatment. An alternative approach was to consolidate operations by forming compact blocks of replanted cocoa, using resistant varieties where available (Thresh *et al.* 1988b) and provide essential logistical support. Ollennu *et al.* (1989) concluded that by following this approach there is little risk of reinfection where scattered outbreaks occur, provided all surrounding cocoa is under regular inspection and treatment. Where mass infection is present, new plantings should allow a cordon-sanitaire of at least 10m from the plot boundary.

21.7.3 Mathematical modeling of CSSV epidemics

The long-term nature and failure of the eradication program attempted for CSSV in Ghana has prompted attempts to model aspects of epidemiology, notably re-infection and disease spread in newly planted areas. Jeger and Thresh (1993) modeled these processes in situations where radial and jump spread were present to varying extents and where a cordon-sanitaire was or was not imposed. The impact of jump spread appeared to be small by contrast to radial spread, but there are few estimates of its frequency or its contribution to a cordon-sanitaire being crossed. A cordon-sanitiare markedly delays re-infection (by whatever process) but subsequent eradication is necessary to restrict spread. The effect of size of outbreak, as noted by Thresh (1959), was also apparent in the importance of a relative rather than absolute width for the cordon-sanitaire. The model however was insufficiently developed to assess where eradication or rouging operations would be sufficient when combined with replanting with healthy cocoa (Jeger and Chan, 1995).

Chan and Jeger (1994) developed an analytical framework for examining rouging and replanting by partitioning trees into healthy, latently-infected, infectious and post-infectious categories. This enabled an analysis of whether rouging could be successful for virus diseases such as citrus tristeza, banana bunchy top and plum pox. With cocoa swollen shoot it was shown that rouging based only on symptom appearance would fail because these appear relatively late relative to epidemic dynamics (Jeger and Chan, 1995). However the lack of detail on the mealybug-virus interactions and mealybug dispersal limited further elaborations. Subsequently Jeger *et al.*

(1998) developed a more comprehensive theoretical model that incorporated these vector characteristics, which generated further questions concerning control through rouging and other type of control options. However, although this analysis has been applied to the vector-borne phytoplasma disease lethal yellowing of coconut (Jeger, 1997), it has not been applied to swollen shoot.

21.8 New developments for control

21.8.1 Vector control with insecticides

Clearly, the control of the mealybug vector would have minimized the spread of the virus by eliminating or decreasing vector populations, but this was not achieved. Trials were conducted in the 1950s with organophosporous systemic insecticides. One was used in a large field trial (Hanna and Hetherington, 1957). Though the trial was successful, it was not pursued as numerous applications of insecticides were necessary and the material used was expensive, had a high mammalian toxicity and caused an undesirable taint of the cocoa beans.

More promising results were obtained by using formicides to control the attendant ants, but the trials were abandoned because the chemicals used had undesirable side effects and led to a big increase in pod-mining insects and other cocoa pests (Thresh and Owusu, 1986). Little further attention has been given to the use of insecticides for routine control of mealybugs on cocoa because these insects are only important as virus vectors and cause little direct yield loss or other forms of damage.

21.8.2 Biological control

Studies carried out in the 1950s at exploiting exotic biological agents to control the CSSV mealybug vector were discouraging despite the identification of several indigenous natural enemies from a nation-wide collection. (Posnette and Strickland, 1949; Strickland 1951a, b: Anon. 1953, Donald, 1956). Since then, the only reported study on mealybug natural enemies was that by Bigger (1976). Interest in beneficial insects was rekindled when Asante *et al.* (1988) reared several unidentified specimens from field samples of cocoa mealybugs. Details of the development of biological control of cocoa mealybugs have been recently reviewed (Padi, 1997). Ackonor and Mordjifa (1999) reported that *Planococcoides njalensis*, the major vector of CSSV hosts several natural enemies but in low numbers and the extent to which they contribute to the normally low population of the pest on cocoa is not certain. They concluded that the efficacy of the beneficial insects is likely to be impaired by their own natural enemies. Padi *et al.* (2002) imported the coccinellid beetle, *Cryptolaemus montrouzieri*, an important predator of mealybugs and successfully reared it in the laboratory. However, one month after release in the field none of the beetles could be recovered.

21.8.3 Breeding for resistance

One of the reasons why swollen shoot disease has caused serious problems in West Africa (Ghana in particular) is that the main cultivar grown is the West African Amelonado, which is both vulnerable and sensitive to CSSV infection. Trees usually die after infection with virulent strains that are prevalent in the region. An obvious means of control would be to introduce cultivars with immunity or high levels of resistance to enable them to grow and yield satisfactorily, even after becoming infected. Such cultivars are not yet available, though the merits of such approach were appreciated from the outset, and a considerable effort has been made over many years to select resistant or tolerant genotypes from those already being grown or from subsequent introductions (e.g. Thresh *et. al.* 1988a).

Some genotypes of Upper Amazon parentage in Ghana are more difficult to infect than Amelonado and they also tend to be less sensitive to infection (Posnette and Todd, 1951). These features are likely to have contributed to the generally successful performance in Ghana of the F3, Upper Amazon and the Series II hybrids (Amazon x Amelonado and Amazon x Trinitario) that have been used wildly, since the 1950s for new plantings and to replace Amelonado trees removed during the eradication campaign (Thresh *et. al.* 1988b). However, the resistance obtained in these hybrids (Legg and Lockwood, 1981) is often insufficient to prevent virus spread. Meanwhile, the most resistant (inter Upper Amazon) hybrids available are supplied to farmers but there is a problem with low yields of these hybrids and the apparent causes are currently being investigated (Adomako and Adu-Ampomah, 2000).

Promising mutants have been obtained from mutation breeding using gamma irradiation at Cocoa Research Institute of Ghana. The mutants are currently being assessed in the field (Adu-Ampomah *et. al.* 1996).

21.8.4 Mild strain protection

The practice of using mild strains of CSSV to protect cocoa plants against the virulent effects of severe strains was demonstrated in Ghana in the 1950s (Posnette and Todd, 1955). However, its practical application in disease control was not assessed. The possible use of this concept in the control of CSSV in Ghana particularly in the Eastern Region was recently revived (Ollennu and Owusu, 1989; Ollennu *et al.* 1996). The ability of mild strains from the Eastern Region to cross-protect against CSSV severe IA has been determined in the glasshouse. Results from field trials indicate that natural infection of CSSV mild strain protected plants by severe strains was slow and involved fewer plants (Ollennu and Owusu, 1997, 2001; Ollennu *et al,* 1999). Some of these plants, which initially produced severe symptoms later, recovered. Field trials in which challenge inoculations were carried out by tissue union, a higher percentage of the trees (c.29.5%) lost their protection compared with c14.0% natural infection (Ollennu and Owusu, 2001, 2002). The mild strain reduced yield by less than 10% (Ollennu and Owusu, 2002). Further investigations are required on the practical usefulness of the protection afforded by the mild strains on farmers' farms. To exploit the potential for mild strain protection arch has to determine best combinations of mild strains and new hybrids giving high quality cocoa as well as adequate protection (Hughes and Ollenu, 1994)

21.8.5 Roguing and phytosanitation

VanderPlank (1948) described the swollen shoot disease, as a crowd disease, which does not spread far and therefore, should be easily controlled by eradication. Posnette (1943) demonstrated that destruction of visibly infected trees was effective in controlling the disease and that the control was more efficient if neighboring contact trees were also removed. Since then eradication has been the principal means of control adopted in Ghana, Nigeria and Togo. This has led to the removal of cocoa trees in these countries, yet control has not been achieved due to several factors including political and socio-economic problems.

In Ghana, attempts to control the disease by grubbing infected as well as neighboring contact trees in 40 years, has resulted in the destruction of about 200 million cocoa trees from about 130,000 hectares of land (Ampofo, 1997). The eradication measures involve physical inspection of cocoa trees and cutting-out all trees with visible infection and their neighboring contact trees. The branches are removed and packed together such that viruliferous mealybug vectors are confined and die as the branches wither. Ollennu *et al.* (1989a) reported that if all known infected trees in the vicinity were removed before replanting, there was little risk of reinfection. They recommended that new plantings in worst affected areas should stop at least 10m short of the plot boundary to provide a "cordon sanitore. They added that plantings should ideally be large and compact to decrease the proportion of trees in the vulnerable peripheral areas.

The eradication measures were reviewed in Ghana (Ollennu *et. al.* 1989b) making it more drastic. All trees with symptoms are to be destroyed together with adjacent symptomless trees to distances of 5, 10 and 15 metres around out-breaks of 1-10, 11-100 and >100 infected trees, respectively. This will ensure the removal of many of the latent and missed infections that would have remained and contributed to further spread.

21.8.6 Cultural control

Much of the spread of cocoa swollen shoot virus occurs over short distances by mealybugs walking between the inter-locking branches of adjacent trees. In Ghana, the common practice of planting to the boundary of the land available, with little or no break between adjacent farms has greatly facilitated the spread of CSSV. Ollennu *et. al.* (1989a) showed that there are steep gradients of infection from the perimeter with many of the infections in the outmost row immediately alongside obvious sources of infection around the perimeter. This emphasizes the importance of plants at the greatest risk in replanting operations. The likely benefits of some form of barrier or mixed cropping systems are therefore being evaluated in Togo and Ghana (Tsatsu *et al.* 1997; Ollennu *et al.*1994). Tsatsu *et al.* (1997) initially used three rows of coffee to surround new cocoa establishment adjacent to old cocoa plantings. In addition, they sprayed the young cocoa with systemic insecticides against the mealybug vector. In later trials they planted kola followed by two rows of coffee. While in Ghana, Ollennu *et al.* (1994) evaluated the effectiveness of cash crops such as kola, oil palm and citrus as barrier between old and new establishments. The objective of these trials was to minimize the early reinfection of newly planted cocoa from old adjacent plantations. The results so far obtained are encouraging.

21.8.7 Integrated management

The Swollen Shoot Virus (CSSV) disease has caused incalculable damage to the cocoa industries in Ghana and Togo. However, despite the extensive eradication measures used since 1946 in Ghana, CSSV is still prevalent. Experience has shown that no single measure on its own will solve the swollen shoot problem and that a number of measures may need to be taken together.

In this regard a new approach towards integrated control of the disease would have to be adopted to achieve success. The strategy would continue to be based primarily on eradication. There is considerable scope for decreasing rates of reinfection in the worst affected areas by planting large blocks away from known sources of infection and by using cocoa varieties that are partially resistant to infection. Other factors that could be integrated may include agronomic factors (planting densities, barrier cropping), breeding for disease resistance, vector control (biological) and molecular biology techniques (mild strains, gentransfers). Additionally as the smallholder cocoa farmer is disinclined to adopt radical new techniques rapidly, recommendations must be compatible with their traditional farm practices.

References

Ackonor, J.B. and Morjifa, D.K. 1999. Parasitism and predation in *Planococcoides njalensis* (Laing). (Homoptera: Pseudococcidae) on cacao in Ghana. *Trop. Agric.* (Trinidad) 76: 269-274.

Adomako, B. and Adu-Ampomah Y. 2000. Reflections on the yield of Upper Amazon cocoa hybrids in Ghana with reference to breeding for cocoa swollen shoot virus resistant varieties. *Cocoa Growers' Bull., 2000,* No. 52, pp. 33-45.

Ackonor, J.B. and Morjifa, D.K. 1999. Parasitism and predation in *Planococcoides njalensis* (Laing). (Homoptera: Pseudococcidae) on cacao in Ghana. *Trop. Agric.* (Trinidad) 76: 269-274.

Adomako, B. and Adu-Ampomah Y. 2000. Reflections on the yield of Upper Amazon cocoa hybrids in Ghana with reference to breeding for cocoa swollen shoot virus resistant varieties. *Cocoa Growers' Bull., 2000,* No. 52, pp. 33-45.

Adomako, D., Owusu, G.K. and Oduro, K.K. 1974. Purification of cocoa necrosis virus from cocoa leaves. *Phytopathology,* 64: 1325-1330.

Adomako, D., Lesemann, D.E., Paul, H.L. and Owuwu, G.K. 1983. Improved methods for the purification and detection of cocoa swollen shoot virus. *Ann. appl. Biol.* 103: 109-116.

Adu-Ampomah, Y., Owusu, G.K., Sackey, S., Padi, B. and Abdul-Karimu, A. 1996. Use of gamma rays to induce mutants resistant to cocoa swollen shoot disease in *Theobroma cacao* L. *Pl. Breeding* 115: 74-76.

Ampofo, S.T. 1997. The current cocoa swollen shoot disease situation in Ghana. *Proc. First Inter. Cocoa Pests and Diseases Seminar, Accra, Ghana,* 6-10 November, 1995. pp 175 – 178.

Anonymous, 1946. Virus research. In: *Rept. West African Cocoa Res. Inst. for 1944/45,* p.12-21.

Anonymous, 1947. Virus research. In: *Rept. West African Cocoa Res. Inst. for 1945/46,* pp.11-25.

Anonymous, 1948. Virus research. Swollen shoot and other diseases of cacao. I. A new cacao virus? In: *Rept. West African Cocoa Res. Inst. for 1947/48,* pp. 11.

Anonymous, 1950. Virus research. In: *Rept. West African Cocoa Res. Inst. for 1948/49,* pp. 12-20.

Anonymous, 1953. Biological control of mealybugs. *Rept. of the Cocoa Res. Inst. of Ghana,* 1950-51 pp 40-42

Asante, S.K., Owusu-Manu, E. and Mordjifa, D.K. 1988. Studies on the natural enemies of cacao mealybug. *Rept. of the Cocoa Res. Inst. Ghana,* 1986/87, pp 33-24

Attafuah, A. 1965. Occurrence of cocoa viruses in wild plant species in Ghana. *Ghana J. Agric. Sci.* 5: 97-101.

Attafuah, A., Blencowe, J.W. and Brunt, A.A. 1963. Swollen shoot disease in the Sierra Leone. *Trop. Agric. Trinidad,* 40: 229-232.

Attafuah, A. and Brunt, A.A. 1960. Virus research: classification and comparison of isolates. In: *Rep. of the West African Cocoa Res. Inst. for 1958/59,* p. 16-17.

Baker, R.E.D. and Dale, W.T. 1947. Notes on virus diseases of cacao. *Annals appl. Biol.,* 34: 60-65.

Bigger, M. 1976. Oscillations of tropical insect populations. *Nature* 259: 207-209.

Bigger, M. 1993a. Time series analysis of variation in abundance of selected cocoa insects and fitting of simple linear predictive models. *Bull. Ent. Res.* 83: 153-169.

Bigger, M. 1993b. Ant-homopteran interactions in a tropical ecosystem. Descriptions of an experiment on cocoa in Ghana. *Bull. Ent. Res.* 83: 475-505.

Blencowe, J.W. 1971. Cocoa and coconuts in Malaysia. *The Planter* 47: 299-303.

Blencowe, J.W., Brunt, A.A., Kenten, R.H. and Lovi, N.K. 1963. A new virus disease of cocoa in Sierra Leone. *Trop. Agric., Trinidad* 40: 233-236.

Bouhida, M., Lockhart, B.E.L. and Olszewski, N.E. 1993. An analysis of the complete sequence of sugarcane bacilliform virus genome infectious to banana and rice. *J. Gen. Virol.* 74: 15-22.

Box, H.E. 1945. Insect transmission of the 'swollen shoot' virus in West Africa. *Nature* 155: 608-609

Brunt, A.A. and Kenten, R.H. 1971. Viruses infecting cacao. *Rev. Plant Pathol.* 50: 591-602.

Brunt, A.A. and Kenten, R.H. 1963. The use of protein in the extraction of cocoa swollen shoot virus from cocoa leaves. *Virology.* 19: 388-392.

Campbell, C.A.M. 1975. The distribution of mealybug vectors of CSSV within trees. *Proc. 4th Conf. West Afric. cocoa entomologists, 9-13 December 1974,* Legon, Ghana, pp67-71.

Campbell, C.A.M. 1983. The assessment of mealybugs (Pseudococcides) and other Homoptera on mature cocoa trees in Ghana. *Bull. Ent. Res.* 73: 137-151.

Campbell, C.A.M. 1984. The influence of overhead shade and fertilizers on the Homoptera of mature Upper-Amazon cocoa trees in Ghana. *Bull. Ent. Res.* 74: 163-174.

Carter, W. 1956. Notes on some mealybugs (Coccidae) of economic importance in Ceylon. *FAO Plant Prot. Bull.* 4: 49.

Chan, M-S. and Jeger, M.J. 1994. An analytical model of plant virus disease dynamics with rouging and replanting. *J. appl. Ecol.* 31: 413-427.

Ciferri, R. 1948. A virus disease of cacao in Columbia and Santa Dominigo. *Review Fac. Mac Agro.Columbia,* 8: 79-84.

Cornwall, P.B. 1958. Movements of the vectors of virus diseases of cacao in Ghana. I. Canopy movement in and between trees. *Bull. Ent. Res.* 49: 613-630.

Cornwall, P.B. 1960. Movements of the vectors of virus diseases of cacao in Ghana. II. Wind movements and aerial dispersal. *Bull. Ent. Res.* 51: 175-201.

Cotterell, G.S. 1943. Entomology. *Rep. Central Cocoa Res. Stat. Tafo, Gold Coast 1938-1942, pp* 46-52.

Dale, W.T. 1962. Diseases and pests of cocoa. Virus diseases. In: Wills JB, ed. *Agriculture and Land Use in Ghana.* London, UK: Oxford University Press, pp 286-316.

Diekman, M. 1999. Options for movement of cacao germplasm in relation to the risk of moving pests. *Pl. Gen. Resources Newsl.* No. 117, 58-59.

Ding, S.W., Mackenzie, A., Torronen, M., Gibbs, A. 1990. Nucleotide sequence of the virion protein gene of cacao yellow mosaic tymovirus. *Nuc. Acids Res.* 18, p. 5886.

Donald, R.G. 1956. The natural enemies of some Pseudo-coccidae in the Gold Coast. *J. West Africa Science Assoc.* 2(1): 48-68.

Dufour, B. 1991. Place et importance des différentes espèces d'insectes dans l'écologie du CSSV (cocoa swollen shoot virus) au Togo. *Café, Cacao, Thé,* 35: 197-204.

Frison, E.A. and Feliu, E. 1989. FAO/IBPGR *Technical Guidelines for the safe movement of cocoa germplasm.* FAO/IBPGR, Rome. 29 pp.

Fütterer, J., Rothnie, H.M., Hohn, T. and Potrykus, I. 1997. Rice tungro bacilliform virus open reading frames II and III are translated from polycistronic pregenomic RNA by leaky scanning. *J. Virol.* 71: 7984-7989.

Fütterer, J., Potrykus, I., Valles-Brau, M.P., Dasgupta, I., Hull, R. and Hohn, T. 1994. Splicing in a plant pararetrovirus. *Virology* 198: 663-670.

Grimsley, N., Hohn, B., Hohn, T. and Walden, R. 1986. Agroinfon , an alternative route for viral infection of plants by using the Ti plasmid. *Proc. Natl. Acad. Sci. USA* 83: 3282-3286.

Guerra-Peraza, O., de Tapia, M., Hohn, T. and Hemmings-Mieszczak, M. 2000. Interaction of the cauliflower mosaic virus coat protein with the pregenomic RNA leader. *J. Virol.* 74: 2067-2072.

Hagen, L.S. 1994. Analyse du génome du 'cacao swollen shoot virus' et mise au point d'une méthode d'infection du cacaoyer. PhD dissertation Université de Paris 6, France 193 pp.

Hagen, L.S., Jacquemond, M., Lepingle, A., Lot, H. and Tepfer, M. 1993. Nucleotide sequence and genomic organization of cacao swollen shoot virus. *Virology* 196: 619-628.

Hagen, L.S., Lot, H., Godon, C., Tepfer, M. and Jacquemond, M. 1994. Infection of *Theobroma cacao* using cloned DNA of cacao swollen shoot virus and particle bombardment. *Phytopathology* 84: 1239-1243.

Hanna, A.D. and Heatherington, W. 1957. Arrest of the swollen-shoot virus disease of cacao in the Gold Coast by controlling the mealybug vectors with systemic insecticide dimefox. *Ann. appl. Biol.* 45: 473-840.

Herzog, E., Guerra-Peraza, O. and Hohn, T. 2000. The rice tungro bacilliform virus gene II product interacts with the coat protein domain of the viral gene III polyprotein. *J. Virology* 74: 2073-2083.

Hoffmann, K., Sackey, S.T., Maiss, E., Adomako, D. and Vetten, H.J. 1997. Immunocapture polymerase chain reaction for the detection and characterization of cacao swollen shoot virus 1A isolates. *Phytopathology* 145: 205-212.

Hoffmann, K., Uhde, C., Lesemann, D.E., Sackey, S.T., Adomako, D. and Vetten, H.J. 1999. Production, characterization and application of monoclonal antibodies to the cacao swollen shoot virus isolate 1A. *J. Phytopathol.* 147: 725-735.

Hohn, T., Corsten, S., Dominguez, D., Fütterer, J., Kirk, D., Hemmings-Mieszczak, M., Pooggin, M., Scharer-Hernandez, N. and Ryabova, L. 2001. Shunting is a translation strategy used by plant pararetroviruses (*Caulimoviridae*). *Micron* 32: 51-57.

Huang, Q. and Hartung, J.S. 2001. Cloning and sequence analysis of an infectious clone of citrus yellow mosaic virus that can infect sweet orange via *Agrobacterium*-mediated inoculation. *J. Gen. Virol.* 82: 2549-2558.

Hughes, J.d'A. and Ollennu, L.A.A. 1993. The virobacterial agglutination test as a rapid means of detecting cocoa swollen shoot virus. *Ann. app. Biol.* 122: 299-310.

Hughes J.d'A. and Ollennu, L.A.A. 1994. Mild strain protection of cocoa in Ghana against cocoa swollen shoot virus – a review. *Pl. Pathol.* 43: 442-457.

Hughes, J.d'A., Adomako, D. and Ollennu, L.A.A. 1995. Evidence from the virobacterial agglutination test for the existence of eight serogroups of cocoa swollen shoot virus. *Ann. appl. Biol.* 127: 297-307.

Jacquot, E., Hagen, L.S., Jacquemond, M. and Yot, P. 1996. The open reading frame 2 product of cacao swollen shoot badnavirus is a nucleic acid-binding protein. *Virology* 225: 191-195.

Jacquot, E., Hagen, L.S., Michler, P., Rohfritsch, O., Stussi-Garaud, C., Keller, M., Jacquemond, M. and Yot, P. 1999. *In situ* localization of cacao swollen shoot virus in agroinfected *Theobroma cacao*. *Arch. Virol.* 144: 259-271.

Jeger, M.J. and Thresh, J.M. 1993. Modelling reinfection of replanted cocoa by swollen shoot virus in pandemically diseased areas. *J. applied Ecol.* 30: 187-196.

Jeger, M.J. and Chan, M-S. 1995. Theoretical aspects of epidemics: uses of analytical models to make strategic management decisions. *Can. J. Pl. Pathol.* 17: 109-114.

Jeger, M.J. 1997. Theoretical considerations in the epidemiology and control of vector-borne diseases of perennial crops, such as lethal yellowing-like diseases of coconut. In: *International workshop on lethal yellowing like diseases of coconut, Elmina, Ghana,* Nov. 1995. (eds. S.J. Eden-Green and F. Ofori). Natural Resources Institute, Chatham, UK. pp 239-246.

Jeger, M.J., Van den Bosch, F., Madden, L.V. and Holt, J. 1998. A model for analysing plant-virus transmission characteristics and epidemic development. *IMA Math. Appl. Med. Biol.* 15: 1-18.

Kenten R.H. 1972. The purification and some properties of cocoa necrosis virus, a serotype of tomato black ring virus. *Ann. appl. Biol.* 71: 119-126.

Kenten, R.H. 1977. Cocoa necrosis virus, Kew, Commonwealth Mycological Institute (CMI/AAB *Descriptions of Plant Viruses* No.173).

Kenten, R.H. and Legg, J.T. 1965. Observations on the purification and properties of cocoa swollen shoot virus. *Ghana J. Sci.* 5: 221-225.

Kenten, R.H. and Legg, J.T. 1968a. Purification and properties of cocoa swollen shoot virus (CSSV) and cocoa mottle leaf virus (CMLV). *Rept. Cocoa Res. Inst. Ghana for 1965/66, pp* 23-24.

Kenten, R.H. and Legg, J.T. 1968b. Purification and properties of cocoa swollen shoot virus. *Rept. Cocoa Res Inst. Ghana for 1966/67, pp* 22.

Kenten, R.H. and Legg, J.T. 1971. Serological relationships of some viruses from cocoa (*Theobroma cacao* L.) in Ghana. *Ann. appl. Biol.* 67: 195-200.

Knight, R. and Tinsley, T.W. 1958. Some histological observations on virus infected *Theobroma cacao L. Ann. appl. Biol.* 46: 7-10.

Laco, G.S. and Beachy, R.N. 1994. Rice tungro bacilliform virus encodes reverse transcriptase DNA polymerase and RNAse H activities. *Proc. Natl. Acad. Sci. USA* 91: 2654-2658.

Legg, J.T. and Agbodjan, F.X. 1969. Swollen shoot disease. R*ept. Of the Cocoa Res. Inst. Ghana for 1967/68, pp* 23-25.

Legg, J.T. and Lockwood, G. 1981. Resistance of cocoa to swollen-shoot virus in Ghana. I. Field trials. *Ann. appl. Biol.* 97: 75-81.

Lockhart, B.E.L., Menke, J., Dahal, G. and Olszewski, N.E. 2000. Characterization and genomic analysis of tobacco vein clearing virus, a plant pararetrovirus that is transmitted vertically and related to sequences integrated in the host genome. *J. Gen. Virol.* 81: 1579-1585.

Lot, H.E., Djiekpor, E. and Jacquemond, M. 1991. Characterization of the genome of cacao swollen shoot virus. *J. General Virol.* 72: 1735-1739.

Mangenot, G., Alibert, H. and Basset, A. 1946. Sur le caractère du "swollen shoot" en Côte d'Ivoire. *Rev. Int. Appl. Agric.* 26: 173-184.

Marmey, Ph., Bothner, B., Jacquot, E., de Kochko, A., Ong, C.A., Yot, P., Siuzdak, G., Beachy, R.N. and Fauquet, C. 1999. Rice tungro bacilliform virus ORF 3 encodes a single 37 kDa coat protein. *Virology* 253: 319-326.

McClure, M.A. 2000. The complexities of genome analysis, the Retroid agent perspective. *Bioinformatics* 16: 79-95.

Muller, E., Jacquot, E. and Yot, P. 2001. Early detection of cacao swollen shoot virus using the polymerase chain reaction. *J. Virol. Methods* 93: 15-22.

Ollennu, L.A.A. and Hughes, J.d'A. 1991. The future of cocoa swollen shoot control in Ghana. *Proc. Int. Cocoa Conf. of the 90s.* Kuala Lumpur, Malaysia 25th to 28th September, 1991.

Ollennu, L.A.A., Hughes J.d'A. and Owusu, G.K. 1996, Mild strain cross-protection of cocoa against cocoa swollen shoot badnavirus. *Trop. Sci.* 36: 116-128.

Ollennu, L.A.A. and Owusu, G.K. 1997. Mild virus strain protection studies with cocoa swollen shoot badnavirus (CSSV) in Ghana. *Proc. First Int. Cocoa Pests and Diseases Sem. Accra, Ghana,* 6-10 November, 1995 pp 204-211.

Ollennu, L.A.A., Owusu, G.K. and Dzahini-Obiatey, H. 1999. Recent studies of mild strain cross-protection with cocoa swollen shoot virus. *J. Ghana Sci. Assoc.* 2 (3): 5-11.

Ollennu, L.A.A. and Owusu, G.K. 2002. Field evaluation of the protective capability of CSSV mild strain NI against severe strains. *Ghana J. Agric. Sci.* (in press).

Ollennu, L.A.A. and Owusu, G.K. 1989. Isolation and study of mild strains of cocoa swollen shoot virus for possible cross protection. In: *Proc. IV Int. Plant Virus Epidemiology Workshop. Montpellier 1989,* pp.119-122.

Ollennu, L.A.A., Owusu, G.K. and Thresh, J.M. 1989. The control of cocoa swollen shoot disease in Ghana. *Coc. Growers Bull.* 46: 26-36.

Owusu, G.K. 1971. Cocoa necrosis virus in Ghana. *Trop. Agric. Trinidad* 48: 133-139.

Owusu, G.K. 1988. Virus diseases in Ghana. In: *Virus diseases in Africa* .Ed. Williams AO, Mbiele AL, Nkuaka N, *OAU/STRC Scientific Publications, Lagos, Nigeria,* pp 9-27.

Owusu, G.K., Asare-Nyako, A., Lovi, N.K. and Halm, B.J. 1985. Association of cocoa necrosis virus infection with mineral deficiency. In: *Ann. Rept. Cocoa Res. Inst. Ghana, for 1979/80-1981/82,* pp 158-159.

Padi, B., and Hollander, J. den. 1996. Morphological variation in *Pianococcoides njalensis* occurring on cocoa in Ghana. *En.a Experiment. Appl.* 79: 317-328.

Padi, B. 1997. Prospects for the control of cacao mealybugs. *Proc. 1st Int. Cocoa Pests and Diseases Sem. Accra, Ghana* November 6-10, 1995, pp 249-263.

Padi, B., Ackonor, J.B., Sarfo, J.E. and Ollennu, L.A.A. 2002. Biological control of mealybug vectors of cocoa swollen shoot virus (CSSV). *Rep. Cocoa Res. Inst. Ghana,* 2000/2001 (in press).

Padwick, G.W. 1956. Losses caused by plant diseases in the colonies. *Phytopath. Papers* 1: 1-60.

Partiot, M., Djiekpor, E.K., Amefia, Y.K. and Bakar, K.A. 1978. Le "swollen shoot" du cacaoyer au Togo: Inventaire preliminaire et premiere estimation des pertes causees par la maladie. *Café Cacao The,* 22: 217-228.

Pfeiffer, P., Hohn, T. 1983. Involvement of reverse transcription in the replication of cauliflower mosaic virus: A detailed model and test of some aspects. *Cell* 33: 781-789.

Pooggin, M.M., Futterer, J., Skryabin, K.G. and Hohn, T. 1999. A short open reading frame terminating in front of a stable hairpin is the conserved feature in pregenomic RNA leaders of plant pararetroviruses. *J. Gen. Virol.* 80: 2217-2228.

Posnette, A.F. 1940. Transmission of 'swollen shoot' disease of cacao. *Trop. Agric. Trinidad* 17: 98.

Posnette, A.F. 1941. Swollen shoot virus disease of cacao. (Review of Research Work to November, 1940). *Trop. Agric. Trinidad* 18: 87-90.

Posnette, A.F. 1943. The diagnosis of swollen shoot disease of cacao. *Trop. Agric. Trinidad* 21: 156-158.

Posnette, A.F. 1944. Swollen shoot in Trinidad. *Trop. Agric. Trinidad* 21: 105-106.

Posnette, A.F. 1947. Virus diseases of cacao in West Africa. I Cacao viruses 1A, 1B, 1C and 1D. *Ann. appl. Biol.* 34: 388-402.

Posnette, A.F. 1951. Virus research at the West African Cocoa Research Institute, Tafo, Gold Coast. *Trop. Agric. Trinidad* 28: 133-142.

Posnette, A.F. 1981. The role of wild hosts in cocoa swollen shoot disease. In: Thresh JM, ed., *Pests, pathogens and vegetation, London,* Pitman, pp. 71-78.

Posnette, A.F. and Palma, M. 1944. Observations on cacao in the Para Penninsula, Venezuela. *Trop. Agric. Trinidad* 21: 130-132.

Posnette, A.F., Robertson, N.F. and Todd, J. Mc.A. 1950. Virus diseases of cacao in West Africa. V. Alternative host plants. *Ann. appl. Biol.* 37: 229-240.

Posnette, A.F. and Strickland, A.H. 1949. Parasitism of mealybug vectors of swollen shoot of cocoa. *Nature* 163: 105-106.

Posnette, A.F. and Strickland, A.H. 1948. Virus diseases of cacao in West Africa III. Technique in insect transmission. *Ann. appl. Biol.* 35: 53-63.

Posnette, A.F. and Todd, J.McA. 1951. Virus diseases of cacao in Africa VIII. The search for virus-resistant cacao. *Ann. appl. Biol.* 38: 785-800.

Posnette, A.F. and Todd, J.Mc.A. 1955. Virus diseases of cacao in West Africa IX. Strain varieties and interference in virus IA. *Ann. appl. Biol.* 43: 433-453.

Purseglove, J.W. 1977. *Tropical Crops. Dicotyledons.* Longman, London 719 pp.

Renaud, R. 1957. Distribution of virus diseases of cocoa in the Ivory Coast. In: *Proc. Cocoa Conf. London, pp* 79-80.

Richert-Pöggeler, K.R. and Shepherd, R.J. 1997. Petunia vein-clearing virus: a plant pararetrovirus with the core sequences for an integrase function. *Virology* 236: 137-146.

Sagemann, W., Paul, H.L., Adomako, D. and Owusu, G.K. 1983. The use of enzyme-linked immunosorbent assay (ELISA) in the detection of cocoa swollen shoot virus (CSSV). *Phytopathol. Z.* 106: 281- 286.

Sagemann, W., Lesemann, D-E., Paul, H.L., Adomako, D. and Owusu, G.K. 1985. Detection and comparison of some Ghanaian isolates of cacao swollen shoot virus (CSSV) by enzyme-linked immunosorbent assay (ELISA) and immunoelectron microscopy (IEM) using an antiserum to CSSV strain 1A. *Phytopathol. Z.* 114: 78-89.

Senagum, M. 1961. Mosaic symptoms of cocoa leaves. *Bull. No. 2 Phytopathology,* University of Gadjah, Mada, Jakarta, pp 3-9.

Steven, W.T. 1936. A new disease of cocoa in the Gold Coast. *Gold Coast Farmer* 5: 122.

Strickland, A.H. 1951a. The entomology of swollen shoot of cocoa – I. The insect species involved, with notes on their biology. *Bull. Ent. Res.* 41: 725-748.

Strickland, A.H. 1951b. The entomology of swollen shoot of cocoa II. The bionomics and ecology of the species involved. *Bul. Ent. Res.* 42: 65-103.

Thorold, C.A. 1975. *Diseases of cacao.* Oxford, U.K: Clarendon Press.

Thresh, J.M. 1958. The spread of virus disease in cacao. *Cocoa Res. Inst. Technical Bull. No. 5:* 36 pp.

Thresh, J.M. 1959. The control of cacao swollen shoot disease in Nigeria. *Trop. Agric.* 36: 35-44.

Thresh, J.M. 1961. Some isolates of virus causing swollen shoot disease of cacao in Nigeria and their inter-relationships. *Ann. appl. Biol.* 49: 340-346.

Thresh, J.M. 1988. Eradication as a virus disease control measure. In: *Control of plant diseases: Costs and benefits.*B.C. Clifford and E. Lester (eds). Blackwell Scientific Publications, Oxford, pp155-194.

Thresh, J.M. and Lister, R.M. 1960. Coppicing experiments on the spread and control of cacao swollen shoot disease in Nigeria. *Ann. appl. Biol.* 48: 65-74.

Thresh, J.M. and Owusu, G.K. 1986. The control of cocoa swollen shoot disease in Ghana: an evaluation of eradication procedures, *Crop Prot.* 51: 41-52.

Thresh, J.M., Owusu, G.L.K. and Ollennu, L.A.A. 1988b. Cocoa swollen shoot: an archetypal crowd disease. *Z. Pfl Krank. Pflschutz* 95: 428-446.

Thresh, J.M., Owusu, G.K., Boamah, A. and Lockwood, G. 1988a. Ghanaian cocoa varieties and swollen shoot virus. *Crop Prot.* 7: 219-231.

Thresh, J.M. and Tinsley, T.W. 1959. The viruses of cacao. *Tech. Bull. West African Cocoa Res. Inst.* 7: 1-32.

Tinsley, T.W. 1953a. The strains of cacao swollen shoot viruses occurring in West Africa. In: *Proc. West Africa Int. Cocoa Res. Conf. Tafo, Gold Coast,*1953, pp 20-21.

Tinsley, T.W. 1953b. Strains of the cacao swollen shoot virus (CSSV), *Ann. Rept. of the West African Cocoa Res. Inst. for 1952/3,* pp 7.

Tinsley, T.W. 1955. The host range of cacao viruses. *Rept. West African Cocoa Res. Inst. for 1954/55,* pp 30-35.

Todd, J. Mc.A. 1951. An indigenous source of swollen shoot disease of cacao. *Nature* 167: 952.

Tsatsu, K., Djiekpor, K.E. and Bekuo, K. 1997. Lutte contre le swollen shoot du cacaoyer au Togo: sélection de matériel résistant et replantation des zones virosées (control of swollen shoot of cocoa in Togo: selection of resistant material and replanting of virus infected areas). *Proc. First Int. Cocoa Pests and Diseases Sem.* Accra, Ghana 6-10 November, 1995, pp 202-203.

Turner, P.D. 1972. Diseases of oil palm and cocoa in North Sumatra. *Document 17 UNDP/FAO Project INS/72/004.*

Tzafrir, I., Ayala-Navarrete, L., Lockhart, B.E.L. and Olszewski, N.E. 1997. The N-terminal portion of the 216-kDa polyprotein of Commelina yellow mottle badnavirus is required for virus movement but not for replication. *Virology* 232: 359-368.

Vander Plank, J.E. 1948. The relation between the size of field and the spread of plant disease into them. Part I, Crowd diseases *Empire J. Exp. Agric.* 16: 134-142.

Williams, C.N., Chew, W.Y. and Rajaratnam, J.A. 1987. *Tree and field crops of the wetter regions of the tropics.* Longman, Harlow, Essex. 262 pp.

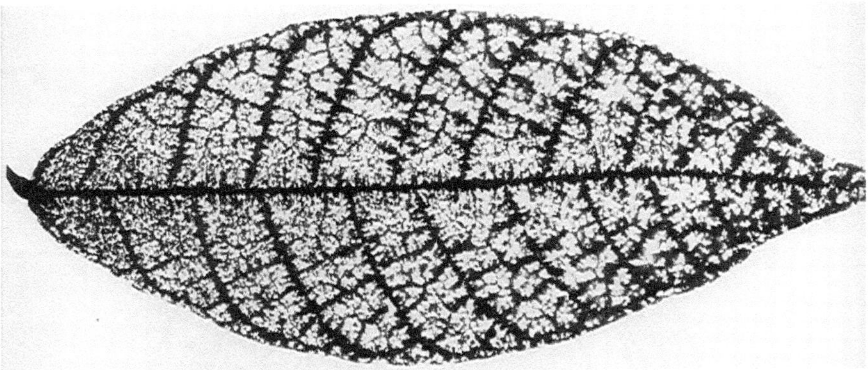

Figure 21.1(a) Leaf symptoms extending over the whole leaf lamina, which results in chlorotic vein clearing or banding which when confined to the secondary veins forms a 'fern leaf' pattern.

Figure 21.1(b) Stem swellings, most conspicuous on rapidly growing shoots, develop a nodes, internodes or apex of stems, which may end in dead tip swelling with further growth restricted to a lateral bud.

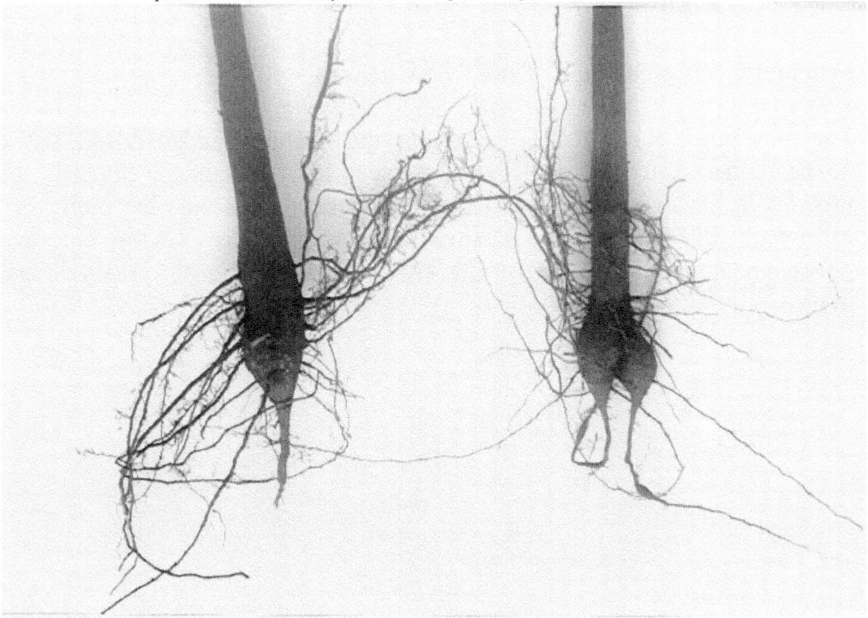

Figure 21.1(c) Many virus isolates inducing swelling on the roots, especially the tap root, usually accompanied by necrosis of the secondary and tertiary roots.

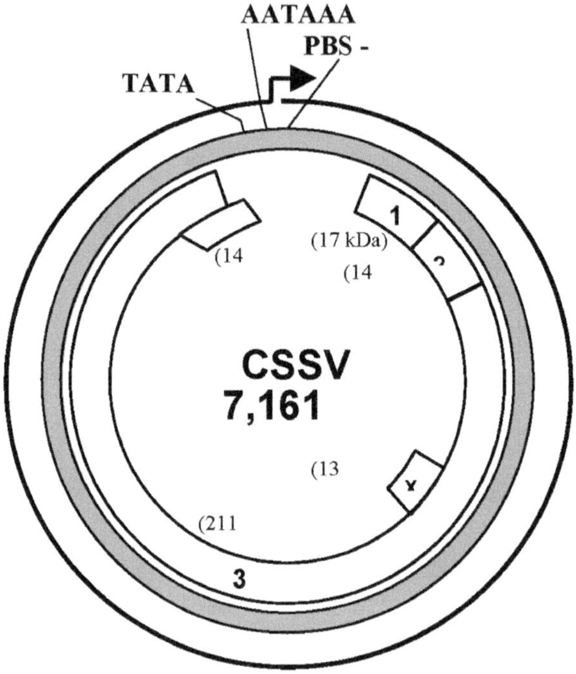

Figure 21.2

Schematic representation of the CSSV genom

The grey inner double circle represents the double-stranded DNA. White boxes represent the five ORFs. The predicted molecular weight of the proteins is stated in parentheses. The outer circle represents the more than full-length RNA transcript. Positions of the putative TATA box and polyadenylation signal and of the tRNAmet$_{init}$ sequence (PBS-1) are indicated.

22. Sugarcane

Grant R Smith[1] and Philippe Rott[2]
[1]David North Plant Research Centre, Bureau of Sugar Experiment Stations, PO Box 86, Indooroopilly, Q 4068, Australia and [2]UMR 385 ENSAM-INRA-CIRAD Biologie et Génétique des Interactions Plante-Parasite, TA 71/09, Avenue Agropolis, 34398 Montpellier Cedex 5, France

22.1 Introduction

Sugarcane (*Saccharum* L. spp. and *Saccharum* interspecific hybrids) and sugarbeet (*Beta vulgaris* L. and *Beta* interspecific hybrids) are the two principal sources of world sugar. Sugarbeet, grown in 49 countries accounts for 28% of world sugar production, while sugarcane grown in 82 countries accounts for the remaining 72% (Anon 2002). In 2000/01, 93.7 million tonnes of sugar was produced from sugarcane. India and Brazil (20.1 and 17.0 m tonnes, respectively) are the two largest cane sugar producers followed by China (5.8), Mexico (5.2) and Thailand (5.2) (Anon 2002).

Sugarcane is a C_4 tropical grass in the family Gramineae (Daniels and Roach, 1987). Contemporary sugarcane cultivars are complex aneuploid polyploid hybrids of species within the genus *Saccharum,* usually *S. officinarum* and *S. spontaneum* with limited infusion of *S. sinsense* and *S. robustum* (Berding *et al.* 1997). The 2n+n chromosome transmission into the first introgression generation (Piperidis and D'Hont, 2001) followed by (apparently random) loss of chromosomes has resulted in the total chromosome number of the modern clones ranging from about 100 to 130.

Recombination between chromosomes derived from the different species has been also observed. Piperidis and D'Hont (2001) established that a range of modern clones contain approximately 80% *S. officinarum,* 10% *S. spontaneum*, and 10% recombinant chromosomes.

Sugarcane is a perennial crop, growing from the stool left in the ground after harvest as a ratoon crop. After a number of ratoons the stool is ploughed out and new 'seed cane' is planted. As sugarcane does not produce true-to-type seed, propagation is via lengths of stalk (setts) cut so that at least one bud is present per piece. Both the planting and harvesting systems present difficulties when diagnosing and eliminating diseases. Many diseases express poorly on the stalks and seed is not available as a method to eradicate seed-borne systemic pathogens. The cutting of stalks for harvest whether mechanically or by hand and the perennial nature of the crop present opportunities for the mechanical transmission of pathogens such as *Sugarcane mosaic, Sorghum mosaic* and *Sugarcane streak mosaic viruses.* Viral diseases represent a significant constraint on sugarcane production around the world, but no single disease predominates. Mosaic is the most widespread disease having been reported from every canegrowing country with the exception of Mauritius and Guyana, whereas striate mosaic is restricted to one region in Australia. Many of the diseases such as Fiji leaf gall or mosaic have caused, or can cause, significant production losses, while for others such as leaf fleck or yellow leaf the data is contradictory and probably reflects genotype response in the different areas. Here, we review the status of the different viral-induced diseases of sugarcane and their causal pathogens (Table 21.1).

Table 22. 1. Virus diseases of sugarcane

Disease	**Causal agent**
Fiji leaf gall (previously Fiji disease)	*Fiji disease virus* Genus *Fijivirus (FDV)*
Leaf fleck	*Sugarcane bacilliform virus* Genus *Badnavirus* (SCBV)
Mild mosaic	*Sugarcane mild mosaic virus* Tentative Genus *Closterovirus* (SCMMV)
Mosaic	*Sugarcane mosaic virus* or *Sorghum mosaic virus* Genus *Potyvirus* (SCMV)
Red leaf mottle	*Peanut clump virus* Genus *Pecluvirus* (PCV)
Streak	*Sugarcane streak virus, Sugarcane streak Egypt virus, Sugarcane streak Mauritius virus* Genus *Mastrevirus* (SSV, SSEV, SSMV)
Streak mosaic	*Sugarcane streak mosaic virus* Tentative Genus *Tritimovirus* (SCSMV)
Striate mosaic	*Sugarcane striate mosaic associated virus* Genus unassigned *(SCSMaV)*
Yellow leaf	*Sugarcane yellow leaf virus* Genus unassigned in Family *Luteoviridae* (SCYLV)

22.2 ***Fiji disease virus*** (FDV) causes Fiji leaf gall disease (previously Fiji disease).The only definitive symptom of Fiji leaf gall is raised whitish enations (galls) on the abaxial side of the leaf laminae and/or midrib. Other symptoms include death of the apical meristem, side-shooting, fan-like appearance of the whole top and bitten appearance of the top leaves. In some cases infection ultimately results in death of the stool (Smith, 2000). The disease is named after the country in which it was first observed in 1886, when the disease was already widespread in commercial sugarcane crops in one district (Daniels *et al.* 1972). The disease has been reported from Australia, Fiji, Indonesia, Madagascar, Malaysia, New Caledonia, Papua New Guinea, Philippines, Samoa, Solomon Islands, Thailand and Vanuatu (Smith, 2000).

Fiji leaf gall is one of the most important diseases of sugarcane. It has been present in Australia since at least 1890, and caused considerable problems before the application of effective disease control measures that reduced it to a minor problem in the 1950s (Toohey and Nielsen, 1972). With the widespread planting of the susceptible cultivar NCo310, there was a re-emergence of Fiji leaf gall in the early to mid 1960s (Egan and Fraser, 1977). NCo310 suffered 50% loss of yield in the main crop and 100% loss by second ratoon. This re-emergence peaked in 1979 in the Bundaberg region of south Queensland where an estimated 70 million stools of diseased cane were present. Eventually, resistant clones such as Q110 and healthy plant sources brought the disease under control. Madagascar has also suffered considerable losses due to epidemics of Fiji leaf gall. The economic impact of the disease via loss of production, early plough out and replacement of dead or infected stools was further compounded by the effects on plant breeding programs and the slowing of overall genetic gain for agronomic and production characteristics to provide disease resistance. During the Bundaberg epidemic, approximately 80% of parents were discarded from the breeding population due to susceptibility to Fiji leaf gall disease. It is considered that it took 10 to 15 years to recover the lost genetic potential in the south Queensland breeding program after the last epidemic (Smith, 2000).

Fiji disease virus is transmitted in infected plant material and by species of the delphacid planthopper genus *Perkinsiella*, including *P. saccharicidia*, *P. vastatrix* and *P. vitiensis*. The virus is transmitted propagatively by adults and second to fifth instar nymphs, and is not mechanically transmissible. The disease is generally limited to members of the genus *Saccharum*, although it has been noted in *Erianthus maximus,* a member of the '*Saccharum* complex,' (Daniels and Roach, 1987) in Fiji. *S. officinarum* and *S. edule* are relatively susceptible, whilst *S. spontaneum* and *S. robustum* clones are relatively resistant. FDV has been experimentally transmitted to maize and sorghum by *P. saccharicidia* nymphs in glasshouse experiments. These plant species have, however, never been naturally observed infected with Fiji leaf gall (Egan *et al.*1989).

In the field, Fiji leaf gall is diagnosed by the presence of the distinctive galls. Diagnosis of asymptomatic plants relies on the molecular assays, or waiting for symptoms to develop. Some high quality antisera have been developed for research purposes (Soo *et al.* 1998). Earlier sera had low titer and limited specificity (Francki *et al.* 1986). Both probes and RT-PCR have been developed for the diagnosis of FDV (Smith *et al.* 1992b, 1994; Smith and Van de Velde, 1994). The RT-PCR test is in routine application in screening germplasm in Australia (James *et al.* 2001).

The virions are icosahedral, with protrusions (A-spikes) at each of the twelve vertices. The double-shelled 70nm particles are unstable *in vitro*, readily losing the outer shell and A spike protrusions to expose the more stable 54 nm sub-viral particles. The sub-viral particles are also icosahedral with prominent B-spike protrusions at each of the vertices. Virions can be purified by extraction in 0.1M glycine pH 8.5 based buffer containing Nonidet-P40, followed by differential and sucrose density gradient centrifugation. Each virion contains ten segments of dsRNA ranging in size from approximately 4.0 kb to 1.8 kb, and a RNA-dependent RNA polymerase capable of transcribing ssRNA from the dsRNA. Some of the segments have been cloned and analysed (Soo *et al.* 1998). There are no recognised strains of FDV, although little work has been conducted on this aspect of the virus.

Control of Fiji leaf gall currently relies on the use of resistant cultivars, planting of disease-free material and inspection and roguing of diseased stools. In some cases, early harvest and the ploughing out of diseased crops are necessary for disease control.

In Australia, there is a significant focus on breeding cultivars with adequate levels of resistance to Fiji leaf gall. While the basis of resistance is unknown, highly resistant agronomically elite cultivars have been successfully bred and selected for commercial production. Transgene-mediated resistance has been demonstrated against FDV in glasshouse trials (McQualter *et al.* 2000, 2001) but is yet to be exploited commercially. For the foreseeable future, programs providing for phytosanitary selection will be indispensable in obtaining healthy planting material. A set of regularly inspected nurseries, separated from the commercial production areas of the industry for the supply of planting material is essential.

22.3 *Sugarcane bacilliform virus* (SCBV) is a recently recognized pathogen of sugarcane (Lockhart and Autrey, 1988). In most sugarcane cultivars, SCBV infection is not associated with any foliar symptoms. In some clones, whitish to yellow chlorotic streaking sometimes striate is evident. The virus is widely distributed throughout the major geographical regions where sugarcane is grown and infects *S. officinarum, S. robustum, S. spontaneum, S. barberi*, commercial sugarcane cultivars, *Erianthus araundinaceus* and *E. ravennae*. SCBV, unlike other Badnavirus species, infects several different host plant genera and species including *Saccharum, Sorghum halapense, Brachiaria sp. Panicum maximum* and *Rottboellia exaltata*, and has been experimentally transmitted to *Oryza sativa, Musa* 'Dwarf Cavendish' and *Sorghum vulgare* (Lockhart and Autrey, 2000b). Only limited and difficult to interpret data on the economic effect of SCBV infection is available.

SCBV is transmitted in infected material and by pseudococcid mealybugs in a number of genera. *Saccharicoccus sacchari* and *Dysmicoccus boninsis*, the pink and grey sugarcane mealybugs respectively, are efficient vectors of the virus (Lockhart and Autrey, 2000b; Lockhart and Olszewski, 1994). *Planococcus citri*, the citrus mealybug which usually does not colonise sugarcane will also transmit SCBV (Lockhart and Autrey, 2000b). SCBV is closely related to *Banana streak virus*, but unlike BSV and a number of other badnaviruses, is not pollen or seed transmitted (Lockhart and Olszewski, 1994; Daniels *et al.* 1995). SCBV is poorly transmitted by mechanical inoculation (Lockhart and Autrey, 2000b). It is suspected to be the cause of yield losses in cultivar Mex57-473 (Lockhart and Autrey, 1988), and a correlation appeared to exist between high viral titer and lower biomass production in three sugarcane clones (Comstock and Lockhart, 1996). Lockhart and Autrey (2000b) also reported an apparent synergistic interaction between SCBV and *Sugarcane mosaic virus*, in that the effects of mosaic infection may be enhanced in cultivars infected with SCBV.

A range of diagnostic tests has been developed for the detection of SCBV. Antiserum-based assays have been developed, but considerable serological variation between SCBV isolates exists. DAS-ELISA is unreliable for screening germplasm due to this variation, whereas a TAS-ELISA test is more dependable (Ndowora, 1998). Particles can be detected by immuno-electron microscopy (IEM) with good sensitivity using partially purified extracts (Lockhart *et al.* 1996) or by immunocapture PCR. Other PCR-based assays using total plant DNA have been developed for the diagnosis of SCBV (Braithwaite *et al.* 1995). However, results from these tests must be treated with caution due to the potential integration of badnaviral sequences into the sugarcane genome similar to that reported for *Musa* (LaFleur *et al.* 1996; Ndowora *et al.* 1999).

The virus is purified from infected leaf tissue using the method described by Lockhart and Autrey (1988). Bacilliform shaped particles are generally

130 by 30 nm in size although elongated (300–500 nm) particles can also be observed. The genome of circular double-stranded DNA is approximately 7.5 kbp, with three open reading frames and a proposed transcription initiation binding site of 19nt, 17 of which are complementary to the plant cytoplasmic tRNA^{met} sequence (Bouhida *et al.* 1993; Geijskes *et al.* 2002). No strains of SCBV are currently recognized, although considerable serological and genomic diversity has been reported (Lockhart *et al.* 1996; Geijskes *et al.* 2002).

SCBV is not eliminated by normal thermotherapy of stalk cuttings, tissue culture or thermotherapy-apical meristem culture (Lockhart and Autrey, 2000b). The use of virus-free seed material and controlling the mealybug vectors are the only viable control options.

22.4 *Sugarcane mild mosaic virus* (SCMMV) has only been found naturally in mixed infections with *Sugarcane bacilliform virus*. SCMMV can be separated from SCBV by passage through clones of *S. officinarum,* which support limited replication of SCBV such as IJ76-391 and IK76-69 (Lockhart and Autrey, 2000a). In most commercial cultivars, there are no distinct symptoms associated with the mixed infection. Ganoo *et al.* (1998) reported narrow leaves, shoot withering and retarded growth of plants infected with both viruses in Mauritius. When SCMMV was separated from SCBV, transient mild mosaic symptoms were found on the young leaves of well-grown plants. The economic importance of SCMMV has not been defined, as cultivars showing poor growth or decline were also infected with SCBV.

SCMMV has been found in *S. officinarum*, *S. barberi*, *S. sinense* and a range of commercial hybrids (Ganoo *et al.* 1998). The virus is spread with infected planting material and by the pink sugarcane mealybug *Saccharicoccus sacchari*. SCMMV has been transmitted by *S. sacchari* to rice (*Oryza sativa*) and *Sorghum halepense*, causing mild mosaic symptoms (Lockhart and Autrey, 2000a). SCMMV can be reliably detected by immuno-electron microscopy (IEM), but as the antisera were prepared from a virion preparation that contained SCBV particles, ELISA does not discriminate between the two viruses (Lockhart and Autrey, 2000a).

The virus is purified from infected leaf tissue by the method described by Lockhart *et al.* (1992). Virions are 1500-1600 nm in length and have the helical characteristics of *Closteroviruses* (Lockhart *et al.* 1996). The genome of SCMMV has not been cloned, and no strains have been recognized so far. SCMMV is not eliminated from sugarcane by standard hot water treatments (Lockhart and Autrey, 2000a).

22.5 Sugarcane mosaic virus and Sorghum mosaic virus

Mosaic is one of the most widely distributed diseases of sugarcane, with only two important cane-growing countries, Guyana and Mauritius, free of the disease (Koike and Gillaspie, 1989). The disease was first noted in 1892 in Java and named "gelestrepenziekte" (yellow-stripe disease). Historically, the disease has been more serious in Louisiana than elsewhere in the world, with many sugarcane infecting isolates of *Sugarcane mosaic virus* (SCMV) and *Maize dwarf mosaic virus* (MDMV) . The probable reason why mosaic is more serious in Louisiana than elsewhere is that mosaic is superimposed on the established diseases *Pythium* root rot and red rot (Koike and Gillaspie, 1989). Estimated losses from mosaic vary considerably. In the period 1918 to 1930, when mosaic was spreading in North America, yield losses of 30-40% and sometimes 60-80% were reported (Koike and Gillaspie, 1989).

McKern *et al.* (1991) provided a new taxonomic system based on peptide profiling of the virus coat protein and split the Sugarcane mosaic virus subgroup into four distinct new species: *Sugarcane mosaic virus*, *Sorghum mosaic virus* (SrMV), *Johnson grass mosaic virus* (JGMV) and *Maize dwarf mosaic virus*. Both the SCMV and SrMV species contain sugarcane-infecting isolates. The original A, B, D and E strains of SCMV were placed in the new SCMV taxon, whilst the original SCMV strains H, I and M were placed in the SrMV taxon. The original SCMV-F strain is now classified as *Sugarcane streak mosaic virus* (see section below). Handley *et al.* (2001) suggested a new taxonomic system for sugarcane infecting potyviruses and that viruses classified in SCMV *sensu lato* should be able to infect sugarcane. This proposal has yet to be more broadly considered by sugarcane pathologists and viral taxonomists.

SCMV infection has been recorded in a wide range of cultivated and wild grasses. However, with the reclassification of the subgroup (McKern *et al.* 1991), the validity of these lists for SCMV *senso lato* and SrMV is questionable. The most obvious symptom of mosaic is a pattern of contrasting shades of green on the leaf. This pattern is the result of varying concentrations of chlorophyll, but the cultivar, growth conditions, temperature and strain of the virus, all contribute to the intensity of the symptom (Koike and Gillaspie, 1989; Grisham, 2000). In some genotypes, sharply defined chlorotic areas develop that are accompanied by necrosis or reddening. The symptoms are best observed on young actively growing leaves, while on older leaves the symptoms tend to fade. Chlorotic symptoms can extend to the leaf sheath and stalk, in particular on *S. officinarum* accessions. Mosaic is relatively easy to diagnose by symptomatology, but pathogen identification is necessary to differentiate the disease from streak mosaic in countries where both diseases are present. ELISA and PCR-based (Smith and Van de Velde, 1994; Yang and Mirkov 1997) assays have been developed for detection. Smith *et al.* (1995a) developed a high titer, highly specific antiserum by expressing the SCMV-A coat protein in *E. coli* as a fusion protein and then purifying the expressed protein prior to injection into a rabbit. This serum could detect all SCMV-SrMV strains. In comparison, the RT-PCR test developed by Smith and Van de Velde (1994) only detected the SCMV-A, B, D and E strains. Yang and Mirkov (1997) used an RFLP supplemented RT-PCR assay to discriminate strains without the need to inoculate differential hosts. The viral coat protein of SCMV-A was first cloned and sequenced (Smith *et al.* 1992a). The variability of the viral coat protein (Smith *et al.* 1996; Handley *et al.* 1988; Alegria *et al.* in press) and other regions of the viral genome (Handley *et al.* 1996) were studied to ascertain the effectiveness of PCR tests to diagnose all isolates and to correlate sequences to host origin.

The viruses have been purified by a number of techniques such as that of Bond and Pirone (1971). SCMV and SrMV show typical potyvirus morphology with flexuous filamentous particles 750 by 13 nm. The virion consists of a single coat protein subunit of 37 kDa surrounding a monopartite ssRNA genome of 9.6 kb (Pickering, 2002). The virus is transmitted by a number of aphid species including *Dactynotus ambrosiae* and *Rhopalosiphum maidis*, and by infected planting material. Whilst SCMV and SrMV are mechanically transmittable, there is negligible spread by farm implements such as knives and harvesters (Koike and Gillaspie, 1989).

Spread of mosaic in a field is usually more rapid in subtropical rather than tropical regions. This is probably the result of large numbers of aphids migrating from dying weeds in the autumn and spring to young, rapidly growing plants when they are most susceptible (Grisham, 2000). The most effective means of control is the use of resistant cultivars. Where virus-resistant germplasm is not available, virus-free planting material can be an effective control measure. Both serial hot water treatment and apical meristem culture have been demonstrated to free germplasm from virus. Timing of planting and subsequent harvest (for growth of the ratoon crop) may assist in reducing virus levels in susceptible genotypes by separating the period of the rapid growth of new plants from the period of high populations of vectors and high vector activity (Grisham, 2000). Transgenic resistance to both SCMV (Joyce *et al.* 1998a; 1998b) and SrMV (Ingelbrecht *et al.* 1999) have been demonstrated but is yet to be commercially exploited.

22.6 *Peanut clump virus*

Red leaf mottle is a relatively minor disease of sugarcane. No yield losses have been reported in commercial fields. The only report of yield loss was in plant cane in yield trials in Senegal (Baudin *et al.* 1989). The disease has been reported from a number of African countries (Burkina Faso, Chad, Congo, Gabon, Niger, Senegal and Sudan) as well as from India (Rott and Chatenet, 2000). The symptoms of red leaf mottle vary with genotype, and include red spots with diffuse edges and no defined shape, chlorotic bands that can appear yellow or rust colored or spotted, and white stripes with wavy edges (Baudin and Chatenet, 1988; Rott, 1996). Sugarcane is a minor host of *Peanut clump virus*: the principal hosts are sorghum and peanut. The disease is soil borne. It can be transmitted through the soil from sugarcane to sugarcane, and sugarcane to groundnut, and in cuttings from symptomatic sugarcane stalks.

The virus is purified from infected leaf tissue by the method described by Baudin and Chatenet (1988). In preparations from sugarcane leaves, rigid rods of 200 to 300 nm in length and 20 nm in width can be observed. The virus is diagnosed in sugarcane by electron microscopy of leaves or roots using the leaf dip technique (Rott and Chatenet, 2000). Various serological techniques (ELISA, dot-blot, tissue-blot) can be used to detect the pathogen in leaves, stalks and roots (Baudin and Chatenet, 1988; Chatenet and Saeed, 1995). Strains of the pathogen infecting sugarcane are not recognized. Two major strains of PCV have been identified in peanut: one from Africa, the other from India. The physical and biological properties of the two strains are identical, but they show considerable variation in their serological properties (Reddy *et al.* 1985). The genome consists of 2 RNA segments of approximately 5.9 (Herzog *et al.* 1994) and 4.5 kb (Manohar *et al*1993).

Sandy soils with a pH of 7 or higher favor transmission. No efficient technique exists to eliminate PCV from infected sugarcane (Rott and Chatenet, 2000).

22.7 Sugarcane streak virus, Sugarcane streak Egypt virus, Sugarcane streak Mauritius virus

Three viral species within the genus *Mastrevirus* cause streak in sugarcane. Leaves of plants infected with one of the three variants of streak show numerous small spots or streaks, which are translucent, narrow and parallel to the veins (Rott and Peterschmitt, 2000). The streaks observed on the young leaves of an infected cutting can be wider but coalesce and can form a mosaic pattern (Storey and Thomson, 1961). Streak was first described on cultivar Uba in Natal in 1914 causing yield losses of about 11%

(Storey, 1925), although it was probably present prior to this being described as yellow stripe (Bock and Bailey, 1989). Streak is found in a number of African countries including Benin, Cape Verde, Ivory Coast, Egypt, Kenya, Malawi, Mozambique, South Africa, Sudan, Uganda and Zimbabwe. It is also found in Madeira, Pakistan, India, Reunion and Mauritius. Commercial sugarcane genotypes, *S. officinarum* and *S. sinense* are susceptible to streak, while *S. barberi* and *S. spontaneum* are relatively resistant. *Cenchrus echinatus* and *Coix lacryma-jobi* are alternative hosts for SSMV in Mauritius, and *C. echinatus* may be a reservoir of SSMV in Mauritius.

Originally, virus isolates causing streak in sugarcane were considered to be different strains of *Maize streak virus* (MSV). The viral particles had twinned quasi-isometric morphology, were transmitted by cicadellid leafhoppers in the genus *Cicadulina,* and were serologically related (Bock and Bailey, 1989). Hughes *et al.* (1993) and Hughes (1994) first proposed that the streak isolates from Mauritius and South Africa be considered distinct *Mastrevirus* species based on results from nucleic acid hybridisation, RFLP analysis and genomic sequence comparisons. The results of Bigarré (*et al.* 1999) following the sequencing of five further streak isolates confirmed that sugarcane streak is caused by three distinct species. The three sugarcane-infecting mastreviruses are transmitted in infected planting material or by species of the leafhopper *Cicadulina*, in particular *C. mbila* and *C. bipunctella* (Bock and Bailey, 1989; Rott and Peterschmitt, 2000).

Purification of the viral particles was described by Bock and Bailey (1989). The *Sugarcane streak mastrevirus* virions show the typical two-joined quasi-isomatric subunits with the characteristic constriction in the middle with a total size of about 30 x 18 nm. The genome is ssDNA of approximately 2.7 kb with the usual two ORFs on the sense and two ORFs on the complimentary strand. Symptoms of streak are useful for diagnosis and are best observed on younger leaves as the streaks become more less obvious and more diffuse in older leaves (Bock and Bailey, 1989). SSMV is milder and induces finer streak symptoms. Antibodies produced against isolates of MSV can be used for the diagnosis of the sugarcane mastreviruses, but due to the variation in the coat protein of the sugarcane infecting viruses, no single antiserum can diagnose all three. Molecular PCR-based assays have been developed for the diagnosis of SSV, SSMV (Rybicki and Hughes, 1990) and SSEV (Shamloul *et al.* 2001).

Resistant cultivars provide the most effective means of control. With the replacement of Uba in the 1935-35 period in South Africa, streak became a rare disease of sugarcane (Bock and Bailey, 1989). Peros *et al.* (1990) demonstrated that virus-free plants could be produced by tissue culture of infected material.

22.8 *Sugarcane streak mosaic virus*

A potyvirus isolated from sugarcane imported from Pakistan into the USA was putatively identified as *Sugarcane mosaic virus*, strain F based on symptomology on indicator hosts (Gillaspie and Mock, 1984; Gillaspie *et al.* 1978). Symptoms are variable depending on genotype. On cultivar CP31-294, infection results in finely divided mosaic mottling with conspicuous green islands without necrosis (Bhargava *et al.* 1972). The host range differed from that of any other known potyvirus and the purified coat protein generated a unique profile after tryptic digestion (Jensen and Hall, 1993a, 1993b). The virus had no serological relationship with SCMV-A or with a range of monocot infecting viruses (Jensen and Hall, 1993a), suggesting that this virus was a novel virus of sugarcane. The 3' 2000 bp of the viral genome was cloned and sequenced. Phylogenetic comparisons revealed that this new virus was most similar to the *Wheat streak mosaic virus* (WSMV) and *Brome streak mosaic virus* (BStMV). At that time, WSMV was a member of the genus *Rymovirus*, but the analysis of Hall *et al.* (1998) revealed that the genus *Rymovirus* contained at least two unique evolutionary lineages. This is acknowledged in the Seventh Report of the International Committee on Taxonomy of Viruses (Anon., 2000) with the creation of the genus *Tritimovirus* with WSMV as the type species and containing BStMV. Based on the similarity of the sequence data from the new sugarcane viruses to WSMV, the name *Sugarcane streak mosaic virus* was proposed by Hall *et al.* (1998), with subsequent reports referring to SCSMV being a member of the genus *Tritimovirus* (Hema *et al.* 1999). Recent results of cloning and sequencing other regions of the viral genome of the Indian isolate of SCSMV (referred to as SCSMV-AP for Andhra Pradesh) revealed low identity at the amino acid level to members of the *Tritimovirus* genus. It was suggested that the Andhra Pradesh isolate SCSMV-AP be classified into a yet undescribed new genus in the family *Potyviridae* (Hema *et al.* 2002).

SCSMV has been reported from sugarcane in Pakistan (Gillaspie and Mock, 1984; Gillaspie *et al.* 1978; Hall *et al.* 1998) and India (Hema *et al.* 1999). It was also recently detected by electron microscopy and RT-PCR in leaves showing mosaic symptoms from Bangladesh, Sri Lanka, Thailand and Vietnam (Chatenet M., personal communication). The virus is mechanically transmittable (Hall *et al.* 1998; Hema *et al.* 1999) to *Sorghum bicolor*. Under natural conditions it is transmitted via infected setts (Hema *et al.* 2002). No insect vector has been identified. If it is a *Tritimovirus,* it should be persistently transmittable by an eriophyid mite species. The virus can be purified using methods described by Hall *et al.* (1998) or Hema *et al.* (1999). The virion is a flexuous filament 900 x 15 nm in size comprised of coat protein subunits of 40 kDa. The genome is monopartite ssRNA of approximately 10 kb. Approximately half the genome (3' portion) has been cloned and sequenced and that segment shows the organisation typical of a member of the family *Potyviridae.*

22.9 Sugarcane striate mosaic associated virus

The disease is restricted to areas within the Burdekin River district in north Queensland, Australia, occurring in distinct patches of up to one ha in size, within fields (Croft and Randles, 2000). The disease is associated with shallow or sandy soil areas in a field. Similar symptoms of unknown cause have also been reported from India (Nayudu *et al.* 1971). The disease does not appear to significantly spread from these areas, and reappears whenever a susceptible genotype is planted to the area. Infected plants are yellowish and severely stunted, often dying or not ratooning, resulting in large bare patches in a field. Short chlorotic striations appear on glasshouse plants grown at 25-30°C, while longer and broader chlorotic striations can appear on the leaf sheath and stems. As the disease is so restricted, it's effect is minimal to the overall Australian sugar industry, but can cause significant losses within a field. The death of patches of cane in otherwise productive fields often leads to weed infestation, early plough out, and replanting.

The virus can be purified using the method described by Thompson (2001). Sugarcane striate mosaic associated virus has a slightly flexuous filamentous particle of 950 by 50 nm. The RNA genome is 8,146 nt in length and contains five open reading frames. Phylogenetic analysis of the genome places SCSMaV between the genera *Foveavirus* and *Carlavirus*, with the absence of a sixth open reading frame precluding assignment with Carlaviruses, while the coat protein is approximately half the size of that of *Apple stem pitting virus*, type member of the *Foveavirus* genus. Genome analysis further revealed that SCSMaV is most similar to *Cherry green ring mottle virus*, but the two viruses are morphogically and biologically distinct. SCSMaV may represent a new taxon of plant viruses (Thompson and Randles, 2001).

The fine chlorotic striations on the leaves are the best diagnostic symptom, but at temperatures higher than 30°C these symptoms are generally not present (Croft and Randles, 2000). Extraction of dsRNA reveals a characteristic 9 kbp band (Choi, 1997; Choi and Randles, 1997; Choi *et al.* 1999). A more specific assay based on the genomic sequence of SCSMaV is being developed (Croft and Randles, 2000). The main method of control is the use of disease-free planting material and resistant genotypes.

22.10 *Sugarcane yellow leaf virus*

Yellow leaf syndrome (YLS) has been associated with a range of abiotic and biotic causes over the past forty years. Yellow wilt in Africa (Ricaud, 1968) and autumn decline in Brazil (Hughes 1964) are probably earlier reports of the syndrome. The symptoms re-emerged in 1988 when a leaf-yellowing symptom of sugarcane was first noted in Hawaii (Schenck, 1990). YLS was subsequently reported from Brazil, other countries of central and South America, mainland USA, Australia, Mauritius, Réunion Island and various countries on the African continent (Ulian and Sanguino, 1994; Fors, 1994, Comstock *et al*, 1994; Smith *et al.* 1995b; Anon 1995; Rassaby *et al.* 1999; Bailey *et al.* 1996; Zvoutete, 1997). The adaxial surface of the leaf midrib of plants showing YLS becomes yellow and sometimes reddened or red flecked (Comstock *et al.* 1994,;Vega *et al.* 1997). Sugarcane yellows phytoplasma (SCYP), belonging to the western X group has been associated with YLS symptoms in South Africa and neighbouring countries (Cronjé *et al.* 1998), and the disease caused by this pathogen has been recently renamed leaf yellows. In the rest of the world, *Sugarcane yellow leaf virus* (SCYLV) has been more widely associated with YLS, and the disease caused by this virus is now known as yellow leaf.

SCYLV infects a range of genotypes within the 'Saccharum complex' including the commercial interspecific hybrids and *S. officinarum*, *S. robustum*, *S. spontaneum*, and *S. sinensis.* SCYLV is transmitted by vegetative propagation and by the aphid species *Melanaphis sacchari*, *Rhopalosiphum maidis* and *R. rufiabdominalis,* but not by the sugarcane aphid *Sipha flava* (Lockhart and Cronjé, 2000) The virus is not transmittable by *R. maidis* to sweet corn, sorghum or Johnson grass (Lockhart and Cronjé, 2000). Because of the wide range of biotic and abiotic causes of leaf yellowing in sugarcane, diagnosis by symptomatology is not reliable. DAS-ELISA (Scagliusi and Lockhart, 2000), TBIA (tissue blot immunoassay) and RT-PCR (Comstock *et a.* 1998) systems have all been developed and have proven reliable for the identification of SCYLV. Molecular beacons were also recently used for highly sensitive and specific detection of the virus in plant material and its aphid vector (Gonçalves *et al.* 2002). To date, there is no evidence of strains of this virus.

Particles are purified by extraction in 0.25M sodium-potassium phosphate buffer with Triton X-100, followed by differential centrifugation, sucrose density gradient separation and isopycnic banding (Lockhart and Autrey, 1988). Icosahedral virions are 24 to 29 nm in diameter. There is a single capsid protein and a genomic RNA of approximately 5.9 kb. The viral genome is monopartite and formed by positive-sense single stranded RNA of 5,895-5,898 nucleotides (Moonan *et al.* 2000; Smith *et al.* 2000). It shows a typical *Polerovirus* ORF 0 to ORF 5 genome organization, but nucleotide similarities suggest that at least two independent recombinations have occurred during genome evolution. A sobemo-like ORF 1/2 RNA dependent RNA polymerase (RdRp) block is present with very high similarity to *Potato leaf roll virus* genus *Polerovirus* RdRp, followed by an ORF 3/4 coat protein/movement protein block similar to the *Barley yellow dwarf virus* genus *Luteovirus* gene block. The 5' end of the read-through domain (ORF 5) is most similar to RNA1 of *Pea enation mosaic* genus *Enamovirus* (Moonan *et al.* 2000; Smith *et al.* 2000). The genus *Saccharvirus* in the family *Luteoviridae* was proposed for viruses such as SCYLV with this *Polerovirus-Luteovirus* genomic organisation (Borg *et al.* 1999), but is yet to be accepted.

Up to twenty percent yield loss has been reported from Brazil in the widely grown cultivar SP71-6163 (Lockhart and Cronjé, 2000), this being a revision of earlier reports of 50% losses (Vega *et al.* 1997). There is also an increase in the brix (total soluble solids) in the leaves (Comstock *et al.* 1994). A number of highly susceptible clones including RB72-454 and H65-7052 have been identified (Lockhart and Cronjé, 2000), as well as plant resistance in clones such as H78-4153. Control strategies are based on the lack of identified alternative plant hosts, suggesting that the virus is principally spread in infected planting material and by the aphid vector. Hence virus-free germplasm and the exploitation of plant resistance in breeding programs are the main control measures. SCYLV has also been successfully eliminated from infected material by apical meristem culture and leaf callus culture (Chatenet *et al.* 2001; Fitch *et al.* 2001; Parmessur, 2002).

22.11 Conclusions

There are a number of serious diseases of sugarcane caused by viruses that affect sugar production worldwide. These viral pathogens represent the four basic genome types (single and double-stranded RNA and DNA) with a variety of insect vectors as the basic mechanism for transmission. There are also a number of diseases of sugarcane with a suspected viral aetiology including Ramu streak (Magarey and Suma, 2000), chlorotic streak (Magarey and Egan, 2000), sembur, sereh and spike (Rott and Comstock, 2000). Based on the number of viruses reported causing diseases in either rice or maize, two other members of the *Poaceae* grown in the tropics and/or sub-tropics, many more viruses await discovery in sugarcane.

Diagnosis of viruses in sugarcane has improved steadily over the past ten or so years with a range of sensitive, specific molecular-based assays replacing or complementing symptomatology and serological-based assays

(Braithwaite and Smith, 1996, 2001; Dietzgen *et al.* 1999; Smith, 2001). In some instances, these molecular assays have been incorporated into routine quarantine systems for germplasm exchange further enhancing the value of the exchange by reducing the risk of exotic disease introduction (James *et al.* 2001; Rott *et al.* 1998). Resistance in cultivars produced by traditional breeding systems is the principal method for control of most sugarcane virus diseases. However, the genetic complexity of sugarcane makes selection for disease resistance for a range of pathogens difficult, (whilst improving agronomic characters such as yield and sugar content) and more than one agronomically elite, advanced genotype have been discarded late in the selection process due to disease susceptibility. Genetically engineered disease resistance to mosaic (Ingelbrecht *et al.* 1999; Joyce *et al.* 1998a, 1998b; Smith *et al.* 1999,) and Fiji leaf gall (McQualter *et al.* 2000, 2001) has already been demonstrated. Transgenic resistance should be achievable against most, if not all, of the virus afflicting sugarcane (Smith and Harding 2001) and remove viral-induced diseases as a constraint to worldwide sugar production.

References

Alegria, O., Royer, M., Bousalem, M., Chatenet, M., Peterschmitt, M., Girard, J-C. and Rott, P. 2002. Genetic diversity in the coat protein coding region of eighty-six sugarcane mosaic virus isolates from eight countries, particularly from Cameroon and Congo. *Arch. Virol.* (in press).

Anon (1995) Yellow leaf syndrome. *Mauritius Sugar Ind. Res. Inst. Annu. Rept 1995*, p. 51.

Anon (2000). Family Potyviridae. In: *Virus Taxonomy. Classification and Nomenclature of Viruses. Seventh Report of the International Committee on Taxonmoy of Viruses.* Van Regenmortel, M.H.V., Fauquet C.M., Bishop, D.H.L., Carstens, E.B., Estes, M.K., Lemon, S.M., Maniloff, J., Mayo, M.A., McGeoch, D.J., Pringle, C.R. and Wickner, R.B. (eds). Academic Press, San Diego, USA. pp 703-724.

Anon (2002). F.O. Lichts World Sugar Statistics 2002 (supplement). *F.O. Lichts World Sugar & Sweetener Yearbook 2002.*

Bailey, R.A., Bechet, G.R. and Cronjé, C.P.R. 1996. Notes on the occurrence of yellow leaf syndrome of sugarcane in southern Africa. *Proc. South Afric. Sugar Technol. Ass.* 70: 3-6.

Baudin, P. and Chatenet, M. 1988. Détection sérologique du PCV, isolat canne à sucre, agent de la marbrure rouge des feuilles. *L'Agronomie Tropicale* 43: 228-235.

Baudin, P., Sene, A. and Marion, D. 1989. Effect of peanut clump virus on the yields of two sugarcane varieties. In: *Current Trends in Sugarcane Pathology (Prof. K.S. Bhargava Festschrift).* G.P. Rao, A.G. Gillaspie A.G. Jr, P.P. Upadhyaya, A. Bergamin Filho, V.P. Agnihotri and C.T. Chen (eds). International Books and Periodicals Supply Service, Dehli. pp 141-150.

Berding, N., Moore, P.H. and Smith, G.R. 1997. Advances in breeding technology for sugarcane. In: *Intensive sugarcane production: meeting the challenges beyond 2000.* B.A. Keating and J.R. Wilson (eds). CAB International, Wallingford, UK. pp 189-219.

Bhargava, K.S., Joshi, R.D. and Rishi, N. 1972. Occurrence of strains A and F of sugarcane mosaic virus in Uttar Pradesh (India). In: *Proceedings of the 14th Congress of the*

International Society of Sugar Cane Technologists. M.T. Henderson (ed). Franklin Press, Inc. Baton Rouge, Louisiana, USA. pp 949-954.

Bigarré, L., Salah, M., Granier, M., Frutos, R., Thouvenel, J. and Peterschmitt, M. 1999. Nucleotide sequence evidence for three distinct sugarcane streak mastreviruses *Arch. Virol.* 144: 2331-2344.

Bock, K.R. and Bailey, R.A. 1989. Streak. In: *Diseases of Sugarcane. Major Diseases*. C. Ricaud, B.T. Egan, A.G. Gillaspie Jr and C.G. Hughes (eds). Elsevier Science Publishers B.V. Amsterdam, The Netherlands. pp 323–332.

Bond, W.P. and Pirone, T.P. 1971. Purification and properties of sugarcane mosaic virus strains. *Phytopathology. Z.* 71: 56-65.

Borg, Z., Gibbs, M., Lockhart, B., Braithwaite, K. and Smith G. 1999. Sugarcane yellow leaf virus: a novel recombinant virus from the Luteoviridae. *Proc. XIth International Congress Virol.* VW15.02, p 47.

Bouhida, M., Lockhart, B.E.L. and Olszewski. N.E. 1993. An analysis of the complete sequence of a sugarcane bacilliform virus genome infectious to banana and rice. *J. Gen. Virol.* 74: 15-22.

Braithwaite, K.S., Egeskov, N.M. and Smith, G.R. 1995. PCR detection of sugarcane bacilliform badnavirus in diseased sugarcane. *Plant Dis.* 79: 792-796.

Braithwaite, K.S. and Smith, G.R. 1996. Molecular tools for indexing in sugarcane. Sugarcane Germplasm Conservation and Exchange. In: *Sugarcane Germplasm Conservation and Exchange*. B.J. Croft, C.M. Piggin, E.S. Wallis and D.M. Hogarth (eds). Proceedings No 67. Australian Centre International Agricultural Research, Canberra, Australia. pp 75-78.

Braithwaite, K.S. and Smith, G.R. 2001. Molecular-based diagnosis of sugarcane virus diseases. In: *Current Trends in Sugarcane Pathology Vol. II. Virus and Phytoplasma Diseases.* G.P. Rao, R.E. Ford, M. Tosic and D.S. Teakle (eds). Science Publishers, Inc. Enfield, New Hampshire, USA. pp117-208.

Chatenet, M. and Saeed, I. 1995. First report of sugarcane red leaf mottle virus in Sudan. *Plant Dis.* 79: 321.

Chatenet, M., Delage, C., Ripolles, M., Irey, M., Lockhart, B.E.L. and Rott, P. 2001. Detection of *Sugarcane yellow leaf virus* in quarantine and production of virus-free sugarcane by apical meristem culture. *Plant Dis.* 85: 1177-1180.

Choi, Y.G. 1997. The etiology of sugarcane striate mosaic disease. PhD Thesis, The University of Adelaide, Australia..

Choi, Y.G. and Randles, J.W. 1997. Microgranular cellulose improves dsRNA recovery from plant nucleic acid extracts. *Biotechniques* 23: 610-611.

Choi, Y.G., Croft, B.J. and Randles, J.W. 1999. Identification of sugarcane striate mosaic-associated virus by partial characterisation of its double-stranded RNA. *Phytopathology* 89: 877-883.

Comstock, J.C. and Lockhart, B.E.L. 1996. Effect of sugarcane bacilliform virus on biomass production of three sugarcane cultivars. *Sugar Cane* 4 : 12-15.

Comstock, J.C., Irvine, J.E. and Miller, J.D. 1994. Yellow leaf syndrome appears on the United States mainland. *Sugar J.* March 1994: 33-35.

Comstock, J.C., Irey, M.S., Lockhart, B.E.L. and Wang, Z.K. 1998. Incidence of yellow leaf syndrome in CP cultivars based on polymerase chain reaction and serological techniques. *Sugar Cane* 4: 21-28.

Croft, B.J. and Randles, J.W. 2000. Striate mosaic. In: *A guide to sugarcane diseases*. P. Rott, R.A. Bailey, J.C. Comstock, B.J. Croft and A.S. Saumtally (eds.) Cirad Publications Service, Montpellier. pp. 265-267.

Cronjé, C.P.R., Tymon, A.M., Jones, P. and Bailey, R.A., 1998. Association of a phytoplasma with a yellow leaf syndrome of sugarcane in Africa. *Ann. App. Biol.* 133: 177-186.

Daniels, J., Husain, A.A. and Hutchinson, P.B. 1972. The control of sugar cane diseases in Fiji. In: *Proceedings of the 14th Congress of the International Society of Sugar Cane Technologists.* M.T.Henderson (ed). Franklin Press, Inc. Baton Rouge, Louisiana, USA. pp. 1007-1014.

Daniels, J. and Roach, B.T. 1987. Taxonomy and Evolution. In: *Sugarcane Improvement through Breeding.* D.J.Heinz (ed). Developments in Crop Science 11. Elsevier, Amsterdam, The Netherlands. pp 7-84.

Daniels, J., Thomas, J.E. and Smith, M. 1995. Seed transmission of banana streak virus confirmed. *Infomusa* 4: 7.

Dietzgen, R.G., Thomas, J.E., Smith, G.R. and Maclean, D.J. 1999. PCR-based detection of viruses in banana and sugarcane. *Curr. Topics Virol.* 1: 105-118.

Egan, B.T. and Fraser, T.K. 1977. The development of the Fiji disease epidemic at Bundaberg. *Proc. Qld. Soc. Sugar Cane Technol.* 44: 43-48.

Egan B.T., Ryan C.C., Francki R.I.B., 1989. Fiji disease. In: *Diseases of Sugarcane. Major Diseases.* C. Ricaud, B.T. Egan, A.G. Gillaspie Jr and C.G. Hughes (eds). Elsevier Science Publishers B.V, Amsterdam, The Netherlands. pp 263–287.

Fitch, M.M.M., Lehrer, A.T., Komor, E. and Moore, P.H. 2001. Elimination of *Sugarcane yellow leaf virus* from infected sugarcane plants by meristem tip culture visualized by tissue blot immunoassay. *Plant Pathol.* 50 : 676-680.

Fors, A.L. 1994. Una anormalidad de la cana de azucar que se asemeja al "sindrome de la hoja amarilla" observada en Venezuela y El Salvador. *Sugar J.* 57: 14-15.

Francki, R.I.B., Ryan, C.C., Hatta, T., Rohozinski, J. and Grivell, C.J. 1986. Serological detection of Fiji disease virus antigens in the planthopper *Perkinsiella saccharicida* and its inefficient ability to transmit the virus. *Plant Pathol.* 35: 324-328.

Ganoo, S., Autrey, L.J.C. and Lockhart, B.E.L. 1998. Distribution of sugarcane mild mosaic virus in sugarcane germplasm in Mauritius. *Sugar Cane* 2: 22–24.

Geijskes, R.J., Braithwaite, K.S., Dale, J.L., Harding, R.M. and Smith, G.R. 2002. Sequence analysis of an Australian isolate of Sugarcane bacilliform badnavirus. *Arch. Virol.* 147: 2393-2404.

Gillaspie, A.G. and Mock, R.G. 1984. Sugar cane mosaic virus. A survey of strains from sugarcane with a set of sorghum inbred lines. *Sugar Cane* 2: 1-3.

Gillaspie, A.G., Mock, R.G. and Smith F.F. 1978. Identification of Sugarcane mosaic virus and characterization of strains of the virus from Pakistan, Iran and Cameroon. In: *Proc. 16th Congresss Int. Soc. Sugar Cane Technol. Vol 1.* F.S. Reis and J. Dick (eds). Impres, Sao Paulo, Brasil. pp 347-355.

Gonçalves, M.C., Klerks, M.M., Verbeek, M., Vega, J. and van den Heuvel, J.F.J.M. 2002. The use of molecular beacons combined with NASBA for the sensitive detection of *Sugarcane yellow leaf virus*. *European J. Plant Pathol.* 108: 401-407.

Grisham, M.P. 2000. Mosaic. In: *A guide to sugarcane diseases.* P. Rott, R.A. Bailey, J.C. Comstock, B.J. Croft and A.S. Saumtally (eds.) Cirad Publications Service, Montpellier. pp 249-254.

Hall, J.S., Adams, B., Parsons, T.J., French, R., Lane L.C. and Jensen, S.G. 1998. Molecular cloning, sequencing and phylogenetic relationships of a new potyvirus: Sugarcane streak mosaic virus and a reevaluation of the classification of the potyviridae. *Mol. Phylogen.. Evol.* 10: 323-332.

Handley, J.A., Smith, G.R., Dale. J.L. and Harding R.M. 1996. Sequence diversity in the NIb coding region of eight sugarcane mosaic potyvirus isolates infecting sugarcane within Australia. *Arch. Virol.* 141: 2289-2300.

Handley, J.A., Smith, G.R., Dale, J.L. and Harding, R.M. 1998. Sequence diversity in the coat protein coding region of twelve sugarcane mosaic potyvirus isolates from Australia, USA and South Africa. *Arch. Virol.* 143: 1145-1153.

Handley, J.A., Smith, G.R. and Harding, R.M 2001. Current concepts of the taxonomy and variation of sugarcane-infecting potyviruses. In: *Current Trends in Sugarcane Pathology Vol. II. Virus and Phytoplasma Diseases*. G.P. Rao, R.E. Ford, M. Tosic and D.S. Teakle (eds). Science Publishers, Inc. Enfield, New Hampshire, USA. pp73-81.

Hema, M., Joseph, J., Gopinath, K., Sreenivasulu, P. and Savithri, H.S. 1999. Molecular characterization and interviral relationships of a flexuous filamentous virus causing mosaic disease of sugarcane (*Saccharum officinarum* L.) in India. *Arch. Virol.* 144: 479-490.

Hema, M., Sreenivasulu, P. and Savithri, H.S. 2002. Taxonomic position of sugarcane streak mosaic virus in the family *Potyviridae*. *Arch. Virol.* 147: 1997-2007.

Herzog, E., Guilley, H., Manohar, S.K., Dollet, M., Richards, K., Fritsch, C. and Jonard, G. 1994. Complete nucleotide sequence of peanut clump virus RNA 1 and relationships with other fungus-transmitted rod-shaped viruses. *J. Gen. Virol.* 75: 3147-3155.

Hughes, C.G. 1964. Genetical effects. In: *Sugar-Cane Diseases of the World Vol 2.* C.G. Hughes, E.V. Abbott and C.A. Wismer (eds). Elsevier Publishing Company, Amsterdam. pp. 267-296.

Hughes, F.L. 1994. Sugarcane streak. In: *Current Trends in Sugarcane Pathology (Prof. K.S. Bhargava Festschrift).* G.P. Rao, A.G. Gillaspie A.G. Jr, P.P. Upadhyaya, A. Bergamin Filho, V.P. Agnihotri and C.T. Chen (eds). Inter. Books Periodicals Supply Service, Dehli. pp 131-140.

Hughes, F.L., Rybicki, E.P. and Kirby, R. 1993. Complete nucleotide sequence of sugarcane streak monogeminivirus. *Arch. Virol.* 132: 171-182.

Ingelbrecht, I., Irvine, J. and Mirkov, T.E. 1999. Post-transcriptional gene silencing in transgenic sugarcane: dissection of homology-dependent virus resistance in a monocot that has a complex polyploid genome. *Plant Physiol.* 119: 1-11.

James, A.P., Fowler, A.R., Hekmeijer, S., Braithwaite, K.S., Whittle, P.J.L. and Smith, G.R. 2001. Implementation of molecular assays for the routine screening of quarantined germplasm. In: *Proc. 24th International Soc. Sugar Cane Technol. Congress Vol. II.* D M Hogarth (ed). ASSCT, Brisbane, Australia. pp 604-606.

Jensen, S.G. and Hall, J.S. 1993a. Characterization of a krish-sorghum infecting potyvirus. *Sorghum Newsl.* 34: 17.

Jensen, S.G. and Hall, J.S. 1993b. Identification of a new sorghum infecting species of potyvirus from sugarcane. *Phytopathology.* 83: 884.

Joyce, P.A., McQualter, R.B., Bernard, M.J. and Smith, G.R. 1998a. Engineering for resistance to SCMV in sugarcane. *Acta Hort.* 461: 385-391.

Joyce, P.A., McQualter, R.B., Handley, J.A., Dale, J.L., Harding, R.M. and Smith, G.R. 1998b. Transgenic sugarcane resistant to sugarcane mosaic virus. *Proc. Australian Soc. Sugar Cane Technol.* 20: 204-210.

Koike, H. and Gillaspie, A.G. Jr. 1989. Mosaic. In: *Diseases of sugarcane. Major diseases.* C. Ricaud, B.T. Egan, A.G. Gillaspie Jr and C.G. Hughes (eds.) Elsevier Science Publishers B.V., Amsterdam. pp. 301-322.

LaFleur, D.A., Lockhart, B.E.L. and Olszewski, N.E. 1996. Portions of the banana streak badnavirus genome are integrated in the genome of its host *Musa* sp. *Phytopathology* 86: S100.

Lockhart, B.E.L. and Autrey, L.J.C. 1988. Occurrence in sugarcane of a bacilliform virus related serologically to banana streak virus. *Plant Dis.* 72: 230–233.

Lockhart B.E. and Autrey L.J.C. 2000a. Mild mosaic. In: *A guide to sugarcane diseases.* P. Rott, R. Bailey, J.C. Comstock, B. Croft and S. Saumtally (eds). CIRAD/ISSCT, Montpellier, France. pp 245-248.

Lockhart B.E. and Autrey L.J.C. 2000b. Sugarcane bacilliform virus. In: *A guide to sugarcane diseases.* P. Rott, R. Bailey, J.C. Comstock, B. Croft and S. Saumtally (eds). CIRAD/ISSCT, Montpellier, France. p. 268-272.

Lockhart, B.E.L. and Cronjé, C.P.R. 2000. Yellow leaf syndrome. In: *A guide to sugarcane diseases.* P. Rott, R. Bailey, J.C. Comstock, B. Croft and S. Saumtally (eds). CIRAD/ISSCT, Montpellier, France. pp. 291-295

Lockhart, B.E.L. and Olszewski, N.E. 1994. Badnavirus group. In: *The Encyclopedia of Virology Vol 1.* R.G. Webster and A. Granoff (eds). Acad. Press, New York. pp. 139-143.

Lockhart, B.E.L., Autrey, L.J.C. and Comstock, J.C. 1992. Purification and serology of sugarcane mild mosaic virus, a mealybug-transmitted closterolike virus. *Phytopathology* 82: 691-695.

Lockhart, B.E.L., Irey, M.J. and Comstock, J.C. 1996. Sugarcane bacilliform virus, sugarcane mild mosaic virus and sugarcane yellow leaf syndrome. In: *Sugarcane Germplasm Conservation and Exchange.* B.J. Croft, C.M. Piggin, E.S. Wallis and D.M. Hogarth (eds). ACIAR Proceedings No 67. pp108-112.

Magarey, R.C. and Egan, B.T. 2000. Chlorotic streak. In: *A guide to sugarcane diseases.* P. Rott, R. Bailey, J.C. Comstock, B. Croft and S. Saumtally (eds). CIRAD/ISSCT, Montpellier, France. pp 278-283.

Magarey, R.C. and Suma, S. 2000. Ramu streak. In: *A guide to sugarcane diseases.* P. Rott, R. Bailey, J.C. Comstock, B. Croft and S. Saumtally (eds). CIRAD/ISSCT, Montpellier, France. pp 289 290.

Manohar, S.K., Guilley, H., Dollet, M., Richards, K. and Jonard, G. 1993. Nucleotide sequence and genetic organization of peanut clump virus RNA 2 and partial characterization of deleted forms. *Virology* 195: 33-41.

McKern, N.M., Shukla, D.D., Toler, R.D., Jensen, S.G., Tosic, M., Ford, R.E., Leon, O. and Ward, C.W. 1991. Confirmation that the sugarcane mosaic virus subgroup consists of four distinct potyviruses by using peptide profiles of coat proteins. *Phytopathology* 81: 1025-1029.

McQualter R.B., Harding, R.M., Dale, J.L. and Smith G.R. 2000. Virus derived transgenes from Fiji disease fijivirus confer resistance to Fiji disease in transgenic sugarcane plants. *Proc. 6th ISSCT Pathol. Workshop*, Cha-am, Thailand, July 2000.

McQualter, R.B., Harding, R.M., Dale, J.L. and Smith, G.R. 2001. A gene encoding a putative structural protein confers transgenic resistance in sugarcane to Fiji disease fijivirus. *Proc. Plant and Animal Genome IX.* W112, p35, San Diego, January 2001.

Moonan, F., Molina, J., Mirkov, T.E., 2000. Sugarcane yellow leaf virus: an emerging virus that has evolved by recombination between luteoviral and poleroviral ancestors. *Virology* 269: 156-171.

Nayudu, M.V., Sastry, K.S. and Sarma Y.R. 1971. Striate mosaic of sugarcane in India. *Sugarcane Pathologists' Newsl.* 7: 24.

Ndowora, T. R. C. 1998. Ph.D. Dissertation, University of Minnesota, St Paul..

Ndowora, T., Gahal, D., LaFleur, D., Harper G., Hull, R., Olszewski, N.E. and Lockhart, B.E.L. 1999. Evidence that badnavirus infection in Musa can originate from integrated pararetroviral sequences. *Virology* 255: 214-220.

Parmessur, Y., Aljanabi, S., Saumtally, S. and Dookun-Saumtally, A. 2002. *Sugarcane yellow leaf virus* and sugarcane yellows phytoplasma: elimination by tissue culture. *Plant Pathol.* 51 : 561-566.

Peros, J-P., Bonnel, E. and Reynaud, B. 1990. In vitro culture of sugarcane infected with maize streak virus (MSV). *Plant Cell, Tiss. Org. Cult.* 23: 145-149.

Pickering, L. A. (2002). Genome organisation of Sugarcane mosaic virus. PhD thesis submitted October 2002. The University of Queensland, Brisbane, Australia..

Piperidis, G. and D'Hont, A. (2001). Chromosome composition analysis of various Saccharum interspecific hybrids by genomic in situ hybridisation (GISH). In: *Proc. 24th International Soc. Sugar Cane Technol. Congress Vol. II.* D.M. Hogarth (ed). ASSCT, Brisbane, Australia. pp 565-566.

Rassaby, L., Girard, J-C., Irey, M.S., Lockhart, B.E.L. and Rott, P. 1999. Survey of sugarcane yellow leaf syndrome in Réunion Island. *Sugar Cane* 5: 16-18.

Reddy, D.V.R., Robinson, D.J., Roberts, I.M. and Harrison, B.D. 1985. Genome properties and relationships of Indian peanut clump virus. *J. Gen. Viro*l. 66: 2011-2016.

Ricaud, C. 1968. Yellow wilt of sugarcane in eastern Africa. *Sugarcane Pathologists' Newslett.* 1: 45-48.

Rott, P. 1996. Red leaf mottle. In: *Sugarcane Germplasm Conservation and Exchange.* B.J. Croft, C.M. Piggin, E.S. Wallis and D.M. Hogarth (eds.). ACIAR Proc. No 67, p 125.

Rott, P. and Chatenet, M., 2000. Red leaf mottle. In: *A guide to sugarcane diseases.* P. Rott, R. Bailey, J.C. Comstock, B. Croft and S. Saumtally (eds). CIRAD/ISSCT, Montpellier, France. pp 255-258.

Rott, P. and Comstock, J.C. 2000. Minor diseases of uncertain etiology. In: *A guide to sugarcane diseases.* P. Rott, R. Bailey, J.C. Comstock, B. Croft and S. Saumtally (eds). CIRAD/ISSCT, Montpellier, France. pp 296-297.

Rott, P. and Peterschmitt, M. 2000. Streak. In: *A guide to sugarcane diseases.* P. Rott, R. Bailey, J.C. Comstock, B. Croft and S. Saumtally (eds). CIRAD/ISSCT, Montpellier, France. pp 259-264.

Rott, P., Bousquet, J.F., Chatenet, M., Darroussat, M.J., Muller, M. and Ripolles, M. 1998. A quarantine unit in Montpellier (France) for safe movement of sugarcane germplasm. In: *Proc. 7th International Congress of Plant Pathology,* Edinburgh, Scotland 9-16 August 1998. Vol. 3, Abstract 4.6.1.

Rybicki, E.P. and Hughes F.L. 1990. Detection and typing of maize streak virus and other distantly related geminiviruses of grasses by polymerase chain reaction amplification of a conserved viral sequence. *J. Gen. Virol.* 1990: 2519-26.

Shamloul, A.M., Abdallahb, N.A., Madkourb, M.A. and Hadidi, A. 2001. Sensitive detection of the Egyptian species of sugarcane streak virus by PCR-probe capture hybridization (PCR-ELISA) and its complete nucleotide sequence. *J. Virol. Meth.* 92: 45-54.

Scagliusi, S.M. and Lockhart, B.E.L., 2000. Transmission, characterization and serology of a luteovirus associated with yellow leaf syndrome of sugarcane. *Phytopathology* 90: 120-124.

Schenck, S., 1990. Yellow leaf syndrome – a new disease of sugarcane. *Report of the Hawaiian Sugar Planters Association Experiment Station for 1990,* pp 38.

Smith, G.R. 2000. Fiji disease. In: *A guide to sugarcane diseases.* P. Rott, R. Bailey, J.C. Comstock, B. Croft and S. Saumtally (eds). CIRAD/ISSCT, Montpellier, France. pp 239-244.

Smith, G.R. 2001. Progress and potential in sugarcane molecular pathology. In: *Proc. 24th International Soc. Sugar Cane Technol. Congress Vol. II.* D M Hogarth (ed). ASSCT, Brisbane, Australia. pp 601-603.

Smith, G.R. and Harding, R.M. 2001. Genetic engineering for virus resistance in sugarcane. In: *Current Trends in Sugarcane Pathology Vol. II.* Virus and Phytoplasma Diseases. G.P. Rao, R.E. Ford, M. Tosic and D.S. Teakle (eds). Science Publishers, Inc. Enfield, New Hampshire, USA. pp335-365.

Smith, G.R. and Van de Velde, R., 1994. Detection of sugarcane mosaic virus and Fiji disease virus in diseased sugarcane using the polymerase chain reaction. *Plant Dis*. 78: 557-561.

Smith, G.R., Borg, Z., Lockhart, B.E.L., Braithwaite, K.S. and Gibbs, M.J., 2000. Sugarcane yellow leaf virus: a novel member of the Luteoviridae that probably arose by inter-species recombination. *J. Gen. Virol.* 81: 1865-1869.

Smith, G.R., Clarke, M.L., Van de Velde, R. and Dale, J.L., 1994. Chemiluminescent detection of Fiji disease virus in sugarcane with biotinylated DNA probes. *Arch. Virol.* 136: 325-334.

Smith, G.R., Ford, R., Bryant, J.D., Gambley, R.L., McGhie, T.K., Harding, R.M. and Dale, J.L. 1995a. The expression, purification, and use as an antigen of recombinant sugarcane mosaic virus coat protein. *Arch. Virol.* 140: 1817-1831.

Smith, G.R., Ford, R., Frenkel, M.J., Shukla, D.D. and Dale, J.L. 1992a. Transient expression of the coat protein of sugarcane mosaic virus in sugarcane protoplasts and expression in *Escherichia coli*. *Arch. Virol.* 125: 15-23.

Smith, G.R., Fraser, T.A., Braithwaite, K.S. and Harding, R.M. 1995b. RT-PCR amplification of RNA from sugarcane with yellow leaf syndrome using luteovirus group-specific primers. *Proc. 10th Australasian Plant Pathol. Soc. Conference* pp.84.

Smith, G.R., Handley, J.A., Harding, R.M., Dale, J.L., Gambley, C.F. and Braithwaite, K.S. 1996. Variability in sugarcane bacilliform and mosaic viruses and consequences for diagnosis. *Proc. XXII Congress International Soc. Sugar Cane Technol. Vol 2.* J.C. Autrey (ed). pp 390-396.

Smith, G.R., Joyce, P.A., Nutt, K.A., McQualter, R.B., Taylor, G.O. and Allsopp, P.G. 1999. Evaluation of transgenic plants engineered for resistance to mosaic and canegrubs. In: *Proc. XXIII Congress International Soc. Sugar Cane Technol. Vol 2.*. V. Singh and V. Kumar (eds). pp 291-297.

Smith, G.R., Van de Velde, R. and Dale, J.L., 1992b. PCR amplification of a specific double-stranded RNA region of Fiji disease virus from diseased sugarcane. *J. Virol. Meth.* 39: 237-246.

Soo, H.M., Handley, J.A., Maugeri, M.M., Burns, P., Smith, G.R., Dale, J.L. and Harding, R.M., 1998. Molecular characterisation of Fiji disease fijivirus genome segment 9. *J. Gen. Virol.* 79: 3155-3161.

Storey, H.H. 1925. Streak disease of sugar-cane. *South African. Dept. Agric. Bulletin* 39.

Storey, H.H. and Thomson, G.M. 1961. Streak disease. In: *Sugar-Cane Diseases of the World, Vol. 1.* J.P. Martin, E.V. Abbott and C.G. Hughes (eds). Elsevier Publishing Company, Amsterdam, The Netherlands. pp 461–476.

Toohey, C.L. and Nielsen, P.J. 1972. Fiji disease at Bundaberg. Proc.Qld. Soc. Sugar Cane Technol. 39: 191-196.

Thompson, N. 2001. Sugarcane striate mosaic associated virus: RNA sequence and genome organisation, taxonomy and detection. PhD thesis, Adelaide University, Australia.

Thompson, N. and Randles, J.W. 2001. The genome organisation and taxonomy of Sugarcane striate mosaic associated virus. *Arch. Virol.* 146: 1441-51.

Ulian, E.C. and Sanguino, A. 1994. Yellow leaf syndrome in Brazil. *Proc. 4th International Soc. Sugar Cane Technol. Pathology Workshop*. Brisbane, Australia.

Vega, J., Scagliusi, S.M.M., Ulian, E.C., 1997. Sugarcane yellow leaf disease in Brazil: evidence of association with a luteovirus. *Plant Dis.* 81: 21-26.

Yang, Z.N. and Mirkov, T.E. 1997. Sequence and relationships of sugarcane mosaic and sorghum mosaic virus strains, and development of RT-PCR based RFLPs for strain discrimination. *Phytopathology* 87: 932-939.

Zvoutete, P. 1997. The history and distribution of yellow leaf syndrome in the Southeast lowveld of Zimbabwe. Proc. 5th International Soc. Sugar Cane Technol. Pathology Workshop. Umhlanga Rocks, South Africa.

Fig. 22.1 Sugarcane streak mosaic

23. Coconut and other palm trees

23.1 Viroid Diseases

Maria Judith B. Rodriguez
Philippine Coconut Authority- Albay Research Center, Banao, Guinobatan, Albay 4503, Philippines

23.1.1 Introduction

Coconut is popularly known in the tropics as the "tree of life." Largely grown by smallholders, a wide variety of food and non-food products can be derived from its various parts. The main producers are the Philippines, Indonesia, India, Sri Lanka, Papua New Guinea and the Pacific Islands. It is also a locally important crop in at least 30 countries in the Americas, in 22 countries along the coasts of East and West Africa, and in the islands of the Indian Ocean (Persley 1992).

In the Philippines, about 300 million coconut palms are planted over 4.09 million hectares of farmland. Of the total production, 80% goes to foreign markets while the remaining 20% is used domestically. Around 3.4 million families are directly or indirectly dependent on the industry for their livelihood.

A major constraint to coconut production worldwide is the prevalence of lethal diseases (some of unknown etiology) that are killing millions of palms and precluding new plantings in some areas. Among these diseases are those caused by viroids – the cadang-cadang disease in the Philippines and tinangaja in Guam. Viroids are molecular parasites of higher plants composed of naked, single-stranded, covalently closed circular, low molecular weight RNAs (Diener 1979). Both cadang-cadang and tinangaja lead to premature decline and death of the host. However, intensive study has been done only on cadang-cadang as tinangaja is not considered economically important in Guam where no coconut industry exists.

23.1.2 Cadang-Cadang Disease

The term "cadang-cadang" meaning dead or dying in the Bicol dialect of the Philippines is used to refer to a premature decline and death of coconut palms. No epidemic of cadang-cadang disease is known outside the Philippines.

The first serious outbreak of cadang-cadang was reported in 1931 on San Miguel Island, Albay, Philippines (Ocfemia 1937). This was followed by observation of the disease at increasing distances from the Bicol Peninsula with intervening water or disease-free coconut areas (Price 1971). By the early 80's, it had spread widely in southern Luzon and neighboring areas with approximately 30 million coconut palms killed (Zelazny *et al.* 1982). At present, the disease continues to spread but at a slow pace. New disease outbreaks are still confined within the central eastern Philippines. Mindanao, the country's top coconut producing region which generates 55% of the total coconut production, is free from the disease.

23.1.2.1 Symptoms and Epidemiology

A palm infected with cadang-cadang shows a series of symptoms (Fig. 23.1.1A) starting with the bearing of smaller, more spherical and scarified nuts. Then, yellow spots develop in the leaves. The spots become numerous as the disease progresses giving the lower two-thirds of the crown a yellowish appearance. Inflorescences become necrotic, infertile and nut production eventually ceases. Fibers remain attached to the bases of the fronds rather than break off as in a healthy palm. Frond production and size gradually decline, leaflets become brittle and death of the palm follows (Randles *et al.* 1998). The time between the appearance of the first symptoms and death of the palm ranges from three to more than 15 years and is, on the average, about 10 years (Zelazny and Niven 1980). The early stage lasts an average of 2 years in 19-30 year-old palms but up to 3.75 years in older palms. The medium stage lasts for an average of just over 2 years. The late stage until death averages about 5 years.

The cadang-cadang disease is rarely detected in young palms. It is usually observed only after palms reach 10 years of age and incidence increases approximately linearly for up to 40 years. Incidence then usually remains constant for older palms (Zelazny and Pacumbaba 1982). The disease has a scattered and apparently random distribution. The rate of spread is slow and gradual ranging from about 0.1% to 1% in low and high incidence areas, respectively (Randles *et al.* 1988). The outward advance of the disease is about 500m p.a. There is no evidence that it originated at one point (Zelazny 1979). The lack of a specific pattern of disease increase does not allow the source of infection in plantation to be inferred (Randles *et al.* 1992).

An obvious new symptom type first appeared in 1987 in a significant number of artificially inoculated palms (Rodriguez 1993, 1998). Its main feature was the great loss of leaf lamina giving the palm a "broomed" appearance associated with severe stunting of the palm which led to premature death (Fig.23.1.1C,D). The basal fibers also remained attached to the fronds as in a typical cadang-cadang infected palm. The severe brooming symptom was not specific to a particular type of coconut. It was observed 3-7 years after inoculation in approximately 12% of the 1,787 palms screened. Occasional observations of similar palms in plantations suggest that brooming occurs naturally.

23.1.2.2 Isolation, Transmission and Host Range of Coconut Cadang-Cadang Viroid (CCCVd)

The detection of two small disease-associated RNAs by Randles in 1975 provided the initial clue to the etiology of cadang-cadang. Characterization of pure preparations by electron microscopy (Imperial *et al.* 1981), nucleotide sequencing (Haseloff *et al.* 1982; Rodriguez 1993) and transmission experiments that demonstrated the infectivity of these RNAs (Randles *et al.* 1977; Rodriguez *et al.* 1995), finally proved that cadang-cadang is caused by a viroid. It was named the coconut cadang-cadang viroid or CCCVd.

Partial isolation of CCCVd was done by preparing a low molecular weight nucleic acid extract from infected plant tissue. Further purification was achieved by electrophoresis that separates the circular viroids from the contaminating cellular linear RNAs (Schumacher *et al.* 1983; Rivera-Bustamante *et al.* 1986). In a denatured state, the circular viroids have much lower electrophoretic mobility than linear RNAs of the same size.

The usual leaf rubbing/pin-pricking inoculation techniques did not work for CCCVd. By using a hand-primed high-pressure injector, however, a 70 to 95% transmission was observed within 2 years after inoculation when 1 to

7 day-old sprouts that emerged from trimmed seednuts were inoculated twice at the base with 0.8 to 8.0 ug of CCCVd per sprout (Fig.23.1.1B) (Imperial *et al.* 1986). The inoculum was prepared by blending chopped leaflets (from the early stage palms) in sodium sulfite solution, followed by precipitation of nucleic acids with polyethylene glycol, then serial extraction with a phenol-chloroform mixture and finally, precipitation of CCCVd with ethanol. This inoculum can be stored at –20°C as precipitates in ethanol for as long as 3 years without any reduction in infectivity. The efficiency of the above mechanical transmission of CCCVd was found to be influenced by factors related to the inoculum, age of test plant, and mode and site of inoculation.

As to the natural mode(s) of cadang-cadang transmission, evidences were obtained that transmission through pollen and seed can occur but at a very low level (Pacumbaba *et al.* 1994; Bonaobra *et al.* 1998). F1 progenies of healthy palms pollinated with diseased pollen contained the viroid and exhibited disease symptoms 6 years after germination. Also, there were indications that the disease could be transmitted through contaminated farm tools such as bolo and harvesting scythe (Manalo *et al.* 2000). Experimental transmission of the disease using insects was unsuccessful (Zelazny *et al.* 1982).

The natural and experimental hosts of cadang-cadang identified so far are all members of the palm family (Randles *et al.* 1980; Imperial *et al.* 1985). They are *Elaies guineensis* (oil palm), *Corypha elata* (buri palm), *Areca catechu* (betel nut palm), *Adonidia merillii* (Manila palm), *Chrysalidocarpus lutescens* (palmera*), Oreodoxa regia* (royal palm), *Ptycosperma macarthuri* (Macarthur palm) and *Phoenix dactylifera* (date palm).

23.1.2.3 Screening of Coconuts for Resistance to CCCVd

Using the efficient high-pressure inoculation technique, 93 cultivars consisting of talls, dwarfs and hybrids were screened for resistance to CCCVd (Bonaobra *et al.* 1998; Orolfo *et al.* 2000). The infected palms manifested the typical symptoms of cadang-cadang. No absolutely resistant population was identified but ranked data on relative susceptibility to CCCVd (a range of 20-100%) could be useful in choosing appropriate materials for replanting purposes. The populations with susceptibility lower than 50% are mostly hybrids, two are among the 15 coconut hybrids being recommended by the Philippine Coconut Authority. These are MRD x BAYT and MRD x PYT with yields of 6 and 4 tons of copra per ha per year, respectively. The LAGT population which is the prevalent type of coconut planted in the cadang-cadang affected areas is one of the most susceptible cultivars tested (with 100% infection) both in screening trials by artificial inoculation and by natural exposure to the disease.

23.1.2.4 Molecular Structure of CCCVd

CCCVd with 246 nucleotides is the smallest known viroid. In addition to this basic form, at least three types of sequence variants have been recognized by sequencing isolates from individual palms and from fronds of different ages within a single palm (Haseloff *et al.* 1982; Keese *et al.* 1988). These include an insertion of cytosine at position 197 of the central conserved (CC) domain to give a 247-nucleotide sequence variant, duplications of the variable (V) and terminal (T2) domains of 41, 50, 55 or 100 nucleotides and mutations of bases adjacent to the boundaries of some of the partial duplications. Detected concurrently with the monomers are their respective dimers which are covalently linked forms of the monomers, incorporating the same sequence variations.

The above variations in CCCVd were correlated with the stages of disease development as manifested by typical symptoms. A systemic progression was observed for the commonly detected variants (Imperial *et al.* 1981; Imperial and Rodriguez 1983). The small 246/247 variants (with their corresponding dimers) appear early in infection but as the disease progresses, the large 296/297 variants (also with their corresponding dimers) arise and eventually dominate the viroid population until the palm dies. More specifically, if infection starts with variant 246, 247 appears next followed by 296, then 297. If variant 247 appears first, only 297 replaces it. All these variants, containing the basic 246 nucleotides, are infectious and associated with disease progression but not with symptoms.

The frequency of occurrence of 246 in the early stage palms is greater than that of 247 as it is the more active infecting agent responsible for disease spread than the latter (Rodriguez *et al.* 1995). With 296 or 297 as the inoculum in the field, the disease could be stable for as long as 6 years before new infection could be observed.

The unusual, severe brooming symptom was found to be associated with new types of mutation (Rodriguez and Randles 1993). Polyacrylamide gel electrophoresis (PAGE) and molecular hybridization analyses of 136 palms with brooming symptom showed that all of them have electrophoretic forms distinct from the detected variants of CCCVd associated with disease progression (Rodriguez 1993, 1998). Whereas four bands representing viroids with 246, 247, 296 and 297 nucleotides were observed in extracts from palms with typical cadang-cadang symptoms, the palms with brooming symptom showed a total of 11 additional or replacement bands. On 20% PAGE, three of these bands had electrophoretic mobilities between CCCVd variants 247 and 246; two, between 246 and 297; three, between 297 and 296, and three, above 296 (Fig. 23.1.1E). To determine whether these electrophoretic forms are mutants of CCCVd, the isolates were first purified by bi-directional PAGE and then amplified by polymerase chain reaction (PCR) via their cDNAs. Cloning of the PCR products had to be done since most of the isolates contained more than one electrophoretic form making it difficult to relate a specific form to a particular symptom as well as to analyze data. Fluorescent dye primer cycle sequencing was performed on 28 full-length and 19 half-length clones. With the basic CCCVd 246 as control, the same result was obtained when compared with that of the direct RNA sequencing method used by Haseloff *et al.* (1982). Thus, PCR artifact was ruled out.

Nine sequence variants associated with the brooming symptom were obtained. Four of them have 247 nucleotides, but differ in sequence at positions 197 (C→AU, UG or UU) and 216 (U→A). Two are 248 in size with substitution of C at position 197 with either AU or UU and addition of A at position 87 (Fig. 23.1.2A,B). Three large variants (294, 303 and 304) have partial sequence duplications in the V and T2 domains and mutations at the boundaries of the duplicated sequences (Fig. 23.1.2C). In two of these large variants, no mutation was observed in the left-hand portion of the native molecules. The third large variant, however, which was isolated singly from a palm with brooming but with mild stunting had a U→A substitution at position 216 in the pathogenicity (P) domain of the left-hand structure. The direct causal relationship between the mutations and brooming can be determined only by inoculating the mutant clones to coconut palms. This will establish what particular type or combination of mutations causes the severe symptom.

Position 197 of the central conserved region seemed vulnerable to mutations with the resulting variants having different mobilities in PAGE. Although a C→CC addition or C→UU substitution appeared not to affect secondary structure, it caused the CCCVd molecule to migrate with markedly different electrophoretic mobility from the basic CCCVd 246. With a C→AU or C→UG substitution, a U-loop was replaced with a base pair. This site may be important in controlling the pathogenicity of CCCVd. The mutations may have an effect on the switching event predicted to occur during replication in which the native structure is converted into a stem loop (Keese *et al.* 1988).

23.1.2.5 Molecular Diagnostics for CCCVd

Since cadang-cadang develops slowly in coconut palms, diagnosis by symptoms is unreliable particularly on a single observation. Hence, there is a need for the application of molecular techniques in disease diagnosis. The success in detecting CCCVd depends largely upon which part of the palm is to be assayed for early detection (Rodriguez 1999). Non-destructive sampling, convenience in the collection and ease in macerating the tissues should also be considered. CCCVd can be detected earlier in leaves than in primary root tips, although it is easier to collect and macerate the latter tissues. In the very early stage of infection, CCCVd is detectable in young but not in old leaf fronds of the coconut crown. For large number of samples, extraction is done by grinding the sample tissues individually in a disposable plastic bag that contains the extraction buffer. Partial purification of the extract involves a phenol and/or chloroform deproteinisation step (Rodriguez *et al.* 1989).

Detection of CCCVd in leaf isolates by electrophoresis is based on its size and thermodynamic properties. Optimum resolution of CCCVd and its sequence variants is obtained by increasing the concentration of polyacrylamide from the standard 5% to 20% under non-denaturing conditions (Imperial and Rodriguez 1983). The nucleic acid bands are visualized by staining with silver nitrate. This PAGE assay is suitable for detection of CCCVd in leaf samples where the concentration of CCCVd is usually above 50 ng and where it is necessary to determine the CCCVd variant present which is indicative of the stage of infection (Rodriguez *et al.* 1989, 1998; Namia *et al.* 1998).

For more sensitive and specific CCCVd detection, dot-blot or Northern-blot hybridization is performed (Rodriguez, 1999). Total nucleic acid sample is isolated, purified to a certain degree and then transferred to a supporting membrane filter before being hybridized with a probe specific to CCCVd. The procedures for sample and probe preparation, type of membrane filter and buffer conditions used during transfer and hybridization all influence the detection of CCCVd. Due to the hazards in handling radioisotopes and their limited shelf-life, radioactive label is replaced with digoxigenin (DIG). Different types of DIG probes can be prepared. These include synthetic oligonucleotides of defined sequences specific to CCCVd and ss cDNA or ss cRNA full-length probes. With the sequencing of CCCVd, oligonucleotide probes (OPs) with lengths of 20-32 nucleotides were designed and tested (Rodriguez 1993). However, with the cloning of ds cDNA of full-length CCCVd in recombinant DNA plasmid that contains highly specific promoter sequences (SP6/T7) for in vitro transcription, it is convenient and economical to prepare large quantities of ss cRNA full-length probe (M.J.B. Rodriguez, unpublished). The Boehringer DIG RNA labeling and chemiluminescent or colorimetric detection kits can be used. High stringent hybridization is necessary to prevent non-specific binding. The detection limit is approximately 10 pg.

A one-tube, reverse transcription-polymerase chain reaction (RT-PCR) analysis was also developed for CCCVd detection (Rodriguez 1993). Since the starting material is RNA, the viroid has to be reverse transcribed first to cDNA and this lengthens the whole process. Thus, a system was designed where all the components necessary for both reverse transcription and PCR are combined in a single tube, and the two reactions carried out sequentially in a single thermal cycling program. The antisense DNA primers selected for use in the PCR also served to prime specifically for the reverse transcription. Three sets of primers have been designed resulting in amplification of either the left or right half-length of the total CCCVd or full-length copy. The priming sites are at the central conserved and pathogenicity regions of CCCVd. Electrophoresis and ethidium bromide staining visualize the PCR products.

23.1.3 Tinangaja Disease

According to farmers in Guam, the word "tinangaja" means becoming smaller. The tinangaja disease of coconuts in Guam was first reported by Weston (1917). A second report in 1961 named the disease "infectious yellow mottle decline" and described it as closely resembling cadang-cadang (Reinking, 1961). Like cadang-cadang, tinangaja primarily affects palms that are 25-30 or more years old. The leaflets of tinangaja-affected coconuts show chlorotic spots very similar to those observed in cadang-cadang. However, there is dissimilarity in nut symptom which is used for convenience as the principal diagnostic character. In the early stage, cadang-cadang induces the production of smaller-than-normal, rounded nuts scarified around the equator while no such symptom has been reported for tinangaja. Instead the formation of crinkled, mummified nuts with no kernel inside has been observed (Rodriguez and Carpio 1987; Boccardo *et al.* 1981; Holmes 1962; Reinking 1961; Weston 1918). These mummified nuts occur in the middle stage of the disease, which seems to last the longest. Palms die soon after nut production ceases, whereas in cadang-cadang, sterility is induced much earlier and the late stage (no nuts) can last for years (Zelazny and Niven 1980).

Available information on Guam soils (USDA Soil Conservation Services draft report on Soil Survey of the Territory of Guam) indicated that diseased palms are found regardless of soil units /classes.

Information on insect surveys done in low and high incidence areas was gathered and compared with those done in the Philippines (Rodriguez and Carpio 1987). In both countries, the insect populations found mostly in coconut palms belong to the family *Derbidae*, although of different species.

23.1.3.1 Molecular Structure of Coconut Tinangaja Viroid (CTiVd)

Viroid-like RNAs were detected in nucleic acid preparations of tinangaja-infected palms (Boccardo *et al.* 1981). Comparative electrophoretic analysis in 5% polyacrylamide gels showed that low-molecular-weight RNAs with the same apparent mobilities as the cadang-cadang viroid were present in palms with tinangaja symptoms. Thus, a common etiology was suggested between the two diseases while the dissimilarity in nut symptom was ascribed to varietal difference between the coconut palm hosts in Guam and the Philippines.

To validate the suggested common identity of the causal agents of the two diseases, isolates of tinangaja-associated viroid-like RNAs were sequenced by Keese *et al.* (1988). Two sequence variants (with corresponding dimers) were obtained, both are 254 nucleotide residues in size but differ in sequence at two positions. In comparison with other viroids, the sequences and proposed secondary structure of CTiVd show a high degree of homology with CCCVd (64%). This extensive but not complete homology between CTiVd and CCCVd could then account for the difference in nut symptom (Randles *et al.* 1994).

23.1.3.2 Molecular Diagnostics for CTiVd

Tinangaja is not considered an economically important disease in Guam. Because of the interchanges among the Marianas Islands, however, the disease (if not contained) poses a serious threat to the neighboring coconut growing islands. For mapping the distribution of CTiVd in the field, there is a need for an efficient diagnostic procedure.

A method of extracting leaf tissue followed by agarose gel electrophoresis was developed and used to detect the viroid in the field (Hodgson *et al.* 1998). Confirmation of the identity of CTiVd and detection below the threshold of agarose gel electrophoresis was achieved either by diagnostic oligonucleotide probe (DOP) hybridization assay or by RT-PCR with the oligonucleotide probe as one of the two PCR primers. RT-PCR was not substantially more sensitive than DOP-hybridization assay.

23.1.4 Control Approaches: Prospects

The cadang-cadang disease has caused prohibition in the international trade of Philippine coconut products such as fresh young coconut, desiccated coconut, coconut in powder and sliced form, by-products of coconut fibers, nata de coco and Makapuno. The main reasons for the export ban were apparently fear of health hazard and the possible transport of the pathogen. The availability of a sensitive, specific, rapid and economical diagnostic method is, therefore, most important in determining whether or not the products or the source palms are contaminated. It should also be noted that cadang-cadang is present only in certain areas in the Philippines. Since Mindanao in southern Philippines is free from cadang-cadang, coconut products from this island should be spared from trade restriction.

To further address the problem, a comprehensive cadang-cadang containment scheme based on research and strict enforcement of quarantine measures should become operative to prevent entry of contaminated plant materials/products into disease-free areas. The availability of a molecular assay should enable "early detection-diseased palm removal-replanting" as a feasible and practical control strategy considering that cadang-cadang has a long incubation period and a low rate of dissemination. Early detection is of prime importance as attempts to eradicate cadang-cadang failed because only the late stage palms were removed. In addition, decontamination of farm tools should be observed in the field. To maintain plantation production, replanting with tolerant high-yielding cultivars and hybrids should be considered as well as intercropping with cash crops to augment the income of farmers.

The detection of sequence variants of CCCVd may also provide an approach to disease control. Change(s) in sequence on certain site(s) in the viroid molecule can be induced so as to modify pathogenicity and obtain mild isolates. These mild strains may then be tested for cross protection against severe forms of the disease.

As cadang-cadang and tinangaja belong to the same group of viroids aside from having the same host, prospects and recommendations can apply to both.

References

Boccardo, G., Beaver, R.G., Randles, J.W. and Imperial, J.S. (1981). Tinangaja and bristletop, coconut diseases of uncertain etiology in Guam and their relationship to cadang-cadang disease of coconut in the Philippines. *Phytopathology* 71: 1104-1107.

Bonaobra III, Z.S., Baylon, G.B., Orense, O. and Rodriguez, M.J.B. (1998). The role of pollen and seed in transmission of CCCVd. Pp. 67-70 in: *Report on ACIAR-funded research on viroids and viruses of coconut palm and other tropical monocotyledons, 1985-1993.* D. Hanold and J.W. Randles, eds. ACIAR Working Paper No. 51 (also published at http://www.aciar.gov.au/viroids/viroids.htm).

Bonaobra III, Z.S., Rodriguez, M.J.B., Estioko, L.P., Baylon, .G.B, Cueto, C.A. and Namia, M.T.I. (1998). Screening of coconut populations for resistance to CCCVd using coconut seedlings. Pp. 77-83 in: *Report on ACIAR-funded research on viroids and viruses of coconut palm and other tropical monocotyledons, 1985-1993.* D. Hanold and J.W. Randles, eds. ACIAR Working Paper No. 51 (also published at http://www.aciar.gov.au/viroids/viroids.htm}.

Diener, T.O. (1979). *Viroids and Viroid Diseases*. John Wiley and Sons, New York.

Haseloff, J., Mohamed, N.A. and Symons, R.H. (1982). Viroid RNAs of cadang-cadang disease of coconuts. *Nature* 299: 316-321.

Hodgson, R.A.J, Wall, G.C. and Randles, J.W. (1998). Specific identification of coconut tinangaja viroid for differential field diagnosis of viroids in coconut palm. *Phytopathology* 88 : 774-781.

Holmes, F.O. (1962). The Guam disease of coconut palms. *FAO Plant Prot Bull* 10: 25-28.

Imperial, J.S., Bautista, R.M. and Randles, J.W. (1985). Transmission of the coconut cadang-cadang viroid to six species of palm by inoculation with nucleic acid extracts. *Plant Pathology* 34: 391-401.

Imperial, J.S., Olfato, R.B. and Rodriguez, M.J.B. (1986). Studies on the factors affecting cadang-cadang transmission by mechanical inoculation to coconut seedlings. *PCA-ARDB Annual Report.*

Imperial, J.S., and Rodriguez, M.J.B. (1983). Variation in the coconut cadang-cadang viroid: Evidence for single base addition with disease progress. *The Philippine J. of Crop Science* 8(2): 87-91.

Imperial, J.S., Rodriguez, M.J.B. and Randles. J.W. (1981). Variation in the viroid-like RNA associated with cadang-cadang disease: Evidence for an increase in molecular weight with disease progress. *The J. of Gen. Virology* 56: 77-85.

Keese, P., Osorio-Keese, M.E. and Symons, R.H. (1988). Coconut tinangaja viroid: Sequence homology with coconut cadang-cadang viroid and other potato spindle tuber viroid related RNAs. *Virology* 162: 508-510.

Keese, P., Visvader, J.E. and Symons, R.H. (1988). Sequence variability in plant viroid RNAs. Pp. 71-98 in: *RNA Genetics, Variability of RNA Genomes*, Vol. III. E Domingo, JJ Holland and P Ahlquist, eds. CRC Press, Boca Raton, FL.

Manalo, G.G , Estioko, L.P. and Rodriguez, M.J.B. (2000). Studies on the identification of the natural mode(s) of transmission of the coconut cadang cadang viroid (CCCVd). *PCA-ARDB Annual Report.*

Namia, M.T.I., Rodriguez, M.J.B. and Randles, J.W. (1998). Diagnosis of cadang-cadang by rapid polyacrylamide gel electrophoresis. Pp. 43-46 in: *Report on ACIAR-funded research on viroids and viruses of coconut palm and other tropical monocotyledons, 1985-1993.* D Hanold and JW Randles, eds. ACIAR Working Paper No. 51 (also published at http://www.aciar.gov.au/viroids/viroids.htm).

Ocfemia, G.O. (1937). The probable nature of cadang-cadang disease of coconuts. *Philippine Agriculturist* 26: 338-340.

Orolfo, M.B., Estioko, L.P. and Rodriguez, M.J.B. (2000). Screening of coconut populations for resistance to coconut cadang-cadang viroid (CCCVd). *PCA-ARDB Annual Report.*

Pacumbaba, E.P., Zelazny, B., Orense, J.C. and Rillo, E.P. (1994). Evidence for pollen and seed transmission of the coconut cadang-cadang viroid in *Cocos nucifera*. *Phytopathology*. 142: 37-42.

Persley, G.J. (1992). *Replanting the Tree of Life: Towards an International Agenda for Coconut Palm Research*. C.A.B. International, UK.

Price, W.C. (1971). Cadang-cadang of coconut – A review. *Plant Science* 3: 1-13.

Randles, J.W. (1975). Association of two ribonucleic acid species with cadang-cadang disease of coconut palm. *Phytopathology* 65: 163-167.

Randles, J.W., Boccardo, G. and Imperial, J.S. (1980). Detection of the cadang-cadang associated RNA in African oil palm and buri palm. *Phytopathology* 70: 185-189.

Randles, J.W., Boccardo, G., Retuerma, M.L. and Rillo, E.P. (1977). Transmission of the RNA species associated with cadang-cadang of coconut palm, and insensitivity of the disease to antibiotics. *Phytopathology* 67: 1211-1216.

Randles, J.W., Hanold, D, Pacumbaba, E.P. and Rodriguez, M.J.B. (1992). Cadang-cadang disease of coconut palm. Pp. 277-295 in: *Plant Diseases of International Importance.* AN Mukhopadhyay, J Kumar, HS Chaube and US Singh, eds. Prentice Hall Inc., New Jersey.

Randles, J.W., Hanold, D., Pacumbaba, E.P. and Rodriguez, M.J.B. (1998). Cadang-cadang disease of coconut palm – an overview. Pp. 11-27 in: *Report on ACIAR-funded research on viroids and viruses of coconut palm and other tropical monocotyledons, 1985-1993.* D Hanold and JW Randles, eds. ACIAR Working Paper No. 51 (also published at http://www.aciar.gov.au/viroids/viroids.htm).

Randles, J.W., Rodriguez, M.J.B. and Hanold, D. (1994). Coconut diseases caused by nonfungal agents: Cadang-cadang and tinangaja. Pp. 28-29 in: *Compendium of Tropical Fruit Diseases.* RC Ploetz, GA Zentmyer, WT Nishjima, KG Rohrrbach and HD Ohr, eds. APS Press, U.S.A.

Randles, J.W., Rodriguez, M.J.B. and Imperial, J.S. (1988). Cadang-cadang disease of coconut palm. *Microbiological Science* 5(1); 18-22.

Reinking, O.A. (1961). Yellow mottle decline in the territory of Guam. *Plant Disease Rep* 45: 599-604.

Rivera-Bustamante, R.F., Gin, R. and Semancik, J.S. (1986). Enhanced resolution of circular and linear molecular forms of viroid and viroid-like RNA by electrophoresis in a discontinuous-pH system. *Anal Biochem* 156: 91-95.

Rodriguez, M.J.B. (1993). *PhD Thesis*, University of Adelaide, South Australia.

Rodriguez, M.J.B. (1998). Brooming, a severe form of cadang-cadang disease. Pp. 99-114 in: *Report on ACIAR-funded research on viroids and viruses of coconut palm and other tropical monocotyledons, 1985-1993.* D Hanold and JW Randles, eds. ACIAR Working Paper No. 51 (also published at http://www.aciar.gov.au/viroids/viroids.htm).

Rodriguez, M.J.B. (1999). Detection and diagnosis of coconut cadang-cadang. Pp. 221-226 in: *Current Advances in Coconut Biotechnology.* C Oropeza, JL Verdeil, GR Ashburner, R Cardeña and JM Santamaria, eds. Kluwer Academic Publishers, The Netherlands.

Rodriguez, M.J.B. and Carpio, C.B. (1987). *Survey report on the tinangaja disease and coconut varieties/populations in Guam, USA.* 21pp.

Rodriguez, M.J.B., Estioko, L.P. and Manalo, G.G. (1995). Infectivity studies on the commonly detected sequence variants of the coconut cadang-cadang viroid (CCCVd) in relation to disease spread. *The Philippine J. of Science* 124(1): 1-13.

Rodriguez, M.J.B., Ignacio-Namia, M.T.R. and Estioko, L.P. (1989). A mobile diagnostic laboratory for the cadang-cadang disease of coconut. *The Philippine J. of Coconut Studies* 14(2): 21-23.

Rodriguez, M.J.B., Namia, M.T.I. and Estioko, L.P. (1998). The mobile diagnostic laboratory. Pp. 47-52 in: *Report on ACIAR-funded research on viroids and viruses of coconut palm and other tropical monocotyledons, 1985-1993.* D Hanold and JW Randles, eds. ACIAR Working Paper No.51 (also published at http://www,aciar.gov.au/viroids/viroids.htm).

Rodriguez, M.J.B. and Randles, J.W. (1993). Coconut cadang-cadang viroid (CCCVd) mutants associated with severe disease vary in both the pathogenicity domain and the central conserved region. *Nucleic Acids Res* 212 (11): 2771.

Schumacher, J., Randles, J.W. and Riesner, D. (1983). A two-dimensional electrophoretic technique for the detection of circular viroids and virusoids. *Anal Biochem* 135: 288-295.

Weston, W.H. Jr (1918). Report on the plant disease situation in Guam. Pp 45-62 in: *Guam Agric Exp Stn Rep* 1917. 89 pp.

Zelazny, B. (1979). Distribution and spread of the cadang-cadang disease of coconut palm. *Acta Phytopath Acad Sci Hung* 14: 115-126.

Zelazny, B. and Niven, B.S. (1980). Duration of the stages of cadang-cadang disease of coconut palm. *Plant Disease* 64: 841-842.

Zelazny, B. and Pacumbaba, E.P. (1982). Incidence of cadang-cadang of coconut palm in the Philippines. *Plant Disease* 66: 547-549.

Zelazny, B., Randles, J.W., Boccardo, G. and Imperial, J.S. (1982). The viroid nature of the cadang-cadang disease of coconut palm. *Scientia Filipinas* 2: 46-63.

Legends to Figures

Figure 23.1.1 (**A**) Typical symptoms of cadang-cadang: nuts become smaller, more spherical and scarified around the equator (left), inflorescences become necrotic (middle) and stipules remain attached to bases of the fronds (right). (**B**) CCCVd- inoculation of a 3-day old coconut sprout using a hand-primed high-pressure injector. (**C**) The unusual brooming syndrome observed in a CCCVd-inoculated coconut palm: severe stunting, much shorter frond and reduced leaf laminae. (**D**) Leaflets from a brooming palm (top) compared with those from cadang-cadang infected (middle) and healthy palms (bottom). (**E**) 20% non-denaturing PAGE assay of extracts from palms showing varying intensities of the brooming symptom (nos. 1-5 and a mixture of their isolates in lane no.6; M denotes CCCVd marker consisting of the commonly detected variants 246, 247, 296 and 297).

Figure 23.1.2 (**A**) The primary and secondary structure of CCCVd 246 with demarcation according to the proposed terminal (T1 and T2), pathogenicity (P), central conserved (CC) and variable (V) domains of Keese *et al.* (1988); the sites where mutations associated with the brooming symptom occur are denoted by red asterisks. (**B**) Mutations at positions 87, 197 and 216 and their putative effects on secondary structure; the substituted or added nucleotides are shown in red letters. (**C**) Partial sequence duplications in the V and T2 domains (numbering according to that of CCCVd 246) and mutations at the boundaries of the duplicated sequences (highlighted with red asterisks) in isolates from palms with brooming symptom; the duplicated sequences vary in positions and lengths (51, 53 and 57 nts).

Figure 23.1.1

Figure 23.1.2

23. Coconut and other palm trees

23.2 Foliar Decay Disease in Vanuatu

D. Hanold[1], J.P. Morin[2], J.P. Labouisse[3] and J.W. Randles[1]
[1]Waite Campus, University of Adelaide, Glen Osmond, South Australia; [2]CIRAD-CP, Montpellier, France; [3]CIRAD-CP, Centre Agronomique de Recherche et de Formation du Vanuatu, Saraoutou, Espiritu Santo, Vanuatu.

23.2.1 Introduction

In 1962, the French and British governments decided to establish a coconut research station in the Condominium of the New Hebrides under the direction of a French research organisation, the Institut de Recherches pour les Huiles et Oléagineux (IRHO). At that time, coconut was the mainstay of the economy of this archipelago and represented 78% of export earnings. The station was set up in the Saraoutou area, about 11 kilometres north from Luganville on the island of Espiritu Santo. After independence of Vanuatu in 1980, the station was named Vanuatu Agricultural Research and Training Center (VARTC) in 1994.

The main objective of the research at the station was the raising of coconut productivity through agronomic and genetic improvement. It was decided to develop high-yielding hybrids suitable for the region to replace the ageing plantations of coconut palms of local varieties. A number of dwarf and tall varieties were introduced and planted into the station genebank, but they quickly started to succumb to a previously unseen 'wilt' of unknown etiology. Malayan Red Dwarf (MRD) was found to be the most sensitive and generally started to show symptoms 18 months after planting, with up to 90% mortality after 7 years (Calvez *et al.* 1980). Subsequently, the symptoms also appeared on introduced varieties on other islands of Vanuatu while the local Vanuatu Tall (VTT) variety remained unaffected. Some of the most productive hybrids based on introduced varieties were seriously affected, but a number of lower-yielding VTT hybrids showed varying degrees of tolerance and were therefore used as a compromise in the establishment of breeding programmes (Calvez *et al.* 1985).

The new 'wilt' was described (Calvez *et al.* 1980) and initially named New Hebrides disease. Later it was also called foliar decay induced by *Myndus taffini* (FDMT), foliar decay disease (FDD) and it is now known as coconut foliar decay (CFD). CFD had the characteristics of an infectious disease, and initially it was suspected that a pathogen might have been introduced unwittingly together with the new germplasm for the station. CFD posed a serious threat to the national coconut improvement programme. Thus a control strategy was urgently needed, and as a prerequisite the search for a cause and mode of spread commenced (Dollet and de Taffin 1978).

Fungi, bacteria or nematodes could not be implicated in the disease. In 1981, the plant-hopper *Myndus taffini* was shown to be the vector of CFD (Julia, 1982). Symptoms could be transmitted to MRD seedlings after only short exposure to small numbers of insects. Tetracycline treatment similar to the one known to ameliorate lethal yellowing of coconut palm caused by a phytoplasma did not have any effect on CFD symptoms. Thus a virus-like organism was considered the most likely cause (Julia *et al.* 1985).

A small circular single-stranded (ss) DNA was shown to be associated with CFD (Randles *et al.* 1986, 1987). A diagnostic test for the disease was developed based on hybridisation assays using cloned sequences of the unique CFD DNA as probes. With this method it was shown that (a) the symptomless VTT contained CFD DNA and was susceptible to infection by the causal agent; it was therefore tolerant rather that resistant to CFD and thus a putative pathogen was likely to be local rather than introduced; (b) *Hibiscus tiliaceus* (locally called burao), the breeding host of the vector which is a common forest tree and used widely as a hedging plant in the region, was not infected with the disease agent; and (c) CFD DNA was only about half the size of geminivirus DNA and not typical of any other groups of viruses known at the time (Hanold *et al.* 1988). The viral nature of the causal agent was confirmed when unusual 20nm icosahedral particles were found to co-purify with CFD DNA. Particle morphology supported the hypothesis that a novel virus was associated with CFD and it was named *Coconut foliar decay virus* (CFDV; Randles and Hanold, 1989). Subsequently, CFDV distribution in the host and the vector was studied (Randles *et al.* 1992).

The nucleotide sequence of the CFDV DNA was determined and it was confirmed that the virus represented a new taxonomic group (Rohde *et al.* 1990). It has now been tentatively assigned to the genus *Nanovirus*, however it differs from the other members of this group in a number of characteristics including its plant-hopper rather that aphid vector, its monopartite genome and the size of the capsid protein (Randles *et al.* 2000).

The cause, vector and disease cycle of CFD were determined in less than 10 years. Effective management strategies for CFDV are now in place in Vanuatu which include control directed at the insect vector *M. taffini*, removal of its breeding host *H. tiliaceus*, and through breeding for tolerant lines of coconut palms. The story of CFD is an example of how effective and efficient an integrated research approach based on collaboration of expert scientists and continuous funding can be for solving a new disease problem.

23.2.2 Distribution, crop losses and host range

The occurrence of CFDV is limited to Vanuatu and probably determined by the distribution of its specific vector *M. taffini*. Other species of the genus *Myndus* were identified in the South Pacific region, but because CFD has not been found in other countries it is assumed that they do not transmit CFDV (Wilson, 1988). Analysis of a number of samples collected from coconut palms throughout the Pacific and South-East Asia during a survey for coconut cadang-cadang related pathogens (Hanold and Randles 1991) did not provide any definite positive results for CFDV DNA in areas outside Vanuatu (D. Hanold and J.W. Randles, unpublished).

The coconut palm is a vital part of the social and economic structure of some of the poorest areas in the world (Persley 1992) and is often referred to as the 'Tree of Life'. General data on crop losses are not available for CFDV and the susceptibility of varieties varies considerably with MRD being the most susceptible (Calvez *et al.* 1980; Randles *et al.* 1992). It is estimated that, over all varieties, ca. 50% of introduced germplasm seed planted at Saraoutou station was lost due to CFDV over the years. CFDV tolerant hybrids have resulted in considerable improvement in yield over unimproved VTT, which on average produces 1 to 2 tonnes of copra per hectare annually. The most promising hybrids appear to be VTT x Rennell Island Tall (RIT) for which an average yield of 2.6 tonnes per hectare has been recorded, and Vanuatu Red Dwarf (VRD) x VTT with an average annual yield of 3.4 tonnes per hectare (Calvez *et al.* 1985; Morin 1994; J.P. Labouisse, unpublished). However, the performance of these improved lines is still considerably lower than what could have been expected for hybrids used in other Pacific areas which can yield as much as 6.5 tonnes of copra per hectare under favourable conditions (Persley, 1992). Thus, as CFDV is limited to Vanuatu where the local cultivars are resistant to this disease, the real future loss for the country is a serious loss in production potential rather than a loss of trees. In addition, germplasm from Vanuatu cannot be distributed with safety due to the possibility of transmission of CFDV or its vector with seed, eliminating a potential source of export earnings for the country (Persley 1992).

The known natural host range of CFDV is limited to *Cocos nucifera* on which the disease was first reported. *C. nucifera* has also been experimentally infected in insect-proof cages by exposure to *M. taffini* collected from CFD trees (Julia 1982). The highly susceptible MRD variety is used as an indicator plant for epidemiology and insect transmission studies. Mechanical inoculation to other members of the *Arecaceae* or herbaceous hosts has not been attempted. *H. tiliaceus,* the breeding host of the vector, does not appear to be a host of CFDV (J.W. Randles and E. Wefels, unpublished).

23.2.3 Symptoms and transmission

Symptoms were first described on MRD (Calvez *et al.* 1980). In the field, yellowing symptoms first appear in several leaflets of fronds 7 to 11 positions down in the crown from the unopened spear leaf. More general yellowing of the fronds follows and the petioles of affected fronds develop lateral necrosis. Fronds die prematurely, hanging from the petiole downward through the canopy. Other fronds become yellow and die as they reach the central whorl of the crown, so that affected palms characteristically show a normal apex, several yellowish fronds, then several young dead fronds hanging down through green older fronds. The trunk generally narrows. Susceptible cultivars die between 1 and 2 years after symptoms appear. Spathes at the base of yellow fronds rot if the subtending frond dies, or they produce an inflorescence that develops normally but which bears fewer nuts than normal. While MRD and its hybrids show these specific symptoms (Fig. 23.2.1), responses in other palm cultivars may vary (Calvez *et al.* 1980). Seedlings that are experimentally infected by insects may show similar symptoms on fronds 3 to 5 at 6 to 11 months after inoculation (Randles *et al.* 1986), or they may show progressive yellowing and necrosis depending on the variety. Remission of symptoms has been observed on individual palms of susceptible tall and dwarf varieties 1-2 years after symptoms developed. These palms then appear to remain disease-free even if exposed to high infection pressure in the field (Anon. 1990; Randles *et al.* 1992).

CFDV is transmitted in the field by the plant-hopper vector *Myndus taffini (Homoptera Cixiidae*). The female lays eggs on *Hibiscus tiliaceus* (burao) roots where the larvae develop. Adults feed on coconut leaves, but can survive for several days on indigenous forest palms (Julia, 1982). *M. taffini* has been reared in the laboratory providing a base for research into virus acquisition and transmission (Morin, 1993; 1994). The minimum inoculation time is around one hour and the virus is retained for a period of 2-3 days being transmitted in a persistent or semi-persistent manner (Julia *et al.* 1985). CFDV has been detected in vector insects in the abdomen rather than the head (Randles *et al.* 1992). When the isolated guts of adult *M. taffini* were probed for CFDV DNA, the signal was found to be associated with the full length of the gut of several females (Fig. 23.2.2). Semipersistently transmitted viruses are commonly retained selectively in the gut so this observation would be consistent with such a mode of transmission. After insect inoculation, CFDV DNA could be detected in the host at 6-8 months and 1-4 months before symptom development (Randles *et al.* 1992).

The virus has not been transmitted by mechanical inoculation (Dollet and de Taffin 1978), and graft transmission experiments cannot be carried out since coconut palms will not graft. Although CFDV DNA was detected in embryos, no virus transmission through seed has been demonstrated so far. However, husks contain CFDV, and movement of nuts may allow transfer of virus to new sites. It will be necessary to determine whether vectors can acquire virus from green husk to decide whether there are quarantine risks from the movement of nuts (Randles *et al.* 1992; Persley 1992). It should be noted that VTT nuts have been exported on several occasions to research stations in other countries (Ivory Coast, Philippines, Brazil, Tanzania, Ghana, Jamaica) before seed export was banned, without any expression of the disease in these countries. No CFDV DNA was detectable in pollen samples (Randles *et al.* 1992).

23.2.4 Diagnosis

Diagnosis of CFD by symptoms is unreliable because of the variation in responses by different palm cultivars (Calvez *et al.* 1980). Only MRD palms show characteristic symptoms of the disease but are unsuitable as biological indicator plants since mechanical transmission appears to be unsuccessful. Virions have not been seen in sap extracts (Randles and Hanold 1989) and thus electron microscopy cannot be used for CFDV diagnosis.

The only available test for diagnosis is via the viral DNA, but it is vital that sampling and extraction protocols are optimised to obtain reliable results (Randles *et al.* 1992). Infection can be recognised in coconut palms by gel electrophoresis (Randles *et al.* 1986, 1999). CFDV DNA migrates as a single band in denaturing polyacrylamide gel electrophoresis (PAGE), but generally as two or three bands in non-denaturing PAGE (Randles *et al.* 1987). Routine diagnosis by this method lacks proof that bands have nucleotide sequence similarity to CFDV. Therefore, gel blotting and hybridisation assays for the DNA should be done in conjunction with PAGE or agarose gel electrophoresis. Dot blot hybridisation assays may also be suitable (Hanold *et al.* 1988, Randles *et al.* 1992).

For hybridisation assays, ^{32}P labelled probes made from either clones of partial CFDV sequences (Hanold *et al.* 1988) or by transcription of near full-length cDNA from purified CFDV DNA have been used, the latter being more sensitive (Rohde *et al.* 1990; Randles *et al.* 1992). Variation in signal strength may be due to differences in CFDV DNA concentration in tissue, sample preparation, and assay procedures. A non-radioactive probing system was also developed using digoxygenin (DIG) labelled cRNA. This alkaline phosphatase based colorimetric assay required more rigorous sample purification to prevent development of a non-specific endogenous reaction with the non-radioactive probing system (Randles *et al.* 1992). We have previously observed apparent endogenous phosphatase activity in coconut nucleic acid extracts which was highly stable and interfered with colorimetric detection methods (D. Hanold and J.W. Randles, unpublished).

Hybridisation tests are more sensitive than PAGE assays for coconut palms, and probably have broader applicability to other species for which the PAGE nucleic acid preparation procedure developed for coconut palms may not be appropriate.

When the development of a diagnostic assay based on polymerase chain reaction (PCR) was attempted, extraction procedures had to be optimised to remove enzyme inhibitors. A pair of oligonucleotide primers representing the putative stem-loop region of the CFDV DNA was used, however complete primer extension could not be achieved even under high stringency conditions, presumably due to the secondary structure of the target molecules. Thus a protocol for unilateral PCR was developed for routine use (Randles *et al.* 1999).

CFDV DNA was found to be distributed unevenly in whole plants and in an apparently non-systematic pattern. Thus, multiple sampling is needed for reliable diagnosis. Sampling should include tissue from a number of the younger fronds, and root sampling should be combined with leaf sampling, particularly where new areas are being surveyed for CFDV incidence (Randles *et al.* 1992).

23.2.5 Purification and characteristics of the virus

CFDV occurs in very low amounts in diseased coconut palms, but in the absence of known alternate hosts it was necessary to develop purification methods appropriate to this host for both CFDV DNA and particles. A small single-stranded DNA extractable only by optimised purification methods from diseased trees was shown to be associated with CFD. Despite its small size, it had low electrophoretic mobility in 5% PAGE and it could only be detected by silver staining due to its low concentration in the host tissue (Randles *et al.* 1986). Examination of nucleic acid spreads by electron microscopy indicated a mean Mr of around 4.3 x 10^3. Analysis of the DNA by sedimentation gave an S value of 12S to 15S, which was consistent with it having a molecular weight below that of *Geminivirus* DNA (16S at pH 7.0). Variations in secondary structure were suggested to explain the broad distribution of DNA in both zonal and isopycnic density gradient centrifugation. Electron microscopy, 2D PAGE and a polynucleotide kinase assay were used to show that molecules are circular (Randles *et al.* 1987).

Cloning of CFDV DNA was attempted at first to determine whether its nucleotide sequence was represented in either the DNA of the host plant or the DNA of a *Geminivirus* infecting the grass *Digitaria sanguinalis* in the plantation understorey (Hanold *et al.* 1988). Southern blot hybridisation studies showed that the sequences in the cloned fragments were uniquely associated with CFD-infected but not with healthy palms and showed no homology with the *D. sanguinalis Geminivirus* DNA. CFDV DNA was, therefore, unique supporting the conclusion that CFDV was a member of a previously undescribed taxonomic group. To obtain sequence data, CFDV DNA was purified, amplified by the polymerase chain reaction, and subcloned. Its sequence was established by analysis of overlapping subgenomic cDNA clones (Rohde *et al.* 1990).

Unusual 20nm icosahedral particles which co-purify with the CFDV DNA were isolated and were considered to be the particles of CFDV (Randles and Hanold 1989). Smaller spherical particles of 12-15 nm in diameter were also detected in lower density fractions from both healthy and diseased plants and a few were commonly seen in preparations of the 20nm particles. The 20nm particles were exclusively associated with gradient fractions of Nycodenz® in the density range 1.27 - 1.30 g/ml. Aliquots of fractions were assayed by PAGE (Randles *et al.* 1987) or dot blot hybridisation (Hanold *et al.* 1988) showing that they contained CFDV DNA. The number of particles seen was approximately proportional to the amount of CFDV DNA detected in the respective fraction. CFDV particles could be extracted from fresh or frozen leaves. Virions are 20.0 ± 0.4 nm (n = 212) in diameter with isometric nucleocapsids. In electron microscopy, some stained particles appeared hexagonal in outline, and shadowed particles also showed an angular outline which would be consistent with an icosahedral structure of 12 vertices and 20 faces (T = 1 symmetry). A surface capsomere arrangement was not obvious. Virions have a density of 1.21-1.25 g/ml in Nycodenz® and 1.30 g/ml in caesium sulphate and are found in leaves, roots and meristem (Randles and Hanold 1989).

The particles contain possibly one molecule of circular single stranded DNA. The detection of two or three bands in non-denaturing PAGE which have identical electrophoretic mobility in the second denaturing dimension of 2D PAGE suggests that under 'native' conditions the DNA occurs as two or three conformers which differ in secondary structure (Randles *et al.* 1987). The CFDV DNA is 1291 nucleotides long containing open reading frames for six putative proteins with an Mr larger than 5 x 10^3. One of these (ORF1, Mr 33.4 x 10^3) codes for a leucine-rich protein with the nucleoside triphosphate-binding motif GXGKS (glycine-x-glycine-lysine-serine) and may possibly participate in virus replication. The putative viral protein encoded by ORF3 (Mr 6.4 x 10^3) is a positively charged arginine-rich protein with homology to the capsid protein of nuclear polyhedrosis virus, and may represent the CFDV coat protein. CFDV DNA can form a stable stem structure of 10 GC base pairs subtending a loop sequence, which in one orientation closely resembles the motif TAATATTAC conserved in a similar structural arrangement within the *Geminivirus* group (Rohde *et al.* 1990). The only other significant sequence similarity occurs in a segment of 52 nucleotides, which has 70% homology to a DNA containing *Badnavirus*. Repeated stem-loop structures upstream from this segment are consistent with the CFDV sequence having promotor activity (Rohde *et al.* 1995; Randles *et al.* 1999). Cloning of the putative viral replication protein was carried out from palm extracts by PCR using primers specific to ORF 1. The cloned inserts varied from the published sequence (Rohde *et al.* 1995) in several nucleotides indicating a possible variation between viral genomes extracted from different host plants. *In vitro* expression of the protein was unsuccessful (Cowan, 1997).

All previously described plant viruses with ssDNA had been placed in the *Geminiviridae.* The CFDV DNA is significantly smaller than *Geminivirus* DNA, but is close to the size of the DNA of *Porcine circovirus* (PCV; Tischer *et al.* 1982) which sediments at 12.7S, has a contour length of 633 µm, 1.76 kilobases, and a Mr of 5.8 x 10^3. The co-purification of the particles with the disease-specific DNA and their exclusive association with diseased palms led to the conclusion that CFD is a virus disease and that the particles described were those of CFDV. The infectivity of the particles remains to be tested, thus they have not yet been proven to be the etiological agent of CFD. They are similar in size and symmetry to half-geminate (Hatta and Francki 1979) particles and to *Subterranean clover stunt virus* (SCSV) particles (Chu and Helms, 1987). They differ from these other plant viruses in that their single-stranded circular DNA is intermediate in size between that of geminiviruses (ca 2,700 nucleotides) and SCSV (900-1,000 nucleotides). They most closely resemble PCV (Tischer *et al.* 1982).

CFDV, therefore, appears to represent a new taxonomic group of plant viruses and has been classified as a tentative member in the genus *Nanovirus* (Randles *et al.* 2000).

23.2.6 Epidemiology

In the field, gradients of disease occur in susceptible host varieties from the borders adjacent to native forest into the plantation (Fig. 23.2.3). *M. taffini* breeds on the roots of a tree species (*Hibiscus tiliaceus*) which predominates in the bordering forest (Julia, 1982) and the presence of this species adjacent to and within plantations provides foci for the plant-hopper. Vector insects commonly settle on coconut fronds at the junction of leaflet with rachis where *in situ* studies have shown that virus is likely to be available, and thus they may spread the virus between host plants in this manner. The frequent detection of virus in coconut roots raises an alternative possibility that the soil-inhabiting larvae of *Myndus* could acquire virus from or infect coconut roots during their pre-adult stage. CFDV has not been detected in the dicotyledon host of *Myndus* larvae, *H. tiliaceus*, but the frequently occurring proximity of coconuts to *Hibiscus* could allow acquisition of virus by larvae from roots of infected coconut palm growing adjacent to roots of *Hibiscus*. The association of CFDV DNA with the abdominal areas and isolated gut of *Myndus* is consistent with accumulation of virus in the gut (Randles *et al.* 1992; Fig. 23.2.2). Further research is needed to determine whether the virus may be acquired by phloem feeding, whether it accumulates or replicates in the insect and/or its eggs and whether it has a semi-persistent, circulative, or propagative association with its vector.

CFDV can be detected in leaves of seedlings within 6 months of controlled inoculation with *Myndus*, and symptoms appear 1-4 months later. Palms infected by natural exposure in the field show CFDV in most tissues sampled within 4-7 months of symptoms first appearing. Palms assayed within about 8 months of inoculation just after the first appearance of symptoms generally had higher concentrations of CFDV DNA than palms infected by natural exposure in the field. The location of the virus in vascular bundles and phloem in particular may account for the low recovery of virus from tissue (Randles *et al.* 1992). CFDV DNA is present in young unaffected fronds, and both within and adjacent to the necrotic region at the early stages of lesion development. It was not detected within the necrotic zone of older fronds of the same tree, but it was detected in the non-necrotic zones either side of older lesions. These observations would be consistent with CFDV being associated with and causing the necrosis, then becoming degraded after the development of necrosis. No reason is known for the localisation of necrosis on petioles in a particular part of the crown. The yellowing of leaflets may be associated with cytopathic effects in phloem and reduced carbohydrate translocation (Randles *et al.* 1992).

Neither *H. tiliaceus*, the breeding host of *M. taffini*, or geminivirus-infected *D. sanguinalis* contains detectable sequences of CFDV DNA, and neither therefore could be implicated as a virus source. Other species are yet to be assayed as potential natural sources of CFDV. The lack of homology between the *D. sanguinalis Geminivirus* DNA and CFDV DNA is in agreement with the observation that the two disease agents have different vectors (*Nesoclutha declivata* and *M. taffini* respectively (Julia and Dollet 1989). Disease-free trees, including tolerant cultivars like the symptomless VTT, and palms in disease remission generally contain CFDV DNA in leaflets and rachis, and a large latent reservoir of virus can, therefore, exist in apparently disease-free coconut growing areas (Hanold *et al.* 1988; Randles *et al.* 1992).

Detection of CFDV DNA independently of symptomatology allows a more reliable evaluation of epidemiology to be undertaken.

23.2.7 Control strategies

Control of CFDV is mainly based on planting tolerant lines of the host. Three types of planting material are available for farmers in Vanuatu. The 'Improved VTT', obtained from several cycles of mass selection, has an average annual yield of 2.2 tonnes of copra per hectare and good copra content of the fruit (195g per nut). It is completely tolerant to the disease and can be easily multiplied by farmers. The VRD x VTT hybrid line, tolerant to CFDV, is produced in seed gardens at Saraoutou station and has been released to farmers between 1986 and 1996. In spite of its high-yielding potential (3.4 tonnes copra per hectare were recorded in station trials), the dissemination of VRD x VTT was discontinued due to its low rate of germination in the nursery, a low content of copra (154g/nut) and the frequent dropping of still immature bunches. The VTT x RIT hybrid is the most promising one with an average yield of 2.6 tonnes and high copra content of 237g/nut. The first lines of this hybrid showed a very slight susceptibility to CFDV. The tolerance has been recently improved by using, as female parent, several self-pollinated progenies of RIT which express no symptoms of the disease (J.P. Labouisse, unpublished). A number of VTT hybrids with introduced dwarf varieties achieve better yields, but are slightly sensitive to CFDV. However, they may be used under controlled conditions including removal of burao plants within a radius of several hundred metres from the plantings and a protective outer row of VTT to reduce invasion of vector insects. An exceptionally high rate of remission of 92% has been recorded for the hybrid Cameroon Red Dwarf (CRD) x RIT, and testing of a number of other hybrids is still in progress. Pure lines derived from individual trees of MRD and Malayan Yellow Dwarf (MYD) with markedly reduced sensitivity have also been established and are being assessed (Morin, 1994).

A most interesting observation is that all MRD plants exposed to the vector in insect cages and subsequently showing symptoms in the nursery, recovered after planting in the field and were asymptomatic and bearing normally 8 years after cage inoculation. At the same time, all uninoculated controls subsequently showed the disease after transplanting to the field and many of them had died. Thus the percentage of palms going into remission after cage inoculation in the nursery was considerably higher than the remission rate occurring in field infected palms of the same variety (Morin, 1994). The reason for this phenomenon is unclear, but it may be useful for assessing long-term effects of inoculation on yield. It may also provide a base for investigating the practical feasibility of a routine inoculation procedure for higher yielding sensitive varieties in the nursery.

References

Anonymous 1990. *IRHO-CIRAD Saraoutou Research Station 1989* Activities Rept. Republic of Vanuatu, Minis. Agric., Fish. Forestry.

Calvez, C., Julia, J.F. and de Nuce de Lamothe, M. 1985. Improvement of coconut in Vanuatu and its importance for the Pacific region. *Oléagineux* 40: 477-490.

Calvez, C., Renard, J.L. and Marty, G. 1980. Tolerance of the hybrid coconut Local x Rennell to New Hebrides disease. *Oléagineux* 35: 443-451.

Chu, P.W.G. and Helms, K. 1987. Purification and characterisation of a novel single-stranded DNA virus associated with subterranean clover stunt disease in Australia. *Proc. 7th Intern. Congress Virology*. p. 233.

Cowan, M.R. 1997. Isolation and cloning of the putative replication protein gene of Coconut foliar decay virus. BSc(Hons) *Thesis. Flinders Univ., South Australia.*

Dollet, M. and de Taffin, G. 1978. Progress report on the virological study of the New Hebrides coconut disease. *IRHO-CIRAD Saraoutou Res. Stat.* 1978 Activities Rept.. Republic of Vanuatu, Minis. Agric., Fish. Forestry.

Hanold, D. and Randles, J.W. 1991. Detection of coconut cadang-cadang viroid-like sequences in oil and coconut palm and other monocotyledons in the south-west Pacific. *Ann. Appl. Biol.* 118: 139-151.

Hanold, D., Langridge, P. and Randles, J.W. 1988. The use of cloned sequences for the identification of coconut foliar decay disease-associated DNA. *J. Gen. Virol.* 69: 1323-1329.

Harrison, B.D. 1985. Advances in geminivirus research. *Annu. Rev. Phytopath.* 23: 55-82.

Hatta, T. and Francki, R. 1979. The fine structure of chloris striate mosaic virus. *Virology* 92: 428-435.

Julia, J.F. 1982. Myndus taffini (Homoptera Cixiidae), vector of foliar decay of coconuts in Vanuatu. *Oléagineux* 37: 409-414.

Julia, J.F., Dollet, M., Randles, J.W. and Calvez, C. 1985. Foliar decay of coconut by Myndus taffini (FDMT): New results. *Oléagineux* 40: 19-27.

Julia, J.F. and Dollet, M. 1989. Nesoclutha declivata, Homoptera Cicadellinidae, vector of the Digitaria Streak Virus (Geminivirus) in Vanuatu. *J. Phytopath.* 127: 42-48.

Morin, J.P. 1993. Elevage et biologie de Myndus taffini Bonfils (Homoptera : Cixiidae) vecteur du deperissement foliaire du cocotier au Vanuatu. *Proc. A.N.P.P. Troisieme Conf. Intern. Ravageurs en Agric.*. Montpellier 1993. pp. 1375-1383.

Morin, J.P. 1994. Le Dépérissement Foliaire du Cocotier. IRHO-CIRAD Saraoutou Research Station, Final Activities Rept. Republic of Vanuatu, Minis. Agric., Fish. Forestry. Cirad doc. 230/94.

Persley, G. 1992. Replanting the Tree of Life. Toward an International Agenda for Coconut Palm Research. Technical Advisory Committee of the Consultative Group on International Agricultural Research, TAC Secretariat, FAO, Rome, Italy.

Randles, J.W. 1998. Coconut foliar decay disease and identification of its viral agent. In: Report on ACIAR-funded research on viroids and viruses of coconut palm and other tropical monocotyledons 1985-1993. (eds) D. Hanold and J.W. Randles. ACIAR Monograph No. 45. Austral. Centre Intern. Agric. Res., Canberra, ACT. http://www.aciar.gov.au/publications/monographs/45//index.htm

Randles, J.W. and Hanold, D. 1989. Coconut foliar decay virus particles are 20-nm icosahedra. *Intervirology* 30: 177-180.

Randles, J.W., Chu, P.W.G., Dale, J.L., Harding, R., Hu, J., Katul, L., Kojima, M., Makkouk, K.M., Sano, Y., Thomas, J.E. and Vetten, H.J. 2000. Genus *Nanovirus*. In *Virus Taxonomy- Classification and Nomenclature of Viruses*. (eds) M.H.V. van Regenmortel, C.M. Fauquet, D.H.L. Bishop, E.B. Carstens, M.K. Estes, S.M. Lemon, J. Maniloff, M.A. Mayo, D.J. Mc Geoch, C.R. Pringle and R.B. Wickner. Academic Press, San Diego. pp 303-309.

Randles, J.W., Hanold, D. and Julia, J.F. 1987. Small circular single-stranded DNA associated with foliar decay disease of coconut palm in Vanuatu. *J. Gen. Virol.* 68: 273-280.

Randles, J.W., Julia, J.F., Calvez, C. and Dollet, M. 1986. Association of single-stranded DNA with the foliar decay disease of coconut palm in Vanuatu. *Phytopathol.* 76: 889-894.

Randles, J.W., Miller, D.C., Morin, J.P., Rohde, W. and Hanold, D. 1992. Localisation of coconut foliar decay virus in coconut palm. *Ann. Appl. Biol.* 121: 601-617.

Randles, J.W., Wefels, E., Hanold, D., Miller, D.C., Morin, J.P. and Rohde, W. 1999. Detection and diagnosis of coconut foliar decay disease. In: *Current Advances in Coconut Biotechnology* (eds) C. Oropeza, J. Verdeil, G.R. Ashburner, R. Cardeña and J. Santamaria. Vol. 35, *Current Plant Science and Biotechnology in Agriculture*. pp. 247-258. Kluwer Academic Publishers, Dordrecht/Boston/London.

Rohde, W., Becker, D. and Randles, J.W. 1995. The coconut foliar decay virus promoter directs phloem-specific reporter gene expression in transgenic tobacco. *Plant Mol. Biol.* 27: 623-628.

Rohde, W., Randles, J.W., Langridge, P. and Hanold, D. 1990. Nucleotide sequence of a circular single-stranded DNA associated with coconut foliar decay virus. *Virology* 176: 648-651.

Tischer, J., Gelderblom, H., Vetterman, W. and Koch, M.A. 1982. A very small porcine virus with circular single-stranded DNA. *Nature* 295: 64-66.

Wilson, M.R. 1988. The genus *Myndus* (Hemiptera: Cixiidae) in the Solomon Islands and Vanuatu and its relation to foliar decay of coconut palms in Vanuatu. *Bull. ent. Res.* 78: 519-526.

FIGURE LEGENDS

Fig. 23.2.1. Symptoms of CFD on naturally infected MRD hybrids: a) late stage with collapse of affected fronds; b) necrosis on petioles.

Fig. 23.2.2. Distribution of CFDV DNA in the gut of three adult females of *Myndus taffini*. Insects collected from diseased trees at Saraoutou, Vanuatu, were dissected, their gut removed and stained briefly with aqueous toluidine blue for ease of manipulation. The gut was extended and applied to a nylon membrane (upper panel). After crosslinking, the membrane was hybridized with ^{32}P-labelled CFDV DNA probe and autoradiographed (lower panel). In this trial, 4/17 female and 0/10 male adults showed the distribution of radioactive signal along the isolated gut.

Fig. 23.2.3. Arial view of a variety trial at Saraoutsu Research Station, showing both the gradients of disease incidence from the native forest into the planting, and the variation in susceptibility of different varieties. Note particularly the two rows of tolerant VTT adjacent to the forest – marked by an arrow. The other dense plantings next to the forest are replanted areas.

23.2.2

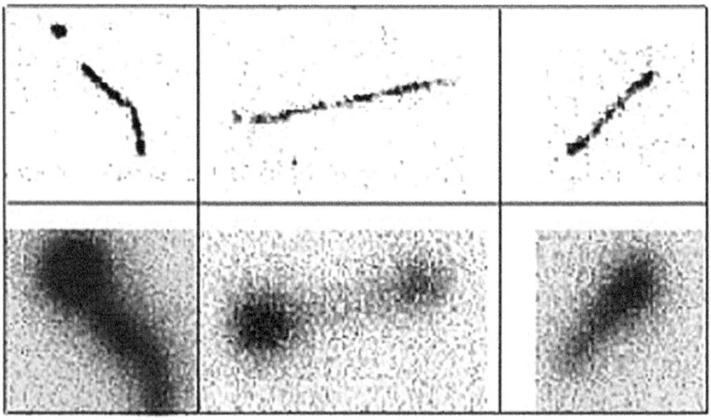

23.3 Lethal yellowing of coconut and lethal declines of palms

J. H. Tsai and N. A. Harrison
Fort Lauderdale Research and Education Center, University of Florida, IFAS, Fort Lauderdale, Florida 33314-7719

23.3.1 Introduction

Of the 2800 species within the Arecaceae, coconut palm (*Cocos nucifera* oil palm (*Elaeis guineensis*) and date palm (*Phoenix dactylifera*) are the most important plantation crops in the world. Coconut palm represents 'the tree of life' for millions of people in the tropics. It is estimated that about 500 million coconut palms are grown annually on 6.5 million hectares of land throughout the tropics. They provide sources of shelter, drink, fiber, oil, food and clothing (Ohler, 2002). Many islands in the tropics are inhabitable only because of the availability of coconut palms (Oliver, 1961). In addition, coconut palms constitute one of the vital components of the tropical forest ecosystems of Africa and South America (Moore, 1973).

In Florida, coconut palms are grown for their aesthetic qualities as they impart all of the magic of the tropical atmosphere. Since 1965, most of Florida's 1 million 'Atlantic Tall' coconut palms have been killed by a pandemic disease called lethal yellowing (LY). LY is a rapidly spreading and highly devastating disease affecting coconut and 35 additional palm species in Florida, Jamaica, and several other Caribbean countries as well as neighboring regions of Central America. LY was first reported in Jamaica (Fawcett, 1891). It had remained on the western part of the island until the 1960's. Over half of 4.3 million 'Atlantic Tall' coconut palms had succumbed to LY in ensuing years. To date, the once prevalent 'Jamaican Tall' coconut palms have virtually disappeared from both locations. Similar lethal declining diseases of coconut palms are also endemic in East and West Africa and Southeast Asia (Warokka, 1999). Collectively, these diseases pose a great threat to global coconut production.

23.3.2 Symptomatology

The symptoms of LY on 'Atlantic Tall' coconut palms are unique and diagnostic for the disease. On fruit-bearing coconut palms the symptoms include: (1) a premature nut fall, accompanied by the appearance of a brown or black water-soaked mark on the calyx of the fallen nuts; (2) a necrosis of inflorescences in newly open and unopened spathes; (3) the appearance of a yellow flag-leaf in the middle of the crown, followed by a progressive yellowing of successively younger leaves; (4) root rot; (5) the decay of the spear leaf and heart; and (6) the toppling of the tree's crown, leaving a trunk that resembles a telephone pole. The entire syndrome normally takes 3 to 6 months from the onset of symptoms until death of the tree. The symptoms on young coconut palms include leaf yellowing, a decaying spear leaf and heart, and root rot. Other LY-affected palms show similar symptoms, for example, *Caryota mitis*, *Corypha elata, Pritchardia* spp., *Syagrus schizophylla, Trachycarpus fortunei, Dictyosperma album,* and *Hyophorbe verschaffeltii*; whereas other palms, such as *Adonidia merrillii, Chrysalidocarpus cabadae, Borassus flabellifer,* and *Phoenix* spp., do not exhibit leaf yellowing. Instead the leaves turn brown, with water-soaked marks appearing along the pinnae. As foliar discoloration progresses, browning extends to the rest of the fronds. On Adonidia palms, discolored leaves often break at the joint of the midrib and the leaf base. The spear leaves remain on top of the dead trees. On *Pritchardia* palms and *P. dactylifera*, death of the spear leaf always occurs early on in the foliar discoloration phase (McCoy *et al.* 1983).

No single symptom is considered diagnostic of LY disease. Premature nutfall and foliar discoloration may be induced by such abiotic factors as nutritional deficiencies. Other biotic factors such as fungal, nematode, uniflagellate protozoan and palm weevil can also induce symptoms, which in their advanced stages closely resemble those of LY. Nonetheless, the appearance and characteristic syndrome development adequately describe and identify LY disease.

23.3.3 Epidemiology

The typical spread of LY first involves a local center of infection on one or two palms. Later, other infected palms appear in a random pattern surrounding initial disease foci from which further local spread occurs. The other type of spread is a long-distance spread (jump spread), which is followed by the pattern of local spread. The distance over which jump spread occurs can be as far as 70 km (Carter, 1966). McCoy (1976) identified and recorded the locations of nearly 20,000 LY-infected palms and their locations in Dade County and, by using proportional analysis, concluded that the apparent infection rate was 0.21 logits per month.

LY disease involves a prolonged latent (incubation) period. The time from initial infection to first appearance of symptoms in young coconut palms has been estimated at 114-262 days (Dabek, 1975). Recently, an LY specific polymerase chain reaction (PCR) assay has been used to determine at what point during the latent period the LY phytoplasma reached detectable levels. Monitoring 26 symptomless, nonbearing Jamaican tall coconut palms monthly for evidence of LY infection, achieved this. These palms were also visually monitored on a daily basis for the onset of foliar symptoms. By this approach, it was determined that the pathogen could be detected within this highly susceptible ecotype between 47-57 days prior to the symptom appearance (Harrison *et al.* 1997).

Embryos excised from seeds of ten coconut palms with primary LY symptoms were tested for carrying phytoplasma by LY-specific PCR. At least one positive embryo was detected among both aborted and attached seeds collected from each of the five test palms. Overall, 21% (12/57) of the total embryos tested were found to contain the LY phytoplasma (Harrison *et al.* 1997).

23.3.4 Host range

LY has a wide range of hosts as at least 36 palm species originating from North Africa, South America, the Western Pacific and Southeast Asia are affected (Table 23.3.1) (McCoy *et al.* 1983). Also, the arborescent monocot *Pandanus utilis* has also been identified as a nonpalm host of LY (Thomas and Donselman, 1979). The susceptibility of palm species to LY varies greatly, ranging from highly susceptible to slight susceptible. Palms considered as highly susceptible to LY include *Arenga engleri*, *Cocos nucifera*, *Corypha elata*, *Phoenix dactylifera*, *Pritchardia affinis* Becc., *P. pacifica*, *P. thurstonii* and other *Pritchardia* spp. The moderately susceptible species include *Syagrus schizophylla*, *Borassus flabellifer*, *Caryota mitis*, *Dictyosperma album* , *Hyophorbe verschaffeltii*, *Lantania* spp., *Phoenix canariensis*, *Trachycarpus fortunei*, and *Adonidia merrillii* (=*Veitchia merrillii*). Three species are known to be slightly susceptible to LY: *Chrysalidocarpus cabadae*, *Phoenix reclinata* and *Veitchia* spp.

The following 16 palm species are known hosts of LY also but their numbers are too few to determine their relative susceptibility: *Aiphanes lindeniana.*, *Allagoptera arenaria*, *Gaussia attenuata*, *Howea belmoreana*,

Livistona chinensis, *Nannorrhops ritchiana, Dypsis decaryi*, *Phoenix sylvestris*, *P. remota*, *Ravenea hildebrandtii*, *Caryota rumphiana*, *Chelyocarpus chuco*, *Crysophila warseceniczii*, *Howea forsteriana*, *Livistona rotundifolia*, *P. rupicola* (McCoy *et al.* 1983, Harrison *et al.* 1999).

23.3.5 Etiology

The first observations of phytoplasmas (Sears and Kirkpatrick, 1994) previously known as mycoplasmalike organisms or MLOs associated with LY infected coconut palms was obtained by transmission electron microscopy (TEM) of ultrathin sections (Plavsic-Banjac *et al.* 1972). In sieve tube elements, cells of these obligate polymorphic mollicutes are generally 142-295 nm in diameter and may vary from 1-16 μm in length.

From extensive TEM surveys of various LY-affected species, it was found that overall phytoplasma concentrations were quite low and unpredictably localized in tissues of most symptomatic palms. The pathogen could be most readily found in unemerged phloem-rich leaf bases surrounding the apical meristem and, to a lesser extent, in inflorescences and tertiary roots. However, in most mature tissues, phytoplasma concentrations were found generally below detectable levels (Thomas, 1979, Waters and Hunt, 1980).

In southeast Florida both syndromes of coconut lethal yellowing and lethal decline diseases of numerous other palm species were first recognized in the 1970s (Thomas, 1979). As with coconut lethal yellowing, phytoplasmas were implicated by means of TEM as the probable etiological agents of these lethal decline diseases. This evidence coupled with the appearance on some palm species of symptoms superficially similar to those of coconut lethal yellowing indicated that these diseases were probably all caused by the same pathogen (McCoy *et al.* 1983). The subsequent transmission of lethal yellowing agent by field-collected *Myndus crudus* adults (Howard *et al.* 1983) and symptom remission on palms following antibiotic injections (McCoy,1972) strongly supported a phytoplasma etiology. Since LY phytoplasma, like all other phytoplasmas, remains unculturable *in vitro* in cell free media this precludes culture techniques for use in etiological determination.

Further evidence of a phytoplasma association with LY of coconut palm and other palm species was obtained by DNA hybridization assays using cloned genomic DNA fragments of the LY agent as probes (Harrison *et al.* 1992, 1997). The development of primers from an uncharacterized genomic DNA sequence of the LY phytoplasma enabled specific and reliable detection of phytoplasmas in other palms affected by lethal decline diseases in the western Caribbean region and in Africa. Comparative studies further demonstrated that the LY and lethal declining phytoplasmas were very similar but not genetically identical (Harrison *et al.* 1994a, b).

23.3.6 Development and utility of molecular techniques

Improved methods for extraction of phytoplasmas from host tissues (Kirkpatrick *et al.* 1995) together with the advent of recombinant DNA technology has greatly facilitated the cloning of LY phytoplasma DNA. Eleven fragments of chromosomal DNA of LY phytoplasma strains infecting *Adonidia merrillii* and *Trachycarpus fortunei* (Harrison *et al.* 1992, 1997, 1999) have been cloned and identified. Used individually as ^{32}P-labeled probes in dot hybridizations at high stringency, four of five probes derived from *A. merrillii* consistently hybridized to DNA extracts derived from LY-affected palms only. At moderate stringency, all probes hybridized with DNA of other phytoplasmas that occur in Florida. In addition to Manila palms, probes detected the presence of LY phytoplasma DNA in DNA samples extracted from heart tissues of LY-diseased true date (*Phoenix dactylifera*), cliff date (*P. rupicola*), Chinese fan (*Livistona chinensis*), and five coconut (*Cocos nucifera*) palm cultivars. Probes also hybridized to DNA from symptomatic *Caryota rumphiana* and *L. rotundifolia*, two palm species previously not known to be affected by LY disease (McCoy *et al.* 1983, Harrison *et al.* 1992). Cloned LY phytoplasma DNA used as molecular probes in DNA hybridization assays have proved to be sensitive and specific and also provided a quantitative means to measure the phytoplasma populations in infected palm tissues (Harrison *et al.* 1992, 1994b). Probes have been used successfully to identify the LY agent in the non-palm host *Pandanus utilis* and to assess genetic variation among the pathogen infecting 11 ornamental palm species in Florida (Harrison *et al.* 1999). Initial use of diagnostic DNA probes for detection of LY phytoplasma within the planthopper vector, *M. crudus* has been less ideal; over 5000 *M. crudus* adults tested individually, only two showed weak positives. This outcome was due to a low LY incidence among the coconut palms from which the test insects were collected rather than the lack of sensitivity of this technique (Harrison *et al.* 1995). Subsequently, when planthoppers were collected and assayed from symptomatic palms at the mid-foliar yellowing stage of disease development the numbers of phytoplasma positive insects was substantially higher. Dot hybridization signals of varying intensities were obtained for 39 of 1321 (29%) insects (Harrison et *al.* 1997). DNA amplification by PCR has been used successfully to detect phytoplasma in lethal yellowing diseased palms in Florida. For PCR, a pair of oligonucleotide primers (LYF1- 5'-cat att tta ttt cct ttg caa tct g-3' nd LYR1-, 5'-tcg ttt tga taa tct ttc att tga c-3')was synthesized according to partial sequences was synthesized according to partial sequences of a cloned anonymous 1.3 kb fragment of LY phytoplasma-specific genomic DNA isolated from diseased *Trachycarpus fortunei*. A DNA product of about 1 kb was specifically amplified by PCR in reaction mixtures containing template DNA derived from heart, inflorescence or leaf tissues of LY-affected palms. PCR performed for 35 cycles was sufficient to consistently amplify the same DNA product from hearts of 11 other species comprising 30 symptomatic palms. Similar reliable detection of the LY agent in palm inflorescence spikelets was also achieved after 35 cycles of PCR. When template DNA was derived from tissues of the most immature emerging leaf, a 40-cycle reaction was necessary for consistent foliar detection of the LY pathogen in all coconut palms including palms with earliest visible symptoms of disease (Harrison *et al.* 1994b). The chosen LY primer pair (LYF1 and LYR1) was specific and discerned the difference between Florida LY phytoplasma and phytoplasmas associated with African coconut lethal decline diseases (Harrison *et al.* 1994a, Tymon *et al.* 1997, 1998). Furthermore, LY specific PCR has enabled the detection of the LY agent in embryos excised from aborted nests (5/34) collected from coconut palms at the onset of disease symptoms (Harrison *et al.* 1995). PCR has also been used to demonstrate the occurrence of the LY phytoplasma in *M. crudus* adults. A total of 1321 adults collected from seven LY symptomatic palms were tested. Of these, 105 (7.9%) individuals yielded PCR positives although two palms only contributed 84% of positives (Harrison *et al.* 1997). PCR has also been used to study the acquisition of the LY MLO by *M. crudus* and non-LY vector, *Peregrinus maidis* (Ashmead). Of 60 *M. crudus* adults tested after a 3-day acquisition access feeding on LY symptomatic coconut fronds, 13 individuals had acquired detectable numbers of the LY phytoplasma. Of 60 *P. maidis* assayed in a similar manner, 31 (51.7%) tested positive for phytoplasma acquisition. The sensitivity of PCR in detection LY phytoplasma in *M. crudus* with the LY specific PCR was demonstrated. Of 94 adults collected from a young coconut palm showing mid-stage foliar symptoms, 36 were positive for LY phytoplasma using as little as 1/50th of the total DNA extracted from each insect as template during a 40-cycle reaction (Harrison *et al.* 1995).

23.3.7 Prevention and control of LY

The currently recommended program for LY control is through an integrated management program including eradication, quarantine, chemotherapy and resistant palms. The removal of infected palms is considered an important means of disease control especially in the initial stage of LY outbreaks. Quarantine measures imposed by several states in the

USA are important to prevent the introduction of the LY agent or its vector from the infected area into disease-free areas. Although antibiotic treatment with oxytetracycline cannot cure LY, it can prolong the longevity of LY infected palms in the landscape while resistant palms are being replanted. Although this measure is not economically feasible for a large coconut plantation or for the poor farmers, nonetheless to date several municipalities like Palm Beach and Naples, Florida have been able to preserve the majestic tall coconut palms in their landscape since the inception of the program. At present, host resistance offers the best solution to disease control but this approach has been slowed because of only a limited choice of proven resistant cultivars. The resistant cultivar 'Malayan Dwarf' and hybrid coconut 'Maypan' (Malayan Dwarf x Panama Tall) have been widely planted in Jamaica and Florida to replace tall-type cultivars lost to LY with desirable results.

23.3.8 Conclusions

LY is a limiting factor in coconut production in the Western Hemisphere. New LY incidences have been continually reported from both old and new localities. For many years, LY research has been hampered by several factors including the ability of this disease to spread and kill palms rapidly; inability to culture the LY phytoplasma *in vitro*. In addition, the generally low titer of phytoplasma in palm tissues, absence of means to maintain the disease in palms, lack of alternate host(s) for LY agent and unequivocal proof of LY vector(s) . Most critical and essential factor is lack of a diagnostic tool to efficiently and correctly detect and assay the LY phytoplasma in plants and insects. Though the development of diagnostic DNA probes have greatly improved detection and identification of the phytoplasma associated with LY disease. But the requirement for use of a labeled radioisotope and unusually low titre of LY phytoplasma in most palm tissues has severely restricted in their use by many laboratories. The convenience and extreme sensitivity of LY specific PCR have made it possible to routinely detect LY infection of various hosts; to study phylogenetic relationship of LY phytoplasma and other similar disease; and to screen plants for interstate and international movements. More importantly it would provide an accurate and practical tool to search for alternate hosts and other homopteran vector(s) of LY phytoplasma which otherwise have previously been relied on the conventional technology.

References

Ahrens, U., Seemüller, E., Kirkpatrick, B.C., Gao, J-L. and Harrison, N. 1992 Detection and differentiation of plant pathogenic MLOs using PCR and MLO-specific 16S rRNA primers. *IOM Lett*. 2: 25.

Carter, W. 1966. Susceptibility of coconut palm to lethal yellowing disease. *Nature* (London) 212 (5059): 320.

Dabek, A. J. 1975. The incubation period, rate of transmission and effect on growth of coconut lethal yellowing disease in Jamaica. *Phytopathol. Z.* 84: 1-9.

Deng, S. and Hiruki, C. 1991. Amplification of 16S rRNA genes from culturable and non-culturable mollicutes. *J. Microbiol. Meth.* 14: 53-61.

Eden-Green, S. J. 1997. History, world distribution and present status of lethal yellowing like *diseases of palms. Pp 9-25. In:* "Proc. Intern. Workshop Lethal *Yellowing-like Diseases of Coconut*". S.J.Eden-Green, and F. Ofori (eds.) Nat. Resour. Inst., UK.

Howard, F. W. and Thomas, D.L.. 1980. Transmission of palm lethal decline to *Veitchia merrillii* by planthopper, *Myndus crudus. J. Econ. Ent.* 73: 715-717.

Fawcett, W. 1981. Rep. Coconut Disease at Montego Bay. *Bull. Bot. Dept.* Jamaica 23: 2.

Harrison, N. A., Bourne, C.M., Cox, R.L.,Tsai, J.H. and Richardson, P.H.. 1992. DNA probes for detection of mycoplasmalike organisms associated with lethal yellowing of palms in Florida. *Phytopathology* 82: 216-224.

Harrison, N. A., I Cordova, Richardson, P. and DiBonito, R. 1999. Detection and diagnosis of lethal yellowing. *In: "Current Advances in Coconut Biotechnology"* Oropeza, C., J. L. Verdeil, G. R. Ashburner, R. Cardeña, and J. M. Santamaria (eds.) pp. 183-196. Kluwer Academic Publishers Dordrecht, Netherlands.

Harrison, N.A. and Oropeza C. 1997. Recent studies on detection of lethal yellowing disease phytoplasmas in the Americas. *In*: *Proc. Intern. Workshop Lethal Yellowing-Like Diseases of Coconut.* S J Eden-Green, F Ofori (eds.) pp. 221-234, Nat. Resour. Inst. UK.

Harrison, N.A., Griffiths, H. M., Carpio, M, L. and Richardson, P.A. 2001. Detection and characterization of an elm yellows (16SrV) group phytoplasma infecting Virginia creeper plants in southern Florida. *Plant Disea.* 85: 1055-1062.

Harrison, N. A, Narváez M, Almeyda H, Cordova I, Carpio M. L and Oropeza C. 2002a. First report of group 16SrIV phytoplasmas infecting coconut palms with leaf yellowing symptoms on the Pacific coast of Mexico. *BSPP New Dis. Rept*, Vol. 5. htpp://www.bspp.org.uk/ndr/

Harrison, N. A., Richardson, P.A., Jones, P., Tymon, A.M., Eden-Green, S. and Mpunami, A.A. 1994a. Comparative investigation of mycoplasmalike organisms associated with Caribbean and African coconut lethal decline diseases by DNA hybridizations and PCR assays. *Plant Dis*. 78: 507-511.

Harrison, N. A., P. A. Richardson, J. B. Kramer and Tsai, J.H. 1994b. Detection of the mycoplasma-like organism associated with lethal yellowing disease of palms in Florida by polymerase chain reaction. *Plant Path.* 43: 998-1008.

Harrison, N. A., P. A. Richardson and Tsai, J.H. 1995. Detection and diagnosis of lethal yellowing: Conventional methods and molecular techniques. *Pp. 79-91. In: "Lethal Yellowing: Research and Practical Aspects"* Oropeza, C., F. W. Howard and G. R. Ashburner (eds.). Kluwer Acad. Publ.

Harrison N. A., Womack, M. and Carpio, M.L. 2002b. Detection and characterization of a lethal yellowing (16SrIV) group phytoplasma in Canary Island date palms affected by lethal decline disease in Texas. *Plant Dis.* 86: 676-681.

Innis, M. A., and Gelfand, D. H. 1990. Optimization of PCRs. *In*: *PCR Protocols: A Guide to Methods and Applications.* M. A. Innis, D. H. Gelfand, J.J. Sninsky, and T. J. White, (eds.) pp. 3-12, Academic Press, Inc. San Diego, USA.

Kirkpatrick, B. C., N. A. Harrison, I.-M. Lee, H. Neimark, and Sears, B.B.. 1995. Isolation of mycoplasma-like organism DNA from plant and insect hosts. *In*: *"Molecular and Diagnostic Procedures in Mycoplasmology. Volume 1, Molecular Characterization"*

S. Razin, and J. G. Tully (eds.) pp. 105-115 Academic Press, San Diego, USA.

McCoy, R. E. 1972. Remission of lethal yellowing in coconut palm treated with tetracycline antibiotics. *Plant Dis. Reptr.* 56: 1019-1021.

McCoy, R. E. 1976. Comparative epidemiology of the lethal yellowing, Kaincope, and cadang-cadang diseases of coconut palm. *Plant Dis. Reptr.* 60: 498-502.

McCoy, R. E., F. W. Howard, J. H. Tsai, H. M. Donselman, D. L. Thomas, H. G. Basham, R. A. Atilano, F. M. Eskafi, L. Britt, and Collins, M.E. 1983. Lethal yellowing of palms. *Univ. of Florida, IFAS Agric. Exp. Sta. Tech. Bull.,* No. 834, 100 pp.

Moore, H. E. 1973. Palms in the tropical forest ecosystem of Africa and South America. *In*: "*Tropical Forest Ecosystems in Africa and South America: A Comparative Review*", B. J. Meggers, E. S. Ayensu, and W. D. Duckworth (eds.). pp. 63-68. Smithsonian Institution Press, Washington, D.C. 350 pp.

Ohler, J. G. 2002. Modern Coconut Management: Palm Cultivation and Products. *Ecoport eArticle*, http://ecoport.org/EP.exe$EAFull?ID=127

Oliver, P. L. 1961. *The Pacific Islands*. Doubleday and Co., Garden City, New York. 456 pp.

Plavsic-Banjac, B, P. Hunt, and Maramorosch, K. 1972. Mycoplasma-like bodies associated with lethal yellowing disease of coconut palms. *Phytopathology* 62: 298-299.

Sears, B. B., and Kirkpatric, B. C. 1994. Unveiling the evolutionary relationships of plant-pathogenic mycoplasma-like organisms. *Am. Soc. Microbiol. News* 60: 307-312.

Smart, C.D., B. Schneider, C.L. Blomquist, L.J. Guerra, N.A. Harrison, U. Ahrens, K-H. Lorenz, E. Seemüller and Kirkpatrick, B. C. 1996. Phytoplasma-specific PCR primers based upon sequences of the 16-23S rRNA spacer region. *Appl. Environ. Microbiol.* 62: 2988-2993.

Thomas, D. L. 1979. Mycoplasmalike bodies associated with lethal declines of palms in Florida. *Phytopathology* 69: 928-934.

Thomas, D. L. and Donselman, H.M. 1979. Mycoplasma-like bodies and phloem degeneration associated with declining Pandanus in Florida. *Plant Dis. Reptr.* 63: 911-916.

Tymon, A. M., P. Jones, and Harrison, N.A. 1997. Detection and differentiation of African coconut phytoplasmas: RFLP Analysis of PCR-amplified 16Sr DNA and DNA hybridization. *Ann. appl. Biol.* 131: 91-102.

Tymon, A. M., P. Jones, and Harrison, N.A. 1998. Phylogenetic relationships of coconut phytoplasmas and the development of specific oligonucleotide PCR Primers. *Ann. appl. Biol.* 132: 437-452.

Warokka, J. S. 1999. The impact and etiology of coconut wilt diseases in Indonesia. Pp. 39-42. *In*: *"Proc. Intern. Workshop Impact Lethal Dis. Coconut caused by Phytoplasmas in South East Asia.* APCC Publishers, Jakarta, Indonesia.

Waters, H. and Hunt, P. 1980. The *in vivo* three-dimensional form of a plant mycoplasma-like organism by the analysis of serial ultrathin sections. *J. Gen. Microbiol.* 116: 111-131.

Table 23.3.1 Listing of palm species susceptible to lethal yellowing disease in Florida.

Scientific name	Common name	Region of origin
Adonidia merrillii	Christmas palm	Western Pacific
Aiphanes lindeniana	Ruffle palm	Caribbean
Allagoptera arenaria	Seashore palm	South America
Arenga engleri	Dwarf sugar palm	Southeast Asia
Borassus flabellifer	Palmyra palm	India
Caryota mitis	Clustering fishtail palm	Southeast Asia
Caryota rumphiana	Giant fishtail palm	Southeast Asia
Chelyocarpus chuco	-------	South America
Cocos nucifera	Coconut palm	Western Pacific
Corypha elata	Buri palm	India
Crysophila warsecewiczii	Rootspine palm	Central America
Cyphophoenix nucele	-------	Western Pacific
Dictyosperma album	Princess palm	Madagascar
Dypsis cabadae	Cabada palm	Madagascar
Dypsis decaryi	Triangle palm	Madagascar
Gaussia attenuata	Puerto Rican Gaussia palm	Caribbean
Howea belmoreana	Belmore sentry palm	Western Pacific
Howea forsteriana	Kentia palm	Western Pacific
Hyophorbe verschafeltii	Spindle Palm	Madagascar
Lantania lontaroides	Latan palm	Madagascar
Livistona chinensis	Chinese fan palm	China
Livistona rotundifolia	Footstool palm	Southeast Asia
Nannorrhops ritchiana	Mazari palm	Asia Minor
Phoenix canariensis	Canary Island date palm	Canary Islands
Phoenix dactylifera	Edible date palm	North Africa
Phoenix reclinata	Senegal date palm	Africa
Phoenix rupicola	Cliff date palm	India
Phoenix sylvestris	Silver date palm	India
Pritchardia affinis	Kona palm	Hawaii
Pritchardia pacifica	Fiji Island fan palm	Western Pacific
Pritchardia remota	Remota loulu palm	Hawaiian Islands
Pritchardia thurstonii	Thurston palm	Western Pacific
Ravenea hildebrantii	Hildebrandt's palm	Madagascar
Syagrus schizophylla	Arikury palm	South America
Trachycarpus fortunei	Chinese windmill palm	China
Veitchia arecina	-------	Western Pacific

24. Graft-transmissible diseases of citrus

Characteristics of the pathogens, economic impact, and management strategies

Richard F. Lee* and Moshe Bar-Joseph**
*University of Florida, Citrus Research and Education Center, Lake Alfred, FL 33850, USA:
**Volcani Centre, Agricultural Research Organization, Bet Dagan 50250 Israel

24.1 Characteristics of the pathogens

24.1.1 *Citrus tristeza virus* Genus *Closterovirus* (CTV)

Tristeza, meaning sadness in Spanish or Portuguese, was the name first given to a decline of trees on sour orange rootstock in South America. Tristeza was later shown to be caused by *Citrus tristeza virus* (CTV), a member of the closterovirus group within the *Closteroviridae*. Seedling yellows disease is a synonym used for tristeza. CTV is probably the most destructive citrus virus in the world. There have been documented losses of about 30 million trees on sour orange rootstock in Brazil and Argentina in the 1940s and 1950s, of 6.6 million in Venezuela in the 1980, and an estimated 10 million trees in Florida and other Caribbean Basin countries, all these losses were due to the introduction of *Toxoptera citricida*, commonly called the brown citrus aphid (BrCA), which is the most efficient vector of CTV (Rocha-Pena *et al.* 1995). The BrCA has been increasing its geographical area throughout the Caribbean Basin since 1989 and was first found in Florida in November 1995 (Halbert, 1998). Even without the BrCA, CTV has killed about 10 million trees Spain (Cambra *et al.* 1988) and millions more which are not well documented in Israel, Florida, California and in other areas (Bar-Joseph *et al.* 1989).

CTV is present in most citrus areas of the world. It has been reported in the African region, the Eastern Asian region, the Eurasian region, the Mediterranean region, the North American region and the Pacific region (Bar-Joseph and Lee, 1989). If the trees are planted on a CTV tolerant rootstock, such as *Poncirus trifoliata,* citranges, citrumelo, or mandarin, the appearance may not be noticeable unless diagnostic methods are used to detect the virus.

CTV causes several different diseases, depending on the strain of CTV. There are five major disease symptoms: 1) Mild, causing little or no noticeable symptoms on commercial citrus and only slight symptoms of vein flecking, leaf cupping, and stem pitting on very susceptible hosts such as Mexican lime, *Citrus excelsa*, or *Citrus macrophylla.* 2) Decline strains which cause decline and death of trees on sour orange rootstock (Figure 24.1A). Decline strains of CTV can reduce fruit size (Figure 24.1B). Sweet orange, grapefruit, mandarin scions on sour orange rootstock are susceptible to CTV decline. Usually tiny pinholes are produced on the inner surface of a bark patch, which is removed from the sour orange rootstock, and the sour orange will have corresponding tiny bristles projecting from the wood. Often there is a bulge immediately above the budunion, however in areas of Florida the trees are killed by decline strains of CTV so rapidly that the bulge does not develop, instead a brown line forms at the budunion a few minutes of exposure to the air after the patch is removed. 3) Seedling yellows (SY) is a stunting of the plant and chlorosis or yellowing of the leaves, which occurs on sour orange, lemon, and grapefruit following inoculation with some strains of CTV. SY has been used as a presumed marker for identification of severe strains (Dodds *et al.* 1987). Usually SY is a greenhouse disease occurring on indicator plants, however if growers topwork grapefruit or lemon scions onto their trees which are infected with SY strains of CTV, SY symptoms can be expressed in the field. 4) Orange stem pitting (OSP) strains of CTV will stem pit sweet orange varieties and 5) Grapefruit stem pitting (GSP) strains of CTV stem pit grapefruit (Figure 24.1C). Stem pitting is considered the most severe symptom of CTV; this aspect cannot be controlled by growing the scions on a CTV tolerant rootstock. Stem pitting strains reduce fruit size, fruit quality, and overall yield (Rocha-Pena *et al.* 1995). There have been instances where CTV strains will severely stem pit rootstock varieties, which are normally considered to be CTV tolerant. For example, the Capao Bonito isolate in Brazil causes severe stem pitting on Rangpur lime rootstock (Targon *et al.* 2000), strains of CTV causing stem pitting on rough lemon have been reported (Lee *et al.* 1994; S.M. Garnsey, personal communication), and some strains even stem pit rootstocks such as Cleopatra mandarin (Manjunath, *et al.* 1995).

CTV infects all *Citrus* species and *Aeglopsis chevalieri, Afraegle paniculata, Pamburus missiones*, and *Passiflora gracilis* (Bar-Joseph and Lee, 1989).

Molecular Properties

CTV has a positive-sense single-stranded RNA genome encapsidated in flexuous particles about 2000 nm in length (Bar-Joseph and Lee, 1989). The virions contain two capsid proteins (CP) arranged in a "rattlesnake" structure: a 25 kDa CP that encapsidates ~95% of the particle and a 27 kDa minor CP that encapsidates ~5% at one terminus (Febres et al, 1996). The entire sequence of the T36 isolate of CTV was reported by Karasev *et al* 1995.

Genomic Organization and Replication Strategies

CTV is a member of the monopartite genus Closterovirus in the family *Closteroviridae* (Bar-Joseph and Lee, 1989). The genomic RNA (gRNA) of this virus contains from 19,226 to 19,302 nt, depending on the isolate (Karasev *et al.* 1995; Mawassi *et al.* 1996; Suastika *et al.* 2001; Vives *et al.* 1999; Yang *et al.* 1999b), which occur as 12 open reading frames (ORFs), potentially encoding 19 protein products (Karasev, 2000; Karasev *et al.* 1995) (Figure 24.2).

Transmission

The virus is transmitted by grafting and mechanically by a knife-cut and slashes inoculation (Bar-Joseph and Lee, 1989; Garnsey *et al.* 1977; Garnsey and Muller, 1988). It also is vectored by several aphid species in a semipersistent manner with the aphid retaining the ability to transmit the virus for up to 24-48 hrs after acquisition (Bar-Joseph and Lee, 1989; Raccah *et al.* 1989). The most efficient vector for CTV is *T. citricida* (Bar-Joseph and Lee, 1989; Yokomi *et al.* 1994). Other aphid vectors of CTV are *Aphis gossypii* , *Aphis spiraecola* and *Toxoptera aurantii* . *A. gossypii* is the second most efficient vector after *T. citricida.* Epidemiology studies have shown that *A. gossypii* generally transmits to trees, which are 8-10 trees away from the source tree, while *T. citricida* transmits to the nearest 1 or 2 trees. When both vectors are present, the incidence of CTV is greatly accelerated with *A. gossypii* spreading the longer distances and *T. citricida* filling in the spaces (Gottwald *et al.* 1996). *A. spiraecola* may at times be an important vector because of the high populations which can be reached. *T. aurantii* can transmit some isolates of CTV, mostly the more severe isolates, but does not transmit Florida isolates of CTV (R. Yokomi and K. L. Manjunath, personal communication).

Methods of detection

The first method developed for diagnosis of CTV was biological indexing on Mexican lime. While quick methods have been developed for the detection of CTV in infected tissue, the biological activity of an isolate of CTV can only be determined by biological indexing on a battery of different citrus hosts. At the minimum, the hosts should include Mexican lime as a susceptible indicator plant for universal detection, sour orange as an indicator for SY, sweet orange on sour orange rootstock as an indicator for decline strains, grapefruit as an indicator GSP strains, and sweet orange as an indicator for OSP strains (Garnsey et al, 1987). Mexican lime will show vein flecking when infected with mild strains, severe strains will cause leaf cupping, interveinal chlorosis, and pronounced vein flecking and occasionally vein corking on the upper surface of the leaves. It usually takes 3-6 months for the Mexican lime index. Sour orange seedlings will be stunted and display chlorosis or yellowing in the leaves with SY strains of CTV. Non-SY strains do not have a pronounced effect of the growth of sour orange seedlings. The SY index usually required 6-9 months. The sweet orange on sour orange indicator plants will display stunting, chlorosis, and reduced vigor if decline strains of CTV are present, but they rarely die under greenhouse conditions. This index requires 12-15 months under ideal growing conditions. The stem pitting evaluations usually require 12-15 months before rating, severe stem pitting strains of CTV often cause reduced growth on the sweet orange and/or grapefruit indicator plants. Leave chlorosis is often expressed in grapefruit indicator plants infected with GSP strains of CTV. All indicator plants except the Mexican lime are maintained to one growing shoot. When the plants are cut back, the length and weight of the flush removed are recorded.

CTV may be detected by staining thin sections cut from tissue with Azura A, which stains the inclusion bodies present in CTV-infected tissue. They may be observed using a light microscope (Christie and Edwardson, 1977).

Several serological methods have been developed for the detection of CTV (Rocha-Pena and Lee, 1991). SDS-immunodiffusion was one of the early methods developed, presently for large-scale assays ELISA is the method of choice, a modification of this procedure is the dot immunobinding assay on nitrocellulose or equivalent membrane, or a direct tissue immunoblot. A monoclonal antibody has been selected which reacts preferentially with severe (decline) isolates in Florida (Permar *et al.* 1990), more recently an ELISA system was developed for selective detection of CTV strains causing sweet orange stem pitting (Nikolaeva *et al.* 1999). Serologically specific electron microscopy is useful, especially when monitoring the effect of additives on purification (Lee *et al* 1987). Staining with fluorescent-labeled antibodies to detect inclusion bodies in CTV infected tissue also has been used (Brlanksy *et al.* 1988).

Reverse transcriptase polymerase chain (RT-PCR) assay has become the method of choice for ultra-sensitive detection of CTV. RT-PCR has been used to detect CTV present in single aphids (Mehta *et al.* 1997). Modifications of the RT-PCR have resulted in the development of some procedures, which are useful to predict quickly the biological activity of an unknown isolate of CTV (Niblett *et al.* 2000; Lopez *et al.* 1998; Hilf *et al.* 2000).

24.1.2 *Citrus psorosis virus*

Psorosis was first described in 1898 by Swingle and Webber, and in the early 1930s was the first citrus disease proven to be graft transmissible (Fawcett, 1934). It was present almost universally in old-line budwood sources before the development of certification programs in the early 1950s (Roistacher 1993). It is considered the most important disease of citrus in Argentina and Uruguay where the disease has a natural, unknown means of spreading from tree to tree.

Psorosis was present in many old-line bud sources, and thus, the movement of budwood distributed it almost worldwide especially before the nature of graft transmissible pathogens was understood.

Psorosis symptoms include bark scaling, with the wood beneath the scaling remaining alive. If a cross section is cut through a limb having bark scaling, gum impregnation of the wood is apparent. The leaves will show flecking and vein clearing under cool weather conditions, such as the spring growth flush. The tips of new flush growth often become necrotic and die. In the literature, psorosis A is described as a disease having psorosis-like leaf and foliar symptoms, but little bark scaling occurring on the limbs and trunk whereas psorosis B had both leaf and foliar symptoms plus a lot of bark scaling with symptoms also occurring commonly on the fruit. Wallace (1957) demonstrated that both psorosis A and B were strains of the same virus by preinoculating psorosis A into indicator plants, later when psorosis B was inoculated into the same plants, psorosis A protected against the display of bark scaling commonly expressed for psorosis B when inoculated by itself.

Citrus ringspot virus (CRSV) used to be considered a different virus, although it was accepted as psorosis-like based on production of psorosis-like symptoms on indicator plants. CRSV commonly produces ringspot-like symptoms on the leaves of infected plants and on fruit. CRSV isolates produce necrotic local lesions when upon mechanical inoculation of the older, fully expanded leaves of *Chenopodium quinoa* or *Phaseolus vulgaris* cv. Red Kidney. Some isolates of CRSV produce bark scaling on citrus hosts whereas other isolates seldom if ever produced bark scaling (Garnsey and Timmer, 1988).

Psorosis virus infects all *Citrus* species and produces typical symptoms on sweet orange, grapefruit and mandarin varieties. Other varieties and *Citrus* species may remain asymptomatic, but they are latent hosts for the virus. Additionally there are several herbaceous hosts: *C. quinoa, P. vulgaris cv.* Red Kidney are common diagnostic hosts, other hosts reported are *Catharanthus roseus, Chenopodium amaranticolor, Crotalaria spectabilis, Cucumis sativus, Gomphrena globosa, Helianthus annuus, Nicotiana benthamiana, N. clevelandii, N. megalosiphon, N. rustica, N. tabacum, Petunia hybrida, Pisium sativum, Poncirus trifiliata, Sesamum indicum,* and *Zinnia elegans* (Timmer *et al.* 1978)

Molecular properties

In 1988 Derrick et al reported the purification of CRSV, isolate 4E, which had two components, both components were required for infectivity. A 48 kdalton protein was isolated from purified CRSV, and a monoclonal antibody was made against this protein. The resultant antibody was very isolate specific and could not be used for broad scale detection of other psorosis isolates. Following N-terminal sequencing of the 48 Kd protein, enough genome sequence was obtained to enable a RT-PCR assay to be developed with amplified the capsid protein of CRSV and other isolates of psorosis (Barthe *et al.* 1998). Based on molecular comparisons (Gracia *et al.* 1997; Barthe *et al.* 1998; Legarreta *et al.* 2000), it appears that psorosis A, psorosis B and CRSV are all strains of psorosis, but that concave gum, impietratura, and cristacortis, all of which produce similar symptoms in indicator plants, are distinct viruses, and they are not members of the psorosis group.

Means of spread

The long distance spread of psorosis and psorosis-like viruses has been by movement of infected nursery material. The virus may be mechanically transmitted to citrus and herbaceous hosts. Clearly a natural means of spread occurs in Argentina and Uruguay, but the means of this spread has never been identified (Roistacher 1993). It most likely is an insect although it could be by pollen or other means.

Detection

Antisera have been developed in Italy (Garcia *et al.* 1997), which permits serological detection of psorosis by ELISA methodology. RT-PCR is probably the method of choice (Garcia *et al.* 1997; Barthe *et al.* 1998; Legarreta *et al.* 2000).

CRSV isolates produce necrotic local lesions on the inoculated leaves on Chenopodium quinoa, but fully expanded leaves must be inoculated. If younger leaves are inoculated, no lesions will develop. On Dweet tangor, sweet orange, or grapefruit indicator plants, psorosis, including CRSV, produces a shock symptom on the first flush which emerges after inoculation; the tip of the flush often necrosis and dies. The leaves usually show pronounced vein clearing and flecking and/or transient oak leaf patterns. Bark scaling on the stem and twigs of the indicator plants may or may not be expressed, depending on the isolate. Concave gum, which produces symptoms on the same indicator plants, usually shows oak leaf patterns and little vein clearing and flecking, and the shock symptoms are now expressed.

24.1.3 Citrus viroids including Citrus exocortis viroid and Cachexia

. The first viroid to be described from citrus was citrus exocortis viroid (CEV). The discover of CEV subsequently lead to the finding of several additional viroids which have been grouped into five viroid groups based on size plus CEV (Duran-Vila *et al.* 1988). CEV produces the strongest symptoms on indicator plants of any of the citrus group of viroids. Citrus viroids seldom are found naturally occurring as single viroids in infected trees, rather they usually occur as a mixture of two or more viroids. Thus, it is hard to say, based on field observations and biological indexing, that CEV or any one of the other viroids produces a given symptom. There has been some work done to establish a correlation between a known viroid and symptoms expressed, but there is need for more of this correlation between specific viroids and combinations of viroids and the expression of symptoms on susceptible hosts from different citrus areas.

Citrus viroids are almost universally distributed and present in all citrus areas of the world. CEV is most destructive to sweet orange, grapefruit, or mandarin on *Poncirus trifoliata* rootstock, or a rootstock containing *P. trifoliata* as one of the parents of a hybrid, such as citranges or citrumelos. Affected trees on these rootstocks show dwarfing and stunting, bark scaling and cracking occurs on the rootstock portion of the tree, and productivity is reduced with small fruit and reduced yield (Figure 24.1 D).

Citron plants infected with CEV display severe leaf epinasty and rugosity with the midrib showing browning and cracking on the underside, especially at the leaf tip and the petiole (Figure 24.1 E). On lemon plants infected with CEV, bark shelling and cracking occur, and often the leaves have a blotchy mottle.

Tomato variety Rutgers displays leaf epinasty, rugosity and stunting when infected with CEV. *Gynura aurantiaca*, velvet plant, is used as the principal herbaceous indicator, it shows leaf epinasty and rugosity upon inoculation with CEV.

Cachexia, also known as xyloprososis, is caused by a Group II viroid. While cachexia, by itself, will not cause obvious problems to trees on *P. trifoliata* rootstock, or citranges or citrumelos, it does cause stunting, dwarfing, and very poor growth of trees on mandarin or *Citrus macrophylla* rootstocks. Gum pockets and pitting occur in the bark over the rootstock of these rootstocks.

Gum pocket disease on grapefruit trees propagated on *P. trifoliata* rootstock has been associated with a group III viroid in Swaziland (Marais *et al.* 1996).

All citrus species may be infected with citrus viroids including CEV and cachexia. Herbaceous hosts include tomato, *Petunia hybrida*, *Gynura saramentosa*, *Gynura aurantiaca*, *Chyrsanthemum morifolium*, *Daucus carota* (carrot), *Solanum melongena* (eggplant), *Brassica campestris* (turnip), *Vica faba* (broad bean), *Tagetes patula* (French marigold), and *Vitis* species (Sanger, 1988).

Molecular properties

The viroids are small species of circular RNA, which are highly base paired. The largest citrus viroid at 371 bases is CEV, and the smallest is a Group IV viroid with about 235 bases. While the viroids are circular, they are also infectious if they are nicked and appear in their linear form.

Means of spread

The long distance distribution has been by movement of infected nursery material. Because they are very easily mechanically transmitted, it is not uncommon for a viroid infection to spread to surrounding trees unless great care is taken to disinfest clippers and cutting tools between trees.

Detection

Biological indexing may be used for the detection of citrus viroids (Duran-Vila *et al.* 1993). All viroids should be biologically indexed under warm growing conditions, 30 C night – 38 C day. Etrog citron is a commonly used indicator plant especially for CEV. For cachexia indexing, Parson's special mandarin is grafted onto rough lemon or other vigorous rootstock. Once the Parson's special mandarin bud is growing off, the inoculum is grafted into the rootstock. After a few months, the typical gumming and pitting will occur at the budunion if cachexia is present. Orlando tangelo may also be used as a biological indicator, but this is a longer-term index taking 2-3 years. Scion trees are grafted onto Orlando tangelo as a rootstock, these trees are inoculated with the source to be tested, and bark patches removed at internals beginning 18 months after inoculation. Severe isolates of cachexia are easy to diagnose, mild isolates producing slighter symptoms are harder to detect using biological indexing. *Gynura aurantiaca* and Rutgers tomatoes are commonly used as herbaceous indicator plants for CEV, the other viroids usually do not express symptoms.

Viroids may be detected by purification and electrophoresis on polyacylamide gels. There have been several extraction methods reported. For best results, the viroid extraction should be made from Etrog citron tissue. Commonly the viroid is affinity purified using CF-11 cellulose powder or by portioning into the 2.0 M lithium chloride soluble fraction. Two dimensional polyacylamide gel electrophoresis is performed, the first gel is a native, non-denaturing gel. The viroid band is removed with a portion of the first dimension gel, and then the viroid is run in a denaturing gel, achieved by use of SDS and urea or by heated return gel electrophoresis (Asia *et al,* 1998). The circular forms of the viroids are retarded on the denatured gel and are well separated from host nucleic acids, the gel is silver-stained and the viroids may be identified by their respective rate of migration on the gel. CEV, being the largest viroid, migrate the least distance whereas the smallest viroids, Group IV, are the smallest and migrate the greatest distance.

RT-PCR assays have been developed (Gillings *et al.* 1988; Yang *et al.* 1992). Hybridization assays have been used to detect CEV and other viroids using various labels, such as biotin, digoxigenin, or radioactive labeled (Palacio *et al.* 2000). Viroid group specific primers are available for RT-PCR of each viroid group, and multiplex RT-PCR assays have been developed which enable simultaneous detection of several viroid groups at once.

24.1.4 Citrus variegated chlorosis (CVC)

Citrus variegated chlorosis (CVC), caused by a strain of *Xylella fastidiosa*, was first found in northern Sao Paulo state and in Minas Gerais in 1987 (Lee *et al.* 1991). The disease spread rapidly in Sao Paulo state and to other growing regions in Brazil through movement of infected nursery material. It soon became one of the economically most important diseases of citrus in Brazil with annual losses now estimated at $100 million per year (Donaldio and Moreira, 1998). Strains of *X. fastidiosa* cause other important diseases such a Pierce's disease of grapevine, phony peach, leaf scorch diseases of almond, coffee, oak, plum and sycamore (Hopkins, 1989; Purcell and Hopkins, 1996). Coffee leaf scorch (CLS) disease was first reported in Brazil, and the CVC strain of *X. fastidiosa* is closely related to the CLS strain (Beretta *et al.* 1996). Inoculation of the CVC strain into coffee results in CLS-like symptoms, but the CLS strain does not cause CVC symptoms when inoculated into citrus.

CVC is found in throughout the citrus areas of Brazil, in Argentina, Paraguay (Donaldio and Moreira, 1998). Recently it was reported to be present in Costa Rica, Central America (Moriera *et al.* 2003).

The symptoms of CVC usually begin with a zinc deficiency-like chlorosis appearing on one sector of the trees; trees become stunted upon infection (Figure 24.2 A). The leaves develop gummy raised lesion on their underneath side with a corresponding yellow chlorosis appearing on the upper surface of the leaf (Figure 24.2 B). As the symptoms spread, the new leaves are small and tend to point upward, twig dieback occurs, the fruit size is greatly reduced, and the fruit has a hard rind (Figure 24.2 C). The sugar content of the fruit is higher than in non-affected fruit, and the fruit ripen earlier. Once infected with CVC the trees have reduced growth and are rendered nonproductive in three years. Younger trees are more susceptible to CVC that trees which are 10 years of age or older. Symptom expression and incidence of CVC appear to be greater in warmer climates (Lee *et al.* 1991; Donaldio and Moreira, 1998).

CVC infects most species of citrus, but symptoms are displayed most on sweet orange cultivars. Lemons, limes, mandarins, mandarin hybrids such as Murcott and Sunburst, kumquats, trifoliate orange and grapefruit usually do not show symptoms of CVC but allow some multiplication of the bacterium (Donaldio and Moreira, 1998).

Molecular properties

The entire genomes of three strains of *Xylella fastidiosa* have been sequenced: CVC (Simpson *et al*, 2000), Pierce's disease of grapevine, and Oleander leaf scorch.

Means of spread

The disease is easily graft transmitted if xylem tissue is included in the bud chip used for propagation. The long distance spread of CVC is by movement of infected nursery material.

The CVC bacterium is transmitted by several species of sharpshooters which are common in Brazilian citrus areas. The efficiency varies among species. Ultrastructure studies have shown that the bacterium may attach to the inside of the stylet, pump organ (cybarium) and pre-cybarium of the sharpshooter. The sharpshooter loses the ability to transmit *X. fastidiosa* whenever a molt occurs. Once an adult acquires *X. fastidiosa*, they retain the ability to transmit for life. At least 11 species of sharpshooter have been identified to transmit CVC in Brazil (Donaldio and Moreira, 1998). The most important vectors in Brazil are *Acrogonia terminalis*, *Dilobopterus costalimai*, and *Oncometopia fascialis*, other common vectors present are *Sonesimia grossa*, *Hortensia similis*, *Ferrariana* sp. and *Molomea* sp. The glassy winged sharpshooter, *Homoladisca coagulate,* present in the southeast United State and now in California has been shown to be capable of transmitting CVC (R. H. Brlansky and V. Damsteegt, personal communication).

Detection

Diagnosis of CVC in the field can be confused with other decline diseases of citrus. CVC infected trees will take up water by the syringe injection test while blight infected trees do not. Diagnostic field symptoms are the small fruit having high sugar content, the gummy lesion on the underside of the leaves, and the small, pointy leaves at the top of the tree. Laboratory detection methods include serological assays, culturing the bacterium, microscopy, and PCR (Donaldio and Moreira, 1998). The problem with most of these detection methods is that it is difficult to differentiate between the CVC strain and other strains of *X. fastidiosa*. PCR methods have been reported which allow specific detection of the CVC strain.

24.1.5 Huanglongbing (Citrus greening disease)

Citrus greening is referred to also as Huanglongbin, yellow shoot, likubin, leaf mottling, and vein-phloem degeneration (da Graca, 1991; Garnier and Bove, 1993). The disease is caused by a systemic phloem-inhabiting bacterium, *Candidatus* Liberobacter. In thin sections under the transmission electron microscope (TEM), this fastidious bacterium is pleomorphic having a three-layer envelope. There are two forms of greening, each form has a similar host range but they differ in the temperature under which they express strongest symptoms. The African form, *Candidatus* L. africanus causes symptoms under cool conditions while the Asian form, *Candidatus* L. asiaticus causes symptoms under warm conditions. The Asian forms, in general, express stronger symptoms than African forms.

HLB is one of the most destructive diseases of citrus. Once established, the management of the disease to achieve continued production of citrus is difficult and expensive. The HLB bacterium is spread by citrus psyllids (McClean and Oberholzer, 1965). When the psyllid vector is present and under ideal climatic conditions, HLB can rapidly decimate productive citrus plantings. If trees are infected while young, they often have no fruit production.

Present in most areas of Africa, Asia, Southeast Asia, Arabian Peninsula, Indonesia and the Philippines.

Symptoms of Asian greening include leaf chlorosis and canopy dieback, the early symptoms usually appear only on one sector or branch of the tree (Figure 24.2D). The chlorosis spreads, often resembling a zinc deficiency symptom. Twig dieback occurs, and the affected trees decline to a non-productive state. Fruit is small, lopsided, with the basal end often remaining green, and the seeds are usually aborted. The fruit has a bitter taste. The symptoms of the African form of greening are similar except they are expressed during cooler weather, and a blotchy mottle symptom is common on the leaves, especially young flush leaves grown under cool temperature (Oberholzer *et al.* 1965; da Graca, 1991).

The HLB bacterium infects nearly all citrus species, cultivars and hybrids and some citrus relatives. Sweet orange, mandarin and mandarin hybrids are most susceptible; lemons, grapefruit and pummelos are moderately affected; and Mexican lime, trifoliate orange, citranges and citrumelos are more tolerant, often expressing foliar symptoms but little twig dieback (da Graca, 1991). Periwinkle and tobacco are experimental hosts, infected by use of dodder (Gao *et al.* 1993).

The bacterium causing greening has not been cultured, and Koch's postulates have not convincingly been fulfilled although there are reports of bacterium having the traits of greening being cultured (Garnet *et al.* 1985).

Means of spread

Greening is graft transmissible. The distribution of the greening bacterium within an infected tree can be irregular so not all buds will contain the bacterium or transmit the disease. The more phloem tissue that is included in the inoculum, the greater the probability of graft transmission when transmitting by grafting. Transmission from plant to plant and to experimental species has been done by the use of dodder (Gao *et al.* 1993).

Greening bacterium is transmitted by psyllids. *Trioza erytreae* occurs in Africa, Yemen, islands in the Indian Ocean, and is associated with the African form of greening. *Diaphorina citri*, commonly called the Asian citrus psyllids, is better adapted to warm humid climates and occurs in Asia, the Indian Subcontinent, Saudi Arabia, Reunion, Mauritius, South America, and more recently in several Caribbean islands, areas in Central America and Florida. Both psyllids have been shown to transmit either form of greening, Asian or African. The bacterium is transmitted in a persistent manner with a latent period occurring after the psyllid acquires the bacterium before the insect is able to transmit. The bacterium multiplies in the psyllid vector. Because of the latent period, most psyllids capable of transmitting greening are either late-stage nymphs or adults. The psyllids remain capable of transmitting greening for the duration of their life once the bacterium has been acquired. Citrus species is the primary host for feeding of both psyllid species; other hosts include *Clausena anisata*, *Vepris lanceolata* and *Murraya paniculata* and a few other species of Rutaceae. The psyllids prefer to feed on young flush tissue. Hosts, which are vigorous and always flushing, such as lemon, lime, and *M. paniculata* are ideal hosts to look at for presence of psyllids.

Detection

The quickest and most reliable method for detection and diagnosis of greening is by PCR using symptomatic tissue as the source (Jagoueix *et al.* 1994: Harakava *et al.* 2000). Biological indexing is often difficult because severe strains of citrus tristeza virus (CTV) occur in most areas where greening occurs, and the symptoms of CTV can mask the presence of greening. Cool greenhouse conditions, 24-27 C, are required for symptom development of African forms of greening, while warm conditions, 30-37 C, are required for symptom development of Asian forms of greening. Asymmetric growth of young leaves, especially on greenhouse indicator plants, is indicative of presence of greening, in addition to a blotchy mottle leaf symptom for African forms and zinc deficiency-like symptoms for Asian forms. Grapefruit and lemon plants are good indicator plants, showing good symptoms under greenhouse conditions. Mandarin indicator plants may be useful if severe strains of CTV are present as the mandarins are tolerant of CTV and most of the symptoms being expressed are due to greening. From studies using *in planta* cultures in the quarantine facility, Beltsville, MD, dweet tangor has proven to be a sensitive indicator plant (M. Hooker and R. Lee, unpublished). Other methods of assay which have been reported are by the use of transmission electron microscopy (Moll and Martin, 1973), a method not useful for a large number of samples, serological detection (Sdoodee and Garnett, 1994) and assays for presence of gentisic acid, a method which is not as accurate and sensitive as PCR, but it does not require elaborate equipment (Hooker *et al.* 1993).

24.1.6 Witches' broom disease of lime (WBDL)

In the early 1980s a devastating disease of lime, now known as witches' broom disease of lime (WBDL), occurred in the Liwa area in Batina, Oman (Figure 24.3E) (Garnier *et al.* 1991). This disease was never reported elsewhere before that time. WBDL is caused by a mycoplasma-like organism know by the *Candidatus* name Phytoplasma aurantifolia (Zreik *et al* 1995)

Up until the early 1990s, citrus ranked as the second most important crop in Oman, after dates. The area used for citrus production was around 2500 hectares. Ninety percent of the citrus grown is a small-fruited acid lime (*Citrus aurantifolia*) known locally as Omani lime. The Omani limes, either fresh or dried and pulverized, are an important food seasoning throughout the Near East. Farms affected with WBDL showed an infection rate of 1.2-3.3% in 1993. This rate increased rapidly and by 1996 the infection rate in surveyed farms was almost 100%. In 1998 the export of limes was ceased and the price of lime increased markedly (Bove, 1986).

There are other reports of phytoplasmas producing witches' brooms in citrus:

In India: a witches' broom in acid lime shows initial symptoms of highly proliferated shoots and shortened internodes with small chlorotic leaves. The leaves drop prematurely and infected twigs become distorted, resembling a broom-like appearance. In advanced stages the infected branches develop dieback. Infected branches do not produce any flowers or fruit. Symptom development is restricted to a single branch while the rest of the branches have normal vegetation and fruiting. Also in Nagpur mandarin (*C. reticulata*) the symptoms are of two different types. The first symptom has the whole plant showing excessive sprouting with small, thick leaves. The branches are very flexible with no flowering or fruiting. The second type of symptoms are similar to those described above in acid lime with the development of highly proliferating shoots with small chlorotic leaves which become distorted, followed by a dieback that give a dry broom-like appearance (D. Ghosh, personal communication). A similar witches' broom has been observed in mandarin in Oman.

In Jamaica a phytoplasma-like disorder has been seen. The disorder was first observed on a 20-year-old Cleopatra mandarin tree. Galls were formed, followed by bud proliferation forming witches-broom like structures but without the usual pronounced internode shortening. As high as five percent incidence of this disorder has been observed on citrumelo and mandarin rootstocks and on Valencia sweet orange budlings in nursery situations. The disorder also has been observed on Clementine trees and older Swingle seed source trees (Figure 24.2 F). The broad host range and rapid spread makes this disorder a threat to citrus production (R. Lee, unpublished).

WBDL occurs in Oman, United Arab Emirates. A disease with all the symptoms associated with WBDL has been reported in southeast Iran and in the Nagpur region of India, but there is no molecular evidence that these diseases having similar symptoms are caused by the same phytoplasma.

The production of witches-brooms symptoms is the main characteristic of WBDL infections. The brooms are compact with very small, pale green leaves. In early symptom trees, there are only a few witches'-brooms with the rest of the canopy looking normal. The first witches' brooms to form are often in the interior of the tree. Additional witches-brooms develop over time (Figure 24.2 E). In the advanced stages of the disease, the leaves of the older witches-brooms die, but some dry leaves remain attached for a period of time before they finally fall off. The disease is lethal in 5 - 10 years after the first witches-broom is found. All ages of trees appear to be susceptible to WBDL.

Small-fruited acid lime (*Citrus aurantifolia*), Palestine sweet lime (*C. limetta*) and sweet limetta (*C. limettoides*) are naturally infected and show symptoms. Experimentally, WBDL has been transmitted by graft inoculation to: *C. aurantifolia, C. excelsa, C. hystrix, C. ichangensis, C. karna, C. macrophylla,* Etron citron (*C. medica*), Meyer lemon (*C. limon*), Rangpur lime (*C. limonia*), rough lemon (*C. jambhiri*), and Troyer citrange (*Poncirus trifoliata* x *C. sinensis*) (Bove *et al.* 1996).

WBDL is caused by a mycoplasma-like organism know by the *candidatus* name Phytoplasma aurantifolia. The phytoplasma has not been cultured. The phytoplasma is closely related to the phytoplasmas of alfalfa, sesame, and sunhemp phyllodies based on sequencing of the 16S rDNA genes (Zreik *et al.* 1995).

Means of spread

WBDL has been shown to be graft transmissible. Recently there have been reports that WBDL is transmitted through seed collected from WBDL affected trees. The WBDL is thought to be transmitted by the leafhopper *Hishimonus phycitis*, but this has not been verified by transmission tests. WBDL may be transmitted by dodder.

Detection

The most common method of detection of WBDL is by PCR amplification of the 16S rDNA gene. Primer pairs are available which enable selective amplification of a region of the 16S rDNA gene (Khan *et al.* 2003). TEM examination of fixed tissue, immunofluorescence using monoclonal antibodies, and the use of labeled DNA hybridization probes have also been used for detection (Bove *et al.* 1993)

24.1.7 Citrus blight (CB)

Citrus blight, also called young tree decline and sandhill decline in Florida and declinio in Brazil, is a chronic decline disease of unknown etiology. It appears in trees on susceptible rootstock between 6-8 years of age when the tree is bearing, but the tree does not usually die. The disease causes a plugging of the xylem vessels.

Citrus blight occurs in most citrus areas having a warm, humid climate, but it has not been reported from countries having a Mediterranean-like climate. There are reports of blight from Florida, Texas, Louisiana, Hawaii, throughout the countries in the Caribbean Basin and Central America,

Colombia, Argentina, Brazil, South Africa, and Australia (Timmer *et al.* 2000).

The first visual symptom of blight is the development of a slight grayish color change of the canopy (Figure 24.1 F). Symptoms are preceded by an increase in zinc in the phloem, then in the wood of the tree. Often one sector of the tree will develop zinc deficiency-like symptoms in the leaves. In the northern hemisphere this sector is usually in the southwest quadrant of the tree while in the southern hemisphere, the first sector to show zinc deficiency symptoms is in the northwest quadrant. The tree then begins to decline with twig dieback, reduced fruit size, high sugar content in the fruit, delayed blossom set, and the leaves at the top of branches become small and point upward. Within 1-2 seasons, the tree is rendered non-productive but the tree does not die.

Citrus blight affects most commercial varieties on rootstocks as well as seedling trees which are bearing fruit. In the field some rootstocks are very susceptible to blight, such as rough lemon, Carrizo citrange, Rangpur lime, *Poncirus* trifoliate and Volkamer lemon while other rootstocks have a good field tolerance, such as sour orange, Cleopatra mandarin, and Swingle citrumelo.

Citrus blight is a disease of unknown etiology, but it has been shown to be transmitted by grafting roots from a blighted tree onto the roots of a healthy tree, but the disease has not been transmitted by any above ground parts of the tree (Timmer *et al.*. 1987; Derrick and Timmer, 2000).

Means of spread

Citrus blight has been shown to be graft transmitted using roots from blighted trees to graft to roots of healthy trees. With some rootstocks such as Cleopatra mandarin and sour orange, which normally show field tolerance for blight, blight will spread down the row tree to tree. This is presumed to be due to natural root grafting between trees, as herbicide applied to the stump of a tree, which has been sawed off, will result in phytotoxicity showing up 1-2 trees down the row. The first occurrence of blight in a grove is randomly distributed, but subsequent spread tends to be 1-2 trees away from the blighted tree (Timmer *et al.*. 2000).

Detection

Because blighted trees have amorphous plugs in their xylem, a water injection test is often used in the field to diagnose blight (Lee *et al.* 1984). Healthy trees will take up 10 ml of water within 30 sec when using a 30 ml syringe to push the water into a 1/8th inch diameter hole in the tree trunk, blight trees will not take up water. Blight may also be diagnosed by the high zinc content in the wood, presence of the amorphous plugs in the xylem vessels, and the presence of blight-associated proteins in leaves and roots using serological methods (Wutscher *et al.* 1977; Derrick and Timmer, 2000).

24.2 Economic impact case studies

It is difficult to get good yield loss data due to the presence of graft-transmissible pathogens. Losses become apparent to the growers who realize that the inputs into their grove are exceeding the income, and the reduced vigor and declining trees are apparent. However, comparable control blocks are usually lacking because the trees are a different scion variety, different rootstock, or under a different management regime. The value of proactive programs, such as mandatory citrus certification programs, comes from maintaining productivity or even increasing productivity over a period of time due to the introduction of superior clones which are free of graft-transmissible pathogens. The value of these proactive programs is difficult to calculate because the potential losses don't occur and cannot be accurately estimated.

24.2.1 Citrus viroids

Low vigor and stunting were apparent in three to five year old plantings of Valencia sweet orange trees grafted onto Rangpur lime or Carrizo citrange rootstock, depending on the grove, in Belize. The budwood for these trees had been imported into Belize from outside the country on the assumption that the budline was superior to clones of this variety already available within Belize. Several groves of Valencia of the same age and same rootstock had been propagated on the same rootstocks, but using a locally available clone of Valencia which was free of citrus viroids. All the groves were under the same management regime. The yield loss was readily apparent and production cost records and yields from similar groves with three groves infected with citrus viroids and three groves which were not affected were analyzed. The groves were all planted at a tree density of 225 trees per hectare. At the end of the eight year period when the viroid infected block was removed, the total loss was US$5,147 per hectare. For the grove planted with the viroid free budwood, the total profit at the end of eight years was US$10,898 per hectare (Roistacher *et al.*. 1996).

In an earlier study (Roistacher *et al.*. 1991), an experimental planting of navel sweet orange was designed to test the effect of citrus viroids on tree growth and productivity on susceptible (Troyer citrange) and tolerant (sour orange) rootstock. Citrus viroid isolates tested included citrus viroid (CV) CV-IIa alone; CVs-IIa and IIIb; CVs- Ia, IIa, IIIb; and CEV with CVs-IIa, and IIIb. The trees were planted in the Central Valley of California, 10 replicates of each treatment and planted in a randomized block. Yield and tree measurements were made for 10 years, and the trees size measurements represented the average size of all trees in the treatment at the end of 10 years, the yield was expressed as average yield in Kg over the 10 year period for each treatment.

The effect of the various viroid combinations on the tree volume after 10 years of growth on the two different rootstocks is shown in Figure 24.4 A. The effect of the viroid combinations on the 10 year accumulative average yield is shown in Figure 24.4 B. CEV, Cvds- IIa and IIIb together reduced the canopy volume on Troyer rootstock by 59.2% while Cvds- Ia, IIa, and IIIb together and IIa and IIIb together reduced the canopy volume by 49.7% and 43.6%, respectively. It is interesting to note that the canopy volume of the trees on sour orange rootstock was reduced by 36.1%, a significant reduction, especially considering that sour orange is a viroid-tolerant rootstock. Considering the average yield over the 10 year period, CEV with Cvds-IIa and IIIb together reduced yield by 34.2%, Cvds-Ia, Ib, and IIIb by 21%, and Cvds-IIa and IIIb by 19% for the trees on Troyer citrange rootstock. CEV with Cvds-IIa and IIIb reduced yield by 21.6% on sour orange rootstock, a significant reduction in yield. Fruit quality was not reduced by the presence of CEV and the other citrus viroids.

24.2.2 Citrus tristeza virus

It is very difficult to determine yield losses due to stem pitting strains of CTV because in the areas where severe stem pitting, there are no healthy trees for comparison. Some estimates may be made however, by examining the yield and performance data from mild strain cross protection experiments. A good example is the summary of 20 years production from mild strain cross-protected Marsh grapefruit trees in Australia by Broadbent *et al.* (1991). *Toxoptera citricida* was present in Australia at the time the trials were planted in the field. After 20 years, at the location having the milder climate there was no breakdown in trees, which had been inoculated with mild isolates while no marketable fruits were produced by trees which had been inoculated with a severe stem pitting CTV isolate. At the warmer location all trees were smaller at the end of 20 years. The trees inoculated with the severe CTV isolate began declining and showing symptoms in only the last few years of the trial; the cumulative yield loss for the trees inoculated with the severe strain ranged from 19.4% to 20.3. However, the fruit size was greatly reduced with 79.1% of the fruit being size class 125 or greater compared to 22.5% and 14.4% for mild isolates. At the mild climate location, the cumulative yield for the trees inoculated with the severe isolate ranged from 52.9% to 44.2% when compared against the yield of trees inoculated with the mild isolates. The fruit size was also greatly reduced with 76.3% of the fruit being size 125 or greater compared to 23% and 17.8% from trees inoculated with mild strains.

In South Africa, Marsh grapefruit trees were propagated on rough lemon rootstock and then inoculated with the Nartia mild isolate of CTV that was used widely in South Africa for several years as an universal mild isolate for cross protection (Marais, 1994; Marais *et al.*1996). When the propagated trees were planted into the field in Nkwaleni Valley, Northern Natal Province, about 25% of them were severely affected by CTV stem pitting. It is thought these plants were probably infected with severe strains of CTV by aphids feeding on the rough lemon liners before the trees were propagated. From the field planting, 10 trees were selected which had mild stem pitting, and 10 trees were selected with had severe stem pitting but yet the tree canopy was vigorous and healthy appearing. Yield data was collected from these selected trees for seven years. The average yield per year over the seven-year period was 221.6 kg per tree for the mild stem pitting trees and 178 kg per tree for the severe stem pitting trees. This is an average yield reduction of 20%. A total of 33.3% of the fruit from the mild stem pitting affected trees was size 40 or smaller compared to 72.6% of the fruit from the severe stem pitting affected trees.

Toxoptera citricida, the efficient vector of CTV, has expanded it's geographical area to the Caribbean Basin beginning in the mid-1980s. *T. citricida* was first found in Jamaica in July 1995, and although it was known that decline strains of CTV were present in the country, there had been no outbreaks of tree losses on sour orange rootstock due to CTV. In 1997, the first outbreak of tree losses was found on sour orange rootstock in the Bog Walk Valley, Jamaica (Lee *et al.* 2002). Based on past performance, the yield losses due to CTV were documented from 1997 through 2001. At one location of 541 acres, a total loss of 489,751 boxes of fruit (standard 90 pound boxes) were lost while at a second location of 618 acres, a total loss of 525,384 boxes of fruit were lost. By November 2001, the total losses for the two locations totalled US$4,413,630, using US$4.35 per box of fruit. These losses did not include the cost of removing the dead trees and replanting with new plants.

24.2.3 Huanglongbing (Citrus greening)

HLB is perhaps one of the most difficult citrus diseases to control. In areas where HLB is endemic, it is rapidly spread by the psyllid vectors. At two locations in India, 67% of the certified HLB-free trees became infected five years after planting (Azzaro *et al.* 1993). A method for living with greening has been developed in South Africa where HLB was first reported in South Africa in 1928/1929 (Buitendag and von Broembsen, 1993). This disease was very severe at elevations greater than 700 m between the years

1932-1936, 1939-1946, and again in 1958. Between the late 1960s and early 1970s, three major citrus production areas were abandoned due to HLB. By 1975, it was estimated that 4 million of 11 million trees in South Africa were infected with HLB (Buitendag and von Broembsen, 1993). Subsequently, a method was developed to live with greening and to grow citrus productively. This method includes planting healthy trees in the field, reduction of inoculum by pruning symptomatic foliage from trees, reducing psyllid populations by removing hosts which are attractive to psyllid breeding and use of systemic insecticides in the citrus trees to reduce psyllid populations.

In Thailand, HLB was first reported by Schwarz *et al.* (1974), who noted that two years appeared to pass between infection of trees and the debilitation of the trees. The trees were not killed, but rendered nonproductive. Marcotting was a common practice in Thailand, and since most trees were infected with HLB, the marcotted cutting was also infected. Su and Tolley (1992) observed that the average life span of a citrus grove was 8 years. They recommended implementation of a clean stock program and citrus certification programs. Roistacher (1994) observed trees declining due to HLB at 5 years of age and even younger. He estimated that between 10-15% of the tangerine trees were removed from productivity annually due to HLB. Grenzebach (1994) studies the economics of producing citrus in Thailand. He calculated the financial returns from an average citrus farm. He noted that if the citrus planting was healthy appearing when the grove was planted, it was about 8 years old before noticeable decline due to HLB occurred and by 12 years, the grove was ready to remove. Individual trees were removed as they declined, so the groves contained trees of several ages ranging from the date of planting to recent replants. Grenzebach used 12.5 ton/ha as the average national production in the country based on data made available by the Department of Agricultural Extension. From his analysis, yields reached 22.75 ton/ha in years 5 through 8, then declined to 6.5 ton/ha by the 12 year due to HLB infections. Roistacher reanalyzed the data from Grenzebach, and based on his personal observations, realized that some plantings went to the field already infected with HLB, and in many instances the decline due to HLB began much earlier than the 8 years used by Grenzebach. A graphic representation of the profit and losses for 6, 8, 10 and 12 years in Thailand are shown in Figure 24.5. Based on the interpolation of Grenzenbach's data, it is estimated that a grove must live for 10 years to make a profit, and if the grove could live and be productive for 20 years, a cumulative profit of about $125,000 would be realized (Roistacher, 1996).

24.3 Management Strategies

For control of graft transmissible pathogens of citrus, the starting point is to be able to plant a healthy tree. This is best accomplished through a mandatory citrus certification program. In addition to providing a cost effective method to enable all growers in a region to plant healthy trees free of graft transmissible pathogens, it also ensures that the grower has planted citrus germplasm of the highest possible quality, which has a proven record of performance.

A citrus certification program is composed of three distinct and different programs: 1) quarantine, 2) clean stock, and 3) certification (Lee *et al.* 1999; Navarro, 1993). These three programs must be integrated to provide for a functional, effective certification program.

1) Quarantine Programs for safe introduction of selected horticultural germplasm. It is often desirable to move citrus species and varieties between different citrus areas for commercial and/or scientific purposes. Uncontrolled introduction of such budwood carries a risk of introduction of new pests and pathogens. Provision has to be made to import such budwood safely to reduce the risk to the importing industry or country. This is usually accomplished by careful introduction through quarantine stations, which perform procedures for the safe importation of the germplasm. There are two main approaches to importing citrus budwood under quarantine conditions: the classic method of propagation of the imported material in an isolated condition until the material can be freed of graft-transmissible pathogens by thermotherapy and/or shoot tip grafting, indexed to verify their pathogen-free status, and then released to the clean stock program for subsequent use in the citrus industry. The second method maintains the imported budwood in tubes. Imported budwood is immediately surface sterilized, placed in culture media in test tubes, buds are forced, shoot tip grafting is performed, and the resulting plants are then indexed to verify freedom from pathogens. This approach is now used in Spain and California, and offers the advantage of more rapid introduction. As an example, one budstick from Texas was introduced into Spain in January 1988, and in May 1989, 10,300 healthy buds were released to citrus nurseries (Roistacher, 1994).

2) Clean Stock Programs for identification and production of desirable sources, and their maintenance as pathogen-free propagating stock.

In a citrus certification scheme, it is desirable to recover healthy plants from the local varieties or cultivars. These varieties may be best adapted for the local climate, soils, and market. Clean stock programs are usually carried out by research institutions with the joint participation of scientists in horticulture, virology, and tissue culture, and in close communication with the Ministry of Agriculture or Commissioner of Agriculture. A clean stock program consists of six steps: 1) selection of mother trees of local cultivars, 2) indexing of the mother trees, 3) recovery of pathogen-free plants by shoot tip grafting and/or thermotherapy, 5) indexing of the recovered plants, 5) horticultural evaluation of the healthy plants, and 6) maintenance of healthy plants under protected conditions with recurring indexing to ensure they have not become contaminated again. When selecting mother trees, selection of individual trees of the different cultivars should be made according to documented horticultural criteria, such as superior production, higher fruit color, early or late ripening, etc. The presence or absence of pathogens should not be a deciding factor; the therapy treatment will rid the selection from the pathogens.

3) Certification Programs, for maintenance and distribution of virus-free propagating materials for commercial use, guarantee the sanitary status and trueness-of-type of the propagating material during the process of commercial propagation through the nurseries. Additionally, they also control the horticultural quality of nursery plants. These programs have legal regulations governing the different steps of nursery operations and require periodic indexing and inspection of trees of the different blocks used for nursery propagations. .

Certification programs have to be adapted to the specific situation of each citrus area. Organization of the citrus industry, pests and diseases present, and sources of funding so the program will be sustainable are important considerations when developing a citrus certification scheme. Certification programs only give a guarantee for those pathogens, which are actually included and tested for in the program.

The propagation scheme allows for concentration of resources of the expensive and time consuming indexing for freedom of graft-transmissible pathogens, verification of horticultural trueness-of-type, and selection for highest horticultural quality to be focused on the foundation trees, yet the benefits of this effort is realized by all the resulting propagations. Certification programs must have the support of the growers and members of the industry, and funds must be available to sustain operation of the program.

These programs are the foundation on which to build any efforts to reducing damage caused by diseases and pests, and for the improvement and continued productivity of a citrus industry.

References

Asai M, Ohara T, Takahashi T, Saito S, Tanaka K. 1998. Detection of viroids in fruit trees by return gel electrophoresis. *Res. Bull. Plant Protec. Service*, Japan 34:99-102.

Azzaro, A., Leocata, S., and Polizzi, G. 1993. Greening spread in two locations of Uttar Pradesh (India). In: *Proc. 12th Conf. IOCV, IOCV, Riverside*. pp. 443-444.

Bar-Joseph, M., and Lee, R. 1989. Citrus tristeza virus. *CMI/AAB Descriptions Plant Viruses* **353**. Assoc. Appl. Biol., Wellesbourne, UK. p7
Bar-Joseph, M., Marcus, R., and Lee, R.F. 1989. The continuous challenge of citrus tristeza virus control. *Annu. Rev. Phytopathol.* 27:291-316.
Barthe, G.A., Ceccardi, T.L., Manjunath, K.L., and Derrick, K.S. 1998. Citrus psorosis virus: nucleotide sequencing of the coat protein gene and detection by hybridization and RT-PCR. *J. Gen. Virology* 79:1531-1537.
Beretta, M.J.G., Harakava, R, Chagas, C.M., Derrick, K.S., Barthe, G.A., Ceccardi, T.L., and Lee, R.F. 1996. First report of *Xylella fastidiosa* in coffee. *Plant Dis.* 80:821.
Bove, J.M. 1986. Witches' broom disease of lime. *FAO Plant Prot. Bull.* 34:217-218.
Bove, J.M., Navarro, L., Bonnet, P., Zreik, L., and Garnier, M. 1996. Reaction of citrus cultivars to graft-inoculation of phytoplasma aurantifolia-infected lime shoots. In: *Proc. 13th Conf. IOCV.* IOCV, Riverside. pp. 249-251.
Bove, J.M., Zreik, L., Danet, J. L., Bonfils, J., Mjeni, A. M. M., and Garnier, M. 1993. Witches' broom disease of lime trees: monoclonal antibody and DNA probes for the detection of the associated MLO and the identification of a possible insect vector. In: *Proc. 12th Conf. IOCV.* IOCV, Riverside. pp. 342-348.
Brlansky, R.H., Lee, R.F., and Garnsey, S.M. 1988 In situ immunofluorescence for the detection of citrus tristeza virus inclusion bodies. *Plant Dis.* 72: 1039-1041.
Broadbent, P., Bevington K.B., and Cotte, B.G. 1991. Control of stem pitting of grapefruit in Australia by mild strain protection. In: *Proc. 11th Conf. IOCV.* IOCV, Riverside. pp 64-70.
Buitendag, C.H., and von Broembsen, L.A. 1993. Living with citrus greening in South Africa. In: *Proc. 12th Conf. IOCV.* IOCV, Riverside. pp 269-273.
Cambra, M., Serra, J., Villalba, D., and Moreno, P. 1988. Present situation of the citrus tristeza virus in the Valencian Community. In: *Proc 10th Conf. IOCV.* IOCV, Riverside. pp 1-7
da Graca, J.V. 1991. Citrus greening disease. *Annu. Rev. Phytopathol.* 29:109-136.
Derrick, K.S., Brlanksy, R.H., da Graca, J.V., Lee, R.F., Timmer, L.W., and Nguyen, T.K. 1988. Partial characterization of a virus associated with citrus ringspot. *Phytopathology* 78:1293-1301.
Derrick, K.S., Lee, R.F., Brlansky, R.H., Timmer, L.W., Hewitt, B.G., and Barthe, G.A. 1990. Proteins associated with citrus blight. *Plant. Dis.* 74:168-170.
Derrick, K.S. and Timmer, L.W. 2000. Citrus blight and other diseases of recalcitrate etiology. *Annu. Rev. Phytopathol.* 38:181-205.
Dodds, J.A., Jarupat, T., Lee, J.G. and Roistacher, C.N. 1987. Effects of strain, host, time of harvest, and virus concentration on double-stranded RNA analysis of citrus tristeza virus. *Phytopathology* 77: 442-447.
Donaldio, L.C., and Moriera, C.S. (Eds.). 1998. *Citrus variegated chlorosis.* Fundecitrus, Bebedouro, SP. Brazil. 166 pp.
Duran-Vila, N., Pina, J.A., and Navarro, L. 1993. Improved indexing of citrus viroids. In: *Proc. 12th Conf. IOCV.* IOCV, Riverside. pp. 202-211.
Duran-Vila, N., Roistacher, C.N., Rivera-Bustamante, R., and Semancik, J.S. 1988. A definition of citrus viroid groups and their relationship to the exocortis disease. *J. Gen. Virol.* 69:3069-3080.
Christie, R.G., and Edwardson, J.R. 1977. Light and electron microscopy of plant virus inclusions. *Florida Agricultural Experiment Stations Monograph Series* 9, 150 pp.
Fawcett, H.S. 1934. Is psorosis of citrus a virus disease? *Phytopathology* 24:659-668.

Febres, V.J., Ashoulin, L., Mawassi, M., Frank, A., Bar-Joseph, M., Manjunath, K.L., Lee, R.F. and Niblett, C.L. 1996. The p27 protein is present at one end of citrus tristeza virus particles. Phytopathology 86:1331-1335.

Gao, S., Garnier, M., and Bove, J.M. 1993. Production of monoclonal antibodies recognizing most strains of the greening BLO by in vitro immunization with an antigenic protein purified from the BLO. In: *Proc. 12th Conf. IOCV*. IOCV, Riverside. pp. 244-249.

Garcia, M.L., Sanchez de la Torre, M.E., Dal Bo, E., Djelouah, K., Rouag, N., Luisoni, E., Milne, R.G., and Grau, O. 1997. Detection of citrus psorosis-ringspot virus using RT-PCR and DAS-ELISA. *Plant Pathol.* 46:830-836.

Garnett, H.M. 1985. Isolation of the greening organism. *Citrus Subtrop. Fruit J.* 611:4-6.

Garnier, M., and Bove, J.M. 1993. Citrus greening disease and the greening bacterium. In: *Proc. 12th Conf. IOCV*. IOCV, Riverside. p. 212-219.

Garnier, M., Zreik, L, and Bove, J.M. 1991. Witches' broom, a lethal mycoplasmal disease of lime trees in the Sultanate of Oman and the United Arab Emirates. *Plant Dis.* 75:546-551.

Garnsey, S.M., Gonsalves, D., and Purcifull, D.E. 1977. Mechanical transmission of citrus tristeza virus. *Phytopathology* 67: 965-968.

Garnsey, S. M., Gumpf, D.J., Roistacher, C.N., Civerolo, E.L., Lee, R.F., Yokomi, R.K. and Bar-Joseph, M. 1987. Towards a standardized evaluation of the biological properties of citrus tristeza virus. *Phytophylactica* 19:151-157.

Garnsey, S.M., and Muller, G.W. 1988. Efficiency of mechanical transmission of citrus tristeza virus. In: *10th Proc. Conf. IOCV*. IOCV, Riverside. pp. 46-54.

Garnsey, S.M. and Timmer, L.W. 1988. Local lesion isolate of ringspot virus induces psorosis bark scaling. In: *Proc. 10th Conf. IOCV*. IOCV, Riverside. pp. 334-339.

Gillings, M.R., Broadbent, P., and Gollnow, B.I. 1988. Biochemical indexing for citrus exocortis viroid. In: *Proc. 10th Conf. IOCV*. IOCV, Riverside. pp. 178-187.

Gottwald, T.R., Garnsey, S.M., Cambra, M., Irey, M., Moreno, P., and Borbon, J. 1996. Differential effects of *Toxoptera citricida* vs *Aphis gossypii* on temporal increase and spatial patterns of spread of citrus tristeza. In: *Proc. 13th Conf. IOCV*. IOCV, Riverside. pp. 120-129.

Grenzebach, E. 1994. Integrated pest management in selected fruit trees. Report No. 42 on a short term consultancy mission: Feb. 27-March 12, 1994 on behalf of Deutche Gesellschaft Für Technische Zusammenarbeit (GTZ). 55 pp.

Halbert, S.E., Evans, G.A., and Clinton, D.C. 1998. Establishment of *Toxoptera citricida* in Florida. In: *Aphids in Natural and Managed Ecosystems.* (Nieto Nafria, J.M. and Dixon, A.F.G., Eds.). Universidad de León, León, Spain. pp. 547-554.

Harakava, R., Marais, L.J., Ochasan, J., Manjunath, K.L., Febres, V.J., Lee, R.F., and Niblett, C.L. 2000. Improved sensitivity in the detection and differentiation of citrus huanglongbing bacteria from South Africa and the Philippines. In: *Proc. 14th Conf. IOCV*. IOCV, Riverside. pp.195-200.

Hilf, M.E, and Garnsey, S.M. 2000. Characterization and classification of citrus tristeza virus isolates by amplification of multiple molecular markers. In: *14th Proc. Conf. IOCV*. IOCV, Riverside. pp. 18-27.

Hooker, M.E., Lee, R.F., and Civerolo, E.L. 1993. Reliability of gentisic acid, a fluorescent marker, for diagnosis of citrus greening disease. *Plant Dis.* 77:174-180.

Hopkins, D.L. 1989. *Xylella fastidiosa*: xylem-limited bacterial pathogen of plants. *Annu. Rev. Phytopatol.* 27:271-290.

Hutton, R.J., Broadbent, P., and Bevington, K.B. 1999. Viroid dwarfing for high density citrus plantings. *Hortic. Rev.* 24:277-317.

Jagoueix, S., Bove, J.M., and Garnier, M. 1994. The phloem-limited bacterium of greening disease of citrus is a member of the ά subdivision of the proteobacteria. *Int. J. Syst. Bacteriol.* 44:397-386.

Jagoueix, S. Bove, J.M., and Garnier, M. 1996. PCR detection of the two liberobacter species associated with greening disease of citrus. *Mol. Cell. Probes* 10:43-50.

Karasev, A. V. 2000. Genetic diversity, and evolution of closteroviruses. *Annu. Rev. Phytopathol.* 38:293-324.

Karasev, A. V., Boyko, V.P., Gowda, S., Nikolaeva, O.V., Hilf, M.E., Koonin, E.V., Niblett, C.L., Cline, K., Gumpf, D.J., Lee, R.F., Garnsey, S.M., Lewandowski, D.J., and Dawson, W.O. 1995. Complete sequence of the citrus tristeza virus RNA genome. *Virology* 208:511-520.

Khan, I.A., Lee, R.F., and Hartung, J. 2003. Confirming seed transmission of witches' broom disease of lime. Proc. 8th International Congress of Plant Pathology, 2-7 February, 2003, Christchurch, New Zealand.

Lee, R. F., Baker, P.S., and Rocha-Pena, M.A. 1994. The citrus tristeza virus (CTV). *International Institute of Biological Control*, Berks, UK 197 pp.

Lee, R.F., Derrick, K.S., Beretta, M.G.J., Chagas, C.M., and Rosetti, V. 1991. Citrus variegated chlorosis: a new destructive disease of citrus in Brazil. *Citrus Industry* Oct 1991:12-13, 15.

Lee, R. F., Garnsey, S.M., Brlansky, R.H., and Goheen, A.C. 1987. A purification procedure for enhancement of citrus tristeza virus yields and its application to other phloem-limited viruses. *Phytopathology* 77:543-549.

Lee, R. F., Lehman, P.S., and Navarro, L. 1999. Nursery practices and certification programs for budwood and rootstocksIn: *Citrus Health Management*, APS. APS, St. Paul, MN. .pp. 35-46.

Lee, R.F., Marais, L.J., Timmer, L.W., and Graham, J.H. 1984. Syringe injection of water into the trunk: a rapid diagnostic test for citrus blight. *Plant Dis.* 68:511-513.

Lee, R. F., McConnell, P., Manjunath, K.L., Cevik, B., Nikolaeva,O.V.,Dekkers, M.G.H., and Niblett, C.L. 2002. The *Citrus tristeza virus* epidemic in Bog Walk Valley, Jamaica. In: *Proc. 15th Conf. IOCV.* IOCV, Riverside. pp. 95-101.

Legarreta, G.G., Garcia, M.L., Costa, N., and Grau, O. 2000. A highly sensitive heminested RT-PCR assay for the detection of citrus psorosis virus targeted to a conserved region of the genome. *J. Virol. Meth.* 84:15-22.

Lopez, C., Ayllon, M.A., Navas-Castillo, J., Guerri, J., Moreno, P., and Flores, R. 1998. Molecular variability of the 5 '- and 3 '-terminal regions of citrus tristeza virus RNA. *Phytopathology* 88: 685-691.

Manjunath, K.L., Hooker, M., Pappu, H.R., Pappu, S.S., Powell, C.A., Bar-Joseph, M., Niblett, C.L., and Lee, R.F. 1995. Synthesis of a functional single chain antibody against citrus tristeza closterovirus (CTV) in bacteria. In: *The Programme and Abstracts, 13th Conf. of IOCV.*, Fuzhou, China, Nov 16-23, 1995. pp 39.

Marais, L.J. 1994. Citrus tristeza virus and its effect on the southern African citrus industry. *Citrus Industry* 75(6):58-60.

Marais, L.J., Lee, R.F., Breytenbach, J.H.J., Manicom, B.Q., and van Vuuren, S.P. 1996. Association of a viroid with gum pocket disease of trifoliate orange. In: *Proc. 13th Conf. IOCV.* IOCV, Riverside. pp. 236-244.

Marais, L.J., Marais, M.L., and Rea, M. 1996. Effect of citrus tristeza stem pitting on fruit size and yield of Marsh grapefruit in Southern Africa. In: Proc. *13th Conf. IOCV.* IOCV, Riverside. pp. 163-167.

Mawassi, M., Mietkiewska, E., Gofman, R., Yang, G., and Bar-Joseph, M. 1996. Unusual sequence relationships between two isolates of citrus tristeza virus. *J Gen Virol* **77**:2359-2364.

McClean, A.P.D., and Oberholzer, P.C.J. 1965. Citrus psylla, a vector of the greening disease of sweet orange. *S. Afr. J. Agric. Sci.* 8:297-298.

Mehta, P., Brlansky, R.H, Gowda, S. and Yokomi, R.K. 1997. Reverse-transcription polymerase chain reaction detection of citrus tristeza virus in aphids. *Plant Dis.* 81:1066-1069.

Moll, J. N., Martin, M. M. 1973. Electron microscope evidence that citrus psylla (Trioza erytreae) is a vector of greening disease in South Africa. *Phytophylactica* 5:41-44

Moreira, L., Rodríguez, C.M., Godoy, C., Villalobos, W., Saborío-Rodríguez, G., Montero-Astúa, M., González-Herrera, A., Sánchez F., and Rivera, C. 2003 *Xylella fastidiosa* In Costa Rica. *Abstr. Pan American Plant Disease Conf., April 5-10, 2003, South Padre Island, Texas*. 279 pp.

Navarro, L. 1993. Citrus sanitation, quarantine and certification programs. In: *Proc. 12th Conf. IOCV.* IOCV, Riverside. pp. 383-391.

Niblett, C. L., Genc, H., Cevik, B., Halbert, S., Brown, L., Nolasco, G., Manjunath, K.L., Febres, V.J., Pappu, H.R., and Lee, R.F. 2000. Progress on strain differentiation and the epidemiology of citrus tristeza virus. *Virus Res.* 71: 97-106.

Nikolaeva, O.V., Karasev, A.V., Garnsey, S.M., and Lee, R.F. 1999. Serological differentiation of the citrus tristeza virus isolates causing stem pitting in sweet orange. *Plant Dis.* 82: 1276-1280.

Oberholzer, P.C.J., van Staden, D.F.A., and Basson, W.J. 1965. Greening disease of sweet orange in South Africa. In: *Proc. 3rd Conf. IOCV.* Univ. of Florida Press, Gainesville. pp 213-219.

Ochoa, F.M., Cevik, B., Febres, V., Niblett, C.L., and Lee, R.F. 2000. Molecular characterization of Florida citrus tristeza virus isolates with potential use in mild strain cross protection. In: *14th Proc. Conf.IOCV.* IOCV, Riverside. pp. 94-102.

Palacio, A., Foissac, X., and Duran-Vila, N. 2000. Indexing of citrus viroids by imprint hybridization: comparison with other detection methods. In: *Proc. 14th Conf. IOCV.* IOCV, Riverside. pp. 294-301.

Permar, T.A., Garnsey, S.M., Gumpf, D.J., and Lee, R.F. 1990. A monoclonal antibody that discriminates strains of citrus tristeza virus. *Phytopathology* 80:224-228.

Purcell, A.H. and Hopkins, D.L. 1996. Fastidious xylem-limited bacterial plant pathogens. *Annu. Rev. Phytopathol.* 34:131-151.

Raccah, B., Roichtacher, C.N., and Barbagallo, S. 1989. Semipersistent transmission of viruses by vectors with special emphasis in citrus tristeza virus. *Adv. Dis. Vector Res.* 6:301-340.

Rocha-Pena, M.A. and Lee, R.F. 1991. Serological techniques for detection of citrus tristeza virus. *J. Virol. Meth.* 34:311-331.

Rocha-Pena, M.A., Lee, R.F., Lastra, R., Niblett, C.L., Ochoa-Corona, F.M.,, Garnsey, S.M., and Yokomi, R.K. 1995. Citrus tristeza virus and its aphid vector *Toxoptera citricida.* Threats to citrus production in the Caribbean and Central and North America. *Plant Dis.* 79:437-445.

Roistacher, C.N. 1994. Rapid multiplication of citrus from a single plant. In: *Proc. International Society of Citriculture: Vol. 1. Taxonomy, breeding and varieties, rootstocks and propagation, plant physiology and ecology*: 7th International Citrus Congress, Acireale, Italy, 3-13 March, 1992. pp. 309-312.

Roistacher, C.N. 1993. Psorosis – A review. In: *Proc. 12th Conf. IOCV*. IOCV, Riverside. pp. 139-154.

Roistacher, C. N. 1996. The economics of living with citrus diseases: Huanglongbing (Greening) in Thailand. In: *Proc. 13th Conf. IOCV*. IOCV, Riverside. pp. 279-285.

Roistacher, C.N., Canton, H., and Reddy, P.S. 1996. The economics of living with citrus viroids in Belize. In: *Proc. 13th Conf. IOCV*. IOCV, Riverside. pp. 370-375.

Roistacher, C.N., Pehrson, J.E., and Semancik, J.S. 1991. Effect of citrus viroids and the influence of rootstocks on field performance of Navel orange. In: *Proc. 11th Conf. IOCV*. IOCV, Riverside. pp. 234-239.

Sanger, H.L. 1988. Viroids and viroid diseases. *Acta Horti*. 234:79-87.

Schwarz, R.E., Knorr, L.C., and Prommintara, M. 1974. Citrus greening disease in Thailand. FAO Technical Document No. 93, 14 pp. FAO, Rome.

Sdoodee, R. and Garnett, H. 1994. Detection of citrus greening bacterium by immunoblotting. *Songklanakarin J. Sci. Technol*. 16(3): 291-300.

Simpson, A.J., Reinach, F.C., Arruda, P., Abreu, F.A., Acencio, M., Alvarenga, R., Alves, L.M., Araya, J.E., Baia, G.S., Baptista, C.S., Barros, M.H., Bonaccorsi, E.D., Bordin, S., Bove, J.M., Briones, M.R., Bueno, M.R., Camargo, A.A., Camargo, L.E., Carraro, D.M., Carrer, H., Colauto, N.B., Colombo, C., Costa, F.F., Costa, M.C., Costa-Neto, C.M., Coutinho, L.L., Cristofani, M., Dias-Neto, E., Docena, C., El-Dorry, H., Facincani, A.P., Ferreira, A.J., Ferreira, V.C., Ferro, J.A., Fraga, J.S., Franca, S.C., Franco, M.C., Frohme, M., Furlan, L.R., Garnier, M., Goldman, G.H., Goldman, M.H., Gomes, S.L., Gruber, A., Ho, P.L., Hoheisel, J.D., Junqueira, M.L., Kemper, E.L., Kitajima, J.P., Krieger, J.E., Kuramae, E.E., Laigret, F., Lambais, M.R., Leite, L.C., Lemos, E.G., Lemos, M.V., Lopes, S.A., Lopes, C.R., Machado, J.A., Machado, M.A., Madeira, A.M., Madeira, H.M., Marino, C.L., Marques, M.V., Martins, E.A., Martins, E.M., Matsukuma, A.Y., Menck, C.F., Miracca, E.C., Miyaki, C.Y., Monteriro-Vitorello, C.B., Moon, D.H., Nagai, M.A., Nascimento, A.L., Netto, L.E., Nhani, A. Jr., Nobrega, F.G., Nunes, L.R., Oliveira, M.A., de Oliveira, M.C., de Oliveira, R.C., Palmieri, D.A., Paris, A., Peixoto, B.R., Pereira, G.A., Pereira, H.A. Jr., Pesquero, J.B., Quaggio, R.B., Roberto, P.G., Rodrigues, V., de M Rosa, A.J., de Rosa, V.E. Jr., de Sa, R.G., Santelli, R.V., Sawasaki, H.E., da Silva, A.C., da Silva, A.M., da Silva, F.R., da Silva, W.A. J.r, da Silveira, J.F., Silvestri, M.L., Siqueira, W.J., de Souza, A.A., de Souza, A,P., Terenzi, M.F., Truffi, D., Tsai, S.M., Tsuhako, M.H., Vallada, H., Van Sluys, M.A., Verjovski-Almeida, S., Vettore, A.L., Zago, M.A., Zatz, M., Meidanis, J,. Setubal, J.C. 2000. The genome sequence of the plant pathogen *Xylella fastidiosa*. The *Xylella fastidiosa* Consortium of the Organization for Nucleotide Sequencing and Analysis. Nature 406: 151-157.

Su, H-J, and Tolley, I.S. 1992. First consulting visit to Thailand. *Reports No. 10 and 11 on behalf of Dutche Gesellschaft Für Technische Zusammenarbeit (GTZ)*. 49 pp.

Suastika, G., Natsuaki, T., Terui, H., Kano, T., Ieki, H., and Okuda, S. 2001. Nucleotide sequence of citrus tristeza virus seedling yellows isolate. *J. Gen Plant Pathol* 67:73-77.

Targon, M.L.P.N., Machado, M.A., Muller, G.W.,Manjunath, K.L., and Lee, R.F. 2000. Sequence of the coat protein gene of the severe Capao Bonito citrus tristeza virus complex. In: *Proc. 14th Conf. IOCV*. IOCV, Riverside. pp. 418.

Timmer, L.W., Garnsey, S.M., and Graham, J.H. (Eds). 2000. *Compendium of Citrus Diseases, sec. edition*. APS Press, St. Paul, MN. 92 pp.

Timmer. L.W., Garnsey, S.M. and McRitchie, J.J. 1978. Comparative symptomatology of Florida and Texas isolates of citrus ringspot virus on citrus and herbaceous hosts. *Plant Dis. Reptr*. 62:1054-1058.

Timmer, L.W., Lee, R.F., Brlansky, R.H., Graham, J.H., Albrigo, L.G., Derrick, K.S., and Tucker, D.P.H. 1992. The infectious nature of citrus blight. *Proc. Fla. State Hortic. Soc.* 105:21-26.

Vives, M.C., Rubio, L., Lopez, C., Navas-Castillo, J., Albiach-Marti, M.R., Dawson, W.O., Guerri, J., Flores, R., and Moreno, P. 1999. The complete genome sequence of the major component of a mild citrus tristeza virus isolate. *J. Gen Virol* 80:811-816.

Wallace, J.M. 1957. Virus-strain interference in relation to symptoms of psorosis disease of citrus. *Hilgardia* 27:223-245.

Wutscher, H.K., Cohen, M. and Young, R.H. 1977. Zinc and water-soluble phenolic levels in the wood for diagnosis of citrus blight. *Plant Dis. Reptr.* 61:572-576.

Yang, G., Mawassi, M., Gofman, R., Gafny, R., and Bar-Joseph, M. 1997. Involvement of a subgenomic mRNA in the generation of a variable population of defective citrus tristeza virus molecules. *J. Virol.* 71: 9800-9802.

Yang. X., Hadidi, A., and Garnsey, S.M. 1992. Enzymatic cDNA amplification of citrus exocortis and cachexia viroids from infected citrus hosts. *Phytopathology* 82:279-285.

Yokomi, R. K., Lastra, R., Stoetzel, M.B., Damgsteet, V.D., Lee, R.F., Garnsey, S.M., Rocha-Pena, M.A., and Niblett, C.L. 1994. Establishment of the brown citrus aphid *Toxoptera citricida* (Kirkaldy) (Homoptera: *Aphididae*) in Central America and the Caribbean basin, and its transmission of citrus tristeza virus. *J. Econ. Entomol.* 87:1078-1085.

Zreik, L., Carle, P., Bove, J.M., and Garnier, M. 1995. Characterization of the mycoplasmalike organism associated with witches' broom disease of lime and propositioln of a Candidatus taxon for the organism, "*Candidatus* Phytoplasma aurantifolia". *Int. J. Syst. Bacteriol.* 45:449-453.

Figure captions:

Figure 24. 1. Symptoms expressed by *Citrus tristeza virus* (CTV), citrus viroids, and citrus blight. A. Hamlin sweet orange tree on sour orange rootstock undergoing decline due to decline strains of CTV. B. Fruit size of Hamlin fruit from the tree illustrated in A, compared to fruit from a non-declining tree, immediately behind the declining tree in A. Depending on the time of year, amount of fruit already set, and length of time of the tree to completely die due to decline strains of CTV, the fruit size may be significantly reduced. C. Stem pitting due to CTV on twigs from a Marsh grapefruit tree showing pits with gumming beneath due to stem pitting strains of CTV. Stem pitting due to CTV can occur on sweet orange also. Fruit size and quality is reduced on trees affected with stem pitting strains of CTV. D. Citrus exocortis induced bark scaling on *Poncirus trifoliata* rootstock, note that the bark scaling occurs on the rootstock only and not on the scion. E. Variation of symptoms due to citrus viroids expressed on the viroid indicator plant Etrog citron clone 861. The plant on the right, labeled "H" is healthy, the rest of the plants were inoculated from field sources containing various mixtures of citrus viroids. F. A Valencia sweet orange tree on rough lemon rootstock showing the symptoms of citrus blight (CB), a disease of unknown etiology which may be graft transmitted using root grafting. The trees affected with CB first develop an off color green appearance, often followed by zinc deficiency-like chlorosis on the southwest quadrant of the tree (Northern hemisphere). Canopy dieback then occurs and the tree declines over a period of several years. Fruit size and yield are reduced, blossom set is delayed 2-3 weeks compared to healthy trees.

Figure 24.2. Symptoms expressed by fastidious prokaryotes causing diseases in citrus. A. Citrus variegated chlorosis (CVC), caused by *Xylella fastidiosa,* causes a general stunting in affected trees, and the tree takes on a yellow or chlorotic appearance. The tree in the foreground is CVC infected while the tree immediately behind is growing normally. B. With CVC, leaf symptoms begin with an interveinal chlorosis which is apparent on both sides of the leaf. Gummy slightly raised lesions are then formed on the underside of the leaf in the areas which are chlorotic. C. CVC reduces fruit size, the fruit have hard rinds, and the fruit ripen early and have a higher than normal sugar content. The Natal sweet orange fruit on the left came from a tree apparently free of CVC, the fruit on the right came from a CVC infected tree. D. Yellow shoot symptom on a Valencia tree infected with an African strain of Huanglongbing (HLB), commonly called citrus greening. The sectored branches showing general chlorosis are common on early symptomatic HLB affected trees. E. Witches' broom of lime (WBDL) from Oman. The witches' brooms start to develop in the inner parts of the tree canopy, then gradually spread and appear throughout the tree's canopy. The WBDL affected trees eventually die. The WBDL in Oman and surrounding countries is caused by *Candidatus* Phytoplasma aurantifolia. F. A 20 year old Swingle citrumelo tree in Jamaica showing witches' broom symptoms. A phytoplasma was identified from samples taken from this tree (R. Lee, unpublished).

Figure 24.3. Representation of the gene expression and genome organization of citrus tristeza virus. Open reading frames are shown as boxes and the putative domains on ORF 1a and 1b are separated by lines. PRO: papain-like proteases 1 and 2; RdRp: RNA-dependent polymerase. The genomic and subgenomic RNAs are shown by solid lines, the size, in kilobases, is indicated by the scale at the top. Defective-RNA strategy is shown by dashed lines (Ochoa-Corona *et al.* 2000).

Figure 24.4. The effect of different citrus viroid combinations on the canopy area at the end of 10 years (Fig. 6 A) and the cumulative 10 year yield (Fig. 6 B) of Naval sweet orange on Troyer citrange, a viroid susceptible rootstock, and sour orange, a viroid tolerant rootstock. The canopy area is expressed as cubic meters, the yield is expressed as kilograms of fruit. Data from Roistacher *et al.* 1991.

Figure 24.5. Huanglongbing (HLB) is a limiting factor of citrus production in Thailand. Grenzenback (1984) and Roistacher (1996) studied the cost of producing citrus in Thailand, calculated the financial returns from several average citrus farms, and extrapolated the profit or loss in US$ per hectare depending on the life span of the block. The rate of spread of HLB was about 10% per year, the average life span was eight years with a range of five to twelve years. The extrapolated profit/loss for 6, 8, 10, and 12 years of age are illustrated in A, B, C, and D, respectively. The short life span groves coincided with planting trees which were already infected with HLB. This example highlights the need to plant healthy plants.

24.1A
24.1B
24.1C
24.1D
24.1E
24.1F

24.2A
24.2B
24.2C
24.2D
24.2E
24.2F

Figure 24.3.

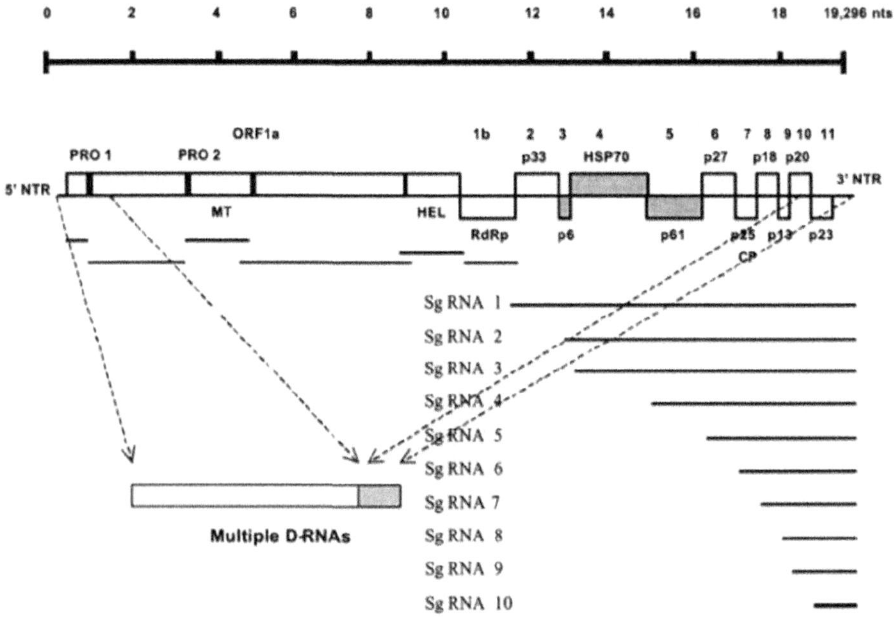
0
2
4
6
8
10
12
14
16
18
19,296 nts
ORF1a
1b
2
3
4
5
6
7
8
9
10
11
PRO 1
PRO 2
p33
HSP70
p27
p18
p20
5' NTR
3' NTR
MT
HEL
RdRp
p6
p61
p25
p13
p23
CP
Sg RNA 1
Sg RNA 2
Sg RNA 3
Sg RNA 4
Sg RNA 5
Sg RNA 6
Sg RNA 7
Sg RNA 8
Sg RNA 9
Sg RNA 10
Multiple D-RNAs

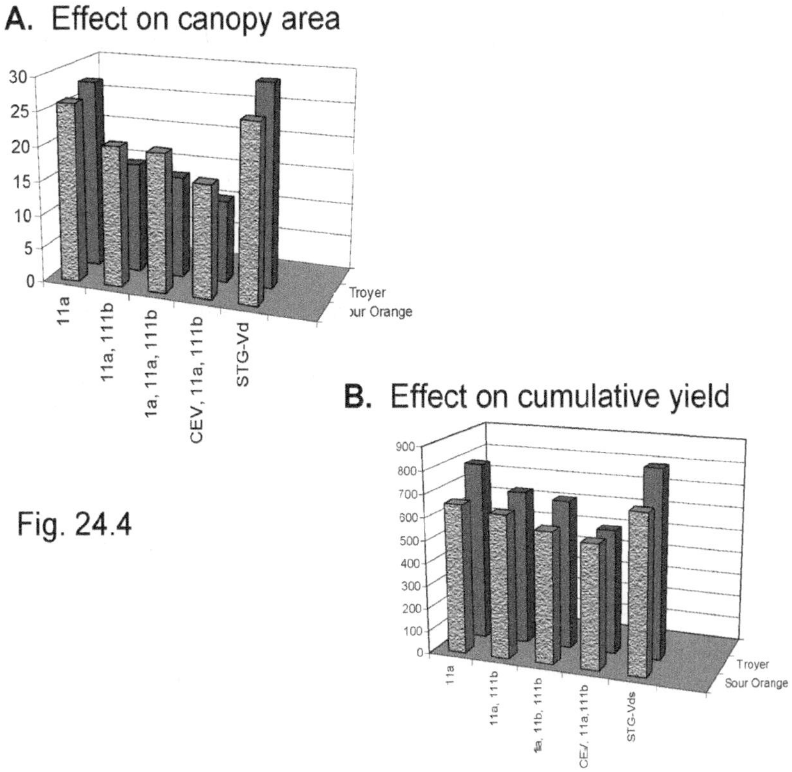

Fig. 24.4

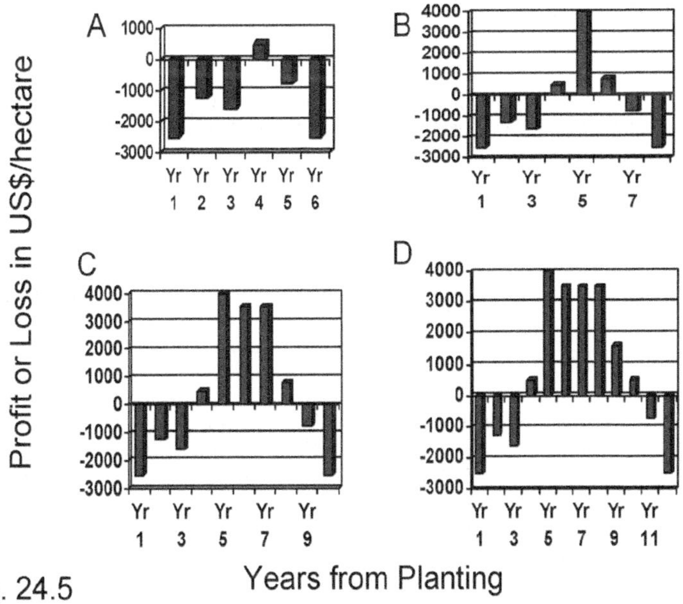

Fig. 24.5

25. Tomato

Y. Antignus
Department of Virology, the Volcani Center, ARO, P. O. B. 6, Bet Dagan, 50250, Israel

25.1 Introduction

The cultivated tomato *Lycopersicon esculentum* has achieved high popularity thanks due to its versatility in fresh or processed forms. Tomato is of 'New World' origin and wild species arc found in the Andean region, Peru, and adjacent areas. The wild cherry tomato, *L. esculentum* var. *cerasiforme*, is common in tropical and subtropical America and is probably the progenitor of the cultivated tomato. There are more than 4,000 varieties of tomato, ranging from the small, marble-size cherry tomato to the giant Ponderosa that can weigh more than 1.5 kg (Rick, 1978).

Plant virus diseases have become a limiting factor in tomato production worldwide. However, information from developing countries on the economic damage caused by viruses is scarce. In the USA, the estimated losses caused by whitefly-transmitted geminiviruses reach about 20% of tomato production, but in the Dominican Republic, Cuba, Mexico, Guatemala, Honduras, Nicaragua, Costa Rica, Venezuela and Brazil the damage is much greater, ranging between 30 and 100% of the yield. The losses in the Dominican Republic during 1989-1995 were estimated at $50 million (Polston and Anderson, 1997).

In African countries, a total of 11,694,000 metric tons of tomatoes were produced on an area of 602,744 ha with an average yield of 19.4 t/Ha. The main producers were Egypt (189,102 ha) and Nigeria (126,000 ha), which together include 50% of the total production area (FAOSTAT). As most of the production is in open fields, plants are exposed to infection by tomato yellow leaf curl virus (TYLCV), which results in severe epidemics of the disease (personal observations). In 1988, losses from this disease in Egypt were estimated at 32,000 ha (Nakhla *et al.* 1993).

Developing countries in Asia produced 43,332,000 metric tons of tomatoes, grown on 1,822,021 ha with an average yield of 23.8 t/ha.

The production in South America is around 6,000,000 metric tons, harvested from an area of 152,813 ha, with an average yield of 39.2 t/ha (FAOSTAT).

This chapter will deal with the important virus and viroid diseases affecting tomato, with emphasis on those occurring in developing countries. The principles of integrated pest management (IPM) to reduce damage due to virus will be outlined. The viruses will be grouped according to their insect vectors, as the interaction between the virus and its vector is a major factor in the control strategy to be selected.

25.2 The Main Viruses

25.2.1 Whitefly-transmitted viruses

Four types of damage are caused by whiteflies *Bemisia tabci* : depletion of the plant nutrient, plant toxication as the insect feeds, plant contamination by honey dew and sooty molds, and damage caused by transmission of plant viruses belonging to several taxonomic families (Byrne *et al.*, 1990; Duffus, 1987).

25.2.1.1 Begomoviruses, family *Geminiviridae*

All 'New World' begomoviruses have a bicomponent genome, whereas some of the 'Old World' ones have a single genome. All group members are transmitted by the whitefly, *Bemisia tabaci.*

Tomato yellow leaf curl virus Genus ***Begomovirus*** (TYLCV)

The TYLCV virion has a geminate shape and was purified and characterized by Czosnek *et al.* (1988). TYLCV has a single partite genome of circular ssDNA, about 2.8 kb in size that is encapsidated in a bisegmented virion about 18 x 20 nm (Rybicki *et al.* 2000). The genome has six open reading frames (ORFs) organized bidirectionally on both strands of the replicative form of the genomic DNA (Lazarowitz, 1992). The two transcriptional units are separated by an intergenic region of *c.* 300 nucleotides that is involved in replication and transcription. The viral capsid protein encoded by V1 is responsible for movement and interaction with the whitefly vector. V2 and C4 are associated with symptom expression and virus movement. C1 encodes the replication protein (Rep). The C2 ORF is a transcriptional activator protein for the coat protein promoter; and the C3 ORF interacts with the Rep protein and enhances viral DNA accumulation (Jupin *et al.* 1995; Wartig *et al.* 1997).

TYLCV infects tomato in tropical and subtropical regions. Disease symptoms in tomato include upward curling of leaflet margins, degeneration of the apical part of the plant, and yellowing (Fig. 25.1)

Although 13 plant species in six botanical families are susceptible to the virus, in nature TYLCV infects *Cynanchum acutum*, *Datura stramonium*, *Malva parviflora* (Cohen and Antignus, 1994) and *Mercurialis ambigua* (Sanches-Campos *et al.* 2000). Among the cultivated crops other than tomato, TYLCV-Is has been detected in common bean (*Phaseolus vulgaris*), and Lisianthus (*Eustoma grandiflora*) (Cohen *et al.* 1995). TYLCV-Sar was found in naturally growing *D. stramonium* and *Solanum nigrum* (Bedford *et al.*, 1998), *Solanum luteum* (Sanchez-Campos *et al.*, 2000) and *Euphorbia* spp. (Moriones and Navas-Castillo, 2000).

TYLCV, like other begomoviruses, is transmitted by *B. tabaci* in a persistent circulative manner. The virus is retained in the insect for periods ranging from weeks to lifetime, and its transmissibility is retained through molts. It has been hypothesized that the transmission process consists of two phases: first, acquisition that involves passage of the virus through the gut wall into the haemocoel; and then the passage of the virus from the haaemocoel into the salivary gland. It has also been suggested that the viral coat protein is responsible for virus-vector specificity and that this specificity resides in the haemocoel-salivary gland barrier (Cohen and Antignus, 1994; Hunter *et al.* 1998; Rossel *et al.* 1999). The specificity of geminivirus transmission may result from an intimate relationship between exposed areas of the CP and certain insect receptors (Noris *et al.* 1998).

The first description of the disease caused by TYLCV in Israel was in 1939 from the Jordan Valley, and the disease spread later to all parts of Israel. It became a limiting factor for tomato production, and susceptible varieties could no longer be grown in the open (Cohen and Antignus, 1994). Two strains of the virus, TYLCV-Is (Navot *et al.* 1991) and TYLCV-Is [Mild] (Antignus and Cohen, 1994) were characterized, both differing in the severity of symptoms they show in tomato. The virus was also reported from Jordan (Makkouk, 1978), and a disease incidence of 93-100% was recorded in the open field in fall-grown tomatoes (Al-Musa, 1982). TYLCV was reported from Saudi Arabia and, as in other Mediterranean countries, the peak of the epidemic was during summer and autumn, when the environmental conditions are optimal for virus spread (Mazyad *et al.* 1979). An isolate with high sequence homology to TYLCV-Is was found in the northern part of Saudi Arabia, and a second isolate with 82% homology to TYLCV was described in the southern part of the country (Hong and Harrison, 1995). A TYLCV isolate was also described in the Republic of Yemen (Bedford *et al.* 1994), and samples collected in Lebanon, Syria and Turkey were found to be closely related to TYLCV-Is (Czosnek and Latterot, 1997).

Viruses of the TYLCV complex were reported from the Far East. TYLCV-Taiwan and TYLCV-China possess a single genomic component, as does TYLCV-Is, but their relationships with TYLCV-Is are relatively distant (Czosnek and Latterot, 1997). Two Japanese TYLCV isolates have a sequence similarity of over 95% with TYLCV-Is Mild (Kato *et al.* 1998).

TYLCV isolates from Thailand and those from India possess two genomic components. Interestingly, the A component of TYLCV Thailand is capable on its own to cause symptomatic infections in tomato, but the symptoms of TYLCV-Th are more severe when both A- and B- DNAs are introduced into the plant (Rochester *et al.* 1994). Tomato leaf curl virus (ToLCV) isolates from northern India have a bipartite genome (Padidam *et al.* 1995), and they share only 45-47% sequence identity with the monopartite ToLCV isolates (ToLCV-Ban1-4) of southern India (Muniyappa *et al.* 2000).

TYLCV is widely spread in African countries. The Egyptian isolate of TYLCV was nearly identical to TYLCV-Is, with >92% identity (Nakhla *et al.* 1993). However TYLCV from Nigeria and Tanzania (TYLCV-Ng and TYLCV-Tz, respectively) have more distant phylogenic relationships with TYLCV-Is (Moriones and Navas-Castillo, 2000). TYLCV has also been reported from other African countries viz., Mali, Senegal, Burkina Faso and Morocco (Czosnek and Laterrot, 1997; Monci *et al.* 2000).

TYLCV has been known in the Iberian Peninsula since 1992, when TYLCV-M was responsible for the first outbreaks in Spain (Noris *et al.* 1994). There it spread to all the main vegetable-producing regions and became a limiting factor in tomato production. TYLCV-Is and the Sardinian isolate TYLCV-Sar were also reported as the cause of disease outbreaks in Spain. However, a displacement of the latter by the former was later observed, probably because of the more efficient vectoring of TYLCV-Is by the local biotypes of *Bemisia tabaci*, and the existence of a bridge crop (beans) which supports TYLCV-Is between tomato-growing seasons (Navas-Castillo *et al.* 1997; Sanchez-Campos, 1999). TYLCV-Is also invaded the southern parts of Portugal (Louro *et al.* 1996), and TYLCV-Is and TYLCV-Sar were recently observed in the Canary Islands (Monci *et al.* 2000).

Since 1990 tomato producing areas in Florida, the Caribbean, Mexico, Central America, Venezuela and Brazil have suffered from high incidences of whitefly-borne geminiviruses, with severe economic consequences. TYLCV-Is was unknowingly introduced to the Dominican Republic on transplants from Israel (Polston *et al.* 1994; Polston and Anderson, 1997). TYLCV-Is was subsequently identified in Jamaica, Cuba (Gonzalez and Valdez, 1995; McGlashan*et al.* 1994; Ramos *et al.* 1996) and Florida (Polston *et al.* 1999).

A phylogenetic tree of TYLCV isolates shows seven branches of the recognized species for which CP sequences are available (Moriones and Navas-Castillo, 2000). These are TYLCV Thailand (TYLCV-Th), TYLCV China (TYLCV-Ch), TYLCV Sardinia (TYLCV-Sar), TYLCV Tanzania (TYLCV-Tz), TYLCV Nigeria (TYLCV-Ng), TYLCV South Saudi Arabia (TYLCV-SSA) and TYLCV-Israel (TYLCV-Is).

In subtropical regions, TYLCV epidemics have seasonal patterns, characterized by outbreaks during late summer and autumn, which is when the populations of *B. tabaci* reach their peak (Cohen and Antignus, 1994; Ioannou, 1987). Epidemiological studies of TYLCV in the Jordan valley demonstrated that the autumn peak in the whitefly population coincided with the peak of disease incidence in trap plants. The migration of whiteflies from the banks of the Jordan river where *Cynanchum acutum,* a host plant of TYLCV and *B. tabaci* is very common, was monitored by dusting these plants with fluorescent dust. Labeled whiteflies were trapped within the main tomato production area 7 km away from the dusting point (Cohen *et al.* 1988). In areas where tomato crops are grown throughout the year (in greenhouses) the continuous availability of source plants supports the infection cycle without the need of any intermediate hosts.

The large number of strains forming the TYLCV complex is the consequence of a variation in viral proteins and regulating sequences in the genome of members of the complex. The evolution of new strains may be attributed to two mechanisms:

1. Transfer of infected plant material or viruliferous *B. tabaci* individuals *via* plant trade or human movement (Polston and Anderson, 1997); and

2. Recombination between an invading virus and local isolates, which is not rare among begomoviruses, and which would provide additional sources of variation in virus pathogenicity (Moriones, 1997, Navas Castillo *et al.* 2000; Zhou *et al.* 1997).

This evolutionary process is manifested in the displacement of less virulent strains by more efficient strains with a broader host range, better transmission efficiency by the vector or a higher replication capacity in the host. This hypothesis is supported by reports from Spain indicating the displacement of TYLCV-Sar by TYLCV-Is (Navas-Castillo *et al.* 1997; Sanchez-Campos *et al.,* 1999) and the disappearance of TYLCV-Is [Mild] that was collected in field in the early sixties (Antignus and Cohen, 1994), to be replaced by TYLCV-Is (Navot *et al.* 1991).

Begomoviruses infecting tomato in Africa

Tomato curly stunt virus was described in South Africa. The symptoms are similar to those caused by TYLCV. Yield losses were close to 100% when infection was at an early stage. Sequencing data indicate that the causal agent is a distinct *Begomovirus*, closely related to the TYLCV cluster (Pietersen *et al.* 2000).

Begomoviruses infecting tomato in the United States and Mexico

The first description of tomato mottle virus (ToMoV) came from Florida (Kring *et al.* 1991). Disease symptoms include chlorotic mottle, leaf curling, stunting, and reduced fruit size and number. Outbreaks of the disease coincided with the emergence of the B biotype of *B. tabaci.* The virus spread from Florida to other states in the USA, including Puerto Rico where it became endemic. The disease incidence in the open field sometimes reached 100%, but the use of the systemic insecticide imidacloprid reduced the spread of the disease significantly (Polston and Anderson, 1997).

Texas pepper virus (TPV) causes mosaic, leaf curling and stunting. Epidemics of the disease occur regularly in tomato and pepper in Mexico (Torres-Pacheco *et al.* 1996) Geminiviruses that infect tomato in Mexico often occur in complexes. Chino del tomate virus (CdTV) was first noticed during 1970-1971 on cultivated tomatoes grown in the coastal area of Sinaloa, Mexico (Brown *et al.* 2000; Polston and Anderson, 1997). Symptoms include severe leaf curling, vein thickening, yellow mosaic, stunting, and reduced fruit set

Begomoviruses infecting tomato in Central and South America

Major tomato diseases were reported from Guatemala in 1987, El Salvador and Costa Rica in 1988, Honduras in 1989, and Panama in 1991. Additional geminiviruses with temporary designation have been identified in Guatemala, Honduras and Nicaragua (Polston and Anderson, 1997). Tomato golden mosaic virus (TGMV), a whitefly-borne geminivirus has been a problem in Brazil since the 1960s. Recent surveys however, failed to show its presence, probably because it had been displaced by other geminiviruses. Newly emerging geminiviruses were described in several states of Brazil, coincidentaly with the establishment of *B. tabaci,* biotype B in many crops (Polston and Anderson, 1997).

Tomato yellow mosaic virus (TYMV) is a limiting factor in tomato production in Venezuela (Lastra and Uzcategui, 1975). The virus induces bright yellow mosaic in tomato. In 1997, a strain of potato yellow mosaic virus (PYMV) was identified in affected tomatoes (Guzman *et al.* 1997).

Control

The persistent mode of transmission of TYLCV and other begomoviruses by *B. tabaci* makes chemical control of the disease attractive, but the large whitefly populations make this option inefficient. Moreover, frequent applications of insecticides lead to the development of resistance in the whitefly population (Horowitz *et al.* 1994).

Interference with the epidemiological cycle of TYLCV can be accomplished by eradication of host plants that serve as virus reservoirs. However, in most cases the natural host plants of TYLCV are not known. In a relatively isolated region in Cyprus, the spread of TYLCV was efficiently reduced by eradication of inoculum sources in over- wintered tomato in the spring, before the emergence of adult whiteflies (Ioannou, 1987). In isolated growing areas surrounded by desert that supports the growth of only a limited number of plant species, a 'crop-free period' between growing seasons may help to reduce both whitefly populations and inoculum sources significantly, and thus help to diminish disease spread (Ucko *et al.* 1998).

Cultural methods that interfere with the 'vision behavior' of *B. tabaci* have been developed. It was found that whiteflies are attracted to yellow plastic mulches, on which they stay and are killed later by the heat that is reflected from the plastic (Cohen and Melamed-Madjar, 1978). Mulching of crop rows with straw or sawdust gave similar results (Cohen and Berlinger, 1986). It was also shown that soil mulching with aluminum foil and orange plastic films reduced the number of whiteflies on the canopy of tomato plants, though the protective effect faded out when the foliage covered the plastic (Csizinszky *et al.* 1995).

Cucumbers grown between tomato rows in the open field served as bait plants to protect tomatoes from the spread of TYLCV (Al-Musa, 1982). Physical barriers that prevent contact between the whitefly vector and the plant can be used both in the open field and in greenhouses. Tomatoes for the fresh market are grown in Israel either in greenhouses protected with fine net (50-mesh) screens or, during the summer, in screenhouses made from the same nets. Perforated polyethylene or polypropylene sheets (Agryl) can be stretched over the crop to serve as 'floating barriers', thus preventing the landing of the whitefly vector on the crop (Cohen and Berlinger, 1986).

Spectrally modified polyethylene, polyvinylchloride and polycarbonate are used as cladding materials for greenhouses and at the same time act as efficient filters of UV irradiation. It was shown recently in Israel that elimination of UV from the light spectrum by the above-mentioned cladding materials dramatically reduced the influx of insect pests, including whiteflies, and efficiently inhibited the spread of TYLCV in tomato (Antignus, 2000; Antignus *et al.* 1996, 2001).

The best approach to control TYLCV is through the use of resistant tomato varieties. During recent decades many breeding programs have been initiated, that have yielded cultivars with various levels of resistance, by incorporating resistance or tolerance from wild *Lycopersicon* species into cultivated tomatoes. Inheritance seems to be complicated because of the multigenic nature of resistance. Another limitation to this approach is the linkage of resistance with undesired horticultural traits. Some of the available commercial cultivars had high tolerance to the virus, but their quality was not satisfactory (Moriones and Navas-Castillo, 2000).

25.2.1.2 Criniviruses

These viruses have filamentous particles of about 900 nm in size and a divided genome comprising two ssRNA segments (Klassen *et al.* 1996). Both segments are needed for infectivity and are encapsidated separately. The genome ORFs encode the following protein products: papain-like protease, methyl transferase, helicase, RNA polymerase, heat shock protein, coat protein and coat protein analog (Martelli *et al.* 2000).

Criniviruses are transmitted semipersistently by one or more of the following whitefly species: *B. tabaci* biotypes A and B, *Trialeurodes vaporariorum* (greenhouse whitefly) and *Trialeurodes abutilonea* (banded wing whitefly). Some members of this group of viruses can be transmitted by more than one whitefly. There is a positive correlation between the length of the feeding period and the probability of virus acquisition. Preliminary evidence indicates that the virus accumulates in the cibarium of the whitefly (Wisler and Duffus, 2001). A retention period of 1-9 days has been reported for some of the Criniviruses (Wisler and Duffus, 2001).

Tomato infectious chlorosis virus Genus ***Crinivirus*** (TICV)

The virus has been found in California, Italy, Spain and Taiwan. (Wisler and Duffus, 2001). Virus epidemics occur both in open-field and greenhouse tomatoes. Affected plants show interveinal yellowing, necrosis and yield losses, but when the disease appears at the end of the season there is relatively little damage. Additional host plants are: tomatillo (*Physalis ixocarpa*), potato (*Solanum tuberosum*), artichoke (*Cynara scolymus*), lettuce, (*Lactuca sativa*) and petunia (*Petunia hybrida*). The virus is transmitted efficiently by the greenhouse whitefly *Trialeurodes vaporariorum* but not by *B. tabaci* The greenhouse whitefly acquired TICV during a minimum access feeding period of 1 h, with maximum transmission efficiency at 48 h of inoculation feeding access. The persistence of the virus in the vector is 3-4 days (Wisler and Duffus, 2001).

Tomato chlorosis virus Genus ***Crinivirus*** (ToCV)

This virus has a wide host range, including weeds and ornamentals. *Datura stramonium* and *Solanum nigrum* are among its natural hosts. The symptoms in tomato are similar to those described for TICV: infected plants have an irregular chlorotic mottle that develops first on lower leaves and gradually advances toward the growing point; interveinal yellow areas on leaves also develop red and brown necrotic flecks. No visual symptoms develop on fruits, but fruit size and numbers are reduced.

Tomato chlorosis virus is transmitted by four whitefly vectors: *T. vaporariorum, Trialeurodes abutilonea,* and *B. tabaci* biotypes A and B. *T. abutilonea* and biotype B of *B. tabaci* have been found to be vectors with relatively higher transmission efficiencies. This multiplicity of vectors for a distinct virus is unique among whitefly transmitted viruses.

The virus occurs in the USA, Italy, Portugal, Morocco, Spain, Canary Islands, Taiwan, and South Africa (EPPO/CABI, 2001), reportedly causing heavy losses in Spain and Morocco (Navas-Castillo *et al.* 2000).

Control

Tomato transplants should be grown in insect-proof structures and protected during transportation to the site of planting. Control of the whitefly vector can reduce virus infection. However, there are indications that *B. tabaci* has developed resistance to all groups of insecticides. A rational IPM program should include rotation of insecticides, covering of greenhouses with UV-absorbing films, use of 50-mesh screens, and the introduction of natural enemies to reduce whitefly populations. At present no cultivars resistant to TICV or ToCV are available.

25.2.2 Aphid-transmitted viruses

Among insects, aphids are the most successful exploiters of higher plants and about 3700 spp. (of aphids) have been described. Aphids transmit about 50% of the 600 described viruses (Eastop, 1977).

25.2.2.1 Potyviruses

The genus potyvirus is the largest in the *Potyviridae* family. Members of the group have flexous filament-like virions with a modal size of 680-900 nm. These viruses contain a single molecule of linear positive sense ssRNA, about 9.7 kb in size with a poly (A) tract at the 3' end and a genome- linked protein (Vpg) covalently linked at the 5' terminus. Genome expression is by the formation of a polyprotein that is processed by specific virus-encoded proteases to produce the individual proteins. Some of these proteins accumulate in the nucleus or cytoplasm to form inclusion bodies (Berger *et al.*, 2000). These viruses are transmitted in a non-persistent mode by as many as 200 aphid species, of which *Myzus persicae* is the most efficient (Buchen-Osmond, 1996). This type of vector-virus interaction is characterized by a very short retention of the virus in the vector feeding apparatus, lack of a measurable latent period in the vector, loss of the virus when the vector molts, low level of specificity for aphid species, and increased transmissibility following preacquisition fasting. A virus-encoded protein, the helper component (HC), is involved in virus transmission (Raccah *et al.* 2001).

Potato virus Y Genus ***Potyvirus*** (PVY)

The virus occurs world wide but has a narrow host range that includes *Solanum tuberosum, Capsicum* spp. *Nicotiana* spp. and *L. esculentum* (Buchen-Osmond, 1996). Symptoms in tomato are variable and generally consist of faint mottling and slight distortion, but some virus strains can induce irregular necrotic lesions (Rosner *et al.* 2000).

Tomato in Florida, Australia, Taiwan, Argentina, France, Italy, Morocco and Tunisia are sometimes affected by PVY (Rosner *et al.* 2000; Thomas and McGrath, 1988; Zitter, 1993). Inoculum sources can be found in cultivated crops such as potato or, in subtropical regions, in weeds such as *Solanum nigrum, S. atropurpureum* and *Physalis.*, that serve as oversummering reservoirs (Zitter, 1993).

Tobacco etch virus (TEV)

The natural host range includes, *Cassia abtusifolia, Capsicum annum, D. stramonium, Nicotiana tabacum and L. esculentum.* Infected tomato plants show intense mottling, puckering and rugosity of the leaves. Early infection results in severe plant stunting. The fruits are mottled and remain small (Purciful, 1996; Zitter, 1993).

Tobacco etch virus is limited to North and South America. Severe epidemics of the disease in tomato have been reported from Venezuela (Lastra and Uzcategui, 1975).

Control

The nonpersistent nature of transmission makes the use of chemical control inefficient in reducing the spread of the disease. Frequent applications of mineral oil can retard the spread of the epidemic and can reduce the damage. Some cultural methods aim to prevent contact between the plant and the vector. Soil mulching with reflective materials, mass trapping of aphids with sticky yellow sheets, and planting of barrier plants have been found successful in reducing infection rates of non-persistently transmitted aphid-borne viruses (Loebenstein and Raccah, 1980). At present no commercial tomato cultivars are resistant to PVY and TEV are available (Zitter, 1993).

25.2.2.2 Cucumoviruses

Cucumber mosaic virus Genus ***Cucumovirus*** (CMV)

The virus has a very wide host range among both wild and cultivated plants; it is widespread in tomato in the temperate regions (Francki and Habili, 1996). Diseased tomato plants have a stunted bushy appearance, but the most characteristic symptom is the drastic narrowing of the leaf blades, to form a shoestring-like leaf (Fig. 25.2) (Francki and Habili, 1996).

The viral RNA genome consists of three functional ssRNA segments (RNA 1, 2, and 3), all of which are required for infection. Each segment is packed separately in an isometric particle 28 nm in diameter. The subgenomic coat protein messenger RNA is a fourth RNA species (RNA 4): a fifth RNA satellite component, designated CARNA 5 (CMV-associated RNA 5) modulates symptom severity in infected plants. Some of the satellite RNAs induces severe necrotic symptoms, whereas others may eliminate the typical symptoms (Roosinck *et al.* 2000).

On the basis of serological relationships, peptide mapping of the viral coat protein and the nucleic acid hybridization analysis, all but one strain of CMV have been allocated to one of two major subgroups (subgroup I and subgroup II). Subgroup I is common in North America while subgroup II is prevalent in the tropics.

The virus is transmitted by more than 60 aphid species in a non-persistent stylet-borne manner. It is easily transmitted mechanically in the laboratory, but not by human workers because of its instability (Zitter, 1993).

Control

Control strategies are very much the same as those used for aphid-borne viruses. It is recommended to eradicate virus reservoir plants near the crop and to spray mineral oils frequently, especially during the early stages of plant growth. No sources of resistance to the virus have been identified (Zitter, 1993). New biotechnological approaches based on the use of cross-protection, coat protein transgene-mediated protection and satellite RNA have been described and show promising results. Tomato plants harboring a truncated CMV replicase gene were found to be immune to CMV infection under field conditions (Gal-On *et al.* 1998). Genetically engineered tomatoes expressing the coat protein gene of CMV remained free of symptoms under extreme disease pressure in the open field (Murphy *et al.*1997). Transgenic tomatoes expressing a satellite RNA of CMV reacted in mild symptoms to infection with CMV and their yield was 25%-130% higher compared to non-transformed parental lines infected with the virus (Stommel et *al.*1998).

25.2.3 Thrips transmitted viruses

Of the 5000 species of thrips, only 10 species, all in the family Thripidae, are vectors of plant viruses. Most of these species are polyphagous and reproduce on a broad range of hosts. Viruses belonging to four plant virus families are transmitted by thrips (Ulman *et al.*, 1997).

Tomato spotted wilt virus Genus ***Tospovirus*** (TSWV)

The virus occurs worldwide and was reported from Europe, Asia, Africa, North America, Central America and the Caribbean, South America and Oceania (EPPO website). Destructive outbreaks of TSWV have occurred in France and Spain, in both protected and outdoor tomato and pepper crops (EPPO/CABI, 1997). TSWV is a production-limiting factor in some tomato growing regions of Spain (Moriones *et al.* 1998). In Georgia, USA, a yield loss of 5-20% was recorded in tomato crops infected with the virus (http://www.Agcomm.uga.edu/ymw/tswv.htm). The disease has become very important in Argentina and Brazil, where 100% damage has been recorded (Berling *et al.* 1990).

The virus particles are spherical, 70-110 nm in size and display surface projections of glycoproteins which are embedded in a lipid bilayer envelope. The viral genome consists of three segments of negative or ambisense ssRNA, designated L, M and S. The genome has four structural proteins: two external glycoproteins (G1, G2), a nucleocapsid protein (N), and a large transcriptase protein (L) (Elliot *et al.* 2000).

TSWV has a wide host range. In tomato the virus induces symptoms, which vary according to cultivar, plant age, and nutritional and environmental conditions. Symptoms in tomatoes include leaf bronzing, necrotic spots and streaks on leaves, leaf petioles, stems and unripened fruits. In ripe fruits a mosaic of pale red, yellow and red pigmentation is observed (Gibbs, 1996).

The virus is transmitted and spread in nature by several species of thrips: they include *Thrips tabaci, T. setosus, Frankliniella occidentalis, F. fusca, F. intonsa, F. schultzi, and Scirtothrips dorsalis.* The virus is acquired only during the larval stages, and virus inoculation is achieved by shallow feeding on leaf epidermal cells. For transmission, the first-instar nymphs acquire the virus (van de Wetering *et al.* 1992) and, subsequently, the active adults spread the virus. The virus has persistent, circulative relationships with the vector, which retains infectivity for 22-30 days after acquisition. There is accumulating evidence that the virus multiplies within the insect vector, but transovarial transmission has not been observed (Wijkamp *et al.* 1996).

Control

Control is usually through reduction of the thrips population by application of insecticides, combined with eradication of inoculum sources in the neighborhood of the crop. Although chemical control is possible, thrips populations develop resistance when insecticides are repeatedly used. It is important, therefore, to use insecticides with different active substances at each application. A single chemical treatment is unable to kill eggs, pre - pupal and pupal stages that are hidden in leaf tissue or soil. Therefore, when TSWV infections are observed and a thrips population is present, a sequential series of at least three insecticide applications are recommended, to provide a complete eradication of the pest. The combination of chemical control with cultural means (IPM) offers a better alternative. Such a program was employed in Hawaii (Cho *et al.* 1989). It included: use of virus-free planting material; regular applications of insecticide on propagated seedling areas; reduced in-field cultivation, to avoid the transfer of thrips from infected sources; keeping field plots with high disease incidence fallow (for 3-4 weeks), to allow thrips to emerge from infected crop debris and disperse from the field; soil fumigation with etham-sodium (Vapam) or 1,3-dichloropene (Telone) to eliminate thrips from the soil and plant debris; planting non-host plant barriers on the field border; avoidance of sequential planting; phytosanitation to remove inoculum sources from abandoned fields and weeds (Cho *et al.* 1989). In Louisiana (USA), aluminum mulches reduced the numbers of trapped thrips by 33-68% and the incidence of TSWV in tomato plots by 60-78% (Greenough *et al.* 1990).

Resistance to TSWV has been found in several *Lycopersicon* species and resistant varieties are available today but the question is how long this resistance will last before being overcome by new strains of the virus (Roca *et al.* 1997).

25.2.4 Non-vectored viruses

Tomato mosaic virus Genus ***Tobamovirus*** (ToMV)

Tomato mosaic virus and tobacco mosaic virus (TMV) are pathogens of tomato. The virus particles are rigid rods measuring about 200 x 18 nm, and the virus has a positive-sense RNA genome that encodes at least four proteins (Lewandowski, 2000).

The natural host of the virus is *L. esculentum*, and *Nicotiana* spp. *Datura* spp .and *Physalis* spp. are among its diagnostically susceptible hosts (Huttinga and Rast, 1996). The characteristic symptoms are mottled light and dark green areas on the leaves and unregular color development on the fruit (Fig. 25.3). The aucuba mosaic strain causes a bright yellow mottling. Plants that are infected at an early stage form curled leaves with reduced size (fernleaf). Symptoms depend on virus strain, the host cultivar and the light intensity At high temperatures the foliar symptoms may be masked, and tomato varieties with resistance genes Tm2 or $Tm2^2$ react to the virus with necrosis on the fruit (Zitter, 1993).

The virus is readily transmissible by sap, and can be spread by workers' hands, or contaminated tools and clothing; and it can persist in soil and plant debris or in a free form for long time periods (Broadbent, 1963).

ToMV is seed-borne, in the testa and sometimes in the endosperm. The transfer of seedlings from the growing trays to soil can wound the root system resulting in a high proportion of infected plants.

Virus-free seed batches should be used to produce healthy transplants. Seed treatments with 10% Na_3Po_4 for 30 min or heating for 2-4 days at 70^0C were found effective in eliminating the contaminating virus from seed. Before transplanting greenhouse tomatoes the internal space of the greenhouse should be sprayed with 10% Na_3Po_4 or 0.3% sodium hypochlorite, for disinfections. Phytosanitation is an important factor in reducing disease incidence, and seedlings should not be transplanted into fields where root debris of an infected former tomato crop remain (Zitter, 1993).

Pepino mosaic virus Genus ***Potexvirus*** (PeMV)

Pepino mosaic virus has flexuous filamentous particles, 470-580 nm in size, with a single molecule of positive-sense ssRNA. The RNA has five major open reading frames (Brunt *et al.* 2000).

PeMV is naturally present in Peru in wild *Lycopersicon* spp., cultivated tomato and pepino (*Solanum muricatum*). Symptoms in tomato include yellow spots on leaves, mild interveinal chlorosis and, sometimes minor leaf malformations. Fruits may show discoloration. The severity of PeMV symptoms in tomato varies according to climatic conditions and tomato cultivar. In most cases losses do not exceed 5%.

The virus is transmitted by contact with contaminated hands, tools, clothing and direct plant-to-plant contact. Seedborne transmission is unlikely.

Control

Eradication of inoculum sources from both cultivated plants and weeds is essential. The disease spreads very rapidly by contact and therefore, personal hygiene is important to reduce dissemination. Regular monitoring of young tomato crops is vital to ensure that infected plants will be removed immediately, when first signs of the disease are detected.

25.2.5 Viroid Diseases

Viroids are the smallest known agents of plant infectious diseases. Most viroids have genome sizes ranging from 246 to 399 nucleotides, and they consist of highly structured, single-stranded circular RNA molecules. These molecules lack a capsid protein and detectable messenger RNA activity. Taxonomically, viroids are grouped in two families, the larger of which is the *Pospiviroidae,* which includes several viroids that were found to cause disease in field- and greenhouse-cultivated tomato plants (Flores *et al.* 1988). *Tomato planta macho viroid* (TPMVd) (Galindo *et al.* 1982), *Tomato apical stunt viroid* (TASVd) (Walter, 1987), *Potato spindle tuber viroid* (PSTVd) (Puchta *et al.* 1990) and a strain of *Citrus exocortis viroid* (CEVd) (Mishra *et al.* 1991) have been reported to infect tomatoes naturally. Host reactions to these viroid diseases include stunting, apical proliferation, apical leaf narrowing and yellowing, leaf crinkling, brittleness of leaf tissue, and leaf necrosis.

The TPMVd disease has been recorded only in Mexico, where it occurs in regions with mean temperatures above 22^0C. A distinct strain of CEVd, designated CEVd-t was found to cause the bunchy top disease of tomatoes in India (Mishra *et al.* 1991). TASVd was biologically characterized in the Ivory Coast, but no data on its epidemiology or on the economic damage were reported (Walter, 1987). An Indonesian strain of TASVd, which elicits severe symptoms in tomatoes, was sequenced by Candresse *et al.* (1987). A variant of TASVd-Indonesia was found to cause severe losses to greenhouse-grown tomato plants in Israel (Fig. 25.4) (Antignus *et al.* 2002).

Control

One of the primary strategies for controlling viroid diseases is to prevent the introduction of viroid-contaminated material into the growing sites. The viroid RNA is highly stable and can withstand extreme conditions, therefore it remains on contaminated surfaces for a long time. The spread of viroids is fast and is facilitated by mechanical wounding of the plant tissue. Also, transmission *via* contaminated hands or tools is very common if inoculum sources exist within the field. Equipment can be disinfected with 0.25% sodium or calcium hypochlorite, but heat treatment or detergents may not be very effective. The best strategy to fight viroid diseases would be to use resistant cultivars, but such varieties are not available at present (Singh and Dhar, 1998).

25.3 Concluding Remarks

Viruses impose serious limitations on tomato cultivation in all countries, and more so in developing countries, which usually lack appropriate infrastructure that could support the efficient exchange of information between institutions and farmers. The key factors in any efficient control program are reliable identification of the viral pathogen and understanding its natural dissemination mechanism. Unfortunately, many developing countries lack the resources to address these situations.

Integrated pest management (IPM) has been recognized and adopted internationally as the preferred plant protection strategy. The implementation of IPM in developing countries requires an efficient education system that teaches farmers the principles of this approach, and the use of IPM tools as an alternative to the application of pesticides. A reason for optimism regarding the control of devastating viral diseases of plants in developing countries is the success already achieved in introducing resistance in plants, including tomato ((Brunetti *et al.* 2001), by genetic engineering techniques.

References

Al-Musa, A. 1982. Incidence, economic importance, and control of tomato yellow leaf curl in Jordan. *Plant Dis*. 66: 561-563.

Antignus, Y. and Cohen, S. (1994). Complete nucleotide sequence of an infectious clone of a mild isolate of tomato yellow leaf curl virus (TYLCV) *Phytopathology* 84: 707-712

Antignus, Y. 2000. Manipulation of wavelength-dependent behavior of insects: an IPM tool to impede insects and restrict epidemics of insect-borne viruses. *Virus Res*. 71: 213-220.

Antignus, Y., Mor, N., Ben–Joseph, R., Lapidot, M. and Cohen, S. 1996. UV-absorbing plastic sheets protect crops from insect pests and from virus diseases vectored by insects. *Environ. Entomol*. 25: 919-924.

Antignus, Y., Lapidot, M. and Cohen, S. 2001. Interference with ultraviolet vision of insects to impede insect pests and insect-borne plant viruses. In: *Virus-Insect-Plant Interactions,* pp. 331-350. K. F. Harris, O. P. Smith, J. E. Duffus (eds). Academic Press, San Diego, USA.

Antignus, Y., Lachman, O., Pearlsman, M., Gofman R. and Bar-Joseph, M. 2002. A new disease of greenhouse tomatoes in Israel caused by a distinct strain of tomato apical stunt viroid (TASVd). *Phytoparasitica* 30: 502-510.

Bedford, I.D., Briddon, R.W., Jones, P., Alkaff, N. and Markham, P.G. 1994. Differentiation of three whitefly transmitted geminiviruses from the Republic of Yemen. *Eur. J. Plant Pathol.* 100: 243-257.

Bedford, I.D., Kelly, A., Banks, G.K., Briddon, R.W., Cenis, J.L. and Markham, P.G. 1998. *Solanum nigrum*: an indigenous weed reservoir of tomato yellow leaf curl geminivirus in Southern Spain. *Eur. J. Plant Pathol.* 104: 221-222.

Bellows, T. S., Perring, T. M., Gill, R. J. and Headrick, D. H. 1994. Description of a species of *Bemisia* [Homoptera, Aleurodidae]. *Ann. Entomol. Soc. Am.* 87: 195-206

Berger, P.H., Barnett, O.W., Brunt, A.A., Colinet, D., Edwardson, J.R., Hammond, J., Hill, J.H., Jordan, R.L., Kashiwazaki, S., Makkouk, K., Morales, F.J., Rybicki, E., Spence, N., Ohki, S.T., Uyeda, I., van Zaayen, A. and Vetten, H.J. 2000. Family Potyviridae. In: *Virus Taxonomy, Seventh Report of The International Committee on Taxonomy of Viruses,* pp. 703-724. M.H.V. van Regenmortel, C.M. Fauquet, D.H.Y.L. Bishop, E. Carstens, M. Estes, S. Lemon, J. Maniloff, M.A. Mayo, D. McGeoch, C. Pringle and R. Wickner (eds), Academic Press, New York.

Berling, A., Llanas-Bousquet, W, Malezieux, S. and Gebre Selassie, K. 1990. Tomato spotted wilt virus. Connaitre le probleme pour enrayer l'epidemie. *Phytoma* 422: 46-50.

Broadbent, L. 1963. The epidemiology of tomato mosaic. III. Cleaning virus from hands and tools. *Ann. Appl. Biol.* 52:225-232.

Brown, J. K., Ostrow, K. M., Idris, A. M. and Stenger, D. C. 2000. Chino del tomate virus: Relationships to other begomoviruses and identification of A-component variants that affect symptom expression. *Phytopathology* 90:546-552.

Brunetti, A., Tavazza, R., Noris, E., Lucioli, A., Accoto, G. P. and Tavazza, M. 2001. Transgenically expressed T-Rep of tomato yellow leaf curl Sardinia virus acts as a trans – dominant – negative, mutant, inhibiting viral transcription and replication. *J. Virol.* 2001: 10573-10581.

Brunt, A.A., Foster, G.D., Martelli, G.P. and Zavriev, S.K. 2000. Genus Potexvirus. In: *Virus Taxonomy*, Seventh Report of The International Committee on Taxonomy of Viruses, pp. 975-980. M. H. V van Regenmortel, C. M. Fauquet, D. HY. L. Bishop, E. Carstens, M. Estes, S. Lemon, J. Maniloff, M.A. Mayo, D. McGeoch, C. Pringle and R. Wickner (eds), Academic Press, New York.

Buchen-Osmond, C. 1996. Potato Y potyvirus. In*: Viruses of Plants, Descriptions and Lists from the VIDE Database*, pp. 1033-1038. A. A., Brunt, K. Crabtree, M. J., Dallwitz, A. J., Gibbs and L. Watson (eds), CAB International, Cambridge, UK.

Byrne, D.N., Bellows, T.S. and Parella, M.P. 1990. Whiteflies in agricultural systems. In: *Whiteflies: their Bionomics, Pest Status and Management*, pp. 227-261. D. Gerling (ed.). Intercept, Andover, Hants, U. K.

Candresse, T., Smith, D. and Diener, T.O. 1987. Nucleotide sequence of full length infectious clone of the Indonesian strain of tomato apical stunt viroid (TASVd*). Nucl. Acids Res.* 15: 10597.

Cho, J.J., Mau, F.L., German, T.L., Hartman, R.W., Yudin, L.S., Gonsalves, D. and Providenti, R. 1989. A multi-disciplinary approach to management of tomato spotted wilt virus in Hawaii. *Plant Dis.* 73: 375-383.

Cohen, S. and Antignus, Y. 1994. Tomato yellow leaf curl virus a whitefly-borne geminivirus of tomatoes. In *: Adv. Dis. Vector. Res..* K.S. Harris (ed.) Vol. 10 pp. 259-288. Springer-Verlag, New York.

Cohen, S. and Berlinger, M.J. 1986. Transmission and cultural control of whitefly-borne viruses. *Agric. Ecosyst. Environ.* 17: 89-97.

Cohen, S. and Melamed-Madjar, V. 1978. Prevention by soil mulching of the spread of tomato yellow leaf curl virus transmitted by *Bemisia tabaci* (Gennadius) (Homoptera: Aleurodidae) in Israel. *Bull. Entomol. Res.* 68: 465-470.

Cohen, S., Kern, J., Harpaz, I. and Ben-Joseph, R. 1988. Epidemiological studies of tomato yellow leaf curl virus (TYLCV) in the Jordan river, Israel. *Phytoparasitica* 16: 259-270.

Cohen, J., Gera, A., Ecker, R., Ben Joseph, R.,Perlsman, M., Gokkes, M., Lachman, O. and Antignus, Y. 1995. Lisianthus leaf curl a new disease of Lisianthus caused by tomato yellow leaf curl virus. *Plant Dis.*79: 416-420.

Csizinszky, A.A., Schuster, D.J. and Kring, J.B. 1995. Color mulches influence yield and insect pest populations in tomatoes. *J. Amer. Soc. Hort. Sci.* 120: 778-784.

Czosnek, H. and Laterrot, H. 1997. A worldwide survey of tomato yellow leaf curl viruses. *Arch. Virol.* 142 :1391-1406.

Czosnek, H., Ber, R., Antignus, Y., Cohen, S., Navot, N. and Zamir, D. 1988. Isolation of tomato yellow leaf curl virus, a geminivirus. *Phytopathology* 78: 508-512.

Defranc, Q., D'hondt, M. and Russo, M. 1985. Tomato yellow leaf curl in Senegal. *Phytopathol. Z.* 112: 153-160.

Duffus, J. 1987. Whitefly transmission of plant viruses. In : *Current Topics in Vector Research*, pp. 74-91. F. H. Kerry (ed), Springer-Verlag, New York.

Eastop, V.F. 1977. Worldwide importance of aphids as virus vectors. In*: Aphids as Virus Vectors*, pp. 3-62, K. F. Harris, and K. Maramorosch (eds), Academic Press, New York.

Elliot, R.M., Bouloy, M., Calisher, C.H., Goldbach, R., Moyer, J.T, Nichol, S.T., Pettersson, R., Plyusnin, A. and Schmaljohn, C.S. 2000. Family Bunyaviridae. In: *Virus Taxonomy, Seventh Report of The International Committee on Taxonomy of Viruses,* pp. 599-621. M. H. V. Van Regenmortel, C. M. Fauquet, D. HY. L. Bishop, E. Carstens, M. Estes, S. Lemon, J. Maniloff, M.A. Mayo, D. McGeoch, C. Pringle and R. Wickner (eds), Academic Press, New York

EPPO/CABI 1997. Tomato spotted wilt tospovirus In *: Quarantine Pests for Europe* (2nd ed), pp. 1379-1387. CAB International, Wallingford, UK.

Flores, R., Randles, J.W., Bar-Joseph, M. and Diener, T.O. 1988. A proposed scheme for viroid classification and nomenclature. *Arch. Virol.* 143: 623-629.

Francki, R.I.B. and Habili, N. 1996. Cucumber mosaic cucumovirus. In: *Viruses of Plants, Descriptions and Lists from the VIDE Database,* pp. 477-483. A. A., Brunt, K. Crabtree, M. J., Dallwitz, A. J., Gibbs and L. Watson (eds), CAB International, Wallingford, UK.

Friscmuth, T. and Stanely, J. 1993. Strategies for the control of geminivirus diseases. *Sem. Virus Virol.* 4: 329-337. Galindo, A.J., Smith, D.R. and Diener, T.O. 1982. Etiology of planta macho, a viroid disease of tomato. *Phytopathology* 72: 49.

Gal-On, A., Wolf, D., Wang, Y., Faure, J-E., Pilowski, M. and Zelcer, A. 1998. Transgenic resistance to cucumber mosaic virus in tomato: blocking of long-distance movement of the virus in lines harboring a defective viral replicase gene. *Phytopathology* 88: 1101-1107.

Gibbs, A.J. 1996. Tomato spotted wilt tospovirus. In: *Viruses of Plants*, *Descriptions and Lists from the VIDE Database*, pp. 1312-1315. A. A., Brunt, K. Crabtree, M. J., Dallwitz, A. J., Gibbs and L. Watson (eds), CAB International, Wallingford, UK.

Gonzalez, G. and Valdez, S. 1995. Virus del encrespamiento Amarillo de las hojas del tomate (TYLCV) en Cuba. *CEIBA* (Honduras) 36: 103.

Greenough, D.R., Black, L.L. and Bond, W.P. 1990. Aluminium-surfaced mulch: an approach to control of tomato spotted wilt virus in Solanaceous crops. *Plant Dis.*. 74: 805-808.

Guzman, P., Arredondo, C.R., Emmatty, D., Portillo, R.J. and Gilbertson, R.L. 1997. Partial characterization of two whitefly-transmitted geminiviruses infecting tomatoes in Venezuela. *Plant Dis*. 81: 312.

Hong, Y.G. and Harrison, B.D. 1995. Nucleotide sequences from tomato leaf curl viruses from different countries: evidence for three geographically separate branches in the evolution of the coat protein of whitefly transmitted geminiviruses. *J. Gen. Virol.* 76: 2043-2049.

Horowitz, A.R., Forer, G. and Isshaaya, I. 1994. Managing resistance in *Bemisia tabaci* in Israel with emphasis on cotton. *Pestic. Sci.* 42: 113-122.

Hunter, W.B., Hiebert, E., Tsai, J.H. and Polston, J.E. 1998. Location of geminiviruses in the whitefly *Bemisia tabaci* (Homoptera : Aleurodidae). *Plant Dis.* 82: 1147-1157.

Huttinga, H. and Rast, A.Th.B. 1996. Tomato mosaic tobamovirus. In: *Viruses of Plants, Descriptions and Lists from the VIDE Database*, pp. 1241-1244. A. A., Brunt, K. Crabtree, M. J., Dallwitz, A. J., Gibbs and L. Watson (eds), CAB International, Wallingford, UK.

Ioannou, N. 1987. Cultural management of tomato yellow leaf curl disease in Cyprus. *Plant. Pathol.* 36: 367-373.

Jupin, I., Hericourt, F., Benz, B. and Gronenborn, B. 1995. DNA replication specificity of TYLCV geminivirus is mediated by the amino-terminal 116 amino acids of the Rep protein. *FEBS Letters* 362: 116-120.

Kato, K., Onuki, M., Fuji, S. and Hanada, K. 1998. The first occurrence of tomato yellow leaf curl in tomato (*Lycopersicon esculentum* Mill.) in Japan. *Ann. Phytopathol. Soc. Jpn.* 64: 552-559.

Klassen, A., Boeshore, M.L., Kooning, E.V., Tian, T. and Falk, B.W. 1996. Genome structure and phylogenetic analysis of lettuce infectious yellows virus, a whitefly transmitted, bipartite closterovirus. *Virology* 208: 99-110.

Kring, J.B., Schuster, D.J., Price J.F. and Simone, G.W. 1991. Sweet potato whitefly-vectored geminivirus on tomato in Florida. *Plant Dis.* 75: 1186.

Lastra, J.R. and de Uzcategui, R.C. 1975. Viruses affecting tomatoes in Venezuela. *Phytopathol. Z.* 84: 253-258.

Lazarowitz, S.G. 1992. Geminiviruses: Genome structure and gene function. *Crit. Rev. Plant Sci.* 11: 327-349.

Lewandowski, D.J 2000. Genus Tobamovirus. In: *Virus Taxonomy, Seventh Report of The International Committee on Taxonomy of Viruses,* pp. 889-894. M.H.V. Van Regenmortel, C.M. Fauquet, D.H.Y.L. Bishop, E. Carstens, M. Estes (eds), S Academic Press, New York.

Loebenstein, G. and Raacah, B. 1980. Control of non-persistently transmitted aphid-borne viruses. *Phytoparasitica* 8: 221-235.

Louro, D., Noris, E., Veratti, F. and Accoto, G.P. 1996. First report of tomato yellow leaf curl virus in Portugal. *Plant Dis.* 80: 1079.

Makkouk, K.M. 1978. A study on tomato viruses in the Jordan valley with special emphasis on tomato yellow leaf curl. *Plant Dis. Reptr.* 62: 259-262.

Martelli, G.P., Agranovsky, A.A., Bar-Joseph, M., Boscia, D., Candresse, T., Couts, R.H.A., Dolja, V.V., Falk, B.W., Gonsalves, D., Jelkman, W., Karasev, A.V., Minafra, A., Murant, A., Namba, S., Niblett, C.L., Vetten, H.J. and Yoshikawa, N. 2000. Genus *Crinivirus* In: *Virus Taxonomy, Seventh Report of The International Committee on Taxonomy of Viruses*, pp. 949-952. M.H.V. van Regenmortel, C.M. Fauquet, D.H.Y.L. Bishop, E. Carstens, M. Estes, S. Lemon, J. Maniloff, M.A. Mayo, D. McGeoch, C. Pringle and R. Wickner (eds), Academic Press, New York.

McGlashan, D., Polston, J.E. and Bois, D. 1994. Tomato yellow leaf curl geminivirus in Jamaica. *Plant Dis.* 78: 1219.

Mazyad, H.M., Omar, F., Al-Taher, K. and Salha, M. 1979. Observations on the epidemiology of tomato yellow leaf curl disease on tomato plants. *Plant Dis. Reptr.* 63: 695-698.

Mishra M.D., Hammond R.W., Owens, R.A., Smith D.R. and Diener, T.O. 1991. Indian bunchy top disease of tomato plants is caused by a distinct strain of citrus exocortis viroid. *J. Gen. Virol.* 72: 1781-1785.

Monci, F. and Navas-Castillo, J., Cenis, J.L., Lacasa, A., Melloul, A. and Moriones, E. 2000. Spread of tomato yellow leaf curl virus Sar from the Mediterranean basin: Presence in the Canary Islands and Morocco. *Plant Dis*. 84: 490.

Moriones, E. 1997. Natural recombination between tomato yellow leaf curl virus-Is and tomato leaf curl virus. *J. Gen. Virol.* 81: 2797-2801

Moriones, E. and Navas-Castillo, J. 2000. Tomato yellow leaf curl virus, an emerging virus complex causing epidemics worldwide. *Virus Res.* 71: 123-134.

Moriones, E., Aramburu, J., Riudavets, J., Arno, J. and Lavina, A. 1998. Effect of plant age at time of infection by tomato spotted wilt tospovirus on the yield of field grown tomato. *Eur. J. Pl. Pathol.* 104: 295-300.

Muniyappa, V., Venkatesh, H. M., Ramappa, H.K., Kulkarni, R.S., Zeidan, M., Tarba, C.Y., Ganim, M. and Czosnek, H. 2000. Tomato leaf curl virus from Bangalore (ToLCV-Ban4) sequence comparison with Indian ToLCV isolates, detection in plants and insects, and vector relationships. *Arch. Virol.* 145: 1583-1598.

Murphy, J., Sikora, E., Lovins, K., Andrianifahnana, M. 1997. Transgenic tomato withstands CMV epidemics in North Alabama. www.ag.auburn.edu./aaes/communications/highlights/fall97/html

Nakhla, M.K., Mazyad, H.M. and Maxwell, D.P. 1993. Molecular characterization of four tomato yellow leaf curl virus isolates from Egypt and development of detection methods. *Phytopath. Med.* 32: 163-173.

Nakhla, M.K., Maxwell,D.P., Martinez, R.T., Carvalho, M.G. and Gilbertson, R.L. 1994. Occurrence of the Eastern Mediterranean strain of tomato yellow leaf curl geminivirus in the Dominican Republic. *Phytopathology* 84: 1072.

Navas-Castillo, J., Sanchez-Campos, S., Diaz, J.A., Saez-Alonso, E. and Moriones, E. 1997. First report of tomato yellow leaf curl virus-Is in Spain: coexistence of two different geminiviruses in the same epidemic outbreak. *Plant Dis*. 81: 1461.

Navas-Castillo, J., Camero, R., Bueno, M. and Moriones, E. 2000. Severe yellowing outbreaks in tomato in Spain associated with infections of tomato chlorosis virus. *Plant Dis*. 84: 835-837.

Navot, N., Picherski, E., Zeidan, M., Zamir, D. and Czosnek, H.1991. Tomato yellow leaf curl virus : a whitefly transmitted geminivirus with a single genomic component. *Virology* 185: 151-161.

Noris, E., Hidalgo, E., Accotto, G.P. and Moriones, E. 1994. High similarity among tomato yellow leaf curl virus isolates from the West Mediterranean Basin: the nucleotide sequence of an infectious clone from Spain. *Arch. Virol.* 135: 165-170.

Noris, E., Vaira, A.M., Caciagli, P., Masenga, V., Gronenborn, B. and Accotto, G.P. 1998. Amino acids in capsid protein of tomato yellow leaf curl virus that are crucial for systemic infection, particle formation, and insect transmission. *J. Virol.* 72: 10050-10057.

Padidam, M., Beachy, R.N. and Fauquet, C. 1995. Classification and identification of geminiviruses using sequence comparison. *J. Gen. Virol.* 76: 249-263.

Pietersen, G., Idris, M.A., Kruger, K., Brown, J.K. 2000. Tomato curly stunt virus, a new begomovirus of tomato within the Tomato yellow leaf curl virus-IS cluster in South Africa. *Plant Dis*. 84: 810.

Polston, J.E. and Anderson, P.K. 1997. The emergence of whitefly-transmitted geminiviruses in tomato in the Western Hemisphere. *Plant Dis*. 81: 1358-1369.

Polston, J.E., Bois, C., Serra, C.A. and Cocepcion, S. 1994. First report of a tomato yellow leaf curl –like geminivirus from tomato in the Western Hemisphere. *Plant Dis*. 78:831.

Polston, J.E., McGovern, R.J. and Brown, L.G. 1999. Introduction of tomato yellow leaf curl virus in Florida and implications for the spread of this and other geminiviruses of tomato. *Plant Dis.* 83: 984-988.

Puchta, H., Herold, T., Verhoven, K., Roenhorst, A., Ramm, K., Schmidt-Puchta, W. and Sanger, H.L. 1990. A new strain of potato spindle tuber viroid (PSTVd-N) exhibits major sequence differences as compared to all other PSTVd strains sequenced so far. *Pl. Mol. Biol.* 15: 509-511.

Purciful, D.F., 1996. Tobacco etch potyvirus. In: *Viruses of Plants, Descriptions and Lists from the VIDE Database*, pp. 1241-1244. A.A. Brunt, K. Crabtree, M.J., Dallwitz, A.J. Gibbs and L. Watson (eds), CAB International, Wallingford, UK.

Raccah, B., Huet, H. and Blanc, S. 2001. Potyviruses. In: *Virus-Insect-Plant Interactions*, pp. 181-206. K.F. Harris, O.P. Smith and J.E. Duffus (eds), Academic Press, San Diego, USA.

Ramos, P.L.O., Guerra, V., Dorestes, N., Ramirez, N., Rivera-Bustamente, R. and Oramas, P. 1996. Detection of TYLCV in Cuba. *Plant Dis.* 80: 1208.

Rick, C.M. 1978. The tomato. *Sci. Am.* 239: 76-87.

Roca, E., Aramburu, J. and Moriones, E. 1997. Comparative host reaction and *Frankliniella occidentalis* transmission of different isolates of tomato spotted wilt tospovirus from Spain. *Plant Pathol.* 46: 407-415.

Rochester, D.E., DePaulo, J.J., Fauquet, C.M. and Beachy, R.N. 1994. Complete nucleotide sequence of the geminivirus tomato yellow leaf curl virus, Thailand isolate. *J. Gen. Virol.* 75: 477-485.

Roossink, M.J., Bujarski, J., Ding, S.W., Hajimorad, R., Hanada, K., Scott, S. and Tousignant, M. 2000. Genus *Cucumovirus*. In: *Virus Taxonomy, Seventh Report of The International Committee on Taxonomy of Viruses* pp. 929-930. M.H.V. van Regenmortel, C.M. Fauquet, D.H.Y.L. Bishop, E. Carstens, M. Estes, S. Lemon, J. Maniloff, M.A. Mayo, D. McGeoch, C. Pringle and R. Wickner (eds), Academic Press, New York.

Rosner, A. Lachman, O., Pearlsman, M., Maslenin, L., and Antignus, Y. 2000. Molecular characterization and differential diagnosis of a necrotic PVY isolate in tomato. *Ann. Appl. Biol.* 137: 253-257.

Rosell, R.C., Torres-Jerez, I. and Brown K.J. 1999. Tracing geminivirus-whitefly transmission pathway by polymerase chain reaction in whitefly extracts, saliva, hemolymph and honeydew. *Phytopathology* 89: 239-246.

Rybicki , E.P., Briddon, R.W., Brown, J.K., Fauquet, C.M., Maxwell, D.P., Stanely, J., Harrison, B.D., Markham, P.G., Bisaro, D.M. and Robinson, D. 2000. Family Geminiviridae. In: *Virus Taxonomy, Seventh Report of The International Committee on Taxonomy of Viruses,* pp. 285-297. M.H.V. van Regenmortel, C.M. Fauquet, D.H.Y.L. Bishop, E. Carstens, M. Estes, S. Lemon, J. Maniloff, M.A. Mayo, D. McGeoch, C. Pringle and R. Wickner (eds), Academic Press, New York.

Sanchez-Campos, S., Navas Castillo, J., Camero, R., Soria, C., Diaz, J.A. and Moriones, E. 1999. Displacement of tomato yellow leaf curl virus (TYLCV)-Sr by TYLCV-Is in tomato epidemics in Spain. *Phytopathology* 89: 1038-1043.

Sanchez-Campos, S., Navas Castillo, J., Monci, F., Diaz, J.A. and Moriones, E. 2000. *Mercurialis ambigua* and *Solanum luteum*: Two newly discovered natural hosts of tomato yellow leaf curl Geminivirus. *Eur. J. Pl. Pathol.* 106: 391-394.

Singh, R.P. and Dhar, A.K. 1998. Detection and management of plant viroids. In: *Plant Virus Disease Control*, pp. 428-447. A. Hadidi, R. K., Ketharpal and H. Koganezawa (eds), APS Press N.Y.

Stommel, J. R., Tousignant, M.E., Wai, T., Pacini, R., and Kaper, J. M. 1998. Field test results of transgenic tomato plants expressing viral satellites for resistance to cucumber mosaic virus. http:/www.nal.usda.gov/ttic/tekran/data/000007/48/0000074885.html

Thomas, J.E. and McGrath, D.J. 1988. Inheritance of resistance to potato virus Y in tomato. *Aus. J. Agr. Res.* 39: 475-479.

Torres-Pacheco, I., Garzon-Tiznado, J. A., Brown, J. K., Becerra-Flora, A., and Rivera-Bustamente, R. F. 1996. Detection and distribution of geminiviruses in Mexico and the Southren United States. *Phytopathology* 86: 1186-1192.

Ucko, O., Cohen, S. and Ben-Joseph, R. 1998. Prevention of virus epidemics by a crop-free period in the Arava region of Israel. *Phytoparasitica* 26: 313-321.

van de Wetering, F., Goldbach, R., and Peters, D. 1992. Tomato spotted wilt tospovirus ingestion by first instar larvae of *Frankliniella occidentalis* is a prerequisite for transmission. *Phytopathology* 86: 900-905.

Walter, B. 1987. Tomato apical stunt In: *The Viroids*, pp. 321-327. T. O, Diener (ed), Plenum Press, New York.

Wartig, L., Kehyr-Pour, A., Noris, E., De Kouchkovsky, F., Jouanneau, F., Gronenborn, B. and Jupin, I. 1997. Genetic analysis of the monopartite tomato yellow leaf curl geminivirus: roles of V1, V2, and C2 ORFs in viral pathogenesis. *Virology* 228: 132-140.

Wijkamp, I., van de Wetering, F., Goldbach, R., and Peters, D. 1996. Transmission of tomato spotted wilt virus by *Frankliniella occidentalis*: median acquisition and inoculation access period. *Ann. Appl. Biol.* 129: 303-313.

Wisler, G.C. and Duffus, J.E. 2001. Transmission Properties of whitefly-borne criniviruses and their impact on virus epidemiology. In : *Virus-Insect-Plant Interactions*, pp. 293—306. K . F. Harris, O. P. Smith and J. E. Duffus (eds), Academic Press, New York.

Zhou, X., Liu, Y., Calvert, L., Munoz, C., Otim-Nape, G.W., Robinson, D. J. and Harrison, B.D. 1997. Evidence that DNA-A of a geminivirus associated with severe cassava mosaic disease in Uganda arisen by interspecific recombination. *J. Gen. Virol.* 78: 2101-2111.

Zitter, T.A. 1993. Tomato mosaic virus and tobacco mosaic virus. In: *Compendium of Tomato Diseases*, pp. 39. J. B. Jones, J. P. Jones, R. E. Stall, T. A. Zitter (eds). APS Press, New York.

Legends to figures:

Fig. 25.1 – Typical symptoms induced by *Tomato yellow leaf curl virus* in tomato cv. Daniela (photo . G. Loebenstein)

Fig. 25.2 – Shoestring leaves induced by *Cucumber mosaic virus* in tomato cv. Fakulta 38

Fig. 25.3 – Typical symptoms induced by *Tobacco mosaic virus* on tomato fruits cv. Hazera 84

Fig. 25.4 - *Tomato apical stunt viroid* induced symptoms on cherry tomatoes.

25.1
25.2
25.4
25.3

26. Cucurbits

Hervé Lecoq
INRA, Station de Pathologie Végétale, Domaine Saint Maurice, BP 94, 84143 Montfavet cedex, France.

26.1 Introduction

Cucurbits are important vegetable crops in developing countries where they account for more than 50% of the total fresh vegetable production (Anonymous, 2002). About 40 different cucurbit species are cultivated, but in terms of production, the 3 major crops are watermelon (*Citrullus lanatus*), cucumber (*Cucumis sativus*) and melon (*Cucumis melo*) (Table 26.1). Another group of economically important cucurbits includes different squash, pumpkin or marrow types (*Cucurbita pepo, Cucurbita moschata* and *Cucurbita maxima*). These species are cultivated world wide as a diversity of cultivars and landraces adapted to the local uses and consumers' demands. The botanical centres of origin of melon and watermelon are thought to be in the Sudano-Sahelian region of Africa, that of cucumber in the foothills of the Himalayas and squash in Central and South America (Robinson and Deckers-Walters, 1997).

Other species have a more restricted cultivation area. Bitter gourd (*Momordica charantia*), snake gourd (*Trichosanthes cucumerina*) and wax gourd (*Benincasa hispida*) are mostly grown in Asia and fluted pumpkin (*Telfairia occidentalis*) in Africa. Although distributed world wide chayote (*Sechium edule*) is particularly popular in Central America and in the Caribbean (Robinson and Deckers-Walters, 1997).

Cucurbits are mostly grown for food production. Generally they are eaten as mature or immature fruits, fresh or cooked. Seeds may be used as snacks (watermelon, squash and melon) or for oil production (squash). Leaves or shoots may be used as greens (chayote and fluted pumpkin). In arid climatic conditions, watermelon serves as a means to store clean water during the dry season. Some cucurbits are grown for medicinal or utilitarian uses (bottle gourd and vegetable sponge) or as ornamentals (Robinson and Deckers-Walters, 1997).

Developing countries produce 85% of the world cucurbit production and are 14 among the 16 major producers of these different commodities (Table 26.1). China, Turkey, India, Iran and Egypt are the main cucurbit producers (Anonymous, 2002). During the 1991-2001 period, there was a 234% increase in cucurbit production in developing countries, probably due to improved production methods, improved germplasm and better disease and insect control. This might be related to an increase in the local consumers' demand but also, for some countries, to the development of important exports to developed countries (Taylor and Brant, 2002).

Table 26.1: Production and cultivated areas of the major world producers of cucurbit crops (data for 2001, FAO source (Anonymous, 2002))

Watermelon World production: 77.5 10^6 t			Cucumber World production: 31.2 10^6 t		
Country	% world production	Area (10^3 ha)	Country	% world production	Area (10^3 ha)
China	68.9	1716	China	57,9	1200
Turkey	5.2	140	Turkey	5	59
Iran	2.7	90	Iran	4.2	60
Egypt	2,2	57	USA	3.5	65
Melon World production: 21 10^6 t			Pumpkin and gourd World production: 16.2 10^6 t		
Country	% world production	Area (10^3 ha)	Country	% world production	Area (10^3 ha)
China	38.5	400	India	24	400
Turkey	8.5	115	China	23	245
USA	5.6	52	Ukraine	5.5	48
Iran	4.7	70	Egypt	4.4	39

Cucurbit crops in developing countries are affected by at least 39 well-characterized viruses. This viral diversity may be a consequence of the genetic and ecological diversity of their hosts. Indeed, cucurbits are grown in a great variety of climatic, environmental and agricultural conditions. This variety of environments may provide more or less favourable conditions for specific viruses or for their vectors.

Typical viral symptoms in cucurbits fall within three major categories (Blancard *et al.* 1994):

- mosaics on leaves sometimes associated with leaf size reduction, rugosity, lacinations or enations ; fruits may develop a range of discolorations and deformations altering their quality,
- yellows affect mainly older leaves and occasionally the entire plant ; fruit production may be reduced but fruit quality is generally not affected,
- necrosis either as necrotic spots on leaves or as generalised necrosis. Fruits may also develop necrotic symptoms and not reach maturity.

Often, in the fields, more than one virus infects the plant, leading to a combination of these symptoms. Therefore, virus identification is very difficult based on the symptoms only.

Virus diseases may have two effects on cucurbit production: they may induce a quantitative reduction of the gross production and/or they may affect the fruit quality due to mosaic or discoloration on the fruits, deformation or internal necrosis. While the yield reduction will affect equally all producers, the effect on quality may have a differential economic impact according to the target market for the farmer. Indeed, local markets generally tolerate some quality defects that will enable the farmer to sell part of his virus-infected produce while export markets will generally require only top quality products. In this case, farmers may have to downgrade their products to local markets and may get much lower prices for their crops.

26.2 The Main Viruses

The complex viral pathosystem affecting cucurbit crops in the world and in developing countries is better known thanks to the improvement of virus diagnostic methods. The development of the sodium dodecyl sulfate immunodiffusion test (SDS-ID) (Purcifull *et al.* 1988) and of the double antibody sandwich enzyme linked immunosorbent assay (DAS-ELISA) (Clark and Adams, 1977) have greatly simplified virus diagnosis. The availability of commercial diagnostic kits has allowed extensive surveys for virus diseases in many developing countries. It has led to the identification of the major viruses occurring in a given area and has revealed some 'emerging' virus diseases (Table 26.2).

Table 26.2: Results of surveys conducted by DAS-ELISA for cucurbit viruses in developing countries.

Country and year	Sample size	Virus tested	Virus and percentage infection
Brazil	621	5	PRSV (49%), ZYMV (25%), ZLCV

997-99 (1)			(8%), CMV (7%), WMV (5%)
;osta Rica 988-91 (2)	1058	4	PRSV (65-100%), CMV (0-55%), WMV (0-60%), ZYMV (0-30%)
ran 000-02 (3)	215	10	CABYV (62%), WMV (61%), ZYMV (40%), CMV (24%), CGMMV (7%), WmCSV (6%), PRSV (1%), MWMV-ZYFV-MRMV (0%)
Iartinique 997-98 (4)	228	8	ZYMV (74%), PRSV (44%), CMV (1%), SqNV (1%), CABYV-WMV-SqMV-WmCSV (0%)
Iauritius 998 (5)	73	7	ZYMV (62%), PRSV (60%), CMV (8%), CABYV (8%), WMV-SqMV-WmCSV (0%)
Iepal 989-94 (6)	157	7	PRSV (66%), MWMV (10%), ZYMV (8%), CABYV (6%), CMV (3%), WMV-SqMV (0%)
)man 994-98 (7)	389 and 143	5 and 8	ZYMV (70%), WMV (65%), PRSV (29%), CMV (27%), SqMV (9%), TRSV (10%), ToRSV (10%), TSWV (3%)
'uerto Rico 001-02 (8)	320	7	ZYMV (69%), PRSV (60%), SqMV (30%), WMV (20%), TSWV (20%), Begomoviruses (5%), CMV (0%)
iouth Africa 997-98 (9)	Not given	6	ZYMV (50%), MWMV (24%), WMV (22%), PRSV-SqMV (0%)
iudan 992-2002 (10	519	10	CABYV (53%), ZYMV (51%), WmCSV (34%), MWMV (17%), SqMV (12%), CMV (8%), PRSV (4%), MRMV (1%), MNSV-WMV (0%)
Taiwan 1986-87 (11)	1493	5	ZYMV (53%), PRSV (36%), CGMMV (8%), CMV (7%), WMV (1%)
Turkey 1991-99 (11)	187	9	ZYMV (69%), CABYV (57%), WMV (56%), CMV (22%), PRSV (5%), SqMV (1%), MWMV-CGMMV (0%)

(1) Yuki *et al.* 2000; (2) Rivera *et al.* 1993; (3) Bananej and Lecoq, unpublished; (4) Lecoq, Rys and Desbiez, unpublished; (5) Ganoo, Saumtally, Autrey and Lecoq, unpublished; (6) Dahal *et al.* 1997; (7) Zouba *et al.* 1997; (8) Carrasco and Wessel-Beaver, 2002; (9) Cradock *et al.* 2001; (10) Lecoq *et al.* 2002; (11) Hseu *et al.* 1987; (12) Yilmaz and Lecoq, unpublished.
For acronyms of virus names see Tables 26.3, 26.4 and 26.5

Surveys using DAS-ELISA (or other specific diagnostic methods) can only reveal the occurrence of viruses that are looked for (i.e. those for which an antiserum is used). This can occasionally lead to miss important viruses in mixed infection or inducing atypical symptoms (yellowing, necrosis) and for which no antiserum is yet available. Therefore, in some cases no virus can be detected in samples, which have clear symptoms of virus infection (Hseu *et al.* 1987). The development of 'generic' tests, i.e. allowing the detection of all viruses belonging to a specific genus, including those not yet characterised, may in part reduce this limitation (Jordan, 1992; Rojas *et al.* 1993; Gibbs and Mackenzie, 1997).

From such surveys and other field observations, it appears that some cucurbit viruses are widely spread and cause important yield losses in many developing countries (Table 26.3) while others may cause severe epidemics only in restricted geographical areas (Table 26.4), and some have only a limited economic impact (Table 26.5).

Table 26.3 : Major viruses infecting cucurbits in developing countries. (c) indicates that DAS-ELISA diagnostic kits are commercially available

Virus	Means of spread	Distribution in developing countries	Detection methods
Begomovirus			
Squash leaf curl virus - SLCV	whiteflies	Central America, Caribbeans	PCR, ELISA
Watermelon chlorotic stunt virus - WmCSV	whiteflies	Red Sea region, Iran	PCR, ELISA
Crinivirus			
Cucurbit yellow stunt disorder virus - CYSDV	whiteflies	Middle East, Africa, Mexico	RT-PCR
Cucumovirus			
Cucumber mosaic virus - CMV	aphids	World wide	ID, ELISA(c)
Polerovirus			
Cucurbit aphid borne yellows virus - CABYV	aphids	World wide	ELISA
Potyvirus			
Moroccan watermelon mosaic virus - MWMV	aphids	Africa	ID-SDS, ELISA(c)
Papaya ringspot virus - PRSV	aphids	World wide, tropical and subtropical	ID-SDS, ELISA(c)
Watermelon mosaic virus - WMV	aphids	World wide temperate	ID-SDS, ELISA(c)
Zucchini yellow mosaic virus - ZYMV	aphids, seeds	World wide	ID-SDS, ELISA(c)

Table 26.4 : Viruses causing local epidemics in cucurbits in developing countries. (c) indicates that DAS-ELISA diagnostic kits are commercially available.

Virus	Means of spread	Distribution in developing countries	Detection methods
Begomovirus			
Melon chlorotic leaf curl virus - MCLCV	whiteflies	Guatemala	PCR
Squash leaf curl virus-China - SqLCV-C	whiteflies	China	PCR
Squash yellow mottle virus - SqYMoV	whiteflies	Costa Rica	PCR
Tomato leaf curl virus - ToLCV	whiteflies	India, Pakistan, Thailand	PCR
Comovirus			
Squash mosaic virus - SqMV	beetles, seeds, contact	World wide	ID, ELISA(c)
Ipomovirus			
Cucumber vein yellowing virus - CVYV	whiteflies	Middle East	RT-PCR
Potyvirus			
Melon vein-banding mosaic virus - MVBMV	aphids	Taiwan	ID-SDS, ELISA
Telfairia mosaic virus - TeMV	aphids, seeds	Nigeria	ID-SDS, ELISA
Zucchini yellow fleck virus - ZYFV	aphids	Mediterranean Basin	ID-SDS, ELISA
Tobamovirus			
Cucumber green mottle mosaic virus - CGMMV	contact, seeds	Asia	ELISA(c)
Tospovirus			
Watermelon bud necrosis virus - WBNV	thrips	India	RT-PCR
Watermelon silver mottle virus - WSMV	thrips	Taiwan	ELISA
Zucchini lethal chlorosis virus - ZLCV	thrips	Brazil	ELISA

Due to the abundance of literature on cucurbit viruses, as far as possible only general reviews are referred, from which more detailed references can be obtained. Unless otherwise stated, information on individual viruses were obtained from Lovisolo, 1980, Blancard *et al.* 1994, Brunt *et al.* 1996, Provvidenti, 1996, Lecoq *et al.* 1998 and van Regenmortel *et al.* 2000. Information on resistance to viruses infecting cucurbits was obtained from Provvidenti, 1993 and Lecoq *et al.* 1998.

26.2.1 The Genus *Potyvirus*

Several potyviruses cause severe economical losses in cucurbit crops in developing countries. Diagnosis for a specific potyvirus is generally based on serological methods (ID-SDS, DAS-ELISA) (Purcifull *et al.* 1988; Hseu *et al.* 1987). A generic test detected almost all potyviruses by using a commercial monoclonal antibody (Jordan, 1992). Generic potyvirid RT-PCR primers permit the amplification of the 3'-non-translated region, the coat protein gene and the 3'-part of the polymerase gene of all potyviruses (Gibbs and Mackenzie, 1997). Observation of viral inclusions in light microscopy is a reliable and cheap generic diagnostic method since all potyviruses induce typical fibrous inclusions with the Orange-Green staining technique (Christie and Edwardson, 1986). Some potyviruses (such as PRSV) can also produce amorphous inclusions.

Potyviruses are efficiently transmitted in a non-persistent manner by many different aphid species. Acquisition and transmission may occur during very short probes (less than a minute) and aphids retain infectivity for a few hours. Experimentally, potyviruses are easily mechanically transmitted. They have flexuous particles *c.* 700-800 nm long and 12 nm wide. Their genome is a unique single stranded RNA molecule of about 10-12 kb. It is translated in a polyprotein, which is subsequently cleaved by virus-encoded proteases in various functional proteins, including a coat protein of approximately 34-36 kDa.

Cultural practices intended to delay the spread of aphid-borne viruses can control potyviruses, but their efficiency is generally limited. This includes careful weeding to remove virus and vector reservoirs and space management to avoid planting new crops close to fields with old infected crops. The use of barrier crops and intercropping susceptible and non-host crops may contribute to limit virus spread (Akkawi *et al.* 1986). When available, the use of plastic mulch and floating row covers can significantly delay virus spread (Cradock *et al.* 2002, El-Zammar *et al.* 2001; Orozco-Santos *et al.* 1995). Insecticides are generally not efficient in preventing directly potyvirus spread, but have beneficial effects by reducing direct damages caused by aphid infestations. Oil sprays may reduce potyvirus transmission by aphid vectors. However, to be efficient, they require frequent applications of the whole plant foliage with specific equipments (Astier *et al.* 2001).

26.2.1.1 *Papaya ringspot virus* (PRSV)

Papaya ringspot virus (PRSV) formerly known as *Watermelon mosaic virus 1* (WMV1), is mostly limited to tropical or sub-tropical areas. PRSV has been reported as the major virus infecting cucurbits in many regions including Brazil, Central America and the Caribbeans, India, and Nepal. It seems to be of minor importance in other regions such as the Middle East, Eastern Africa and China (Table 26.2). PRSV causes severe mosaic, blisters and deformations on leaves and fruits of melon, cucumber, squash and other cultivated cucurbit species. Some strains of PRSV induce systemic necrosis in some melon cultivars (Quiot-Douine *et al.* 1989). Losses can be high due to the severe reduction in fruit yield and to altered fruit quality. PRSV host range is mostly restricted to the *Cucurbitaceae*; therefore virus sources are either cucurbit crops - in regions where cucurbits are grown all year round - or wild cucurbit species that can serve as reservoirs of the virus.

PRSV strains have been differentiated on a host range basis: the papaya strain (PRSV-P) infects papaya and cucurbits and the watermelon strain (PRSV-W) infects cucurbits and not papaya. Analysis of the coat protein gene of PRSV isolates showed geographical structuring of viral populations, suggesting that the virus evolved independently in different countries from local isolates (Bateson *et al.* 2002).

Resistance to PRSV has been identified in melon, cucumber and various *Cucurbita* species from germplasm collections, but PRSV resistant commercial cultivars are still few. An efficient control of PRSV has been achieved in papaya by cross protection using a mild mutant. Mild variants of PRSV-W have been isolated in Brazil and are presently evaluated for their protective effects in field conditions (Rezende *et al.* 1999).

26.2.1.2 *Watermelon mosaic virus* (WMV)

Watermelon mosaic virus (WMV) was formerly known as *Watermelon mosaic virus 2*, (WMV2). WMV and PRSV were once thought to be strains of the same virus, but now are recognised as distinct entities based on serological, molecular and biological properties. WMV has a worldwide distribution and is prevalent in Mediterranean and other temperate regions while it is not or only rarely detected in tropical areas. WMV is one of the major cucurbit viruses in the Middle East, Iran, China and North Africa (Table 26.2).

Symptoms of WMV are variable according to the cultivar. WMV induces very severe vein banding, green mosaic, deformation and plant stunting in some melon (Figure 26.1.A), cucumber or squash cultivars while it is almost symptomless in others. Symptoms in watermelon are generally inconspicuous. WMV has a wide host range; it infects experimentally plants in at least 16 families. Besides cucurbits, WMV can infect naturally pea, vanilla and many weed species that can be reservoirs of the virus during the winter season.

Various levels of resistance to WMV have been identified in cucumber, melon, watermelon and squash. However, some of them appear to be partial or strain-specific. Very few WMV resistant commercial cultivars are presently available. A good level of resistance has been obtained in squash and melon using the coat protein mediated resistance approach (Tricoli *et al.* 1995). Commercial transgenic squash cultivars resistant to WMV are presently available but their use by farmers depends on local regulations as well as on consumers' acceptance.

26.2.1.3 *Zucchini yellow mosaic virus* (ZYMV)

Zucchini yellow mosaic virus (ZYMV) was the first of a series of 'emerging' cucurbit viruses that have been threatening the cucurbit industry since the early 1980s. First identified in Italy and France, ZYMV spread within a decade to the major cucurbit producing regions world wide including many developing countries. Even now, ZYMV is reported to colonise new regions or islands.

ZYMV induces extremely severe mosaic, blisters, enations or filiformism on leaves of most cultivated cucurbit species (Figure 26.1.B). It causes also mosaic and distortions on mature fruits, making them unmarketable. Early infections generally lead to a complete yield loss. ZYMV host range includes several species outside the cucurbits in 11 families (Desbiez and Lecoq, 1997).

A number of ZYMV variants have been described. Some strains may induce a lethal wilting in melons possessing the *Fn* gene. Others induce symptoms very similar to those caused by PRSV or WMV. ZYMV seed transmission is rare or erratic, although it could well be the means by which the virus spread throughout the world (Desbiez and Lecoq, 1997). Recently, it was shown that ZYMV and PRSV could be acquired from melon fruits by aphids, in natural conditions. This suggests possible virus dissemination over long distances by fruits exported for human consumption (Lecoq *et al.* 2003).

Resistance to ZYMV has been found in cucumber, melon, watermelon, *Cucurbita moschata* and wild *Cucurbita* species. However, resistance in melon and zucchini squash appear to be easily overcome, while that in cucumber seems durable. A good level of resistance has been achieved in squash and melon using the coat protein mediated resistance approach (Tricoli *et al.* 1995). Commercial transgenic ZYMV resistant squashes are now grown in the USA. One variant of ZYMV inducing very mild symptoms has been shown to be efficient in protecting plants against severe forms of the virus in different developing countries (Turkey, Lebanon, Sudan, Tonga, Taiwan) (Lecoq and Raccah, 2001). It is now used as a commercial cross protecting agent in different parts of the world.

26.2.1.4 *Moroccan watermelon mosaic virus* (MWMV)

Moroccan watermelon mosaic virus (MWMV) was first described in Morocco as a strain of WMV2 (Fischer and Lockhart, 1974). Serological and molecular studies revealed later that it is a distinct member of the potyvirus group (Lecoq *et al.* 2001). MWMV and closely serologically related strains are widespread in the African continent (Morocco, Algeria, Sudan, Niger, Cameroon, South Africa) where it may be one of the most damaging cucurbit viruses (Van der Meer *et al.* 1987, Quiot-Douine *et al.* 1989, Lecoq *et al.* 2001). MWMV may also occur in Nepal (Table 26.2). Symptoms are very severe in cucumber, squash and watermelon including mosaic and deformations in leaves and fruits. In most melon cultivars symptoms are systemic necrotic spots often followed by a complete collapse and necrosis of the plant. MWMV has biological, serological and molecular relationships with PRSV, and probably belongs to the same potyvirus subgroup (Quiot-Douine *et al.* 1989). Resistance to MWMV has been observed in cucumber and melon.

26.2.1.5 Other potyviruses

Other potyviruses may occur in cucurbit crops in more restricted geographical regions. *Zucchini yellow fleck virus* (ZYFV) infects the cucurbit weed *Ecballium elaterium* in the Mediterranean Basin. Occasionally, ZYFV may cause a severe mosaic disease in cucumber and watermelon and yellow flecks and fruit deformation in zucchini squash. *Telfairia mosaic virus* (TeMV) causes severe mosaic in fluted pumpkin in Nigeria. It has not yet been reported naturally occurring in other cucurbits or countries (Shoyinka *et al.* 1987). TeMV may be seed transmitted in fluted pumpkin. *Melon vein banding mosaic virus* (MVBMV) is a tentative new potyvirus species causing mosaic and vein banding in melon, squash and cucumber crops in Taiwan (Huang *et al.* 1993).

26.2.2 *Cucumber mosaic virus* (CMV)

Cucumber mosaic virus, Genus *Cucumovirus* (CMV) was the first virus reported to infect cucurbits in the beginning of the century. CMV is still probably one of the most common viruses infecting cucurbits world wide, especially in temperate or Mediterranean climatic conditions. In tropical or subtropical regions, CMV is less frequent in cultivated cucurbits although it may be found in weeds or other crops nearby (Quiot *et al.* 1983; Table 26.2). CMV causes severe diseases in most cucurbits (except in watermelon where it is rare) and with subsequent severe economical losses. CMV has a broad host range, infecting more than 775 different species in 86 different families. CMV induces plant stunting and mosaic, size reduction and distortion of leaves in melon, cucumber and squash. Symptoms on fruits are discolorations and occasionally deformations. A strong synergistic effect is observed when CMV is in mixed infection with potyviruses, leading to more severe mosaic symptoms and pronounced stunting.

CMV is efficiently transmitted by more than 60 aphid species in a non-persistent manner. Diagnosis can be easily done within one to few days through biological tests (local or systemic infections of *Vigna unguiculata*, systemic infection of *Nicotiana tabacum*). Classical serological methods such as SDS-ID or DAS-ELISA can also be used. CMV is a small isometric virus with particles *c.* 29 nm in diameter which has a divided genome constituted of three single stranded RNA molecules of 3.4, 3.1 and 2.2 kb, respectively. Two major sub-groups (type I and type II) have been differentiated according to serological and molecular properties. DAS-ELISA diagnostic kits are commercially available for CMV, including monoclonal antibodies differentiating strains of the two subgroups. A diversity of strains were described differing in symptomatology, host range, virulence towards resistance genes and aphid transmissibility.

Differences in the type of CMV spread may occur in different cucurbit species. CMV epidemics were observed to be rapid in melon crops while they are often slow and limited to a few plants in zucchini squash crops grown nearby.

Cultural practices intended to delay the spread of aphid-borne viruses (see *Potyvirus* section) can be applied to control CMV. Various levels of resistance to CMV are present in some cucumber, melon and squash commercial cultivars. Biotechnology has provided new opportunities to diversify resistance mechanisms to CMV. Significant levels of resistance have been obtained in squash, melon and cucumber using the coat protein mediated resistance (Tricoli et al. 1995).

26.2.3 *Cucurbit aphid borne yellows virus* (CABYV)

Cucurbit aphid borne yellows virus, Genus *Polerovirus* (CABYV) is the first *Luteoviridae* described infecting naturally cucurbits and one of the several viruses inducing yellowing symptoms in these crops (Lecoq *et al.* 1992). CABYV is extremely wide spread, and probably one of the most common virus infecting cucurbits in developing countries. It has been detected whenever it was looked for except in the Caribbeans (Table 26.2). CABYV can significantly reduce cucurbit yields by reducing fruit setting, although it does not affect the quality of the fruits produced on infected plants. CABYV has a moderately wide host range, infecting cucurbits and weeds, that are efficient alternative hosts. CABYV is phloem-limited in plants. Typical symptoms in cucumber, melon, squash and watermelon are yellowing and thickening of the older leaves (Figure 26.1.C). There is a wide range of symptom intensity in susceptible cultivars, varying from a yellowing limited to only a few older leaves to a complete discoloration of the plants. A synergism was observed when CABYV is in mixed infection with potyviruses: CABYV multiplication rate is higher and the yellows symptoms are generally more severe.

CABYV is transmitted in a persistent manner by a few aphid species (*Aphis gossypii*, *Myzus persicae*). Diagnosis is by DAS-ELISA. The virus has small isometric particles *c.* 25 nm in diameter with a single stranded RNA genome of approximately 5.7 kb. The virus genome codes for 6 proteins, including two structural proteins, a coat protein of 23 kDa and a readthrough protein of 76 kDa involved in aphid transmission.

Prevention of virus spread can be achieved by reducing vector population, abundance of reservoirs as well as by the use of plastic mulches that repels aphid vectors. Resistance has been found in cucumber, melon and squash (Lecoq, 1999).

26.2.4 The Genus *Begomovirus*

Begomoviruses emerged as important cucurbit pathogens in developing countries only in recent years. Several new species have been reported in Mediterranean and sub-tropical regions of both the New and the Old world during the last five years. Diagnosis for a specific begomovirus is generally done either by nucleic acid hybridisation or by PCR with specific primers. Serological diagnosis is also possible but it is generally less specific because of the cross reactivity of polyclonal antisera. Serotyping with monoclonal antibodies can help to distinguish between different begomovirus species (Swanson *et al.* 1992). Begomovirus generic diagnosis can be achieved either serologically or using degenerate PCR primers (Rojas *et al.* 1993). Begomoviruses have geminate particles measuring 18 nm x 30 nm containing a single structural protein, the coat protein of 28-34 kDa. The genome consists of two circular single-stranded DNA molecules (components A and B) of 2.5-2.8 kb in size.

Begomoviruses are transmitted by *Bemisia tabaci* in a persistent manner. The acquisition period lasts several hours, followed by a latent period of 12 or more hours. After that, the whiteflies can generally transmit the virus for the rest of their lives. Some begomoviruses are transmitted mechanically, but with difficulties.

Symptoms in infected plants are generally distinctive with bright yellow mosaic or mottle and leaf curling, severe stunting, fruit distortion, and marked yield reduction in susceptible cultivars. Intensification of the cucurbit production with the development of irrigated areas, higher plant densities, extended periods of cultivation and increase in whitefly populations are factors that were probably responsible for the rapid emergence of cucurbit begomoviruses (Brown *et al.* 2002).

26.2.4.1 *Squash leaf curl virus* (SLCV)

The original isolate of *Squash leaf curl virus* (SLCV) described in California has a restricted host range, and infects bean and squash only. Other isolates later described present a high biological and genetic polymorphism, some having an expanded host range which includes melon, cucumber and experimentally several non-cucurbit hosts (Brown *et al.* 1995). SLCV causes reduction in leaf size, curling, vein thickening, enations and plant stunting in squash, yellow mosaic, mottle and curling in watermelon and only a mild mosaic in melon and cucumbers. SLCV is prevalent in Mexico, Central America and the Dominican Republic.

Because of its restricted natural host range, implementing host-free periods and removal of weed reservoirs of the virus and the whitefly vector can control SLCV. Efficient vector control may also contribute to limit virus spread as well as the use of plastic mulches and floating row covers (Hilje *et al.* 2001). Tolerance and resistance to SLCV has been described in several cucurbit species including *Cucurbita moschata*, *C. ficifolia* and *C. pepo*.

26.2.4.2 *Watermelon chlorotic stunt virus* (WmCSV)

First described in Yemen, *Watermelon chlorotic stunt virus* (WmCSV) is now causing devastating epidemics in other countries of the Red Sea region and Iran (Jones *et al.* 1988, Kheyr-Pour *et al.* 2000, Bananej *et al.* 2002). Symptoms are very severe in watermelon, melon and zucchini squash. Once infected, watermelon plants develop yellow leaves with very short internodes (Figure 26.1.D). This gives a typical appearance to the plants with old green leaves in the centre and completely yellow stunted young shoots at the periphery. Infected plants produce only a few misshapen fruits. In melons, symptoms are a bright yellow mosaic or mottle. In zucchini squash, mosaic, leaf deformations and stunting are generally observed.

Tolerances to WmCSV have been observed in *Citrullus lanatus* landraces from Sudan and resistance, in several melon cultivars.

26.2.4.3 Other begomoviruses

Several begomoviruses have recently been reported in different parts of the world. *Melon chlorotic leaf curl virus* (MCLCV) causes patchy foliar chlorosis, leaf curling and reduced fruit sets in export melon crops in Guatemala (Brown *et al.* 2001). *Squash yellow mottle virus* (SqYMoV) induces leaf deformation and yellow mottling in squash in Costa Rica (Karkashian *et al.* 2002). *Cucurbit leaf curl virus* (CuLCV) is occurring in Northern Mexico infecting several cucurbit crops. CuLCV and SLCV can exchange DNA components while in mixed infection, yielding reassortant viruses with new biological properties. If this phenomenon occurs in natural conditions, it may lead to a broader biological diversity of begomoviruses in the future, in regions where two species may occur in mixed infections (Brown *et al.* 2002).

A yellow leaf disease of melon and wax gourd in Thailand (Samretwanish *et al.* 2000), a mosaic disease of bittergourd in India (Khan *et al.* 2002) and a leaf curl disease of watermelon in Pakistan (Mansoor *et al.* 2000) have been associated with the presence of strains of *Tomato leaf curl virus-India.* A leaf curl disease of squash in China was associated to a new virus species, SLCV-China (Hong *et al.* 1995). A virus related to SLCV was also reported in Saudi Arabia (Al-Shahwan *et al.* 2002).

26.2.5 *Cucurbit yellow stunting disorder virus* (CYSDV)

A yellowing disease of cucurbit crops induced by *Cucurbit yellow stunting disorder virus*, Genus *Crinivirus* (CYSDV) was first detected in the United Arab Emirates in 1982 (Hassan and Duffus, 1991). It is wide spread now in the Mediterranean Basin (Jordan, Lebanon, Turkey, Saudi Arabia, Egypt) and Africa (Morocco, Senegal, Sudan) (Rubio *et al.* 1999; Desbiez *et al.* 2000). It has been recently observed in Mexico. CYSDV may induce significant yield reductions, but generally it does not affect the quality of the fruits. CYSDV infects only cucurbits including melon, cucumber, watermelon and squash (Celix *et al.* 1996). Symptoms in melon and cucumbers are pronounced interveinal yellowings, generally limited to the older leaves but occasionally found on all leaves (Figure 26.1.E). The yellowing symptoms may be confused with physiological or nutritional disorders, natural senescence as well as infection by other viruses such as CABYV (Figure 26.1.C). For this reason, CYSDV infections are difficult for growers to recognize. Milder yellowing symptoms are generally observed in squash and watermelon.

CYSDV is efficiently transmitted by the B biotype and relatively inefficiently by the A biotype of *B. tabaci* in a semi-persistent mode. CYSDV is acquired within a few hours and may persist in the vector up to nine days (Wisler *et al.* 1998). CYSDV diagnoses rely so far either on nucleic acid hybridisation techniques or on the use of RT-PCR with specific primers. Primers are generally designed either on the coat protein gene or on a gene coding for an analog of the heat shock protein, HSP70, which is relatively conserved in all *Closteroviridae* (Rubio *et al.* 1999). CYSDV has highly flexuous particles *c.* 800-900 nm long. It has a linear, positive sense single stranded RNA genome, which is divided into two molecules of approximately 8 and 9 kb. Criniviruses express their large genome *via* polyprotein processing, translational frameshifting, and production of multiple subgenomic RNAs. Particles are encapsidated mainly by a major coat protein of 28 kDa.

CYSDV is difficult to control. Prevention mostly relies on whitefly control. Epidemics of CYSDV are particularly severe in protected crops due to the rapid succession of crops and to the high whitefly populations that allow virus transfer from one crop to another. Sources of resistance to CYSDV have been identified in cucumber, melon and watermelon.

26.2.6 Other Important Cucurbit Viruses

Some viruses infecting cucurbits in developing countries may be widely distributed but are only seldom of economic importance. Among those, some are seed borne as *Squash mosaic virus*, Genus *Comovirus* (SqMV) and *Cucumber green mottle mosaic virus*, Genus *Tobamovirus* (CGMMV).

SqMV has isometric particles of *c.* 30 nm and a divided genome consisting of 2 single stranded RNA molecules. It is transmitted by beetles, by contact and also by the seeds. Seed transmission is probably the major means of virus conservation from one crop to another, especially in regions where farmers produce their own seeds. Symptoms on leaves include mosaic, vein banding and deformations on melon or squash leaves. Detection of SqMV in seed lots is possible by DAS-ELISA, but does not correlate well with the actual seed transmission rates. Indeed, SqMV seed transmission occurs only through embryo infection and DAS-ELISA can detect the virus in teguments or papery layers of individual seeds while their embryos are not infected. Production of virus-free seeds will prevent early contaminations and limit subsequent virus spread.

CGMMV has rod shaped particles *c.* 300 nm long and 18 nm in diameter with a single stranded RNA genome. It is easily transmitted mechanically (handling of the plants, leaf contacts), through soil or substrate contamination and by the seeds in cucumber, watermelon, melon and bottle gourd. CGMMV seed contamination seems to be mostly external; the virus can be eliminated from cucumber seeds by dry heat treatment at 70° C during 3 days without impairing seed germination rates. CGMMV induces mosaic of varying intensities on the leaves; fruits are often reduced in size and deformed. CGMMV control includes the use of virus-free seeds and disinfection of stakes and pruning tools. CGMMV can be easily detected by standard serological tests.

Other viruses cause severe epidemics at regional level mainly because of an increase in their vector prevalence.

Cucumber vein yellowing virus, Genus *Ipomovirus* (CVYV) has rod-shaped particles *c.* 750 nm long. It belongs to the *Potyviridae* and is transmitted by *B. tabaci* in a semi-persistent manner. CVYV is common in the southeastern part of the Mediterranean Basin where it infects several cucurbit species. CVYV causes severe vein clearing on young leaves and mosaic and yellowing on older leaves of melon and cucumber (Figure 26.1.F). Symptoms are inconspicuous in watermelon and squash. Several commercial cucumber cultivars are tolerant to CVYV.

Tospoviruses are gaining importance in cucurbit crops, especially in Asia (India, Taiwan) and South America (Brazil). Tospoviruses have quasi-spherical enveloped particles 75-100 nm in diameter, containing three linear single stranded RNA molecules. Tospovirus genomic organization is complex: the large RNA (L-RNA) has a negative polarity while the two others (S-RNA and M-RNA) use ambisense coding strategies. Tospoviruses are transmitted in a circulative manner by different *Thrips* species. *Watermelon silver mottle virus* (WSMV) is transmitted by *Thrips palmi* and occurs in Taiwan. In watermelon, it induces a severe stunting, mottling, yellow spotting and leaf deformation, tip necrosis and dieback. Fruits are malformed and show silver mottling (Yeh and Chang, 1995). *Zucchini lethal chlorosis virus* (ZLCV) host range is limited to cucurbits. It causes severe mosaic, leaf deformation and stunting in zucchini squash in Brazil (Bezerra *et al.* 1999). *Watermelon bud necrosis virus* (WBNV) causes leaf mottling, yellowing, shortened internodes and necrosis and dieback of buds in watermelon in India (Jain *et al.* 1998).

26.2.7 Minor Cucurbit Viruses

Viruses infecting cucurbits in developing countries with only minor economic incidence are listed in Table 26.5.

Table 26.5 : Cucurbit viruses with only minor economical incidence in developing countries. (c) indicates that DAS-ELISA diagnostic kits are commercially available.

Virus	Means of spread	Distribution in developing countries	Detection methods
Carmovirus			
Cucumber soil-borne virus - CuSBV	fungi	Lebanon	ID, ELISA
Melon necrotic spot virus - MNSV	fungi, seeds, contact	Iran	ID, ELISA(c)
Closterovirus			
Beet pseudo yellows virus - BPYV	whiteflies	Turkey	RT-PCR
Crinivirus			
Lettuce infectious yellows virus - LIYV	whiteflies	Mexico	RT-PCR, ELISA
Curtovirus			
Beet curly top virus - BCTV	leaf-hoppers	Middle East	PCR
Ipomovirus			
Squash yellow leaf curl virus - SYLCV	whiteflies	Sultanate of Oman	
Nepovirus			
Tobacco ringspot virus - TRSV	nematodes seeds	Middle East	ID, ELISA(c)
Tomato ringspot virus - ToRSV	nematodes	World wide	ID, ELISA(c)
Nucleorhabdovirus			
Eggplant mottle dwarf virus - EMDV	leaf-hoppers	Mediterranean Basin	ELISA
Ourmiavirus			
Ourmia melon virus - OuMV	unknown	Iran	ID, ELISA
Potexvirus			
Papaya mosaic virus - PapMV	contact	Mexico	ELISA(c)
Tobamovirus			
Zucchini green mottle mosaic virus - ZGMMV	contact, seeds	Korea	ELISA
Tospovirus			
Tomato spotted wilt virus - TSWV	thrips	World wide	ELISA(c)
Tymovirus			
Chayote mosaic virus -	unknown	Costa Rica	ID,

ChMV			ELISA
Melon rugose mosaic virus - MRMV	unknown, seeds	Red Sea region	ID, ELISA
Unassigned			
Cucumber leaf spot virus - CLSV	fungi, seeds	Middle East	ID, ELISA
Squash necrosis virus - SqNV	fungi	Brazil, Martinique	ID, ELISA

26.3 Conclusions

Viruses that infect cucurbits in developing countries constitute a complex and changing pathosystem. Crops are often infected by several viruses, sharing the same vector types (i.e. aphids or whiteflies). In regions where several crop cycles are grown in a year, the viral pathosystem composition may vary according to the season in relation to the vector prevalence (i.e. aphid-borne viruses are generally frequent during mild and wet seasons while whitefly-borne viruses mainly occur during the hot and dry seasons). In addition to seasonal or regional variations, the occurrence of 'new' viruses can also change a pathosystem composition. As an illustration, 16 'new' viruses infecting cucurbits were described during the 1980-1998 period, some of which are now widespread and cause major yield reductions (Lecoq *et al.* 1998). From 1998 to 2002, 7 'new' cucurbit viruses -mostly begomoviruses or tospoviruses- were reported mainly from developing countries.

Among the 'new' viruses, some have probably been present in cultivated cucurbits for a long time, but remained unnoticed because their symptoms were masked by other viruses or previously associated to other causes (i.e. nutritional or physiological disorders). This applies probably to CABYV that causes yellowing symptoms. The world wide distribution of CABYV was in fact established only after a DAS-ELISA was developed to detect it.

Other 'new' viruses may represent typical 'emerging' viruses. Some became suddenly wide-spread because they followed the rapid increase and dissemination of their natural vectors. Typical of this situation are the begomoviruses transmitted by *B. tabaci* (Brown *et al.* 2002). The situation of ZYMV is probably different. This potyvirus, transmitted by many aphid species in a non-persistent manner, was described in the mid 1970s as occurring locally in Northern Italy. Within a decade, ZYMV spread to all the major cucurbit growing areas in the world including many developing countries. The rapid spread of ZYMV occurred by means, which are still unknown and cannot be related to an extension of its aphid vectors which were already prevalent wherever cucurbits were grown (Desbiez and Lecoq, 1997).

Some viruses still restricted geographically, have the potential to become more widely spread if they are introduced to new areas and have a good ecological fitness.

There is still a need for increased research on virus diagnosis. There are major viruses such as the begomoviruses and the criniviruses for which simple, cheap and reliable serological diagnostic methods have to be developed. International cooperation for production and exchange of antisera or molecular probes is essential in order to conduct regular surveys to update the local virus pathosystem compositions. An efficient international cooperation and the development of simple diagnostic methods (SDS-ID, DAS-ELISA) contributed greatly to the identification of ZYMV on a world-wide basis within a few years (Desbiez and Lecoq, 1997). The development of commercially available DAS-ELISA kits is also a step forward for a more accurate evaluation of the virus situation in developing countries although their specificity and cost could be limiting factors (Vegas *et al.* 1998).

The diversity of viruses infecting cucurbits has important consequences on the choice of control strategies. It will depend on the local virus pathosystem, on the agro-ecosystem, on the technical and economical farmer's potential as well as on the target market. Some methods are applicable to control all viruses (use of virus-free seeds or healthy seedlings, elimination of virus or vector reservoirs, barrier crops). Others will be efficient for viruses having the same type of vector (plastic mulching, insect-proof nets, floating row covers), and some will be specific for only one virus (cross protection, resistance). Some methods can be successfully applied in more traditional farming systems still common in many developing countries (barrier crops, intercropping) but would not be compatible with intensive production systems.

The use of virus resistant cultivars is probably the easiest way to control viral diseases at the farmer's level. Although the number of virus resistant cultivars adapted to developing countries is increasing, they are still lacking for many cultivated species and local landraces. In recent years biotechnology has emerged as a new and promising alternative for creating virus resistant cultivars. However, their use in developing countries will depend on local regulations allowing their cultivation, on consumers' acceptance as well as on the patent issue that may increase significantly the cost of the seeds.

References

Akkawi, M., Al-Musa, A.M., Sharaf, N. and Mansour, A. 1986. Control of mosaic affecting squash in Jordan. 1) Effect of intercropping corn, pepper, and eggplants in squash fields on the incidence of mosaic diseases. *Dirasat* 13:157-163.

Al-Shahwan, I.M., Abdalla, O.A. and Al-Saleh, M.A. 2002. Squash leaf curl virus and other begomoviruses in Saudi Arabia. *Dirasat* 29:28-36.

Anonymous, 2002. FAOSTAT, Food and Agriculture Organization of the United Nations, HtmlResAnchor http://apps.fao.org/.

Bananej, K., Ahoonmanesh, A. and Kheyr-Pour, A. 2002. Host range of an Iranian isolate of Watermelon chlorotic stunt virus as determined by whytefly-mediated inoculation and agroinfection and its geographical distribution. *J. Phytopathology* 150:423-430.

Bateson, M.F., Lines, R.E., Rewill, P., Chaleeprom, W., Ha, C.V., Gibbs, A.J. and Dale, J.L. 2002. On the evolution and molecular epidemiology of the potyvirus *Papaya ringspot virus*. *J. Gen. Virol.* 83: 2575-2585.

Bezerra, I.C., Resende, R. de O., Pozzer, L., Nagata, T., Kormelink, R. and De Avila, A.C. 1999. Increase of Tospoviral Diversity in Brazil with the identification of two new tospovirus species one from chrysanthemum and one from zucchini. *Phytopathology* 89:823-830.

Blancard, D., Lecoq, H. and Pitrat, M. 1994. *A colour atlas of cucurbit diseases*. Manson Publishing. London, UK.

Brown, J.K., Idriss, A.M., Alteri, C. and Stenger, D.C. 2002. Emergence of a new cucurbit-infecting begomovirus species capable of forming viable reassortants with related viruses in the Squash leaf curl virus cluster. *Phytopathology* 92:734-742.

Brown, J.K., Idriss, A.M. and Rogan, D. 2001. Melon chlorotic leaf curl virus, a new begomovirus associated with *Bemisia tabaci* infestations in Guatemala. *Plant Dis.* 85:1027.

Brown, J.K., Wendt, K. and Wyatt, S.D. 1995. Genetic variability of Squash Leaf Curl Virus isolates by DNA hybridization and component-specific polymerase chain reaction. In: *Cucurbitaceae '94* G. Lester and G. Dunlap (eds.). Gateway Printing, USA pp. 5-11.

Brunt, A.A., Crabtree, K., Dallwitz, M.J., Gibbs, A.J. and Watson, L. 1996. *Viruses of Plants*. CAB International. Wallingford, UK.

Carrasco, L.P. and Wessel-Beaver, L. 2002. Survey of cucurbit viruses in Puerto Rico. In: *Cucurbitaceae 2002*, D.N. Maynard (ed), ASHS Press, Alexandria, Virginia, USA pp. 259-264.

Celix, A., Lopez-Sese, A., Almarza, N., Gomez-Guillamon, M. and Rodriguez-Cerezo, E. 1996. Characterization of cucurbit yellow stunting disorder virus, a *Bemisia tabaci*-transmitted Closterovirus. *Phytopathology* 86:1370-1376.

Christie, R.G. and Edwardson, J.R. 1986. Light microscopy techniques for detection of plant virus inclusions. *Plant Dis.* 70:273-279.

Clark, M.F. and Adams, A.N. 1977. Characteristics of microplate method of enzyme-linked immunosorbent assay for the detection of plant viruses. *J. Gen. Vir.* 34:475-483.

Cradock, K.R., Da Graca, J.V. and Laing, M.D. 2002. Studies on the cultural control of virus diseases in zucchini crops. *S. Afr. J. Sci.* 98:225-227.

Cradock, K.R., Laing, M.D. and Da Graca, J.V. 2001. Survey of aphid-borne cucurbit viruses in KwaZulu-Natal, South Africa, and their control using cultural methods. *Phytopathology* 92 (Suppl.):S19.

Dahal, G., Lecoq, H. and Albrechtsen, S.E. 1997. Occurrence of papaya ringspot potyvirus and cucurbit viruses in Nepal. *Ann. appl. Biol.* 130:491-502.

Desbiez, C. and Lecoq, H. 1997. Zucchini yellow mosaic virus. *Plant Path.* 46:809-829.

Desbiez, C., Lecoq, H., Aboulama, S. and Peterschmitt, M. 2000. First report of Cucurbit yellow stunt disorder virus in Morocco. *Plant Dis.* 84:596.

El-Zammar, S., Abou-Jawdah, Y. and Sobh, H. 2001. Management of virus diseases of squash in Lebanon. *J. Plant Path.* 83:21-25.

Fischer, H.U. and Lockhart, B.E.L. 1974. Serious losses in cucurbits caused by Watermelon mosaic virus in Morocco. *Plant Dis. Reptr* 58:143-146.

Gibbs, A. and Mackenzie, A. 1997. A primer pair for amplifying part of the genome of all potyvirids by RT-PCR. *J. Virol. Methods* 63:9-16.

Hassan, A.A. and Duffus, J.E. 1991. A review of a yellowing and stunting disorder of cucurbits in the United Arab Emirates. *Emir. J. Agric. Sc.* 2:1-16.

Hilje, L., Costa, H.S. and Stansly, P.A. 2001. Cultural practices for managing *Bemisia tabaci* and associated viral diseases. *Crop Prot.* 20:801-812.

Hong, Y., Wang, X., Tian, B. and Cai, J. 1995. Chinese squash leaf curl virus: a new whitefly-transmitted geminivirus. *Science China* 38:179-186.

Hseu, S.H., Huang, C.H., Chang, C.A., Yang, W.Z., Chang, Y.M. and Hsiao, C.H. 1987. The occurrence of five viruses in six cucurbits in Taiwan. *Plant Prot. Bull.* 29:233-244.

Huang, C.H., Chang, L. and Tsang, J.H. 1993. The partial characterization of Melon Vein-banding Mosaic Virus, a newly recognized virus infecting cucurbits in Taiwan. *Plant Path.* 42:100-107.

Jain, R.K., Pappu, H.R., Pappu, S.S., Krishna Reddy, M. and Vani, A. 1998. Watermelon bud necrosis tospovirus is a distinct virus species belonging to serogroup IV. *Arch. Virol.* 143:1637-1644.

Jones, P., Sattar, M.H.A. and Al Kaff, N. 1988. The incidence of virus disease in watermelon and sweetmelon crops in the People Democratic Republic of Yemen and its impact on cropping policy. *Aspects appl. Biol.* 17:203-207.

Jordan, R. 1992. Potyviruses, monoclonal antibodies and antigenic sites. *Arch. Virol.* Supplementum 5:81-95

Karkashian J.P., Maxwell, D.P. and Ramirez, P. 2002. Squash yellow mottle geminivirus: a new cucurbit-infecting geminivirus from Costa Rica. *Phytopathology* 92 (Suppl.):S125

Khan, J.A., Siddiqui, M.K. and Singh, B.P. 2002. The association of begomovirus with bitter melon in India. *Plant Dis.* 86:328.

Kheyr-Pour, A., Bananej, K., Dafalla, G.A., Caciagli, P., Noris, E., Ahonmanesh, A., Lecoq, H. and Gronenborn, B. 2000. Watermelon chlorotic stunt virus from the Sudan and Iran: sequence comparison and identification of a whitefly-transmission determinant. *Phytopathology* 90:629-635.

Lecoq, H. 1999. Epidemiology of Cucurbit aphid-borne yellows virus. In: *The Luteoviridae.* H.G. Smith and H. Barker (eds). CAB International. Wallingford, U.K. pp.243-248.

Lecoq, H., Bourdin, D., Wipf Scheibel, C., Bon, M., Lot, H., Lemaire, O. and Herrbach, E. 1992. A new yellowing disease of cucurbits caused by a luteovirus, cucurbit aphid borne yellows virus. *Plant Path.* 41:749-761.

Lecoq, H., Dafalla, G., Desbiez, C., Wipf-Scheibel, C., Delécolle, B., Lanina, T., Ullah, Z. and Grumet, R. 2001. Biological ad molecular characterization of Moroccan watermelon mosaic virus and a potyvirus isolate from Sudan. *Plant Dis.* 85:547-552.

Lecoq, H., Dafalla, G., Desbiez, C., Wipf-Scheibel, C. and Kheyr-Pour, A. 2003. A 10-year survey (1993-2002) of cucurbit viruses in Sudan. *J. Plant Dis. Protect.* 110:68-69.

Lecoq, H., Desbiez, C., Wipf-Scheibel, C. and Girard, M. 2003. Potential involvement of melon fruits in the long distance dissemination of cucurbit potyviruses. *Plant Dis.* (*in press*).

Lecoq, H. and Raccah, B. 2001. Cross-protection: interactions between strains exploited to control plant virus diseases. In: *Biotic Interactions in Plant-Pathogen Associations*. M.J. Jegger and N.J. Spence (eds) CAB International. Wallingford, UK.

Lecoq, H., Wisler, G. and Pitrat, M., 1998. Cucurbit viruses: the classics and the emerging. In: *Cucurbitaceae '98*. J.D. McCreight (ed), ASHS Press, Alexandria, Virginia, USA. pp. 126-142.

Lovisolo, O. 1980. Virus and viroid diseases of cucurbits. *Act. Horticult.* 88:33-82.

Mansoor, S., Khan, S.H., Hussain, M., Mushtaq, N., Zafar, Y. and Malik, K.A. 2000. Evidence that watermelon leaf curl disease in Pakistan is associated with Tomato leaf curl virus-India, a bipartite begomovirus. *Plant Dis.* 84:102.

Orozco-Santos, M., Perz-Zamora, O. and Lopez-Arriaga, O. 1995. Effect of transparent mulch on insect populations, virus diseases, soil temperature, and yield of cantaloup in a tropical region. *N. Z. J. Crop Hort. Sc.* 23:199-204.

Provvidenti, R. 1993. Resistance to viral diseases of cucurbits, In: *Resistance to Viral Diseases of Vegetables: Genetics and Breeding.* M. M. Kyle (ed.). Timber Press. Portland, USA. pp. 8-43.

Provvidenti, R. 1996. Diseases caused by viruses. In: *Compendium of Cucurbit Diseases.* T. A. Zitter, D. L. Hopkins and C. E. Thomas (eds.). APS St Paul, USA. pp. 37-45.

Purcifull, D.E., Simone, G.W., Baker, C.A. and Hiebert, E. 1988. Immunodiffusion tests for six viruses that infect cucurbits in Florida. *Proc. Flor. St. Hort. Soc.* 101:400-403.

Quiot, J.B., Labonne, G. and Quiot-Douine, L. 1983. The comparative ecology of cucumber mosaic virus in Mediterranean and tropical regions, In: *Plant Virus Epidemiology*. R. T. Plumb and J. M. Tresh (eds.). Blackwell Scientific Publications. Oxford, UK. pp. 177-183.

Quiot-Douine, L., Lecoq, H., Quiot, J.B., Pitrat, M. and Labonne G. 1989. Serological and biological variability of isolates related to strains of papaya ringspot virus. *Phytopathology* 80:256-263.

Rezende, J.A.M., Pacheco, D.A., and Iemma, A.F. 1999. Effects of cross protection with mild strains of PRSV-W on 'Menina Brasileira' squash. *Pesq. Agropec. Bras.* 34:1481-1489.

Rivera, C., Villalobos, W., Sanchez, M.V., Zumbado, C. and Rodriguez C.M. 1993. Identification and distribution of melon-infecting viruses and their vectors in two provinces of Costa-Rica. *Turrialba* 43:210-215.

Robinson, R.W. and Deckers-Walters, D.S. 1997. *Cucurbits.* CAB International, Wallingford, UK.

Rojas, M.R., Gilbertson, R.L., Russell, D.R. and Maxwell, D.P. 1993. Use of degenerate primers in the polymerase chain reaction to detect whitefly transmitted geminiviruses. *Plant Dis.* 77:340-347.

Rubio L., Soong J., Kao, J. and Falk, B.W. 1999. Geographic distribution and molecular variation of isolates of three whitefly-borne closteroviruses of cucurbits: Lettuce infectious yellows virus, Cucurbit yellow stunting disorder virus and Beet pseudo-yellows virus. *Phytopathology* 89:707-711.

Samretwanich, K., Chiemsombat, P., Kittipakorn, K. and Ikegami, M. 2000. Yellow leaf disease of cantaloupe and wax gourd from Thailand caused by Tomato leaf curl virus. *Plant Dis.* 84:200.

Shoyinka, S.A., Brunt, A.A., Phillips, S., Lesemann, D.-E., Thottappilly, G. and Lastra, R. 1987. The occurrence, properties and affinities of Telfairia mosaic virus, a potyvirus prevalent in *Telfairia occidentalis* (*Cucurbitaceae*) in South Western Nigeria. *J. Phytopath.* 119:13-24.

Swanson, M.M., Brown, J.D., Poulos, B.T. and Harrison, B.D. 1992. Genome affinities and epitope profiles of whitefly transmitted geminiviruses from the Americas. *Ann. Appl. Biol.* 121:285-296.

Taylor, M.J. and Brant, J. 2002. Trends in world cucurbit production 1991 to 2001. In: *Cucurbitaceae 2002*, D.N. Maynard (ed), ASHS Press, Alexandria, Virginia, USA. pp. 373-379.

Tricoli, D., Carney, K., Russell, P., McMaster, J., Groff, D., Hadden, K., Himmel, P., Hubbard, J., Boeshore, M., Reynolds, J. and Quemada, H. 1995. Field evaluation of transgenic squash containing single or multiple virus coat protein gene constructs for resistance to cucumber mosaic virus, watermelon mosaic virus 2, and/or zucchini yellow mosaic virus. *Bio/Technology* 13:1458-1465.

Van der Meer, F.W. and Garnett H.M. 1987. Purification and identification of a South African isolate of Watermelon mosaic virus-Morocco. *J. Phytopath.* 120:255-270.

Van Regenmortel, M.H.V., Fauquet, C.M., Bishop, D.H.L., Carstens, E., Estes, M.K., Lemon, S., Maniloff, J., Mayo, M.A., McGeoch, D.J., Pringle, C.R. and Wickner, R. 2000. *Virus taxonomy. Classification and nomenclature of viruses. VIIth report of the International Committee of Taxonomy of Viruses.* Academic Press, San Diego, USA.

Vegas, A., Gonzalez, A., Trujillo, G. and Pino, I. 1998. Difficulties in the serological diagnosis of attenuated strains of the papaya ringspot virus. *Fitopatol. Venez.* 11:40-44.

Wisler, G., Duffus, J., Liu, H. and Li, R. 1998. Ecology and epidemiology of whitefly-transmitted closteroviruses. *Plant Dis.* 82:270-280.

Yeh, S.D. and Chang, T.F. 1995. Nucleotide sequence of the N gene of watermelon silver mottle virus, a proposed new member of the genus tospovirus. *Phytopathology* 85:58-64.

Yuki, V.A., Rezende, J.A.M., Kitajima, E.W., Barroso, P.A.V., Kuniyuki, H., Groppo, G.A. and Pavan, M.A. 2000. Occurrence, distribution, and relative incidence of five viruses infecting cucurbits in the State of Sao Paulo, Brazil. *Plant Dis.* 84:516-520.

Zouba, A., Khan, A.J., Lopez, M. and Al-Maqbaly, Y.M. 1997. Survey of virus diseases of cucurbits in the Batinah region of the Sultanate of Oman. *Arab J. Pl. Prot.* 15:43-46.

Figure 26.1. (A) Vein banding, blisters and leaf deformation caused by *Watermelon mosaic virus* in melon. (B) Severe stunting, mosaic and leaf deformation caused by *Zucchini yellow mosaic virus* in zucchini squash. (C) Yellowing of older leaves caused by *Cucurbit aphid borne yellows virus* in melon. (D) Severe yellow mosaic caused by *Watermelon chlorotic stunt virus* in watermelon. (E) Yellowing of older leaves caused by *Cucurbit yellow stunting disorder virus* in melon. (F) Vein clearing caused by *Cucumber vein yellowing virus* in cucumber.

27. Other vegetables

Amaranthus, Chayote, Colocasia and Xanthosoma; Eggplant, Lablab, Okra, Onion, Pea and Sweet pepper.

Anupam Varma and Bikash Mandal
Advanced Centre for Plant Virology, Indian Agricultural Research Institute, NewDelhi 110012, India

27.1 Introduction

Viral diseases are a major constraint in improving productivity of crop plants in general and vegetable crops in particular. The losses caused by the viral diseases are greater in the tropics and semitropics, which provide ideal conditions for the perpetuation of not only the viruses but also their vectors and secondary hosts. Estimates of the losses caused by viral diseases in vegetables are not available, but viral infections frequently lead to even abandoning of crops due to high incidence of viruses (Varma and Malathi, 2003). Here, we briefly describe viral diseases of some common vegetables like eggplant, garlic, onion, pea and sweet pepper, and some uncommon vegetables like Amaranthus, chayote, Colocasia and Xanthosoma, Lablab, and okra.

27.2 Amaranthus

More than 70 species of *Amaranthus* are known, some of which are cultivated for grains and/or green leaf vegetable in the tropic and semi-tropic regions of Africa, Asia and Central and South America. *A. blitum, A. cruentus, A. dubious, A. hybridus, A. hypochondriacus, A. lividis, A. thunberghii, A. tricolor* and *A. tristis* are commonly grown. The young leaves of all *Amaranthus* spp. are edible. *Amaranthus* leaf protein, as well as grain protein are rich in lysine. The leaves are also good source of micronutrients and vitamins. Some viral diseases of minor importance have been reported to affect *Amaranthus,* which may also alter the nutritive value of the leaves.

Amaranthus leaf mottle virus Genus ***Potyvirus*** (AmLMV)

AmLMV causes leaf mottling, blistering and growth reduction in *A. deflexus* in Italy (Lovisolo and Lisa, 1979). It is transmitted by aphids *Macrosiphum euphorbiae* and *Myzus persicae* in a non-persistent manner and spreads in Italy, Morocco and Spain. Diagnostic systemic hosts of the virus are *A. caudatus, A. hybridus, A. tricolor, Gomphrena globosa* and *Nicotiana clevelandii*, which develop systemic leaf chlorosis. Diagnostic local lesion hosts are *Chenopodium amaranticolor, C. quinoa* (both of these develop systemic mottling) and *Phaseolus vulgaris* (Lisa and Lecoq, 1984). AmLMV particles have a modal length of 780 nm which are serologically related to potyviruses, *Bean yellow mosaic virus,* Lisianthus virus, *Peanut mottle virus, Plum pox virus* and *Zucchini yellow mosaic virus.*

Amaranthus mosaic virus Genus ***Potyvirus*** (AmMV)

AmMV, infects *A. lividus* and *A. viridis* in North India causing systemic mottle, mosaic, occasional vein yellowing, and stunting (Phatak, 1965), and *A. caudatus* and *A. tricolor* in South India causing systemic mosaic and puckered leaves (Govindaswamy *et al.* 1967). A similar virus disease has also been reported on *A. hybridus* from Nigeria (Taiwo, 1988). AmMV is transmitted efficiently by aphids in a non-persistent manner. It has a restricted host range and produces diagnostic systemic mosaic and mottling in *Amaranthus* spp. except *A. spinosus*, which is reported to be immune to AmMV. Diagnostic insusceptible species are: *C. album, C. amaranticolor, N. glutinosa, N. tabacum* and *Solanum nigrum.* The particles of AmMV are potyvirus-like measuring 700-760 nm (Singh *et al.* 1972; Taiwo, 1988). The Nigerian isolate does not react with AmLMV antiserum in immunodiffusion tests (Taiwo, 1988); further comparative studies are required to determine relationship between the two viruses.

Cucumber mosaic virus Genus ***Cucumovirus*** (CMV)

CMV infects *A. hypochondriacus* in Japan (Ohashi and Kamiunten, 1994), and *A. hypochondriacus* and *A. tricolor* in India (Raj *et al.* 1997). CMV is transmitted by aphids in a non-persistent manner, and have 28 nm isometric particles.

27.3 Chayote

The chayote (*Sechium edule*), also known as vegetable pear, is a perennial, monoecious climber with tuberized roots and slender branching stem which climbs on trees and bears up to 500 fruits per plant. Unlike other cucurbits chayote fruits are single seeded and germinate viviparously. The chayote is cultivated on a commercial scale in Costa Rica, Guatemala, the Dominican Republic and Mexico, and in some parts of India. Two viruses are known to cause diseases in chayote.

Chayote mosaic virus Genus ***Tymovirus*** (ChMV) causes mosaic disease of the chayote in Costa Rica. It induces characteristic mosaic symptoms in cucubitaceous hosts and causes asymptomatic infection in other hosts. ChMV is serologically related to tymoviruses, *Andean potato latent virus* (APLV) and *Eggplant mosaic virus* (EMV). Phylogenetic analysis of the coat protein gene and replicase protein gene also shows close relationship of ChMV with APLV and EMV (Bernal *et al.* 2000). ChMV can be easily detected by serology.

Tomato spotted wilt virus Genus ***Tospovirus*** (TSWV) causes vein clearing, irregular chlorotic spots and leaf deformation on chayote in Brazil (Silveira-Junior, 1985). TSWV infection in chayote can be detected by ELISA and PCR.

Other viruses also cause diseases of chayote. A *Potyvirus,* transmitted by *Aphis craccivora, A. gossypii* and *M. persicae* spreads in southern India. *Chayote yellow mosaic virus*, Genus *Begomovirus*, transmitted by whitefly, *Bemisia tabaci*, spreads in Nigeria and north-eastern India. The virus causes yellow mosaic (Fig. 27.5), enations and leaf distortion.

27.4 Colocasia and Xanthosoma

Several members of family Araceae are grown worldwide as vegetables; of these, the cocoyam, taro or dasheen (*Colocasia esculenta)* and the tannia, yautia or new cocoyam (*Xanthosoma sagittifolium*) are most common. The usable parts in these species are the subterranean tubers, which, contain between 15 and 39 per cent carbohydrates, 2 to 3 per cent protein and have a nutritional value comparable to potato. Young leaves and petiole (leaf stalk) are also used as green vegetables. *Xanthosoma* is more popular in Latin America and Africa, and *Colocasia* in Asia. Both these crops are affected by three different viruses, which infect plants singly or in mixed infection (Shaw *et al.* 1979).

Colocasia bobone disease virus Genus ***Nucleorhabdovirus*** (CBDV)
The virus is reported to occur in *C. esculenta* from the Solomon Islands, South Pacific (James *et al.* 1973) and from Papua New Guinea. It causes

chlorotic streaks (Fig. 27.8), malformation, leaf rolling and leaf thickening soon after infection; later the infected plants recover. *Tarophagus proserpina* (Delphacidae) transmits CBDV in a persistent manner, but the virus is not transmitted by mechanical inoculation or through seed or pollen. The virus develops characteristic symptoms in *C. esculenta* and leaf chlorosis in *Philodendron selloum.* It is not known to infect *Xanthosoma* or any other plant species. Large Rhabdovirus-like particles of 300-335 x 50-55 nm with a distinct core of 9 nm diameter are associated with the disease (James *et al.* 1973). The particles have been found in the perinuclear space, but not within the inner nuclear space.

Dasheen bacilliform virus Genus ***Badnavirus*** (DBV)

'Alomae' disease of *Colocasia* (Gollifer and Brown, 1972) occurs in Cook Islands, Fiji, the Solomon Islands, Vanuatu and Western Samoa. It causes early foliar symptoms, similar to 'bobone' disease, but is followed by systemic necrosis and death of plants. Apparently the plants with one large corm (also referred as 'male' plants) are more susceptible than the plants with one small central corm and many cormels (also referred as 'female' plants). DBV is always associated with the disease together with CBDV and occasionally with DsMV (Kenten and Woods, 1973). DBV is transmitted by *Planococcus citri* (Pseudococcoidae), but not by mechanical inoculation or grafting. The particles of DBV are bacilliform, 125 nm in length and 28 nm wide, and are found in phloem sieve tubes, cytoplasm and cell vacuoles. The virus can be detected by ELISA and PCR.

Dasheen mosaic virus Genus ***Potyvirus*** (DsMV)

The virus was first reported in *Zantedeschia elliottina* (Zettler *et al.* 1970). It is widespread in the tropics and sub-tropics affecting a broad range of aroid plants. In *Colocasia* and *Xanthosoma,* DsMV causes mosaic and chlorotic feathering. DsMV is transmitted by *M. persicae, Aphis craccivora* and *A. gossypii* in a non-persistent manner. The virus is also transmissible by sap inoculation, but not by seed or pollen. DsMV particles are 750 nm long filaments, and contain 5% RNA and 95% protein. Virus particles are found in mesophyll, epidermis and probably phloem; typical pinwheel type inclusions are also formed. It is serologically related to *Blackeye cowpea mosaic virus, Tobacco etch virus* and *Araujia mosaic virus.* Attempts to eliminate the virus by heat treatment of *Xanthosoma* corms have not been successful, but virus-free plants of *Colocasia* and *Xanthosoma* can be produced through tissue culture (Zettler and Hartman, 1980). However, somaclonal variation in tissue culture raised plants has been observed (Liu *et al.* 1982). In Costa Rica, a good system has been developed for supplying growers with virus free planting material of *Xanthosoma* originating from virus-free cultivation of stem tips grown *in vitro.* Use of virus-free planting material is the most remunerative control operation in *Xanthosoma* cultivation (Monge *et al.* 1987).

27.5 Eggplant

Eggplant (*Solanum melongena*), the most widely used solanaceous vegetable next to tomato, is also commonly known as brinjal and aubergine. Africa, China, India, Malaysia and Myanmar are the major centers of variation in eggplant. Various eggplants are grouped in three botanical varieties; the round, or egg shaped type belong to *S. melongena* var. *esculentum*, the long and slender type to var. *serpentinum* and dwarf plant type to var. *depressum*. Several viruses are known to cause diseases of economic importance in eggplant in various parts of the world (Table 27.1).

Eggplant mild mottle virus Genus ***Carlavirus*** (EMMV)

EMMV spreads in the Middle East causing mild mottle (Khalil *et al.* 1982). EMMV is transmitted by sap inoculation, but not by seed or *M. persicae*. It infects also *C. amaranticolor* and *N. tabacum*. The filamentous particles are 693 nm long with thermal inactivation point (TIP) of 70^0C.

Eggplant mottled dwarf virus Genus ***Nucleorhabdovirus*** (EMDV)

EMDV spreads in Algeria, Greece, Iran, Italy, Jordan, Morocco, Portugal, Tunisia and Turkey. The virus causes vein yellowing, chlorotic mottling, leaf crinkling and stunting and the plants do not bear fruits (Martelli and Hamadi, 1986). The natural host range of the virus includes *Hibiscus rosa-sinensis, Lycopersicon esculentum, Solanum nigrum, S. sodomaeum, Lonicera sp.* and *Withania somnifera*. The virus is sap and graft transmissible. Vector of the virus is not known, but *S. sodomaeum* and *W. somnifera* are important perennial sources of infection (Brunt *et al.* 1996; Musa and Lockhart, 1990). The virus induces typical chlorotic local lesions followed by systemic chlorosis in *N. clevelandii, N. glutinosa* and *N. tabacum* and local lesions in *C. amaranticolor*. The TIP of the virus is 54^0C. Particles of the virus are rhabdo-like 220-230 x 80-90 nm with an obvious axial canal of 18 nm; pitch of the basic helix is 4.5 nm. The genome codes for five proteins. Virus particles are found in vascular parenchyma and in the perinuclear space of leaf tissue (Brunt *et al.* 1996).

Brinjal mild mosaic virus (BrMMV), **Eggplant green mosaic virus** (EGMV), **Eggplant severe mottle virus** (ESMoV), ***Pepper veinal mottle virus*** (PVMV) and ***Potato virus Y*** (PVY) all Genus ***Potyvirus***

BrMMV causes mild leaf chlorosis of eggplant under natural conditions in India; the virus is sap transmissible and its host range is restricted to family Solanaceae. TIP of the virus is 65^0C and it has filamentous particles 675-760 nm long and 13 nm wide with sedimentation co-efficient of 168 S. Serologically it is not related to PVY and *Tobacco etch virus* (TEV; Naqvi and Mahmood, 1976). EGMV and ESMoV spread in Nigeria. EGMV is sap transmissible and causes green mosaic disease of eggplant. Its host range is restricted to the Solanaceae; TIP is 60^{0}C; virus particles are flexuous filaments of 820 nm with obscure axial canal and basic helix; typical pinwheel inclusions are formed in infected cells (Ladipo *et al.* 1988a). ESMoV causes severe systemic mottling, blistering and distortion of leaves and abnormal serration of the leaf margins; it is transmitted by sap inoculation and by *A. craccivora* and *M. persicae* in a non-persistent manner; host range of the virus is restricted to Solanaceae and Leguminosae; TIP of the virus is between 55-65°C; infected cells contain typical pinwheel inclusions (Ladipo *et al.* 1988b). PVMV causes mosaic, puckering and stunting of eggplants in Nigeria. The host range of PVMV isolate from eggplant is restricted to Solanaceae. It is transmitted by *M. persicae*, in a non-persistent manner, but not through seeds (Igwegbe and Waterworth, 1982). PVY spreads in India causing prominent mosaic disease. It is transmitted through seeds (up to 40%) and by *M. persicae, A. gossypii* and *A. craccivora* in a non-persistent manner (Surendran *et al.* 2000). A strain of PVY, designated as Eggplant mottle virus (EMoV), causes leaf mottling (Fig. 27.2) and spreads in northern India. Nucleotide sequences of CP gene and the 3'-untranslated region of EMoV show high degree of sequence homology with various PVY strains and isolates (Bhat *et al.* 1999).

Eggplant mottled crinkle virus Genus ***Tombusvirus*** (ECMV)

EMCV spreads in India and Lebanon. It causes mottling and crinkling of the leaves. The virus can be maintained and propagated in *N. clevelandii*, and assayed in *C. amaranticolor* and *N. glutinosa*. The isometric particles of 28-30 nm in diameter have sedimentation coefficient of 132 S. In infected cells particles are found in cytoplasm and chloroplasts (Makkouk *et al.* 1981). The virus could be eliminated from infected tissue culture raised plants by incorporating virazole in the culture medium (Raj *et al.* 1991).

Eggplant mosaic virus Genus ***Tymovirus*** (EMV)

EMV (Gibbs and Harrison, 1973) causes mosaic disease of eggplant, tobacco and tomato in Argentina, Brazil, Trinidad and Tobago; and of common weed *Solanum seaforthianum* in Venezuela. The virus in transmitted by sap inoculation, seed and Coleopteran beetle, *Epitrix* sp. The virus induces characteristic chlorotic or necrotic local lesions followed by systemic infection in *C. amaranticolor, C. quinoa, Cucumis sativus, N. clevelandii* and *N. glutinosa*. The TIP of the virus is between 80 to 90°C. It has isometric particles 30 nm in diameter with conspicuous surface structure. Complete genome sequence is available in the GenBank (accession I04374). EMV particles are found in cytoplasm, nuclei, chloroplasts, mitochondria and vacuoles of infected cells of all the parts of host plants; crystalline inclusions are also formed in cell vacuoles.

Other viruses

Alfalfa mosaic virus (AlMV) causes chlorotic vein banding, leaf distortion and stunting in eggplant in India, where it is transmitted by *A. gossypii,* in a non-persistent manner (Vyanjane and Mali, 1981). *Cucumber mosaic virus* (CMV) infects eggplant in China, India, Israel, Italy, Iraq and

Table 27.1 Viruses causing diseases of Eggplant

Virus genus	Virus species*	Disease	Status**
Alfamovirus	*Alfalfa mosaic virus*	Chlorotic vein banding	Mn
Carlavirus	Eggplant mild mottle virus (EMMV)	Mild leaf mottling	Mn
Cucumovirus	*Cucumber mosaic virus* (CMV)	Mosaic	Mn
Fabaviurs	*Broad bean wilt virus* (BBWV)	Necrotic leaf spot	Mn
Nepovirus	*Tobacco ringspot virus* (TRSV)	Ringspot	Mn
Nucleorhabdovirus	*Eggplant mottled dwarf* virus (EMDV)	Vein yellowing	Mn
Potyvirus	Brinjal mild mosaic virus (BrMMV)	Mild mosaic	Mn
	Eggplant green mosaic vius (EGMV)	Green mosaic	Mn
	Eggplant severe mottle virus (ESMoV)	Severe mottle	Mn
	Pepper veinal mottle virus (PVMV)	Mottle	Mn
	Potato virus Y (PVY)	Mosaic	Mj
Tobamovirus	*Tobacco mosaic virus* (TMV)	Necrotic mosaic	Mn
Tombusvirus	*Eggplant mottled crinkle* virus (ECMV)	Mottling and crinkling	Mn
Tospovirus	*Tomato spotted wilt virus* (TSWV)	Deformation and curling	Mn
Tymovirus	*Eggplant mosaic virus* (EMV)	Mosaic	Mn

*Names of recognized species are in italicized. ** Mn= disease of minor importance; Mj= disease of major importance.

Japan, causing mosaic disease of minor importance. *Broad bean wilt virus* (BBWV) causes necrotic spot disease of eggplants in Japan. Virus particles of the virus are distributed in vacuole and cytoplasm of epidermal, mesophyll and phloem parenchyma cells (Morita *et al.* 1974). *Tobacco ringspot virus* (TRSV), is reported to infect eggplant in India. It is transmitted to the extent of up to 10% through the seeds and causes ringspot disease (Sastry and Nayudu, 1976). *Tobacco mosaic virus* (TMV), causes necrotic mosaic disease in eggplant in Egypt, India, Israel, Turkey and Nigeria. In Turkey, TMV was detected in 7.4% seeds of eggplant (Cicek and Yorganci, 1991), but its role in the spread of TMV is not known. *Tomato spotted wilt virus* (TSWV), causes deformation and leaf curling of the top leaves, bronzing and necrosis of older leaves, stunting and fruit deformation in greenhouse in Italy (Betti, 1992).

27.6 Lablab

Lablab [*Lablab purpureus* (syn. *Dolichos lablab*)], also known as hyacinth beans, is a popular vegetable in India, South East Asia, Egypt and the Sudan. Lablab is a climbing or erect annual or short- lived perennial with pubescent, trifoliate leaves, white or purple flowers, and produces flat or inflated pods, 5-20 cm long and 1-5 cm wide. Each pod bears 3-6 ovoid seeds; with a long white aril around a third of the seed. Both pods and seeds are used as vegetables. Viral diseases are of minor importance in this crop.

Cucumber mosaic virus Genus ***Cucumovirus*** (CMV) and ***Soybean mosaic virus*** Genus ***Potyvirus*** (SbMV) were found to cause mosaic disease of lablab in Nanjing, China (Zhou and Chen, 1990).

Dolichos yellow mosaic virus Genus ***Begomovirus*** (DoYMV) causes yellow mosaic disease of Lablab, throughout India (Capoor and Varma, 1950). The virus is transmitted by whitefly, *B. tabaci*, in a persistent manner. Host range of DoYMV is restricted to *L. purpureus*. DoYMV reacts with monoclonal antibodies to *African cassasva mosaic virus* (ACMV) and *Indian cassava mosaic virus* (ICMV) (Harrison *et al.* 1991).

Sunn-hemp mosaic virus Genus ***Tobamovirus*** (SHMV) causes mosaic and enation disease of minor importance in India (Kassanis and Varma, 1975). The virus has been transmitted by spider mite *Tetranychus ludeni* (Rajagopalan, 1974).

27.7 Okra

Okra is a herbaceous annual vegetable grown from seed in the tropics and sub-tropics. It is also known as 'Ladies finger', 'Bhindi', 'Bhendi', 'Bendi', 'Gombo'. Three species, *Abelmoschus esculentus*, *A. manihot* and *A. moschatus*, are cultivated in various parts of the world. All the three species are endemic in West Africa. Of these, *A. esculentus* is grown most widely. Tender green fruits are used as fresh vegetable. Okra is rich in vitamins, minerals and fibres, and also has medicinal properties. The roots and stems of okra are also used for clearing sugarcane juice by the 'jaggery' industry. Viral diseases are a major constraint in okra cultivation.

Bhendi yellow vein mosaic virus (BYVMV) and ***Okra yellow vein mosaic virus*** (OYVMV) both Genus ***Begomovirus***

Yellow vein mosaic (Fig. 27.4) and leaf curl (Fig. 27.3) diseases caused by begomoviurses are most serious (Varma and Malathi, 2003). Two different begomoviruses, i.e. BYVMV and OYVMV, have been recognized to cause yellow vein mosaic disease. Plants affected by this disease have reduced growth, develop bright yellow vein symptoms and produce fewer yellowish un-marketable fruits. BYVMV is transmitted by *B. tabaci* and by grafting but not by sap inoculation (Varma, 1952). The germinate particles (18 x 30 nm) are serologically related to ACMV and ICMV and hybridze strongly with probes to DNA-A of ICMV.

BYVMV, isolated from southern India is a single component (DNA-A) *Begomovirus*, but it requires a small satellite-like DNA-β component to cause typical yellow vein mosaic disease of okra (Jose and Usha, 2003). BYVMV is widespread in the Indian sub-continent, but has not been observed so far in any other okra growing region of the world. In Pakistan, OYVMV is also associated with the disease (Zhou *et al.* 1998); DNA-A of the two viruses share 88% sequence identity. Good sources of resistance to BYVMV are available. Resistance in okra was reported to be controlled by a single dominant gene, two complementary dominant genes, two duplicate dominant genes and dependent on gene dosage with incomplete dominant gene action (Ali *et al.* 2000). Yellow-coloured polythene mulch has been shown to delay infection upto 69 days after sowing compared to 28 days in control plots. Soil application of insecticide methyl phosphoro-dithioate 10G (Furatox-10G) at 15kg/ha followed by 4 foliar spray of Metosystox (demeton-S-methyl) 25 EC at 0.03% at 15 days intervals from the sowing date reduced BYVMV incidence by nearly 58% and enhanced yields (Khan and Mukhopadhyay, 1985). Application of extracts of some plants and neem oil reduced infection by BYVMV in pot experiments (Pun *et al.* 2000). Their efficacy under field conditions remains to be seen.

Cotton leaf curl Gezira virus (CLCuGV), ***Cotton leaf curl Multan virus*** (CLCuMV), **Okra leaf curl virus** (OkLCuD) and **Okra leaf curl India virus** (OkLCuIV) all Genus ***Begomovirus***

Okra leaf curl disease (OkLCuD), caused by begomoviruses, is endemic in West Africa and also spreads in India and Brazil. The infected plants develop leaf vein thickening and enations, crinkling and curling, remain stunted and produce fewer and smaller fruits. The disease is caused by begomoviruses transmitted by *B. tabaci* in a persistent manner. OkLCuD was first observed in northern India in 1980 on germplasm accessions introduced from West Africa. Gradually the disease spread to region and now it is a major problem (Varma and Malathi, 2003). Four begomoviruses, i.e. CLCuGV, CLCuMV, OkLCuV and OkLCuIV, have been found associated with OkLCuD in different geographical regions; *Cotton leaf curl Gezira virus* (CLCuGV) has been found in Egypt and Sudan, *Cotton leaf curl Multan virus* (CLCuMV) in Pakistan, Okra leaf curl virus (OkLCuV) in the Ivory Coast and Okra leaf curl India virus (OkLCuIV) in India. These viruses are best detected by PCR.

Okra mosaic virus Genus ***Tymovirus*** (OkMV)

Okra mosaic virus causes mosaic disease of okra in West Africa and Brazil. Apart from okra, natural host range of the virus includes *Abutilon hirtum, Corchorus olitorius, Hibiscus rosa-sinensis, H. sabdariffa, Indigofera spictata, Malvastrum coromandelianum* and *Sida* spp. Symptoms in these hosts vary; generally OkMV causes vein cholosis, vein banding and mosaic. The virus is transmitted by chrysomelid beetle *Podagrica sjostedti, P. uniformis* and *Syagrus calcaratus* (Lana and Ajibola-Taylor, 1976), and by sap. TIP of the virus is 80°C. It has isometric particles of 28 nm diameter which sediment as top and bottom components having sedimentation coefficient of 42 S (T) and 106 S (B) (Givord and Koenig, 1974). Virus particles are found in roots and leaves; the nuclei of infected cells contain large number of empty particles and double membrane vesicles are formed on the peripheries of chloroplasts (Lesemann, 1977). OkMV isolates from Ivory Coast and Nigeria are considered related but distinct strains (Bozarth *et al.* 1977). Treatment with a synthetic pyrethroid and aqueous neem solution has been shown to reduce disease incidence (Atiri *et al.* 1991).

27.8 Onion

Onion (*Allium cepa*) is under cultivation for more than 5000 years. It is mainly grown for its large mature bulbs and to a lesser extent for immature bulbs and leaves (shallots and green onion). Two types of onions, 'fresh onions' and 'storage onions' are grown. A large number of varieties of both types are grown all over the world. In Japan welsh onion (*A. fistulosum*) is also popular. Compared to garlic (*A. sativum*), onion has fewer virus disease problems, as unlike vegetative propagation of garlic, onion is grown from seed and the viruses affecting onion are not transmitted through seeds. However, like garlic, virus free clones of onion can also be obtained by meristem culture.

Onion mite-borne latent virus Genus ***Allexivirus*** (OMbLV), causes usually symptomless infection but occasionally induces faint mottle and more virulent isolates induce diffuse yellow striping. The virus is transmitted by *Aceria tulipae* and by mechanical inoculation but not by seed. OMbLV induces chlorotic local lesions in *Chenopodium morale*, but does not infect *C. quinoa* (Van Dijk *et al.* 1991). The particles of the virus are helical flexuous 775 nm long filaments.

Garlic latent virus Genus ***Carlavirus*** (GarLV), causes symptomless infection in Welsh onion in Japan (Sako *et al.* 1994).

Leek yellowstripe virus (LYSV), ***Onion yellow dwarf virus*** (OYDV) and ***Shallot yellow stripe virus*** (SYSV) all Genus ***Potyvirus***.

LYSV occurs in Iran (Shahraeen 1998) and SYSV occurs in Welsh onion in China (Liu *et al.* 1996). The CP gene and 3' non-translated region sequences of LYSV share 95 and 89% identity with the type strain of SYSV but differ from those of OYDV by > 25 and > 67%, respectively (Vlugt *et al.* 1999). Welsh onion varieties having dark coloured leaves or thicker wax layer on the leaves are more resistant to SYSV (Liu, 1999). OYDV is the most widespread *Potyvirus* affecting onion. It is reported to occur in Argentina, Brazil, Chile, India, Indonesia, Iraq, Japan, Kazakhanistan, the Netherlands, Spain, USA and Yugoslavia; probably distributed worldwide. It is a major problem in onion seed crop in India (Ahlawat and Varma, 1997). The leaves and flower stalks of the affected plants develop mosaic mottling, short chlorotic streaks (Fig. 27.1), general yellowing and stunting. The virus is efficiently transmitted by *M. persicae,* in a non-persistent manner. It is easily transmitted by sap but not by contact, seed or pollen. The particles of OYDV are flexuous, 772-823 nm in length. It can be efficiently detected by dot-immunobinding assay in extracts of infected onion leaves. The isolates of OYDV from onion and garlic are serologically closely related (Conci *et al.* 1999).

In Japan, mixed infection by GarLV and OYDV is common in Welsh onion. In crops grown from meristem-tip cultured plants within one season in field, 38% plants got infected with OYDV and 10% with both OYDV and GarLV (Fukami and Ishii, 1991).

Tobacco rattle virus Genus ***Tobravirus*** (TRV) and **Iris yellow spot virus** Genus ***Tospovirus*** (IYSV), cause diseases in onion (Uhde *et al.*1998; Pozzer *et al.* 1999). IYSV causes 'eye spots' on the leaves and flower stalks leading to flower abortion in onion. IYSVcould become a serious problem if it further spreads in onion crops.

27.9 Pea

The crop was originally grown for use as a grain legume and used as dry seeds. However, in the modern times, it is more often picked green before maturity and eaten fresh. Pea is rich in protein, carbohydrate, vitamin A and C, calcium and phosphorus. Pea contains about 22% protein, therefore, it is considered as an important source of non-meat protein. Nearly 20 viruses are known to cause diseases in Pea (Table 27.2).

Pea streak virus (PeSV) and ***Red clover vein mosaic virus*** (RCVMV), both Genus ***Carlavirus***

Plants infected with PeSV develop necrotic lesions or streak; the symptoms may persist under field conditions but soon disappear under glasshouse conditions (Brunt *et al.* 1996). RCVMV induces mosaic and symptoms persist both under field and glasshouse conditions (Varma, 1970).

Both the viruses, occur in North America and parts of Europe, and (a) are transmittied by aphids in a non-persistent manner, (b) have similar host range, particle morphology (flexuous filaments measuring 610 – 700 nm in lenth and 12-15 nm in diameter, having helical symmetry with a pitch of about 3.4 nm), biophysical (TIP of 55ºC) and biochemical properties, (c) are serologically related to each other and (d) are not known to be transmitted by seed of *P. sativum*. The genome of PeSV has been partially sequenced (AF354652).

Broad bean stain virus (BBSV), ***Broad bean true mosaic virus*** (BBTMV), ***Pea green mottle virus*** (PGMV), ***Pea mild mosaic virus*** (PMiMV) all Genus ***Comovirus***

BBSV causes systemic mottle and necrosis in pea. The virus is transmitted by coleopteran beetles, *Apion vorax* and *Sitona* spp., and through seeds. The Czechoslovakian isolates F1 and VsM are more virulent than the isolate Kow 60, which causes symptomless infection. However, Kow 60 is transmitted through up to 46% seeds compared to up to 10% seed transmission of isolates F1 and VsM (Musil and Kowalska, 1993). BBSV is serologically distantly related to *Cowpea mosaic virus* (CPMV), *Glycine mosaic virus* (GMV), *Red clover mottle virus* (RCMV) and *Squash mosaic virus* (SqMV) (Brunt *et al.*1996). PMiMV causes inconspicuous chlorosis or symptomless infection in pea in New Zealand and is transmitted through seeds (Clark, 1972). It is serologically related to *Cowpea severe mosaic virus.* BBTMV occurs in pea in Poland and PGMV occurs in Czechoslovakia.

Cucumber mosaic virus Genus ***Cucumovirus*** (CMV), causes a severe disease in and around Gorakhpur, India (Rao *et al.* 1995). The disease is characterized by mosaic mottling, puckering of leaves and plant stunting.

Bean leaf roll virus (BLRV) and ***Pea enation mosaic virus –1*** (PEMV-1) both Genus ***Enamovirus,*** and ***Pea enation mosaic virus-2*** Genus ***Umbravirus*** (PEMV-2)

BLRV widely occurs in peas causing leaf rolling and chlorosis. *A. pisum* and *Macrosiphum euphorbiae,* transmit the virus efficiently in a persistent manner and it is retained after moulting; *M. persicae* also transmits the virus but only occasionally; apterous females of *A. pisum* are more efficient vectors than the alate forms (Thottapilly, 1969). The virus is not transmitted through seeds, but it has been detected in pea seeds (Musil *et al.* 1981).

Enation mosaic (Fig. 27.7) disease of pea is widespread in Canada, China, Europe (Denmark, France, Germany, Poland, Sicily, UK, Yugoslavia), Iran, and USA. A complex of PEMV-1 and PEMV-2 causes the disease. *Acyrthosiphon pisum* and *M. persicae* transmit the PEMV complex in a persistent manner. PEMV-1 is readily transmitted mechanically but aphid transmissibility can be lost after several mechanical passages, as a result of changes in 33 kDa protein encoded immediately downstream of coat protein (CP) gene (Demler *et al.* 1997). The two viruses show mutualism and help in biological success of each other (Murant *et al.* 2000). PEMV-2 lacks ORF for CP and depends on PEMV-1 for encapsidation and aphid transmission, but it (PEMV-2) is essential for systemic spread of PEMV-1 in phloem and mesophyll tissues. Transformed peas expressing PEMV-CP show resistance to infection by PEMV and may provide economic management of the disease (Chowrira *et al.* 1998). A marked locus AdH-1 and two DNA markers linked to the gene conferring resistance to PEMV in peas have been identified (Yu *et al.* 1995). These will be useful in breeding resistant varieties.

Bean yellow mosaic virus (BYMV), ***Clover yellow vein virus*** (ClYVV), ***Lettuce mosaic virus*** (LMV), ***Passion fruit woodiness virus*** (PWV), and ***Pea seed-borne mosaic virus*** (PSbMV) all Genus ***Potyvirus***

BYMV is distributed world wide and it causes tip necrosis and mosaic (Fig. 27.6) diseases in pea in East Africa, Europe, Indian subcontinent and Middle East. Good sources of resistance to BYMV in pea are available. BYMV isolates from pea are easily transmitted by sap inoculation and *A. craccivora* and *M. persicae,* in a non-persistent manner (Eldin *et al.* 1983). The TIP of BYMV is 65^0C and it has filamentous 750 nm long particles. The virus forms typical pinwheel inclusions.

ClYVV causes necrosis in pea. It differs from BYMV in its ability to infect white clover, to produce local lesions on cucumber cotyledons and to induce extensive nucleolar enlargements in the host cells (Bos *et al.* 1977). LMV causes a disease of minor importance in Japan and China. An Australian strain of PWV infects peas causing light to moderate mosaic, chlorotic spots, vein chlorosis and stunting. Pea varieties resistant to BYMV are also shown to resist PWV, but resistance to the two viruses is conferred by different genetic factors. A single recessive gene confers resistance to PWV in pea (Provvidenti and Niblett, 1994).

Table 21.2: Viruses causing diseases of Pea

Virus Genus	**Virus Species***	**Disease**	**Status****
Carlavirus	*Pea streak virus* (PeSV)	Streak	Mn
	Red clover vein mosaic virus (RCVMV)	Mosaic	Mn
Comovirus	*Broad bean stain virus* (BBSV)	Motlle and necrosis	Mn
	Broad bean true mosaic virus (BBTMV)	Mosaic	Mn
	Pea green mottle virus (PGMV)	Motlle	Mn
	Pea mild mosaic virus (PMiMV)	Chlorosis	Mn
Cucumovirus	*Cucumber mosaic virus* (CMV)	Severe mosaic	Mn
Enamovirus	*Bean leaf roll virus* (BLRV)	Leaf roll	Mj
	Pea enation mosaic virus –1 (PEMV-1)	Leaf enation	Mj

Umbravirus	*Pea enation mosaic virus-2* (PEMV-2)	Leaf enation	Mj
Fabavirus	*Broad bean wilt virus* (BBWV)	Severe mosaic	Mn
Potexvirus	*White clover mosaic virus* (WClMV)	Mosaic	Mn
Potyvirus	*Bean yellow mosaic virus* (BYMV) (Syn: Pea mosaic virus)	Tip necrosis; mosaic	Mn
	Clover yellow vein virus (ClYVV) (Syn: Pea necrosis virus)	Necrosis	Mn
	Lettuce mosaic virus (LMV)	Vein leaning	Mn
	Passion fruit woodiness virus (PWV)	Mosaic	Mn
	Pea seed-borne mosaic virus (PSbMV)	Mosaic	Mj
Tobravirus	*Pea early-browning virus* (PEBV)	Browning	Mj
Tymovirus	*Plantago mottle virus* (PlMoV)	Mottle	Mn

*Names of recognized species are in italicized. ** Mn= disease of minor importance; Mj= disease of major importance.

PSbMV is the most widely distributed and serious pathogen of pea causing transitory vein clearing, adaxial folding of leaflets, rosetting of branches, stunting, malformation of flowers and production of small pods with few distorted seeds. The virus is transmitted by sap inoculation, by seed and by aphids. Seed transmission of the virus in pea is up to 100%, which plays an important role in disease epidemiology (Khetarpal and Maury, 1987). The extent of seed transmission depends on virus isolate, pea variety, time of infection and environmental condition. The virus has flexuous particles of 770 nm with basic helix pitch of 3.4 nm, and sedimentation coefficient of 154 S. Serologically it is related to BYMV. PSbMV is readily detected in leaves and seeds of pea by ELISA and DIBA (Varma *et al.* 1991).

Different pathotypes of PSbMV have been identified on the basis of resistance conferred by recessive genes *sbm-1*, *sbm-2*, *sbm-3* and *sbm-4*. Gene *sbm-1* confers resistance to pathotype P-1, *sbm-2* and *sbm-3* to pathotype L (also called P-2) and *sbm-4* to pathotype P-4. Some isolates of pathotype P-1 are more frequently transmitted by seed than pathotype P-4 (Johansen *et al.* 1996), and *vice versa* for some other isolates (Ali and Randles, 1998). Quality control of pea seed is important for managing the disease caused by PSbMV (Masmoudi *et al.* 1994). In varieties like Century, the virus is self limiting. Use of such varieties would minimize the losses caused by PSbMV (Zimmer and Lamb, 1993). The best approach is to use virus-free seeds, which can be produced in areas where PSbMV does not spread (Varma *et al.* 1991).

PSbMV is a serious problem in utilizing pea germplasm. However, a practical approach has been developed to eliminate the virus from pea germplasm (Hampton *et al.* 1993). Good sources of resistance to PSbMV are available. These have been used to develop peas resistant not only to PSbMV but also to PEMV (Baggett *et al.* 1994). The lines carrying *sbm-1* show good degree of resistance even under heavy inoculum pressure under field conditions (Khetarpal *et al.* 1990).

Pea early-browning virus Genus ***Tobravirus*** (PEBV)

Systemic necrotic spots, streaks, mosaic, leaf deformation and stunting of pea plants is caused by PEBV in Europe and the Middle East. The virus is transmitted by sap inoculation, by seed and by several species of nematode *Trichodorus*. The virus induces local lesions on the inoculated leaves of *C. amaranticolor* and *C. sativus*, and local lesions followed by systemic infection in *Phaseolus vulgaris*, *P. sativum*, *N. clevelandii* and *N. glutinosa*; in *Tetragonia tetragonioides* it causes characteristic systemic necrotic concentric rings. The TIP of the virus is 74-78^{0}C, and its particles are rod-shaped with two predominant lengths, 215 or 207 nm and 105 nm and width of 21 nm; sedimentation coefficient of the particles are 280 S and 210 S respectively. The ss RNA genome is bipartite with larger RNA-1 of 7.9 kb and the smaller RNA-2 of 4.1 kb (Brunt *et al.* 1996). RNA-2 encodes for four proteins including the coat protein. All these proteins are important for the transmission PEBV by the vector nematode, but deletion of gene encoding 29kDa protein abolish vector transmission (MacFarlane *et al.* 1996). Similarly seed transmission of PEBV is regulated by RNA-1 gene encoding 12kDa protein (Wang *et al.* 1997). *N. benthamiana* plants transformed with the 3'proximal portion (54 kDa) of the gene encoding the 201kDa putative replicase protein of PBEV resisted infection by PEBV (MacFarlane and Davies, 1992).

Other viruses

Broad bean wilt virus (BBWV) causes severe moaic, leaf malformation and stunting of pea in Egypt (Kishtah *et al.* 1978). It also occurs in China; the Chinese isolate is shown to belong to serotype II group of BBWV (Xu *et al.* 1988). *White clover mosaic virus* (WClMV) causes mosaic disease of pea in Japan (Matsumoto *et al.* 1991). *Plantago mottle virus* (PlMoV) causes conspicuous leaf veinal chlorosis mottle and necrosis, in USA (Provvidenti and Garnett, 1976).

27.10 Sweet pepper

Sweet pepper, (*Capsicum annuum*), also known as bell pepper, originated in tropical South America. It is cultivated in Central and South America, Europe and Asia. Fruits vary in colour ranging from green, yellow, red, purple, black, brown, white and orange. The immature fruits are green in colour. It is an excellent source of vitamin A, vitamin C and calcium. Several viruses are known to affect sweet pepper cultivation (Table 21.3). The most common of these are CMV, PVY and TMV.

Cucumber mosaic virus (CMV) and ***Tomato aspermy virus*** (TAV) both Genus ***Cucumovirus***

CMV spreads widely in sweet pepper, causing severe losses in production. The younger leaves of affected plants curl downwards, develop mottling, remain small in size, and show distortion. The infected plants are stunted due to shortening of internodes and the fruits also develop mottling. The anther lobes of virus infected plants show histopathological features and are partially filled with non-functional pollen (Awasthi and Singh, 1974). The virus is efficiently transmitted by a variety of aphids. In greenhouse cultivation, non-vectored soil transmission has also been demonstrated (Pares and Gunn, 1989).

Inoculation of plants with an attenuated strain of CMV effectively reduced damage by severe strain under field conditions in Japan. However some damage is also caused by the attenuated strain (Takeuchi *et al.* 1998). In India, use of sunflower, sesame, sorghum and pearl millet as barrier crops reduced the incidence of CMV in sweet pepper (Deol and Rataul, 1978). Decrease in CMV incidence on sweet pepper sprayed with mineral oils has also been shown (Loebenstein *et al.* 1970).

TAV causes necrotic mosaic in sweet pepper in Japan. The isolate of TAV from sweet pepper is serologically related to another isolate of TAV from petunia, but not to CMV and *Peanut stunt virus*. Association of a satellite-like RNA, designated 'TAVRNA5', which attenuates the virus-induced symptoms has also been observed (Natsuaki *et al.* 1994).

Potato virus Y (PVY), ***Pepper mottle virus*** (PepMoV), ***Pepper veinal mottle virus*** (PVMV) and ***Pepper severe mosaic virus*** (PepSMV) all Genus ***Potyvirus***

Several potyviruses cause diseases in sweet pepper (Table 21.4). Most important of these is PVY, which is widely distributed in sweet pepper causing leaf chlorosis, mosaic and deformation. The virus is easily transmissible by sap inoculation and by aphids in a non-persistent manner. Several strains of PVY infect sweet pepper in different parts of the world. *Nicandra physaloides* and *Solanum nigrum* have been identified as important reservoirs of PVY in South Africa (Budnik *et al.* 1996). Spray with insecticide oil every ten days after transplanting delays infection in pepper by PVY and improves production (Loebenstein *et al.* 1970). Several sources of resistance to different strains of PVY have been identified. In *C. annuum* var. Criollo de Morellos 334, resistance to PVY is governed by a single dominant gene (Bioteux *et al.* 1996) whereas, in some other varieties of sweet pepper it is governed by a recessive gene (Sharma *et al.* 1989).

PepMoV spreads in El Salvador, India and the USA, causing severe systemic leaf mottling in pepper. It is transmitted by aphids in a non-persistent manner but not by seed. Host range of PepMoV is mainly restricted to Solanaceous plants, but it also induces necrotic local lesions in *C. amaranticolor*. The TIP of PepMoV is 50-55^{0}C. Its particles are 737 nm long flaxuous filaments. Serologically it is related to PVY, PVMV and TEV but not to Bidens mosaic virus (BMV), *Lettuce mosaic virus* (LMV) and *Turnip mosaic virus* (TuMV) (Brunt *et al.* 1996).

PVMV spreads in West Africa and South Africa causing mottle and foliar rugosity. It is transmitted by aphids is a non-persistent manner. Host range of the virus is restricted to 3-9 families. However, the host range of Nigerian isolate appears to be restricted to sweet pepper (Lana *et al.* 1975). TIP of PVMV is 55-60^{0}C and it has flexuous flamentous particles of 770 nm length with a sedimentation coefficient of 155 S. It is serologically related to BYMV, ClYVV, *Datura shoestring virus*, PVY, *Tobacco etch virus* (TEV) and *Watermelon mosaic virus* (WMV), but not to *Henbane mosaic virus*, LMV, PSbMV and TuMV (Brunt *et al.* 1996; Brunt and Kenten, 1972).

PepSMV spreads in Argentina. It is transmitted by aphids in a non-persistent manner. TIP of PepSMV is 85-90^{0}C, which is unusual for a *Potyvirus*. Particles of PepSMV are 761 nm long. Serologically it is distantly related to PVY and TEV, but not to PepMoV, PVMV and TuMV (Feldman and Garcia, 1977).

Some provisional potyviruses have also been reported to infect sweet pepper. Pepper Indian mottle virus spreads in Punjab, India (Sandhu and Chohan, 1979) and Pepper mild mosaic virus spreads in Venezuela (Ladera *et al.* 1982). The biological and bio-physical properties of these viruses resemble potyviruses.

Table 27.3: Viruses affecting sweet pepper

Virus genus	Virus species*	Disease	Status **
Alfamovirus	*Alfalfa mosaic virus* (AMV)	Yellow fleck	Mn
Begomovirus	*Pepper golden mosaic virus* (PepGMV)		
	Pepper huasteco yellow vein virus (PHYVV)	Yellow vein	Mn
	Pepper leaf curl Bangladesh virus (PepLCBV)	Leaf curl	Mn
	Pepper leaf curl virus (PepLCV)	Leaf curl	Mn
	Pepper mild tigre' virus (PepMTV)	Mild leaf curling	Mn
	Tomato yellow leaf curl virus (TYLCV)	Chlorosis and curling	Mn
Cucumovirus	*Cucumber mosaic virus* (CMV)	Mosaic	Mj
	Tomato aspesmy virus (TAV)	Necrotic mosaic	Mn
Curtovirus	*Beet curly top virus* (BCTV)	Curly top	Mj
	Tomato pseudo-curly top virus (TPCTV)	Curly top	Mn
Luteovirus	Pepper vein yellow virus (PeVYV)	Vein yellowing	Mn
Polerovirus	Potato leafroll virus (PLRV)	Leafroll	Mn
Potexvirus	*Potato virus X* (PVX)	Mottle	Mn
Potyvirus	Pepper mild mosaic virus (PepMMV)	Mosaic	Mn
	Pepper mottle virus (PepMoV)	Leaf mottle	Mn

	Pepper severe mosaic virus (PepSMV)	Mosaic	Mn
	Pepper vein banding virus (PVBV)	Mosaic	Mn
	Pepper veinal mottle virus (PVMV)	Leaf mottle	Mn
	Potato virus Y (PVY)	Mosaic	Mn
	Tobacco etch virus (TEV)	Mosaic	Mn
Tobamovirus	*Pepper mild mottle virus* (PMMoV)	Chlorosis	Mn
	Tobacco mosaic virus (TMV)	Mosaic	Mn
	Tomato mosaic virus (ToMV)	Necrotic spots on fruits	Mn
Tobravirus	*Pepper ringspot virus* (PepRSV)	Ringspot	Mn
	Tobacco rattle virus (TRV)	Ringspot	
Tombusvirus	*Tomato bushy stunt virus* (TBSV)	Necrosis	Mn
Tospovirus	*Groundnut ringspot virus* (GRSV)	Ringspot	Mn
	Tomato chlorotic spot virus (TCSV)	Mottle	Mn
	Tomato spotted wilt virus (TSWV)	Mosaic, stem necrosis	Mn

*Names of recognized species are in italicized. ** Mn= disease of minor importance; Mj= disease of major importance.

Pepper mild mottle virus (PMMoV), ***Tobacco mosaic virus*** (TMV), ***Tomato mosaic virus*** (ToMV), all Genus ***Tobamovirus***

PMMoV causes mild chlorosis, stunting, fruit mottling and malformation. It is transmitted by contact between plants, by seed and by contaminated soil (Pares and Gunn, 1989). It spreads in Argentina, Australia, Europe, Japan and the USA. PMMoV has a wide host range and infects many Solanaceous species but not tomato and *N. glauca*. TIP of the virus is 95^0C and its particles are rod-shaped 312 nm long and 18 nm wide. Particles are found in mesophyll and phloem cells forming crystalline inclusions. It is serologically related to several tobamoviruses (Brunt *et al.* 1996). Coat protein of PMMoV is the elicitor of *L3* resistance gene-mediated hypersensitive response of *Capsicum* spp. (Gilardi *et al.* 1998).

TMV infection in pepper is widespread. It causes serious damage to greenhouse grown sweet peppers (Betti *et al.* 1984). In var. Soroksari, TMV infection reduced yields by 78.5% under field conditions and 52% under glasshouse conditions (Jasnic, 1979). Strains P1 and P1, 2 which can infect cultivars carrying *L1* resistance gene, and strain P1, 2, 3 which can overcome *L3* resistance gene, and TMV-pepper (TMV-P) strain occur in China (Sakata *et al.* 1997). Yellow mosaic strain is reported to occur in India (Capoor and Sawant, 1987). In Japan TMV-U strain has been identified which differs serologically from the ordinary, tomato and pepper strains of TMV (Nagai *et al.* 1987). Mild isolate of TMV-pepper strain protects plants from infection by severe isolate and improves yields by 24%. Cross protection, however, is not effective, when plants are also infected by other viruses (Goto *et al.* 1977).

ToMV causes necrotic spots on stem and fruits and the disease symptoms are more severe in plants grown at 24^0C than those grown at 18^0C

(Schuerger and Hammer, 1995). The virus is transmitted through contaminated seeds (Ragozzino and Angelaccio, 1981) and contaminated tools and soil. Sweet pepper vars Anaheim, TMR23 and Rio Grande Gold are most resistant to ToMV (Pategas *et al.* 1989); vars KKg-W3, Titan, Poivron and Gunzsulina are also good sources of resistance (Kostova, 1982). Dipping pruning shears in 10% trisodium phosphate provided best management practice to check spread of ToMV in hydroponically grown sweet pepper (Pategas *et al.* 1989).

Tobacco rattle virus Genus ***Tobravirus*** (TRV)

TRV spreads in sweet pepper in Brazil, China, and Italy. Strain TRV-PRSM induces ring spot and yellow mosaic symptoms. It infects 25 species of 5 families by sap inoculation, has TIP of 82-85^{0}C and particles of two different lengths measuring 160-210 nm and 40-80 nm (Tian *et al.* 1995).

Tomato spotted wilt virus Genus ***Tospovirus*** (TSWV)

TSWV affected sweet pepper plants develop leaf necrosis (Fig. 27.9), ring spots, mild mosaic, leaf distortion and stem necrosis. TSWV spreads in sweet pepper in Brazil, Taiwan and USA Virus particles form clusters between membranes surrounded by dense viroplasm (Avilla *et al.* 1991). *Groundnut ringspot virus* (GRSV) also occurs in sweet pepper causing minor disease.

Other viruses

Several begomoviruses have been identified to cause diseases of minor importance in pepper (Table 27.4), but these could emerge as serious pathogens (Varma and Malathi, 2003). Most of these viruses cause interveinal and marginal chlorosis and upward curling of leaves (Fig. 27.10) of bell pepper (Reina *et al.* 1999). Pepper vein yellow virus (PeVYV), a tentative species of family *Luteovirus* causes vein yellowing and leaf roll in bell pepper in Japan. PeVYV is not transmitted by sap inoculation or by seed, but aphids, *A. gossypii* and *M. persicae* transmit it, in a persistent manner. Host range of the virus is limited, it has isometric particles of about 25 nm diameter and serologically it is not related to *Potato leaf roll virus, Tobacco necrotic dwarf virus, Barley yellow dwarf virus* or *Beet western yellows virus*. The particles of the virus are found in phloem cells (Yonaha *et al.* 1995).

Curtovirus, BCTV, causes epidemic in Capsicum crops. Sources of resistance to BCTV in *C. annuum, C. chacoense, C. chinense* and *C. frutescens* have been identified (Bosland, 2000). A cryptic virus has also been found in sweet pepper (Arancibia *et al.* 1995).

Mixed infection by viruses in sweet pepper is common. In Zambia, 80-100% plants were found infected with AMV, 60% with TMV and PVY, 15-60% with CMV and 30-40% with PMMoV (Ndungru and Kapooria, 1999). Treatment of *C. annuum* plants with a leaf-coating compound combined with insecticide application every two weeks increased yield and average fruit weight significantly; although the incidence of aphid-transmitted viruses was not significantly affected, there was a reduction in the number of aphid infested plants (Green and Chou, 1991). Spraying mineral oil also decreases infection by viruses and increases yields (Kiss *et al.* 1988). Mulching with polyethylene plastic sheet, reflective surface stripping and pepper/marigold (*Tagetes erecta*) inter cultivation reduced the incidence of aphid- and whitefly-transmitted viruses in pepper. Economically, inter cultivation of pepper-marigold in a 2:1 ratio was most effective with a marginal return rate 233.7% (Chew *et al.* 1995). Intercropping of *C. annuum* with maize or sorghum also delayed the onset of viral diseases in Mexico (Guigon-Lopez and Gonzalez-Gonzalez, 1999). Growing *Arachis pintoi* as a ground cover with *C. annuum* reduced whitefly and *Begomovirus* incidence and improved production in Honduras (Rafie *et al.* 1999). Polyethylene covers are useful in reducing the incidence of PVY and CMV particularly in areas of high virus pressure (Avilla *et al.* 1997). In glasshouse grown crops, use of anti-insect net completely protected plants from infection by CMV and PVY, but not by TSWV; infection by TSWV was more in plants, which received higher doses of nitrogen compared to plants grown in soils without nitrogen (Camele *et al.* 2000).

Resistance

Sources of resistance to different viruses are available (Green and Kim, 1994), which can be effectively utilized for developing agronomically suitable varieties of *C. annuum*.

Resistance in *C. frutescence* accession 'BG2814-6' to CMV is apparently controlled by two major genes; this accession and *C. annuum* resistant var. Perennial, which is a major source of CMV resistance, appear to share one or more CMV resistance genes (Grube *et al.* 2000).

Capsicum vars Florida VR-2, Tambel 1, Tambel-2, Tam Mild Chile 1 etc. are shown to be resistant to not only PVY but also TEV and TMV (Debrot *et al.* 1986). Pepper var. Dempsay also resists infection by PVY, PepMoV and *Xanthomonas campestris* pv. *vesicatoria* (Lana *et al.* 1975). *C. annuum* vars Florida VR2 and Perennial are susceptible to PVMV, but a double haploid line resourced from F1 hybrid between these two varieties was completely resistant to PVMV due to complementation of two recessive genes, each one coming from one of the parents. The gene from var. Florida VR2 was identical or tightly linked to *pvr 22*, which confers complete resistance to PVY (Caranta, 1996). Uses of resistant cultivars provide potential means of checking spread of TEV in pepper (Padgett *et al.* 1990). *C. annuum* var. Avelar resists infection by PepMoV but the resistance is not effective in plants co-infected with PepMoV and CMV (Guerini and Murphy, 1999).

A high degree of field resistance to TSWV has been found in *C. baccatum*, *C. chinense* and *C. annuum* (Boiteux *et al.* 1993). The resistance, however, varies from isolate to isolate of the virus. The *C. chinense* accessions CMPH 275 and PI159236 are almost immune to the Brazilian isolate TSWV-BsB but are susceptible to isolate TSWV-SP. Similarly, *C. chinense* accessions PI152225 and PI159236, were found resistant to 59 of the 61 TSWV isolates from Southern Louisiana, USA, but susceptible to the two other isolates (Hobbs *et al.* 1994); a single dominant gene determines resistance to TSWV in these accession (Moury *et al.* 1997).

27.11 Concluding remarks

Viruses belonging to more than 25 genera cause diseases in the selected common and uncommon vegetales included in this chapter. The number of viruses affecting common vegetables is greater, which is expected as these vegetables (eggplant, pea, onion and sweet pepper) are grown widely under diverse agroclimatic conditions. Almost all the vegetables included in this chapter are affected by potyviruses, some of which are a major constraint in vegetable production. Cucumoviruses are the next most commonly occurring viruses followed by carla-, tospo-, begomo-, tobamo- and tobraviruses. Tospo and begomoviruses are fast emerging in new crops and new areas.

Of the various approaches available for the integrated management of plant viral diseases (Varma, 1993), avoidance of sources of infection, modification of cultural practices and use of resistant varieties are important approaches for the management of viral diseases of vegetable crops. Attempts have been made to produce viruse-free planting material of some of the vetatively propagated vegetables, but the production of virus-free true-seed has remained largely neglected. In general, the viruses causing diseases of the uncommon vegetables included in this chapter have received relatively little attention. There is a need to characterize various viruses, study their epidemiology, and develop and apply appropriate management practices so that the full productivity potential can be achieved.

Acknowledgements

The authors thank Mr. S.S. Sohrab, Ms. Deepika Mehra and Ms. Pallavi Chakravarty for their ungrudging help in the preparation of this manuscript, and CAB International for the picture of bobone disease of colocasia.

References

Ahlawat, Y.S. and Varma, A. 1997. Serological detection of a mixed viral infections in onion seed crop and possible measures for its management. *Indian Phytopath.* 50: 151-153.

Ali, A. and Randles, J.W. 1998. The effects of two pathotypes of pea seed-borne mosaic virus on the morphology and yield of pea. *Australasian Pl. Pathol.* 27: 226-233.

Ali, M., Hossain, M.Z. and Sarker, N.C. 2000. Inheritance of yellow vein mosaic virus (YVMV) tolerance in a cultivar of okra (Abelmoschus esculentus (L.) Moench). *Euphytica* 111: 205-209.

Arancibia, R.A., Valverde, R.A. and Can, F. 1995. Properties of a cryptic virus from pepper (Capsicum annuum). *Pl. Pathol.* 44: 164-168.

Atiri, G.I., Ivbijaro, M.F. and Oladele, A.D. 1991. Effects of natural and synthetic chemicals on the incidence and severity of okra mosaic virus in okra. *Tropical Agr.* 68: 178-180.

Avilla, A.C.de, Pena, L., Kitajima, E.W., Resende, R.de.O., Diaz-Mugica, M.V., Diaz-Ruiz, J.R., Peters, D. and De-Avilla, A.C. 1991. Characterization of tomato spotted wilt virus (TSWV), isolated from Capsicum annuum L. in the Canary Islands. *Phytopath. Mediterranea* 30: 23-29.

Avilla, C., Collar, J.L., Duque, M., Perez, P. and Fereres, A. 1997. Impact of floating rowcovers on bell pepper yield and virus incidence. *Hort. Sci.* 32: 882-883.

Awasthi, D.N. and Singh, B.P. 1974. Histopathological studies of reproductive organs in Capsicum annuum infected with cucumber mosaic virus. *Indian Phytopath.* 27: 218-22.

Baggett, J.R., Kasimor, K. and Hampton, R.O. 1994. OSU 663, OSU 668, and OSU 677 pea breeding lines resistant to pea seedborne mosaic virus. *Hort. Sci.* 29: 337-338.

Bernal, J.J., Jimenez, I., Moreno, M., Hord, M., Rivera, C., Koenig, R. and Rodriguez-Cerezo, E. 2000. Chayote mosaic virus, a new tymovirus infecting Cucurbitaceae. *Phytopathology* 90: 1098-1104.

Betti, L. 1992. Tomato spotted wilt virus on eggplant in Sicily (Italy). *Phytopath. Mediterranea* 31: 119-120.

Betti, L., Rubbini, M., Tanzi, M. and Canova, A. 1984. Research on TMV of Capsicum. II. Analysis of the aggressiveness of the virus and susceptibility of the host. *Colture Protette* 13: 49-52.

Bhat, A.I., Varma, A., Pappu, HR., Rajamannar, M., Jain, R.K. and Praveen, S. 1999. Characterization of a potyvirus from eggplant (Solanum melongena) as a strain of potato virus Y by N-terminal serology and sequence relationships. *Pl. Pathol.* 48: 648-654.

Boiteux, L.S., Cupertino, F.P., Silva, C., Dusi, A.N., Monte-Neshich, D.C., Vlugt, R.A.A.van-der and Fonseca, M.E.N. 1996. Resistance to potato virus Y (pathotype 1-2) in Capsicum annuum and Capsicum chinense is controlled by two independent major genes. *Euphytica* 87: 53-58.

Boiteux, L.S., Nagata, T., Dutra, W.P. and Fonseca, M.E.N. 1993. Sources of resistance to tomato spotted wilt virus (TSWV) in cultivated and wild species of Capsicum. *Euphytica* 67: 89-94.

Bos, L., Lindsten K and Maat, D.Z. 1977. Similarity of clover yellow vein virus and pea necrosis virus. *Neth. J. Pl. Pathol.* 83: 97-108.

Bosland, P.W. 2000. Sources of curly top virus resistance in Capsicum. *Hort. Sci.* 35: 1321-1322.

Bozarth, R.F., Lana, A.O., Koenig, R. and Reese, J. 1977. Properties of the Nigerian and Ivory Coast strains of the okra mosaic virus. *Phytopathology* 67: 735-737

Brunt, A.A. and Kenten, R.H. 1972. *Pepper veinal mottle virus.* CMI/AAB Descr. Pl. Viruses No. 104 p. 6.

Brunt, A.A., Crabtree, K., Dallwitz, M.J., Gibbs, A.J. and Watson, L. 1996. *Viruses of Plants.* CAB International, Wallingford. pp.1484.

Budnik, K., Laing, M.D., Graca, J.V.-da and Da-Graca, J.V. 1996. An epidemiological study of viruses of pepper (Capsicum annuum L.) in KwaZulu-Natal, South Africa. *African Pl. Protection* 2: 9-14.

Camele, I., Candido, V., Miccolis, V. and Rana, G.L. 2000. Influence of nitrogen fertilization and water stress on incidence of virus infections in sweet pepper culture under glasshouse. *Atti. Giornate Fitopatologiche*, Perugia, 16-20 aprile, 2000, volume primo: 51-54.

Capoor, S.P. and Sawant, D.M. 1987. Studies on bell pepper yellow mosaic virus. *Indian Phytopath.* 40: 443-450.

Capoor, S.P. and Varma, P.M. 1950. A new virus disease of *dolichos lablab. Current Sci.*19: 248-249.

Caranta, C., Palloix, A., Gebre-Selassie, K., Lefebvre, V., Moury, B. and Daubeze, A.M. 1996. A complementation of two genes originating from susceptible Capsicum annuum lines confers a new and complete resistance to pepper veinal mottle virus. *Phytopathology* 86: 739-743.

Chew, M.Y.I., Zavaleta, M., Delgadillo, S.F., Valdivia, A.R., Pena-Martinez, M.R. and Cardenas S.E. 1995. Evaluation of control strategies for virus diseases of pepper (Capsicum annuum L.). *Fitopatologia* 30:7484.

Chowrira, G.M., Cavileer, T.D., Gupta, S.K., Lurquin, P.F. and Berger, P.H. 1988. Coat protein-mediated resistance to pea enation mosaic virus in transgenic Pisum sativum L. *Transgenic Res.* 7: 265-271.

Cicek, Y. and Yorganci, U. 1991. Studies on the incidence of tobacco mosaic virus on certified seed of tomato, pepper and eggplant in Aegean region. *J. Turkish Phytopath.* 20: 57-68.

Clark, M.F. 1972. Purification and some properties of a Comovirus group virus isolated in New Zealand. *New Zealand J. Agric.*15: 846-856.

Conci, V.C., Helguera, M., Nome, S.F. 1999. Serological and biological comparison of onion yellow dwarf virus from onion and garlic in Argentina. *Fitopatol. Bras.* 24: 73-75.

Debrot, E.A., Morales, F., Anzola, D., Perlasca, A. and Betancourt, L.A. 1986. Results of trials for virus control in peppers (Capsicum annuum L.) using resistant varieties. *Agronomia Tropical* 32: 249-260.

Demler, S.A., Rucker-Feeney, D.G., Skaf, J.S., Zoeten, G.A.-de and De-Zoeten, G.A. 1997. Expression and suppression of circulative aphid transmission in pea enation mosaic virus. *J. Gen. Virology* 78: 511-523.

Deol, G.S. and Rataul, H.S. 1978. Role of various barrier crops in reducing the incidence of cucumber mosaic virus in chilli, Capsicum annuum Linn. *Indian J. Entomology* 40: 261-264.

Eldin, A.S.G., El-Kady, M.A. and El-Amrety. A.A. 1983. Pea mosaic virus (PMV) strain of bean yellow mosaic virus isolated from pea. *Egyptian J. Phytopathol.* 13: 23-28.

Feldman, J.M. and Garcia, O. 1977. Pepper severe mosaic virus: A new Potyvirus from pepper in Argentina. *Phytopath. Z.* 89: 146-160.

Fukami, M. and Ishii, K. 1991. Occurrence of garlic latent virus (GLV) and onion yellow dwarf virus (OYDV) on seedless Welsh onions "Bozushirazu" in Chiba Prefecture, and influence of their infection on their yield. *Bull. Chiba Prefectural Agril. Experiment Station* 32: 9-17.

Gibbs, A.J. and Harrison, B.D. 1973. Eggplant mosaic virus. *CMI/AAB Descr. Pl. Viruses* No. 124, p. 4.

Gilardi, P., Garcia-Luque, I. and Serra, M.T. 1998. Pepper mild mottle virus coat protein alone can elicit the Capsicum spp. L3 gene-mediated resistance. *Molec. Pl. Microbe Interact.* 11: 1253-1257.

Givord, L. and Koeing, R. 1974. Okra mosaic virus. *CMI/AAB Descr. Pl. Viruses* No. 128 p. 4.

Gollifer, D.E. and Brown, J.F. 1972. Virus diseases of Colocasia esculenta in the British Solomon Islands. *Plant Dis. Reptr.* 56: 597-599.

Goto, H., Itai, T. and Sato, S. 1977. Selection of attenuated viruses for control of sweet pepper [Capsicum annuum] mosaic disease caused by tobacco mosaic virus pepper strain and cucumber mosaic virus and their effects. *Bull. Oita Prefectural Agril. Res. Center* 27: 79-122.

Govindasamy, C.V., Marappan, V., Kandaswamy, T.K. and Padmanabham, C. 1967. "Mosaic disease of Amaranthus in Coimbatore. (Madras State). *Madras Agric. J.* 54: 35-36.

Green, S.K. and Chou, J.C. 1991. Studies on management practices to reduce aphid-transmitted viruses and their vectors in pepper (Capsicum annuum L.). *Pl. Protection Bull.* Taipei 33: 364-375.

Green, S.K and Kim, J.S. 1994. Sources of resistance to viruses of pepper (Capsicum spp.): a catalog. *Technical Bullt. Asian Veg. Res. Dev. Center.* No. 20, p.64.

Grube, R.C., Zhang, Y.P., Murphy, J.F., Loaiza-Figueroa, F., Lackney, V.K., Provvidenti, R. and Jahn, M.K. 2000. New source of resistance to cucumber mosaic virus in Capsicum frutescens. *Plant Dis*. 84: 885-891.

Guerini, M.N. and Murphy, J.F. 1999. Resistance of Capsicum annuum 'Avelar' to pepper mottle potyvirus and alleviation of this resistance by co-infection with cucumber mosaic cucumovirus are associated with virus movement. *J. Gen. Virology* 80: 2785-2792.

Guigon-Lopez, C. and Gonzalez-Gonzalez, P.A. 1999. Integrated control of viruses of pepper (Capsicum annuum, L.) in the South of Chihuahua. *Rev. Mexic. Fitopatologia* 17: 8-16.

Hampton, R.O., Kraft, J.M. and Muehlbauer, F.J. 1993. Minimizing the threat of seedborne pathogens in crop germ plasm: elimination of pea seedborne mosaic virus from the USDA-ARS germ plasm collection of Pisum sativum. *Plant Dis*. 77: 220-224.

Harrison, B.D., Muniyappa, V., Swanson, M.M., Roberts, I.M. and Robinson, D.J. 1991. Recognition and differentiation of seven whitefly-transmited geminiviruses from India and their relationship to African cassare mosaic and Thailand mung bean yellow mosaic viruses. *Annl. appl. Biol*. 118: 299-308.

Hobbs, H.A., Black, L.L., Johnson, R.R. and Valverde, R.A. 1994. Differences in reactions among tomato spotted wilt virus isolates to three resistant Capsicum chinense lines. *Plant Dis*. 78:1220.

Igwegbe, E.C.K. and Waterworth, H.E. 1982. Proportion and serology of a strain of pepper veinal mottle virus isolated from eggplant (Solanum melongena L.) in Nigeria. *Phytopath Z*. 103: 9-12

James, M., Kenten, R.H. and Woods, R.D. 1973. Virus-like particles associated with two diseases of Colocasia esculenta (L.) Schott in the Solomon Islands. *J. Gen. Virology* 21: 145-153.

Jasnic, S. 1979. Effect of tomato mosaic virus on capsicum yield. *Savremena Poljoprivreda* 27: 335-341.

Johansen, I.E., Keller, K.E., Dougherty, W.G. and Hampton, R.O. 1996. Biological and molecular properties of a pathotype P-1 and a pathotype P-4 isolate of pea seed-borne mosaic virus. *J. Gen. Virology* 77: 1329-1333.

Jose, J. and Usha, R. 2003. Bhendi yellow vein mosaic disease in India is caused by association of a DNA β satellite with a begomovirus. *Virology* 305: 310-317.

Kassanis, B. and Varma, A. 1975. Sunn-hemp mosaic virus, *CMI/AAB Descr. Pl. Viruses* No. 153.

Kenten, R.H. and Woods, R.D. 1973. Viruses of Colocasia esculenta and Xanthosoma saggitifolium. *PANS* 19: 38-41.

Khalil, J.A., Nelson, M.R. and Wheeler, R.E. 1982. Host range, purification, serology and properties of a Carlavirus from eggplant. *Phytopathology* 72: 1064.

Khan, M.A. and Mukhopadhyay, S. 1985. Effect of different pesticide combinations on the incidence of yellow vein mosaic virus disease of okra (Abelmoschus esculentus) and its whitefly vector Bemisia tabaci Genn. *Indian J. Virology* 1: 147-151.

Khetarpal, R.K. and Maury, Y. 1987. Pea seed-borne mosaic virus: a review. *Agronomie* 7: 215-224.

Khetarpal, R.K., Maury, Y., Cousin, R., Burghofer, A. and Varma, A. 1990. Studies on resistance of pea to pea seed borne mosaic virus and new pathotypes. *Ann. appl. Biol*. 116: 297-304.

Kishtah, A.A., Russo, M., Tolba, M.A. and Martelli, G.P. 1978. A strain of broad bean wilt virus isolated from pea in Egypt. *Phytopath. Mediterranea* 17: 157-164.

Kiss, E., Kajati, I., Kolber, M., Basky, Z. and Nasser, M.A.K. 1988. The effect of Atplus 411 F on virus infection of red pepper and bell pepper. *Meded. Facu. Landbouwwetenschappen, Rijksuniversiteit Gent* 53: 479-486.

Kostova, D. 1982. Susceptibility of some Capsicum varieties to tomato strains of tobacco mosaic virus. *Gradinarska I Lozarska Nauka* 19: 95-103.

Ladera, P., Lastra, R. and Debrot, E.A. 1982. Purification and partial characterization of a Potyvirus infecting pepper in Venezuela. *Phytopathology* 83: 514-521.

Ladipo, J.L., Lesemann, D.E. and Koenig, R. 1988a. Further studies on a virus causing a green mosaic disease of eggplant in Nigeria. *J. Phytopathol.* 121: 159-165.

Ladipo, J.L., Lesemann, D.E. and Koenig, R. 1988b. Host ranges, serology and cytopathology of eggplant and tomato strains of eggplant severe mottle virus, a new potyvirus from Nigeria. *J. Phytopathol.* 122: 359-371.

Lana, A.O. and Ajibola-Taylor, T. 1976. The insect transmission of an isolate of okra mosaic virus occurring in Nigeria. *Ann. appl. Biol.* 82: 361-364.

Lana, A.O., Gilmer, R.M., Wilson, G.F. and Shoyinka, S.A. 1975. Identification and characterisation of a severe strain of pepper veinal mottle virus isolated from pepper (Capsicum annuum and C. frutescens) in Nigeria. *Nigerian J. Pl. Protection* 1: 76-81.

Lesemann, D.E. 1977. Virus group specific and virus specific cytological alteration induced by members of the tymovirus group. *Phytopath. Z.* 90: 315-336.

Lisa, V. and Lecoq, H. 1984. Amaranthus leaf mottle virus. *CMI/AAB Descr. Pl. Viruses* No. 282 p. 4.

Liu, H.M., Zhang, Q.P. and Li, J.R. 1996. Studies on the characteristics of welsh-onion yellow stripe virus. *J. Shandong Agric. University* 27: 303-310.

Liu, H.M., Zhang, Q.P. and Wei, Y.Y. 1999. Identification of resistance and primary screening of resistant resources to Welsh onion virus. *J. Shandong Agric. University* 30: 31-36.

Liu, L.J., Licha, M., Baella, D. and Rosa.E. 1982. Variations in morphology and mosaic virus resistance in plantlets of taniers (Xanthosoma sp.) via tissue culture. *Phytopathology* 72: 990.

Loebenstein, G., Alper, M. and Levy, S. 1970. Field test with oil sprays for the prevention of aphid-viruses in peppers. *Phytopathology* 60: 212-215.

MacFarlane, S.A. and Davies, J.W. 1992. Plants transformed with a region of the 201-kilodalton replicase gene from pea early browning virus RNA1 are resistant to virus infection. *Proc. Natl. Acad. Sci. USA* 89: 5829-5833.

MacFarlane, S.A., Wallis, C.V. and Brown, D.J.F. 1996. Multiple virus genes involved in the nematode transmission of pea early browning virus. *Virology* 219: 417-422.

Makkouk, K.M., Koenig, R. and Lesemann, D.E. 1981. Characterization of a tobamoviurs isolated from eggplant. *Phytopathology* 71: 572.

Martelli, G.P. and Hamadi, A. 1986. Occurrence of eggplant mottled dwarf virus in Algeria. *Pl. Pathol.* 35: 595-597

Masmoudi, K., Duby, C., Suhas, M, Guo, J.Q., Guyot, L., Olivier, V., Taylor, J. and Maury, Y. 1994. Quality control of pea seed for pea seed borne mosaic virus. *Seed Sci.Technol.* 22: 407-414.

Matsumoto, T., Ohkawa, T., Hasegawa, S. and Inoue, M. 1991. Occurrence of virus disease of pea caused by white clover mosaic virus in Akita Prefecture. *Ann. Rep. Soci. Pl. Protec. North Japan* 42: 48-52.

Monge, M., Arias, O. and Ramirez, P. 1987. Obtencion de plantas de tiquisque blanco (Xanthosoma sagittifolium), de tiquisque morado (Xanthosoma violaceum) y de nampi (Colocasia esculenta) libre de virus por medio del cultivo in vitro de apices. *Agron. Costar* 11:71-79.

Morita, K., Arai, K., Doi, Y. and Hora, K. 1974. Necrotic leaf spot, a new virus disease of eggplant caused by broad bean wilt virus. *Ann. Phytopathol. Soci. Japan* 40: 288-294.

Moury, B., Palloix, A., Selassie, K.G. and Marchoux, G. 1997. Hypersensitive resistance to tomato spotted wilt virus in three Capsicum chinense accessions is controlled by a single gene and is overcome by virulent strains. *Euphytica* 94: 45-52.

Murant, A.F., Robinson, D.J., Taliansky, M.E., deZoeten, G.A., Falk, B.W., Gibbs, M.J., Pietersen, G. and Waterhouse, P.M. 2000. Family Luteoviridae. In: *Virus Taxonmy*. Van Regenmortel et al. (eds), Academic Press, San Diego. pp. 775-790.

Musa, A.M. and Lockhart, B. 1990. Occurrence of eggplant mottled dwarf virus in Jordan. *J. Phytopathol.* 128: 283-287.

Musil, M. and Kowalska, C. 1993. Transmission of three isolates of broad bean stain virus through seeds of some pea cultivars of the Czechoslovak and Polish assortment. *Ochrana Rostlin* 29: 17-21.

Musil, M., Leskova, O. and Rapi, J. 1981. Effects of some factors on the transmission of the pea leaf roll mosaic virus by seed. *Biologia Czechoslovkia* 36: 889-896.

Nagai, Y., Choi, Y.M. and Tochihara, H. 1987. TMV-U, a new strain of tobacco mosaic virus isolated from sweet pepper. *Ann. Phytopathol. Soci. Japan* 53: 540-543.

Naqvi, Q.A. and Mahmood, K. 1976. Studies on brinjal mild mosaic virus. *Indian Phytopath.* 29: 138-142.

Natsuaki, T., Wang, W.Q. and Okuda, S. 1994. Characterization of chrysanthemum mild mottle virus isolates from sweet pepper (Capsicum annuum) and petunia (Petunia X hybrida). *Bull. College Agr. Utsunomiya Univ.* 15: 11-19.

Ndunguru, J. and Kapooria, R.G. 1999. Identification and incidence of virus diseases of Capsicum annuum in the Lusaka Province of Zambia. *Bull. OEPP* 29: 183-189.

Nono-Womdim, R., Gebre-Selassie, K., Palloix, A., Pochard, E. and Marchoux, G. 1993a. Study of multiplication of cucumber mosaic virus in susceptible and resistant Capsicum annuum lines. *Ann. appl. Biol.* 122: 49-56.

Ohashi, M. and Kamiunten, H. 1994. Mosaic disease of grain amaranth (Amaranthus hypochondriacus L.) caused by cucumber mosaic virus (CMV). *Ann. Phytopathol. Soci. Japan* 60: 119-121.

Padgett, G.B., Nutter, F.W. Jr., Kuhn, C.W. and All, J.N. 1990. Quantification of disease resistance that reduces the rate of tobacco etch virus epidemics in bell pepper. *Phytopathology* 80: 451-455.

Pares, R.D. and Gunn, L.V. 1989. The role of non-vectored soil transmission as a primary source of infection by pepper mild mottle and cucumber mosaic viruses in glasshouse-grown Capsicum in Australia. *J. Phytopathol.* 126: 353-360.

Pategas, K.G., Schuerger, A.C. and Wetter, C. 1989. Management of tomato mosaic virus in hydroponically grown pepper (Capsicum annuum). *Plant Dis.* 73: 570-573.

Phatak, H.S. 1965. Mosaic disease of amaranthus-A new record for India. *Current Sci.* 34: 645-646.

Pozzer, L., Bezerra, I.C., Kormelink, R., Prins, M., Peters, D., Resende, R.de.O. and Avilla, A.C.de. 1999. Characterization of a tospovirus isolate of iris yellow spot virus associated with a disease in onion fields in Brazil. *Plant Dis.* 83: 345-350.

Provvidenti, R. and Granett, A.L. 1976. Occurrence of plantago mottle virus in pea, Pisum sativum, in New York State. *Ann. appl. Biol.* 82: 85-89.

Provvidenti, R. and Niblett, C.L. 1994. Inheritance of resistance to a strain of passionfruit woodiness virus in pea (Pisum sativum L.). *Hort. Sci.* 29: 901-902.

Pun, K.B., Sabitha-Doraiswamy, Jeyarajan, R, Doraiswamy-S. 2000. Screening of virus inhibitory chemicals and neem products against okra yellow vein mosaic virus. *Indian Phytopath.* 53: 95-96.

Rafie, A., Diaz, J. and McLeod, P. 1999. Effects of forage groundnut in reducing the sweetpotato whitefly and associated gemini virus disease in bell pepper in Honduras. *Tropical Agr.* 76: 208-211.

Ragozzino, A. and Angelaccio, C. 1981. Browning of *Capsicum annuum* fruits and stems caused by a strain of tomato mosaic virus. *Phytopath. Mediterranea* 20: 51-55.

Raj, S.K., Amin-ud-din, Aslam, M. and Singh, B.P. 1991. Elimination of eggplant mottled crinkle virus using virazole in explant cultures of Solanum melongena L. *Indian J. Exp. Biology* 29: 594-595.

Raj, S.K., Aminuddin, Singh, B.P. and Pal, M. 1997. Characterization of a cucumber mosaic virus isolate causing leaf crinkle and severe mosaic of Amaranthus in India. *Canadian J. Pl. Pathol.* 19: 97-100.

Rajagopalan, K. 1974. First record of spider mite Tetranychus ludeni Zacher transmitting dolichos enation mosaic virus. *Current Sci.* 43: 488-489.

Rao, G.P., Singh, M., Shukla, K. and Gupta, S.N. 1995. A new strain of cucumber mosaic virus causing mosaic disease of pea (Pisum sativum L.) in India. *Indian J. Virology* 11: 43-48.

Reina, J., Morilla, G. and Bejarano, E.R. 1999. First report of Capsicum annuum plants infected by tomato yellow leaf curl virus. *Plant Dis.* 83: 1176.

Sakata, Y., Chen, Y., Chen, H., Zhao, J. and Xu, Q. 1997. Identification of strains of tobacco mosaic virus (TMV) isolated from Capsicum in southcentral China. *J. Japanese Soci. Hort. Sci.* 66: 99-104.

Sako, I., Takami, T., Nakasone, W., Osaki, T. and Inouye, T. 1994. Occurrence of garlic latent virus and onion yellow dwarf virus in seedless Welsh onion (cv. *Bozushirazu*) and influence of virus reinfection on their yield. *Proc. Kansai Pl. Protec. Soci.* 36: 21-27.

Sandhu, K.S. and Chohan, J.S. 1979. Studies on the characterization of a mottle disease of chilli (Capsicum annuum) in the Punjab. *Indian J. Mycol. Pl. Pathol.* 9: 177-182.

Sastry, K.S. and Nayudu, M.V. 1976. Ringspot symptoms of eggplant, incited by tobacco ring spot virus. *Phytopath. Mediterranea* 15: 60-62.

Schuerger, A.C. and Hammer, W. 1995. Effects of temperature on disease development of tomato mosaic virus in Capsicum annuum in hydroponic systems. *Plant Dis.* 79: 880-885.

Shahraeen, N., Bananej, K., Ahoonmanesh, A. and Lesemann, D.E. 1998. Identification of three elongated viruses from Allium species (onion and garlic) in Tehran province. *Iranian J. Pl. Pathol.* 34: 117-118.

Sharma, O.P., Sharma, P.P. and Chowfla, S.C. 1989. Inheritance of resistance to potato virus Y in garden pepper (Capsicum annuum L.). *Euphytica* 42: 31-33.

Shaw, D.E., Plumb, R.T. and Jackson, G.V.H. 1979. Virus diseases of taro (Colocasia esculenta) and Xanthosoma sp. in Papua New Guinea. *Papua New Guinea Agric. J.* 30: 71-97.

Silveira-Junior, W.G., Avilla, A.C. and Munoz, J.O. 1985. Chayote (Sechium edule Sw), a new host of tomato spotted wilt (vira-cabeca) virus. *Fitopatol. Bras.* 10: 661-665.

Singh, S., Varma, A., Chenulu, V.V. and Verma, V.S. 1972. Pigweed mosaic virus. *Phytopath. Z.* 75: 82-85.

Surendran, M., Shanmugam, V., Kamalakannan, A. and Srinivasan, R. 2000. Transmission studies of brinjal mosaic virus. *Intern. J. Tropical Agr.* 18: 123-131.

Taiwo, M.A. 1988. Studies on a virus disease of Amaranthus hybridus L. in Nigeria. *Intern. J. Tropical Plant Dis.* 6: 195-200.

Takeuchi, S., Kawaike, N. and Kawada, Y. 1998. Control of cucumber mosaic virus by application of attenuated strain in open fields of green pepper (Capsicum annuum L.). *Bull. Kochi Agril. Res. Center* 7: 1-9.

Thottappilly, G. 1969. Investigations on leaf-roll virus of pea and its vectors. I. Transmission tests with different species of aphids that had developed on pea plants infected with the virus. *Phytopathol. Z.* 64: 327-337.

Tian, W.H. and Zhang, S.P. 1995. Identification of tobacco rattle virus on sweet pepper. *J. Hebei Agril. University* 18(4): 32-35.

Uhde, K., Koenig, R. and Lesemann, D.E. 1998. An onion isolate of tobacco rattle virus: reactivity with an antiserum to hypochoeris mosaic virus, a putative furovirus, and molecular analysis of its RNA 2. *Arch. Virol.* 143: 1041-1053.

Van Dijk, P., Verbeek, M. and Bos, L. 1991. Mite-borne virus isolates from cultivated Allium species, and their classification into two new rymoviruses in the family Potyviridae. *Neth. J. Pl. Pathol.* 97: 381-399.

Varma, A. 1970. Red clover vein mosaic virus. *CMI/AAB Descr. Pl. Viruses* No. 22 p. 4.

Varma, A. 1993. Integrated management of plant viral diseases. In: *Crop Protection and Sustainable Agriculture* (CIBA Foundation Symposium 177). D. J. Chadwick and J. Marsh (eds). Chichester, John Wiley & Sons. pp. 140-157.

Varma, A. and Malathi, V.G. 2003. Emerging geminivirus problems: A serious threat to crop production. *Ann. appl. Biol.*142: 145-164.

Varma, A., Khetarpal, R.K., Vishwanath, S.M., Dhirendra, K., Maury, Y., Sharma, B., Tyagi, M.C. and Kumar, D. 1991. Detection of pea seed-borne mosaic virus in commercial seeds of pea, and germplasm of pea and lentil. *Indian Phytopath.* 44: 107-111.

Varma, P.M. 1952. Studies on the relationship of the Bhendi yellow-vein mosaic virus and its vector, the white-fly (Bemisia tabaci Gen.). *Indian J. Agric. Sci.* 22: 75-91.

Vlugt, R.A.A.-van-der, Steffens, P., Cuperus, C., Barg, E., Lesemann, D.E., Bos, L. and Vetten, H.J. 1999. Further evidence that shallot yellow stripe virus (SYSV) is a distinct potyvirus and reidentification of Welsh onion yellow stripe virus as a SYSV strain. *Phytopathology* 89: 148-155.

Vyanjane, N.T. and Mali, V.R. 1981. Occurrence of alfalfa mosaic virus on brinjal in India. *Indian Phytopath.* 34: 164-168.

Wang, D.W., MacFarlane, S.A. and Maule, A.J. 1997. Viral determinants of pea early browning virus seed transmission in pea. *Virology* 234: 112-117.

Wetter, C., Dore, I. and Bernard, M. 1987. Bell pepper mottle virus, a distinct tobamovirus infecting pepper. *J. Phytopathol.* 119: 333-344.

Xu, Z.G., Cockbain, A.J., Woods, R.D. and Govier, D.A. 1988. The serological relationships and some other properties of isolates of broad bean wilt virus from faba bean and pea in China. *Ann. appl. Biol.* 113: 287-296.

Yonaha, T, Toyosato, T., Kawano, S. and Osaki, T. 1995. Pepper vein yellows virus, a novel luteovirus from bell pepper plants in Japan. *Ann. Phytopathol. Soci. Japan* 61: 178-184.

Yu, J., Gu, W.K., Provvidenti, R. and Weeden, N.F. 1995. Identifying and mapping two DNA markers linked to the gene conferring resistance to pea enation mosaic virus. *J. American Soci. Hort. Sci.* 120: 730-733.

Zettler, F.W., Foxe, M.J., Hartman, R.D., Edwardson, R.D. and Christie, R.G. 1970. Filamentous viruses infecting dasheen and other araceous plants. *Phytopathology* 60: 983-987.

Zettler, F.W. and Hartman, R.D. 1980. Evaluation of dieffenbachia maculata 'perfection' plants free of Dasheen mosaic virus. *Acta Hort.* 110: 259.

Zhou, G.Y. and Chen, Y.X. 1990. Identification of staple viruses of Dolichos in Nanjing. *Chinese J. Virology* 6: 233-238.

Zhou, X.P., Liu, Y.L., Robinson, D.J. and Harrison, B.D. 1998. Four DNA-A variants among Pakistani isolates of cotton leaf curl virus and their affinities to DNA-A of geminivirus isolates from okra. *J. Gen. Virology* 79: 915-923.

Zimmer, R.C. and Lamb, R.J. 1993. Amplification and spread of pea seed-borne mosaic virus in field-grown peas. *Canadian J. Pl. Pathol.* 15: 17-22.

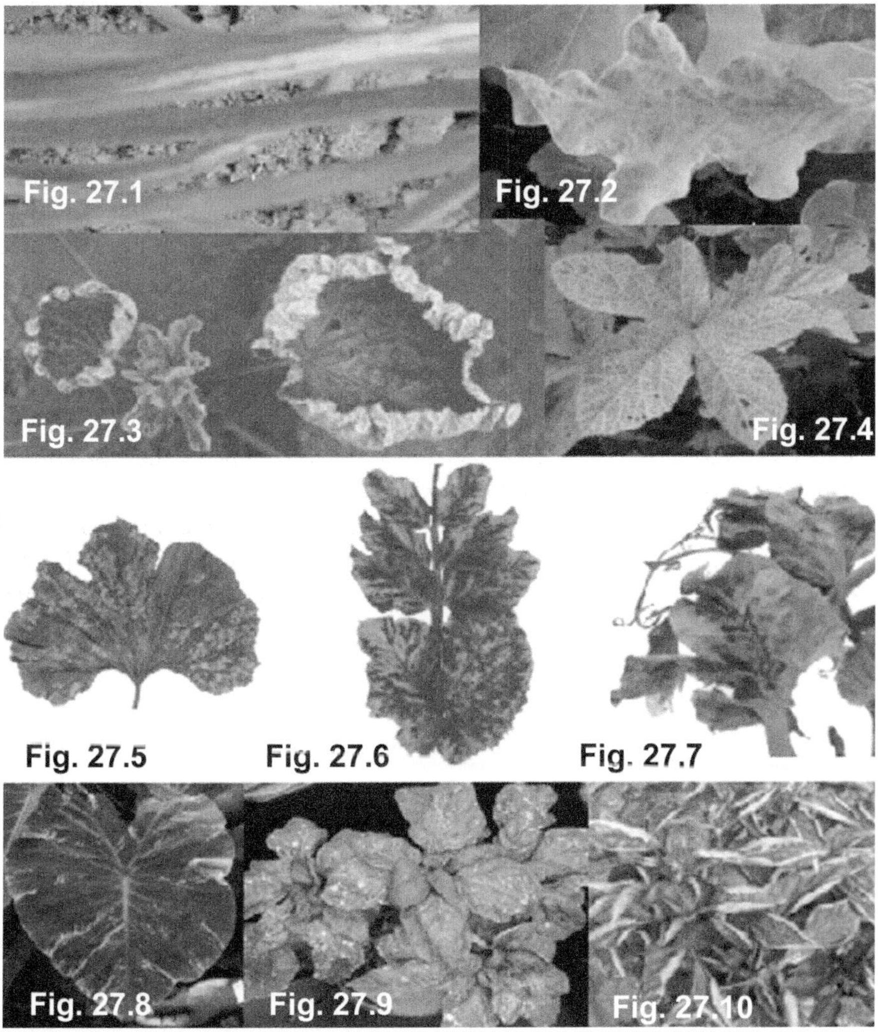

Fig. 27.1 Yellow dwarf mosaic disease of onion. **Fig. 27.2** Mosaic disease of eggplant caused by PVY. **Fig. 27.3** Okra leaf curl disease. **Fig. 27.4** Okra yellow vein mosaic disease. **Fig. 27.5** Mosaic disease of chayote. **Fig. 27.6** Mosaic disease of pea. **Fig. 27.7** Eanation mosaic disease of pea. **Fig. 27. 8** Bobone disease of colocasia. **Fig. 27.9** Symptoms induced by TSWV on sweet pepper. **Fig. 27.10** Leaf curl disease of sweet pepper.

28. Tospoviruses

Tospoviruses: a threat to the intensive agriculture in the tropics

D. Peters
Laboratory of Virology, Wageningen University, Binnenhaven 11, 6709 PD Wageningen, The Netherlands, dick.peters@wur.nl.

28.1 Introduction

Tospoviruses are among the most damaging plant viruses. A disease in tomato described as "spotted wilt" was observed in 1915 in Australia (Brittlebank, 1919). Its viral etiology was established fifteen years later (Samuel *et al.* 1930). TSWV (see Table 28.1 for the name of a virus species when a siglum is used) is now widely known as the causal agent of diseases in many crops and cultures. Diseases like dahlia oak leaf, cyclamen ringspot, tomato carcova, tobacco kromnek, pineapple yellow spot, pine apple side rot, tomato bronze leaf, Kat river disease, makhorka tip chlorosis and 'vira cabeça' are caused by TSWV. For years, TSWV was classified as the sole member of a monotypic virus group.

Application of molecular techniques and of sera against the nucleocapsid protein resulted in the detection of various new viruses with properties identical to those of TSWV (Table 28.1). These viruses now recognized or tentative new species were distinguished from TSWV by host range, vector transmission, and amino acid sequences of the capsid protein and serological properties of the latter. They are classified in one genus, *Tospovirus,* of the family Bunyaviridae. This family consists of four other genera, the viruses of which infect animals. For several decades, TSWV was the only known devastating tospovirus in temperate and subtopic regions. The discovery of GBNV, WBNV and PYSV in India, the occurrence of GBNV and PYSV in South-East Asia, the presence of WSMoV in Japan and Taiwan, the wide distribution of GRSV and TCSV as well as TSWV on the South American continent, and the recent detection of TSWV in Kenya demonstrate that tospoviruses are omnipresent in the tropics. The present chapter in this book aims at summarizing the present knowledge of tospoviruses while emphasizing the various techniques used in tospovirus studies.

28.2 Symptoms and diseases

Most plant species susceptible to a tospovirus react with a systemic infection. Locally infected plants will usually not be noticed in the field. PYSV may form an exception. The large number of yellow spots produced on the leaves of some plants may arise from a high infection pressure.

Tospoviruses can induce a wide variety of symptoms. The plants may respond locally with necrotic lesions or chlorotic spots. Systemic symptoms may consist of wilting leaves, stem and top necrosis, leaf deformation, spotting, streaking, mosaic, mottling, vein yellowing ring spots, line patterns, yellow netting on the leaves, and flower color breaking. Several host species tend to develop more severe symptoms on one side of the plant resulting in asymmetric growth. The severity of symptoms may vary with age, nutritional status as well as growth governing factors such as light intensity and temperature. The various strains of TSWV are probably the main cause of symptom variation within a plant species or cultivar. This will also hold for the other tospovirus species. The incidence of strains responsible for the variation in symptoms may depend on the vector populations prevailing in the different locations and seasons, and the strains in the natural vegetation.

Those parts of plants and fruits that are mature prior to infection will not show symptoms. Symptoms will appear on those parts of the plants that develop after infection or were not mature when the plant became infected. This phenomenon may give the observer, knowing the growth rate and development of the plant, an approximate indication when the plant got infected.

Tomato. This is probably the most often attacked crop by TSWV. The type of symptoms produced depends on the tomato cultivar, age of the plant, and conditions at growth. A slight thickening of the veins and mottling of the younger leaves, and cessation of plant growth are the first symptoms to be observed. The young leaves exhibit also a tendency to curl downwards and inwards or become distorted, whereas also top necrosis may occur (Fig. 28. 1a). The color of the leaves will gradually bronze and can finally become dark purple. This bronze or purple color may cover the leaf completely. Green, yellow or light red spots occur on the fruits. They may later develop into necrotic spots. Occasionally, symptoms are restricted to the fruits, most likely due to a late infection of the plants by thrips.

Sweet pepper. The first chlorotic spots appear on the leaves five days after inoculation. The first systemic symptoms follow a few days later. The latter consists in a development of rugosity and a chlorosis of the smaller veins. This chlorosis leads to the larger veins and to a mottling or ring pattern, or a mosaic with necrotic spots on the leaf blade (Fig. 28.1b). The leaves will curl and the plants become stunted. Leaf and stem necrosis will develop and results eventually in plant death. The infection will develop at one side more rapidly than on the other, so that not all tips become infected at the same moment. INSV causes similar symptoms in peppers. More often, the symptoms are restricted to one leaf and that part of the stem just above that leaf.

Groundnut. GBNV spreads systemically in infected plants. The first symptoms appear usually on newly formed leaves as chlorotic spots (Fig. 28.1d) that may develop into chlorotic and necrotic rings. The leaflets become flaccid and droop, whereas the terminal bud necrotizes. Early infections induce necrosis of all buds. The plants become stunted and axillary shoots are common, giving the plant a bushy appearance with a necrotic center. Because of the great variation in symptoms on plants of the same cultivar, the disease has been described as groundnut mosaic, groundnut rosette, bunchy top, chlorosis, ring mottle, bud blight and ring mosaic (Reddy, 1988). Seeds from early-infected plants are small and shriveled, and their testae show red, brown or purple mottling.

Tobacco. A wide range of symptoms is observed in the field depending on the tobacco variety, developmental stage, plant robustness and weather conditions before and during the infection. Local yellow or necrotic lesions that initially resemble those caused by *Peronospora tabacina* will develop on the infected leaf. A systemic infection appears after 10-15 days and is characterized by the appearance of brown necrotic rings or lines in apical leaves usually forming large necrotic areas and patterns. Older leaves may develop some necrotic spots or severe necrosis around the veins (Fig. 28.1c). Necrotic striation is often observed on one side of the stem causing a characteristic top bending. Milder symptoms, such as mottling, mosaic or vein yellowing and leaf malformation occur at high temperatures, whereas at lower temperatures necrotic lesions and patterns prevail. Infection results in a severe stunting, that often causes death of the seedlings when infection occurs in the seedbeds.

Lettuce. Early symptoms of TSWV infection are usually visible at one side of the plant and consist of chlorotic and brown patches or lines on the leaves (Fig. 28.1e). These symptoms develop rapidly into severe necrotic spots. The symptoms extend to the heart of the plant. This response gives the impression that the plant has been infected by a bacterial or fungal pathogen.

Potato. TSWV has been reported to infect potatoes in Brazil, Iran, Portugal and Tasmania. These infections are usually restricted to a small number of plants. Spread by tubers from infected plants seems to be limited or absent, probably due to the severe necrosis often found in the tubers. (Fig. 28.1f) An unusual infection characterized by extensive stem and petiole necrosis has been described in India (Jain *et al.* 2000). This disease has been encountered in the northwest and the central plains of India. This virus, designated potato stem necrosis virus, reacted with polyclonal antiserum to the N protein of GBNV and WBNV.

28.3 Host range

TSWV can infect 1090 plant species of the 1200 known to be susceptible to tospoviruses (Table 28.2). These 1200 species belong to more than 100 families. The large host range of TSWV consists of plant species grown as agricultural and horticultural crops, naturally infected weeds and experimentally infected hosts. Almost half of the plant species susceptible to TSWV are found in the Asteraceae (236) and Solanaceae (195). Realizing that these families number more than 12000 and 6000 species, respectively, the list of species susceptible to TSWV will be much longer. Since the plant species known to be susceptible to GRSV and TCSV do not essentially differ from those of TSWV, it is likely that these viruses have a much larger host range than the presently listed species (Peters, 2003). The host range of INSV forms the second largest host range with 251-recorded susceptible species, mainly used in the horticultural industry. Although much attention has already been paid to GBNV and WSMoV, their host range is not so widely explored as that of TSWV. The host range of the more recently discovered tospoviruses, like CaCV, CSNV, IYSV, MYSV, PCFV, PYSV, WBNV, and ZLCV seems to be rather limited.

28.4 Thrips and transmision

Tospovirus are transmitted by thrips (Thysanoptera) belonging to the genera *Thrips* and *Frankliniella* of the family *Thripidae*. All tospovirus species, except TSWV, GRSV and TCSV, are transmitted by one thrips species (Table 28.1). The cosmopolitan species, *T. tabaci,* has been known to vector TSWV for several decades (Linford, 1932; Sakimura, 1963). However, the competence of some populations to transmit this virus has been challenged in the last decade and appears to be more complex than supposed. *T. tabaci* is, so far known, the only vector of TSWV in the tobacco industry on the Balkan peninsula (Chatzivassiliou *et al.* 2002), whereas *F. fusca* is the most abundant and effective vector in the tobacco industry in the USA (Eckel *et al.* 1996). Recent studies showed that *T. tabaci* arrhenotokous populations (i.e. populations of males and females), preferring tobacco as host, are efficient transmitters, whereas arrhenotokous populations, preferring other hosts like onion and leek, are poor transmitters. Thelykotous populations, i.e. populations of only females, do not transmit at all (Chatzivassiliou *et al.* 2002; Wijkamp *et al.* 1995). Zawirska (1976) distinguished two *T. tabaci* populations, differing in host preference and some morphological traits. This author proposed that this species consisted of two subspecies, i.e. *T. tabaci ssp. tabaci* and *T. tabaci ssp. communis*. A phylogenetic analysis of mitochondrial cytochrome oxidase I sequences showed that *T. tabaci* populations from tobacco differed indeed from the populations on leek and onion (Chatzivassiliou, personal communication).

The vector status of some other thrips species is also subject of discussion. Transmission of GBNV by *Scirtothrips dorsalis* can be discounted (Palmer *et al.* 1990). Transmission of PCFV by adults of *S. dorsalis* (Chen and Chiu, 1996) can be questioned as only second larval instars had access to the virus. Instars of this age do not acquire virus or very poorly (van de Wetering *et al.* 1996). *F. bispinosa* was shown to transmit TSWV (Webb *et al.* 1998). Since it is not known whether this species reproduce on any host of TSWV, its significance as vector in the field has still to be established. Transmission of ZLCV by *F. zucchini* has been concluded from the abundance of this species on zucchini plants (Nakahara and Monteiro, 1999), but needs to be confirmed.

To transmit tospoviruses larvae have to acquire the virus from infected plant material (Fig. 28.3). This ability rapidly decreases with the age at which the larvae acquire (defined by ingesting followed by replication) the virus (van de Wetering *et al.* 1996). The adults, males as well as females, transmit. Second instars, close to pupation, will also transmit as shown for INSV, IYSV and TSWV (Cortês *et al.* 1998; Wijkamp *et al.* 1993) after being hatched on an infected plant. The thrips remain infectious for their life span after becoming viruliferous. Transovarial transmission has not been demonstrated.

The transmission kinetics has been described by various parameters. Values of 5 or 10 min have been reported for the minimum access acquisition and inoculation periods. These values, obtained with a few thrips specimens out of a much larger population, show that the virus can be acquired and inoculated in very short periods. These short access periods suggest that the virus is acquired and inoculated from and to superficially located plant tissues as could be expected from the ease with which the tospoviruses can usually be inoculated to some plant species.

Measurements of the median acquisition access periods (AAP_{50}), the median latent period (LP_{50}) and the median inoculation access period (IAP_{50}) are more informative as they reflect the transmission competence of a population and not of a few individuals. The AAP_{50}s and IAP_{50}s are in the order of 30 min to 2 h (Chatzivassiliou *et al.* 2002; Wijkamp *et al.* 1996).

LP_{50} studies showed that a part of the adults, who become viruliferous, would already transmit as second instar larvae when the virus was acquired by newborn larvae on infected plants (Wijkamp *et al.* 1993). The LP_{50}s are temperature dependent. Fifty percent of the larvae transmitted TSWV in 170 h when kept at 20°C and in 84 h when kept at 27°C. These LPs were somewhat shorter for INSV (Wijkamp *et al.* 1993). The length of the larval LP_{50}s is slightly, but inversely, correlated with the efficiency with which the larvae transmitted the virus (Wijkamp *et al.* 1993).

Intra- as well as interspecies specificities exist in the transmission between tospoviruses and thrips species. Transmission of a given tospovirus is often restricted to one or a few thrips species (Table 28.2). However, all *Frankliniella* species, known to transmit tospoviruses, also transmit TSWV. Transmission of this virus by *F. zucchini* needs to be confirmed. The ability to transmit depends on the thrips population, but also on the virus isolate. Vector populations differ greatly in their capacity to transmit (Sakurai *et al.* 2002; Chatzivassiliou *et al.* 1999, van de Wetering *et al.* 1999, 1and Peters, 2002) and the various TSWV isolates are also transmitted with different efficiencies (Chatzivassiliou *et al.* 2002).

Thrips can be reared on various hosts. *F. occidentalis* can be reared on *Phaseolus vulgaris* pods in glass jars closed with a lid in which a hole is drilled. This hole has to be closed using screen with 80 holes per cm. Fresh pods are added every third or fourth day, while the dried out ones are discarded. Some pollen has to be added to support the population development. Small pieces of corrugated or other rough (tissue) paper at the bottom provide places for pupation. A small cup of water closed with Parafilm may be added to the jars. To remove the thrips possibly infesting the fresh pods, they are thoroughly washed, dried and stored at 25°C for 5 days and again washed before use.

T. tabaci can also be reared on *P. vulgaris* pods without pollen. Rapidly developing colonies of this species are obtained on segments of the inner part of the pseudostem of leek. The humidity in the jars needs attention as fungi can occasionally infect leek segments.

Different host plants can be used in transmission test of tospoviruses by thrips. Since thrips can readily be recovered from leaf disks placed in small vials (Eppendorf tubes), acquisition of the virus and transmission to leaf disks has to be preferred over placing thrips on test plants (seedlings) in a cage. After the IAP, the disks are floated on water for the development of the infection. Infection can directly be read from the local lesions produced on *Petunia x hybrida* leaf disks within two or three days after inoculation (Fig. 28.1f). Dark-flowering cultivars are preferred as they produce more pronounced lesions. The susceptibility increases considerably when the plants are kept in dark for a number of days. Disks of plants reacting systemically to infection by tospoviruses have to be monitored for infection by ELISA after a longer incubation period on water.

Thrips can be reared or maintained on leaf disks or segments on 1.5% agar in a Petridish closed with clingfilm.

28.5 Purification

Purification of tospoviruses is essential to produce antisera and study their molecular properties. Purification of these viruses meets often with many difficulties. *Nicotiana rustica* is the favoured host to propagate tospoviruses. Some, like INSV and IYSV, cannot be propagated on this species. *N. benthamiana*, probably susceptible to all tospoviruses, is then recommended. This species gave, however, less satisfactory results than *N. rustica*. The quantity of leaf material harvested from a *N. rustica* plant is several times larger than that from *N. benthamiana*. Plants of both species should be inoculated when they are in their fifth or sixth leaf stage. *Chenopodium quinoa* has been used for the extraction of WSMoV dsRNA (Chu *et al.* 2001a) and purification of PCFV nucleocapsids (Chu *et al.* 2001b). Systemically infected leaves of cucumber were used to extract MYSV nucleocapsids (Kato *et al.* 2000).

Maintenance of TSWV by mechanical transfer in solanaceous hosts can readily result in the formation of defective isolates (Inoue-Nagata *et al.* 2000). These isolates consist of four RNA strands, of which one is generated by a deletion in the L RNA. Defective isolates are difficult to purify and are not or only poorly transmissible by thrips (Nagata *et al.* 2000). To avoid defective isolates, inocula should be prepared from plants either infected by thrips or inoculated with samples of stored material transferred once or a few times mechanically. Plant material, infected with fresh isolated virus, should preferably be stored after a first propagation in several vials in liquid nitrogen, at -80° or -20°C. Samples remain also infectious for large periods when properly stored after being dried over $CaCl_2$ in a refrigerator or by lyophilization. The infectivity of stored samples has regularly to be checked for infectivity.

A successful procedure to purify tospoviruses is given in the scheme below:

Harvest systemically infected *N. rustica* or *N. benthamiana* leaves showing clear, but not necrotic, symptoms one or two days after the development of the first systemic symptoms. The purification will be less successful when necrotic symptoms appear on the leaves.

Homogenize the leaves for 10 sec at low speed in a blender with 4 ml extraction buffer per g leaf material. The extraction buffer consists of 0.1 M KH_2PO_4 (300 ml), 0.1 M K_2HPO_4 (500 ml) and 1 g of Na_2SO_3, pH 7.0.)

Strain the homogenate through two layers of cheesecloth (boiled in water with some EDTA).

Centrifuge the filtrate for 10 min at 5000 rpm (Sorvall GSA rotor) to sediment the virus. Almost all virus precipitates in this first low speed centrifugation. After resuspension of these pellets the virus will precipitate at higher centrifugal forces.

Resuspend the pellet in 1 ml resuspension buffer per gram leaf material by gently stirring until the pellets are completely dissolved (30-60 min). This buffer consists of 0.01 M KH_2PO_4 (300 ml) and 0.01 M K_2HPO_4 (500 ml), pH 7.0.

Centrifuge the suspension for 15 min at 8000 *g* (= 9000 rpm Sorvall SS34 rotor) to remove the debris.

Carefully decant the supernatant (use a 25 ml pipette or syringe) to a Ti 45 tube (Beckman Ultracentrifuge) or removed it by vacuum extraction. Make sure not to collect any pelleted material!

Centrifuge for 30 min at 100,000 *g* (29,000 rpm, Beckman Ti45).

Resuspend carefully with a paintbrush the pellet in 2.5 ml resuspension buffer per 100 g leaf material.

Prepare one or two 5-40% sucrose gradients (dissolved in resuspension buffer) in Sw28 tubes, layer the virus suspension (1 ml or a smaller volume) on the gradient(s), and centrifuge for 45 min at 24,000 rpm in a Sw28 rotor (Beckman centrifuge).

Collect the 1-cm-wide milky virus band using a syringe with a bended needle and dilute the collected band at 1:1 in resuspension buffer.

Concentrate the virus by centrifugation for 30 min at 24,000 rpm (R30 rotor, Beckman Ultracentrifuge) and resuspend the pellet in 25 µl H_2O/100 g leaf material, which yield 10 µg/µl or more virus.

Completely pure preparations of some tospoviruses or isolates of TSWV are sometimes hard to obtain. Preparations of purified nucleocapsids can be obtained to a higher degree of purity and are therefore more suited to produce powerful and specific antisera and for molecular studies. The procedure to be used is given below:

Cut approximately 40 gram of systemically infected leaf material in smaller pieces and homogenize them in a blender for 7 sec in 8 ml extraction buffer/g leaf material. This buffer consists of 0.1 M Tris-HCl, 0.01 M EDTA, 0.01 M Na_2SO_3, and 0.1% cysteine, pH 8.0. In the purification of

MYSV nucleocapsids 2-mercaptoethanol and l-ascorbic acid was added Kato *et al.* (2000).

Strain the homogenate through cheesecloth, centrifuge the filtrate for 10 min at 3000 rpm (Beckman JA) at 4° C.

Centrifuge the supernatant for 40 min at 30,000 rpm (17,000 *g*) in a Beckman centrifuge.

Suspend the green pellets in resuspension buffer (2.5 ml/g leaf material), while gently stirring for 40-60 min in the cold room. This buffer consists of 0.01 M Tris-HCl, 0.01 M EDTA, 0.01 M Na_2SO_3, 0.1% cysteine, 0.01 M glycine and 1% Nonidet P40, pH 7.9.

Centrifuge the suspension for 10 min at 3000 rpm. Layer the supernatant on a 30 ml 30 % sucrose foot filling half of the tube and centrifuge for 70 min at 39,000 rpm in a Beckman Ti45 rotor. The sucrose has to be dissolved in resuspension buffer.

Dissolve the light green or yellow, but transparent, pellets in 2 ml 0.01M citrate buffer, pH 6.0.

Centrifuge the suspension for 5 min at 8000 rpm in an Eppendorf centrifuge and then on 20-40% sucrose gradient in 0.01 M citrate buffer for 2.5 h at 36,000 rpm in a Sw41 rotor.

Collect the opalescent bands with a Pasteur pipette, concentrate the nucleocapsids at 40,000 rpm for 60 min, and dissolve the pellet in water.

Analyse the purity of the preparation and the presence of the nucleocapsid protein on a SDS gel.

28.6 Morphology, classification, properties and replication strategy

Morphology The genus *Tospoviruses*, named after TSWV consists of 13 recognized and tentative species (Elliott *et al.* 2000). CaCV has recently been described (McMichael *et al.* 2002). The particles, like all bunyaviruses, are spherical or pleomorphic, 80-120 nm in diameter and display two glycoproteins (G1 and G2) on an approximately 5-nm thick lipid envelope, enclosing three nucleocapsids (Fig. 28.4). These nucleocapsids are formed by three single-stranded, linear RNA segments, denoted L (large), M (medium), and S (small), enwrapped by the nucleocapsid (N) protein and a few RNA-dependent RNA polymerase proteins. These genomic RNAs have conserved 5'- and 3'- terminal sequences of which the ultimate 8 nucleotides are completely complementary (5'-AGAGCAAU...AUUGCUCU-3'). These terminal nucleotides combine in such a way that the segments are non-covalently closed pseudo-circles (Fig. 28.4). The M (~ 4820 nt) and S RNA (~2920 nt) segments are ambisense. The M RNA encodes for the precursor of the G1 and G2 proteins in the non-viral sense and a nonstructural (NSm) protein on the viral sense strand. This protein has a function in the cell-to-cell movement of the virus. The S RNA encodes the N protein in the non-viral sense and a nonstructural (NSs) protein on the viral sense strand. This protein prevents silencing of the viral RNA replication (Bücher *et al.* 2003). The NSs protein can readily be detected in the salivary glands of viruliferous thrips (Wijkamp *et al.* 1993) and in the cells of infected plants in paracrystalline structures (Kitajima *et al.* 1992). A UA-rich intergenic region forming a large energy-rich hairpin separates the genes on the viral and viral-complementary strands of the M and S RNA. The RNA-dependent RNA polymerase is encoded by the negative L RNA (~9000 nt).

Classification.As pointed out earlier, several species have been classified as tospoviruses on the basis of the amino acid sequence identity of the nucleocapsid proteins, vector specificity, serology and host range. Viruses with an amino acid identity lower than 90% are considered to be different species. The recognized and tentative species (Elliot *et al.* 2000, see also Table 28.1) can serologically be differentiated by antisera to the N protein. Some cross reactivity has been shown between TSWV, TCSV and GRSV and between GBNV, WSMoV and CaCV. No cross reactivity has found between these and the other tospoviruses, which do also not mutually show any cross-reaction. TSWV, TCSV and GRSV, and GBNV WSMoV and CaCV are considered to be members of two different serogroups, denoted serogroup I and IV (Jain *et al.* 1998), respectively. Grouping of the tospoviruses in serogroups is rather impractical, as each of the other tospoviruses, thus far identified, form a single member of a serogroup. Referring to TSWV serogroup and WSMoV serogroup in stead of serogroup I and IV would be more useful.

Physical properties Tospoviruses possess properties by which they can readily be distinguished from other plant viruses. Due to their dimensions, they are well discernable, although not readily, in dip preparations for electron microscopy. Extracts from infected plants kept for some hours at room temperature lose their infectivity. These viruses are heat-inactivated at 56 °C within 10 min and sensitive to lipid solvents and detergents. Their instability is a first indication that the infecting virus might be a tospovirus. Loss of infectivity in a one-day old extract from plants with a possible virus infection might point to a tospovirus infection. The RNA is not infectious, but nucleocapsids are after their release from the virus by a detergent. An envelope-deficient TSWV isolate, which does not envelop its nucleocapsids, can infect plants as efficient as complete virus (Resende *et al.* 1991).

Replication strategy Tospovirus infection starts with the introduction of the virus in a cell either by thrips feeding or mechanical inoculation. Upon this entry the virus will fuse with acidic lysosomes, after which the membrane is removed. The set-free nucleocapsids will be released in the cytoplasm. The virus-specific RNA-dependent RNA polymerase (L protein), transcribes (primary transcription) the negative-sensed parts (the L, G1/G2 precursor and N gene) of the genome into mRNAs (Fig. 28.4). The synthesis of these mRNAs starts by snatching the cap from non-virus-specific mRNAs, followed by the transcription of virus-specific sequences. These viral mRNAs are translated into proteins (Duysings *et al.* 1999). The synthesis of N protein results in a switch from the ongoing mRNA synthesis into the synthesis of full-length complementary copies of the infecting RNA (primary replication). These RNA products are encapsidated by the N protein, whereas the produced mRNAs are not. The viral complementary strands serve as templates for (secondary) transcription, resulting in the NSs and NSm mRNAs (Fig. 28.4). The product of the NSm mRNAs enables the spread of virus in the form of nucleocapsids to neighboring cells, where the transcription and replication will start again. After transporting the infecting material to other cells, the infection in the first cell will be completed in a process of secondary transcription and replication. A membrane, on which the G1 and G2 proteins are anchored, will envelop the nucleocapsids in the Golgi complex (Kikkert *et al.* 1999).

28.7 Diagnosis and identification

A small number of plant species have frequently been used in the detection of almost all tospoviruses (Table 28.2). No definite identification can be made as the slight differences in symptom expression on these plant species makes identification almost impossible. Moreover, the symptoms produced are often similar to those evoked by viruses of other families. Some of the plant species listed can be used to remove or separate co-infecting viruses from a mixed infection. TSWV will be separated from *Tobacco mosaic virus* in a mixed infection in *N. glutinosa*, as the latter virus produces only local lesions, whereas TSWV gives a systemic infection. TSWV can be separated from *Potato virus Y* by using *Datura stramonium*, which is not susceptible to this potyvirus. The occurrence of mixed infections can be demonstrated by incubation of an extract for 24 h at room temperature and its inoculation to *N. benthamiana* or any other obvious host. Inoculation of this extract to petunia before incubation is indicative of a tospovirus infection.

Enzyme linked immunosorbent assay (ELISA) is probably the best method to detect tospoviruses in plants and to identify the infecting virus up to species level. The different tospovirus species can well be distinguished by using antisera prepared to the nucleocapsid proteins. The nucleocapsid antisera to all tospoviruses, except GRSV, TCSV and WSMoV, do not show any or only an extremely faint reaction with other tospoviruses. TSWV antisera give slight positive reactions with GRSV and TCSV in reciprocal tests and GBNV with WSMoV. The double-antibody sandwich ELISA is the most used technique. Different formats can be used since large virus concentrations occur in most infected plants and the anti-nucleocapsid protein sera have often high titers. The duration of coating, sample and conjugate incubation steps can be arbitrarily chosen between 2h at 37°C or room temperature, or one night at 4 °C. Readings of 2.000 to 3.000 can be obtained after substrate incubation for 1 h with well-prepared anti-N protein sera and 1:3 or 1:10 w/v diluted extracts from infected plants. Incorporation of polyvinylpyrrolidone in the coating and sample buffer will be helpful in reducing the background, if necessary.

Detection of the virus in thrips is more cumbersome. Several methods have been published for the detection of TSWV in thrips, ranging from ELISA followed by enzymatic amplification of the reaction (Wijkamp *et al.* 1996), nucleic acid blots (Rice *et al.* 1990), immunological squash blots (Aramburu *et al.* 1996), PCR (Mason *et al.* 2003) to real time fluorescent RT-PCR (Boonham *et al.* 2002). Most of these advanced techniques require well-developed laboratories and well-furnished budgets. Some of these methods have been used in experimental studies, but most of them have not been used on a large scale to monitor epidemics or to analyze infection pressure in the prevailing thrips populations. In field studies, a distinction has to be made between adults with a virus load resulting from acquisition of virus after hatching on an infected plant and subsequent replication, or those that ingested virus as adult on infected plants. This problem can be circumvented by starving adults, or placing them on healthy leaves for a day to deplete ingested virus from the intestinal tract (Aramburu *et al.* 1996). All viruliferous thrips are ELISA positive, but there is no correlation between the ELISA values and the number of local lesions produced on petunia leaf disks (Wijkamp and Peters, 1993). Analyzing adults that acquired the virus as newborn larvae showed that adults could be divided in three groups of response. One group of positive adults that transmits TSWV, one positive group that fails to transmit, and the third group does not give a positive response and fails to transmit (van de Wetering *et al.* 1996).

Several universal primer sets have been designed to develop molecular techniques for identification and characterization of tospoviruses in infected plants or thrips. Degenerate primers designed from conserved regions of the L genes of TSWV, INSV, WSMoV and GBNV were used to amplify the corresponding regions of L genes in total RNA extracts of plants infected with five tospoviruses (Chu *et al.* 2001a). The results obtained demonstrate that the infecting virus can be identified up to the level of genus, but not to species level. This primer set may be of great value in analyzing those L genes that have still not been sequenced. Another approach was made in the characterization of a novel tospovirus, probably identical to MYSV (Cortez *et al.* 2001). First strand cDNA was synthesized using the primer J13 (5'dCCCGGATCCAGAGCAAT) containing the conserved 8 terminal nucleotides. Full-length S cDNA was then amplified using J13 in 3-min extension cycles. Complete N and NSs protein genes were produced using J13 and 5'dCACTGGATCCTTTTGTTTTTGTTTTTTG. This primer (UHP) was designed on the base of the highly homologous sequence within the intergenic region of the S and M RNAs (Cortez *et al.* 2001).

28.8 Control

Various methods have been advocated to control tospoviruses. The viruses have to be introduced in the crop by thrips from outside, as the viruses are not seed transmitted and most important crops, susceptible to tospoviruses, are grown from seeds. The omnipresence of vector and virus hosts in native environments, surrounding crops and screenhouses, makes control of the virus extremely difficult. Control of thrips can be achieved by application of insecticides in the field, but the spread of virus seems less affected by these applications than the thrips population.

The first infected plants might be rogued. However, this is not often a useful strategy as the appearance of an infected plant in a young crop is usually a sign that more infections will follow either from outside or inside spread. The first spread from a primarily infected plant in a crop may occur approximately two and a half-week after the initial infection. The first larvae may acquire the first virus approximately eight days after infection of the plant. These larvae need another nine or ten days before they will spread the virus as adults.

Hygienic measures like removal of weeds in a wide area around a crop may relieve the infection pressure. Keeping fields bare, including weeding and tilling, may affect the survival of the pupae in a negative sense. Additional measures can be taken to control the virus in greenhouses. Roguing of infected plants and frequent control for infections and thrips should be routine. To detect infections in a greenhouse, petunia plants have been used to bait TSWV (Allen and Matteoni, 1991), while yellow or blue sticky traps could monitor thrips. The use of virus free stock material, airlocks at entrances of screenhouses, thorough removing of plant material after harvest and, as the pupae may survive for some time, delay of planting the new crop may reduce the inside infection pressure.

A risk assessment index to predict TSWV infection in groundnut crops is successfully applied in extension work to farmers in Georgia, USA. The index is determined by the sum of several weighted factors like the susceptibility of the groundnut variety, planting date, plant density, at-plant application of insecticides (phorate seemed to suppress infection), history of the infection, row pattern and tillage. The total sum of points to be allocated range from 35 to 155. Low risk may be expected when the sum is lower than 65, a moderate risk with a sum ranging between 70-110, and a high risk when more than 115 points are scored (Culbreath *et al.* 2003).

Host resistance to tospoviruses. Since hypersentivity or other forms of natural resistance to tospoviruses is not known in most plant species, breeding for virus resistance is a cumbersome approach. Crosses made between resistant accessions of *C. chinense* PI152225 or PI159236 and susceptible *C. annuum* resulted in some accessions giving a hypersensitive response after infection with TSWV. This hypersensitivity is strongly influenced by temperature and physiological conditions. A hypersensitive response is obtained in PI152225 on 20% of the plants with two true leaves and 45% with four true leaves were inoculated with TSWV and kept continuously at 33 °C (Roggero *et al.* 2002). Breeding for resistance in tomato resulted in the discovery of a single dominant gene, *Sw-5*. Several tomato lines have now been released with this resistance (Stevens *et al.* 1992). TSWV resistance breaking strains are encountered in the field in pepper cultivars showing resistance to TSWV conferred by Tsw. Such strains have also been found in tomato carrying the *Sw-5* gene (Latham and Jones, 1998; Thompson and van Zijl, 1996). Resistance to TSWV has also been found in some accessions of crops like, chrysanthemum and lettuce.

Non-convential strategies for tospovirus resistance. As natural sources of resistance to tospoviruses are limited, novel forms of resistance have to be developed. Transgenic expression of the TSWV N and NSm gene are effective against TSWV (Gielen *et al.* 1992; Prins *et al.* 1995), but not to the related TCSV and GRSV (de Haan *et al.* 1992). However, simultaneous expression of the three N genes of TSWV, TCSV and GRSV, inserted at a single site, resulted in resistance to these viruses. This transgenic resistance is not only induced by the complete N gene, but also by translational defective copies of this gene, and segments as small as 110 nucleotides (de Haan *et al.* 1992; Pang *et al.* 1998). Sequences of other genes were not effective in inducing resistance in transgenic plants (Prins and Goldbach, 1998).

Host resistance to thrips. Resistance to thrips may add to the control of TSWV epidemics. Significant levels of this resistance have been found in cabbage, chrysanthemum, groundnut, peanut, pepper and tomato. Reduced spread of GBNV in the groundnut accession Robut 33-1 was attributed to the lower numbers of thrips infesting plants of this accession (Reddy *et al.* 1983). Studies with thrips-resistant and susceptible pepper accessions, which are both fully susceptible to TSWV, showed that infection of thrips-resistant plants was restricted to inoculation by invading thrips. In populations with plants of both accessions, infection of resistant plants was significant delayed compared with the development of the infection among the susceptible plants. These results show that thrips resistance will provide a significant protection to TSWV infections, even when the crop is fully susceptible to the virus (Maris *et al.* 2003). Besides the absence of any population development, changes in the feeding behaviour may also affect the transmission of TSWV to thrips resistant accessions. A disturbed feeding behaviour of *F. occidentalis* and *T. tabaci* was observed on resistant cucumber and leek, respectively, resulting in a lower number of penetrations and reduced feeding times (Harrewijn *et al.* 1996). The effect of the difference in feeding behaviour on the spread remains to be studied.

28.9 Conclusion.

This chapter describes the increase of knowledge on tospoviruses in the last one and a half decade. These studies were unexpectedly ignited by the almost worldwide expansion of *F. occidentalis* in the 1980 and 1990s to greenhouse crops in the temperate climatic zones, and to field and screen-houses in the subtropics. TSWV was present wherever this thrips species settled. In its wake, this thrips species also introduced INSV in the horticultural industry of these regions. *T. palmi* seems now to be in a process of expansion. This expansion will be more into the direction of countries with a tropical climate than into the temperate regions. The incidence of diseases caused by viruses, like WSMoV and GBNV, seems apparently to be on the increase. Other viruses transmitted by *T. palmi* are at present of local significance. However, there is no reason to believe that they will not follow their vector. With this in view, it can be concluded that the importance of tospoviruses will increase in the coming decade in the tropics.

References

Allen, W.R. and Matteoni, J.A. 1991. Petunia as indicator plant for use of growers to monitor for thrips carrying the tomato spotted wilt virus in greenhouses. *Plant Dis.* 75: 78-82.

Aramburu, J. Riudavets, J., Arno, J., Lavina, A. and Moriones, E. 1996. Rapid serological detection of tomato spotted wilt virus in individual thrips by squash-blot assay for use in epidemiology studies. *Plant Pathol.* 45: 367-374.

Boonham, N., Smith, P., Walsh, K., Tame, J., Morris, J., Spence, N., Bennison, J. and Barker, I. 2002. The detection of *Tomato spotted wilt virus* (TSWV) in individual thrips using real time fluorescent RT-PCR (Taqman). *J. Virol. Meth.* 101: 37-48.

Brittlebank, C.C. 1919. Tomato diseases. *J. Agric. Victoria* 27: 213-235.

Bücher, E., Sijen, T., de Haan, P., Goldbach, R. and Prins, M. 2003. Negative-strand tospoviruses and tenuiviruses carry a gene for a suppresssor of gene silencing at analogous genomic positions. *J. Virol.* 77: 1329-1336.

Chatzivassiliou, E., Nagata, T., Peters, D. and Katis, N.I. 1999. The transmission of tomato spotted wilt tospovirus (TSWV) by *Thrips tabaci* Lind. (Thysanoptera: Thripidae) populations originating from leek. *Plant Pathol.* 48: 700-706.

Chatzivassiliou, E., Peters, D. and Katis, N.I. 2002. The efficiency by which *Thrips tabaci* populations transmit *Tomato spotted wilt virus* depends on their host preference and reproductive strategy. *Phytopathology* 92: 603-609.

Chen, C.C. and Chiu, R.J. 1996. A tospovirus infecting peanut in Taiwan. *Acta Hortic.* 431: 57-67.

Chu, F-H., Chao, C-H., Chung, M-H, Chen, C-C. Yeh, S-D. 2001a. Completion of the genome sequence of *Watermelon silver mottle virus* and utilization of degenerate primers for detecting tospoviruses in five serogroups. *Phytopathology* 91: 361-368.

Chu, F-H., Chao, C-H., Peng, Y-C., Lin, S-S., Chen, C-C. and Yeh, S-D. 2001b. Serological and molecular characterization of *Peanut chorotic fan-spot*, a new species of the genus *Tospovirus. Phytopathology* 91: 856-863.

Cortês, I., Livieratos, J., Derks, A., Peters, D. and Kormelink, R. 1998. Molecular and serological characterization of iris yellow spot virus, a new and distinct tospovirus species. *Phytopathology* 88: 1276-1282.

Cortez, I., Saaijer, J., Wongjkaew, K.S., Pereira, A.-M., Goldbach, R., Peters, D. and Kormelink, R. 2001. *Arch. Virol.* 146: 265-278.

Culbreath, A. K., Todd, J.W. & Brown, S.L. 2003. Epidemiology and management of tomato spotted wilt virus. *Annu. Revs. Phytopathol.* 41 (in press).

de Haan, P., Gielen, J.J.L., Prins, M., Wijkamp, I.G., van Schepen, A., Peters, D., Grinsven, Q.J.M. and Goldbach, R. 1992. Characterization of RNA-mediated resistance to tomato spotted wilt virus in transgenic tobacco plants. *Biotechnology* 10: 133-1137.

Duysings, D., Kormelink, R. and Goldbach, R. 1999. Alfalfa mosaic virus RNAs serve as cap donors for tomato spotted wilt virus transcription during co-infection of *Nicotiana benthamiana. J. Virol.* 73: 5172-5175.

Eckel, G.S., Cho, K., Walgenbach, J.F., Kennedy, G.G. and Moyer, J.W. 1996. Variation in thrips species composition in field crops and implications for tomato spotted wilt epidemiology in North Carolina. *Entom. Exp. Appl.* 78: 19-29.

Elliott, R.M., Bouloy, M., Calisher, C.H., Goldbach, R., Moyer, J.T., Nichol, S.T., Petterson, R., Plyusnin, A. and Schmalljohn, C.S. 2000. Bunyaviruses. In: *Virus Taxonomy*. Seventh Rept. Intern. Com. Taxonomy of Viruses (Eds. M.H.V. van Regenmortel *et al.*), Academic Press, San Diego/London, pp 599-621.

Gielen, J.J.L., de Haan, P., Kool, A.J., Peters, D., van Grindsven, M.Q.J.M. and Goldbach, R.W. 1992. Engineered resistance to tomato spotted wilt virus, a negative-strand RNA virus. *Biotechnology* 9: 1363-1367.

Harrewijn, P., Piron, P.G.M. and Mollema, C. 1996. Electrically recorded probing behaviour of thrips species on optimal and suboptimal hosts. *Entom. Exp. Appl.* 80: 43-45.

Inoue-Nagata, A.K., Kormelink, R., Nagata, T., Kitajima, E.W., Goldbach, R. and Peters, D. 1997. Effects of temperature and host on the generation of tomato spotted wilt virus defective interfering RNAs. *Phytopathology* 87: 139-144.

Jain, R.K., Pappu, H., Pappu, S., Krishna Reddy, M. and Vani, A. 1998. Watermelon bud necrosis tospovirus is a distinct virus species belonging to serogroup IV. *Arch Virol.* 143: 1637-1644.

Jain, R.K., Paul Khurana, S.M., Gourgopal, R., Hegde, V., Singh, R.A. and Varma, A. 2000. Serological and molecular characterization of the tospovirus associated with potato stem necrosis disease. In: *Proc. Global Conf. Potato, New Delhi, India.* Vol 1: 456-458.

Kato, K., Hanada, K. and Kameya-Iwaka, M. 2000. *Melon yellow spot virus*: A distinct species of the genus *Tospovirus* isolated from melon. *Phytopathology* 90: 422-426.

Kikkert, M., van Lent, J., Storms, M., Bodegom, P., Kormelink, R. and Goldbach, R. 1999. *Tomato spotted wilt virus* particle morphogenesis in plant cells. *J. Virol.* 73: 2288-2297.

Kitajima, E.W., de Ávila, A.C., Resende, R. de O., Goldbach, R.W. and Peters, D. 1992. Comparative cytological and immunogold labelling studies on different isolates of tomato spotted wilt virus. J. submicrosc. *Cytol. Pathol.* 24: 1-14.

Latham, L.J., and Jones, R.A.C. 1998. Selection of resistance breaking strains of tomato spotted wilt tospovirus strains. *Ann. appl. Biol.* 133: 385-402.

Linford, M.B.1932. Transmission of the pineapple yellow spot virus by *Thrips tabaci. Phytopathology* 22: 301-324.

Maris, P.C., Joosten, N.N., Goldbach, R.W. and Peters, D. 2003. Restricted spread of *Tomato spotted wilt virus* in thrips resistant pepper (*Capsicum annuum*). *Phytopathology* (submitted).

Mason. G., Roggero, P. and Tavela, L. 2003. *Tomato spotted wilt virus* in its vector *Frankliniella occidentalis* by reverse transcription-polymerase chain reaction. J. Virol. Meth. 109: 69-73.

McMichael, L.A., Persley, D.M. and Thomas, J.E. 2002. A new tospovirus serogroup IV species infecting capsicum and tomato in Queensland, Australia. *Austral. Plant Pathol.* 31: 231-239.

Nagata, T., Inoue-Nagata, A.K., Prins, M., Goldbach, R. and Peters, D. 2000. Impeded thrips transmission of defective tomato spotted wilt virus isolates. *Phytopathology* 90: 454-459.

Nagata, T. and Peters, D. 2001. An anatomical perspective of tospovirus transmission. In: *Virus-Insect-Plant Interactions* (K.F.Harris, O.P.Smith and J.E. Duffus, eds), Academic Press, San Diego. 51-63 pp.

Nakahara, S. and Monteiro, R.C. (1999). *Frankliniella zucchini* (Thysanoptera: Thripidae), a new species and vector of tospovirus in Brazil. *Proc. Entomol. Soc.* Washington 101: 290-294.

Palmer, J.M., Reddy, D.V.R., Wightman, J.A. and Ganga Rao, G.V. 1990. New information on the thrips vectors of tomato spotted wilt virus in groundnut crops in India. *Int. Arachis Newslett.* 7: 24-25.

Pang, S.-Z., Nagpala, P., Wang, M., Slightom, J.L. and Gonsalves. D. 1998. Resistance to heterologous isolates of *Tomato spotted wilt virus* in transgenic tobacco expressing its nucleocapsid protein gene. *Phytopathology* 82: 1223-1229.

Peters, D. 2003. A collated list of plant species susceptible for tospoviruses. (Unpublished, if information required, request dick.peters@wur.nl).

Prins, M., de Haan, P., Luyten, R., van Veller, M., van Grindsven, M.Q.J.M. and Goldbach, R.W. 1995. Broad resistance to tospoviruses in transgenic tobacco plants expressing three tospoviral nucleoprotein gene sequences. *Molec. Plant-Microbe Interac.* 8: 85-91.

Prins, M. and Goldbach, R. 1998. The emerging problem of tospovirus infection and nonconventional methods of control. *Trends in Microbiol.* 6: 31-35.

Reddy, D.V.R, Amin, P.W., McDonald, D. and Ghanekar, A.M. 1983. Epidemiology and control of groundnut bud necrosis and other disease of legume crops in India caused by tomato spotted wilt virus. In: *Plant Virus Epidemiology*. R.T. Plumb and J.M., Thresh (eds.), Blackwell Sci. Publ., Oxford, pp 93-102.

Reddy, D.V.R. 1988. Groundnut. In: *Virus Diseases*. P.S. Reddy (ed.), Indian Counc. Res., New Delhi, India, pp 508-525.

Resende, R. de O., de Haan, P., de Ávila, A.C., Kitajima, E.W. Goldbach, R. and Peters, D. 1991. Generation of envelope and defective interfering RNA mutants of tomato spotted wilt virus by mechanical passage. *J. Gen. Virol.* 72: 2375-2383.

Rice, D.J., German, T.L., Mau, R.F.L. and Fujimoto, F.M. 1990. Dot blot detection of tomato spotted wilt virus RNA in plant and thrips tissues by cDNA clones. *Plant Dis.* 74: 274-276.

Roggero. P., Pennazio, S., Masenga, V. and Tavella, L. 2002. Resistance to tospovirus in pepper. Thrips and Tospviruses: *Proc. 7th Int. Symp. Thysanoptera.* (R.Marullo and L.Mound, eds.), pp 105-110.

Sakurai, T., Inoue, T. and Murai, T. 2002. Intraspecific interaction in transmission of TSWV by *Frankliniella occidentalis* results from distinct virus accumulation. *Proc. 7th Int.* Symp. *Thysanoptera.* (R.Marullo and L.Mound, eds.) pp 51-57.

Sakimura, K.1963. The present status of thrips-borne viruses. In: *Biological Transmission of Disease Agents.* (K.Maramorosch, ed.), Academic Press, New York, pp 33-40.

Samuel, G., Bald, J.G., and Pittman, H.A. 1930. Investigations of "spotted wilt" of tomatoes. *Commonw. Austr. Counc.Sci. Ind. Res. Bull.* 44, 64pp.

Stevens, M.R., Scott, S.J. and Gergerich, R.C. 1992. Inheritance of a gene for resistance to tomato spotted wilt virus (TSWV) from *Lycopersicum peruvianum* Mill. *Euphytica* 59: 9-17.

Thompson, G.J. and van Zijl, J.J.B. 1996. Control of tomato spotted wilt virus in South Africa. *Acta Hortic.* 431: 379-384.

van de Wetering, F., Goldbach, R. and Peters, D. 1996. Tomato spotted wilt virus ingestion by first instar larvae of *Frankliniella occidentalis* is a prerequisite for transmission. *Phytopathology* 86: 900-905.

van de Wetering, F., van der Hoek, M., Goldbach, R.W., Mollema, C. and Peters, D. 1999. Variation in tospovirus transmission between populations of *Frankliniella occidentalis* (Thysanoptera: Thripidae). *Bull. Entomol. Res.* 89: 579-588.

Webb, S., Tsai, J. and Mitchell, F. 1998. Bionomics of *Frankliniella bispinosa* and its transmission of tomato spotted wilt virus. In: *Recent Progress in Tospovirus and Thrips Research.* (D.Peters and R.Goldbach, eds), pp 67.

Wijkamp, I. and Peters, D. 1993. Determination of the median latent period of two tospoviruses in *Frankliniella occidentalis*, using a novel leaf disk assay. *Phytopathology* 83: 986-991.

Wijkamp. I., van Lent, J., Kormelink, R., Goldbach, R. and Peters, D. 1993. Multiplication of tomato spotted wilt virus in insect vector, *Frankliniella occidentalis. J. Gen. Virol.* 74: 341-349.

Wijkamp, I., Almarza, N., Goldbach, R. and Peters, D. 1995. Distinct levels of specificity in thrips transmission of tospoviruses. *Phytopathology* 85: 1069-1074.

Wijkamp, I., van de Wetering, F., Goldbach, R and Peters. D. 1996. Transmission of tomato spotted wilt virus by *Frankliniella occidentalis*: Median acquisition and inoculation access period. *Ann. appl. Biol.* 129: 303-313.

Zawirska, I. 1976. Untersuchungen über zwei biologische Typen von *Thrips tabaci* Lind. (Thysanoptera, Thripidae) in der VR Polen. *Arch. Phytopathol. Pflanzensch.* 12: 411-422.

Legends to figures:

Fig. 28.1. Infection by tospoviruses. a) top necrosis induced in tomato; b) infection in sweet pepper; c) infection in tobacco; d) mosaic in early infected groundnut; e) early infection in lettuce (notice the necrosis; f) necrosis in potato tubers; g) local lesions on a petunia leaf disk exposed to viruliferous thrips (left) , feeding damage evoked by non-viruliferous thrips (middle), and an unexposed disk (right).

Fig. 28.2. Accumulation of TSWV particles after an infection with a non-defective isolate (left), and of nucleocapsid aggregates induced after infection with an defective isolate (right) as seen in thin section of infected *Nicotiana rustica* plants.

Fig. 28. 3. Life cycle of the thrips *Frankliniella occidentalis* and the periods in which they acquire a tospovirus to become viruliferous and after which they can transmit.

Fig. 28. 4. Electronmicroscopy of *Tomato spotted wilt virus,* its schematic morphology and outline of the genome expression. The virus particles show the lipid membrane with the projections consisting of the G1 and G2 proteins. The core of the schematic particle consists of the three viral nucleocapsids. White: nucleocapsid (N), yellow: RNA-dependent RNA (L) polymerase. The schematic particle is enveloped by two glycoproteins (red =G1, blue is G2. NSs and NSm are the nonstructural proteins. Double arrows indicate replication, where as single arrows indicate transcription and translation. Courtesy of Dr. M. Prins.

Fig. 28.1

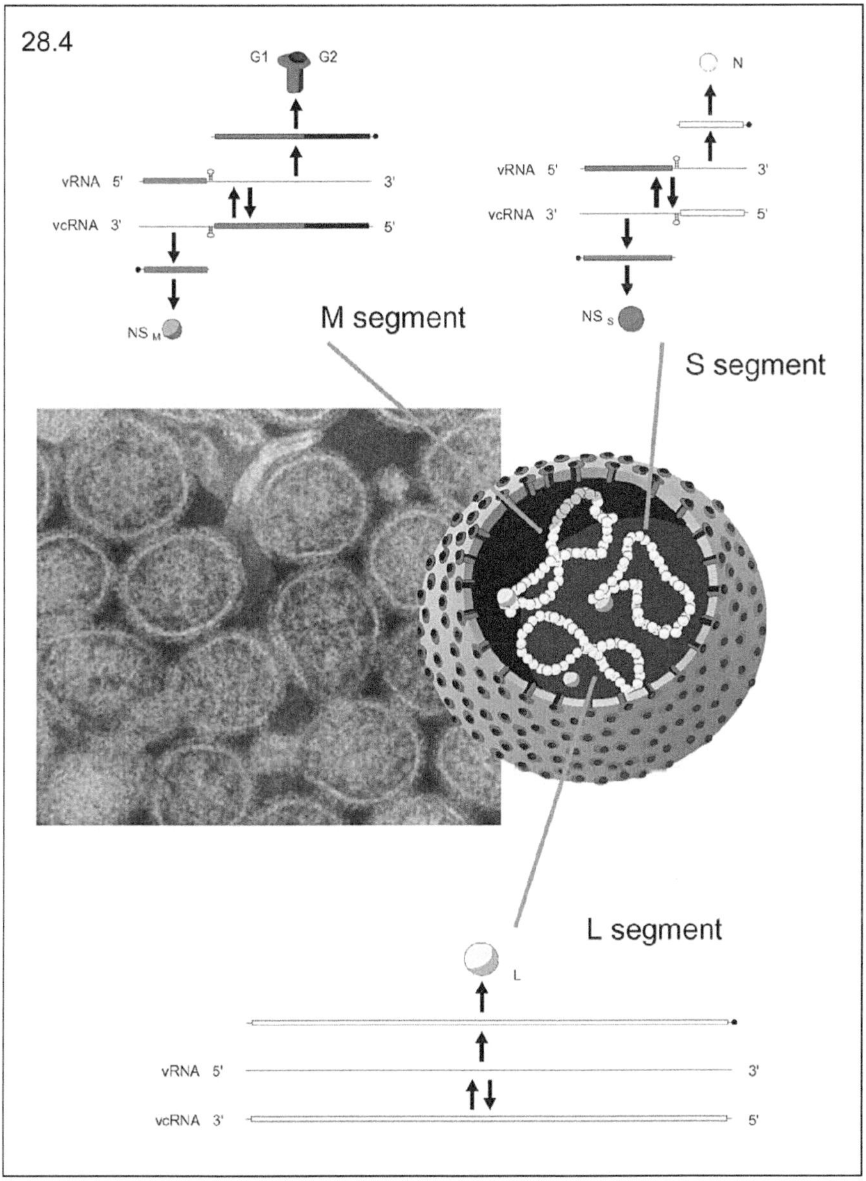
28.4
G1
G2
N
vRNA 5'
3'
vcRNA 3'
5'
NS M
M segment
NS S
S segment
L segment
L

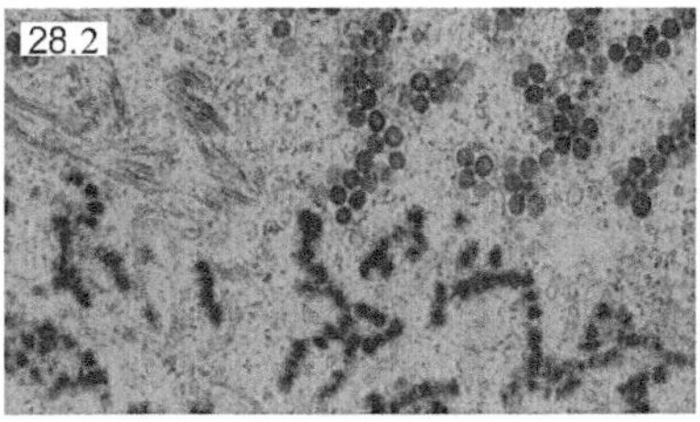
28.2

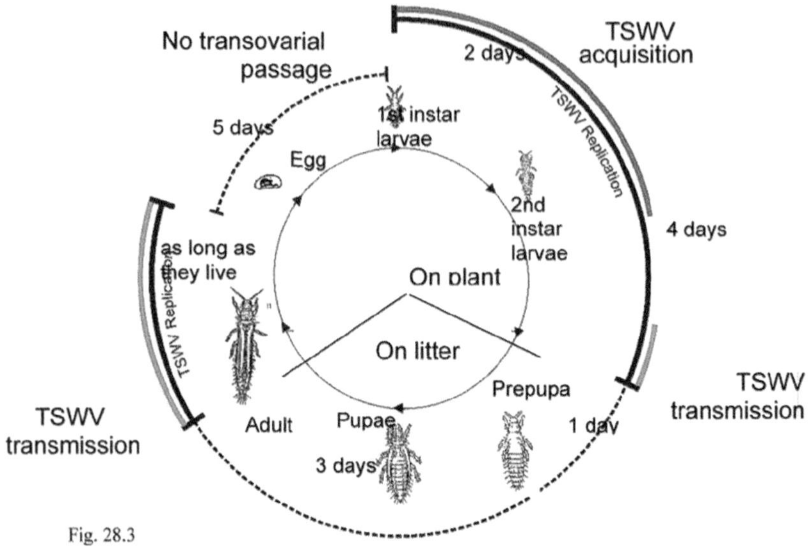
No transovarial passage
TSWV acquisition
2 days
TSWV Replication
5 days
1st instar larvae
Egg
2nd instar larvae
4 days
as long as they live
On plant
TSWV Replication
On litter
TSWV transmission
Prepupa
TSWV transmission
Adult
Pupae
1 day
3 days

Fig. 28.3

Table 28.1. Geographical distribution, host range and vector species of the recognized (italics) and species tentative or proposed (roman) tospovirus species

Tospovirus species	Geographical distribution	Hosts	Vector species
Groundnut bud necrosis virus (GBNV)	India, South-east Asia	Watermelon	*Thrips palmi*
Groundnut ringspot virus (GRSV)	South America, South Africa	Groundnut, tomato	*Frankliniella occidentalis, F. schultzei*
Impatiens necrotic spot virus (INSV)	USA, West and South Europe	Ornamentals, e.g. alstroemeria, begonia, impatiens, gerbera and other asteraceous species	*F. occidentalis*
Peanut yellow spot virus (PYSV)	India, Thailand	Groundnut	?
Tomato chlorotic spot virus (TCSV)	South America	Tomato, sweet pepper	*F. intonsa, F. occidentalis, F. schultzei*
Tomato spotted wilt virus (TSWV)	Worldwide, but is not frequently recorded in African countries, except South Africa	Monocots and dicots, i.e. chrysanthemum, cyclamen, dahlia, gerbera, groundnut, impatiens, lettuce, pea, sweet pepper	*F. bispinosa, F. fusca F. intonsa, F. occidentalis F. schultzei,*
		potato, tobacco, tomato	*T. setosus, T. tabaci*
Watermelon silver mottle virus (WSMoV)	Japan, Taiwan	Watermelon, other cucurbits, tomato	*T. palmi*
Zucchini lethal chlorotic virus (ZLCV)	Brazil	Zucchini (*Cucurbita pepo*)	*F. zucchini?*
Capsicum chlorosis virus (CaCV)	Australia	Chilli, sweet pepper, tomato	?
Chrysanthemum stem necrosis virus (CNSV)	Brazil	Chrysanthemum	*F. occidentalis, F. schultzei*
Iris yellow spot virus (IYSV)	Brazil, Israel, the Netherlands, USA	Iris, leek, onion	*T. tabaci*
Melon yellow spot virus (MYSV)	Taiwan	Melon	*T. palmi*
Peanut chlorotic fan-spot virus (PCFV)	Taiwan	Groundnut	?
Watermelon bud necrosis virus (WBNV)	India	Watermelon	*T. palmi*

Table 28.2. Diagnostically susceptible hosts species used in the identification of tospoviruses

	TSWV	INSV	GBNV	MYSV	GRSV	TCSV	WSMoV	WBNV	CSNV	CaCV	PCFV	PYSV	IYSV	ZLCV
	1089#	251	53	33	32	25	24	22	20	18	18	12	7	3
Plant species														
Arachis hypogaea	+	+	+	-	+	+		+		L	L	L		
Capsicum annuum	+	+	+	-	+	+	+	+	L	+				
Chenopodium quinoa	L	L	+	L	L	L	+	+		L	L	L		
Gomphrena globosa	L	L	+	L	L	L	+	+		L	L			
Cucumis sativus	L	L	+	+	L	L	+	+	L	L				+
Datura stramonium	+	+	+	L	+	+	+	+	L	+	L	L	+	
Emilia sonchifolia	+	+	+	+	+	+		+						
Impatiens spp	+	+			+	+			L					
Lycopersicum esculentum	+	+	+	-	+	+	+	+	+	+		-	-	+
Nicotiana benthamiana	+	+	+	+	+	+		+	L	+	+	L	+	
N. clevelandii	+	+	+	+	+	+	+		L					
N. glutinosa	+	+	+	L	+	+	+	+	L	L	L	L	-	
N. rustica	+	+	+	L	+	+		+	L	L	L		-	
Petunia hybrida	L	L	L	L	L	L	L	L	L	+	L	L	L	
Phaseolus vulgaris	+	L	+	L	L	L	+		L	L	+	L	-	
Pisum sativum	+	+	+	-	+	+						+	-	
Vigna unguiculata	+	L	+		+	+	+	+		L	+	+		

' Number of plant species reported to susceptible.
'+' = reacting locally and systemically; L = local reactions on the inoculated leaves; '-' = no visible reaction after inoculation

29. Cotton

V.G. Malathi, G. Radhakrishnan and A. Varma
Advanced Centre for Plant Virology, Indian Agricultural Research Institute, NewDelhi 110012, India

29.1 Introduction

Cultivation of cotton for the fibre perhaps marked the beginning of human civilization. The term cotton refers to four species of the genus *Gossypium, G. arboreum, G. herbaceum, G. hirsutum* and *G. barbadense*. The chromosome number of *Gossypium* sp. is n=13, and different genome groups from A-F have been identified (Stewart, 2001). Cotton is also commercially important as an oil crop as the cottonseed contains about 17% oil and 24% protein. The produce after oil extraction is concentrated into protein rich oil cake, which is used as livestock feed. China is the largest producer of cotton followed by USA, India, Pakistan and former Soviet Union. However, Israel ranks first in productivity (1706 kg/ha) followed by Australia, Syria, Turkey, Spain and China, all of which produce more than 1000 kg/ha. Low productivity in other parts of the world is mainly due to major biotic stresses, including viral diseases. Cotton cultivation is affected by several begomo-, ilar- and luteoviruses (Table 29.1). Most devastating of these are the begomoviruses causing diseases in major cotton growing areas of the world.

29.2 *Cotton leaf crumple virus* Genus *Begomovirus* (CLCrV)

Cotton leaf crumple disease (CLCrD) has been known to occur for a long time in southwestern United States of America. Earlier the disease was known as "Acromania" or "Crazy top" (Cook, 1924).

Table 29.1 Viruses affecting cotton

Virus genus	Virus species*	Vector	Disease	Geographical distribution
Begomovirus	*Cotton leaf crumple virus* (CLCrV)	*B. tabaci*	Leaf crumple	South-western USA, Mexico, Guatemala
	Cotton leaf curl Alabad virus (CLCuAV)	-do-	Leaf curl and vein enation	Pakistan
	Cotton leaf curl Gezira virus (CLCuGV)	-do-	-do-	Sudan
	Cotton leaf curl Kokhran virus (CLCuKV)	-do-	-do-	Pakistan
	Cotton leaf curl Multan virus (CLCuMV)	-do-	-do-	Pakistan
	Cotton leaf curl Rajasthan virus (CLCuRV)	-do-	-do-	India
	Cotton yellow mosaic virus	-do-	African cotton mosaic	Western-central Africa
	?	-do-	Yellow mosaic	Brazil
	?	-do-	Leaf curl	Philippines
	?	-do-	Infectious variegation	India
	?	-do-	Mosaic	Central America, USA
Ilarvirus	*Tobacco streak virus*	*Thrips palmi*	Leaf necrosis	India, Pakistan, Brazil
Luteovirus	?	*Aphis gossypii*	Reddening of leaf (Anthocyanosis)	Brazil, Thailand, India, Sudan, Philippines
	?	*Aphis gossypii*	Green or bluish colour of leaf	Brazil, Africa, Philippines
	?	-do-	Leaf roll	Thailand
	?	-do-	Bunchy top	Australia

(?) Not yet identified

The disease is now prevalent in Guatemala, Mexico, Arizona, California and Texas. It caused yield reduction of 23-55% in ratoon crop and 11-17% in seeded cotton (Russel, 1982).

CLCrV has a narrow host range and was experimentally transmitted through whitefly inoculation only to a few host species belonging to Malvaceae and Fabaceae (Brown and Nelson, 1986). Hosts of CLCrV include *G. arboreum, G. barbadense, G. herbaceum, G. hirsutum, G. thurberi, Hibiscus cameronii, H. diversifolius, H. palustris, H. sabdariffa, Malva parviflora, Abutilon theophrasti, Althaea officinalis, A. rosea, Castanospermum australe, Glycine max* and *Phaseolus vulgaris*. The characteristics symptoms induced by CLCrV are mosaic, downward or upwards curling of the leaves and veinal hypertrophy, giving rise to crumpled appearance. Minute veinal outgrowths or veinal hypertrophy were also seen on petals. The bolls are malformed and the plants look stunted.

CLCrV is transmitted by the whitefly, *Bemisia tabaci,* in a persistent, circulative manner. The virus is not transmitted through seed or sap. Minimum acquisition access period (AAP) required is one hour (24% transmission); with increasing AAP, transmission efficiency also increases. Whiteflies can transmit the virus for up to 9 days depending on the time given for AAP.

29.2.1 Purification and genome characteristics

CLCrV can be partially purified from infected bean plants according to Brown and Nelson, 1984. Single, paired, chains of geminivirus-like particles were observed by transmission electron microscopy of partially purified preparations. However, visualization of virus particles in infected cotton plants was difficult. The genome of CLCrV consists of DNA A and B components (Nadeem *et al.* 1997) and resembles bipartite New World begomoviruses like *Tomato golden mosaic virus* and *Bean golden mosaic virus*. DNA A component shows characteristic features of a *Begomovirus*.

CLCrV isolates collected from Arizona, California, Coborca,, Mexico and Texas, share > 98% sequence identity in the coat protein gene indicating limited variability between CLCrV isolates. CLCrV can be detected by PCR using primers specific to the middle or core region of coat protein gene or intergenic region.

29.2.2 Epidemiology and control

Spread of CLCrV in southwestern USA coincides with the build up of whitefly population during the warmer summer months. Besides ratooned cotton, weeds like *Malva parviflora* along the irrigation channel may harbour the virus. Some weeds belonging to Solanaceae and Asteraceae, help in the build up of whitefly population. Another important development is the establishment of 'B' biotype of *B. tabaci,* identified as *B. argentifolii,* in cotton-vegetable agro-system. *B. argentifolii* has broader host range and better transmission efficiency than *B. tabaci*, and it may result in an increase in CLCrV incidence. Elimination of cultivation of ratoon cotton crop helped in decreasing the disease incidence. Resistance genes have been identified in *G. barbadense* cv. Cedix from El Salvador and *G. arboreum* lines. Efforts are being made to combine resistance to the virus and its vector as seen in some okra-leaf-like cotton lines NIC 71-23, 71-1655 (Brown, 1992).

29.3 *Cotton leaf curl Alabad virus* (CLCuAV), ***Cotton leaf curl Gezira virus*** (CLCuGV), ***Cotton leaf curl Kokhran virus*** (CLCuKV), ***Cotton leaf curl Multan virus*** (CLCuMV) and ***Cotton leaf curl Rajasthan virus*** (CLCuRV), all Genus ***Begomovirus***

Cotton leaf curl disease (CLCuD) was first reported by Farquharson in 1912 in *G. peruvianum* and *G. vitifolia* in Nigeria. Later in 1923 the disease appeared in Sakel cotton in the Gezira area in Sudan (Jones and Mason, 1926), which developed into epidemic in 1927 and 1928. The disease also had spread in Tanzania, Egypt, Malavi and in all the cotton producing regions north of equator in Africa, except in Benin, Chad, Togo and Upper volta (Brown, 2001). In Pakistan, the disease was first observed in 1960s in a small area in Multan, but in 1987 it assumed epidemic proportion and spread to nearly one million hectares of (*G. hirsutum*) cotton (Hussain and Mahmood, 1988; Hussain *et al.* 1991). In India the disease was first observed in a few isolated plants of *G. barbadense* in 1989 in the experimental fields of Indian Agricultural Research Institute, New Delhi (Anonymous, 1990). It was also observed in some perennial cotton plants in kitchen gardens in Karnataka State, in South India (Nateshan *et al.* 1991). In 1994, the disease appeared in alarming proportion in the northwestern India (Varma *et al.* 1995). The disease affected all *G. hirsutum* varieties but *G. arboreum* varieties were free of infection. By 1997, an area of about 0.2 million ha was estimated to be affected by the disease in northwestern India (Narula *et al.* 1999). The Sudan was worst hit by cotton leaf curl disease in 1950's, and yield losses of up to 40% were reported (Tarr, 1957; 1964). The disease progress was very rapid in susceptible cotton varieties GS83 (3), Dominian Sakel and Bar 14/25, and they are no longer grown in areas prone to the disease. In Pakistan due to CLCuD the national average cotton yield dropped by nearly 30%, resulting in losses of US $5 billion between 1992-97 (Briddon and Markham, 2000). The losses in yield are mainly due to reduction in the number of bolls produced. The disease also affected the quality of the cotton fibre.

Characteristic symptoms of CLCuD (Figs. 29.1–29.5) are vein thickening and vein greening, which is best observed in transmitted light. Later the leaves curl upwards due to an increase in the growth of veinal tissues on the abaxical side of the leaves. The leaves of diseased plants become leathery, brittle and greener than healthy leaves. Similar symptoms are also observed on bracts. The leaves also develop vein enations, which frequently develop into cup shaped, leaf-like outgrowths. In more genotypes of *G. barbadense*, vein yellowing and blister mosaic symptoms are also seen. The severely affected plants have bushy appearance with dark green color, short internodes and are without flowers and bolls. Twisting of the petiole, peduncle, and the youngest parts of the stem is also common; in severe cases buds are either totally suppressed or die shortly after formation causing partial or complete sterility of the plant. In plants affected at early age, reduction of internodal length leads to stunting and reduced flowering/fruiting.

The host-range of the begomoviruses causing CLCuD seems to vary. CLCuGV infects large number of host species in Malvaceae and Tiliaceae, but it is not transmitted to *G. arboreum* and *G. herbaceum*. CLCuAV, ALCuMV and CLCuKV infect several plant species belonging to Fabaceae Malvaceae and Solanaceae, causing vein thickening, vein greening and moderate leaf curling (Harrison *et al.* 1997). The Karnataka isolate from South India infected 24 species in 6 plant families, which include Asteraceae, Euphorbiaceae, Fabaceae, Malvaceae, Oxalidaceae and Solanaceae (Nateshan *et al.* 1996). CLCuRV infects species belonging to Malvaceae, Solanaceae, Asteraceae and Fabaceae (Radhakrishnan *et al.* 2001).

29.3.1 Transmission

The begomoviruses causing CLCuD are tansmitted by the whitefly *B. tabaci,* in a persistent manner but it is not transmitted tranasovarially, by seed, by sap or by soil. Single insects are able to transmit the virus. In experiments, 3h 20min of AAP and 30min of inoculation access period (IAP) were found sufficient for transmission (Kirkpatrick, 1931). The transmission threshold was reported to be 6h 30min. Nateshan *et al.* (1996) showed that a minimum of 1 h AAP and minimum 5 min IAP were required, with a threshold of 8 h for Karnataka isolate. They also found that female *B. tabaci* transmitted this isolate more efficiently than the male whitefly.

29.3.2 Diagnosis and virion characteristics

Typical vein thickening is a diagnostic symptom. Dot blot hybridization and PCR can be effectively used for the diagnosis of CLCuD (Harrison *et al.* 1997; Radhakrishnan *et al.* 2001).

Polyclonal antibody to SqLCV and ICMV detected typical geminate particle in severely infected cotton plants from Rajasthan in ISEM studies (Varma *et al.* 1993). Panels of monoclonal antibody to ACMV/ICMV have been extensively used to study the particle and epitope profile of different cotton leaf curl isolates from Pakistan (Harrison, *et al.* 1997). The Begomoviruses causing CLCuD have two genomic components, DNA-A and a satellite-like DNA molecule approximately 1350 nt size, called as β DNA. DNA A and β DNA together induce typical disease symptoms (Briddon *et al.* 2001; Radhakrishnan *et al.* 2001). In addition to these two components a DNA molecule, 1376 nt in length and resembling the genome component of *Nanovirus* coding for replicase, were detected in leaf curl virus isolates from Pakistan (Mansoor *et al.* 1999). However, the role of the Nanovirus-like component in causing CLCuD is not known. Based on the DNA A complete nucleotide sequence, the leaf curl virus isolates infecting cotton were separated into five different species (Fauquet *et al.* 2003).

DNA A component of all the begomoviruses causing CLCuD show typical features of DNA A component of begomoviruses originating from the old world. All these viruses have intergenic region with nonanucleotide sequence and have six ORFs, two in viral and four in complementary sense. The ORF AV2 characteristic of the old world begomoviruses is present in all these viruses. The satellite or DNA β associated with CLCuD has nonanucleotide sequence TAATATTAC similar to DNA A component. There are four ORFs in viral sense and three in complementary sense coding for peptides of less than 10 kDa. There is only one ORF in the complementary sense, ORF βC1 that codes ~13 kDa protein. This protein is suggested to be involved in cell-to-cell movement and to function as replication enhancer protein (Briddon *et al.* 2001).

CLCuGV shows only 74% sequence identity with CLCuD associated begomoviruses from India and Pakistan (Idris and Brown, 2002). This suggests that the viruses causing CLCuD in the Indian Subcontinent have originated independently and have not been introduced (Briddon and Markham, 2000). These viruses might have emerged as recombinants between bhendi/okra yellow vein and other yet to be identified begomoviruses (Idris and Brown, 2002; Varma and Malathi, 2003).

29.3.3 Epidemiology and control

CLCuD has been endemic in Africa for a long time. Recently it has, emerged as serious constraint in cotton cultivation in the north-western cotton belt of the Indian Sub-continent. Several factors like the emergence of new variants of viruses, increase in vector population, weather, changing cropping systems and introduction of susceptible genotypes have singly or in combination contributed to the emergence of CLCuD in these areas (Varma and Malathi, 2003).

Whitefly populations are greatly influenced by weather conditions. The most important weather parameter is the wind, its strength and direction as seen in the case of CLCuGV in the Sudan. Whiteflies do not fly long distances but are passively wind driven resulting in spread of begomoviruses. The sudden epidemic of CLCuD in India was attributed to the dust storms in north-western region of the Indian Subcontinent (Varma and Malathi, 2003). Appearance of CLCuD in the years (1994, 1995) coincided with northeasterly dust storm, about 2 weeks earlier, which apparently blew in viruliferous whiteflies from neighbouring fields across the border with Pakistan. In both years the incidence was greater in fields adjoining citrus and mango orchards which provided windbreaks for the landing of wind blown whiteflies.

The begomoviruses associated with CLCuD have a wide host range including bean, tobacco, tomato, okra and hollyhock. Some of these plants when grown in the neighbouring fields could provide source of virus for the cotton crops. Indeed, in the Sudan, infected okra crops were a major source of virus infection in cotton. An array of weeds in and around cotton fields were also affected by begomoviruses, but experimental evidence of any weed acting as a source of virus infection of cotton is lacking. Date of sowing and the time of initial infection influence the epidemic development. Incidence of the disease was found to be more in early sown crop (Idris, 1990; Monga, 1998).

Various approaches like the avoidance of source(s) of infection, control of whitefly population and use of resistant cultivars have been used, for minimizing the losses caused by CLCuD. In the Sudan a moratorium on the cultivation of okra before sowing of cotton crop reduced CLCuD incidence. Management of the whitefly is an onerous task, but some chemicals like Ethion and Triazophos, and some biocontrol agents/botanicals like *Paecilomyces lilacinus, Verticillium chlamydosporium,* Nimbokil and neem oil (Singh *et al.* 1994; Khan *et al.* 2000) effectively reduced vector population and CLCuD incidence.

Varietal resistance plays a vital role in managing CLCuD epidemic. CLCuD was first noticed in New Delhi in 1989 in a few highly susceptible lines of *G. barbadense*. A strategy comprising cultivation of *G. arboreum* and resistant varieties of *G. hirsutum* in areas most prone to the disease and avoidance of cultivation of *G. barbadense* and its hybrids in north-west India helped in slowing the spread of CLCuD. Avoidance of *G. barbadense* and use of tolerant varieties have also minimized CLCuD incidence in the Sudan (Nour and Nour, 1964). Similarly cotton varieties that have increased tolerance like CIM 1100, CIM 448 CIM 443 helped in containing the disease in Pakistan (Briddon and Markham, 2001), but in some areas resistance breaking strains have emerged leading to up to 50% CLCuD incidence even in these varieties (Mansoor *et al.* 2003). In India, cotton vars RS875, LRA 5166 and LRK 516 were shown to be resistant (Narula *et al.* 1999). *G. arboreum* is free of virus infestation both in India and Pakistan, indicating presence of high degree of resistance. Attempts to generate transgenic resistance by experiencing coat protein gene and whole or fragments of DNA β of CLCuMV are in progress (Mansoor *et al.* 2001). It is proposed to achieve resistance through RNAi strategy. Orientation of the field avoiding the direction in which wind is blowing or growing wind barriers may also be useful in reducing CLCuD incidence.

29.4 Other Viruses

Some incompletely characterized viruses also cause diseases in cotton. **African cotton mosaic:** The disease is reported to occur throughout Africa and caused as severe epidemic in 1968-1971 in Chad. Affected plants develop small chlorotic pattern on leaves, bigger regular chlorosis giving rise to typical mosaic pattern, vein thickening and leaf deformation, severe mosaic and sterility. The causal virus is transmitted by grafting and by *B. tabaci*. Management of the disease is essentially by cultivation of resistant varieties like EH9 and Y1422.

Cotton anthocyanosis: The disease is also known as reddening disease due to reddening of leaves, which turn almost purple (Fig. 29.6). The disease was first observed in Brazil in 1954, (Costa, 1957) and was also reported from Thailand, Philippines, India and the Sudan. The virus, isolated from the affected plants has isometric particles of 25-30nm; considered to belong to the genus *Luteovirus* (Waterhouse *et al.* 1988). The virus is transmitted by *Aphis gossypii* in a persistent manner. It is transmissible to *G. hirsutum, G. barbadense, G. arboreum* and *Sida micrantha*. Cultivation of resistant lines, and application of insecticide to reduce vector populations, help in controlling the disease.

Cotton blue disease: Cotton blue disease was first reported in the Central African Republic in 1949; currently it also occurs in other countries (Cauquil *et al.* 1993). Yield losses of 15-20% are reported due to reduction in flowering. Characteristics symptoms are dark green to bluish coloured leaves, epinasty and stunting, occasionally accompanied by leaf shoot rolling and leathery texture. The causal virus is transmitted by *A. gossypii*; insect related toxin is also suspected to induce some of the symptoms. Triple hybrid cotton lines (Crosses between *G. hirsutum, G. arboreum* and *G. rainomdii*) have a high degree of resistance. Eradication of infected cotton plants from the field, control of weeds and high plant density help in its management.

29.5 Diseases caused by *Phytoplasma*

Some of the virus-like diseases of cotton are suspected to be caused by *Phytoplasma*. These include flavescence, flower virescence, phyllody, psyllosis and small leaf diseases (Costa, 1923; Cauquil *et al.* 1993). Flavescence disease occurs in theCentral African Republic; the affected plants develop leaf chlorosis and reddening, curved petioles and stunting. Scale insects *Margarodes* sp., that survive in the soil, are suspected to vector the causal agent. Flower virescence and phyllody are serious diseases of cotton in Ivory Coast, Mali, Nigeria and Upper Volta. The disease is transmitted by the leaf hopper *Orosius cellolosus.* Psyllosis disease occurs in Belgium Congo. Affected plants develop scale-like purple leaves forming rosette and remain sterile. Larvae of the psyllid *Paurocephala gossypii* transmitted the disease. Small leaf disease occurs in China, India and Pakistan. In the Indian Subcontinent it affects *G. herbaceum* but not *G. barbadense*.

Acknowledgement

The authors thank Ms. Pallavi Chakravarty and Mr. P.N. Sivalingam for their help in the preparation of this chapter, and The America Phytopathological Society for the picture of Cotton anthocyanosis disease.

References

Anonymous 1990. Cotton leaf curl virus is a whitefly transmitted geminivirus. *Ann. Rept,* IARI. 74-77

Briddon, R.W. and Markham, P.G. 2000. Cotton leaf curl virus disease. *Virus Res.* 71:151-159.

Briddon, R.W., Mansoor, S., Bedford, I.D., Pinner, M.S., Saunders, K., Stanley, J., Zafar, Y., Malik, K.A. and Markham, P.G. 2001. Identification of DNA components required for induction of cotton leaf curl disease. *Virology* 285: 234-243.

Brown, J.K. 1992. Virus diseases of cotton. In: Cotton Diseases (Ed. R.J. Hillocks). *Commonwealth Agricultural Bureaux International,* Oxon, UK, 415 pp.

Brown, J.K. 2001. Virus and Phytoplasma diseases. In: Compendium of cotton diseases Second edition, Eds. Kirkpatrick, T.L. and Rothrock, C.S. *American Phytopathol. Soci.*: 48-57

Brown, J.K. and Nelson, M.R. 1984. Geminate particles associated with cotton leaf crumple disease in Arizona. *Phytopathology,* 74: 987-990.

Brown, J.K., Nelson, M.R. and Lambe, R.C. 1986. Cotton leaf crumple virus transmitted from naturally infected bean from Mexico. *Plant Dis.* 70: 981

Cauquil, J. and Follin, J.D. 1993. Presumed virus and mycoplasma-like organism diseases in subsaharan Africa and the rest of the world. *Coton et Fibers Tropicales* 38: 293-317.

Cook, O.F. 1923. Malformation of cotton plants in Haiti: a new disease named smalling or stenosis, causing abnormal growth and sterility. *J. Heredity* 14: 323-325.

Cook, O.F. 1924. Acromania or 'crazy top', a growth disorder of cotton. *J. Agril. Res.* 28: 803-828

Costa, A. S. 1957. Anthocyanosis, a virus disease in cotton in Brazil. *Phytopathol. Z.* 28: 167-186

Farquharson, C.O. 1912. *Report of a mycologist. Ann. Rep. Agril. Department,* Nigeria.

Fauquet, C.M. and Stanley, J. 2003. Geminivirus classification and nomenclature: progress and problems. *Annl. appl. Biol.* 142: 165-189

Harrison, B.D., Liu, Y.L. Khalid, S., Hameed, S., Otim-Nape, G.W. and Robinson, D.J. 1997. Detection and relationships of cotton leaf curl virus and allied whitefly transmitted geminiviruses occurring in Pakistan. *Ann. Appl. Biol.* 130: 61-75.

Hussain, T. and Mahmood, T. 1988. A note on leaf curl disease of cotton. *Pak. Cottons* 32: 248-251.

Hussain, T., Tahir, M.and Mahmood, T., 1991. Cotton leaf curl virus. *Pak. J. Phytopathol.* 3: 57-61

Idris, A.M. and Brown J.K. 2002. Molecular analysis of cotton leaf curl virus-Sudan reveals an evolutionary history of recombination. *Virus Genes* 24: 249-56.

Idris, A.M. 1990. Cotton leaf curl virus disease in Sudan, *Med. Fac. Lanbow, Rijksunir, Gent,* 55: 1990-2029.

Jones, G.H. and Mason, I.G. 1926. Studies on two obscure diseases of cotton. *Ann. Bot.* 160: 759-72.

Khan, N.U., Abro, H.K., Gul, H., Kumbhar, M.B. and Abdur, R. 2000. Leaf curl virus in American upland cotton in NWFP. *Pak. J. Phytopathol.* 12: 79-86.

Kirkpatrick, T.W. 1930. Leaf curl in cotton. *Nature* 125: 672.

Kirkpatrick, T.W. 1931. Further studies on leaf-curl of cotton in the Sudan. *Bull. Entomol. Res.* 22: 323-363.

Mansoor, S., Amim, I., Iram, S., Hussain, M., Zafar, Y., Malik, K.A. and Briddon, R.W. 2003. The breakdown of rresistance in cotton to cotton leaf curl disease in Pakistan. *New Dis. Rep.* 7: 1-2

Mansoor, S., Hussain, M., Yasir-Ul-Musawaar, Khan, M.A., Iram, S., Zafar, Y., Briddon, R.W., Markham, P.G. and Malik, K.A. 2001. Strategies for broad-spectrum resistance against cotton leaf curl and tomato curl disease in Pakistan. *Assoc. Appld. Biologists, Int. Geminivirus Symposium.* 59

Mansoor, S., Khan, S.H., Bashir, A., Saeed, M., Zafar, Y., Malik, K.A. Briddon, R., Stanley, J. and Markham, P.G. 1999. Identification of a novel circular single-stranded DNA associated with cotton leaf curl disease in Pakistan. *Virology.* 259: 190-199.

Monga, D., Raj, S. and Verma, S.K. 1998. Preliminary investigations on reaction of cotton germplasm to leaf curl and losses caused by the disease. *Paper presented at National Symposium on Diseases of Oil seeds and Pulses* 17-19.

Nadeem, A., Weng, Z., Nelson, M.R. and Xiong, Z. 1997. Cotton leaf crumple virus and cotton leaf curl virus are two distantly related geminiviruses. *Molec. Pl. Pathol. On-line.* http://www.bspp.org.uk/mppol/1997/0612nadeem

Narula, A.M., Monga, D., Chauhan, M.S. and Sheo Raj. 1999. Cotton leaf curl virus disease in India: The challenge ahead. *J. Cotton Res. Dev.* 13: 129-138.

Nateshan, M.M., Muniappa, V., Siwanson, M.M. and Harrison, B.D. 1996. Host range, vector relations and serological relationships of cotton leaf curl virus from Southern India. *Ann. appl. Biol.* 128: 233-244.

Radhakrishnan, G., Malathi, V.G. and Varma, A. 2001. Novel features of cotton leaf curl disease in India. **In**: 3rd *Intern. Geminivirus Symp.* July 24-28, 2001, John Innes Centre, Norwich, Norfolk, UK, 53pp.

Russell, T.E. 1982. Effect of cotton leaf crumple on stub and planted cotton. *Arizona Pl. Pathol. / Cooperative Extension Service Newsletter* 1: 3-5

Singh, J.A.S., Sahi Mann, H.S. and Kapur, S.P. 1994. Studies on whitefly Bemisia tabaci (Genn) transmitted cotton leaf curl virus disease in Punjab. *J. Insect Sci.* 7: 194-198.

Stuart, J.M. (2001). In: Compendium of cotton diseases Second edition, Eds. Kirkpatrick, T.L. and Rothrock, C.S. *American Phytopathol. Soci.*: 1-3

Tarr, S.A.J. (1957). The host range of cotton leaf curl virus in Sudan. *Empire Cotton Growers Review*, 34: 258-262.

Tarr, S.A.J. (1964). Virus Diseases of Cotton. *Miscellaneous Publication* No.18. Commonwealth Mycological Institute, Kew, Surrey, England, 23pp.

Varma, A. 1995. Leaf curl disease of cotton in North West India. *Report of the ICAR committee.* 15p

Varma, A. and Malathi, V.G. 2003. Emerging geminivirus problems: A serious threat to crop production. *Annl. appl. Biol.* 142: 145-164

Varma, A., Malathi, V.G., Handa, A., Ailon, M., Harrison, B.D., Verma, J.P., Singh, R.P., Singh, M., Srivastava, M. and Singh, J. 1993. Occurrence of leaf curl of cotton and okra in Northern India. **In**: 6th *Intern. Cong. Plant Pathology,* July 28-August 6, Montreal, Canada, 17.5.14.

Waterhouse, P.M., Galt, F.E. and Johnston, G.R. 1988. Descriptions of plant viruses: Luteovirus group no. 339. *Assoc. Appl. Biologists*, Kew England

Fig. 29.1 Field symptoms of cotton leaf curl disease (CLCuD).
Fig. 29.2 Diagnostic vein thickening symptom of CLCuD. Fig. 29.3 Severe upward leaf curling due to CLCuD. Fig. 29.4 Yellow vein symptom due to CLCuD.
Fig. 29.5 Blister mosaic (CLCuD). Fig. 29.6 Cotton anthocyanosis.

30. Sunflower

S. Winter[1] and K. Ravi[2]
[1] *DSMZ, Plant Virus Division, c/o BBA, Braunschweig, Germany, and* [2] *Division of Molecular Virology, Mahyco Life Sciences Research Center, Jalna, India*

30.1 Introduction

Sunflower, (*Helianthus annuus*) is an important oilseed crop grown in many countries predominantly for oilseed and for direct consumption as confection seed. The crop is grown in tropical and moderate climates, from India to France, from China to Ukraine and Tanzania.

More than 18 million ha worldwide are currently under sunflower cultivation. The Russian Federation (3.8 Mio. ha), Ukraine (2.4 Mio. ha), Argentina (1.9 Mio. ha), India (1.2 Mio. ha) and China (1.1 Mio. ha) have most acreage under sunflower production (FAOSTAT 2001). In 2001, 21 million metric tons of sunflower seeds were harvested and processed into high quality oils or, used for direct consumption. Sunflower production (yield/ha) can vary significantly. While 1.4 metric tons/ha is the world's average harvest, sunflower seed harvests range from 0.3-0.5 metric tons/ha on marginal lands and 2.5 metric tons/ha in high input agriculture. This demonstrates the potential of existing sunflower cultivars and shows that only a fraction of this potential is utilized in many areas of sunflower cultivation.

Diseases present a major constraint to sunflower production and can lead to significant reductions in yield and seed quality. More than 30 fungal diseases of sunflower are known, but only a few are common pathogens causing considerable damage. Downy mildew, rust, *Sclerotinia* stalk and head rot, *Verticillium* wilt, *Phoma* black stem, and *Alternaria* leaf and stem spot can cause up to 12% yield loss. In contrast, virus diseases of cultivated sunflower (*Helianthus annuus* L.) are not very common and until recently, none of the few virus diseases described for this crop has reached epidemic levels. Natural occurrences of virus diseases in sunflower are summarized in Table 30.1.

Helianthus annuus is an experimental host for many virus species from diverse genera and particularly susceptible to potyviruses. For some diseases of sunflower, a definite virus etiology has been proven, while in many earlier reports on virus-like diseases of sunflower, proof for the causal virus (es) is rather circumstantial and the virus identity remains elusive. This report focuses on the definite viruses occurring in sunflower and the diseases of sunflower with putative virus etiology emphasizing on diseases with economic significance.

30.2 The main viruses

Sunflower chlorotic mottle virus Genus ***Potyvirus*** (SuCMoV)

This virus disease was first found in Argentina during surveys for sunflower diseases in the province Entre Ríos in 1992 and 1993 (Lenardon, 1994). In subsequent inspections, plants with similar symptoms were also found in neighboring provinces of the country both in commercially cultivated crops and in wild sunflower (*Helianthus petiolaris*). Diseased plants showed remarkable symptoms consisting of chlorotic mottling, mosaic and ring spots.

The virus is readily sap transmitted to sunflower and typical chlorotic mottle symptoms appear 8-10 days after inoculation. In inoculated sunflower plants, mottling symptoms coalesce to yellow blotches followed by distortion and stunting of the plant. Severe symptoms with bright yellow mottling on leaves and stunted growth are a result of infections at early stages of plant development.

The virus is mainly infectious to plant species of the genera *Asteracea* causing chlorotic mottle symptoms in *Zinnia Bidens*, *Tagetes* and *Chrysanthemum* spp. Only few of the commonly used virus indicator plants are susceptible to the virus. Inoculation of *Nicotiana occidentalis, Datura metel and Gomphrena globosa* result in systemic infections, while infections of *Chenopodium quinoa* and *C. amaranticolor* show only local lesions in inoculated leaves.

Flexuous rod shaped particles of *ca.*770 nm in length present in leaf sap preparations and typical cytoplasmic laminar aggregates and pinwheel inclusions in infected leaf tissue, classified this virus as a typical potyvirus (Dujovny *et al.* 1998). The sunflower virus is non-persistently transmitted by the green peach aphid *Myzus persicae*, but is not seed transmissible. An antiserum prepared against purified virus preparations showed strong homologous reactions with all geographical isolates of this virus. There is no heterologous reaction with *potato virus Y* or *maize dwarf mosaic virus*, although a weak serological relationship to *Bidens mottle virus*, BiMoV, and to *Bidens mosaic virus*, BiMV, was demonstrated in dot blot immunoassays (Dujovny *et al.* 1998).

Detection of SuCMoV in sunflower is possible by ELISA using the homologous antiserum or, by utilizing the PTY-Mab (Jordan and Hammond, 1991) or, MAb P-3-3H8 (Richter *et al.* 1995) commercially available for group specific detection of potyviruses in an indirect ELISA format (www.AGDIA.com; www.dsmz.de/plvirus).

To clarify the taxonomic position of the sunflower potyvirus, the 3'-terminal nucleotide sequence of the viral RNA was determined (Dujovny *et al.* 2000). This region, comprising the coat protein cistron (CP) and the 3'-non translated region (NTR) is of taxonomic relevance and provides a reliable and practical approach to potyvirus identification and classification (Frenkel *et al.* 1989; Ward *et al.* 1992). The highest sequence relationship of SuCMoV was found with PVY-N strains NTN and NsNr. Based on the established potyvirus phylogeny, (Shukla *et al.* 1994), SuCMoV from sunflower in Argentina can be considered as a distinct virus species, despite striking sequence homologies with PVY sub-group viruses.

The relationship of SuCMoV with the potyviruses from *Bidens*, BiMoV and BiMV (Nagata *et al.* 1995), is also not established. Both potyviruses have a broad host range and can infect sunflower. BiMV was found in commercially grown sunflower in Brazil (Costa and Kitajima, 1966). BiMoV infections result in faint to bright yellow mottle symptoms (Purcifull *et al.* 1971) while BiMV causes severe systemic mosaic symptoms, stunting and leaf distortion respectively (Christie *et al.* 1968). *Bidens* spp. are common wild plants and might serve as reservoirs for SuCMoV or, for BiMoV and BiMV, aiding infection of sunflower.

SuCMoV is so far described only from Argentina, where it occurs in different sunflower growing regions. Despite the distinct symptoms caused by this virus, substantial yield losses are not reported. Nevertheless, infections at early stages in plant development lead to stunted growth and disturbances in flowering and seed setting. The narrow host range reported for SuCMoV and the lack of seed transmissibility makes this virus a relatively limited risk in becoming a serious sunflower pathogen.

Sunflower mosaic virus Genus ***Potyvirus*** (SuMV)

This disease was originally observed on wild sunflower in Texas, USA, and the causal pathogen described as a potyvirus (Arnott and Smith, 1967). The re-occurrence of this disease in 1997 provided the incentive for a further characterization of this virus (Gulya *et al.* 1998).

Sunflower mosaic virus, SuMV, is found on wild sunflower, *H. annuus*, growing in abandoned fields and along roadside ditches in Texas. It does not occur in the major areas of commercial sunflower production, in North and South Dakota of the United States. In 1999, the virus was detected in cultivated sunflower in certain locations in Texas only. SuMV is not reported from other sunflower growing areas, although mosaic symptoms of sunflower are quite common and were described for other viruses found in this crop, *i.e.* CMV (Venugopal *et al.* 1987; Orellana and Quacquarelli, 1968).

SuMV causes mild to severe systemic mosaic symptoms in infected sunflower plants. Aside from the diagnostic hosts *Chenopodium quinoa* and *C. ambrosoides* (LL hosts), the host range of this virus is confined to *Asteracea*, with species of the genera *Helianthus, Zinnia* and *Sanvitalia* which develop systemic mosaic symptoms upon virus infection. There appears to be no resistance to SuMV in sunflower since in resistance trials, all sunflower cultivars tested became systemically infected upon mechanical inoculation with this virus (Gulya *et al.* 2002).

SuMV has at least two aphid vector species, *Myzus persicae* and *Capitphorus elaegeni*, transmitting the virus quite efficiently, although at different frequencies. Seed transmission also is demonstrated for SuMV with transmission rates of up to 12% for some sunflower lines and missing seed transmission for others. Hence, this feature seems to be cultivar specific (Gulya *et al.* 2002).

An antiserum prepared against purified SuMV was used to detect virus in systemically infected sunflower plants, although with very low ELISA readings indicating low virus concentrations in infected plants. Hence, ELISA as the sole means for detection of SuMV in sunflower is not recommended.

SuMV is a typical potyvirus with flexuous rod shaped particle morphology *c.* 720nm in length). Systemic virus infections result in typical cytoplasmic inclusion bodies and laminated aggregates in cytoplasm of infected sunflower cells. The size of the ssRNA genome is approximately 9800 nucleotides and the size of the coat protein is estimated between 34 kDa and 35 kDa. A phylogenetic analysis of the coat protein amino acid sequence (Gulya *et al.* 2002) revealed the closest relationship with *Tobacco etch virus* (74,5% amino acid identity) and, with *Columbian datura virus* and *Endive necrotic mosaic virus*. Based on the criteria for potyvirus discrimination (Shukla *et al.* 1994) SuMV is a distinct species in the genus *potyvirus*.

The potyvirus SuMV causing a mosaic disease of sunflower is so far described only from a geographically limited area in southern Texas, USA. From the limited samples where this virus was found in the field, estimates on yield losses cannot be concluded. High yield reductions were however observed in controlled inoculation experiments with sunflower hybrids (75%- 40% yield loss) and inbred lines (100%-50% yield loss) indicating a potentially severe impact of this virus to sunflower seed production (Gulya *et al.* 2002). However, susceptibility of sunflower to this virus decreases with plant age and successful virus inoculations are limited to the seedling stages only, thus reducing the potential threat to sunflower cultivation.

30.3 Viruses occasionally found in sunflower

Helianthus annuus is an experimental host for many viruses and is susceptible to more than 30 definite virus species. Some of these viruses are also found in naturally infected sunflower. However, their occurrence in natural sunflower infections rather is a reflection of a general virus pressure in a respective environment and with sunflower serving as an alternative or additional host. In general, infections do not reach sizable proportions, especially when other parameters, e.g. vector abundance, are not supporting the spread of virus in the field. Still, the viruses are listed below, since some of them might have a great potential of becoming serious threats to sunflower production, when favorable conditions prevail.

An aphid-transmitted yellowing disease of sunflower occurring in the breeding plots of the Plant Breeding Institute, Cambridge, UK, was attributed to an infection with *Beet western yellows virus,* BWYV (Russell *et al.* 1975). The virus caused leaf thickening and interveinal chlorosis developing into irregular necrotic patches. Transmission experiments with *Myzus persicae* proved that the virus is aphid-transmitted in a persistent manner. Upon aphid inoculations, indicator hosts for BWYV developed symptoms typical of infections with BWYV. The *Polerovirus*, (family*Luteoviridae),* BWYV is endemic in England and has a wide host range including many weed species. This incidence of BWYV infections in sunflower in England exemplifies the inherent risks of new virus disease when crop cultivation expands into areas where vectors and prospective viruses are abundant.

The ubiquitous *Cucumber mosaic virus* (CMV) was described from sunflower in China (Wang *et al.* 1983; Liu and Li, 1988), Japan (Nishimura *et al* 1995) and India (Nagaraju *et al.* 1997). It was also found in research plots in Beltsville, USA, where it caused a mosaic disease in sunflower (Orellana and Quacquarelli, 1968).

Other viruses reported from sunflower are *Tobacco ringspot virus*, found in wild sunflower in Texas, USA (McLean, 1962), *Potato virus Y* , reported from the Czech Republic (Chod *et al.* 1996) and the tospovirus *Tomato spotted wilt virus*, occurring in sunflower fields in the Ukraine (Zakusilo *et al.* 1994). J. Dale described a virus with isometric 28nm particles, with an ssRNA genome of 4 genome segments from *Helianthus annuus* in Australia. The virus resembled members of the *Bromoviridae* and was named sunflower ringspot (?) ilarvirus (Brunt *et al.* 1996). The ilarvirus *Tobacco streak virus*, TSV, was found in 1980 in the garden around the Laboratory of Virology, University of Wageningen, Holland. This virus is a serious pathogen of sunflower causing severe leaf curl symptoms and chlorosis in leaves followed by necrosis and leaf scorch in severe infections (Dijkstra, 1983). Since this first and isolated report of TSV naturally infecting *Helianthus annuus* ornamental plants in Holland, the disease has now reached epidemic proportions in sunflower in India.

30.4 Virus diseases of sunflower in India

Several reports originating from Indian researchers describe virus or virus-like diseases from sunflower. Battu and Pathak (1965) described a mosaic disease of sunflower and found a sap transmissible agent. Gupta and Gupta (1977) isolated 3 viruses from sunflower with mosaic, yellow rings and speckling symptoms. Venugopal *et al.* (1987) and later Nagaraju *et al.* (1997), investigated a mosaic disease of sunflower. However, none of the reports lead to a confirmatory description of the virus associated with the disease, although CMV was mostly discussed as a candidate virus responsible for the diseases.

Sunflower necrosis disease (SND)

A serious disease of sunflower with putative virus etiology was found first in Karnataka in 1997 and from then on in all major sunflower growing regions, Andhra Pradesh, Maharashtra, Karnataka and Tamil Nadu in India (Anon, 1998; 1999). Losses to sunflower during the year 2000 in South India alone were estimated to exceed US $ 120 million. Symptoms of the disease vary with genotype and range from a general chlorosis to mosaic, chlorotic mottling and chlorotic to necrotic blotches on leaves. Stunted growth and stem bending were seen in severely affected plants, general necrosis, spreading to leaf midribs, petioles, stem, flower bracts and head eventually results in dieback (Figures 30.1 and 30.2).

In the past 4 years, the disease has assumed epidemic proportion with infection rates up to 100 % , resulting in a largely suppressed seed formation, which often means the complete loss of the harvest. The incidence of SND recorded in the years 1999-2002 varied in the regions of cultivation. The three southern states Maharashtra, Karnataka and Andhra Pradesh are experiencing very high incidence of sunflower necrosis disease. In sunflower growing regions in and around Jalna and Latur from Maharashtra State a 40-100 % disease incidence was recorded from 1999 onwards. In 2002 disease incidence reached up to 100 % in major sunflower growing regions of Karnataka State. Kurnool and Ananthpur districts of Andhra Pradesh recorded SND incidence up to 50 % during 2000 to 2002 (Ravi unpublished).

Initial studies ascribe the etiology of SND as a putative tospovirus, related to groundnut bud necrosis and watermelon silver mottle virus to SND (Jain *et al.* 2000 and Venkata Subbaiah *et al.* 2000). In contrast to these findings, Ramiah *et al.* (2001) found isometric virus-like particles consistently in necrosis disease affected sunflower leaves and tentatively assigned the virus in sunflower to the genus *Ilarvirus*. The causal agent of SND was then identified as a strain of *Tobacco streak virus* (TSV) a virus previously not reported from sunflower in India (Ravi *et al.* 2001) and finally designated it as TSV-SF (Bhat *et al.* 2002).

TSV-SF shows all characteristic features of a typical subgroup 1 *Ilarvirus,* with serological relationships only found with other TSV isolates. The virus was purified from systemically infected *Chenopodium foliosum* plants (Ravi *et al.* 2001) and an antiserum was subsequently prepared against purified particle preparations. This antiserum (DSMZ-AS0615) proved useful for virus detection in DAS-ELISA, and serological indexing for presence of TSV is recommended for routine diagnosis and virus screening.

TSV, like all *Bromoviridae*, has a tripartite genome, RNA 1, 2, 3, acting as mRNAs. RNA 3 carries the ORF, which codes for the viral coat protein. Sequence comparison of RNA 3 and the coat protein gene provided evidence for the taxonomic position of TSV-SF (Ravi *et al.* 2001; Bhat *et al.* 2002). Approximately 90% identical amino acids in the coat protein genes of TSV-SF and TSV and only a distant sequence relationship with *Hydrangea mosaic virus* (60% amino acid identity) proved the strain affiliation of this virus. From preliminary molecular studies with geographic isolates of TSV-SF, it can be concluded that the sequence variability, at least within the RNA 3 fragment analyzed, is limited to about 3%.

TSV-SF is readily sap transmissible and has a wide host range comprising plants from many families (Brunt *et al.* 1996) including many significant crop species. Table 30.2 lists plant species of agricultural and horticultural importance in India of which many species are susceptible to TSV-SF showing severe systemic symptoms, often resulting in total necrosis and dieback of the whole plant. The peanut stem necrosis disease, reported from Andhra Pradesh in 2000, is also caused by TSV (Reddy *et al.* 2002; see Chapter 16).

Many peanut fields affected by this disease were adjacent to sunflower cultivation areas with SND incidence. Spread of TSV in the field is very rapid, indicating a very efficient means of virus dissemination. TSV is seed borne in some plant species; however, this has not been verified for TSV in sunflower. Transmission through infected pollen seems to be the likely means of virus spread in the field. Virus from pollen of infected plants is efficiently mechanically transmitted through wounding resulting from feeding activity of the *Tysanoptera* vector *Frankiniella schultzei*. A sole vector transmission by *Frankiniella schultzei* was not successful.

Considering the epidemiological parameters for TSV in sunflower and in other major crops in India, it is evident that thisdisease has reached its magnitude most likely because virus transmission patterns have changed. From the number of wild and cultivated plants in India infected with TSV and the widely occurring weed species (i.e., *Parthenium, Xanthium, Tridax*) serving as efficient carriers and source plants of the virus, it is clear that TSV is not a recent introduction to the Indian subcontinent. Because the virus problem and its identification is recent, little is known about the role of *Tysanoptera* vector species in the transmission of TSV both within the field and among different crops.

Still, more descriptive research is needed, to define the relationships of TSV isolates present in different crops and at different locations. This will elucidate the existence of pathotypes of TSV in diverse crops, as was reported by Kaiser *et al.* (1991). TSV found in sunflower, peanut, cotton and okra might then segregate into pathogens with differential capacities to infect a specific host.

Information regarding resistance against TSV in sunflower is very limited and more research is needed to select breeding lines with tolerance/resistance to the virus. Most sunflower hybrids/varieties are susceptible to the disease. Sunflower necrosis disease caused by *Tobacco streak virus* is currently a significant threat to sunflower production.

Sunflower leaf curl disease

Sunflower plants with distinct leaf curl symptoms have been observed in and around Jalna, Maharashtra state, India, during January 2000. The number of plants with pronounced symptoms of the disease incidence was high and the causal agent of the disease was characterized at Mahyco Life Sciences Research Center, Jalna, India,

The most characteristic symptom of the disease is an upward leaf curling of the leaves with prominent vein thickening and formation of enations. Severely infected plants show severe leaf deformation and stunting.

The sunflower leaf curl disease is not mechanically transmissible, but the whitefly, *Bemisia tabaci*, is readily transmitting the disease. Characteristic symptoms of leaf curl, vein thickening and enations were reproduced in sunflower upon whitefly inoculations. *Bemisia tabaci* insects acquired the virus after a minimum acquisition-feeding period of 12 hrs followed by an inoculation feeding period of 12 hours. Typical symptoms of virus infections became visible 17 to 26 days after inoculation. In addition to sunflower, the virus infected tomato *Lycopersicon esculentum* (cultivars Pusa Rubi and Arka Sourab), *Nicotiana clevelandii*, *N. silvestris*, *N. tabacum* var. Xanthi 'nc', *N. tabacum* var. Samsun and in the weed host *Parthenium hysteroporus*. The inoculated plants produced leaf curl and vein thickening symptoms in 11 to 28 days after inoculation.

Symptoms in sunflower, vein thickening and enations indicated a geminivirus etiology. To verify the presence of a geminivirus in leaf curl affected plants and to identify the causative agent of the disease, a PCR approach utilizing oligonucleotide primers deduced from conserved sequences of geminiviruses flanking the entire intergenic region was followed. A *c.* 700 bp PCR fragment consistently amplified from leaf curl infected plants verified the presence of a geminivirus in symptomatic plants. The subsequent sequence analysis of this PCR fragment revealed nucleotide identities of *c.* 89 % with *Tomato leaf curl virus* from India (Ravi unpublished).

Symptoms in sunflower, *Bemisia tabaci* transmission studies and nucleotide sequence analyses provided evidence that the sunflower leaf curl disease in India is caused by a begomovirus related to *Tomato leaf curl virus*.

30.5 Virus diseases of sunflower in Africa

In Africa, the major sunflower cultivation is in north and east Africa and in the Republic of South Africa. In 1970, Martens *et al.* reported virus diseases from sunflower in Kenya, apparently associated with similar viruses but causing different disease phenotypes. Separate names were assigned to the putatively distinct viral entities, sunflower crinkle virus, SuCV and sunflower yellow blotch virus, (SuYBV). In the 1976/1977 growing season, 45-55% of all sunflower plants in Kenya were affected by the sunflower rugose mosaic disease (used as a synonym for SuCV) (Singh, 1979), The affected plants appeared stunted and leaves were markedly reduced in size. Leaf crinkling, severe mottle and leave puckering were the most pronounced symptoms. Severely affected plants yielded few or no seeds at all. Similar disease symptoms were found in sunflower from Tanzania (Fivawo, 1987) showing rugose mosaic and yellow ringspot symptoms on leaves. In 1998, sunflower plants with typical symptoms of bright yellow leaf blotches, concentric rings, line pattern and vein banding (Figure 30.3) were observed in fields in Zambia (Ndunguru and Mwale, 1999). In a survey covering 5 provinces, 80-100% of the plants showed symptoms characteristic of the disease. Improved sunflower hybrids were more severely affected by SuYBD than local varieties. In June 2002 (http://www.scienceinafrica.co.za) a survey in the region revealed a 35% incidence of this disease in sunflower grown in Tanzania and 60% of the plants were infected in sunflower cultivation areas of Zambia. Leaf distortion, irregular vein banding, chlorotic blotch symptoms and stunted growth were characteristic for severe infections. The flower heads were reduced in size and often seed setting was completely aborted. Consequently, the number of flowering heads was reduced sometimes up to 70%.

To date, SuYBD is the most serious disease of sunflower in Africa. Despite the epidemic history of this disease, since more than 30 years, information on the biology and ecology of the causative virus (es) is inadequate. Although it can be speculated that the types of the disease, yellow blotch and leaf crinkle, only reflect the plasticity of symptoms caused by one pathogen, it has still not been established whether a single virus or a virus complex exists in the different locations in East Africa. Consequently, more research is needed to fully characterize the viruses contributing to sunflower virus disease(s) in East Africa.

Evidence for a virus identity of SuCV (syn. Sunflower rugose mosaic virus) and SuYBV (syn. Sunflower yellow ringspot virus) was provided from experiments done by Singh (1979) and Theuri *et al.* (1987). Both diseases were mechanically transmissible to many different plant species from the families *Compositae* (*Asteraceae*), *Solanacea* and *Leguminosae* (soybean and peanut). In sunflower, inoculations with SuCV first resulted in blotch symptoms and later turned into the leaf crinkle (disease), indicating distinct pathogens. Despite earlier reports (Singh, 1979), seed transmission was not proven, but *Aphis gossypii* readily transmitted both diseases. Partial purified virus preparations from infected sunflower, revealed virus-like particles in electron microscopy resembling luteoviruses. In immuno-electron microscopy, BWYV antiserum decorated these virus-like particles, suggesting a luteovirus etiology of the sunflower disease(s) in East Africa. However, the mechanical transmission of the disease (in contrast to BWYV) might indicate presence of a second virus, providing for the mechanical transmissibility while the putative luteovirus provided helper virus functions for aphid transmissibility. Different symptoms obtained in sunflower might result from either unstable transmissions of both partners or presence of only one virus in a particular area.

This resembles the disease situation with groundnut rosette disease of peanuts in Africa, caused by *Groundnut rosette virus* Genus *Umbravirus* (GRV), which depends on its helper luteovirus, *Groundnut rosette assistor virus* (GRAV) for encapsidation and insect transmission. The phenotype of the disease is further influenced by an associated satellite RNA, '*Groundnut rosette virus* satellite RNA'. It can be speculated that the sunflower disease of East Africa is similar, with a luteovirus providing helper functions and a hitherto unknown umbravirus as the disease causing agent. A satellite RNA might furthermore be present and responsible for differences in disease severity and phenotypes.

Several aspects in this disease need attention. It is evident that the virus (es) are endemic in all environments where sunflower is grown in East Africa: Wild plants, i.e. *Tridax procumbens* and *Osteospermum vaillanti* serve as good source plants for both virus and vector. Other hosts, including peanut, can be infected with viruliferous aphids and since groundnut rosette disease and sunflower yellow blotch disease occur in neighboring locations, it is crucial to characterize the viruses in order to establish similarities and relationships between the two diseases. This will undoubtedly improve the diagnosis of diseases as well as efforts in identifying resistance selection and screening programs.

30.6 Concluding remarks

Helianthus annuus is an experimental host for many viruses from different families. However, in nature virus infections of sunflower do not reach epidemic significance. Viruses in sunflower that can be considered endemic are SuCMoV from Argentina, SuMV from Texas and the viruses contributing to the sunflower crinkle and yellow blotch diseases existing in East Africa. The potyvirus found in wild sunflower in Texas is well adapted and present in wild sunflower populations. Although highly infectious to cultivated sunflower at early stages in plant development, older plants become 'resistant' to infection and it is unlikely that these viruses will become serious problems for sunflower production. The virus from sunflower in Argentina, known since 1949 (Traversi, 1949) has only a limited geographic distribution and probably circulates between its wild reservoir hosts and sunflower, reaching epidemiological significance periodically with changing vector populations.

The putative umbravirus(es) causing the serious sunflower disease in Eastern Africa is endemic with wild plants acting as source for virus and insect vectors. Virus infections take place whenever and wherever sunflower is grown, with more uniform infections in plantations of genetically homogenous hybrid cultivars. This virus disease presents a serious threat to sunflower production in Africa and hence introducing sunflower varieties with some degree of resistance is strongly recommended. This however, requires a solid knowledge on the causative pathogen(s) and the forces determining its epidemiology.

The disease caused by *Tobacco streak virus* in India is an example for outbreaks reaching epidemic dimensions when conditions favor high vector populations. This virus together with tospoviruses are highly infectious to diverse crops (sunflower, peanut), and cause sporadic outbreaks of the diseases.

TSV is not a specific sunflower virus and with its wide host range presents a threat also to other crops. Therefore, control measures, including crop management, crop rotation, date of planting, vector, weed control and breeding resistant or tolerant varieties of sunflower have to be developed.

Newly emerging virus disease in sunflower, like the recently identified leaf curl disease, caused by a whitefly-transmitted geminivirus, need to be monitored. Geminivirus diseases are threatening the cultivation of many agricultural and horticultural crops - especially cotton and okra - in India.. As with other crops, virus resistance in sunflower, reached through classical plant breeding or genetic engineering, is the most promising and durable approach.

References

Anon 1998. Annual Report (1998-1999). Directorate of Oilseeds Research, Hyderabad, India.

Anon 1999. Annual Report (1998-1999). Directorate of Oilseeds Research, Hyderabad, India.

Arnott, H.J. and Smith, K.M., 1967. Electron microscopy of virus infected sunflower leaves. *J. Ultrastruct. Res.* 25: 3389-3402.

Battu, A.N. and Phatak, H.C. 1965. Observations on a mosaic disease of sunflower. *Indian Phytopathology*. 18: 317.

Bhat, A.I., Jain, R.K., Kumar, A., Ramiah, M. and Varma, A. 2002 Serological and coat protein sequences studies suggest that necrosis disease of sunflower in India is caused by a strain of *Tobacco streak ilarvirus*. *Arch.Virol.*: 651-658.

Brunt, A.A., Crabtree, K., Dallwitz, M.J., Gibbs, A.J. and Zurcher, E.J. eds. 1996 onwards. `Plant Virus Online: Descriptions and lists from the Vide database. http://biology.anu.edu.au/Groups/MES/vide

Chod, J., Skaloud V. and Jokes M. 1996. Detection of potato Y virus in the connection with sunflower mosaic symptoms. Sb. *UVITZ Ochrana Rostlin* 26: 11-16.

Christie, S.R., Edwardson, J.R. and Zettler, F.W. 1968. Characterization and electron microscopy of a virus isolated from *Bidens* and *Lepidium*. *Plant Dis. Reptr.* 52: 763-768.

Costa, A.S. and Kitajima, E.W. 1966. Virus do mosaice do picão ataca o girasol. O. Estrado de São Paulo, São Paulo. *Suppl. Agric. J.* 605: 12-13.

Dijkstra, J., 1983. Tobacco streak virus in sunflower (Helianthus annuus L). Neth. J. Plant Pathol. 89: 153-169.

Dujovny, G., Sasaya, T., Koganesawa, H., Usugi, T., Shohara, K., Leonardon, S.L. 2000. Molecular characterization of a potyvirus infecting sunflower. Arch. Virol. 145: 2249-2258.

Dujovny, G., Usugi, T., Shohara, K., Leonardon, S.L. 1998. Characterization of a potyvirus infecting sunflower in Argentina. Plant Dis. 82: 470-474.

FAOSTAT 2001, www.fao.org

Fivawo, N.C. 1987. Diseases and pests of sunflower in Tanzania. *TARO Newsletter* 2/3: 6-8.

Frenkel, M.J., Warnd, C.W. and Shukla, D.D. 1989. The use of 3'noncoding nucleotide sequences in the taxonomy of potyviruses: application to watermelon mosaic virus 2 and soybean mosaic virus. *J. Gen. Virol.* 70: 2775-2783.

Gulya, T.J., Berger, P.H., Shiel, P.J., Freeman, T.F., Isakeit, T.A. and Berger, P.H. 1998. Mosaic-causing potyvirus on wild sunflower from southern Texas. (Abstr.) *Phytopathology* 88 (suppl.) : S34.

Gulya, T.J., Shiel, P.J., Freeman, T.F., Jordan R.L. and Isakeit, T.A. and Berger, P.H., 2002. Host range and characterization of *Sunflower mosaic virus*. *Phytopathology* 92: 694-702.

Gupta, K.C. and Gupta, M.N. 1977. Viral diseases of sunflower, new records from India. *Curr. Sci.* 46: 285.

Jain, R.K, Bhat., A.I., Byadgi, A.S., Nagaraju, H.S., Halkeri, A.V., Anahosur, K.H. and Varma A. 2000. Association of a tospovirus with sunflower necrosis disease. *Current Science* 79: 1703-1705.

Jordan, R., and Hammond, J. (1991). Comparison and differentiation of potyvirus isolates and identification of strain-, virus-, subgroup-specific and potyvirus group-common epitopes using monoclonal antibodies. *J. Gen. Virol.* 72: 25-36.

Kaiser, J.W., Wyatt, S.D. and Klein, R.E. 1991. Epidemiology and seed transmission of two tobacco streak virus pathotypes associated with seed increases of legume germplasm in eastern Washington. *Plant Dis.* 75: 256-264.

Lenardon, S. L. 1994. Síntomas de etilogía viral en cultivos de girasol. In: *Enfermendades del Girasol en la Argentina*. pp.99-103. (eds) V.R. Pereyra and A.R. Escande, INTA, Balcare, Buenos Aires, Argentina.

Liu, L. and Li, X. 1988. The geographical distribution of sunflower diseases in China. *Plant Pathol.* 37: 470-474.

Martens, J.W., Ravagan, G. and MacDonald, W.C. 1970. Diseases of sunflower in Kenya. *East African Agric. Forestry J.* 35: 103-106.

McLean, D.M., 1962. Common weeds of tobacco ringspot virus in the lower Rio Grande Valley of Texas. *Plant Dis. Reptr.* 48: 5-7.

Nagaraju, V., Muniyappa V., Singh, S.J. and Virupahshappa, K. 1997. Occurrrence of a mosaic virus disase on sunflower in Karnataka. *Indian Phytopath.* 50: 277-281.

Nagata, T., Inoué, A.K., Dusi, A.N. and Kirajima, E.W. 1995. Bidens mosaic potyvirus newly isolated from pea, its characteristics and serological relationship with other potyviruses. *Fitopatol. Bras.* 20: 473-478.

Ndunguru, J. and Mwale, S. 1999. Putative virus disease of sunflower in Zambia. OEPP/EPPO Bulletin 29: 215-216.

Nishimura, N.T., Kikuchi, S., Tsuda, A. and Tsuchizaki.T. 1995. Cucumber mosaic virus isolated from sunflower (Helianthus annuus L.). Proc. Kanto Tosan Plant Prot. Soc. 42: 151-152.

Orellana, R.G. and Quacquarelli, A.A., 1968. Sunflower mosaic caused by a strain of cucumber mosaic virus. Phytopathology 58: 1439-1440.

Purcifull, D.E., Christie, S.R., Zitter, T.A. and Bassett M.J. 1971. Natural infections of lettuce and endive by Bidens mottle virus. Plant Dis. Reptr. 55: 1061-1063.

Ramiah, M., Bhat, A.I., Jain, R.K., Pant, R.P., Ahlawat, Y.S, Prabhakar K. and Varma, A. 2001.Partial characterization of an isometric virus causing sunflower necrosis disease in India. Plant Dis. 85: 443.

Ravi, K.S., Butgereitt, A., Kitkaru, A.S., Deshmukh, S., Lesemann, D.E. and Winter, S. 2001. Sunflower necrosis disease from India is caused by an ilarvirus related to Tobacco streak virus. Plant Pathol. 50: 1-1.

Reddy, A.S., Prasad Rao, R.D.V.J., Thirumala-Devi,K., Reddy,S.V., Mayo,M.A.,Roberts,I., Satyanarayana, T., Subramaniam,K. and Reddy, D.V.R. 2002. Ocurrence of Tobacco streak virus on peanut (Archis hypogaea) in India. Plant Dis. 86:173-178.

Richter, J., Rabenstein, F., Pröll, E. and Vetten, J. 1995. Use of cross-reactive antibodies to detect members of the Poytiviridae. J. Phytopathol., 143: 459-464.

Russell, G.E., Cook, P.H.L. and Bunting, E.S. 1975. An aphid-transmitted yellowing disease of sunflower. Plant Pathol. 24: 58-59.

Shukla, D.D., Ward, D.D. and Brunt A. 1994. The Potyviridae. CAB International, Wallingford.

Singh, J.P. 1979. Mechanical and seed transmission of the causal agent of sunflower rugose mosaic in Kenya. *Sunflower Newsletter* 3: 13-14.

Sdoodee, R. and Teakle, D.S. 1987. Transmission of tobacco streak virus by *Thrips tabaci*; a new method of plant virus transmission. *Plant Pathol.* 36: 377-380.

Theuri, J.M., Bock, K.R. and Woods, R.D. 1987. Distribution, host range and some properties of a virus disease of sunflower in Kenya. *Tropical Pest Manag.* 33: 202-207.

Traversi, B.A., 1949. Estudio inicial sobre una enferemedad del girasol (*Helianthus annuus* L.) Argentina. *Revta Invest agric.,* B. Aires 3: 345-351.

Venkata Subbaiah, K., Sai Gopal,V.R. and Krishna Reddy, M. 2000. First report of a tospovirus on sunflower (*Helianthus annuus* L.) from India. *Plant Dis.* 84: 1343.

Venugopal, R.R., Madhusudan, T. and Sastry, K.S. 1987. Studies on a mosaic disease of sunflower. *J. Oilseed Res.* 4: 286-288.

Wang, X.F., Tien W.H., Qui, B.S. and Pei, M.Y. 1983. Sunflower ringspot mosaic virus caused by a strain of CMV. *Acta Phytopath. Sinica* 13: 29-36.

Ward, C.W., McKern, N.M., Frenkel, M.J. and Shukla, D.D. 1992. Sequence data as the major criterion for potyvirus classification. In: *Potyvirus Taxonomy* pp.283-297. (ed) Barnett O.W. Springer, Wien New York (Arch. Virol. suppl. 5).

Zakusilo A., Knyaseva, N., Didenkoa, L. and Boyko, A. 1994. Identification of an ukranian virus isolate from *Helianthus annuus* L. with yellow spot mosaic symptoms. *Arch. Phytopath. Plant Protection* 29: 13-19.

Legends to Figures:

Fig.30.1 - Severe symptoms of *Tobacco streak virus* in sunflower from Jalna, Maharasta state, India, infected at early stages of plant development.

Fig. 30.2 - Sunflower plant with severe necrosis symptoms caused by infections of *Tobacco streak virus* in India

Fig. 30.3 - Chlorotic ringspot and mottle symptoms on leaves of sunflower from Zambia, presumably infected with a virus related to sunflower yellow blotch virus.

30.1

30.3

30.2

Table 30.1 Viruses reported from natural infections of sunflower

Virus	**Genus**	**Country**	**Reference**
Cucumber mosaic virus	*Cucumovirus*	China, Japan, India and USA	Orellana and Quacquarelli, 1968 ; Wang *et al.* 1983; Liu & Li, 1988; Nishimura *et al.* 1995; Nagaraju *et al.*, 1997
Sunflower chlorotic mottle virus	*Potyvirus*	Argentina	Lenardon *et al.* 1994 ; Dujovny *et al.* 1998
Tobacco ringspot virus	*Nepovirus*	USA	McLean, 1962
Sunflower mosaic virus	*Potyvirus*	USA	Arnott & Smith, 1967; Gulya *et al.* 1998, 2002
Tomato spotted wilt virus	*Tospovirus*	Ukraine	Zakusilo *et al.* 1994
Tobacco streak virus	*Ilarvirus*	Netherlands,India	Dijkstra,1983; Ravi *et al.* 2001, Ramiah *et al.* 2001
Potato virus Y	*Potyvirus*	Czech Republik	Chod *et al.* 1996
Beet western yellows virus	*Polerovirus*	U.K.	Russsel *et al.* 1974
Diseases with unclear or unconfirmed virus aetiology			
Mosaic disease		India	Gupta and Gupta, 1977; Venugopal *et al.* 1987
Yellow blotch disease	*Polerovirus/* Umbravirus	Kenia, Tanzania	Theuri *et al.* 1987; Ndunguru & Mwale, 1999
Sunflower crinkle virus disease	*Polerovirus/* Umbravirus	Kenya, Tanzania	Theuri *et al.* 1987

Table 30.2 Reactions of selected plant species to inoculations with TSV-SF.

Common name	**Scientific name**	**Systemic symptoms**	**days p.i.**
Sunflower	*Helianthus annuus* L.	CMo, Mo, Ne, die-back	6
Peanut	*Arachis hypogaea* L.	Ne, die-back	7 to 13
Okra	*Abelmoschus esculentus* L.	CMo, Mo	7 to 14
Cotton	*Gossypium arboreum* L.	Cl, NeR, LC	3
Soybean	*Glycine max* (L.) Merr.	NeR, die-back	3 to 6
Cowpea	*Vigna unguiculata* (L.) Walp.	NeR	3 to 9
Brinjal	*Solanum* spp.	Ne	3 to 6
Pumpkin	*Cucurbita pepo*	CMo, Mo	8 to 16
Squash	*Cucurbita maxima*	CMo, Mo	10
Watermelon	*Citrulus vulgaris* (Schrad.)	localised infection Cl	4
Cucumber	*Cucumis sativus* L.	Ch	3 to 4
Parthenium	*Parthenium hysterophorus* L.	Ch	7
Xanthium	*Xanthium strumarium* L.	NeR, VC,	3
Cosmos	*Cosmos sulphureus* Cav.	Cl, Ne	15 to 20
Marigold	*Tagetes tenuifloia* Cav.	Ne	7
Petunia	*Petunia hybrida* Vilm.	Cl, Ne	12

Days p.i. = days after inoculation until first appearance of symptoms. Ch = chlorosis, Cl = chlorotic lesions CMo = chlorotic mottle, Mo = mosaic, Ne = necrosis, NeR = necrotic rings, VC = vein chlorosis, LC = leaf curl.

31. Spices

Chilli pepper, black pepper, cardamom, garlic and ginger.

Anindya Roy [1], H.S.Savithri [1] and R.Usha [2]
[1]*Department of Biochemistry, Indian Institute of Science, Bangalore 560012, India.*
[2]*Department of Plant Biotechnology, School of Biotechnology, Madurai Kamaraj University, Madurai 625021, India.*

31.1 Chilli pepper (*Capsicum* spp.)

Chilli peppers were domesticated in Mexico. Since pre-historic times, red chilli peppers have been used in tropical South America and hot chilli peppers are popular in India, Indonesia, Malaysia, Thailand, Sri Lanka, Mexico and Brazil. India ranks first in chilli production, followed closely by Indonesia and Brazil and about 915,000 ha are under chilli cultivation in India, producing about 1018,000 tons per year (Peter and Nybe, 2002). The peppers are spicy hot due to the presence of capsaicin, an acrid, volatile alkaloid.

31.1.1 ***Pepper vein banding virus*** Genus ***Potyvirus*** (PVBV)

PVBV was first reported in *Capsicum annuum* from Uttar Pradesh, India (Joshi and Bhargava, 1962) and is prevalent in the chilli-growing areas of India, mainly in Karnataka. PVBV infects a wide range of *Solanaceous* plants including bell pepper and chilli. In chilli, systemic infection leads to mottling and leaf malformation, including narrowing and split tips (Ravi *et al.* 1997). The virus is transmitted by *Aphis craccivora* and *Myzus persicae* in a non-persistent manner. PVBV can be purified from infected chilli according to Ravi *et al.* (1997). The virus can be detected by inoculation to *Nicotiana glutinosa* (mosaic, mottling); *N. tabacum* (vein banding, leaf distortion); *Datura metel*; *D. stramonium* (mosaic, mottling, leaf distortion, reduced leaf); (*Lycopersicon esculentum* (vein banding, mosaic, mottling) and by DAS-ELISA (Ravi *et al.* 1997).

Virions are non-enveloped flexuous rods 900 x10nm (Joseph, 1999). The coat protein is 34 kDa in size and has been shown to assemble into virus-like particles when over-expressed in *E.coli* (Joseph and Savithri, 1999). The non-structural protein NIa protease over-expressed in *E.coli* was shown to cleave the coat protein from NIb-Coat protein in *trans* (Joseph and Savithri, 2000).

Virions contain 5-6 % nucleic acid (Joseph, 1999). The genome is made up of one molecule of linear positive-sense single-stranded RNA. The 5' end of the genome has a genome-linked protein (VPg) and poly A tail at the 3′ end (Joseph and Savithri, 1999). A comparison of the 3' untranslated region and the coat protein sequence with other potyviruses shows that PVBV is closely related to *Chilli vein-banding mottle virus* (CVbMV), (Chiemsombat *et al.*1998) with maximum identity of 74.1% and 96.5% respectively. This suggests that PVBV and CVbMV represent strains of the same virus (Joseph and Savithri, 1999).

31.1.2 ***Pepper severe mosaic virus*** Genus ***Potyvirus*** (PeSMV).

It was first reported in *Capsicum annuum*, from San Juan, Argentina. (Feldman and Gracia, 1977). PeSMV is the most severe virus that infects chilli in tropical countries and is prevalent all over the world. PeSMV infects bell pepper and chillies but damage to the plant is more severe in the latter. Common symptoms include necrotic spots and streaks in leaves and stems, followed by chlorosis.

The virus is transmitted by *Aphis craccivora, Myzus persicae* and *Macrosiphon euphorbae* in a non-persistent manner. PeSMV was purified by Feldman and Gracia (1977). The virus can be detected by ELISA using polyclonal antisera against *Tobacco etch virus* (TEV) and *Potato virus Y* as they are serologically related. Presence of 'pinwheel' like structure could be detected in infected plant parts. The virus can be detected by inoculation to test plants such as *Nicandra physalodes, Nicotiana debneyi, N. rustica, N. tabacum* (local lesions, followed by systemic vein-clearing mosaic and/or chlorotic rings); *Chenopodium amaranticolor, C. quinoa* (local lesions; not systemic).

Virions are non-enveloped, flexuous filaments, 761x13 nm. Capsid protein is the only structural protein and cDNA sequence of coat protein gene is known (Rabinowicz *et al.* 1993). Virion nucleic acid is single-stranded, positive sense RNA. The total genome is approximately 10 kb.

31.1.3 *Pepper mild mottle virus* Genus *Tobamovirus.* (PMMV)

PMMV occurs in Brazil, India and Southeast Asia. In chilli it causes mild chlorosis, stunting, small malformed and mottled fruits and necrotic depressions. The virus is transmitted by mechanical inoculation, grafting, contact with infected plants, and through seed. PMMV was purified according to Francki *et al.* (1985), and can be detected by ELISA using polyclonal antisera of *Tobacco mosaic virus.* The virus can also be detected by inoculation to test plants such as *Datura metel* (local necrosis and abscission); *D. stramonium* (small necrotic local lesions, not systemic); *Chenopodium amaranticolor, C. quinoa* (chlorotic local lesions; not systemic); *Nicotiana glutinosa, N. sylvestris* (small necrotic local lesions; not systemic).

Virions are non-enveloped, rigid rods, 312 x 18 nm with an axial canal and of 4 nm in diameter. The nucleic acid is single-stranded, positive sense RNA. Total genome length is 6357 nucleotides (Alonso *et al.* 1991) as determined from the Spanish isolate of PMMV. The viral genome encodes four open reading frames (ORFs), which encode a 126 kDa protein and a read-through 183 kDa protein, a 28 kDa protein, and a 17.5 kDa coat protein.

31.1.4 *Pepper mild tigre virus* Genus *Begomovirus* (PepMTV).

The virus was first reported in *Capsicum annuum*, from Mexico (Brown *et al.* 1989), and occurs in South America, especially in Mexico and in the Indian subcontinent. It infects all cultivated hot pepper varieties and commonly found symptoms are mild yellow leaf curling and mosaic. Alone, the virus induces mild leaf symptoms, but in mixed infection with other whitefly-transmitted viruses, severe crop loss is observed. The virus is transmitted by Bemisia *tabaci* in a non-persistent way.

Purification of PepMTV was described by Brown *et al.* (1989). The virions are geminate; 20 nm in diameter, 30 nm long, their surface capsomer arrangement is not clear. Virus is therefore not always detected by electron microscopy. The nucleic acid is single-stranded DNA. of approximately 2.5kb size. The virus can be detected by PCR, using two partial sequences available in the database (acc. no L27270 and L27271). The virus can be detected by inoculating test plants such as *Lycopersicon esculentum cv.*Pole Boy (systemic yellow mosaic leaf curling).

31.1.5 *Pepper mottle virus* Genus *Potyvirus* (PepMoV)

The virus was first reported from Mexico (Nelson *et al.* 1982), and occurs in South America, Mexico and El Salvador. The virus infects *Capsicum annuum* causing mottling and malformed leaves, *Datura metel Solanum sp.* causing mild mottle. *Datura meteloides, Lycopersicon esculentum* (mild systemic mottling.); (*Nicotiana glutinosa, N. tabacum, Physalis floridana* (systemic leaf mottling); *Chenopodium amaranticolor* (necrotic local lesions; not systemic). The virus is transmitted by *Aphis craccivora, Myzus persicae,*and *Macrosiphon euphorbae* in a non-persistent manner.

Purification of PepMoV was described by Dougherty *et al.* (1985). 'Pinwheel' structures were observed in infected tissues. Polyclonal antisera of Potyviruses, such as potato virus Y, tobacco etch and pepper veinal mottle viruses could be used to detect the virus. Virions are flexuous filaments, similar to that of other *Potyviridae*, with two structural proteins, VPg and coat protein (Hiebert and McDonald, 1973).

Virions contain one molecule of linear single-stranded RNA. Total sequence of cDNA of California isolate of PepMoV has been reported (Vance *et al.* 1992).

31.1.6 *Pepper veinal mottle virus* Genus *Potyvirus* (PVMV)

The disease was first reported in *Capsicum annuum, C. frutescens* and *Petunia hybrida* from Ghana (Konate and Traore,1999). It is transmitted in a non-persistent manner by *Aphis gossypii, A. spiraecola, Myzus persicae* and *Toxoptera citricidus* and also by mechanical inoculation and by grafting. The virus can be diagnosed by inoculation on *Chenopodium amaranticolor, C. oetidum* (local chlorotic or necrotic ringspots) and *Nicotiana clevelandii, N. megalosiphon* (systemic mosaic). The method for purification of the virus was described by Konate and Traore,1999. The virions are filamentous, not enveloped, with a clear modal length of 770 nm and width of 12 nm. The genome consists of single-stranded RNA. The molecular weight of the coat protein is 32-33 kDa.

31.1.7 ***Cucumber mosaic virus*** Genus ***Cucumovirus*** (CMV)

CMV is an important virus that affects chilli cultivation all over the world. It is described in detail under the heading 'Black pepper'.

31.1.8 Resistance

The sources of resistance to viruses of chilli have been compiled by Green and Kim (1994). More recently considerable progress has been made in the understanding of virus resistant genes and their manipulation towards achievement of virus resistance through molecular breeding. Caranta and Palloix (1996) showed that a chilli cultivar 'Indian pepper' (*Capsicum annuum*) showed absolute resistance to potato virus Y strain O (PVYO), Potyvirus E and *Chilli veinal mottle virus* (CVMV) and a partial resistance to *Potato virus Y* pathotype II and I. Resistance to CVMV was found to be due to two independent genes. By using molecular plant breeding technology, location of QTLs responsible for resistance was mapped and was found to be located near the pvr2 and pvr6 loci (Caranta *et al.* 1997).

Virus infection often leads to hypersensitive response (HR). It has been shown that HR to tobamovirus infection, which was mediated by L2 gene in capsicum, requires coat protein and not RNA (Cruz *et al.* 1997). Although plant viruses cause considerable damage to the pepper production and affect the economy of pepper producing third world countries in many ways, the information on exact amount of crop loss due to plant virus infection in developing countries is lacking.

31.2 Black Pepper (*Piper nigrum)*

Black pepper is considered as the 'King of Spices'. India is the leading producer of black pepper with an area of 1,95,050 ha, producing 65,000 tons of black pepper, followed closely by Vietnam, which produced 55,000 tons in 2001 (Peter and Nybe, 2002).

31.2.1 ***Cucumber mosaic virus*** Genus ***Cucumovirus*** (CMV)

CMV in black pepper was first reported by Fajardo *et al.* (1996) and is the only reported virus that causes mosaic disease in black pepper in India, Malaysia, Indonesia and Sri Lanka. (Ravindra, 2000).

CMV has a wide host range and is transmitted non-persistently by aphids, such as *Aphis craccivora, Myzus persicae,* and *Acyrthosiphon pisum.*

The virus has been purified from chilli and black pepper (Kiranmai *et al.* 1997 and Sarma *et al.* 2001). The virus isolated from black pepper cross-reacts with CMV-L, CMV-A and CMV isolated from chilli, banana and brinjal (Sarma *et al.* 2001). Virions are isometric, about 29nm in diameter, with a coat protein of 26.1 kDa.

The nucleic acid isolated from purified virus resolved into four species of molecular weight 1.21, 1.10, 0.87 and 0.37×10^6 Da on 1% agarose gels (Sarma *et al.* 2001). Similar size RNAs were observed with CMV isolated from infected chilli (Kiranmai *et al,* 1997).

ELISA kits for detection of *Pepper mild mottle virus, Pepper mottle virus, and Cucumber mosaic virus* are available from Agadia, Indiana, USA (www.agadia.com). For detection by PCR, diagnostic kits can be obtained from the DSMZ Braunschweig, Germany (www.dsmz.de).

The virus can be detected by inoculation to test plants such as *Chenopodium amaranticolor, C. quinoa* (chlorotic local lesions); *Vigna unguiculata* (necrotic local lesions); *Lycopersicon esculentum* (mosaic, reduction of leaf laminae ("fernleaf") and stunting).

31.2.2 *Piper yellow mottle virus* Genus *Badnavirus* (PYMV)

PYMV was found to be the causal agent of a disease of black pepper in Malaysia, the Philippines, Sri Lanka nd Thailand (Lockhart *et al.* 1997). The disease symptoms include mild to severe chlorotic mottling, vein-clearing, inter-veinal chlorosis , reduction in leaf size and leaf puckering and deformation. The virus was also associated with a disease of betelvine (*P. betel*). The virus from *P.nigrum* and *P betel* could be transmitted to healthy *P. nigrum* both mechanically and by the mealybug *Planococcus citri.* However, no other test plants could be inoculated by either method.

The virus was purified by Lockhart *et al.* (1997). The virions are non-enveloped, bacilliform averaging 30x125 nm in size. The genome is double-stranded DNA. The virus was found to be serologically related to banana streak virus and sugarcane bacilliform virus. Genomic PYMV sequences were amplified by PCR using badnavirus-specific oligonucleotide primers and sequence analysis comparisons of the putative reverse transcriptase domain showed PYMV to be closely related to other mealybug-transmitted badnaviruses. Lockhart *et al.* (1997) raised a polyclonal antiserum against the purified virus. Black pepper infected with PYMV sometimes contained one or more isometric virus-like particles. Therefore Lockhart *et al.* (1997) conclude that PYMV may be only one component of a virus complex infecting blackpepper in Southeast Asia.

31.3 Small cardamom (*Elettaria cardamomum*)

Small cardamom is popularly known as the "Queen of Spices". It is one of the major spice crops in India and Guatemala. Cardamom is a perennial herb of the family *Zingiberaceae.* It grows in the high altitudes of 1000 to 1500 m and. is propagated mainly through its rhizomes. A healthy plant produces 500 g to 2 kg of capsules annually.

31.3.1 *Cardamom mosaic virus* Genus *Macluravirus* (CdMV)

An aphid-transmitted disease affecting small cardamom in South India was first reported by Uppal *et al.* (1945). The disease is known as *katte* (meaning disorder) in India. A similar mosaic disease of cardamom from Guatemala was identified by Gonsalves *et al.* (1986). The disease is found all along the cardamom tract in South India, and all varieties of cardamom are susceptible to the disease. Infected plants show 70-100% reduction in the yield. There was a steep reduction in the annual export of cardamom from 2,383 tons in 1985, to 270 tons in 1986, and 297 tons in 1998 (Varmudi, 2000). The following species of the family *Zingiberaceae* are susceptible to the virus: *Amomum muricatum, Amomum pterocarpum, Amomum cannicarpum* and *Alpinia mutica.* (M.N. Venugopal, personal communication).The disease is characterized by mosaic patches and light green stripes on the leaf lamina and pseudostem (Fig. 31.1), and infected plants are stunted.

Only the banana aphid *Pentalonia nigronervosa* naturally transmits the virus, although under experimental conditions, thirteen other types of aphids could transmit the disease (Rao and Naidu, 1974). The disease does not spread through seeds or by mechanical inoculation. Gonsalves *et al.* (1986) and Saigopal *et al.* (1992) detected the virus by ELISA. Jacob and Usha (2002) have demonstrated the detection of the virus coat protein in western blots using antiserum raised against the virus.

CdMV was purified by Jacob and Usha, (2001). The virions are flexuous rod-shaped particles of 700-720 nm length for the Guatemalan virus (Gonsalves *et al.* 1986) and ~800 nm in length and 11-12 nm in diameter for the Indian virus (Jacob and Usha, 2002). Gonsalves *et al.* (1986) observed three protein bands of 37.5, 32.2 and 29 kDa from the virus, and Jacob and Usha (2001) have made similar observations of 38, 35 and 26 kDa bands and showed that the 26 kDa band to be the trypsin-resistant core of the coat protein.The virion contains a single-stranded RNA genome estimated to be 8.5 kb. Based on the 3' terminal sequence of the genome of CdMV, Jacob and Usha (2001) suggested to classify CdMV as a new member of the *Macluravirus* genus of the *Potyviridae*. Based on symptoms and host range, six distinct strains of CdMV have been identified (M.N.Venugopal, personal communication). The present approach for preventing the spread of the disease is by rouging infected plants. Development of transgenic, virus-resistant cardamom by the pathogen-derived resistance is in progress at Dr. R.Usha's lab in Madurai Kamaraj University, India.

31.3.2 *Cardamom necrosis virus*

The necrosis disease of cardamom was found relatively recently in India, in the Nilgiri hills of Tamil Nadu, and in some isolated pockets in Kerala and Karnataka (Naidu and Thomas, 1994). The necrosis disease though severe, occurred only in small pockets. The affected plants were destroyed in order to prevent the spread of the disease. Therefore, only limited work could be done on the virus.

Infected plants show chlorotic or necrotic patches on the leaves, leaf shredding and distortion of panicles and fruits.

The disease is transmitted by infected rhizomes and not through soil, seed, sap, aphids, thrips or whiteflies (Sridhar *et al.* 1990).

The virus was purified by Saravanakumar *et al.* (1998). Using leaf-dip electron microscopy, Naidu and Saigopal detected filamentous particles of ~650nm (personal communication). However, Saravanakumar *et al.* (1998) found flexuous filaments of ~1550nm and attributed the lengthening to end-to-end aggregation of the necrosis virus or with the co-infecting mosaic virus. The molecular weight of the virus protein was determined to be 30.2 kDa (Saravanakumar *et al.* 1998). No data are available on the size of the genome. Saravanakumar *et al.* (1998) developed a degenerate carlavirus-specific antisense primer for the detection of the viral nucleic acid in northern slot blots.

31.3.3 Cardamom vein clearing disease

Vein clearing disease, also known as *kokke kandhu,* is another severe disease recently reported from the Coorg district of Karnataka (Govindaraju *et al.*, 1994). The causal agent is assumed to be a virus. No information is available about the virus.

31.4 Large cardamom (*Amomum subulatum*)

Large cardamom is grown in the Indian states of Sikkim and West Bengal. It is affected by two diseases, supposedly of viral origin: (1)'Foorkey', transmitted by *Pentalonia nigronervosa* (Varma and Capoor,

1964) and by *Micromyzus kalimpongensis* (Basu and Ganguly ,1968). It infects also small cardamom (Rao, 1972; Rao and Naidu,1973); (2) 'Chirke' disease transmitted by the aphids *Brachycaudus helichrys* and *Rhopalosiphum maidis* (Raychaudhuri and Chatterjee ,1964).
No further data are available on these viruses.

31.5 Garlic (*Allium sativum*)

Garlic is one of the most popular spices in the world. A native of Central Asia, garlic was well known to ancient Egyptians way back in 2600 B.C.

Garlic is a hardy bulbous perennial, of the family *Alliaceae* and is grown all over the world. Among the developing countries, it is grown mainly in Argentina, Brazil, China and Egypt. In India, 114,000 ha are grown yielding 495,000 tons annually (Peter and Nybe, 2002).

Most of the garlic plants cultivated throughout the world are infected by viruses which give rise to mosaic or streak symptoms on their leaves and reduction in bulb and clove size (Ahlawat, 1974).

31.5.1 *Garlic virus X* Genus *Allexivirus* (GVX)

GVX occurs in Korea (Song *et al.* (1998)) and China (Chen *et al.* 2001), causing mosaic or streak symptoms. The virus is transmitted by mites (Chen *et al.* 2001), and may be detected by immunoblotting or with a specific cDNA probe (Song *et al.* 1998). GVX can be purified as described by Nagakubo *et al.* (1994). The deduced size of the coat protein is 26 kDa.

The RNA genome of GVX contains 8106 nucleotides, excluding the 3'-end poly(A) tail. Based on the organization of the open reading frames in the genome, Song *et al.* (1998) conclude that GVX does not belong to the carla or potexviruses but is a separate member of a new and unassigned group of plant viruses that includes *Shallot virus X* and GarV-type viruses. Sumi *et al.* (1999) have classified the GarV-type viruses in the *Allexivirus* genus. Therefore it is likely that GVX from Korea is also a member of the *Allexivirus* genus. Chen *et al.* (2001) have isolated a GVX of 8319 nt (AJ292229) with 90.2% identity with the nucleotide sequence of the Korean isolate (U89243).

31.5.2 *Garlic common latent virus* (?) Genus *Carlavirus* (GCLV)

The virus was first reported from France (Delacolle and Lot, 1981)., and is distributed worldwide. Among the developing countries, it is found in Argentina (Conci *et al.* (1992) and China (Chen *et al.* 2001). The virus occurs in *Allium sativum* (garlic), *A. ampeloprasum* var. holmense (great headed garlic) and *A. porrum* (leek) (Van Dijk, 1993). Infected plants may show weak mosaic symptoms. The virus is transmitted mechanically.

An antiserum prepared against the coat protein, which was expressed in *E.coli,* detected specifically the virus in ELISA, immunoelectron microscopy and in direct tissue blotting immunoassay (Tsuneyoshi and Sumi, 1996). An RT-PCR protocol for detecting the virus was also developed (Tsuneyoshi *et al.* 1998).

The virus was purified as described by Mohamed and Young (1981), and filamentous particles of 650nm were observed in electron micrographs.The deduced molecular weight of the coat protein is 32.8 kDa (Chen *et al.* 2001). The sequences of the RNA genome coding for the coat protein and the nucleic acid binding protein (AB004458) have been reported by Tsuneyoshi *et al.* (1998). Chen *et al.* (2001) have sequenced the complete genome (8363 nt) of a Chinese isolate of the virus (AJ292226), which has 76.4% identity at the nucleotide level with the Korean isolate (Z68502).

31.6 Ginger (*Zingiber officinale*)

The use of the ground or powdered root of ginger is mentioned by Confucius and in the Koran. Ginger is probably the oldest of all today's culinary flavorings for nearly 2,500 years.

The world's main production areas for ginger are Brazil, Costa Rica, India, Thailand and China. The area under ginger cultivation in India is 62,000 ha with an annual production of 186,000 tons (Spices Statistics, 1997).

31.6.1 *Ginger chlorotic fleck virus(?)* Genus ***Sobemovirus*** (GCFV).

The virus was reported from New Zealand and Australia in ginger imported from India, Malaysia, and Mauritius (Thomas, 1986). All infected material was destroyed and there was no evidence that it became established in Australia. Its presence has not yet been confirmed in the countries of origin of the infected ginger.

The virus was inefficiently transmitted between ginger plants by mechanical inoculation and was not transmissible to any of the experimental hosts tested. The natural host range of the virus is not known.

Infected ginger has prominent, chlorotic flecks 1-10 mm long on the leaves parallel to the veins. Symptoms first appear in the younger leaves and persist in these and all subsequently formed leaves. No obvious symptoms occur in infected rhizomes.The virus was not transmitted by the aphids *Myzus persicae, Pentalonia nigronervosa, Rhopalosiphum maidis* or *Rhopalosiphum padi.*

The virus was purified according to Thomas (1986). Particles have an average diameter of 29 nm measured across the parallel sides and 33 nm across opposite apices and the virus particle has a major protein of 29 kDa. The virion contains a single-stranded RNA genome with a major species of molecular weight 1.5×10^{6} and two minor species of molecular weights 0.59×10^{6} and 0.14×10^{6}.Thomas (1986) concludes that GCFV is a possible sobemovirus.

An antiserum produced against the virus detected a number of virus isolates from ginger rhizomes.

Acknowledgements: The help of Dr.Y.S.Rao of Spices Board, Cochin is acknowledged for providing data on ginger and garlic. Ms. Archana Somanath, Mr. T. Jacob and Mr. Phaneeswara Rao are thanked for help in collecting some of the references.

References

Ahlawat, Y.S. 1974. A mosaic disease of garlic in Darjeeling hills. *Science and Culture* 40 : 466-467.

Alonso, E., Garcia-Luque, I., de la Cruz, A., Wicke, B., Avila-Rincon, M.J., Serra, M.T., Castresana, C. and Diaz-Ruiz, J.R. 1991. Nucleotide sequence of the genomic RNA of Pepper mild mottle virus, a resistance-breaking tobamovirus in pepper. *J. Gen. Virol.*72: 2875-84.

Anstead, R. D. 1924. *Rept. Operations Dept. Agric., Madras Presidency for 1923-1924.* pp 41.

Basu, M. and Ganguly,P. 1968. A note on the transmission of 'Foorkey' disease of large cardamom by the aphid *Micromyzus kalimpongensis. Indian Phytopath.* 21:127.

Brown, J.K., Pozo-Campodonico, O. and Nelson, M.R. (1989). A whitefly-transmitted geminivirus from peppers with tigre disease. *Plant Dis.* 73: 610-617

Caranta, C. and Palloix, A. 1996. Identification and comparative mapping of a dominant potyvirus resistance gene-cluster in Capsicum. *Theor.Appl.Gen.* 92: 15-20.

Caranta, C., Lefebvre,V. and Palloix, A.1997. Polygenic Resistance of Pepper to Potyviruses Consists of a Combination of Isolate-Specific and Broad-Spectrum Quantitative Trait Loci. *Mol.Plant.Mic.Inter.* 10: 872-878.

Chen, J., Chen, J. and Adams, M.J. 2001. Molecular characterization of a complex mixture of viruses in garlic with mosaic symptoms in China. *Arch. Virol.* 146: 1841-1853.

Chiemsombat, P., Sae-Ung, N., Attathom, S., Patarapuwadol, S., and Siriwong,P.1998. Molecular taxonomy of a new potyvirus isolated from chilli pepper in Thailand. *Arch. Virol.*143: 1855-1863.

Conci, V., Nome,S.F., and Milne,R.G. 1992. Filamentous viruses of garlic in Argentina. *Plant Dis.* 76: 594-596.

Cruz, A., Lopez, L., Tenllado, F., Diaz-Ruiz, J.R., Sanz, A.I., Vaquero, C., Serra, M.T., Garcia-Luque, I. and De-la-Cruz, A.1997. The coat protein is required for the elicitation of the Capsicum L2 gene-mediated resistance against the tobamoviruses. *Mol.Plant.Mic.Int.*10: 107-113

Delacolle, B. and Lot, H. 1981. Virus diseases of garlic. I. Detection and attempts at characterization by immuno-electron microscopy of a complex of three viruses in different garlic populations attacked by mosaic. *Agronomie* 1: 763 -769.

Deol, G.S. and Rataul, H.S. 1978. Influence of plant age at inoculation by cucumber mosaic virus on disease incidence and yield of chilli.. *Ind. Jl.Pl.Protection*. 6:82-83.

Dougerty,W.G., Allison, R.F., Parks, T.D., Johnston, R.E., Field, M.J. and Armstrong F.B. 1985. Nucleotide sequence of the 3' terminus of Pepper mottle virus RNA. Evidence for alternative mode of capsid protein gene organization. *Virology* 146: 282-285.

Fajardo,T.V.M., Carvalho, M.G. and Maciel-Zambolim,E. 1996. Cucumber mosaic virus detected in leaves of black pepper (Piper nigrum) with the enzyme-linked immunosorbent assay (ELISA). *Fitopatologia Brasileira* 21: 99-104.

Feldman, J.M. and Gracia, O. 1977. Pepper severe mosaic virus : a new poty virus from pepper in Argentina. *Phytopath. Z.* 89: 146-149.

Francki,R.I.B., Milne,R.G. and Hatta,T. 1985. *Atlas of plant viruses* Vol II CRC press Boca Raton FL, 284 pp.

Garzon-Tiznado, JA, I Torres-Pacheco, JT Ascencio-Ibanez, L Herrera-Estrella and RF Rivera-Bustamante. 1993. Inoculation of peppers with infectious clones of a new geminivirus by a biolistic procedure. *Phytopathology*. 83: 514-521.

Gonsalves, D., Trujillo, E. and Hoch, H. C. 1986. Purification and some properties associated with cardamom mosaic, a new member of potyvirus group. *Plant Dis*. 70: 65-69.

Gorsane F, Fakhfakh H, Tourneur C, Marrakchi M, Makni M. 2001. Nucleotide sequence comparison of the 3' terminal region of the genome of pepper vein mottle virus isolates from Tunisia and Ivory Coast. *Arch Virol*. 146(3):611-8.

Govindaraju, C., Venugopal, M. N. & Sudarshan, M. R. (1994). An apprisal of 'Kokkekandu'- a new disease of cardamom and Katte (Mosaic) disease in Karnataka. *Jl. Plantation Crops* 22: 57-59.

Green, S.K and Kim, J.S.1994. Sources of resistance to viruses of pepper (Capsicum spp) a catolog. Tech*nical Bull, No 20. Asian Veg. Res. Devel. Center* publication, Taipei.

Hiebert, E. and McDonald, J.G. 1973. Characterization of some proteins associated with viruses in the potato Y group. *Virology* 56: 349-361.

Hovarth, J. 1978. New artificial hosts and non hosts of plant viruses and separation of viruses-IV, *Acta Phytopath Acd Sci, Huang.,*13:57-73

Jacob,T. and Usha, R. 2001. 3' Terminal sequence analysis of the RNA genome of the Indian isolate of *Cardamom mosaic virus*: A New member of genus *Macluravirus* of *Potyviridae. Virus Genes* 23: 81-88.

Jacob, T. and Usha, R. 2002. Expression of *Cardamom mosaic virus* coat protein in *Escherichia coli* and its assembly into filamentous aggregates. *Virus Res*. 86: 133-141.

Jeyarajan, R., and Ramakrishnan, K. 1969. Potatovirus Y on chilli. *Madras Ag. Jl.* 56:761-766.

Joseph, J. 1999. Molecular characterization of pepper vein banding virus. *PhD Thesis*,Indian Institute of Science, Bangalore, India.

Joseph, J and Savithri, H.S.1999. Determination of 3'-terminal nucleotide sequence of pepper vein banding viral RNA and expression of its coat protein in *E.coli. Arch. Virol.*144: 1679-1687.

Joseph, J and Savithri, H.S. 2000. Mutational analysis of the NIa protease from pepper vein banding potyvirus. *Arch. Virol.*145: 2493-2502.

Joshi, R.D. and Bhargava, K.S.1962. A vein banding mosaic virus disease of chillies. (*Capsicum frutescens* L) *Ind. J. Micol.*.2: 29-34.

Kiranmai, G., Sreenivasulu, P. and Nayudu, M.V. 1997. Characterization of cucumber mosaic cucumovirus isolates naturally infecting three solanaceous vegetable crops in Andhra Pradesh. *Ind. J. Phytopathol.* 50: 416-425.

Konaté, G and Traoré, O.1999. Characterization and distribution of pepper veinal mottle virus in West Africa. *Agricultures.* 8: 132-134.

Kulkarni, C. S. 1924. Mosaic and other related diseases of crops in the Bombay Presidency. *Poona Agric. Coll. Magazine* 16: 6-12.

Lockhart, B.E.L., Kiratiya-Angul, K., Jones, P., Eng, L., De Silva, P., Olszewski, N.E., Lockhart, N., Deema, N. and Sangalang, J. 1997. Identification of *Piper* yellow mottle virus, a mealybug-transmitted badnavirus infecting *Piper* spp. In Southeast Asia. *Eur. J .Pl. Path.* 103: 303-311.

Misra, A., and Jha, A.1972. transmission of chilli mosaic virus through pollen and root contact. *Phyton(Austria)*, 15:103

Mohamed, N.A. and Young, B.R. 1981. Garlic yellow streak virus, a potyvirus infecting garlic in New Zealand. *Ann. appl. Biol.* 97: 65-74.

Nagakubo, T., Kubo, M. and Oeda, K. 1994. Nucleotide sequences of the 3'- regions of two major viruses from mosaic-diseased garlic: Molecular evidence of mixed infection by a potyvirus and a carlavirus: *Phytopathology* 84: 640-645.

Naidu, R. and Venugopal, M. N. 1984. Epidemiology of Katte disease of small cardamom. I:Disease incubation period and role of different host parts as a source of inoculums in relation to disease spread. *Proc. Plantation Crops Symp.* VI: 121-127.

Naidu, R. and Thomas, J. 1994. Viral diseases of cardamom: in *Advances in Horticulture : Plantation and Spice Crops.* Part 2. K.L. Chadha, and P. Rethinam, (eds.) Malhothra Publishing House, New Delhi.

Nambiar, K.K.N. and Sarma, Y.R. (1974). Mosaic disease of ginger. *Arecanut and Spices Bull.* 6: 3-4.

Nariani,T.K. and Sastry, K.S.M.1961. Insect transmission of chilli mosaic virus. *Ind. Phytopath.*15:173-179

Nelson,M.R.,Wheeler,R.E.,and Zitter,T.A. 1982. *Pepper mottle virus, CNI/AAB Description of plant viruses* No.253.

Peter, K.V. and Nybe, E.V. 2002. Dominating global markets. In *The Hindu Surv. Indian Agric.* pp 87-95.

Prasad Rao, R.D.V.J., Chakrabarty, S.K., Rajeswari, R., and Reddy, A.S.1987. Occurrence of tomato spotted wilt virus on cowpea, chilli, and brinjal. *Indian J. Pl. Protec.* 15:117-119.

Rao, D.G. 1972. Katte disease of cardamom and its possible control. In *III Intern Symp. Sub-tropical and Tropical Fruits.* New Delhi.

Rao, D.G. and Naidu, R. 1973. Studies on katte or mosaic disease of cardamom. *J.Plant. Crops.*1:129-136.

Rao, D. G. and Naidu, R. 1974. Additional vectors of 'katte' disease of small cardamom. *Indian J. Hort.* 31: 380-381.

Rabinowicz,P.D., Bravo-Almonacid, F. F. and Mentaberry, A. N. 1993. cDNA sequence of the pepper severe mosaic virus coat protein gene. *Plant Physiol.* 103 : 1023-1027.

Ravi, K. S., (1991) Studies on Pepper vein banding virus and other components of murda syndrome in chilli. *PhD thesis*, University of Agricultural Sciences, Bangalore, India.

Ravi, K. S., Joseph, J., Nagaraju, N., KrishnaPrasad, S., Reddy, H. R. and Savithri, H. S. 1997. Characterization of a pepper vein banding virus of chilli pepper in India. *Plant Dis.* 81: 673-676.

Ravindra, P. N. (Ed). 2000. *Black Pepper:Piper nigrum*, Harwood Academic Publishers. NY.

Raychaudhuri, S.P. and Chatterjee, S.N. 1964. Aphid transmission of a new streak mosaic of large cardamom. *Phytopathology* 54 : 904.

Rodriguez-Cerezo E, Moya A, Garcia-Arenal F. 1989. Variability and evolution of the plant RNA virus pepper mild mottle virus. *J Virol.* 63: 2198-203

Saigopal, D. V. R., Naidu, R. and Thomas, J. 1992. Early detection of Katte disease of small cardamom through ELISA. *J.Plant.Crops* 20 : 73-75.

Saravanakumar, M., Panicker, S. and Usha, R. 1998. Partial characterization and early detection of a virus associated with the necrosis disease of small cardamom in south India. *Indian J. Virol.* 14 : 59-63.

Sarma, Y.R., Kiranmai, G., Sreenivasulu, P., Anandaraj, M.,.Hema, M., Venkataramana, M., Murthy, A.K. and Reddy, D.V.R 2001. Partial characterization and identification of a virus associated with stunt disease of black pepper (*Piper nigrum*) in South India. *Curr.Sci.* 80: 459-462.

So, I.Y. (1980). Studies on ginger mosaic virus. *Korean J. Plant Protection* 19: 67-72.

Song S.I, Song J.T, Kim C.H, Lee J.S, Choi Y.D (1998) Molecular characterization of the garlic virus X genome. *J. Gen. Virol.* 79: 155-159.

Spices Statistics (1997). 3rd edition, Spices Board, Kochi.

Sridhar, V.V., Muthusamy, M. and Naidu, R. 1990. Survey on the occurrence of Nilgiri necrosis- a new virus disease of cardamom. *South Indian Hort.* 38: 165.

Sumi,S., Matsumi, T. and Tsuneyoshi, T. 1999. Complete nucleotide sequences of garlic viruses A and C, members of the newly ratified genus *Allexivirus. Arch. Virol.* 144(9): 1819-1826.

Thomas, J.E. (1986). Purification and properties of ginger chlorotic fleck virus. *Ann. appl. Biol.* 108: 43-50.

Tsuneyoshi, T. and Sumi, S. 1996. Differentiation among garlic viruses in mixed infections based on RT-PCR procedures and direct tissue blotting immunoassays. *Phytopathology* 86: 253-259.

Tsuneyoshi, T., Matsumi, T., Deng, T.C., Sako, I. and Sumi, S. 1998. Differentiation of Allium carlaviruses isolated from different parts of the world based on the viral coat protein sequence. *Arch.Virol* 143: 1093-1107.

Uppal, B. N., Varma, P. M. and Capoor, S. P. 1945. Mosaic disease of cardamom. *Curr. Sci* 14: 208-209.

Vance, V.B., Moore, D., Turpen, T.H., Bracker, A. and Hollowell, V.C. 1992.The complete nucleotide sequence of pepper mottle virus genomic RNA: comparison of the encoded polyprotein with those of other sequenced potyviruses. *Virology* 191:19-30.

Varma, P.M. and Capoor, S.P. 1964. 'Foorkey' disease of large cardamom. *Indian J. Agric.. Sci.* 34:56-62.

Varmudi, V. 2000. Decreasing flavor of Indian cardamom in the international market. *Spice India* April 2000,13-15. Directorate of Economics and Statistics, New Delhi.

Waller, J.M., Lenne, J.M., and Waller, S.J.,(ed). 2002. *Plant Pathologist's Pocketbook*, CAB international, Surrey, UK.

Warren C. E. and Murphy J. F. The complete nucleotide sequence of Pepper mottle virus-Florida RNA. 2003. *Arch Virol.*148 :189-97.

Fig. 31.1 (b) Cardamom leaf with typical mosaic symptoms caused by *Cardamom mosaic virus*. (a) Healthy cardamom leaf.

Table 31. 1 Viruses affecting chilli

No.	**Virus & Distribution**	**Transmission**	**Key References**
A	**SOUTH & CENTRAL AMERICA**		
1	*Pepper ring spot virus* (PRSV) ;Tobravirus Brazil	Sap: Yes Vector: No Seed : NT	Waller *et al,*2002
2	*Pepper mild mottle virus (PMMV)** *Tobamovirus* Argentina	Sap: Yes Vector: No Seed: Yes	Alonso *et al,*1991 Rodriguez-Cerezo *et al,*1989
3	*Pepper mild tigre virus (PepMTV)* *Geminiviridae: Begomovirus* Mexico	Sap: Yes Vector: Bemisia tabaci Seed: No	Brown *et al,*1989
4	*Pepper huasteco virus(PHV)** *Geminiviridae: Begomovirus* Mexico	Sap: Yes Vector: Bemisia tabaci. Seed: NT	Garzon-Tizando *et al,*1993. Torres-Pacheco *et al,* 1993.
5	*Pepper severe mosaic virus (PepSMV)* *Potyviridae: Potyvirus* Argentina	Sap: Yes Vector: *Macrosiphum euphorbae, Aphis craccivora, Myzus persicae* Seed: NS.	Rabinowicz *et al,*1993
6	*Pepper mottle virus* (PMV) *Potyviridae: Potyvirus* El Salvador	Sap: Yes Vector: *Aphis craccivora, Myzus persicae Aphis gossypii* Seed : NT	Warren and Murphy,2003
B	**INDIAN SUBCONTINENT**		
7	*Pepper vein banding virus* (PVBV)* Potyviridae: Potyvirus. India and Sri Lanka.	Sap: Yes Vector: *Aphis craccivora, Myzus persicae Aphis gossypii*	Joseph and Savithri, 1999

		Seed: No.	
8	*Indian chilli mosaic virus* (ICMV) *Potyviridae: Potyvirus* India	Sap: Yes Vector: *Aphis gossypii* Seed: NS.	Nariani and sastry, 1961 Misra and Jha, 1972
9	(*Chilli*)Tomato *spotted wilt virus (TSWV)* *Bunyaviridae:Tospovirus* India	Sap: Yes Vector: *Thrips tabaci,* *Thrips flavus* Seed: NT	Prasad Rao et al, 1987
C	**AFRICA**		
10	*Pepper vein mottle* *Virus* (PVMV)* *Ghana, Kenya, Nigeria, South Africa.*	Sap: Yes Vector: *Aphis spiraecola,* *Myzus persicae* *Aphis gossypii* Seed ; No	Gorsane et al, 2001
11	*Moroccan pepper virus* (Potexvirus) Morocco	Sap: Yes Vector: No Seed : No.	Gallitelli and Ruso,1987
D	**Viruses Distributed Worldwide**		
12	*Tobacco mosaic virus* *Tobamovirus* Causes chilli mosaic	Sap: Yes Vector: No Seed : Yes	Hovarth, 1979
13	*Cucunber mosac virus* *Cucumoviridae* Causes chilli mosaic	Sap: Yes Vector: *Acyrthosiphon pisum* *Myzus persicae* *Aphis gossypii* Seed : No	Deol and Rataul, 1978
14	*Potato virus Y* *Potyviridae: Potyvirus* *Causes chilli mosaic*	Sap: Yes Vector: *Myzus humuli,* *Myzus certus,* *Rhopalosiphum Insetum* Seed : NT	Jeyrajan and Ramakrishnan, 1969

NS= not studied; NT= not tested in chilli

Index